Theory of Martingales

Mathematics and Its Applications (*Soviet Series*)

Volume 49

Theory of Martingales

by

R. Sh. Liptser

and

A. N. Shiryayev

Steklov Institute of Mathematics,
Academy of Sciences of the U.S.S.R.,
Moscow, U.S.S.R.

KLUWER ACADEMIC PUBLISHERS
DORDRECHT / BOSTON / LONDON

Library of Congress Cataloging in Publication Data

Liptser, R. Sh. (Robert Shevilevich)
[Teoriia martingalov. English]
Theory of martingales / by R.Sh. Liptser and A.N. Shiryayev ;
[translated from the Russian by K. Dzjaparidze].
p. cm. -- (Mathematics and its applications. Soviet series ;
49)
Includes bibliographical references.
1. Martingales (Mathematics) I. Shiriaev, Al'bert Nikolaevich.
II. Title. III. Series: Mathematics and its applications (Kluwer
Academic Publishers). Soviet series ; 49.
QA274.5.L5713 1989
519.2'87--dc20 89-15522

ISBN-13: 978-94-010-7600-5 e-ISBN-13: 978-94-009-2438-3
DOI: 10.1007/978-94-009-2438-3

Published by Kluwer Academic Publishers,
P.O. Box 17, 3300 AA Dordrecht, The Netherlands.

Kluwer Academic Publishers incorporates
the publishing programmes of
D. Reidel, Martinus Nijhoff, Dr W. Junk and MTP Press.

Sold and distributed in the U.S.A. and Canada
by Kluwer Academic Publishers,
101 Philip Drive, Norwell, MA 02061, U.S.A.

In all other countries, sold and distributed
by Kluwer Academic Publishers Group,
P.O. Box 322, 3300 AH Dordrecht, The Netherlands.

printed on acid free paper

This is the expanded and revised translation of the original work
ТЕОРИЯ МАРТИНГАЛОВ
Published by Nauka Publishers, Moscow, © 1986

Translated from the Russian by K. Dzjaparidze.

SERIES EDITOR'S PREFACE

'Et moi, ..., si j'avait su comment en revenir, je n'y serais point allé.'

Jules Verne

The series is divergent; therefore we may be able to do something with it.

O. Heaviside

One service mathematics has rendered the human race. It has put common sense back where it belongs, on the topmost shelf next to the dusty canister labelled 'discarded non-sense'.

Eric T. Bell

Mathematics is a tool for thought. A highly necessary tool in a world where both feedback and non-linearities abound. Similarly, all kinds of parts of mathematics serve as tools for other parts and for other sciences.

Applying a simple rewriting rule to the quote on the right above one finds such statements as: 'One service topology has rendered mathematical physics ...'; 'One service logic has rendered computer science ...'; 'One service category theory has rendered mathematics ...'. All arguably true. And all statements obtainable this way form part of the raison d'être of this series.

This series, *Mathematics and Its Applications*, started in 1977. Now that over one hundred volumes have appeared it seems opportune to reexamine its scope. At the time I wrote

> "Growing specialization and diversification have brought a host of monographs and textbooks on increasingly specialized topics. However, the 'tree' of knowledge of mathematics and related fields does not grow only by putting forth new branches. It also happens, quite often in fact, that branches which were thought to be completely disparate are suddenly seen to be related. Further, the kind and level of sophistication of mathematics applied in various sciences has changed drastically in recent years: measure theory is used (non-trivially) in regional and theoretical economics; algebraic geometry interacts with physics; the Minkowsky lemma, coding theory and the structure of water meet one another in packing and covering theory; quantum fields, crystal defects and mathematical programming profit from homotopy theory; Lie algebras are relevant to filtering; and prediction and electrical engineering can use Stein spaces. And in addition to this there are such new emerging subdisciplines as 'experimental mathematics', 'CFD', 'completely integrable systems', 'chaos, synergetics and large-scale order', which are almost impossible to fit into the existing classification schemes. They draw upon widely different sections of mathematics."

By and large, all this still applies today. It is still true that at first sight mathematics seems rather fragmented and that to find, see, and exploit the deeper underlying interrelations more effort is needed and so are books that can help mathematicians and scientists do so. Accordingly MIA will continue to try to make such books available.

If anything, the description I gave in 1977 is now an understatement. To the examples of interaction areas one should add string theory where Riemann surfaces, algebraic geometry, modular functions, knots, quantum field theory, Kac-Moody algebras, monstrous moonshine (and more) all come together. And to the examples of things which can be usefully applied let me add the topic 'finite geometry'; a combination of words which sounds like it might not even exist, let alone be applicable. And yet it is being applied: to statistics via designs, to radar/sonar detection arrays (via finite projective planes), and to bus connections of VLSI chips (via difference sets). There seems to be no part of (so-called pure) mathematics that is not in immediate danger of being applied. And, accordingly, the applied mathematician needs to be aware of much more. Besides analysis and numerics, the traditional workhorses, he may need all kinds of combinatorics, algebra, probability, and so on.

In addition, the applied scientist needs to cope increasingly with the nonlinear world and the extra mathematical sophistication that this requires. For that is where the rewards are. Linear models are honest and a bit sad and depressing: proportional efforts and results. It is in the nonlinear world that infinitesimal inputs may result in macroscopic outputs (or vice versa). To appreciate what I am hinting at: if electronics were linear we would have no fun with transistors and computers; we would have no TV; in fact you would not be reading these lines.

There is also no safety in ignoring such outlandish things as nonstandard analysis, superspace and anticommuting integration, p-adic and ultrametric space. All three have applications in both electrical engineering and physics. Once, complex numbers were equally outlandish, but they frequently proved the shortest path between 'real' results. Similarly, the first two topics named have already provided a number of 'wormhole' paths. There is no telling where all this is leading - fortunately.

Thus the original scope of the series, which for various (sound) reasons now comprises five subseries: white (Japan), yellow (China), red (USSR), blue (Eastern Europe), and green (everything else), still applies. It has been enlarged a bit to include books treating of the tools from one subdiscipline which are used in others. Thus the series still aims at books dealing with:

- a central concept which plays an important role in several different mathematical and/or scientific specialization areas;
- new applications of the results and ideas from one area of scientific endeavour into another;
- influences which the results, problems and concepts of one field of enquiry have, and have had, on the development of another.

A martingale is first of all a kind of restraining strap, occasionally used as part of a horse harness, especially for rather wild ones, to prevent them rearing or throwing their heads back. There is a second nautical meaning of similar flavour, and finally it is a system in gambling which consists in doubling the stake when losing: 'You have not played as yet? Do not do so; above all avoid a martingale if you do' (Thackeray; courtesy of the Shorter Oxford English Dictionary).

In mathematics, a martingale is a random process $\{x(s)\}$ such that $x(s)=[x(t)\,|\,\mathcal{F}(s)]$ almos surely for $s<t$. A martingale represents a mathematical model for the fortune of a player of a fai game without built-in tendencies to lose or win: if the player has fortune $x(s)$ at time s then giver the past up to and including the present he expects his fortune at a later time t to be $x(s)$ again.

The theory of martingales was established by J.L. Doob and apparently it is the first meaning o the word martingale given above (rather than the one with the gambling connotation), whicl inspired the name because of the strong interrelations between the random variables $x(s)$ makin up the process, which the martingale property implies.

At least after the fact of their discovery (or invention, depending on one's ideological back ground) martingales turn out to be such a natural, powerful, and unifying concept and tool that on wonders why they did not appear earlier; one is, so to speak, inevitably led to them by the study o conditional expectations, and they provide a natural unified method for dealing with limit theorem It is thus no strain at all to understand why they so quickly became an indispensable and centr concept in stochastic processes and stochastic calculus.

Thus it is a real pleasure to welcome in this series such a comprehensive modern treatise on th subject of martingale theory and its natural generalizations and applications; a treatise, moreove written by one of the best writing teams in probability.

The shortest path between two truths in the real domain passes through the complex domain.

J. Hadamard

La physique ne nous donne pas seulement l'occasion de résoudre des problèmes ... elle nous fait pressentir la solution.

H. Poincaré

Never lend books, for no one ever retur them; the only books I have in my libra are books that other folk have lent me.

Anatole Fran

The function of an expert is not to be m right than other people, but to be wrong more sophisticated reasons.

David But

Bussum, March 1989

Michiel Hazewink

TABLE OF CONTENTS

PART II

PREFACE

In our earlier monograph[1] "Statistics of random processes", published in 1974, the *martingale theory* has been represented presumably by the square integrable case. Correspondingly, stochastic integration theory has also been described for square integrable martingales only.

During the last decade it became clear that, though the case of square integrable martingales is rather important, it does not meet all demands of intrinsic developments of the martingale theory and its applications. It became apparent that the basic concepts of the martingale theory are the notions of a local martingale and a semimartingale - a stochastic process representable as the sum of a local martingale and a process of locally bounded variation. Stochastic integration theory has been developed first relative to local martingales and then to semimartingales. At the same time the very construction of the stochastic integral $H \cdot X \equiv \int H\, dX$ with respect to a semimartingale X revealed the essence and importance of the predictability concept, and of predictability conditions imposed on an integrand process H.

A new branch of the martingale theory has also been developed, based on the concepts of an integer-valued random measure μ (in particular the jump measure of a process), its compensator ν and a stochastic integral with respect to a "martingale" measure $\mu - \nu$.

In the present monograph we intend to give a sufficiently complete exposition of the *modern theory of martingales with continuous time,* including stochastic calculus with respect to local martingales, semimartingales and "martingale" measures.

A number of necessary facts and assertions of the *general theory of stochastic processes,* already presented in books, are given without proof. This concerns mainly the first chapter.

Semimartingales present a rather wide class of processes which are invariant relative to such operations as a continuous change of a measure, random change of time, restriction of a flow of σ-algebras, etc. Each process with discrete time is a semimartingale. Ito processes, diffusion processes, point (counting) processes belong to the class of semimartingales too.

A *canonical representation* of a semimartingale and a notion of a *triplet* of its predictable characteristics are the important and useful "working" concepts. In the case in which a semimartingale is a process with independent increments, its triplet turns out

[1] See [188-190]. The Russian original of the referred monograph, as well as of the present book, is included in the series of monographs on Probability Theory and Mathematical Statistics by the publishers "Nauka", Moscow (S. D.).

to be nonrandom. In this sense semimartingales present a natural extension of the class of processes with independent increments.

The first part of the monograph (Chapters 1 - 4) consists of the material of the *martingale theory* which concerns the concepts mentioned above and includes the stochastic calculus. The second and third parts (Chapters 5 - 10) may be regarded as an illustrative application of the notions and methods of the martingale theory to limit theorems of type $X^n \to X$, $n \to \infty$, where X^n is a semimartingale and X a sufficiently simply constructed semimartingale (say, a process with independent increments or with conditionally independent increments). The convergence is understood as the weak convergence of finite dimensional distributions or as the weak convergence of measures in Skorohod's space. The method for proving the weak convergence of finite dimensional distributions which we call the *method of stochastic exponentials,* is a natural extension of the method of characteristic functions to the case of dependent variables. It should be emphasized that the verification of the weak convergence of finite dimensional distributions of semimartingales to that of, say, processes with independent increments is reduced by the method of stochastic exponentials to the verification of laws of large numbers for certain functionals defined by the triplets of predictable characteristics of prelimiting and limiting processes.

This method of proving limit theorems, such as the central limit theorem for dependent variables , called often a *martingale method* (because of the use of notions and results of the martingale theory), does not use directly the considerations of weak dependence type, or the reduction to the case of independent random variables. Its essence, as is mentioned above, consists in the reduction to the verification of certain laws of large numbers.

Since processes with discrete time are semimartingales, our "semimartingale scheme" includes also the classical "scheme of series". The reader will see that the martingale methods used lead to new results even for the scheme of series of independent random variables or random variables generating martingale differences.

The reader may get an idea of the concreteness of the presented material by consulting the table of contents. In the historic-bibliographical notes information can be found concerning sources of the various results, as well as additional references concerning the subjects discussed.

In each chapter an autonomous numeration system is used; results referred to from other chapters are indicated by first the chapter number, then the section number, and finally the number of the corresponding assertion. Within each section a single number is used, within a chapter the section number is followed by the number of the assertion.

Each section is followed by problems of various nature. Sometimes it is suggested to prove assertions which are presented without proof in the main text, or to prove certain useful and interesting assertions which enrich the exposition.

The authors express their gratitude to S.E. Kuznetsov who has read the whole manuscript and made essential remarks on the style of the exposition and on the formulation and proof of a number of propositions.

PART I

CHAPTER 1

BASIC CONCEPTS AND THE REVIEW OF RESULTS OF «THE GENERAL THEORY OF STOCHASTIC PROCESSES»

§ 1. Stochastic basis. Random times, sets and processes

1. The considerations of the general theory of stochastic processes relies upon the concept of the stochastic basis.

Definition 1. A complete probability space $(\Omega, \mathcal{F}, \mathsf{P})$, equipped with a nondecreasing family of σ-algebras $\mathbb{F} = (\mathcal{F}_t)_{t \geq 0}$, $\mathcal{F}_s \subseteq \mathcal{F}_t \subseteq \mathcal{F}$, $s \leq t$, which satisfies the ("usual") conditions:
(a) right continuity ($\mathcal{F}_t = \mathcal{F}_{t+}$ with $\mathcal{F}_{t+} = \bigcap_{u > t} \mathcal{F}_u$, $t \geq 0$),
(b) completeness (i.e. $\mathcal{F}_0$ is augmented by the sets from $\mathcal{F}$ of P-null probability) is called a *stochastic basis.*

A stochastic basis introduced this way will be denoted by $(\Omega, \mathcal{F}, \mathbb{F}, \mathsf{P})$ or $(\Omega, \mathcal{F}, (\mathcal{F}_t)_{t \geq 0}, \mathsf{P})$. A family $\mathbb{F} = (\mathcal{F}_t)_{t \geq 0}$ is often called a *flow of σ-algebras* or a *filtration.* By $\mathcal{F}_\infty$ we denote a σ-algebra $\bigvee_t \mathcal{F}_t \equiv \sigma(\bigcup_{t \geq 0} \mathcal{F}_t)$; we suppose also that

$$\mathcal{F}_{t-} = \bigvee_{s < t} \mathcal{F}_s,\ t > 0,\ \mathcal{F}_{0-} = \mathcal{F}_0,\ \mathbb{F}_- = (\mathcal{F}_{t-})_{t \geq 0}.$$

In the general theory of stochastic processes one often encounters a probability space $(\Omega, \mathcal{F}, \mathsf{P})$ which is not necessarily complete and a flow of σ-algebras $\mathbb{F} = (\mathcal{F})_{t \geq 0}$ which not necessarily satisfies the condition of right continuity. In this case a stochastic basis $(\Omega, \mathcal{F}^{\mathsf{P}}, \mathbb{F}^{\mathsf{P}}_{t+}, \mathsf{P})$ is constructed in the following way: $\mathcal{F}^{\mathsf{P}}$ is the completion of $\mathcal{F}$ with respect to the measure P, while

$$\mathcal{F}^{\mathsf{P}}_{t+} = \bigcap_{\varepsilon > 0} (\mathcal{F}_{t+\varepsilon} \vee \mathcal{N})$$

where $\mathcal{N}$ is the system of sets from $\mathcal{F}^{\mathsf{P}}$ of P-measure zero.

2. Definition 2. A random variable $\tau = \tau(\omega)$ given on $(\Omega, \mathcal{F})$ and taking values in $\bar{R}_+ = [0,\infty]$ is called a *random time*. A random time $\tau = \tau(\omega)$ is called a *Markov time* if for each $t \geq 0$

$$\{\omega: \tau(\omega) \leq t\} \in \mathcal{F}_t. \tag{1.1}$$

In case $P(\tau < \infty) = 1$ a Markov time τ is called a *stopping time*.

Further on we will assume that $(\Omega, \mathcal{F}, \mathbb{F}, P)$ is a stochastic basis.

The class of all Markov times will be denoted by $\mathbf{T}$ or $\mathbf{T}(\mathbb{F})$. The assumption of the right continuity of the family $\mathbb{F}$ made above allows us to assert that $\tau \in T(\mathbb{F})$ if and only if for each $t > 0$

$$\{\omega: \tau(\omega) < t\} \in \mathcal{F}_t.$$

If $\tau \in \mathbf{T}(\mathbb{F})$, then we set

$$\mathcal{F}_\tau = \{A \in \mathcal{F} : A \cap \{\tau \leq t\} \in \mathcal{F}_t, \forall t \geq 0\}.$$

It is easily verified that $\mathcal{F}_\tau$ is a σ-algebra, and if $\tau(\omega) \equiv t$ with $t \in R_+ = [0,\infty]$, then $\mathcal{F}_\tau = \mathcal{F}_t$.

If $\tau \in \mathbf{T}(\mathbb{F})$, then by $\mathcal{F}_{\tau-}$ we denote the σ-algebra generated by sets from $\mathcal{F}_0$ and by sets of type $A \cap \{t < \tau\}$ with $A \in \mathcal{F}_t$ and $t > 0$.

A vivid interpretation of the σ-algebras $\mathcal{F}_t$, $\mathcal{F}_\tau$ and $\mathcal{F}_{\tau-}$ is presented as follows: $\mathcal{F}_t$ is the totality of all the events observed before or at time t; $\mathcal{F}_\tau$ before or at time τ; $\mathcal{F}_{\tau-}$ strictly before time τ.

The following properties of Markov times are directly verified:

(1) $\tau \in \mathbf{T} \Rightarrow \mathcal{F}_{\tau-} \subseteq \mathcal{F}_\tau$, τ is $\mathcal{F}_{\tau-}$-measurable;

(2) $\tau \in \mathbf{T} \Rightarrow \tau + t \in \mathbf{T}$, $\forall t \in R_+$;

(3) $\tau, \sigma \in \mathbf{T}$, $\tau(\omega) \leq \sigma(\omega) \Rightarrow \mathcal{F}_{\tau-} \subseteq \mathcal{F}_{\sigma-}$, $\mathcal{F}_\tau \subseteq \mathcal{F}_\sigma$;

(4) $\tau_n \in \mathbf{T}$, $n \geq 1 \Rightarrow \sigma \equiv \inf \tau_n \in T$, $\tau \equiv \sup \tau_n \in \mathbf{T}$, $\mathcal{F}_\sigma = \cap \mathcal{F}_{\tau_n}$;

(5) if $A \in \mathcal{F}_\sigma$, then $A \cap \{\sigma < \tau\} \in \mathcal{F}_{\tau-}$;

(6) if $A \in \mathcal{F}_\infty$, $\tau \in \mathbf{T}$, then $A \cap \{\tau = \infty\} \in \mathcal{F}_{\tau-}$.

3. Definition 3. Each subset of a set

$$\Omega \times R_+ = \{(\omega, t): \omega \in \Omega, t \in R_+\}$$

is called a *random set*.

Let

$$\pi_A = \{\omega: \exists t \in R_+ \text{ such that } (\omega, t) \in A\} \in \mathcal{F}.$$

If $P(\pi_A) = 0$, then it is said that a set A is *evanescent* (or P-*negligible*).

The important examples of random sets are presented by the random intervals

$$[\![\sigma, \tau]\!] = \{(\omega, t): \sigma(\omega) \le t \le \tau(\omega)\},$$
$$[\![\sigma, \tau[\![= \{(\omega, t): \sigma(\omega) \le t < \tau(\omega)\},$$
$$]\!]\sigma, \tau]\!] = \{(\omega, t): \sigma(\omega) < t \le \tau(\omega)\},$$
$$]\!]\sigma, \tau[\![= \{(\omega, t): \sigma(\omega) < t < \tau(\omega)\},$$

which are generated by the random times σ and τ.

The stochastic interval $[\![\tau, \tau]\!] = \{(\omega, t): \tau(\omega) = t\}$ is denoted also by $[\![\tau]\!]$, and is called the *graph* of a random time τ.

Definition 4. A random set A is called *thin* if it has a form $A = \bigcup_n [\![\tau_n]\!]$, where $(\tau_n)_{n \ge 1}$ is a sequence of random times. If $[\![\tau_i]\!] \cap [\![\tau_j]\!] = \emptyset$, $i \ne j$, then $(\tau_n)_{n \ge 1}$ is called a sequence *exhausting* a random set A.

Definition 5. For random set $A \subseteq \Omega \times R_+$ the function

$$D_A = \begin{cases} \inf\{t: (\omega,t) \in A\}, & \text{if } \{t: (\omega,t) \in A\} \ne \emptyset, \\ \infty, & \text{if } \{t: (\omega, t) \in A\} = \emptyset, \end{cases}$$

is called a *debut of the set* A.

4. Let $X = (X_t)_{t \ge 0}$ be a *stochastic process*, that is, a family of random variables $X_t = X_t(\omega)$ defined on $(\Omega, \mathcal{F})$. If for each $t \ge 0$ the random variables X_t are $\mathcal{F}_t$-measurable, then we use also the notations $X = (X_t, \mathcal{F}_t)_{t \ge 0}$ or $X = (X_t, \mathcal{F}_t)$, saying that the process X is $\mathbb{F}$-*adapted* (with respect to the family $\mathbb{F} = (\mathcal{F}_t)_{t \ge 0}$) and writing $X \in \mathbb{F}$.

By C and D we will denote the spaces of real-valued functions $x = x(t)$ defined on $R_+ = [0, \infty]$, which are continuous or, respectively, right continuous having left hand limits. Clearly, $\mathsf{C} \subset \mathsf{D}$. We write $X \in \mathsf{C}$ or $X \in \mathsf{D}$, if for each $\omega \in \Omega$ the trajectories $(X_t(\omega))_{t \ge 0}$ as functions of $t \in R_+$ belong to C or D respectively.

For each process $X \in \mathsf{D}$, one can define the processes

$$X_- = (X_{t-})_{t \ge 0} \quad \text{and} \quad \Delta X = (\Delta X_t)_{t \ge 0}$$

with $X_{0-} = X_0$, $X_{t-} = \lim_{s \uparrow t} X_s$ for $t > 0$ and $\Delta X_t = X_t - X_{t-}$, also

$$X^* = (X_t^*)_{t \ge 0} \text{ with } X_t^* = \sup_{s \le t} |X_s|,$$

$$(\Delta X)^* = ((\Delta X)_t^*)_{t \ge 0} \text{ with } (\Delta X)_t^* = \sup_{s \le t} |\Delta X_s|,$$

and, finally,

$$\sum_s (\Delta X_s)^2 = \left(\sum_{s \le t} (\Delta X_s)^2 \right)_{t \ge 0} .$$

The indicator of a set A we denote by I_A (or I (A)).

For a random time τ, we set

$$\Delta X_\tau = X_\tau I_{\{\tau < \infty\}} - X_{\tau -} I_{\{\tau < \infty\}} .$$

The "stopped" (in time τ) process $(X_{t \wedge \tau})_{t \ge 0}$ is denoted by X^τ.

5. Let $(\Omega, \mathcal{F}, \mathbb{F}, \mathsf{P})$ be a stochastic basis, and B (R), B (R_+), B ([0, t])... be Borel σ-algebras on R, R_+, [0, t]...

Definition 6. A random process $X = (X_t)_{t \ge 0}$ is called *measurable* if the function X (ω, t), defined on $\Omega \times R_+$ by the equality X (ω, t) = $X_t(\omega)$, is $\mathcal{F} \times B\ (R_+)$-measurable, i.e.

$$\{(\omega, t): X(\omega, t) \in A\} \in \mathcal{F} \otimes B\ (R_+) ,\ \forall\ A \in B\ (R).$$

A measurable adapted stochastic process X is called *progressively measurable* if, for each $t \in R_+$ the function X (ω, t), defined on $\Omega \times [0, t]$, is $\mathcal{F}_t \times B\ ([0, t])$-measurable.

A random set $A \in \Omega \times R_+$ is called *measurable*, $\mathbb{F}$*-adapted* or *progressively measurable* if the process $X = I_A$ is measurable, $\mathbb{F}$-adapted or, respectively, progressively measurable.

Two stochastic processes $X = (X_t)_{t \ge 0}$ and $Y = (Y_t)_{t \ge 0}$ are called *stochastically equivalent* if $\mathsf{P}(X_t \ne Y_t) = 0$ for each $t > 0$. (A process Y, stochastically equivalent to a process X, is often called a *modification* or *version* of X.)

Processes X and Y are called *indistinguishable* (or P-indistinguishable), if the random set A = {(ω, t): X $\ne$ Y} is evanescent, i. e. $\mathsf{P}(\pi_A) = 0$. A process X, indistinguishable from a process identically equal to zero, is called *negligible* or *P-negligible*. Further on the expression X = Y or $X \overset{\mathsf{P}}{=} Y$ will indicate that processes X and Y are indistinguishable (with respect to the measure P).

6. Theorem 1. *Let a process* $X \in \mathsf{D} \cap \mathbb{F}$. *Then the random set* $\{\Delta X \ne 0\}$ *is a thin set* : $\{\Delta X \ne 0\} = \cup [\![\tau_k]\!]$ *where* $(\tau_k)_{k \ge 1}$ *is a sequence of Markov times.*

Proof. Define inductively a sequence of random times $(\sigma(n; p))_{n \ge 1,\, p \ge 0}$ by setting $\sigma(n; 0) = 0$ and

$$\sigma(n; p + 1) = \inf \{t > \sigma(n; p): |\Delta X_t| > 2^{-n}\}.$$

It is easily seen that $\sigma(n; p)$, $p \ge 1$ is the debut of the set

$$]\!]\, \sigma(n; p - 1), \infty [\![\cap \{ |\Delta X| > 2^{-n} \}.$$

If $\sigma(n; p-1)$ is a Markov time, then the random set under consideration is progressively measurable (and, in terms of the next section, even optional). Consequently, in virtue of Problem 3 $\sigma(n; p)$ is a Markov time. Next, since $X \in D$, on a finite time interval (for each $\omega \in \Omega$) there exists at most a finite number of jumps that excede 2^{-n} ($|\Delta X| > 2^{-n}$). Hence

$$\lim_{p} \sigma(n; p) = \infty,$$

and this means that

$$\{\Delta X \neq 0\} = \bigcup_{n,\ p} [\![\sigma(n; p)]\!].$$

Problems

1. Show that each left (right) continuous adapted stochastic process is progressively measurable.

2. Let $\tau \in \mathbf{T}$, $A \in \mathcal{F}_\tau$ and

$$\tau_A(\omega) = \begin{cases} \tau(\omega), & \text{if } \omega \in A. \\ \infty, & \text{if } \omega \notin A. \end{cases}$$

Show that $\tau_A \in \mathbf{T}$.

3. Show that under the "usual conditions" debut D_A of each progressively measurable set A is a Markov time.

4. Show that if $\tau \in \mathbf{T}(\mathbb{F})$, then $\tau \in \mathbf{T}(\mathbb{F}_+)$, where $\mathbb{F}_+ = (\mathcal{F}_{t+})_{t \geq 0}$.

5. Prove that $\tau \in \mathbf{T}(\mathbb{F}_+)$ if and only if $\{\tau < t\} \in \mathcal{F}_t$ for each $t \geq 0$.

6. Let $\sigma, \tau \in \mathbf{T}$. Show that $\{\sigma = \tau\}$, $\{\sigma \leq \tau\} \in \mathcal{F}_\tau \cap \mathcal{F}_\sigma$, $\{\sigma < \tau\} \in \mathcal{F}_{\tau-}$, $\{\sigma < \infty\} \in \mathcal{F}_{\sigma-}$, $A \cap \{\tau \leq t\} \in \mathcal{F}_t$ for each $A \in \mathcal{F}_\tau$.

7. Let X be a progressively measurable stochastic process, and $\tau \in \mathbf{T}$. Show that $I_{\{\tau < \infty\}} X_\tau$ is a $\mathcal{F}_\tau$-measurable random variable.

§ 2. Optional and predictable σ–algebras of random sets

1. Consider a stochastic basis $(\Omega, \mathcal{F}, \mathbb{F}, P)$. In the space of random sets $\Omega \times R_+$ several σ-algebras are singled out in a natural way which play an important rôle in the general theory of stochastic processes. One of them is the σ-algebra $\mathcal{F} \otimes B(R_+)$, called the σ-algebra of measurable (random) sets. Two other σ-algebras $\mathfrak{O} = \mathfrak{O}(\mathbb{F})$ and $\mathcal{P} = \mathcal{P}(\mathbb{F})$ of optional and, respectively, predictable sets are defined as follows:

Definition 1. A σ-algebra $\mathfrak{O} = \mathfrak{O}(\mathbb{F})$ of *optional* (or well measurable) *sets* is the smallest σ-algebra, generated by the random sets $[\![0, \tau[\![$ with $\tau \in \mathbf{T}(\mathbb{F})$.

Definition 2. A σ-algebra $\mathcal{P} = \mathcal{P}(\mathbb{F})$ of *predictable sets* is the smallest σ-algebra, generated by random sets $[\![0, \tau]\!]$ with $\tau \in \mathbf{T}(\mathbb{F})$, and $[\![0_A]\!] = \{(\omega,t): \omega \in A, t = 0\}$ where $A \in \mathcal{F}_0$.

Since for each $\tau \in \mathbf{T}$ also $\tau + \frac{1}{n} \in \mathbf{T}$ and $[\![0, \tau]\!] = \bigcap_n [\![0, \tau + \frac{1}{n}[\![$, we have

$$\mathcal{P}(\mathbb{F}) \subseteq \mathfrak{O}(\mathbb{F}). \tag{2.1}$$

Note that in general the inclusion is strict here (Problem 1).

Definition 3. A random process $X = (X_t)_{t \geq 0}$ is called *optional* ($X \in \mathfrak{O}$) or *predictable* ($X \in \mathcal{P}$) if the function $X(\omega, t)$, defined on $\Omega \times R_+$ by the equality $X(\omega, t) = X_t(\omega)$ is $\mathfrak{O}$-measurable or $\mathcal{P}$-measurable respectively.

Another characterization of the σ-algebras $\mathfrak{O}(\mathbb{F})$ and $\mathcal{P}(\mathbb{F})$ is given in the following theorems.

Theorem 1. *An optional σ-algebra $\mathfrak{O}(\mathbb{F})$ coincides with the smallest σ-algebra of random sets $\mathfrak{O}_1(\mathbb{F})$, generated by $\mathbb{F}$-adapted processes taking values in* D.

Theorem 2. *A predictable σ-algebra $\mathcal{P}(\mathbb{F})$ coincides with the smallest σ-algebra of random sets $\mathcal{P}_1(\mathbb{F})$, generated by $\mathbb{F}$-adapted processes taking values in* C.

2. Proof of Theorem 1. By definition $\mathfrak{O}_1(\mathbb{F})$ is the smallest σ-algebra, generated by sets of type $A = \{(\omega,t): Y(\omega,t) \in B\}$ with $B \in B(R)$, while the functions $Y = Y(\omega, t)$, $\omega \in \Omega$, $t \geq 0$ are such that for each $\omega \in \Omega$ the trajectories $(Y(\omega, t))_{t \geq 0} \in$ D, and $Y(\omega, t)$ is $\mathcal{F}_t$-measurable for each $t \geq 0$.

Let $\tau \in \mathbf{T}(\mathbb{F})$. Then the process $Y = I_{[\![0, \tau[\![}$ is an $\mathbb{F}$-adapted process with trajectories in D, and hence $\mathfrak{O}(\mathbb{F}) \subseteq \mathfrak{O}_1(\mathbb{F})$.

Let us prove the reverse inclusion $\mathfrak{D}_1(\mathbb{F}) \subseteq \mathfrak{D}(\mathbb{F})$. To this end consider first a random process $Y = I_H I_{[\![\sigma, \tau[\![}$ with $\sigma, \tau \in \mathbf{T}(\mathbb{F})$, $\sigma \le \tau$, and a set $H \in \mathfrak{F}_\sigma$.

According to Problem 1.2 σ_H and τ_H are Markov times, $[\![\sigma_H, \tau_H[\![= [\![0, \tau_H[\![\setminus [\![0, \sigma_H[\![\in \mathfrak{D}(\mathbb{F})$, and since $Y = I_H I_{[\![\sigma, \tau[\![} = I_{[\![\sigma_H, \tau_H[\![}$, then Y is $\mathfrak{D}(\mathbb{F})$-measurable. It follows from this that any process Y of type $Y = \eta I_{[\![\sigma, \tau[\![}$ where η is a $\mathfrak{F}_\sigma$-measurable random variable and $\sigma, \tau \in \mathbf{T}$, is $\mathfrak{D}(\mathbb{F})$-measurable too. Hence, to prove the inclusion $\mathfrak{D}_1(\mathbb{F}) \subseteq \mathfrak{D}(\mathbb{F})$ we have to show that any process $X \in D \cap \mathbb{F}$ is a pointwise limit of a sequence of processes $(X^n)_{n \ge 1}$ each of which is nothing more then a countable linear combination of processes Y of type $Y = \eta I_{[\![\sigma, \tau[\![}$ considered above.

Thus, let $X = (X_t(\omega))\, D \cap \mathbb{F}$ and $\varepsilon > 0$. Define a sequence $(\tau_n^\varepsilon)_{n \ge 1}$ inductively by setting

$$\tau_1^\varepsilon(\omega) = 0, \dots, \tau_{n+1}^\varepsilon(\omega) = \inf\{t > \tau_n^\varepsilon(\omega) : |X_t(\omega) - X_{\tau_n^\varepsilon(\omega)}(\omega)| > \varepsilon\}, \dots$$

and assuming $\inf \varnothing = \infty$.

Let us show that τ_{n+1}^ε are Markov times. The proof will be carried out by induction. Suppose that $\tau_n^\varepsilon \in \mathbf{T}$, and note that τ_{n+1}^ε is the debut $D_{A_n^\varepsilon}$ of the set

$$A_n^\varepsilon = \{(\omega, t) : |Y_n^\varepsilon| > \varepsilon\}$$

with

$$Y_n^\varepsilon = I_{]\!]\tau_n^\varepsilon, \infty[\![}\,[X - I_{\{\tau_n^\varepsilon < \infty\}} X_{\tau_n^\varepsilon}], \tag{2.2}$$

since

$$\{(\omega, t) : t > \tau_n^\varepsilon(\omega), |X_t(\omega) - X_{\tau_n^\varepsilon(\omega)}| > \varepsilon\} = A_n^\varepsilon.$$

A process X from $D \cap \mathfrak{F}$ is progressively measurable. Therefore $X_{\tau_n^\varepsilon}$ is $\mathfrak{F}_{\tau_n^\varepsilon}$–measurable (Problem 2). It follows from this that the process $I_{]\!]\tau_n^\varepsilon, \infty]\!]} I_{\{\tau_n^\varepsilon < \infty\}} X_{\tau_n^\varepsilon}$ is progressively measurable, and hence the process Y_n^ε is progressively measurable too, by (2.2). It follows then from Problem 1.3 that time τ_{n+1}^ε which coincides with the debut $D_{A_n^\varepsilon}$, is a Markov time.

The right continuity of a process X yields

$$|X_{t^\varepsilon_{n+1}(\omega)}(\omega) - X_{\tau^\varepsilon_n(\omega)}(\omega)| \geq \varepsilon, \quad \omega \in \{\tau^\varepsilon_{n+1}(\omega) < \infty\} \tag{2.3}$$

and

$$|X_t(\omega) - X_{\tau^\varepsilon_n(\omega)}(\omega)| < \varepsilon, \quad (t, \omega) \in [\![\tau^\varepsilon_n(\omega), \tau^\varepsilon_{n+1}(\omega)[\![. \tag{2.4}$$

Denote

$$\tau^\varepsilon(\omega) = \sup_n \tau^\varepsilon_n(\omega)$$

and consider the set $\{\omega: \tau^\varepsilon(\omega) < \infty\}$. Then, due to (2.3) on this set the limit $\lim_{t \uparrow \tau^\varepsilon(\omega)} X_t(\omega)$ does not exist. As the left-hand limits of a process X exist, this yields the equality $\tau^\varepsilon(\omega) = \infty$ for each $\omega \in \Omega$. Thus the process $X^{(\varepsilon)} = (X^{(\varepsilon)}_t(\omega))$ is defined where

$$X^{(\varepsilon)}_t(\omega) = \sum_{n \geq 1} X_{\tau^\varepsilon_n(\omega)}(\omega)\, I_{[\![\tau^\varepsilon_n(\omega),\, \tau^\varepsilon_{n+1}(\omega)[\![}(t).$$

By (2.4) the process $X^{(\varepsilon)}$ converges uniformly to the process X as $\varepsilon \downarrow 0$. Therefore as for the desired sequence $(X^n)_{n \geq 1}$ one can take, for instance, the sequence with $X^n = X^{(1/n)}$.

3. Proof of Theorem 2. Along with the σ-algebras $\mathcal{P}(\mathbb{F})$ and $\mathcal{P}_1(\mathbb{F})$ we introduce the following σ-algebras of random sets that are useful for proving the equality $\mathcal{P}(\mathbb{F}) = \mathcal{P}_1(\mathbb{F})$, as well as for alternative descriptions of the system of predictable sets:

$\mathcal{P}_2 = \mathcal{P}_2(\mathbb{F})$ is the σ-algebra generated by $\mathbb{F}$-adapted processes that are left-hand continuous on $[0, \infty)$;

$\mathcal{P}_3 = \mathcal{P}_3(\mathbb{F})$ is the σ-algebra generated by sets of type $A \times (s, t]$, $A \in \mathcal{F}_{s-}$, $s \leq t$, and $A \times \{0\}$, $A \in \mathcal{F}_0$;

$\mathcal{P}_4 = \mathcal{P}_4(\mathcal{F})$ is the σ-algebra generated by sets of type $A \times (s, t]$, $A \in \mathcal{F}_{s-}$, $s \leq t$, and $A \times \{0\}$, $A \in \mathcal{F}_0$;

$\mathcal{P}_5 = \mathcal{P}_5(\mathbb{F})$ is the σ-algebra generated by sets of type $A \times [s, t)$, $A \in \mathcal{F}_{s-}$, $s \leq t$;

$\mathcal{P}_6 = \mathcal{P}_6(\mathbb{F})$ is the σ-algebra generated by sets of type $]\!]\sigma, \tau]\!]$ with $\sigma, \tau \in \mathbf{T}(\mathbb{F})$, and of type $A \times \{0\}$, $A \in \mathcal{F}_0$.

Clearly $\mathcal{P}_1 \subseteq \mathcal{P}_2$, $\mathcal{P}_3 \subseteq \mathcal{P}_2$. Let us show that $\mathcal{P}_2 \subseteq \mathcal{P}_3$.

Let $X \in \mathcal{P}_2$ (i.e. a process X is measurable with respect to the σ-algebra $\mathcal{P}_2$) and

$$X^n = X_0 I_{[\![0]\!]} + \sum_{k=1}^{\infty} X_{\frac{k}{2^n}} I_{]\!]\frac{k}{2^n}, \frac{k+1}{2^n}]\!]}$$

Since the variables $X_{k/2^n}$ are $\mathcal{F}_{k/2^n}$-measurable, $X^n \in \mathcal{P}_3$. Next, it can be assumed without loss of generality that the process X is left-hand continuous and hence $X^n \to X$ pointwise on $\Omega \times R_+$ as $n \to \infty$. Thus $X \in \mathcal{P}_3$, so that $\mathcal{P}_2 = \mathcal{P}_3$.

Let us show now that $\mathcal{P}_3 \subseteq \mathcal{P}_1$.

Take a set $A \times (s, t]$, $A \in \mathcal{F}_s$, $s \le t$. There can always be found a sequence of continuous functions $\phi_n = \phi_n(u)$, $u \in R_+$ such that $\phi_n(u) = 0$ for $u < s$ and $I_{(s, t]}(u) = \lim_n \phi_n(u)$. Then each of the processes $I_A\phi_n$ is $\mathbb{F}$-adapted and continuous, and hence $I_A\phi_n \in \mathcal{P}_1$. Analogous considerations are applied to a set $A \times \{0\}$, $A \in \mathcal{F}_0$ too. Hence $\mathcal{P}_1 = \mathcal{P}_2 = \mathcal{P}_3$.

Let us verify now the equalities $\mathcal{P}_3 = \mathcal{P}_4 = \mathcal{P}_5$. To this end note that $\mathcal{P}_4$ and $\mathcal{P}_5$ are generated by sets of type

$$A \times (s, t], \quad A \in \bigcup_{r<s} \mathcal{F}_r, \quad s \le t, \quad \text{and} \quad A \times \{0\}, \quad A \in \mathcal{F}_0,$$

and, respectively

$$A \times [s, t), \quad A \in \bigcup_{r<s} \mathcal{F}_r, \quad s \le t, \quad \text{and} \quad A \times \{0\}, \quad A \in \mathcal{F}_0.$$

Clearly

$$I_{(s, t]} = \lim_n I_{[s + \frac{1}{n}, t + \frac{1}{n})} \tag{2.5}$$

and

$$I_{[s, t)} = \lim_n I_{(s - \frac{1}{n}, t - \frac{1}{n}]}. \tag{2.6}$$

This gives $\mathcal{P}_4 = \mathcal{P}_5$. Clearly $\mathcal{P}_4 \subseteq \mathcal{P}_3$ by the definitions, and by (2.5) we have $\mathcal{P}_3 \subseteq \mathcal{P}_4$. Hence $\mathcal{P}_3 = \mathcal{P}_4 = \mathcal{P}_5$.

Next, $\mathcal{P} \subseteq \mathcal{P}_2$ evidently. Let us show that $\mathcal{P}_3 \subseteq \mathcal{P}$. In fact, if $A \in \mathcal{F}_s$ and $s \le t$, then $A \times (s, t] =]\!]s_A, t_A]\!] = [\![0, t_A]\!] \setminus [\![0, s_A]\!] \in \mathcal{P}$, since the times $s_A, t_A \in \mathbf{T}$. Hence $\mathcal{P} = \mathcal{P}_2 = \mathcal{P}_3$ and $\mathcal{P} = \mathcal{P}_1 = \mathcal{P}_2 = \mathcal{P}_3 = \mathcal{P}_4 = \mathcal{P}_5$. Finally, since $]\!]\sigma, \tau]\!] = [\![0, \tau]\!] \setminus [\![0, \sigma]\!]$, then $\mathcal{P}_6 \subseteq \mathcal{P}$, and since $\mathcal{P}_3 \subseteq \mathcal{P}_6$ evidently, then $\mathcal{P} = \mathcal{P}_k$ for each $k = 1, \ldots, 6$. This proves the theorem.

Problems

1. Construct an example, showing that in general the inclusion in (2.1) is strict.

2. Prove that for each progressively measurable (and, in particular, optional) process X and each $\tau \in \mathbf{T}$ a variable $I_{\{\tau < \infty\}} X_\tau$ is $\mathcal{F}_\tau$-measurable.

3. Show that a function $\tau = \tau(\omega)$ taking values in $\bar{R}_+ = [0, \infty]$ will be a Markov time if and only if the process $I_{[\![0, \tau[\![}$ is optional.

4. Let $\tau \in \mathbf{T}$ and let X be a progressively measurable (optional) process. Then the stopped process $X^\tau = (X_{t \wedge \tau}, \mathcal{F}_t)$ is progressively measurable (optional) too.

5. Show that in absence of the right continuity of a flow $\mathbb{F}$ we have

$$\mathcal{P}(\mathbb{F}) = \mathcal{P}(\mathbb{F}_-) = \mathcal{P}(\mathbb{F}_+).$$

6. If $X \in \mathbf{D} \cap \mathbb{F}$, then a process X_- is predictable.

7. If $\tau \in \mathbf{T}(\mathbb{F})$, then $[\![\tau]\!] \in \mathcal{D}(\mathbb{F})$.

8. Construct an example of a Markov time $\tau \in \mathbf{T}(\mathbb{F})$ for which $[\![\tau]\!] \notin \mathcal{P}(\mathbb{F})$.

9. Show that if $\tau \in \mathbf{T}$ and $A = [\![\tau, \infty[\![$, then $\tau = D_A$.

10. If $\tau = \mathbf{T}(\mathbb{F})$, then $]\!]0, \tau]\!] \in \mathcal{P}(\mathbb{F})$.

11. If $\tau \in \mathbf{T}(\mathbb{F})$, then the stochastic intervals $[\![0, \tau[\![$ and $]\!]0, \tau[\![$ simultaneously present or do not present predictable sets.

12. If $\sigma, \tau \in \mathbf{T}(\mathbb{F})$ and a random variable ξ is $\mathcal{F}_\sigma$-measurable, then the process $\xi I_{]\!]\sigma, \tau]\!]} \in \mathcal{P}$.

§ 3. Predictable and totally inaccesssible random times. Classification of Markov times. Section theorems

1. It will be supposed that a stochastic basis $(\Omega, \mathcal{F}, \mathbb{F}, \mathsf{P})$ is given.

Definition 1. A random time $\tau = \tau(\omega)$, taking values in $[0, \infty]$, is called a *predictable time* if its graph $[\![\tau]\!] \in \mathcal{P}$.

Since $\tau = D_{[\![\tau]\!]}$, i.e. τ is the debut of the set $[\![\tau]\!]$, we see that, according to Problem 1.3 and the fact that any predictable set is progressively measurable, a predictable time is a Markov time. Thus, if the class of predictable times is denoted by $\mathbf{T}_p$, then $\mathbf{T}_p \subseteq \mathbf{T}$.

By this inclusion and by the definition of the σ-algebra of predictable sets (§ 2), it is not difficult to verify that time τ is predictable if and only if any of the stochastic intervals

$$]\!]0, \tau[\![, \ [\![0, \tau[\![, \ [\![\tau, \infty[\![\tag{3.1}$$

is a predictable set.

If time $\tau \in \mathbf{T}$ and $t > 0$, then $\tau + t$ serves as an exemple of a predictable time. In fact $\tau + t \in \mathbf{T}$ and $]\!]\tau + t(1-1/n), \tau + t]\!] \in \mathcal{P}$. Hence

$$[\![\tau + t]\!] = \bigcap_n \,]\!]\tau + t\left(1 - \frac{1}{n}\right), \tau + t]\!] \in \mathcal{P}.$$

The simplest, though important properties of predictable times are given in the following theorem.

Theorem 1.

1) *If* $\sigma, \tau \in \mathbf{T}_p$, *then* $\sigma \wedge \tau \in \mathbf{T}_p$, $\sigma \vee \tau \in \mathbf{T}_p$.

2) *If* $\tau_n \in \mathbf{T}_p$, $n \geq 1$, *then* $\sup \tau_n \in \mathbf{T}_p$.

3) *If* $\tau_n \in \mathbf{T}_p$, $n \geq 1$, $\tau \equiv \inf \tau_n$ *and* $\bigcup_n \{\tau = \tau_n\} = \Omega$, *then* $\tau \in \mathbf{T}_p$.

4) *If* $\tau \in \mathbf{T}$, *then a decreasing sequence* τ_n, $n \geq 1$ *of predictable times can be found such that* $\tau = \inf \tau_n$.

5) *If* $\tau \in \mathbf{T}_p$ *and* $A \in \mathcal{F}_{\tau-}$, *then* $\tau_A \in \mathbf{T}_p$.

6) *Let a set* $A \in \mathcal{P}$. *The debut* D_A *will be a predictable time if and only if the random set* $[\![D_A]\!] \cup A \in \mathcal{P}$.

7) *Let* $\sigma \in \mathbf{T}_p$, $A \in \mathcal{F}_{\sigma-}$ *and* $\tau \in \mathbf{T}$. *Then*

$$A \cap \{\sigma \leq \tau\} \in \mathcal{F}_{\tau-}.$$

Proof. Assertions 1) and 2) follow from the equalities

$$[\![0, \sigma \wedge \tau[\![= [\![0, \sigma[\![\cap [\![0, \tau[\![,$$

$$[\![0, \sigma \vee \tau[\![= [\![0, \sigma[\![\cup [\![0, \tau[\![,$$

$$[\![0, \sup \tau_n[\![= \bigcup_n [\![0, \tau_n[\![.$$

3) By the assumption $\bigcup_n \{\tau = \tau_n\} = \Omega$ we have $[\![0, \tau[\![= \bigcap_n [\![0, \tau_n[\![$, and since $[\![0, \tau_n[\![\in \mathcal{P}$, $n \geq 1$, we have $[\![0, \tau[\![\in \mathcal{P}$ too. Hence $\tau \in \mathbf{T}_p$.

4) To prove this assertion it suffices to assume $\tau_n = \tau + 1/n$ and to observe that $\tau_n \in \mathbf{T}_p$, $n \geq 1$ as it was noted above.

5) To prove this assertion we note first that if $A_n \in \mathcal{F}$, $n \geq 1$, then

$$\tau_{\cup A_n} = \inf_n \tau_{A_n}, \quad \tau_{\cap A_n} = \sup_n \tau_{A_n}.$$

Denote $\mathbf{A} = \{A \in \mathcal{F} : \tau_A \in \mathbf{T}_p\}$.

A system of sets $\mathbf{A}$ is closed with respect to the operation $\bigcap_n$ (as $\sup_n \tau_{A_n} \in \mathbf{T}_p$ by 2) as well as with respect to the operation $\bigcup_n$ (by 3) and by the definition of τ_A).

Next, if $A \in \mathbf{A}$, then $[\![\tau, \tau_A[\![\in \mathcal{P}$. Hence

$$[\![0, \tau_{\bar{A}}[\![= [\![0, \infty[\![\setminus [\![\tau, \tau_A[\![\in \mathcal{P}$$

and consequently time $\tau_{\bar{A}} \in \mathbf{T}_p$. Thus the system $\mathbf{A}$ is a σ-algebra.

The system of sets $\mathcal{F}_{\tau-}$ is a σ-algebra, generated by sets from $\mathcal{F}_0$ and by sets of type $A = B \cap \{t < \tau\}$ with $B \in \mathcal{F}_t$, $t > 0$. If $A \in \mathcal{F}_0$, then $[\![0, \tau_A[\![= [\![0, \tau[\![\cup (\bar{A} \times R_+) \in \mathcal{P}$. If $A = B \cap \{t < \tau\}$ with $B \in \mathcal{F}_t$, then $A \in \mathcal{F}_t$ and (since $\mathcal{P} = \mathcal{P}_3$; cf. the proof of Theorem 2.2)

$$[\![0, \tau_{\bar{A}}[\![= [\![0, \tau[\![\cup (A \times (t, \infty)) \in \mathcal{P},$$

i.e. $\bar{A} \in \mathbf{A}$ and hence $A \in \mathbf{A}$.

6) Since $D_A \in \mathbf{T}$, then $[\![0, D_A]\!] \in \mathcal{P}$. But $[\![D_A]\!] = ([\![D_A]\!] \cup A) \cap [\![0, D_A]\!]$ which proves the desired assertion.

7) Let us represent a set $A \cap \{\sigma \leq \tau\}$ in the following way:

$$A \cap \{\sigma \leq \tau\} = \{\sigma_A \leq \tau < \infty\} \cup [A \cap \{\tau = \infty\}].$$

According to 5) we have $\sigma_A \in \mathbf{T}_p$. This yields the predictability of the process $X = I_{[\![\sigma_A, \infty[\![}$ and hence $\{\sigma_A \le \tau < \infty\} = \{X_\tau I_{\{\tau < \infty\}} = 1\} \in \mathcal{F}_{\tau-}$ (cf. Theorem 3 below). Let us verify now that $A \cap \{\tau = \infty\} \in \mathcal{F}_{\tau-}$. Since $A \in \mathcal{F}_\infty = \bigvee_t \mathcal{F}_t$, it suffices to show only that $A \cap \{\tau = \infty\} \in \mathcal{F}_{\tau-}$ for $A \in \mathcal{F}_t$. But

$$A \cap \{t = \infty\} = A \cap \{t < \tau\} \cap \{\tau = \infty\} \in \mathcal{F}_{\tau-}$$

as $A \cap \{t < \tau\} \in \mathcal{F}_{\tau-}$ (by the definition of $\mathcal{F}_{\tau-}$), and $\{\tau = \infty\} \in \mathcal{F}_{\tau-}$ as τ is $\mathcal{F}_{\tau-}$-measurable.

2. The following results present easy corrolaries of the preceding theorem.

Theorem 2. *Let* $\sigma, \tau \in \mathbf{T}$ *and let* ξ *be a* $\mathcal{F}$*-measurable variable. Then*

1) *if* $\tau \in \mathbf{T}_p$, *and* ξ *is* $\mathcal{F}_\sigma$*-measurable, then the process* $\xi I_{]\!]\sigma, \tau[\![}$ *is predictable;*

2) *if* $\sigma \in \mathbf{T}_p$, *and* ξ is $\mathcal{F}_{\sigma-}$*-measurable, then the process* $\xi I_{[\![\sigma, \tau]\!]}$ *is predictable;*

3) *if* $\sigma, \tau \in \mathbf{T}_p$, *and* ξ *is* $\mathcal{F}_{\sigma-}$*-measurable, then the process* $\xi I_{[\![\sigma, \tau[\![}$ *is predictable.*

Proof. Assertion 1) follows from the fact that

$$\xi I_{]\!]\sigma, \tau[\![} = (\xi I_{]\!]\sigma, \tau]\!]}) I_{[\![0, \tau[\![} \in \mathcal{P}$$

(cf. Problem 2.12 and (3.1)). Analogously 3) follows from 2). To prove 2) it suffices to consider the case in which $\xi = I_A$ with $A \in \mathcal{F}_{\sigma-}$.

Then $\xi_{[\![\sigma, \tau]\!]} = I_{[\![\sigma_A, \tau]\!]}$ and $[\![\sigma_A, \tau]\!] = [\![0, \tau]\!] \setminus [\![0, \sigma_A[\![\in \mathcal{P}$ by Assertion 5) of the preceding theorem.

3. A variable $X_\tau I_{\{\tau < \infty\}}$ is $\mathcal{F}_\tau$-measurable for progressively measurable (and, in particular, for optional) processes X and for each $\tau \in \mathbf{T}$ (Problem 2.2). In case of predictable processes X this assertion admits the following strengthening.

Theorem 3. *Let* X *be a predictable process and* $\tau \in \mathbf{T}$. *Then:*

1) *the variable* $X_\tau I_{\{\tau < \infty\}}$ *is* $\mathcal{F}_{\tau-}$*-measurable;*

2) *a stopped process* $X^\tau = (X_{t \wedge \tau})_{t \ge 0}$ *is predictable too.*

Proof. 1) Consider the processes $X = I_{]\!]\sigma, \tilde{\sigma}]\!]}$ and $Y = I_{A \times \{0\}}$, where $\sigma, \tilde{\sigma} \in \mathbf{T}$, $\sigma \le \tilde{\sigma}$, $A \in \mathcal{F}_0$. As $\mathcal{P}_6 = \mathcal{P}$ (cf. the proof of Theorem 2.2), processes of type X and Y generate a σ-algebra $\mathcal{P}$. Hence it suffices to prove the desired assertion only for the processes X and Y introduced above. Evidently, the variable $Y_\tau I_{\{\tau < \infty\}} = I_A I_{\{\tau = 0\}}$ is

$\mathcal{F}_{\tau-}$-measurable. Next

$$X_\tau I_{\{\tau<\infty\}} = I_{\{\sigma<\tau\le\tilde{\sigma}\}} I_{\{\tau<\infty\}} = [I_{\{\tau\le\tilde{\sigma}\}} - I_{\{\tau\le\sigma\}}]\, I_{\{\tau<\infty\}}.$$

Since (Q is the set of rational numbers)

$$\{\tilde{\sigma}<\tau\} = \bigcup_{r\in Q} \{\tilde{\sigma}\le r<\tau\} = \bigcup_{r\in Q} \{\tilde{\sigma}\le r\}\cap\{r<\tau\} \in \mathcal{F}_{\tau-}$$

and τ is $\mathcal{F}_{\tau-}$-measurable, then $X_\tau I_{\{\tau<\infty\}}$ is $\mathcal{F}_{\tau-}$-measurable too.

2) It suffices to prove the desired assertion only for $\mathbb{F}$-adapted continuous processes X. Clearly, a stopped process X^τ is continuous too, and by the representation

$$X_t^\tau = X_t I_{\{t<\tau\}} + X_\tau I_{\{\tau\le t\}}$$

it is $\mathbb{F}$-adapted as well.

4. Let τ be a random time for which there exists a nondecreasing sequence $(\tau_n)_{n\ge1}$ of Markov times such that

a) $\tau(\omega) = \lim_n \tau_n(\omega), \quad \omega\in\Omega;$

b) $\tau_n(\omega) < \tau(\omega), \quad \omega\in\{\tau(\omega)>0\}.$

(A sequence $(\tau_n)_{n\ge1}$ is called *announcing* for τ.) Since $]\!]0,\tau_n]\!] \in \mathcal{P}$, then

$$]\!]0,\tau[\![= \bigcup_n]\!]0,\tau_n]\!] \in \mathcal{P}.$$

Consequently time τ is predictable and $\mathcal{F}_{\tau-} = \bigvee_n \mathcal{F}_{\tau_n}$ (Problem 8). Thus each random time τ admitting an announcing sequence of Markov times $(\tau_n)_{n\ge1}$, is predictable. It turns out that the converse (complicated!) result holds as well.

Theorem 4 ([81], Ch. IV, Theorem 12). *If τ is predictable, then an announcing sequence of predictable times $(\tau_n)_{n\ge1}$ can be found.*

Corollary. *A random time τ is predictable if and only if there exists a sequence of Markov times announcing it.*

According to Definition 4 in § 1, a set A is called thin if it has a form $A = \bigcup_n [\![\tau_n]\!]$ where $(\tau_n)_{n\ge1}$ is a sequence of Markov times. It turns out that under the additional condition of predictability of a set A Markov times τ_n can be chosen to be predictable. In particular, the following theorem holds.

Theorem 5 ([81], Ch. IV, Theorem 17). *Each thin predictable set* A *admits a* *sequence of predictable times* $(\tau_n)_{n \geq 1}$, *exhausting* A, *i.e.*

$$A = \bigcup_n [\![\tau_n]\!], \quad [\![\tau_i]\!] \cap [\![\tau_j]\!] = \emptyset, \quad i \neq j.$$

If a process X is predictable, then the process ΔX is predictable too. Hence, from Theorem 5 and Theorem 1.1 follows the following result.

Theorem 6 ([81], Ch. IV, Theorem 30). *For each predictable process* $X \in D \cap \mathbb{F}$ *the thin set* $\{\Delta X \neq 0\}$ *is exhausted by a sequence of predictable times.*

5. Definition 2. A Markov time τ is called *accessible* if a sequence of predictable times $(\tau_n)_{n \geq 1}$ can be found such that $[\![\tau]\!] \subseteq \bigcup_n [\![\tau_n]\!]$.

In the following definition a new class of Markov times is introduced which in certain sense (cf. Theorem 7 below) is "orthogonal" to all accessible times.

Definition 3. A Markov time σ is called *completely inaccessible* if $P(\sigma = \tau < \infty) = 0$ for each predictable time τ.

The introduction of the notions of accessible and completely inaccessible times is motivated by the following theorem.

Theorem 7 ([81], Ch. III, Theorem 41). *For each Markov time* T *there exists one and only one (up to* P*-negligibility) pair of Markov times* τ *and* σ *such that*

(1) τ *is an accessible time*;

(2) σ *is a completely inaccessible time*;

(3) $[\![T]\!] = [\![\tau]\!] \cup [\![\sigma]\!]$ and $[\![\tau]\!] \cap [\![\sigma]\!] = \emptyset$.

This theorem yields the following result, concerning the structure of trajectories of a process $X \in D \cap \mathbb{F}$.

Theorem 8 ([81], Ch. IV, § 3). *Each process* $X \in D \cap \mathbb{F}$ *can be represented in the form*

$$X = Y + \sum_n \Delta X_{\sigma_n} I_{[\![\sigma_n]\!]} + \sum_m \Delta X_{\tau_m} I_{[\![\tau_m]\!]}$$

where Y *is a left continuous process, while* $(\sigma_n)_{n \geq 1}$ *is a sequence of totally inaccessible and* $(\tau_m)_{m \geq 1}$ *a sequence of predictable times. If, in addition the process* X *is predictable, then all of the variables* ΔX_{σ_n} *are* P*-negligible.*

6. Definition 4. A process $X \in D \cap \mathbb{F}$ is called *left quasicontinuous* if $\Delta X_\tau = 0$ $(\{\tau < \infty\}$; P-a.s.) for each predictable time τ.

In the following theorem the equivalent formulations of this notion are given.

Theorem 9. *Let a process* $X \in D \cap \mathbb{F}$. *Then the following conditions are equivalent*:

(a) X *is left quasicontinuous*;

(b) *jump times of X are exhausted by totally inaccessible Markov times;*

(c) *for each increasing sequence of Markov times* $(\tau_n)_{n\geq 1}$ *with a limit* τ

$$\lim_n X_{\tau_n} = X_\tau \qquad (\{\tau < \infty\};\ \mathsf{P}\text{-a.s.}).$$

Proof. Since there exists a sequence of Markov times exhausting all jump times of X, the equivalence of Assertions (a) and (b) follows from Definitions 3 and 4.

(c) ⇒ (a). Let (c) hold, however X is not left quasicontinuous, that is, there exists a predictable time T with an announcing sequence $(T_n)_{n\geq 1}$ such that $\mathsf{P}(\Delta X_T \neq 0, T < \infty) > 0$. But $\lim_n X_{T_n} = X_{T-}$ on $\{0 < T < \infty\}$, hence $\lim_n X_{T_n} \neq X_T$ on $\{\Delta X_T \neq 0, T < \infty\}$ and this contradicts Assertion (c).

(a) ⇒ (c). Let (a) hold, however (c) fails for a certain increasing sequence $(T_n)_{n\geq 1}$ with $\lim_n T_n = T$. Put

$$S_n = (T_n)_{\{T_n < T\}}, \quad S = T_A, \quad A = \bigcap_n \{T_n < T\}.$$

Then time S is predictable with an announcing sequence $(S_n)_{n\geq 1}$. Obviously,

$$\{\lim_n X_{T_n} \neq X_T,\ T < \infty\} \subseteq \{\Delta X_S \neq 0,\ S < \infty\}.$$

Hence, if (c) fails, then $\mathsf{P}(\Delta X_S \neq 0, S < \infty) > 0$, and this contradicts the assumption.

In the following theorem necessary and sufficient conditions are given for the predictability of a process X.

Theorem 10. *It is necessary and sufficient for the predictability of a process* $X \in \mathsf{D} \cap \mathbb{F}$ *that the following two conditions hold:*

(a) *for each predictable time* τ *the random variable* $X_\tau I_{\{\tau<\infty\}}$ *is* $\mathbb{F}$*-measurable.*

(b) *for each totally inaccessible time* σ *the set* $\{\Delta X \neq 0\} \cap [\![\sigma]\!]$ *is* P*-negligible.*

Proof. Sufficiency. Let $(\tau_n)_{n\geq 1}$ be a sequence of Markov times announcing τ. Then the random variable $X_{\tau-}I_{\{\tau<\infty\}} = \lim_n X_{\tau_n} I_{\{\tau<\infty\}}$ is clearly $\mathcal{F}_{\tau-}$-measurable and by Assumption (a) the variable $\Delta X_\tau = X_\tau I_{\{\tau<\infty\}} - X_{\tau-} I_{\{\tau<\infty\}}$ is $\mathcal{F}_{\tau-}$-measurable too. Hence, according to Assertion 2) of Theorem 2, the process $\Delta X_\tau I_{[\![\tau]\!]}$ is predictable. Thus it follows from Theorem 8 that the process X is predictable.

Necessity. Let X be a predictable process. Then by Assertion 1) of Theorem 3 the variable $X_\tau I_{\{\tau<\infty\}}$ is $\mathcal{F}_{\tau-}$-measurable, that is, (a) holds. Assumption (b) follows from the last assertion of Theorem 8.

7. In the present subsection a complicated, yet important theorem (section theorem) is given which plays a basic rôle in proving results on "uniqueness" (of a type of the existance of a process with that or another property, unique up to indistinguishability).

Theorem 11 (the Section Theorem). *Let* A *be an optional (predictable) random set. For each* $\varepsilon > 0$ *there exists a Markov (respectively predictable) time* τ_ε *such that* $[\![\tau_\varepsilon]\!] \subseteq A$ *and*

$$P(\tau_\varepsilon < \infty) \geq P(\pi_A) - \varepsilon,$$

where $\pi_A = \{\omega: \exists\, t \in R_+ \text{ with } (\omega, t) \in A\}$.

Proof (see [81], Ch. IV, § 2).

A typical example of applying this result is presented by

Theorem 12. *Let X and Y be optional (predictable) processes. For each Markov (predictable) time let*

$$X_\tau = Y_\tau \quad (\{\tau < 0\};\ P\text{-a.s.}).$$

Then the processes X and Y are indistinguishable.

Proof. To be definite suppose that X and Y are optional processes, and consider the set

$$A = \{(\omega, t): X_t(\omega) \neq Y_t(\omega)\}.$$

It suffices to show that $P(\pi_A) = 0$.

Let $P(\pi_A) = 2\delta > 0$. Then by the Section Theorem one can find a time τ_δ such that $P(\tau_\delta < \infty) \geq \delta > 0$, hence $X_{\tau_\delta} \neq Y_{\tau_\delta}$ with a positive probability and this contradicts the assumption $X_{\tau_\delta} = Y_{\tau_\delta}$ $(\{\tau_\delta < \infty\};\ P\text{-a.s.})$.

8. Theorem 13 ([81], Ch. 5, § 2). *Let X be a measurable stochastic process such that* $X \geq 0$ *or* $|X| < c$. *Then there exists one and only one (up to* P*-indistinguishability) process* ${}^0X$, *and one and only one (up to* P*-indistinguishability) predictable process* ${}^pX$ *such that*

$$\mathbf{E}\,[X_\tau I_{\{\tau < \infty\}} \mid \mathcal{F}_\tau] = {}^0X_\tau I_{\{\tau < \infty\}} \quad \forall\, \tau \in \mathbf{T}, \tag{3.2}$$

$$\mathbf{E}\,[X_\tau I_{\{\tau < \infty\}} \mid \mathcal{F}_{\tau-}\,] = {}^pX_\tau I_{\{\tau < \infty\}} \quad \forall\, \tau \in \mathbf{T}_p. \tag{3.3}$$

The processes ${}^0X$ and ${}^pX$ are called the *optional* and *predictable projections* of a process X (with respect to $(\mathbb{F}, P)$).

Clearly, by (3.2) and (3.3) the projections ${}^0X$ and ${}^pX$ satisfy the relations

$$\mathbf{E}\,[X_\tau I_{\{\tau < \infty\}}] = \mathbf{E}\,[{}^0X_\tau I_{\{\tau < \infty\}}] \quad \forall\, \tau \in \mathbf{T}, \tag{3.4}$$

$$\mathbf{E}\,[X_\tau I_{\{\tau < \infty\}}] = \mathbf{E}\,[{}^pX_\tau I_{\{\tau < \infty\}}] \quad \forall\, \tau \in \mathbf{T}_p. \tag{3.5}$$

Conversely, if the optional and predictable processes ${}^{0}X$ and ${}^{p}X$ satisfy the relations (3.4) and (3.5) for the times $\tau \in \mathbf{T}$ and $\tau \in T_p$ respectively, then formulas (3.2) and (3.3) take place.

The assumption made in Theorem 13, concerning the nonnegativity or boundedness of X, can be relaxed. Namely, if X takes values in $[-\infty, \infty]$, then the predictable process ${}^{p}X$, for instance, defined by the equality

$$ {}^{p}X = \begin{cases} {}^{p}(X^{+}) - {}^{p}(X^{-}) & \text{on } \{{}^{p}|X| < \infty\}, \\ +\infty & \text{on } \{{}^{p}|X| = \infty\} \end{cases} $$

takes values in $(-\infty, \infty]$ and possesses the property (3.3). (If $\mathcal{G}$ is a σ-subalgebra of $\mathcal{F}$ and Y is a random variable taking values in $[-\infty, \infty]$, then $\mathbf{E}(Y \mid \mathcal{G})$ is understood as a random variable defined by the equality

$$ \mathbf{E}(Y \mid \mathcal{G}) = \begin{cases} \mathbf{E}(Y^{+} \mid \mathcal{G}) - \mathbf{E}(Y^{-} \mid \mathcal{G}), & \text{if } \mathbf{E}(\,|Y| \mid \mathcal{G}) < \infty, \\ +\infty, & \text{if } \mathbf{E}(\,|Y| \mid \mathcal{G}) = \infty). \end{cases} $$

9. Let us establish the relation between the optional and predictable projections.

Theorem 14. *Let X be a measurable stochastic process,* $X \in D$ *and* ${}^{0}X \in D$. *Then*

$$ {}^{p}(X_{-}) = {}^{0}(X)_{-}. $$

Proof. By Theorem 12 it suffices to show that for each $\tau \in \mathbf{T}_p$

$$ {}^{p}(X_{-})_{\tau} = {}^{0}(X)_{\tau-} \qquad (\{\tau < \infty\};\ \mathrm{P}\text{-a.s.}). $$

Suppose first $|X| \le c$. By the definition of a predictable projection we have

$$ {}^{p}(X_{-})_{\tau} = \mathbf{E}(X_{\tau-} \mid \mathcal{F}_{\tau-}) \qquad (\{\tau < \infty\};\ \mathrm{P}\text{-a.s.}) $$

Let $(\tau_n)_{n \ge 1}$ be a sequence announcing τ. Then (cf. [188], Theorem 1.6)

$$ \mathbf{E}(X_{\tau-} \mid \mathcal{F}_{\tau-}) = \lim_{n} \mathbf{E}(X_{\tau_n} \mid \mathcal{F}_{\tau_n}) \qquad (\{\tau < \infty\};\ \mathrm{P}\text{-a.s.}), $$

and hence by the assumtion ${}^{0}X \in D$

$$ {}^{p}(X_{-})_{\tau} = \lim_{n} {}^{0}(X)_{\tau_n} = {}^{0}(X)_{\tau-} \qquad (\{\tau < \infty\};\ \mathrm{P}\text{-a.s.}). $$

If $X \geq 0$, then ${}^{P}(X_{-} \wedge n) = {}^{0}(X \wedge n)_{-}$, $n \geq 1$, and the desired equality

$$ {}^{P}(X_{-})_{\tau} = {}^{0}(X)_{\tau-} \ (\{\tau < \infty\};\ \mathsf{P}\text{-a.s.}) $$

is obtained by taking the limit as $n \to \infty$.

In the general case the desired result takes place in an evident manner.

Problems

1. If a Markov time τ is accessible and in the same time totally inaccessible, then $\tau = \infty$ (P-a.s.).

2. Let $\tau \in \mathbf{T}$ and $A \in \mathcal{F}_{\tau}$. If τ is an accessible (totally inaccessible) time, then such is time τ_A too.

3. Let $\sigma \in \mathbf{T}_P$, $\tau \in \mathbf{T}$ and $A = \{\tau < \sigma\}$. Then $\sigma_A \in \mathbf{T}_p$.

4. Let $X = W$ be a Wiener process. Show that ${}^{P}X = W$, and deduce from this that the consideration only of bounded times τ in (3.5) does not imply the validity of this property for all $\tau \in \mathbf{T}$.

5. Show that the consideration of the equalities (3.2) and (3.3) for bounded times τ implies their validity for τ which takes values in $[0, \infty]$.

6. Prove that if X is a predictable process and Y a measurable process, then ${}^{P}X = X$ and ${}^{P}(XY) = X\,({}^{P}Y)$.

7. If σ is a totally inaccessible time, then $\mathsf{P}\,(\sigma > 0) = 1$.

8. Show that $\mathcal{F}_{\tau-} = \bigvee_{n \geq 1} \mathcal{F}_{\tau_n}$ if $\tau \in \mathbf{T}_p$ and $(\tau_n)_{n \geq 1}$ is an announcing sequence.

9. Let $f = f(x)$ be a concave function, and let a measurable process X be such that $\mathbf{E}\,|\,X_{\tau}\,| < \infty$ and $\mathbf{E}\,|\,f(X_{\tau})\,| < \infty$ for each $\tau \in \mathbf{T}$. Show that the following Jensen inequality holds

$$ {}^{P}f(X) \geq f({}^{P}X). $$

10. Let $X \in \mathsf{D}$ and $\mathbf{E} \sup_{t \geq 0} |\,X_t\,| < \infty$. Show that a modification of ${}^{0}X$ may be chosen with right continuous trajectories having left hand limits.

11. Let $X \in \mathsf{D} \cap \mathcal{P}$ and $\tau_a = \inf(t: X_t \geq a)$ where $\inf \varnothing = \infty$. Show that $\tau_a \in \mathbf{T}_p$.

§ 4. Martingales and local martingales

1. Let $(\Omega, \mathcal{F}, \mathbb{F}, \mathsf{P})$ be a stochastic basis.

Definition 1. A stochastic process $X = (X_t, \mathcal{F}_t) \in \mathsf{D} \cap \mathbb{F}$ is called a *martingale* (respectively *supermartingale* or *submartingale*) if $\mathbf{E}\,|X_t| < \infty$, $t \geq 0$ and (P-a.s.)

$$\mathbf{E}\,(X_t \mid \mathcal{F}_s) = X_s, \quad s \leq t \tag{4.1}$$

(respectively if $\mathbf{E}\,(X_t \mid \mathcal{F}_s) \leq X_s$ or $\mathbf{E}\,(X_t \mid \mathcal{F}_s) \geq X_s$).

The class of all processes generating a martingale is denoted by $\bar{\mathcal{M}}$ or by $\bar{\mathcal{M}}\,(\mathbb{F})$ or by $\bar{\mathcal{M}}\,(\mathbb{F}, \mathsf{P})$, if it is necessary to point out the flow $\mathbb{F}$ and the measure P with respect to which the given processes are considered. By $\mathcal{M}$ or $\mathcal{M}\,(\mathbb{F})$ or $\mathcal{M}\,(\mathbb{F},\mathsf{P})$ we denote the class of uniformly integrable martingales $X = (X_t, \mathcal{F}_t)$ for which $(X_t)_{t \geq 0}$ is uniformly integrable, i.e.

$$\sup_{t \geq 0} \mathbf{E}\,[\,|X_t|\, I(\,|X_t| > N)] \to 0, \quad N \to \infty. \tag{4.2}$$

Clearly $\mathcal{M} \subseteq \bar{\mathcal{M}}$.

The class of uniformly integrable martingales has an especially simple structure.

Theorem 1 ([188], Ch. 2, Theorem 2.7; [217], Ch. 5, Theorem 21, Ch. 6, Theorem 6).

1) *If* $X = (X_t, \mathcal{F}_t) \in \mathcal{M}$, *then there exists an integrable random variable* X_∞ *such that*

$$X_t \xrightarrow{\text{a.s.}} X_\infty,\ t \to \infty,$$
$$\mathbf{E}\,|X_t - X_\infty| \to 0,\ t \to \infty,$$
$$X_t = \mathbf{E}\,(X_\infty \mid \mathcal{F}_t) \quad (\mathsf{P}\text{-a.s.}),\ t \geq 0,$$

and for each $\sigma, \tau \in \mathbf{T}$

$$X_{\tau \wedge \sigma} = \mathbf{E}\,(X_\sigma \mid \mathcal{F}_\tau) \quad (\mathsf{P}\text{-a.s.}).$$

The process $(X_t, \mathcal{F}_t)_{0 \leq t \leq \infty}$ *is a martingale.*

2) *If* Y *is an integrable random variable, then there exists one and only one* (*up to* P-*indistinguishability*) *martingale* $X = (X_t, \mathcal{F}_t) \in \mathcal{M}$ *such that* (P-a.s.) *for each* $t \geq 0$

$$X_t = \mathbf{E}\,(Y \mid \mathcal{F}_t).$$

3) *If* $(\tau_n)_{n \geq 1}$ *is a sequence of increasing Markov times, then* (P-a.s.)

$$\lim_n X_{\tau_n} = \mathbf{E}\,[X_{\lim_n \tau_n} \mid \bigvee_n \mathcal{F}_{\tau_n}].$$

In particular, if τ *is a predictable time, then*

$$X_{\tau-} = \mathbf{E}(X_\tau \mid \mathcal{F}_{\tau-}).$$

2. Definition 2. Let $\mathcal{K}$ be a certain class of stochastic processes. We say that a process $X = (X_t)_{t \geq 0}$ belongs to a class $\mathcal{K}_{loc}$ if there exists a ("localizing") nondecreasing sequence of Markov times $(\tau_n)_{n \geq 1}$ such that $\lim_n \tau_n = \infty$, and if for each $n \geq 1$ a "stopped" process $X^{\tau_n} = (X_{t \wedge \tau_n})_{t \geq 0}$ belongs to class $\mathcal{K}$.

Remark. Under the "usual conditions" one can require that a localizing sequence $(\tau_n)_{n \geq 1}$ in the definition of $\mathcal{K}_{loc}$ possesses only the property $\lim_n \tau_n = \infty$, P-a.s., since in this case a sequence of Markov times $(\sigma_n)_{n \geq 1}$ with $\sigma_n = \tau_n$ (P-a.s.) can be constructed such that $s_n \uparrow \infty$ for all $\omega \in \Omega$.

By taking $\mathcal{K}$ in the above definition to be the class of uniformly integrable martingales, we arrive at the following definition.

Definition 3. A process $X = (X, \mathcal{F}_t) \in D \cap \mathbb{F}$ is called a *local martingale* if X belongs to the class $\mathcal{M}_{loc}$ where $\mathcal{M}$ is the class of uniformly integrable martingales.

If $X \in \bar{\mathcal{M}}$ and $\tau_n \equiv n$, then a stopped process $X^{\tau_n} = (X_{t \wedge n}, \mathcal{F}_t)_{t \geq 0}$ is constructed very simply:

$$X_{t \wedge n} = \mathbf{E}(X_n \mid \mathcal{F}_t) \quad \text{(P-a.s.)}, \quad t \geq 0,$$

and hence, $X^{\tau_n} \in \mathcal{M}$, $n \geq 1$ (Theorem 1).

Thus the following result holds.

Theorem 2. *Each martingale is a local martingale too and*

$$\mathcal{M} \subseteq \bar{\mathcal{M}} \subseteq \mathcal{M}_{loc}.$$

The following definition is useful for characterizing the class of uniformly integrable martingales within the class of local martingales.

Definition 4. A process $X = (X, \mathcal{F}_t)_{t \geq 0}$ is called a *process of class* $(\mathcal{D})$ (Dirichlet class) if the family of random variables

$$\{X_\tau; \tau \in Y, \tau < \infty\}$$

is uniformly integrable, i.e.

$$\sup \mathbf{E}\left[\,|X_\tau|\, I(|X_\tau| > N)\right] \to 0, \quad N \to \infty,$$

where sup is taken over the class of stopping times.

Theorem 3. *A local martingale* X *belongs to the class* $\mathcal{M}$ *if and only if* $X \in (\mathcal{D})$, *i.e.*

$$\mathcal{M}_{loc} \cap (\mathcal{D}) = \mathcal{M}.$$

Proof. If $X \in \mathcal{M}$, then by Theorem 1 we have $X_\tau = \mathbf{E}(X_\infty \mid \mathcal{F}_\tau)$ for each $\tau \in \mathbf{T}$, where $\mathbf{E}\,|X_\infty| < \infty$. As is well known, the family of random variables $\{\mathbf{E}(X_\infty \mid \mathcal{F}_\tau): \mathbf{E}\,|X_\infty| < \infty, \tau \in \mathbf{T}\}$ is uniformly integrable. Hence, $\mathcal{M} \subseteq (\mathcal{D})$, and by Theorem 2 we have $\mathcal{M} \subseteq (\mathcal{D}) \cap \mathcal{M}_{loc}$. Conversely, let $X \in (\mathcal{D}) \cap \mathcal{M}_{loc}$ and let $(\tau_n)_{n \geq 1}$ be a localizing sequence for X. Since $X^{\tau_n} \in \mathcal{M}$, then

$$X_s^{\tau_n} = \mathbf{E}(X_t^{\tau_n} \mid \mathcal{F}_s), \quad s \leq t,$$

and hence $X_{s \wedge \tau_n} = \mathbf{E}(X_{t \wedge \tau_n} \mid \mathcal{F}_s)$. Since $X \in (\mathcal{D})$, sequences of random variables $(X_{s \wedge \tau_n})_{n \geq 1}$ and $(X_{t \wedge \tau_n})_{n \geq 1}$ are uniformly integrable, and evidently

$$X_{s \wedge \tau_n} \overset{a.s.}{\to} X_s, \quad X_{t \wedge \tau_n} \overset{a.s.}{\to} X_t$$

(by taking into account that $\lim_n \tau_n = \infty$).

This gives $\mathbf{E}\,|X_{t \wedge \tau_n} - X_t| \to 0$ and consequently

$$\mathbf{E}\,|\mathbf{E}(X_{t \wedge \tau_n} \mid \mathcal{F}_s) - \mathbf{E}(X_t \mid \mathcal{F}_s)| \leq \mathbf{E}\,|X_{t \wedge \tau_n} - X_t| \to 0.$$

Taking the limit in $X_{s \wedge \tau_n} = \mathbf{E}(X_{t \wedge \tau_n} \mid \mathcal{F}_s)$ we get $X_s = \mathbf{E}(X_t \mid \mathcal{F}_s)$ (P-a.s.) which gives $X \in \bar{\mathcal{M}}$, and since $X \in (\mathcal{D})$, we have $X \in \mathcal{M}$.

Remark. According to Definition 1 martingales are understood as processes $X \in \mathbb{F}$ with trajectories belonging to the space D. This last presumption (under the right continuity assumption on the family $(\mathcal{F}_t)_{t \geq 0}$) is not restrictive at all because each martingale (in the sense of Definition 1 but without the assumption $X \in D$) admits a modification $\tilde{X} \in \mathbb{F} \cap D$ which is a martingale again. An analogous claim is true also concerning supermartingales, for which the function $(\mathbf{E}X_t)_{t \geq 0}$ is right continuous [217].

Problems

1. Show that $\bar{\mathcal{M}}_{loc} = \mathcal{M}_{loc}$.

2. Let $X \in \mathbb{F} \cap D$, and let the limit $\lim_{t \to \infty} X_t \,(= X_\infty)$ exist P-a.s. Show that it is necessary and sufficient for the inclusion $X \in \mathcal{M}$ that $\mathbf{E}\,|X_0| < \infty$ and $\mathbf{E}\,X_\tau = \mathbf{E}\,X_0$ for each $\tau \in \mathbf{T}$.

3. If $X \in \mathcal{M}_{loc}$, then ${}^pX = X_-$ and ${}^p(\Delta X) = 0$.

4. Show that a nonnegative local martingale is a supermartingale.

5. (The case of a discrete time.) A sequence $X = (X_n, \mathcal{F}_n)_{n \geq 0}$ presents a local martingale if and only if (P-a.s.) for every n

$$\mathbf{E}\,(\,|X_{n+1}|\,|\,\mathcal{F}_n) < \infty, \qquad \mathbf{E}\,(X_{n+1}\,|\,\mathcal{F}_n) = X_n.$$

6. If $X = (X_t, \mathcal{F}_t) \in \mathcal{M}_{loc}$ and $\tau \in \mathbf{T}_p$, then (P-a.s.)

$$\mathbf{E}[\,|X_\tau|\,I(\tau < \infty)\,|\,\mathcal{F}_{\tau-}] < \infty,$$

$$\mathbf{E}\,[X_\tau I(\tau < \infty)\,|\,\mathcal{F}_{\tau-}] = X_{\tau-}\,I(\tau < \infty).$$

7. Let

$$X \in \mathcal{M}_{loc} \quad \text{and} \quad \|X\|_1 = \sup_{\tau \in T} \mathbf{E}\,[\,|X_\tau|\,I(\tau < \infty)] < \infty.$$

Then X can be represented as $X = Y - Z$ where $Y \geq 0$, $Z \geq 0$, $Y \in \mathcal{M}_{loc}$, $Z \in \mathcal{M}_{loc}$ and $\|X\|_1 = \|Y\|_1 + \|Z\|_1 = \mathbf{E}\,[Y_0 + Z_0]$.

8. If $X \in \mathcal{M}_{loc}$ and $\mathbf{E}\sup_{t \geq 0}|X_t| < \infty$, then $\mathbf{E}\,X_\tau = \mathbf{E}\,X_0$ for each $\tau \in \mathbf{T}$.

9. If $X \in \mathcal{M}_{loc}$, $X \geq 0$ and $X_0 = 0$, then $X = 0$.

10. If $X \in \mathcal{M}$ and $\tau \in \mathbf{T}_p$, then $I_{\{\tau < \infty\}}\mathbf{E}\,(\Delta X_\tau\,|\,\mathcal{F}_{\tau-}) = 0$ (P-a.s.).

11. Let $X \in \mathcal{M}_{loc}$, $X \geq 0$ and $\mathbf{T} = \inf\,(t\colon X_t = 0)$ $(\inf \varnothing = \infty)$. Show that the process $XI_{[\![T,\infty[\![}$ is indistinguishable from zero.

§ 5. Square integrable martingales

1. Definition 1. It is said that a martingale $X \in \mathbb{F} \cap D$ is *square integrable* if

$$\sup_{t \geq 0} \mathbf{E}\, X_t^2 < \infty. \tag{5.1}$$

The class of square integrable martingales will be denoted by $\mathcal{M}^2$, or by $\mathcal{M}^2(\mathbb{F})$, or by $\mathcal{M}^2(\mathbb{F}, P)$.

From (5.1) and from La Vallée-Poussin's uniform integrability criterion it follows that $\mathcal{M}^2 \subseteq \mathcal{M}$, i.e. each square integrable martingale is uniformly integrable, and hence by Theorem 4.1 there exists an integrable random variable X_∞ such that $X_t = \mathbf{E}(X_\infty \mid \mathcal{F}_t)$ (P-a.s.). By Fatou's lemma

$$\mathbf{E} X_\infty^2 \leq \underline{\lim}_{t \to \infty} \mathbf{E}\, X_t^2 \leq \sup_{t \geq 0} \mathbf{E}\, X_t^2 < \infty$$

and conversely, for each variable $X_\infty \in L^2(\Omega, \mathcal{F}_\infty, P)$ a process $X = (X_t, \mathcal{F}_t)$ with $X_t = \mathbf{E}(X_\infty \mid \mathcal{F}_t)$ is a square integrable martingale.

The class of locally square integrable martingales, i.e. processes $X \in \mathbb{F} \cap D$ for which a localizing sequence $(\tau_n)_{n \geq 1}$ can be found such that $X^{\tau_n} \in \mathcal{M}^2$, $n \geq 1$, we denote by $\mathcal{M}^2_{loc}$ or by $\mathcal{M}^2_{loc}(\mathbb{F})$ or by $\mathcal{M}^2_{loc}(\mathbb{F}, P)$. Clearly, $\mathcal{M}^2_{loc} \subseteq \mathcal{M}_{loc}$.

By $\overline{\mathcal{M}}^2$ we denote martingales X from $\overline{\mathcal{M}}$, for which $\mathbf{E}X_t^2 < \infty$, $t \geq 0$.

2. If $X \in \mathcal{M}^2$, then as has been noted above, $X_t = \mathbf{E}(X_\infty \mid \mathcal{F}_t)$ (P-a.s.), with $X_\infty \in L^2(\Omega, \mathcal{F}_\infty, P)$. From this it follows that the vector space $\mathcal{M}^2$ (by identifying P-indistinguishable elements) is algebraicly isomorphic with the space $L^2(\Omega, \mathcal{F}_\infty, P)$.

In the space $\mathcal{M}^2$ one can introduce in a natural manner the scalar product

$$(X, Y) = \mathbf{E}\, X_\infty Y_\infty$$

and the norm

$$\| X \|_2 = \sqrt{\mathbf{E} X_\infty^2}$$

that makes $\mathcal{M}^2$ a Hilbert space.

Definition 2. Two local martingales X and Y are called *strongly orthogonal* ($X \perp\!\!\!\perp Y$) if the process XY is a local martingale with $X_0 Y_0 = 0$.

For each process $X = (X_t)_{t \geq 0}$, set

$$X_\infty^* = \sup_{t \geq 0} | X_t |.$$

Definition 3. The (Banach) space $\mathcal{H}^p$, $p \geq 1$, is a space of martingales for which

$$\| X \|_{\mathcal{H}^p} < \infty$$

where

$$\| X \|_{\mathcal{H}^p} = \| X^*_\infty \|_p$$

and $\| \cdot \|$ is the norm in $L^p = L^p(\Omega, \mathcal{F}_\infty, P)$. (In case $p = 1$ we write simply $\mathcal{H}$ instead of $\mathcal{H}^1$.)

For $p = 2$, in particular, the following inequalities take place:

$$\| X_\infty \|_2 \leq \| X^*_\infty \|_2 \leq 2 \| X_\infty \|_2 \tag{5.2}$$

(the first inequality is evident, and the second one is a consequence of Doob's inequality (Theorem 9.2)).

3. In the following theorem the connection is established between the orthogonality ($X \perp Y$) in the sense of the Hilbert space and the strong orthogonality ($X \perp\!\!\!\perp Y$).

Theorem 1. *Let* $X, Y \in \mathcal{M}^2$. *If* $X \perp\!\!\!\perp Y$, *then* $X_\tau \perp Y_\tau$ *for each* $\tau \in \mathbf{T}$. *Conversely, if* $X_0 Y_0 = 0$ *and* $X_\tau \perp Y_\tau$ *for each* $\tau \in \mathbf{T}$, *then* $X \perp\!\!\!\perp Y$.

Proof. From the inequality (5.2) it follows that $X^*_\infty = \sup_{t \geq 0} |X_t|$ and that $Y^*_\infty = \sup_{t \geq 0} |Y_t|$ belongs to $L^2(\Omega, \mathcal{F}_\infty, P)$, and hence $\mathbf{E}|X^*_\infty Y^*_\infty| < \infty$. Since $(XY)^*_\infty \leq X^*_\infty Y^*_\infty$, then $XY \in \mathcal{H}$ (i.e. $\mathbf{E} \sup_{t \geq 0} |X_t Y_t| < \infty$). The orthogonality $X \perp\!\!\!\perp Y$ means that $XY \in \mathcal{M}_{loc}$ and $X_0 Y_0 = 0$. Then by Problem 4.8 we have $\mathbf{E}\, X_\tau Y_\tau = \mathbf{E} X_0 Y_0 = 0$, $\tau \in \mathbf{T}$, i.e. $X_\tau \perp Y_\tau$.

Conversely, for each $\tau \in \mathbf{T}$ we have $\mathbf{E}\, |X_\tau Y_\tau| < \infty$ and $\mathbf{E}\, X_\tau Y_\tau = 0$. Hence $(X_t Y_t)_{t \geq 0}$ is a uniformly integrable martingale (Problem 4.2) and certainly $XY \in \mathcal{M}_{loc}$. By this and the assumption $X_0 Y_0 = 0$ we have $X \perp\!\!\!\perp Y$.

4. It will be proved below that any martingale $X \in \mathcal{M}^2$ admits the decomposition into two components

$$X = X^c + X^d$$

where X^c is a martingale with continuous trajectories, while X^d is a so called purely discontinuous martingale. With this end in view note first that the spaces $\mathcal{M}^2$ and $\mathcal{H}^2$ coincide (Problem 1), and introduce the following

Definition 4. A subspace $\mathcal{K} \in \mathcal{H}^2$ is called *stable* if:

(1) $\mathcal{K}$ is a closed subspace with respect to the norm $\|\cdot\|_{\mathcal{H}^2}$.

(2) $\mathcal{K}$ is closed relative to "stopping", i.e. if $X \in \mathcal{K}$, then $X^\tau \in \mathcal{K}$ for each $\tau \in \mathbf{T}$;

(3) if $X \in \mathcal{K}$ and $A \in \mathcal{F}_0$, then $I_A X \in \mathcal{K}$.

If $\mathcal{K}$ is a stable subspace, then we denote by $\mathcal{K}^\perp$ a set of martingales $Y \in \mathcal{H}^2$ such that $EX_\infty Y_\infty = 0$ for each $X \in \mathcal{K}$.

Theorem 2. *Let $\mathcal{K}$ be a stable subspace of $\mathcal{H}^2$. Then $\mathcal{K}^\perp$ is a stable subspace too, and if $X \in \mathcal{K}$, and $Y \in \mathcal{K}^\perp$, then $X \perp\!\!\!\perp Y$.*

Proof. Let $X \in \mathcal{K}$ and $Y \in \mathcal{K}^\perp$. Then $X^\tau \in \mathcal{K}$ for $\tau \in \mathbf{T}$, and $\mathbf{E}X_\tau Y_\infty = 0$. Hence

$$\mathbf{E}X_\tau Y_\infty = \mathbf{E}\{\mathbf{E}(X_\tau Y_\infty \mid \mathcal{F}_\tau)\} = \mathbf{E}\{X_\tau \mathbf{E}(Y_\infty \mid \mathcal{F}_\tau)\} = \mathbf{E}X_\tau Y_\tau = 0. \quad (5.3)$$

Taking $\tau = 0$ and the process $I_A X$ instead of X, by (5.3) we get $\mathbf{E}[I_A X_0 Y_0] = 0$ and hence $X_0 Y_0 = 0$ (P-a.s.), which together with (5.3) gives $X \perp\!\!\!\perp Y$.

Further, if $A \in \mathcal{F}_0$, then

$$\mathbf{E}[I_A X_\tau Y_\tau] = \mathbf{E}[X_\infty (I_A Y^\tau)_\infty] = 0.$$

Consequently $I_A Y^\tau \in \mathcal{K}^\perp$ for each $Y \in \mathcal{K}^\perp$, $\tau \in \mathbf{T}$ and $A \in \mathcal{F}_0$, i.e. the stability properties 2) and 3) hold. The stability property 1) follows from the definition of $\mathcal{K}^\perp$. Thus the subspace $\mathcal{K}^\perp$ is stable.

Corollary. *Let $\mathcal{K}$ be a stable subspace of $\mathcal{H}^2$. Then each element $M \in \mathcal{H}^2$ admits the unique decomposition*

$$M = M' + M'' \quad (5.4)$$

with $M' \in \mathcal{K}$, and $M'' \in \mathcal{K}^\perp$.

Proof. Let $\mathcal{K}_\infty$ be a closed subspace in $L^2(\Omega, \mathcal{F}_\infty, P)$ generated by the "terminal" variables $\{X_\infty : X \in \mathcal{K}_\infty\}$ with $X_\infty = \lim_{t \to \infty} X_t$ (Theorem 4.1). Define $\mathcal{K}_\infty^\perp$ analogously. Clearly for each $M \in \mathcal{H}^2$ its terminal value M_∞ has a unique orthogonal (in $L^2(\Omega, \mathcal{F}_\infty, P)$) decomposition

$$M_\infty = M'_\infty + M''_\infty, \quad M'_\infty \in \mathcal{K}_\infty, \quad M''_\infty \in \mathcal{K}_\infty^\perp.$$

Setting

$$M'_t = \mathbf{E}(M'_\infty \mid \mathcal{F}_t), \quad M''_t = \mathbf{E}(M''_\infty \mid \mathcal{F}_t)$$

we obtain the desired decomposition $M = M' + M''$.

5. The decomposition (5.4) is especially important in the special case in which $\mathcal{K} =$

$\mathcal{H}^{2,c}$ is a space of martingales from $\mathcal{H}^2$, having continuous trajectories with origin at zero. First let us verify that $\mathcal{H}^{2,c}$ is a stable subspace. The properties (2) and (3) in Definition 4 are evident. The proof of property (1) follows directly from the following lemma.

Lemma 1. *Let* $X^n \in \mathcal{M}^2$, $n \geq 1$ *and* $X \in \mathcal{M}^2$ *be such that* $\| X^n - X \|_{\mathcal{H}^2} \to 0$, $n \to \infty$. *Then there exists a subsequence* (n_k) *such that* $(X_t^{n_k}(\omega))_{t \geq 0}$ *converges to* $(X_t(\omega))_{t \geq 0}$ *uniformly on* $[0, \infty]$.

Proof. As is given

$$\| X^n - X \|^2_{\mathcal{H}^2} = \mathbf{E} \sup_{t \geq 0} | X_t^n - X_t |^2 \to 0, \quad n \to \infty.$$

Choose a subsequence (n_k) such that

$$\mathbf{E} \sup_{t \geq 0} |X_t^{n_k} - X_t|^2 \leq \frac{1}{2^k}, \quad k \geq 1.$$

Then by the Chebyshev inequality

$$\mathsf{P}\left(\sup_{t \geq 0} | X_t^{n_k} - X_t |^2 > \frac{1}{k} \right) \leq \frac{k}{2^k},$$

and by the Borel-Cantelli lemma

$$\mathsf{P}\left(\bigcap_m \bigcup_{k \geq m} \left\{ \sup_{t \geq 0} | X_t^{n_k} - X_t | > \frac{1}{k} \right\} \right) = 0. \tag{5.5}$$

For each ω belonging to the complement of the set

$$\bigcap_m \bigcup_{k \geq m} \left\{ \sup_{t \geq 0} | X_t^{n_k} - X_t | > \frac{1}{k} \right\}$$

the sequence $(X_t^{n_k}(\omega))_{t \geq 0}$ converges to $(X_t(\omega))_{t \geq 0}$ uniformly. Consequently, by (5.5) this convergence takes place P-a.s.

Thus it can be claimed that $\mathcal{K} = \mathcal{H}^{2,c}$ is a stable subspace in $\mathcal{M}^2$. Denote $\mathcal{K}^{\perp}$ by $\mathcal{M}^{2,d}$. Martingales from $\mathcal{M}^{2,d}$ are usually called purely discontinuous. (A motivation for using this term will be given below; see § 7.)

Thus the following theorem holds.

Theorem 3. *Each square integrable martingale* $M \in \mathcal{M}^2$ *admits the following unique decomposition:*

$$M = M^c + M^d$$

where M^c *is a martingale with continuous trajectories and* $M_0^c = 0$, *while* M^d *is a purely discontinuous martingale such that* $M^d \perp\!\!\!\perp M^c$.

Problems

1. Prove that the space $\mathcal{H}^2$ coincides with a set of those martingales M for which

$$\sup_{t \geq 0} EM_t^2 < \infty.$$

2. Let $M \in \mathcal{M}^2$ and $T_n \in \mathbf{T}$, $n \geq 1$, $T_n \uparrow \infty$. Prove that

$$\| M - M^{T_n} \|_{\mathcal{H}^2} \to 0, \ n \to \infty.$$

3. Show that $M^1 + \dots + M^n$, $n \geq 1$ converges to $M \in \mathcal{H}^2$ in the norm of $\mathcal{H}^2$ if M^n, $n \geq 1$ is a sequence of mutually strongly orthogonal martingales with $\sum_{n=1}^{\infty} E(M_\infty^n)^2 < \infty$. If $N \in \mathcal{M}^2$ and $N \perp\!\!\!\perp (M^1 + \dots + M^n)$, $n \geq 1$, then $N \perp\!\!\!\perp M$.

4. Let $X^n \in \mathcal{H}$, $n \geq 1$, $X \in \mathcal{H}$ and $\| X - X^n \|_{\mathcal{H}} \to 0$, $n \to \infty$. Prove that there exists a subsequence (n_k) such that for P- almost every ω $(X_t^{n_k}(\omega))_{t \geq 0}$ converges to $(X_t(\omega))_{t \geq 0}$ uniformly on $[0, \infty]$. (Hint: see Lemma 1.)

5. Show that if $M \in \mathcal{M}_{loc}$ and $| \Delta M | \leq$ const, then the process $M \in \mathcal{M}_{loc}^2$.

§ 6. Increasing processes. Compensators (dual predictable projections). The Doob-Meyer decomposition

1. Along with the notions of a martingale and local martingale, an essential rôle in the theory of stochastic processes is played by the notion of increasing processes.

Definition 1. A stochastic process $A = (A_t)_{t \geq 0}$ of the class $\mathbb{F} \cap D$ is called *increasing* if $A_0(\omega) = 0$ and if for each $\omega \in \Omega$ the trajectories $(A_t(\omega))_{t \geq 0}$ are nondecreasing functions of t.

The collection of such processes will be denoted by $\mathcal{V}^+(\mathbb{F})$ or $\mathcal{V}^+$. Processes $A \in \mathbb{F} \cap D$ with $A_0 = 0$, having for each $\omega \in \Omega$ bounded variation over each interval $[0, t]$, are denoted by $\mathcal{V}$ or $\mathcal{V}(\mathbb{F})$.

We will use V^+ (V) to denote the subspace of the space D which consists of nondecreasing functions $X = (X_t)_{t \geq 0}$ (the differences of nondecreasing functions).

If $A \in V$, then there exists the unique decomposition: $A_t = B_t - C_t$, with $B \in V^+$, $C \in V^+$ and total variation

$$\operatorname{Var}(A)_t = \int_0^t |dA_s| = B_t + C_t. \tag{6.1}$$

Note that as

$$\operatorname{Var}(A) = \operatorname{Var}(A)_- + |\Delta A|$$

and as for B and C one can take the processes

$$B = \frac{1}{2}[\operatorname{Var}(A) + A], \quad C = \frac{1}{2}[\operatorname{Var}(A) - A],$$

the property $A \in \mathcal{V} \cap \mathcal{P}$ implies $\mathcal{P}$-measurability of the processes $\operatorname{Var}(A)$, B and C.

Definition 2. An increasing process $A = (A_t)_{t \geq 0}$ from the class $\mathbb{F} \cap D$ is called *integrable* if $EA_\infty < \infty$, with $A_\infty = \lim_{t \to \infty} A_t$.

The collection of such processes is denoted by $\mathcal{A}^+$. By $\mathcal{A}$ we denote those processes $\mathcal{A} \in \mathcal{V}$ for which $E \operatorname{Var}(A)_\infty < \infty$.

By using "localizing" sequences of Markov times (cf. Definition 2 in § 4) the classes $\mathcal{V}^+_{loc}$, $\mathcal{V}_{loc}$, $\mathcal{A}^+_{loc}$, $\mathcal{A}_{loc}$ can be introduced. Evidently

$$\mathcal{V}^+_{loc} = \mathcal{V}^+, \quad \mathcal{V}_{loc} = \mathcal{V},$$
$$\mathcal{A}^+ \subseteq \mathcal{A}^+_{loc} \subseteq \mathcal{V}^+, \quad \mathcal{A} \subseteq \mathcal{A}_{loc} \subseteq \mathcal{V}.$$

The following technical lemma will be often used in the sequel.

Lemma 1. *If a process* $A \in \mathcal{V} \cap \mathcal{P}$, *then a localizing sequence* $(T_n)_{n \geq 1}$ *of Markov times can be found such that* $\mathrm{Var}\,(A)_{T_n} \leq n$. *In particular*

$$\mathcal{P} \cap \mathcal{V} = \mathcal{P} \cap \mathcal{A}_{loc}.$$

Proof. Since a process A is predictable, then the random set

$$B_n = \{(\omega, t): \int_0^t |\, dA_s(\omega) \,| \geq n\} \in \mathcal{P}, \quad n \geq 1.$$

Let $\tau_n = D_{B_n}$ be the debut of the set B_n. By $A \in D$, the graph $[\![\tau_n]\!] \subseteq B_n$. Hence $[\![\tau_n]\!] \cup B_n \in \mathcal{P}$, and according to Assertion 6) of Theorem 3.1, time τ is predictable. Since $A_0 = 0$, then $\tau_n > 0$ (P-a.s.). According to Theorem 3.4, a Markov time $\sigma_n < \tau_n$ can be found such that $P\left(\sigma_n \leq \tau_n - \frac{1}{n}\right) \leq 2^{-n}$. Set $T_n = \sup_{m \leq n} \sigma_m$. Then $\lim_n T_n = \infty$ (P-a.s.), and

$$\mathrm{Var}\,(A)_{T_n} = \int_0^{T_n} |\, d\,A_s \,| \leq n.$$

Remark. If a process $A \in \mathcal{V}$ possesses bounded jumps ($|\,\Delta A\,| \leq c$), then $A \in \mathcal{A}_{loc}$. This follows from the fact that in the proof given above one can take $T_n = \tau_n$, and then

$$\int_0^{T_n} |\, d\,A_s \,| \leq n + c.$$

2. Let $A \in \mathcal{V}$ and let $H = (H_t(\omega))_{t \geq 0}$ be a measurable stochastic process. Then a new process $H \circ A = (H \circ A)_t(\omega)$, $t \geq 0$ can be defined by using the Lebesgue-Stieltjes integral via the following formula:

$$(H \circ A)_t(\omega) = \begin{cases} \displaystyle\int_0^t H_s(\omega)\, dA_s(\omega), & \text{if } \displaystyle\int_0^t |\, H_s(\omega) \,|\, d\,\mathrm{Var}\,(A)_s(\omega) < \infty, \\ \infty, & \text{if } \displaystyle\int_0^t |\, H_s(\omega) \,|\, d\,\mathrm{Var}\,(A)_s(\omega) = \infty \end{cases} \tag{6.2}$$

(The integration from 0 to t is understood to be carried out on the set (0, t].)

Theorem 1. *Let* X *and* Y *be two nonnegative measurable processes. Suppose that for each Markov time* τ

$$\mathbf{E}\,[X_\tau I\,(\tau < \infty)] = \mathbf{E}\,[Y_\tau I\,(\tau < \infty)]. \tag{6.3}$$

Then for each $A \in \mathcal{V}^+$ *and* $\tau \in \mathbf{T}$

$$\mathbf{E}\int_0^\tau X_s\, dA_s = \mathbf{E}\int_0^\tau Y_s\, dA_s. \tag{6.4}$$

Proof. Let us use the wellknown device reducing the Lebesgue-Stieltjes integral to the Lebesgue integral. Namely, for $t \ge 0$ take

$$\tau_t(\omega) = \inf(s\colon A_s(\omega) > t).$$

Clearly,

$$\{\tau_t(\omega) < s\} = \bigcup_n \{A_{s-1/n}(\omega) > t\}.$$

and $\{A_{s-1/n}(\omega) > t\} \in \mathcal{F}_{s-1/n} \subseteq \mathcal{F}_s$, as $A \in \mathbb{F}$. Hence, $\{\tau_t(\omega) < s\} \in \mathcal{F}_s$, and since $\mathcal{F}_s = \mathcal{F}_{s+}$, then $\{\tau_t(\omega) \le s\} \in \mathcal{F}_s$. In other words, τ_t is a Markov time for each t.

First, let $\tau = \infty$. Then the Lebesgue-Stieltjes integral is equal to

$$\int_0^\infty X_s\, dA_s = \int_0^\infty I\,(\tau_s < \infty)\, X_{\tau_s}\, ds,$$

and by Fubini's theorem

$$\mathbf{E}\int_0^\infty X_s\, dA_s = \mathbf{E}\int_0^\infty I\,(\tau_s < \infty)\, X_{\tau_s}\, ds = \int_0^\infty \mathbf{E}\,[X_{\tau_s} I\,(\tau_s < \infty)]\, ds.$$

Analogously,

$$\mathrm{E}\int_0^\infty Y_s\, dA_s = \int_0^\infty \mathrm{E}\,[Y_{\tau_s} I\,(\tau_s < \infty)]\, ds.$$

Since τ_s is a Markov time, by the conditions of the theorem we have

$$\mathbf{E}\int_0^\infty X_s\, dA_s = \mathbf{E}\int_0^\infty Y_s\, dA_s. \tag{6.5}$$

The general case (6.4) follows from (6.5) by considering the "stopped" process $A^\tau = (A_{\tau \wedge s})_{s \ge 0}$ instead of $A = (A_s)_{s \ge 0}$.

Corollary. *Let* $A \in \mathcal{V}^+$ *and let* M *be a nonnegative uniformly integrable martingale. Then for each* $\tau \in \mathbf{T}$

$$\mathbf{E}\int_0^\tau M_s\, dA_s = \mathbf{E}\,[M_\tau A_\tau]. \tag{6.6}$$

Proof. In the previous theorem take

$$X_t = M_t I_{[\![0,\tau]\!]}(t), \qquad Y_t = M_\tau I_{[\![0,\tau]\!]}(t).$$

For each $\sigma \in \mathbf{T}$

$$\mathbf{E}\,[X_\sigma I(\sigma < \infty)] = \mathbf{E}\,[M_\sigma I_{[\![0,\tau]\!]}(\sigma)\, I(\sigma < \infty)] = \mathbf{E}\,[M_\sigma I(\sigma \le \tau)\, I(\sigma < \infty)].$$

Then by Theorem 4.1

$$\begin{aligned} \mathbf{E}\,[Y_\sigma I(\sigma < \infty)] &= \mathbf{E}\,[M_\tau I_{[\![0,\tau]\!]}(\sigma)\, I(\sigma < \infty)] \\ &= \mathbf{E}\,(\mathbf{E}[M_\tau I(\sigma \le \tau)\, I(\sigma < \infty) \,|\, \mathcal{F}_\sigma]) \\ &= \mathbf{E}\,[M_\sigma I(\sigma \le \tau)\, I(\sigma < \infty)]. \end{aligned}$$

Consequently Condition (6.3) holds, and hence by (6.4)

$$\mathbf{E}\int_0^\tau M_s\, dA_s = \mathbf{E}\int_0^\tau X_s\, dA_s = \mathbf{E}\int_0^\tau Y_s\, dA_s = \mathbf{E}\int_0^\tau M_\tau I_{[\![0,\tau]\!]}(s)\, dA_s = \mathbf{E}\,[M_\tau A_\tau].$$

3. Let us consider the structure of increasing processes $A \in \mathcal{V}^+$. For each $\omega \in \Omega$ a nondecreasing function $(A_t(\omega))_{t \ge 0}$ admits, as is well known from analysis, the unique decomposition as the sum

$$A(\omega) = A_t^c(\omega) + A_t^d(\omega) \tag{6.7}$$

where A^c is a nondecreasing continuous function, while A^d is a purely discontinuous function (i.e. such that the support of the corresponding measure dA^d consists of a countable number of atoms at most).

In the following theorem a more detailed description of processes A is given.

Theorem 2. *Let $A \in \mathcal{V}^+$. Then a sequence of stopping times $(T_n(\omega))$ and a sequence of positive constants $(\alpha_n)_{n \ge 1}$ can be found such that*

$$A_t = A_t^c + \sum_n \alpha_n I(T_n \le t). \tag{6.8}$$

If in addition a process A is predictable, then times T_n can be chosen to be predictable too.

Proof. According to Theorem 1.1 the set $\{\Delta A \ne 0\}$ is thin, i.e.

$$\{\Delta A \ne 0\} = \cup_k [\![\tau_k]\!], \tag{6.9}$$

where $(\tau_k)_{k \ge 1}$ is a sequence of Markov times (being predictable if A is predictable). Hence

$$A_t^d = \sum_n (A_{\tau_n} - A_{\tau_n -})\, I_{[\![\tau_n, \infty[\![}(t).$$

Denote $H_n(\omega) = (A_{\tau_n} - A_{\tau_n -})\, I(\tau_n < \infty)$ and consider the binary decomposition of $H_n(\omega)$:

$$H_n(\omega) = \sum_{k=-\infty}^{\infty} b_{nk}(\omega)\, 2^{-k}$$

with $b_{nk}(\omega)$ taking the values 0 or 1.

Now, put $B_{nk} = \{\omega: b_{nk}(\omega) = 1\}$ ($\in \mathcal{F}_{\tau_n}$), $\tau_{nk} = (\tau_n)_{B_{nk}}$, i.e.

$$\tau_{nk} = \begin{cases} \tau_n, & \omega \in B_{nk}, \\ \infty, & \omega \notin B_{nk}, \end{cases}$$

and let $\alpha_{nk} = 2^{-k}$. Then

$$A_t^d = \sum_{n,k} \alpha_{nk} I(\tau_{nk} \le t),$$

which by the corresponding denumeration can be rewritten in the form

$$A_t^d = \sum_n \alpha_n I(T_n \le t).$$

In case when a process A is predictable, the variables H_n are $\mathcal{F}_{\tau_n -}$-measurable, $B_{nk} \in \mathcal{F}_{\tau_n -}$ and hence τ_{nk} are predictable times (Theorem 3.1, Assertion 5).

4. The notion of the dual predictable projection (compensator) of an increasing process which will be introduced in this subsection presents one of the fundamental notions of the general theory of stochastic processes.

Theorem 3 ([81], Ch. V, Theorem 28). *Let a process* $A \in \mathcal{A}^+_{loc}$. *Then there exists one and only one predictable process* $\tilde{A} \in \mathcal{A}^+_{loc} \cap \mathcal{P}$, *called the dual predictable projection (compensator) of* A, *such that*

$$A - \tilde{A} \in \mathcal{M}_{loc}. \tag{6.10}$$

Condition (6.10) *is equivalent to any of the following three conditions:*

$$\mathbf{E}A_\tau = \mathbf{E}\tilde{A}_\tau \qquad \forall\, \tau \in \mathbf{T}, \tag{6.11}$$

$$\mathbf{E}(H \circ A)_\infty = \mathbf{E}(H \circ \tilde{A})_\infty \quad \forall\, H \ge 0, \quad H \in \mathcal{P}, \tag{6.12}$$

or

$$\mathbf{E}(^{p}H \circ A)_\infty = \mathbf{E}(H \circ \tilde{A})_\infty, \tag{6.13}$$

where H *are nonnegative measurable processes, while* ^{p}H *is the predictable projection.*

Remark 1. Along with $\tilde{A}$, the notation A^p is also frequently used. It is important to stress that in general ${}^{p}A$ does not coincide with A^{p} (Problem 1).

Remark 2. By analogy with the dual predictable projection $\tilde{A} = A^{p}$, defined, for instance by (6.13), the notion of the dual optional projection A^{0} can also be defined by the equality

$$\mathbf{E}\,({}^{0}H \circ A)_{\infty} = \mathbf{E}\,(H \circ A^{0})_{\infty} \tag{6.14}$$

where ${}^{0}H$ is the optional projection of a nonnegative measurable process H.

Corollary 1. *If* $A \in \mathcal{A}_{loc}$, *then there exists the process* $\tilde{A} \in \mathcal{A}_{loc} \cap \mathcal{P}$ *such that* $A - \tilde{A} \in \mathcal{M}_{loc}$. *If* $H \in \mathcal{P}$ *and* $H \circ A \in \mathcal{A}_{loc}$, *then* $H \circ \tilde{A} \in \mathcal{A}_{loc}$ *and*

$$H \circ A - H \circ \tilde{A} \in \mathcal{M}_{loc}. \tag{6.15}$$

Corollary 2. *If* $A \in \mathcal{A}_{loc} \cap \mathcal{P}$, *then* $\tilde{A} = A$. *If* $A \in \mathcal{A}_{loc}$, *then* ${}^{p}(\Delta A) = \Delta\tilde{A}$. *If* $A \in \mathcal{A}_{loc}$ *and* $\tau \in \mathbf{T}$, *then* $(A^{\tau}) = (\tilde{A})^{\tau}$.

Let us establish certain properties of local martingales belonging to class $\mathcal{V}$ ($\mathbb{F}$).

Theorem 4. 1) *Let* $M \in \mathcal{M}_{loc} \cap \mathcal{V}$ ($\mathbb{F}$). *Then* $M \in \mathcal{A}_{loc}$, *and hence*

$$\mathcal{M}_{loc} \cap \mathcal{V} = \mathcal{M}_{loc} \cap \mathcal{A}_{loc}.$$

2) *If* $M \in \mathcal{M}_{loc} \cap \mathcal{A}_{loc} \cap \mathcal{P}$ *and* $M_0 = 0$, *then* $M = 0$.

3) *If* $M \in \mathcal{M}_{loc} \cap \mathcal{A}_{loc}$ *and* $M_0 = 0$, *then* $\tilde{M} = 0$.

Proof. 1) Denote by (τ_n) a localizing sequence for M ($M^{\tau_n} \in \mathcal{M}$), and let $\sigma_n = \inf$ (t: Var $(M)_t \geq n$). Since $M \in \mathcal{V}$, we have $\sigma_n \uparrow \infty$ and $\tau_n \wedge \sigma_n \uparrow \infty$. Then

$$\text{Var}\,(M)_{\tau_n \wedge \sigma_n} \leq n + |\Delta M^{\tau_n}_{\sigma_n}| \leq 2n + |M^{\tau_n}_{\sigma_n}|.$$

But $\mathbf{E}\,|M^{\tau_n}_{\sigma_n}| < \infty$, and hence $M \in \mathcal{A}_{loc}$.

2) Assume, making use of localizing sequences if necessary, that $M \in \mathcal{M} \cap \mathcal{P}$. As $M \in \mathcal{P}$, the thin set $\{\Delta M \neq 0\}$ is exhausted, according to Theorem 3.6, by a sequence of predictable times. Let $\tau \in \mathbf{T}_p$. Then, by Theorem 3.10, $\Delta M_{\tau} I_{\{\tau < \infty\}}$ is a $\mathcal{F}_{\tau-}$-measurable random variable and, by Theorem 4.1, $\mathbf{E}\,(\Delta M_{\tau} \mid \mathcal{F}_{\tau-})\, I_{\{\tau < \infty\}} = 0$ (P-a.s.). Consequently $\Delta M_{\tau} = 0$ and hence $M \in \mathcal{M}^{c} \cap \mathcal{A}$, where $\mathcal{M}^{c}$ is the class of uniformly integrable martingales with trajectories from $\mathbf{C}$. From this it follows, in particular, that M is a locally bounded martingale. Then $M^2 - M \circ M \in \mathcal{M}_{loc}$ (Problem 4). On the other hand, by properties of the Lebesgue-Stieltjes integral (cf. also Ito's formula

below, Ch. 2, § 3) we have $M_t^2 = 2M \circ M_t$, $t \geq 0$, i.e. $M \circ M = M^2 - M \circ M$, and hence $M \circ M \in \mathcal{M}_{loc}$. By this $M^2 \in \mathcal{M}_{loc}$ and hence $M^2 = 0$ (Problem 4.9).

3) Let B and C be processes from A^+_{loc}, involved in the representation $M = B - C$, and let $\tilde{B}$ and $\tilde{C}$ be their compensators. Then $B - \tilde{B}$ and $C - \tilde{C}$ are local martingales (Theorem 3) and

$$\tilde{B} - \tilde{C} = M - (B - \tilde{B}) + (C - \tilde{C}) \in \mathcal{M}_{loc} \cap \mathcal{A}_{loc} \cap \mathcal{P}.$$

Thus, as has been proved above,

$$\tilde{M} = \widetilde{(B - C)} = \tilde{B} - \tilde{C} = 0.$$

5. Let us consider the structure of compensators for certain simple processes.

Example 1. Let $\pi = (\pi_t, \mathcal{F}_t)_{t \geq 0}$ be a Poisson process with a parameter λ. It is directly verified that a process $(\pi_t - \lambda t, \mathcal{F}_t)$ is a martingale, and this implies that the dual predictable projection (compensator) $\tilde{\pi}$ of the Poisson process π is the deterministic function λt. In this regard it is remarkable to observe that the predictable projection is ${}^P\pi = \pi_-$.

Example 2. Let a process $A = (A_t, \mathcal{F}_t)$ be of the form

$$A_t = \alpha I\,(t \geq \sigma)$$

where σ is a totally inaccessible time, $\mathbf{E}\,|\,\alpha\,| < \infty$ and α is $\mathcal{F}_s$-measurable. Let us show that the compensator $\tilde{A}$ is a continuous process.

In fact, it can be assumed that $\alpha \geq 0$. Since σ is a totally inaccessible time, $P(\sigma > 0) = 1$. According to Theorem 3.9 the process A is left quasicontinuous, i.e. for each predictable τ

$$A_\tau = A_{\tau-} \; (\{\tau < \infty\};\ P\text{-a.s.}).$$

If τ is a bounded predictable time, then, by (6.15),

$$0 = \mathbf{E}\,[A_\tau - A_{\tau-}] = \mathbf{E}\int_0^\infty I_{[\![\tau]\!]}(s)\, dA_s = \mathbf{E}\int_0^\infty I_{[\![\tau]\!]}(s)\, d\tilde{A}_s = \mathbf{E}\,[\tilde{A}_\tau - \tilde{A}_{\tau-}].$$

But $\tilde{A}$ is a predictable increasing process and all moments of its jumps are predictable (Theorem 3.8). Thus $\tilde{A}$ is a continuous process (P-a.s.) by (6.16).

Example 3. Again, let a process $A = (A_t, \mathcal{F}_t)$ be of the form $A_t = \alpha I\,(t \geq \sigma)$, but now let σ be a predictable time, $\mathbf{E}\,|\,\alpha\,| < \infty$, $\mathbf{E}\,(\alpha \,|\, \mathcal{F}_{\sigma-}) = 0$. Let us show then that $\tilde{A} = 0$ and, consequently, the process A is a martingale.

In fact, for each bounded predictable process H a variable H_σ is $\mathcal{F}_{\sigma-}$-measurable (Theorem 3.3) and

$$\begin{aligned} \mathbf{E}(H \circ \tilde{A})_\infty &= \mathbf{E}(H \circ A)_\infty \\ &= \mathbf{E}\int_0^\infty H_s\, dA_s \\ &= \mathbf{E}[H_\sigma(A_\sigma - A_{\sigma-})] \\ &= \mathbf{E}[H_\sigma \mathbf{E}(\alpha \mid \mathcal{F}_{\sigma-})] = 0. \end{aligned}$$

As H is arbitrary, we conclude from this that $\tilde{A} = 0$.

6. We conclude this section by formulating the well known Doob-Meyer theorem, a particular case of which is presented by Theorem 3.

Theorem 5 ([81], [217]). 1. *Let* $X \in \mathbb{F} \cap D$ *be a submartingale of class* $(\mathcal{D})$. *Then there exists a unique up to* P*-indistinguishability increasing integrable predictable process* A *such that* $M = X - A$ *is a uniformly integrable martingale. In other words*

$$X = A + M \quad (A \in \mathcal{A}^+ \cap \mathcal{P}, M \in \mathcal{M}). \tag{6.17}$$

2. *Let* $X \in \mathbb{F} \cap D$ *be a submartingale. Then there exists a unique up to* P*-indistinguishability increasing predictable locally integrable process* A *such that* $M = X - A$ *is a local martingale. In other words*

$$X = A + M \; (A \in \mathcal{A}^+_{loc} \cap \mathcal{P}, M \in \mathcal{M}_{loc}). \tag{6.18}$$

Remark. If $X \in \mathcal{A}^+_{loc}$, then X is a submartingale and the assertion of Theorem 3 follows from the Doob-Meyer theorem just formulated.

Concerning the structure of a predictable process A involved in the Doob-Meyer decomposition X = A + M, additional information may be given. For instance the predictable process A in the decomposition of a submartingale X of the class $(\mathcal{D})$ is continuous if and only if a process X is left quasicontinuous (cf. Definition 4 in § 3), which is equivalent in the present case to the fact that

$$\mathbf{E}X_\tau = \mathbf{E}X_{\tau-} \tag{6.19}$$

for each predictable time τ. (Sometimes processes satisfying property (6.19), are called regular.)

If M is a uniformly integrable martingale ($M \in \mathcal{M}$), then $|M|$ is a submartingale of the class ($\mathcal{D}$), and hence $|M| = A + N$ with $A \in \mathcal{A}^+_{loc} \cap \mathcal{P}$, $N \in \mathcal{M}_{loc}$. Further on the process A just defined will be denoted by $|\tilde{M}|$. It is not hard to verify that in case $M \in \mathcal{M}_{loc}$ the following decomposition takes place:

$$|M| = \widetilde{|M|} + N, \tag{6.20}$$

since the process $|M|$ is locally integrable. Furthermore, if $M \in \mathcal{M}_{loc}$, $q > 1$ and the process $|M|^q$ is locally integrable, then $|M|^q$ also can be decomposed:

$$|M|^q = \widetilde{|M|^q} + N \tag{6.21}$$

with

$$\widetilde{|M|^q} \in \mathcal{A}^+_{loc} \cap \mathcal{P}, \quad N \in \mathcal{M}_{loc}.$$

Problems

1. Give an example, showing that A^p is unequal to ${}^pA$.

2. Let $A, B \in \mathcal{A}^+_{loc}$. Show that $\tilde{A} = \tilde{B}$ if and only if any of the following conditions are satisfied:

(1) $\mathbf{E}A_\tau = \mathbf{E}B_\tau \quad \forall\, \tau \in \mathbf{T}$;

(2) $A - B$ is a local martingale.

3. Let $A \in \mathcal{A}^+ \cap \mathcal{P}$ and let M be a nonnegative uniformly integrable martingale. Then for each $\tau \in \mathbf{T}$

$$\mathbf{E}\int_0^\tau M_{s-}\, dA_s = \mathbf{E}\,[M_\tau A_\tau].$$

If $A \in \mathcal{A}^+$ and if the last equality holds for each bounded martingale M, then $A \in \mathcal{P}$.

4. Let $A \in \mathcal{A}^+_{loc}$ and let M be a local bounded martingale. Then the process

$$MA - M \circ A \in \mathcal{M}_{loc}.$$

If in addition $A \in \mathcal{P}$, then

$$MA - M_- \circ A \in \mathcal{M}_{loc}.$$

5. Let $A \in \mathcal{A}^+_{loc} \cap \mathcal{M}_{loc}$. Then $A_t = 0$ (P-a.s.) for each $t > 0$.

6. If $A \in \mathcal{A}^{+}_{loc}$ and $\Delta A \le c$, then a compensatort $\tilde{A}$ can be chosen in such way that $\Delta\tilde{A} \le c$.

7. Let $A \in \mathcal{A}_{loc}$ and let $\tilde{A}$ be a compensator of A. Show that

$$\mathrm{Var}(\tilde{A}) \le \widetilde{\mathrm{Var}(A)}.$$

8. Let $A \in \mathcal{A}_{loc}$ and let $\tilde{A}$ be a compensator of A. Show that $\mathbf{E}\,|\Delta\tilde{A}_T| \le \mathbf{E}\,|\Delta A_T|$ for each $T \in \mathbf{T}_p$.

9. Let X be a nonnegative process and ${}^0X$ and ${}^pX$ its optional and predictable projections. Show that as $A \in \mathcal{V}^{+}(\mathbb{F})$

$$\mathbf{E}\,(X \circ A_\infty) = \mathbf{E}\,({}^0X \circ A_\infty)$$

and as $A \in \mathcal{V}^{+}(\mathbb{F}) \cap \mathcal{P}$

$$\mathbf{E}\,(X \circ A_\infty) = \mathbf{E}\,({}^pX \circ A_\infty).$$

(Hint: use the technique of proving Theorem 1.)

10. Let $A \in \mathcal{A}^{+}_{loc}$, let $\tilde{A}$ be the compensator of A with respect to the family $\mathbb{F}$ and S a stopping time. Show that $B = (B_t)_{t \ge 0}$ with $B_t = A_{S+t} - A_S$ belongs to $\mathcal{A}^{+}_{loc}(\mathbb{F}^S)$, where $\mathbb{F}^S = (\mathcal{F}^S_t)_{t \ge 0}$ with $\mathcal{F}^S_t = \mathcal{F}_{S+t}$, and that the compensator $\tilde{B}$ of the process B with respect to the family $\mathbb{F}^S$ is given by the formula $\tilde{B}_t = \tilde{A}_{S+t} - \tilde{A}_S$.

§ 7. The structure of local martingales

1. Denote by $\mathcal{M}^c_{loc}$ the class of local martingales with trajectories in C, and define the class $\mathcal{M}^d_{loc}$ of local martingales (called purely discontinuous) in the following manner: $M \in \mathcal{M}^d_{loc}$ if M is strongly orthogonal to any bounded local martingale N from the class $\mathcal{M}^c_{loc}$ with $N_0 = 0$, i.e. $MN \in \mathcal{M}_{loc}$ (observe that in the definition of $\mathcal{M}^d_{loc}$ a bounded local martingale N can be replaced by any process N from the class $\mathcal{M}^c_{loc}$ with $N_0 = 0$).

Introduce also the classes:

$$\mathcal{M}^d = \mathcal{M}^d_{loc} \cap \mathcal{M}, \quad \mathcal{M}^{2,d}_{loc} = \mathcal{M}^2_{loc} \cap \mathcal{M}^d_{loc},$$

$$\mathcal{M}^{2,c} = \mathcal{M}^2 \cap \mathcal{M}^c, \quad \mathcal{M}^{2,d} = \mathcal{M}^2 \cap \mathcal{M}^d.$$

The following two decompositions play a fundamental rôle in studying properties of local martingales, and in developing the stochastic integration theory.

Theorem 1 (first decomposition). *Each local martingale* $M \in \mathcal{M}_{loc}$ *admits the representation*

$$M = M^1 + M^2 \tag{7.1}$$

where $M^1 \in \mathcal{M}_{loc} \cap \mathcal{A}_{loc}$ *and* $M^2 \in \mathcal{M}^2_{loc}$.

Theorem 2 (second decomposition). *Each local martingale admits the unique representation*

$$M = M^c + M^d \tag{7.2}$$

where $M_d \in \mathcal{M}^d_{loc}$ *and* $M_c \in \mathcal{M}^c_{loc}$ *with* $M^c_0 = 0$.

Proof of Theorem 1. Assume $M \in \mathcal{M}$, making use of a localizing sequence if necessary. Put

$$\alpha_t = \sum_{s \leq t} |\Delta M_s| \, I(|\Delta M_s| \geq 1). \tag{7.3}$$

Since $M \in D$, $M(\omega)$ has over each finite interval [0, t] a finite number of jumps at most, the size of which equals (in modulus) to one or exceeds it. Let

$$\sigma_n = \inf(s: \alpha_s \geq n \ \text{ or } \ |M_s| \geq n).$$

Then

$$|\Delta M_{\sigma_n}| \le |M_{\sigma_n}| + |M_{\sigma_n -}| \le |M_{\sigma_n}| + n,$$

and hence

$$\alpha_{\sigma_n} \le \alpha_{\sigma_n -} + |\Delta M_{\sigma_n}| \le |M_{\sigma_n}| + 2n.$$

Since the martingale $M \in \mathcal{M}$ is uniformly integrable, we have $\mathbf{E}\alpha_{\sigma_n} < \infty$, and since $\sigma_n \uparrow \infty$, we have $\alpha \in \mathcal{A}^+_{loc}$. From this it follows that the process A with

$$A_t = \sum_{s \le t} \Delta M_s I(|\Delta M_s| \ge 1)$$

belongs to $\mathcal{A}_{loc}$, and hence (Corollary 1 to Theorem 6.3) its compensator $\tilde{A}$ is defined.

Denote

$$N = M - (A - \tilde{A}).$$

Clearly, $N \in \mathcal{M}_{loc}$. Let us show that $|\Delta N| \le 2$.

Again, making use of a localizing sequence if necessary, we can assume that $A \in \mathcal{A}$, and hence $\tilde{A} \in \mathcal{A}$ too. Then $N \in \mathcal{M}$.

For each time $T \in \mathbf{T}$ on $\{T < \infty\}$ we have $\Delta N_T = \Delta M_T - (\Delta A_T - \Delta\tilde{A}_T)$. If T is a totally inaccessible time, then $\Delta\tilde{A}_T = 0$ (Example 2 in § 6). Hence

$$|\Delta N_T| = |\Delta M_T| I(|\Delta M_T| < 1) \le 1.$$

Now if T is a predictable time, then by Theorem 3.13 and by Corollary 2 to Theorem 6.3 we have on the set $\{T < \infty\}$

$$\Delta\tilde{A}_T = \mathbf{E}(\Delta A_T \mid \mathcal{F}_{T-})$$

which, together with the equality $\mathbf{E}(\Delta M_T \mid \mathcal{F}_{T-}) = 0$ gives

$$\Delta N_T = [\Delta M_T - \mathbf{E}(\Delta M_T \mid \mathcal{F}_{T-})] - [\Delta A_T - \mathbf{E}(\Delta A_T \mid \mathcal{F}_{T-})]$$
$$= \Delta(M - A)_T - \mathbf{E}[\Delta(M - A)_T \mid \mathcal{F}_{T-}].$$

We have $|\Delta(M - A)_T| \le 1$, and hence $|\Delta N_T| \le 2$.

Denote $\tau_n = \inf(t: |N_t| \ge n)$. Then on $[0, \tau_n]$ we have $|N_t| \le n + 2$ and $N^{\tau_n} \in \mathcal{M}^2$ (moreover, N^{τ_n} is a bounded martingale). Thus $M = (A - \tilde{A}) + N$ with $A - \tilde{A} \in \mathcal{M}_{loc} \cap \mathcal{A}_{loc}$ and $N \in \mathcal{M}^2_{loc}$, which implies the first decomposition (7.1).

2. Proof of Theorem 2. Let us begin with proving the uniqueness. Let

$$M = M^{ic} + M^{id}, \; i = 1, 2$$

where $M^{ic} \in \mathcal{M}^c_{loc}$ with $M^{ic}_0 = 0$, and $M^{id} \in \mathcal{M}^d_{loc}$.

Then

$$M^{1d} - M^{2d} = M^{2c} - M^{1c},$$

and hence $M^{1d} - M^{2d}$ is strongly orthogonal to itself. Hence $(M^{1d} - M^{2d})^2$ is a nonnegative local martingale equal to zero at zero, and thus $M^{1d} = M^{2d}$ and $M^{1c} = M^{2c}$ (Problem 4.9).

Consider now the first decomposition $M = M^1 + M^2$, where we assume first that $M^1 \in \mathcal{M} \cap \mathcal{A}$ and $M^2 \in \mathcal{M}^2$. According to Theorem 5.3 we have $M^2 = M^{2,c} + M^{2,d}$ with $M^{2,c} \in \mathcal{M}^{2,c}$ and $M^{2,d} \in \mathcal{M}^{2,d}$. Theorem 3 given below implies that the martingale M^1, as belonging to the class $\mathcal{M} \cap \mathcal{A}$, is strongly orthogonal to any bounded (and hence to any) continuous martingale, i.e. $M^1 \in \mathcal{M}^d$.

Thus, $M = M^{2,c} + (M^{2,d} + M^1)$ with $M^{2,c} \in \mathcal{M}^c$ and $M^{2,d} + M^1 \in \mathcal{M}^d$.

The general case in which $M^1 \in \mathcal{M}_{loc} \cap \mathcal{A}_{loc}$ and $M^2 \in \mathcal{M}^2_{loc}$ is reduced to the case just discussed by making use of localizing sequences. In fact, let $M = M^1 + M^2$ with $M^1 \in \mathcal{M}_{loc} \cap \mathcal{A}_{loc}$ and $M^2 \in \mathcal{M}^2_{loc}$, and let $(\tau_n)_{n \geq 1}$ be a sequence localizing M^1 and M^2 at the same time. Then $M^{\tau_n} = M^c(\tau_n) + M^d(\tau_n)$ with $M^c(\tau_n) \in \mathcal{M}^c$ and $M^d(\tau_n) \in \mathcal{M}^d$. By the uniqueness established above as $m > n$ we have

$$M^c_t(\tau_n) = M^c_t(\tau_m), \quad M^d_t(\tau_n) = M^d_t(\tau_m), \quad t \leq \tau_n.$$

Therefore, the processes M^c and M^d are defined with

$$M^c_t = \lim_n M^c_t(\tau_n), \quad M^d_t = \lim_n M^d_t(\tau_n),$$

possessing the desired property:

$$M = M^c + M^d, \quad M^c \in \mathcal{M}^c_{loc}, \quad M^d \in \mathcal{M}^d_{loc}.$$

3. Theorem 3. *Let* $M - M_0 \in \mathcal{M} \cap \mathcal{A}$ *and* $N \in \mathcal{M}$ *with* $N^*_\infty \leq \text{const}$ *and* $N_0 = 0$. *Then*

$$\mathbf{E} M_\infty N_\infty = \mathbf{E} \sum_{t > 0} \Delta M_t \Delta N_t. \tag{7.4}$$

The equality (7.4) *still holds if*

$$\mathbf{E} N^*_\infty \operatorname{Var}(M)_\infty < \infty.$$

2) *If* $M \in \mathcal{M}^2 \cap \mathcal{A}$ *and* $M_0 = 0$, *then*

$$\mathbf{E}M_\infty^2 = \mathbf{E}\sum_{t>0} (\Delta M_t)^2.$$

Corollary. *Each process from* $\mathcal{M} \cap \mathcal{A}$ *is strongly orthogonal to any bounded continuous martingale* N *with* $N_0 = 0$, *and hence it belongs to* $\mathcal{M}^d$.

Proof. 1) Let B and C be processes from $\mathcal{A}^+$ involved in the representation $M - M_0 = B - C$, and let $\tilde{B}$ and $\tilde{C}$ be their compensators. Then by Theorem 6.4 $\tilde{B} - \tilde{C} = M - M_0 = 0$, and hence

$$M - M_0 = (B - \tilde{B}) - (C - \tilde{C}).$$

Therefore Theorem 6.3 implies

$$\mathbf{E}N_- \circ M_\infty = (\mathbf{E}N_- \circ B_\infty - \mathbf{E}N_- \circ \tilde{B}_\infty) - (\mathbf{E}N_- \circ C_\infty - \mathbf{E}N_- \circ \tilde{C}_\infty) = 0 \tag{7.5}$$

and Theorem 6.1 and its corollary (with $X_s = M_\infty$, $Y_s = N_s$) imply

$$\mathbf{E}M_\infty N_\infty = \mathbf{E}N \circ M_\infty. \tag{7.6}$$

From (7.5) and (7.6) it follows that

$$\mathbf{E}M_\infty N_\infty = \mathbf{E}\Delta N \circ M_\infty = \mathbf{E}\sum_{t>0} \Delta N_t \Delta M_t.$$

Observe that (7.5) and (7.6) take place if

$$\mathbf{E}N_\infty^* \operatorname{Var}(M)_\infty < \infty.$$

Hence under this Condition (7.4) holds.

2) Under the condition $\mathbf{E}(\operatorname{Var}(M)_\infty)^2 < \infty$ the desired equation follows from (7.4) with $N = M$. In the general case introduce the Markov times $\tau_k = \inf(t: \operatorname{Var}(M)_t \geq k)$. Since

$$\operatorname{Var}(M)_{\tau_k} \leq k + |\Delta M_{\tau_k}| \leq 2k + |M_{\tau_k}|$$

and $\mathbf{E}(M_\infty^*)^2 \leq 4\mathbf{E}M_\infty^2$ (Theorem 9.2), we have

$$\mathbf{E}(\operatorname{var}(M)_{\tau_k})^2 \leq 8(k^2 + \mathbf{E}M_\infty^2) < \infty.$$

Hence, $\mathbf{E}M_{\tau_k}^2 = \mathbf{E}\sum_{t \leq \tau_k} (\Delta M_t)^2$, and the desired relation is obtained by taking the limit $\lim_k$ (Problem 5.2), since $\tau_k \uparrow \infty$, $k \to \infty$.

The corollary to Theorem 3 takes place since $\mathbf{E}M_\tau N_\tau = 0$ for each $\tau \in \mathbf{T}$, and due to Problem 4.2 we have $MN \in \mathcal{M}$, which gives $M \perp\!\!\!\perp N$ by Definition 2 in § 5.

4. Thus, each martingale M, being at the same time a process of integrable variation, belongs to the class $\mathcal{M}^d$. In the following theorem the structure of such processes is described.

Theorem 4. *If* $M - M_0 \in \mathcal{M} \cap \mathcal{C}$, *then*

$$M_t = M_0 + \left(\sum_{s \leq t} \Delta M_s\right) - \widetilde{\left(\sum_{s \leq t} \Delta M_s\right)}, \tag{7.7}$$

where

$$\widetilde{\left(\sum_{s \leq t} \Delta M_s\right)}_{t \geq 0}$$

is the compensator of the sum of jumps

$$\left(\sum_{s \leq t} \Delta M_s\right)_{t \geq 0}$$

i.e. M *is the compensator of its jumps.*

Proof. Denote

$$N_t = \sum_{s \leq t} \Delta M_s.$$

Then the process $A = M - M_0 - N \in \mathcal{C}$ is continuous and hence predictable. Therefore

$$A = \tilde{A} = (M - M_0) - \tilde{N}.$$

Now $M - M_0 \in \mathcal{C} \cap \mathcal{M}$ and $(M - M_0) = 0$ (Theorem 6.4). Hence $A = - \tilde{N}$, and at the same time $A = M - M_0 - N$. Thus $M = M_0 + N - \tilde{N}$, and this proves (7.7).

5. As it follows from the proof of Theorem 2, to describe the structure of a process M^d involved in the decomposition $M = M^c + M^d$, one needs to study the structure of processes from the class $\mathcal{M}^{2,d}$ as well.

Theorem 5. *Let* $M \in \mathcal{M}^{2,d}$. *Then*

$$M = \lim_n \sum_{k=1}^{n} M^k \tag{7.8}$$

(the limit is understood in the sense of the convergence in $\mathcal{H}^2$*) where*

$$M_t^k = \Delta M_{\tau_k} I(t \geq \tau_k) - \widetilde{\Delta M_{\tau_k} I(t \geq \tau_k)}, \tag{7.9}$$

while $(\tau_k)_{k \geq 1}$ *is a collection of Markov times such that* $[\![\tau_i]\!] \cap [\![\tau_j]\!] = \emptyset$, $i \neq j$,

$$\{\Delta M \neq 0\} = \{(\omega, t): \Delta M_t(\omega) \neq 0\} \subseteq \bigcup_k [\![\tau_k]\!]. \qquad (7.10)$$

Proof. According to Theorem 1.1, the set $\{\Delta M \neq 0\}$ is thin and it is representable in the form $\bigcup_k [\![\sigma_k]\!]$, where the graphs $[\![\sigma_k]\!]$ of Markov times σ_k do not intersect (otherwise one needs to pass from the sets $[\![\sigma_k]\!]$ to the sets $[\![\sigma_k]\!] \setminus \bigcup_{l<k} [\![\sigma_l]\!]$ and their debuts). Next, according to Theorem 3.7 the times σ_k can be considered as accessible or totally inaccessible. If σ_k is an inaccessible time, then (by Definition 2 in § 3)

$$[\![\sigma_k]\!] \subseteq \bigcup_{p=1}^{\infty} [\![\sigma_k^p]\!]$$

where σ_k^p are predictable times, and their graphs do not intersect (by the same reasons as above). Thus the set $\{\Delta M \neq 0\}$ is contained in the set $\bigcup_k [\![\tau_k]\!]$ where $[\![\tau_i]\!] \cap [\![\tau_j]\!] = \varnothing$, $i \neq j$ and τ_k are predictable or totally inaccessible.

Define

$$A_t^k = \Delta M_{\tau_k} I(t \geq \tau_k).$$

If the time τ_k is totally inaccessible, then the compensator $\tilde{A}^k$ of the process A^k is continuous (Problem 2 in § 6).

But if the time τ_k is predictable, then the compensator $\tilde{A}^k$ of the process A^k is equal to zero, since $\mathbf{E}(\Delta M_{\tau_k} \mid \mathcal{F}_{\tau_k-}) = 0$ (Problem 4).

Consequently, in both cases the martingale

$$M^k = A^k - \tilde{A}^k$$

is a continuous process, except at the time τ_k when $\Delta M_{\tau_k}^k = \Delta M_{\tau_k}$.

Put

$$N^k = M^1 + \dots + M^k$$

and note that $N^k \in \mathcal{M}^2 \cap \mathcal{A}$, since $M^i \in \mathcal{M}^2 \cap \mathcal{A}$, $i = 1, \dots, k$ (Problem 1). Furthermore, $\mathbf{E}(\mathrm{Var}(N^k)_\infty)^2 < \infty$ (Problem 3). Therefore $\mathbf{E}M_\infty^* \mathrm{Var}(N^k)_\infty < \infty$, and analogously to the proof of Theorem 3 for $\tau \in \mathbf{T}$ we have

$$\mathbf{E}M_\tau N_\tau^k = \mathbf{E}\sum_{t \leq \tau} \Delta M_t \Delta N_t^k = \mathbf{E}\sum_{t \leq \tau} (\Delta N_t^k)^2.$$

On the other hand

$$\mathbf{E}\,(N_\tau^k)^2 = \mathbf{E} \sum_{t \le \tau} (\Delta N_t^k)^2.$$

Hence, the processes $M - N^k$ and N^k are strongly orthogonal and consequently

$$\begin{aligned} \mathbf{E}M_\infty^2 &= \mathbf{E}\,(N_\infty^k)^2 + \mathbf{E}\,(M - N^k)_\infty^2 \\ &= \sum_{l=1}^{k} \mathbf{E}\,(M_\infty^l)^2 + \mathbf{E}\,(M - N^k)_\infty^2 \\ &= \sum_{l=1}^{k} \mathbf{E}\,(\Delta M_{\tau_l})^2 + \mathbf{E}\,(M - N^k)_\infty^2. \end{aligned}$$

From this it follows that

$$\sum_{l=1}^{\infty} \mathbf{E}\,(\Delta M_{\tau_l})^2 = \sum_{l=1}^{\infty} \mathbf{E}\,(M_\infty^l)^2 \le \mathbf{E}M_\infty^2 < \infty.$$

Since the processes M^i and M^j with $i \ne j$ are strongly orthogonal, $N^k = M^1 + \dots + M^k$ converges (in $\mathcal{H}^2$ sense) to a certain martingale $N \in \mathcal{M}^2$ (Problem 5.3), and by $M - N^k \perp\!\!\!\perp N^k$ the martingales N and M — N are strongly orthogonal (Problem 5.3). Since $\{\Delta M \ne 0\} \subseteq \bigcup_{j \ge 1} [\![\tau_j]\!]$, then $\{\Delta\,(M - N^k) \ne 0\} \subseteq \bigcup_{j \ge k+1} [\![\tau_j]\!]$.

According to Lemma 5.1, a subsequence of the sequence (N^k) can be chosen converging to N uniformly in t for P-almost all ω. This means that the set $\{\Delta\,(M - N) \ne 0\}$ is P-negligible, i.e. $M - N \in \mathcal{M}^{2,c}$. By assumption, $M \in \mathcal{M}^{2,d}$. Hence, M and M — N are strongly orthogonal (Theorem 5.3).

The properties $M \perp\!\!\!\perp M - N$ and $N \perp\!\!\!\perp M - N$ imply $M - N \perp\!\!\!\perp M - N$, i.e. $(M - N)^2 \in \mathcal{M}$. Consequently, $M = M_0 + N$ (Problem 4.9), i.e. $M - M_0$ is the limit (in the sense of the convergence in $\mathcal{H}^2$) of the sums $N^k = M^1 + \dots + M^k$, which present "the sums of compensated jumps".

Corollary 1. *If* $M \in \mathcal{M}^{2,d}$, *then*

$$\mathbf{E}M_\infty^2 = \mathbf{E}M_0^2 + \mathbf{E} \sum_{t > 0} (\Delta M_t)^2. \tag{7.11}$$

Corollary 2. *If* $M \in \mathcal{M}_{loc}^2$, *then for each* $t > 0$

$$\sum_{s \le t} (\Delta M_s)^2 < \infty \quad \text{(P-a.s.)}. \tag{7.12}$$

6. In this and the subsequent subsections we dwell on further important properties of local martingales.

Theorem 6. *If* $M \in \mathcal{M}_{loc}$, *then for each* $t > 0$

$$\sum_{s \le t} (\Delta M_s)^2 < \infty \quad \text{P-a.s.} \tag{7.13}$$

Proof. According to the first decomposition, $M = M^1 + M^2$ with $M^1 \in \mathcal{M}_{loc} \cap \mathcal{A}_{loc}$ and $M^2 \in \mathcal{M}^2_{loc}$ (Theorem 1). Besides $\sum_{s \le t} (\Delta M^2_s)^2 < \infty$ (P-a.s.), $t > 0$ (Corollary 2 to Theorem 5), and as $M^1 \in \mathcal{A}_{loc}$, we have

$$\sum_{s \le t} (\Delta M^1_s)^2 \le \left(\sum_{s \le t} |\Delta M^1_s| \right)^2 \le (\mathrm{Var}\,(M^1)_t)^2 < \infty \quad \text{(P-a.s.)},\ t > 0.$$

Thus the desired assertion takes place by the obvious inequality

$$\sum_{s \le t} (\Delta M_s)^2 \le 2\left[\sum_{s \le t} (\Delta M^1_s)^2 + \sum_{s \le t} (\Delta M^2_s)^2 \right].$$

7. Recall that for $X \in D$

$$X^*_t = \sup_{s \le t} |X_s|, \quad (\Delta X)^*_t = \sup_{s \le t} |\Delta X_s|.$$

Theorem 7. *For each local martingale* M *the processes*

$$M^* \in \mathcal{A}^+_{loc}, \tag{7.14}$$

$$(\Delta M)^* \in \mathcal{A}^+_{loc}. \tag{7.15}$$

Proof. Let (τ_n) be a localizing sequence for M. Put $\sigma_n = \inf\,(t: |M_t| \ge n) \wedge \tau_n$. Since $M^{\sigma_n} \in \mathcal{M}$, we have $\mathbf{E}\,|M_{\sigma_n}| < \infty$, and hence

$$\mathbf{E}M^*_{\sigma_n} \le n + \mathbf{E}\,|M_{\sigma_n}| < \infty,$$

which proves (7.14). The property (7.15) follows from (7.14).

Remark. The proof entails $M^{\sigma_n} \in \mathcal{H}$, $n \ge 1$.

Theorem 8. *For each local martingale* M *the predictable projection*

$${}^p(\Delta M) = 0. \tag{7.16}$$

Proof. By the definition of the predictable projection given in § 3

$${}^p(\Delta M) = \begin{cases} {}^p((\Delta M)^+) - {}^p((\Delta M)^-) & \text{on } \{{}^p|\Delta M| < \infty\}, \\ \infty & \text{on } \{{}^p|\Delta M| = \infty\}. \end{cases}$$

First, let $M \in \mathcal{H}$. Then $\mathbf{E} \sup_{t > 0} |\Delta M_t| \le 2\mathbf{E} \sup_{t \ge 0} |M_t| < \infty$, and according to Theorem 4.1 and Problem 4.10 for each stopping time τ we have (P-a.s.)

$$I_{\{\tau<\infty\}}\,{}^{p}(\Delta M)_\tau = I_{\{\tau<\infty\}}[\,{}^{p}((\Delta M)^+)_\tau - {}^{p}((\Delta M)^-)_\tau]$$
$$= I_{\{\tau<\infty\}}\mathbf{E}\,[(\Delta M_\tau)^+ - (\Delta M_\tau)^- \mid \mathcal{F}_{\tau-}]$$
$$= I_{\{\tau<\infty\}}\mathbf{E}\,(\Delta M_\tau \mid \mathcal{F}_{\tau-}) = 0.$$

Hence, by Theorem 3.13 for each $M \in \mathcal{H}$ the predictable projection ${}^{p}(\Delta M) = 0$.

Now, let $M \in \mathcal{M}_{loc}$. Then by Theorem 7 there exists a localizing sequence $(\tau_n)_{n\geq1}$ such that $M^{\tau_n} \in \mathcal{H}$. As has been proved ${}^{p}(\Delta M^{\tau_n}) = 0$, $n \geq 1$. The stochastic interval $[\![0, \tau_n]\!] \in \mathcal{P}$ (cf. § 2), and hence ${}^{p}(I_{[\![0,\tau_n]\!]}\Delta M) = I_{[\![0,\tau_n]\!]}\,{}^{p}(\Delta M)$. On the other hand $I_{[\![0,\tau_n]\!]}\Delta M = I_{[\![0,\tau_n]\!]}\Delta M^{\tau_n}$. Therefore

$$I_{[\![0,\tau_n]\!]}\,{}^{p}(\Delta M) = I_{[\![0,\tau_n]\!]}\,{}^{p}(\Delta M^{\tau_n}) = 0, \quad n \geq 1.$$

By this $I_{]\!]\tau_n, \tau_{n+1}]\!]}\,{}^{p}(\Delta M) = 0$, $n \geq 1$. Since $\bigcup_{n\geq1} [\![0, \tau_n]\!] = \Omega \times R_+$, we have

$${}^{p}(\Delta M) = I_{[\![0,\tau_1]\!]}\,{}^{p}(\Delta M) + \sum_{n\geq1} I_{]\!]\tau_n, \tau_{n+1}]\!]}\,{}^{p}(\Delta M) = 0.$$

Problems

1. Let $A_t = \Delta M_T I_{\{t\geq T\}}$, $T \in \mathbf{T}$, $M \in \mathcal{M}^2$ and let $\tilde{A}$ be the compensator of A. Show that $A - \tilde{A} \in \mathcal{M}^2 \cap \mathcal{A}$.

2. Let $\tilde{A}$ be the compensator of A as in Problem 1. Show that $\mathbf{E}\,(\mathrm{Var}\,(\tilde{A})_\infty)^2 < \infty$ (Hint: use the inequality (9.49) and Problem 6.7).

3. Let $N = A - \tilde{A}$ with A and $\tilde{A}$ defined in Problem 1. Show that $\mathbf{E}\,(\mathrm{Var}\,(N)_\infty)^2 < \infty$.

4. Let $\tau \in \mathbf{T}_p$, and let α be an integrable random variable. Show that $X \in \mathcal{M}$ with $X_t = (\alpha - \mathbf{E}\,(\alpha \mid \mathcal{F}_{\tau-}))\, I\,(t \geq \tau)$.

5. Let $M \in \mathcal{M}_{loc} \cap \mathcal{A}_{loc}$ and $H \in \mathcal{P}$, $H \geq 0$. Show that

$$\mathbf{E}\,(H \circ \widetilde{\mathrm{Var}\,(M)_\infty}) = \mathbf{E}\,(H \circ \mathrm{Var}\,(M)_\infty) \leq 2\mathbf{E} \sum_{t>0} H_t\,|\,\Delta M_t\,|.$$

6. Give an example of $M \in \mathcal{M}$ and $\tau \in \mathbf{T}$ such that $M \geq 0$, $\mathbf{E}M_\tau < \infty$ and $\mathbf{E}M_{\tau-} = \infty$.

§ 8. Quadratic characteristic and quadratic variation

1. Let M be a square integrable martingale ($M \in \mathcal{M}^2$, $\mathbf{E} \sup_{t \geq 0} M_t^2 < \infty$). Since M is also a uniformly integrable martingale, by Theorem 4.1 we have

$$M_t = \mathbf{E}\,(M_\infty \mid \mathcal{F}_t)\ (\text{P-a.s.}),\ t \geq 0.$$

Hence $M^2 = (M_t^2, \mathcal{F}_t)$ is a submartingale of the class $(\mathcal{D})$ (Problem 1), and by the Doob-Meyer decomposition (6.17) there exists a predictable process from $\mathcal{A}^+$, denoted by $\langle M, M\rangle$ or by $\langle M\rangle$, such that

$$m = M^2 - \langle M\rangle \tag{8.1}$$

is a uniformly integrable martingale (note that by the definition of class $\mathcal{A}^+$ we have $\langle M\rangle_0 = 0$, and hence $m_0 = M_0^2$).

Analogously, if $M \in \mathcal{M}^2_{loc}$, then by the Doob-Meyer decomposition (6.18) again, a predictable increasing process from $\mathcal{A}^+_{loc}$ can be found, denoted by $\langle M, M\rangle$ or $\langle M\rangle$ as above, such that the process $m = M^2 - \langle M\rangle \in \mathcal{M}_{loc}$.

The process $\langle M\rangle$ is called the *quadratic characteristic* (or the *predictable quadratic variation*) of a local martingale M.

Observe that if $M \in \mathcal{M}^c_{loc}$, then $M \in \mathcal{M}^2_{loc}$ (with a localizing sequence $\tau_n = \inf\,(t: |M_t| \geq n)$, $n \geq 1$), and therefore for such martingales the notion of the quadratic characteristic is defined.

With each pair $M, N \in \mathcal{M}^2_{loc}$ the predictable process $\langle M, N\rangle \in \mathcal{A}_{loc}$ can be associated such that

$$MN - \langle M, N\rangle \in \mathcal{M}_{loc}.$$

The process $\langle M, N\rangle$ is called the *mutual quadratic characteristic* or the *predictable quadratic covariation*, and as is simply verified, it can be defined by the formula

$$\langle M, N\rangle = \frac{1}{4}(\langle M+N, M+N\rangle - \langle M-N, M-N\rangle) \tag{8.2}$$

or equivalently by

$$\langle M, N\rangle = \frac{1}{2}(\langle M+N, M+N\rangle - \langle M, M\rangle - \langle N, N\rangle).$$

Note that if $M, N \in \mathcal{M}^2$, then $\langle M, N\rangle \in \mathcal{A}$.

2. Now, let $M \in \mathcal{M}_{loc}$. By the second decomposition (7.2) $M = M^c + M^d$ with $M^c \in \mathcal{M}^c_{loc}$ and $M^d \in \mathcal{M}^d_{loc}$. Denote ($\Delta M_0 = 0$)

$$[M, M]_t = \langle M^c\rangle_t + \sum_{s\le t} (\Delta M_s)^2 \tag{8.3}$$

where $\langle M^c\rangle$ is the quadratic characteristic of the process $M^c \in \mathcal{M}^2_{loc}$ and

$$\sum_{s\le t} (\Delta M_s)^2 < \infty \quad (\text{P-a.s.})$$

by (7.12).

The process $[M, M]$ is called the *quadratic variation* (or the *optional quadratic variation*) of a local martingale M.

If $M, N \in \mathcal{M}_{loc}$, then analogously to the process $\langle M, N\rangle$ the quadratic covariance is defined by the process

$$[M, N] = \frac{1}{4}([M + N, M + N] - [M - N, M - N]) \tag{8.4}$$

or equivalently by

$$[M, N] = \frac{1}{2}([M + N, M + N] - [M, M] - [N, N]).$$

From this and (8.3) it follows that

$$[M, N] = \langle M^c, N^c\rangle + \sum_s \Delta M_s \Delta N_s.$$

3. Let us dwell on a number of properties of the quadratic variation and the quadratic covariation.

Theorem 1. 1) *Let* $M, N \in \mathcal{M}^2_{loc}$. *Then*

$$[M, N] \in \mathcal{A}_{loc}, \tag{8.5}$$

$$MN - [M, N] \in \mathcal{M}_{loc}, \tag{8.6}$$

$$\widetilde{[M, N]} = \langle M, N\rangle, \tag{8.7}$$

$$M \perp\!\!\!\perp N \Leftrightarrow [M, N] \in \mathcal{M}_{loc}, \quad M_0 N_0 = 0. \tag{8.8}$$

2) *Let* $M, N \in \mathcal{M}_{loc}$. *Then*

$$[M, N] \in \mathcal{V}, \tag{8.9}$$

$$[M, M]^{1/2} \in \mathcal{A}^{+}_{loc}, \tag{8.10}$$

and in case $M_0 = N_0 = 0$ *the properties* (8.6) *and* (8.8) *hold.*

Proof. By virtue of the representations (8.2) and (8.4) it suffices to establish the properties (8.5) — (8.7) only in case N = M. Next, by making use of localizing sequences if necessary, it can be assumed that $M \in \mathcal{M}^2$. Then (8.5) follows from (7.11). By Theorem 5.3 we have $M = M^c + M^d$ where M^c and M^d are square integrable martingales and $M^c \perp\!\!\!\perp M^d$. Therefore, according to (7.11) and the definitions of $\langle M^c \rangle$ and [M, M], for each $\tau \in \mathbf{T}$ we get

$$\begin{aligned} \mathbf{E}M_\tau^2 &= \mathbf{E}\,(M_\tau^c)^2 + \mathbf{E}\,(M_\tau^d)^2 \\ &= \mathbf{E}\,\langle M^c\rangle_\tau + \mathbf{E}M_0^2 + \mathbf{E}\sum_{s\le\tau}(\Delta M_s)^2 \\ &= \mathbf{E}M_0^2 + \mathbf{E}\,[M, M]_\tau. \end{aligned}$$

By the equality $[M, M]_0 = 0$ this entails

$$\mathbf{E}\,(M_\tau^2 - [M, M]_\tau) = \mathbf{E}\,(M_0^2 - [M, M]_0).$$

Consequently, $M^2 - [M, M] \in \mathcal{M}$ (Problem 4.2), and hence (8.6) takes place. The property (8.7) follows from the fact that $M^2 - [M, M] \in \mathcal{M}_{loc}$, $M^2 - \langle M\rangle \in \mathcal{M}_{loc}$. In fact, in this case

$$[M, M] - \langle M\rangle \in \mathcal{M}_{loc},\ [M, M] \in \mathcal{A}^{+}_{loc},\ \langle M\rangle \in \mathcal{A}^{+}_{loc} \cap \mathcal{P}$$

(Theorem 6.3).

The property (8.8) follows from (8.6).

2) As in proving (8.5) it suffices to consider N = M. Then (8.9) follows from (7.12) and from the fact that

$$[M, M]_t + \langle M^c\rangle_t + \sum_{s\le t}(\Delta M_s)^2, \quad \langle M^c\rangle \in \mathcal{V}^{+}.$$

The property (8.10) follows from Davis' inequality

$$c\mathbf{E}\,[M, M]_\tau^{1/2} \le \mathbf{E}M_\tau^*, \quad c > 0, \quad \tau \in \mathbf{T}$$

(cf. (9.27)) and from $M^* \in \mathcal{A}^{+}_{loc}$ (cf. (7.14)). The property (8.6) is established by using Ito's formula (Corollary 2 to Theorem 2.3.1), while property (8.8) follows from (8.6).

4. Let $M^n \in \mathcal{M}_{loc}$, $n \ge 1$. Clearly

$$\sum_{n=1}^{N} M^n \in \mathcal{M}_{loc},\ N \ge 1.$$

In the following theorem the question whether

$$\sum_{n=1}^{\infty} M^n \in \mathcal{M}_{loc}$$

is considered.

Theorem 2. *Let* $M^n \in \mathcal{M}_{loc}$, $M_0^n = 0$, $n \geq 1$, *and let the following conditions be satisfied*:

(a) $$[M^n, M^m] = 0, \quad n \neq m,$$

(b) $$\sum_{n=1}^{\infty} [M^n, M^n]^{1/2} \in \mathcal{A}^+_{loc}.$$

Then

$$\sum_{n=1}^{\infty} M^n \in \mathcal{M}_{loc}.$$

If in addition

(b') $$\sum_{n=1}^{\infty} [M^n, M^n]^{1/2} \in \mathcal{A}^+,$$

then

$$\sum_{n=1}^{\infty} M^n \in \mathcal{H}.$$

Proof. Let Conditions (a) and (b') be satisfied. Then $\mathbf{E}\,[M^n, M^n]_\infty^{1/2} < \infty$ and by Davis' inequality (cf. (9.27)) $\mathbf{E}\,(M^n)^*_\infty \leq C\mathbf{E}\,[M^n, M^n]_\infty^{1/2}$, i.e. $M^n \in \mathcal{H}$.

Consider the process

$$\sum_{n=m}^{m+p} M^n.$$

Clearly, under the assumption (b') we have

$$\sum_{n=m}^{m+p} M^n \in \mathcal{H},$$

and besides, by (a) and Davis' inequality (9.27) we have

$$\mathbf{E}\left(\sum_{n=m}^{m+p} M^n\right)_\infty^* \le C\mathbf{E}\sum_{n=m}^{m+p} [M^n, M^n]_\infty^{1/2}.$$

Therefore

$$\lim_{m,\,p} \mathbf{E}\left(\sum_{n=m}^{m+p} M^n\right)_\infty^* = 0.$$

Thus $\sum_{n=1}^{N} M^n$ converges (by virtue of Cauchy's criterion) uniformly in L_1 (and hence also uniformly in probability) to the limit

$$\sum_{n=1}^{\infty} M^n \text{ with } \mathbf{E}\left(\sum_{n=1}^{\infty} M^n\right)_\infty^* < \infty.$$

Besides,

$$\mathbf{E}\left(\sum_{n=1}^{\infty} M_\infty^n \mid \mathcal{F}_t\right) = \sum_{n=1}^{\infty} M_t^n, \quad t \in R_+.$$

Choose a subsequence (N^k) such that

$$\sum_{n=1}^{N^k} M^n \to \sum_{n=1}^{\infty} M^n$$

uniformly with probability one. Then

$$\sum_{n=1}^{\infty} M^n \in D,$$

and hence

$$\sum_{n=1}^{\infty} M^n \in \mathcal{M}_{loc}.$$

But by Davis' inequality (9.27) and by the assumptions (a) and (b') we have

$$\mathbf{E}\left(\sum_{n=1}^{\infty} M^n\right)_\infty^* \le C\mathbf{E}\sum_{n=1}^{\infty} [M^n, M^n]_\infty^{1/2} < \infty$$

implying

$$\sum_{n=1}^{\infty} M^n \in \mathcal{H}.$$

Now let Conditions (a) and (b) be satisfied. Then there exists a localizing sequence $(\tau_k)_{k \geq 1}$ such that

$$\mathbf{E} \sum_{n=1}^{\infty} [M^n, M^n]_{\tau_k}^{1/2} < \infty, \quad k \geq 1.$$

Denote $M^{n,k} = (M^n)^{\tau_k}$. For each $k \geq 1$ Conditions (a) and (b') are satisfied, and as was proved

$$\sum_{n=1}^{\infty} M^{n,k} \in \mathcal{H}, \quad k \geq 1.$$

Over the stochastic interval $[\![0, \tau_k]\!]$ we have

$$\sum_{n=1}^{\infty} M^{n,k} = \sum_{n=1}^{\infty} M^n.$$

Consequently, by virtue of $\bigcup_{k \geq 1} [\![0, \tau_k]\!] = \Omega \times R_+$ we have

$$\sum_{n=1}^{\infty} M^n \in \mathcal{M}_{loc}.$$

5. Theorem 3. *Let* X *be an optional process. For the existence of a unique (up to* P*-indistinguishability) process* $M \in \mathcal{M}^d_{loc}$ *with* $M_0 = 0$ *and with the property*

$$\Delta M = X,$$

it suffices that

(1) $$^{p}X = 0,$$

(2) $$\left(\sum_s X_s^2 \right)^{1/2} \in \mathcal{A}^+_{loc}$$

where

$$\sum_s X_s^2 = \left(\sum_{s \leq t} X_s^2 \right)_{t \geq 0}$$

Proof. Let

$$\mathbf{E}\left(\sum_{s\geq 0} X_s^2\right)^{1/2} < \infty.$$

Then $\mathbf{E}\,|X_0| < \infty$ evidently. By Condition (1) we have

$$X_0 = \mathbf{E}\,(X_0 \mid \mathcal{F}_{0-}) = ({}^p X)_0 = 0 \quad \text{(P-a.s.)}.$$

Now, observe that the set $\{X \neq 0\} \in \mathcal{D}$, and for P- almost all ω each of its sections is at most countable, by the assumption

$$\mathbf{E}\left(\sum_{s\geq 0} X_s^2\right)^{1/2} < \infty.$$

Therefore by Theorem 33 in [81], Ch. 6, this set presents a union of an at most countable collection of mutually disjoint graphs of Markov times. Thus, up to a P-negligible set

$$\{X \neq 0\} \subseteq \left(\bigcup_{q\geq 1} [\![S_q]\!]\right) \cup \left(\bigcup_{r\geq 1} [\![T_r]\!]\right),$$

where S_q are totally inaccessible times, while T_r are predictable times with mutually disjoint graphs. Note that by the assumption

$$\mathbf{E}\left(\sum_{s\geq 0} X_s^2\right)^{1/2} < \infty$$

the random variables X_{S_q} and X_{T_r} are integrable. Relate to the random variables X_{S_q} and X_{T_r} the processes M^q and N^r by assuming

$$M_t^q = X_{S_q}\, I\,(t \geq S_q) - A_t^q, \quad N_t^r = X_{T_r}\, I\,(t \geq T_r),$$

where $A^q = (A_t^q)_{t\geq 0}$ is the compensator of the process $X_{S_q}\, I\,(t \geq S_q))_{t\geq 0}$. Evidently $M^q \in \mathcal{M}$. Let us show that $N^r \in \mathcal{M}$. According to Problem 3 from § 6 it suffices to show that $\mathbf{E}\,(X_{T_r} \mid \mathcal{F}_{T_r-}) = 0$. The last equality takes place in virtue of (1) (as ${}^pX = 0$) and the definition (3.3) of the predictable projection. Also, observe that $\Delta N^r = X_{T_r}$ and $\Delta M^q = X_{S_q}$ (by the continuity of the process A^q; see Problem 2 in § 6). This entails

$$\mathbf{E}\left(\sum_{q\geq 1} [M^q, M^q]_\infty + \sum_{r\geq 1} [N^r, N^r]_\infty\right)^{1/2} < \infty.$$

If any of the processes M^q, N^r is denoted by Y^i, then evidently $[Y^i, Y^j] = 0$, $i \neq j$. Therefore, by Theorem 2

$$\sum_{q \geq 1} M^q + \sum_{r \geq 1} N^r$$

presents the process $M \in \mathcal{H}$ we are looking for.

If Condition (2) is satisfied, then there exists a localizing sequence $(\tau_k)_{k \geq 1}$ such that

$$E\left(\sum_{s \leq \tau_k} X_s^2\right)^{1/2} < \infty.$$

Therefore processes $M^k \in \mathcal{H}$, $k \geq 1$, can be constructed in such a way that $M^k = (M^k)^{\tau_k}$ and $\Delta M^k = X$ over the stochastic interval $[\![0, \tau_k]\!]$. Consequently, by the equality $\bigcup_{k \geq 1} [\![0, \tau_k]\!] = \Omega \times R_+$ we have

$$M = M^1 + \sum_{k \geq 1} (M^{k+1} - M^k) \in \mathcal{M}_{loc} \text{ and } \Delta M = X.$$

Let a process $\hat{M} \in \mathcal{M}^d_{loc}$, $\hat{M} = 0$ be such that $\Delta\hat{M} = X$. Then $M - \hat{M} \in \mathcal{M}^d_{loc}$ with $(M - \hat{M})_0 = 0$ and $\Delta(M - \hat{M}) = 0$, i.e. $M - \hat{M} \in \mathcal{M}^c_{loc} \cap \mathcal{M}^d_{loc}$. Thus $M = \hat{M}$.

Let $M = (M^1, \ldots, M^k)$, where $M^i \in \mathcal{M}^2_{loc}$, $i = 1, \ldots, k$, i.e. M is a vector-valued local square integrable martingale. By definition the matrix $\langle M \rangle$ with the elements $\langle M^i, M^j \rangle$, $i, j = 1, \ldots, k$, is called the *quadratic characteristic of the vector-valued martingale* M. If $M = (M^1, \ldots, M^k)$ and $M^i \in \mathcal{M}_{loc}$, $i = 1, \ldots, k$, then by definition the *quadratic variation* $[M, M]$ of this vector-valued local martingale is the matrix with the elements $[M^i, M^j]$, $i, j = 1, \ldots, k$.

6. Let us dwell here on one more property of the quadratic characteristic $[M, M]$ of a local martingale M.

Theorem 4. *Let* $M \in \mathcal{M}_{loc}$ *with* $M_0 = 0$ *and let* $[M, M]$ *be its quadratic characteristic. For each* $t > 0$

$$\sum_{j=0}^{n-1} (M_{t^n_{j+1}} - M_{t^n_j})^2 \xrightarrow{P} [M, M]_t, \tag{8.11}$$

where $0 \equiv t^n_0 < t^n_1 < \ldots < t^n_n \equiv t$, $n \geq 1$ *is a sequence of partitions of the interval* $[0, t]$ *such that*

$$\lim_n \max_{0 \le j \le n-1} (t^n_{j+1} - t^n_j) = 0.$$

Proof. We establish first the assertion of the theorem in two particular cases.

α) Let $M \in \mathcal{M}^2_{loc}$. Denote

$$f^n_j(s) = I_{]t^n_j, t^n_{j+1}]}(s)(M_{s-} - M_{t^n_j})$$

and define

$$f^n(s) = \sum_{j=0}^{n-1} f^n_j(s).$$

By Ito's formula (Ch. 2, § 3) we get

$$(M_{t^n_{j+1}} - M_{t^n_j})^2 = 2f^n_j \cdot M_{t^n_{j+1}} + [M, M]_{t^n_{j+1}} - [M, M]_{t^n_j},$$

where $f^n_j \cdot M = (f^n_j \cdot M_t)_{t \ge 0}$ is the stochastic integral with respect to a local martingale (see Ch. 2). In view of the equality $f^n_j \cdot M_{t^n_{j+1}} = f^n_j \cdot M_t$, we get

$$\sum_{j=0}^{n-1} (M_{t^n_{j+1}} - M_{t^n_j})^2 - [M, M]_t = 2f^n \cdot M_t. \tag{8.12}$$

Therefore it suffices to show that

$$\text{P-}\lim_n f^n \cdot M_t = 0.$$

Since $M \in \mathcal{M}^2_{loc}$, then

$$f^n \cdot M = (f^n \cdot M_t)_{t \ge 0} \in \mathcal{M}^2_{loc}$$

and its quadratic characteristic $\langle f^n \cdot M\rangle$ is defined by the formula

$$\langle f^n \cdot M\rangle_t = (f^n)^2 \circ \langle M\rangle_t$$

(see Theorem 2.2.2 and Definition 5 in Ch. 2, § 2). In accordance with Problem 1.9.2 $f^n \cdot M_t \to 0$ in probability, provided $(f^n)^2 \circ \langle M\rangle_t \to 0$ in probability.

To prove the last relation introduce Markov times T^ε_k, $k = 0, 1, \ldots$ by setting $T^\varepsilon_0 = 0$ and

$$T^\varepsilon_k = \inf(t > T^\varepsilon_{k-1} : (M_t - M_{T^\varepsilon_{k-1}})^2 + \langle M\rangle_t - \langle M\rangle_{T^\varepsilon_{k-1}} \ge \varepsilon), \ k \ge 1$$

with $\varepsilon > 0$.

Let A_n^ε be a set of those points ω, for which each of the intervals $]t_j^n, t_{j+1}^n]$ contains at most one point of the sequence $(T_k^\varepsilon)_{k\geq 1}$. Obviously,

$$\lim_n P(\Omega \setminus A_n^\varepsilon) = 0.$$

Hence, it suffices to show that

$$\lim_{\varepsilon\to 0} \overline{\lim_n}\ P((f^n)^2 \circ \langle M\rangle_t \geq a, A_n^\varepsilon) = 0, \quad \forall a > 0. \tag{8.13}$$

Let $\omega \in A_n^\varepsilon$. If the interval $]t_j^n, t_{j+1}^n]$ does not contain any point of the sequence T_k^ε, $k \geq 1$, then

$$\int_{t_j^n}^{t_{j+1}^n} (f^n(s))^2\, d\langle M\rangle_s \leq 2\varepsilon\, [\langle M\rangle_{t_{j+1}^n} - \langle M\rangle_{t_j^n}]. \tag{8.14}$$

If the interval $]t_j^n, t_{j+1}^n]$ contains one point of the sequence T_k^ε, $k \geq 1$, say T_1^ε, then

$$\int_{t_j^n}^{t_{j+1}^n} (f^n(s))^2 d\langle M\rangle_s = \int_{t_j^n}^{T_1^\varepsilon} (f^n(s))^2 d\langle M\rangle_s + \int_{T_1^\varepsilon}^{t_{j+1}^n} (f^n(s))^2 d\langle M\rangle_s$$

$$\leq 2\varepsilon\, [\langle M\rangle_{T_1^\varepsilon} - \langle M\rangle_{t_j^n}] + 2\, [\int_{T_1^\varepsilon}^{t_{j+1}^n} ((M_{u-} - M_{T_1^\varepsilon})^2 + (M_{T_1^\varepsilon} - M_{t_j^n})^2)\, d\langle M\rangle_u]$$

$$\leq 12\varepsilon\, [\langle M\rangle_{t_{j+1}^n} - \langle M\rangle_{t_j^n}] + 4\varepsilon\, (\Delta M_{T_1^\varepsilon})^2. \tag{8.15}$$

By (8.14) and (8.15) we get the following estimate for $(f^n)^2 \circ \langle M\rangle_t$:

$$(f^n)^2 \circ \langle M\rangle_t \leq 12\varepsilon\, (\langle M\rangle_t + [M, M]_t),$$

which gives the desired relation for (8.13).

β) Let $M \in \mathcal{M}_{loc} \cap \mathcal{A}_{loc}$. The representation (8.12) takes place in this case as well. The stochastic integral $f^n \cdot M_t$ coincides here with the Lebesgue-Stieltjes integral (see Definition 1 in Ch. 2, § 2). Hence

$$|f^n \cdot M_t| \leq |f^n| \circ \mathrm{Var}(M)_t$$

i.e. it suffices to show that $|f^n| \circ \mathrm{Var}(M)_t \to 0$ in probability as $n \to \infty$.

To prove the last relation we define Markov times (S_k^ε), $k = 0, 1, \dots$ by setting $S_0^\varepsilon = 0$ and

$$S_k^\varepsilon = \inf (t > S_{k-1}^\varepsilon : | M_t - M_{S_{k-1}^\varepsilon} | + \mathrm{Var}(M)_t - \mathrm{Var}(M)_{S_{k-1}^\varepsilon} \geq \varepsilon), \ k \geq 1$$

with $\varepsilon > 0$.

Let B_n^ε be a set of those points ω, for which each of the intervals $]t_j^n, t_{j+1}^n]$ contains at most one point of the sequence S_k^ε, $k \geq 1$. Obviously

$$\lim_n \mathsf{P} (\Omega \setminus B_n^\varepsilon) = 0.$$

Hence it suffices to show that

$$\lim_{\varepsilon \to 0} \lim_n \mathsf{P} (| f^n | \circ \mathrm{Var}(M)_t \geq a, B_n^\varepsilon) = 0, \quad \forall a > 0. \tag{8.16}$$

Let $\omega \in B_n^\varepsilon$. If the interval $[t_j^n, t_{j+1}^n]$ does not contain any point of the sequence S_k^ε, $k \geq 1$, then

$$\int_{t_j^n}^{t_{j+1}^n} | f^n (s)) \, d \,\mathrm{Var}(M)_s \leq 2\varepsilon [\mathrm{Var}(M_{t_j^n}].$$

If the interval $]t_j^n, t_{j+1}^n]$ contains one point of the sequence S_k^ε, $k \geq 1$, say S_l^ε, then

$$\int_{t_j^n}^{t_{j+1}^n} | f^n (s) | \, d \,\mathrm{Var}(M)_s = \int_{t_j^n}^{S_l^\varepsilon} | f^n (s) | \, d \,\mathrm{Var}(M)_s + \int_{S_l^\varepsilon}^{t_{j+1}^n} | f^n (s) | \, d \,\mathrm{Var}(M)_s$$

$$\leq 2\varepsilon [\mathrm{Var}(M)_{S_l^\varepsilon} - \mathrm{Var}(M)_{t_j^n}] + \int_{S_l^\varepsilon}^{t_{j+1}^n} (| M_{u-} - M_{S_l^\varepsilon} | + | M_{S_l^\varepsilon} - M_{t_j^n} |) \, d \,\mathrm{Var}(M)_s$$

$$< 3\varepsilon [\mathrm{Var}(M)_{t_{j+1}^n} - \mathrm{Var}(M)_{t_j^n}] + \varepsilon | \Delta M_{S_l^\varepsilon} |.$$

This and (8.17) give

$$| f^n | \circ \mathrm{Var}(M)_t \leq 4\varepsilon \,\mathrm{Var}(M)_t$$

and hence the relation (8.16) takes place..

We will use now the fact that a local martingale M admits the decomposition

$$M = M^1 + M^2$$

with $M^1 \in \mathcal{M}_{loc} \cap \mathcal{A}_{loc}$ and $M^2 \in \mathcal{M}^2_{loc}$ (Theorem 1.7.1).

As has been proved

$$\sum_{j=0}^{n-1} (M^i_{t^n_{j+1}} - M^i_{t^n_j})^2 \xrightarrow{P} [M^i, M^i]_t, \quad i = 1, 2.$$

On the other hand by the definition of the quadratic characteristic we have

$$[M, M] = [M^1, M^1] + [M^2, M^2] + 2[M^1, M^2].$$

Therefore the desired assertion takes place, provided

$$\sum_{j=0}^{n-1} [M^1_{t^n_{j+1}} - M^1_{t^n_j}] [M^2_{t^n_{j+1}} - M^2_{t^n_j}] \xrightarrow{P} [M^1, M^2]_t. \tag{8.18}$$

To prove (8.18) we note

$$[M^1, M^2]_t = \sum_{0 < s \le t} \Delta M^1_s \Delta M^2_s = \Delta M^2 \circ M^1_t.$$

Define the function

$$g_n(s) = \sum_{j=0}^{n-1} I_{]t^n_j, t^n_{j+1}]}(s) [M^2_{t^n_{j+1}} - M^2_{t^n_j}].$$

Then

$$g_n \circ M^1_t = \sum_{j=0}^{n-1} [M^1_{t^n_{j+1}} - M^1_{t^n_j}] [M^2_{t^n_{j+1}} - M^2_{t^n_j}].$$

Thus the relation (8.18) takes place, provided $(\Delta M^2 - g_n) \circ M^1_t \to 0$ in probability as $n \to \infty$.

Since

$$|(\Delta M^2 - g_n) \circ M^1_t| \le |\Delta M^2 - g_n| \circ \operatorname{Var}(M^1)_t,$$

it suffices to show that

$$|\Delta M^2 - g_n| \circ \operatorname{Var}(M^1)_t \to 0$$

in probability as $n \to \infty$.

The last relation takes place in virtue of

$$\lim_n g_n(s) = \Delta M^2_s, \quad |g_n(s)| \le 2 \sup_{u \le s} |M^2_u|$$

and Lebesgue's dominating convergence theorem.

Problems

1. Show that for $M \in \mathcal{M}^2$ the process M^2 is a submartingale of the class $(\mathcal{D})$.

2. Let $M \in \mathcal{M}^2_{loc}$. Show that

$$\Delta\langle M\rangle_\tau = \mathbf{E}\,((\Delta M_\tau)^2 \mid \mathcal{F}_{\tau-}), \qquad \tau \in \mathbf{T}_p.$$

3. Let $A \in \mathcal{A}^+_{loc}$, and let $\tilde{A}$ be the compensator of A. If $A^c = 0$, $\Delta A = 0$ or 1, then $M = A - \tilde{A} \in \mathcal{M}^2_{loc}$ and $\langle M\rangle = (1 - \Delta\tilde{A}) \circ \tilde{A}$.

4. Let π be a Poisson process with the intensity λ. Show that the local square integrable martingale $M = (\pi_t - \lambda t)_{t \geq 0}$ has the quadratic characteristic $\langle M\rangle_t = \lambda t$.

5. If $M \in \mathcal{M}^2$, $M_0 = 0$, then $\mathbf{E}M_\tau^2 = \mathbf{E}\,[M, M]_t = \mathbf{E}\,\langle M\rangle_\tau$, $\tau \in \mathbf{T}$.

6. If $M \in \mathcal{M}^2_{loc}$, $M_0 = 0$ and if τ is a stopping time, then $\mathbf{E}M_\tau^2 \leq \mathbf{E}\,[M, M]_\tau = \mathbf{E}\,\langle M\rangle_\tau$.

7. If $M \in \mathcal{M}^2_{loc}$, $M_0 = 0$ and $\tau \in \mathbf{T}$, then $\mathbf{E}\,[M, M]_\tau = \mathbf{E}\,\langle M\rangle_\tau \leq \mathbf{E}\,(M^*_\tau)^2$.

8. Show that for $M \in \mathcal{M}^c_{loc}$ the process $\langle M\rangle \in \mathcal{C}$.

9. Show that for $M, N \in \mathcal{M}^2_{loc}$

$$\operatorname{Var}(\langle M, N\rangle) \leq \frac{1}{2}\,(\langle M\rangle + \langle N\rangle).$$

10. Show that for $M, N \in \mathcal{M}^2_{loc}$

$$\langle M, N\rangle \leq \langle M\rangle^{1/2} \langle N\rangle^{1/2},$$

and for $M, N \in \mathcal{M}_{loc}$

$$[M, N] \leq [M, M]^{1/2} [N, N]^{1/2}.$$

11. Show that for $M \in \mathcal{M}^2_{loc}$

$$[M, M] - \langle M\rangle \in \mathcal{M}^d_{loc}.$$

12. Show that for $M \in \mathcal{M}^2_{loc}$

$${}^p((\Delta M)^2) = \Delta\,\langle M\rangle.$$

13. Let $M \in \mathcal{M}^2_{loc}(\mathbb{F})$, $M_0 = 0$, let S be a stopping time, and let $\mathbb{F}^S = (\mathcal{F}^S_t)_{t \geq 0}$ with $\mathcal{F}^S_t = \mathcal{F}_{S+t}$, $\hat{M}_t = M_{S+t} - M_S$. Show that

$$\hat{M} = (\hat{M}_t)_{t \geq 0} \in \mathcal{M}^2_{loc}(\mathbb{F}^S) \text{ and } \langle \hat{M} \rangle_t = \langle M \rangle_{S+t} - \langle M \rangle_S.$$

14. Let M be a vector-valued local square integrable martingale and let $\langle M \rangle$ be its quadratic characteristic (a matrix). Show that $\langle M \rangle_t$ and $\langle M \rangle_t - \langle M \rangle_s$, $s \leq t$, are symmetric nonnegative definite matrices.

15. Let M be a vector-valued local martingale, and let [M, M] be its quadratic variation (a matrix). Show that $[M, M]_t$ and $[M, M]_t - [M, M]_s$, $s \leq t$, are symmetric nonnegative definite matrices.

16. Let $M \in \mathcal{M}_{loc}$ and $\tau \in T$. Then $[M^\tau, M^\tau] = [M, M]^\tau$. If $M \in \mathcal{M}^2_{loc}$, then $\langle M^\tau \rangle = \langle M \rangle^\tau$.

17. Let $M, N \in \mathcal{M}^2_{loc}$ with $M_0 = N_0 = 0$ and let processes M and N be independent. Show that $\langle M, N \rangle = 0$. (Hint: utilize Theorem 2.3.2.)

§ 9. Inequalities for local martingales

1. Doob's inequality.

Theorem 1. *Let* T *be a Markov time* ($T \le \infty$).

1) *If* M *is a uniformly integrable martingale* ($M \in \mathcal{M}$), *then for each* $a > 0$

$$P(M_T^* \ge a) \le \frac{1}{a} E\, I(M_T^* \ge a)\,|M_T| \le \frac{1}{a} E\,|M_T| \tag{9.1}$$

with $M_t^* = \sup_{s \le t} |M_s|$.

2) *If* M *is a locally integrable martingale* ($M \in \mathcal{M}_{loc}$) *with* $M_0 = 0$, *then*

$$P(M_T^* \ge a) \le \frac{1}{a} E\,|\widetilde{M}|_T. \tag{9.1'}$$

3) *If M is a local square integrable martingale* ($M \in \mathcal{M}^2_{loc}$) *with* $M_0 = 0$, *then*

$$P(M_T^* \ge a) \le \frac{1}{a^2} E\,[M, M]_T = \frac{1}{a^2} E\,\langle M\rangle_T. \tag{9.2}$$

Proof. 1) Let $\tau = \inf(t: |M_t| > a) \wedge T$, with $a > 0$. According to Problem 8 we have $P(M_T^* > a) \le P(|M_\tau| \ge a)$.

Consequently

$$P(M_T^* > a) \le \frac{1}{a} E\,|M_\tau|\, I(|M_\tau| \ge a). \tag{9.3}$$

In virtue of Jensen's inequality $|E(M_T \mid \mathcal{F}_\tau)| \le E(|M_T| \mid \mathcal{F}_\tau)$ and the equality $M_\tau = E(M_T \mid \mathcal{F}_\tau)$ we get

$$E\,|M_\tau|\, I(|M_\tau| \ge a) \le E\,[E(|M_T| \mid \mathcal{F}_\tau)\, I(|M_\tau| \ge a)] = E\,|M_T|\, I(|M_\tau| \ge a).$$

By this

$$E\,|M_\tau|\, I(|M_\tau| \ge a) \le E\,|M_T|\, I(M_T^* \ge a)$$

(cf. Problem 9) and consequently according, to (9.3),

$$P(M_T^* > a) \le \frac{1}{a} E\,|M_T|\, I(M_T^* \ge a). \tag{9.4}$$

Let a_k, $k \ge 1$ be a sequence of numbers such that $0 < a_k < a$, $k \ge 1$ and $a_k \uparrow a$. Then by the inequality (9.4)

$$P(M_T^* > a_k) \le \frac{1}{a_k} E\,|M_T|\, I(M_T^* \ge a_k), \quad k \ge 1. \tag{9.5}$$

The assertion (9.1) follows from (9.5) by taking the limit as $k \to \infty$.

2) Let $(a_k)_{k\geq 1}$ be a sequence of numbers such that $0 < a_k < a$, $a_k \uparrow a$, where a_k is a continuity point for the distribution function of the random variable M^*_T, and let $(T_n)_{n\geq 1}$ be a localizing sequence for M. By 1) we have

$$P(M^*_{T\wedge T_n} > a_k) \leq \frac{1}{a_k} \mathbf{E}\, |M_{T\wedge T_n}|, \quad k \geq 1, \ n \geq 1.$$

By the decomposition (6.20) (Ch. 1) for $|M|$ we have

$$\mathbf{E}\,|M_{T\wedge T_n}| = \mathbf{E}\,\widetilde{|M|}_{T\wedge T_n}, \quad n \geq 1.$$

Hence

$$P(M^*_{T\wedge T_n} > a_k) \leq \frac{1}{a_k} \mathbf{E}\,\widetilde{|M|}_{T\wedge T_n} \leq \frac{1}{a_k} \mathbf{E}\,\widetilde{|M|}_T.$$

This implies the desired assertion by taking the limit $\lim\limits_k \lim\limits_n$.

3) According to Problem 8 we have $P(M^*_T > a) \leq P(M^2_\tau \geq a^2)$. Therefore, by Chebyshev's inequality

$$P(M^*_T > a) \leq \frac{1}{a^2} EM^2_\tau.$$

If a_k, $k \geq 1$, is a sequence introduced in the course of proving Assertion 1), then by taking the limit as $k \to \infty$ in

$$P(M^*_T > a_k) \leq \frac{1}{a_k^2} EM^2_\tau, \quad k \geq 1,$$

we arrive at the inequality

$$P(M^*_T \geq a) \leq \frac{1}{a^2} EM^2_\tau. \tag{9.6}$$

For $M \in \mathcal{M}^2$ we have $EM^2_\tau = \mathbf{E}\,[M, M]_\tau$ (Problem 8.5), which together with the inequality $[M, M]_\tau \leq [M, M]_T$ gives the desired inequality (9.2), in virtue of (9.6).

In the general case $M \in \mathcal{M}^2_{loc}$ consider a localizing sequence (T^n), $n \geq 1$, for which $M^{T^n} \in \mathcal{M}^2$, $n \geq 1$. Then, as has been proved already,

$$P(M^*_{T\wedge T^n} > a_k) \leq \frac{1}{a_k^2} \mathbf{E}\,[M, M]_{T\wedge T^n} = \frac{1}{a_k^2} \mathbf{E}\,\langle M\rangle_{T\wedge T^n} \tag{9.7}$$

where a sequence a_k, $k \geq 1$, introduced in the course of Assertion 1) is such that a_k presents a continuity point for the distribution function of the random variable M_T^*. Then the desired inequality (9.2) follows from (9.7) by taking the limit $\lim_k \lim_n$.

Corollary (Métivier's inequality). *If* T *is a predictable Markov time* ($T \leq \infty$) *and* $M \in \mathcal{M}^2_{loc}$, *then*

$$P(M^*_{T-} \geq a) \leq \frac{1}{a^2} E[M, M]_{T-} = \frac{1}{a^2} E\langle M\rangle_{T-}, \quad a > 0. \tag{9.8}$$

Proof. Let (T^n), $n \geq 1$, be an announcing sequence for T, and a sequence a_k, $k \geq 1$, introduced in the course of proving Assertion 1) in Theorem 1 is such that a_k presents a continuity point for the distribution function of the random variable M^*_{T-}. By Theorem 1

$$P(M^*_{T^n} > a_k) \leq \frac{1}{a_k^2} E[M, M]_{T^n} = \frac{1}{a_k^2} E\langle M\rangle_{T^n}.$$

The desired inequality (9.8) follows from this by taking the limit $\lim_k \lim_n$.

If $M \in \mathcal{M}_{loc}$, then

$$P(M^*_{T-} \geq a) \leq \frac{1}{a} E\widetilde{|M|}_{T-}. \tag{9.8'}$$

The inequality (9.8') is proved analogously.

Theorem 2. *Let* T *be a Markov time* ($T \leq \infty$), $p > 1$ *and let* M *be a uniformly integrable martingale* ($M \in \mathcal{M}$) *with* $E|M_T|^p < \infty$. *Then*

$$E(M_T^*)^p \leq \left(\frac{p}{p-1}\right)^p E|M_T|^p. \tag{9.9}$$

If $M \in \mathcal{M}_{loc}$ *with* $M_0 = 0$ *and* $|M|^p$ *is a locally integrable process, then*

$$E(M_T^*)^p \leq \left(\frac{p}{p-1}\right)^p E\widetilde{|M|}^p_T. \tag{9.9'}$$

Proof. First, suppose $E(M_T^*)^p < \infty$. By Theorem 1

$$P(M_T^* \geq a^{1/p}) \leq a^{-1/p} E I(M_T^* \geq a^{1/p})|M_T|.$$

Therefore

$$\mathbf{E}\,(M_T^*)^p = \int_0^\infty P\,(M_T^* \geq a^{1/p})\,da \leq \mathbf{E}\,|\,M_T\,|\int_0^{(M_T^*)^p} a^{-1/p}da$$

$$= \frac{p}{p-1}\mathbf{E}\,|\,M_T\,|\,(M_T^*)^{p-1}.$$

By using Hölder's inequality with $p' = \dfrac{p}{p-1}$ and $q' = p$, from this we get

$$\mathbf{E}\,(M_T^*)^p \leq \frac{p}{p-1}(\mathbf{E}\,|\,M_T\,|^p)^{1/p}(\mathbf{E}\,(M_T^*)^p)^{\frac{p-1}{p}}.$$

Under the assumption $\mathbf{E}\,(M_T^*)^p < \infty$, the solution of this inequality with respect to $\mathbf{E}\,(M_T^*)^p < \infty$ gives the desired inequality (9.9).

To prove (9.9) in the general case introduce Markov times $T_k = \inf\,(t\colon |\,M_t\,| \geq k) \wedge T$. Then

$$M_{T_k}^* \leq M_{T_k-}^* + |\,\Delta M_{T_k}\,| \leq 2M_{T_k-}^* + |\,M_{T_k}\,| \leq 2k + |\,M_{T_k}\,|.$$

By this

$$(M_{T_k}^*)^p \leq 2^{p-1}[(2k)^p + |\,M_{T_k}\,|^p]. \tag{9.10}$$

Next

$$|\,\mathbf{E}\,(M_T\,|\,\mathfrak{F}_{T_k})\,|^p \leq \mathbf{E}\,(|\,M_T\,|^p\,|\,\mathfrak{F}_{T_k})$$

and $\mathbf{E}\,(M_T\,|\,\mathfrak{F}_{T_k}) = M_{T_k}$. Therefore

$$\mathbf{E}\,|\,M_{T_k}\,|^p \leq \mathbf{E}\,|\,M_T\,|^p. \tag{9.11}$$

From (9.10) and (9.11) it follows that $\mathbf{E}\,(M_{T_k}^*)^p < \infty$. Thus, as has been proved already

$$\mathbf{E}\,(M_{T_k}^*)^p \leq \left(\frac{p}{p-1}\right)^p \mathbf{E}\,|\,M_{T_k}\,|^p \leq \left(\frac{p}{p-1}\right)^p \mathbf{E}\,|\,M_T\,|^p,$$

and the desired inequality (9.9) follows from this by taking the limit as $k \to \infty$, in virtue of Fatou's lemma.

Let $(T_n)_{n\geq 1}$ be a localizing sequence for M and $|\,M\,|^p$. Then by (9.9)

$$\mathbf{E}\,(M_{T\wedge T_n}^*)^p \leq \left(\frac{p}{p-1}\right)^p \mathbf{E}\,|\,M_{T\wedge T_n}\,|^p.$$

In virtue of (6.21) (Ch. 1) we have

$$\mathbf{E}\,|M_{T\wedge T_n}|^p = \mathbf{E}\,|\widetilde{M}|^p_{T\wedge T_n}.$$

Hence

$$\mathbf{E}\,(M^*_{T\wedge T_n})^p \le \left(\frac{p}{p-1}\right)^p \mathbf{E}\,|\widetilde{M}|^p_{T\wedge T_n} \le \left(\frac{p}{p-1}\right)^p \mathbf{E}\,|\tilde{M}|^p_{T}.$$

This implies the inequality (9.9') by Fatou's lemma.

2. The Lenglart - Rebolledo inequality.

Theorem 3. *Let* X *and* Y *be nonnegative processes from the class* $D \cap \mathbb{F}$, *and* $X_0 = Y_0 = 0$, $Y \in \mathcal{V}^+$. *Let* Y *dominate the process* X (*shortly*: X ⦃ Y) *in the sense that for each stopping time* τ

$$\mathbf{E}X_\tau \le \mathbf{E}Y_\tau. \tag{9.12}$$

Then for each Markov time T *and all numbers* $a > 0$, $b > 0$

$$\mathsf{P}\,(\sup_{t\le T} X_t \ge a) \le \frac{1}{a}\mathbf{E}\,[Y_T \wedge (b + \sup_{t\le T}\Delta Y_t)] + \mathsf{P}\,(Y_T \ge b). \tag{9.13}$$

If, in addition the process Y *is predictable, then*

$$\mathsf{P}\,(\sup_{t\le T} X_t \ge a) \le \frac{1}{a}\mathbf{E}\,[Y_T \wedge b] + \mathsf{P}\,(Y_T \ge b). \tag{9.14}$$

Proof. First, suppose $T < \infty$. Denote

$$\alpha = \inf\,(t\colon X_t > a), \quad \beta = \inf\,(t\colon Y_t \ge b)$$

and observe that

$$\mathsf{P}\,(\sup_{t\le T} X_t > a) = \mathsf{P}\,(\sup_{t\le T\wedge\beta} X_t > a,\ \beta > T) + \mathsf{P}\,(\sup_{t\le T} X_t > a,\ \beta \le T)$$

$$\le \mathsf{P}\,(\sup_{t\le T\wedge\beta} X_t > a) + \mathsf{P}\,(\beta \le T). \tag{9.15}$$

Let us evaluate the terms on the right-hand side of (9.15). We have

$$\mathsf{P}\,(\beta \le T) = \mathsf{P}\,(Y_T \ge b). \tag{9.16}$$

Further, according to Problem 8,

$$\{\sup_{t\le T\wedge\beta} X_t > a\} \subseteq \{X_{\alpha\wedge T\wedge\beta} \ge a\}.$$

Due to Chebyshev's inequality by this and (9.12) we get

$$\mathsf{P}\,(\sup_{t\le T\wedge\beta} X_t > a) \le \mathsf{P}\,(X_{\alpha\wedge T\wedge\beta} \ge a) \le \frac{1}{a}\mathbf{E}X_{\alpha\wedge T\wedge\beta} \le \frac{1}{a}\mathbf{E}Y_{\alpha\wedge T\wedge\beta}. \tag{9.17}$$

Observe that

$$Y_{\alpha\wedge T\wedge\beta} \le Y_{T\wedge\beta} = Y_T \wedge Y_\beta = Y_T \wedge Y_{T\wedge\beta} = Y_T \wedge [Y_{(T\wedge\beta)-} + \Delta Y_{T\wedge\beta}]$$

$$\le Y_T \wedge [Y_{\beta-} + \sup_{t\le T} \Delta Y_t]$$

$$\le Y_T \wedge [b + \sup_{t\le T} \Delta Y_t].$$

By this and (9.17)

$$P(\sup_{t\le T\wedge\beta} X_t > a) \le \frac{1}{a} \mathbf{E}\,[Y_T \wedge (b + \sup_{t\le T} \Delta Y_t)].$$

Let a_k, $k \ge 1$, be a sequence of numbers such that $0 < a_k < a$, $k \ge 1$ and $a_k \uparrow a$. Then, as has been proved already,

$$P(\sup_{t\le T} X_t > a_k) \le \frac{1}{a_k} \mathbf{E}\,[Y_T \wedge (b + \sup_{t\le T} \Delta Y_t)] + P(Y_T \ge b), \quad k \ge 1, \tag{9.18}$$

and by taking in this inequality the limit as $k \to \infty$ we arrive at the inequality (9.13).

To prove (9.14) note that since $Y \in \mathcal{V}^+ \cap \mathcal{P}$, time β is predictable. Let $(\beta_n)_{n\ge 1}$ be an announcing sequence of Markov times for β. Then by (9.15) we have

$$P(\sup_{t\le T} X_t > a) \le P(\sup_{t\le T\wedge\beta_n} X_t > a) + P(\beta_n \le T). \tag{9.19}$$

An analogous to (9.17) inequality is presented by

$$P(\sup_{t\le T\wedge\beta_n} X_t > a) \le \frac{1}{a} \mathbf{E} Y_{\alpha\wedge T\wedge\beta_n}.$$

But $Y_{\alpha\wedge T\wedge\beta_n} \le Y_{T\wedge\beta_n} = Y_T \wedge Y_{\beta_n} \le Y_T \wedge b$, and hence

$$P(\sup_{t\le T\wedge\beta_n} X_t > a) \le \frac{1}{a} \mathbf{E}\,[Y_T \wedge b]. \tag{9.20}$$

Let us evaluate $P(\beta_n \le T)$. We have

$$P(\beta_n \le T) = P(\beta_n \le T, \beta - \beta_n > \varepsilon) + P(\beta_n \le T, \beta - \beta_n \le \varepsilon)$$

$$\le P(\beta - \beta_n > \varepsilon) + P(\beta \le T + \varepsilon)$$

$$= P(\beta - \beta_n > \varepsilon) + P(Y_{T+\varepsilon} \ge b).$$

Let a_k, $k \ge 1$, be a sequence of numbers introduced above in Subsection 1 ($0 < a_k < a$, $a_k \uparrow a$) and let b_j, $j \ge 1$, be a sequence of numbers with the properties: $0 < b_j < b$, $j \ge 1$, $b \uparrow b$ where b_j are continuity points for the distribution function of the random variable Y_T. Then

$$P(\beta_n \le T) \le P(\beta - \beta_n > \varepsilon) + P(Y_{T+\varepsilon} > b_j), \quad j \ge 1. \tag{9.21}$$

By (9.19) - (9.21) we get

$$P(\sup_{t\le T} X_t > a_k) \le \frac{1}{a_k} \mathbf{E}\,[Y_T \wedge b] + P(\beta - \beta_n > \varepsilon) + P(Y_{T+\varepsilon} > b_j).$$

The desired inequality (9.14) is obtained by taking the limit $\lim_k \lim_{\varepsilon \to 0} \lim_n$.

Let us establish now the validity of the inequalities (9.13) and (9.14) for an arbitrary Markov time T (not necessarily finite).

Let L_T denote the right-hand side of any of the inequalities (9.13) or (9.14). Let $T_n = T \wedge n$. Then

$$P(\sup_{t\le T_n} X_t \ge a) \le L_{T_n} \le L_T. \tag{9.22}$$

Observe that $\sup_{t\le T_n} X_t \uparrow \sup_{t\le T} X_t$. Choose a sequence a_k, $k \ge 1$, such that $a_k \uparrow a$ and a_k are continuity points for the distribution function of the random variable $\sup_{t\le T} X_t$. Evidently

$$P(\sup_{t\le T_n} X_t > a_k) \le P(\sup_{t\le T_n} X_t \ge a) \le L_T.$$

By this we have

$$P(\sup_{t\le T} X_t > a_k) = \lim_n P(\sup_{t\le T_n} X_t > a_k) \le L_T$$

and

$$P(\sup_{t\le T} X_t \ge a) = \lim_k P(\sup_{t\le T} X_t > a_k) \le L_T,$$

which prove the inequalities (9.13) and (9.14) for an arbitrary Markov time $T \le \infty$.

3. A generalization of the Lenglart - Rebolledo inequality (9.14).

Theorem 4. *Let* X *and* Y *be such as in Theorem 3, and let* Z *be an increasing process with* $Z_0 = 0$, *belonging to the class* $\mathbb{F} \cap \mathrm{D}$ *and such that*

$$\Delta Y \le Z_-.$$

Then for each Markov time T *and numbers* $a > 0$, $b > 0$

$$P(\sup_{t\le T} X_t \ge a) \le \frac{1}{a} \mathbf{E}\,[Y_T + Z_T) \wedge b] + P(Y_T + Z_T \ge b). \tag{9.23}$$

Proof. Denote

$$\tau = \inf(t\colon X_t > a) \wedge T, \quad \sigma = \inf(t\colon Y_t + Z_t \ge b) \wedge T.$$

Let $T < \infty$. Then by taking into consideration Problem 8 we have

$$\begin{aligned} P(\sup_{t\le T} X_t > a) &\le P(X_\tau \ge a) \le P(X_\tau \ge a, \sigma = T) + P(\sigma < T) \\ &\le P(X_{t\wedge\sigma} \ge a) + P(\sigma < T) \\ &\le \frac{1}{a} \mathbf{E} X_{\tau\wedge\sigma} + P(\sigma < T) \\ &\le \frac{1}{a} \mathbf{E} Y_\sigma + P(\sigma < T). \end{aligned}$$

If $0 < a_k < a$, $a_k \uparrow a$, then by taking the limit $\lim_k$ in the inequalities

$$P(\sup_{t \le T} X_t > a_k) \le \frac{1}{a_k} \mathbf{E} Y_\sigma + P(\sigma < T), \quad k \ge 1,$$

we get

$$P(\sup_{t \le T} X_t \ge a) \le \frac{1}{a} \mathbf{E} Y_\sigma + P(\sigma < T). \tag{9.24}$$

Next,

$$Y_\sigma = Y_{\sigma-} + \Delta Y_\sigma \le Y_{\sigma-} + Z_{\sigma-} \le b \wedge (Y_{\sigma-} + Z_{\sigma-}) \le b \wedge (Y_T + Z_T),$$

which together with (9.24) proves the desired assertion for $T < \infty$. In the general case $T \le \infty$ the proof is completed in the same manner as in Theorem 3.

4. Inequalities for the moments of order $0 < p < 2$.

Theorem 5. *Let* T *be a Markov time* ($T \le \infty$) *and* $0 < p < 2$.

1) *If* $M \in \mathcal{M}^2_{loc}$, $M_0 = 0$, *then*

$$\mathbf{E}(M_T^*)^p \le \frac{4-p}{2-p} \mathbf{E} \langle M \rangle_T^{p/2}. \tag{9.25}$$

2) *If* $M \in \mathcal{M}^c_{loc}$, $M_0 = 0$, *then*

$$\frac{2-p}{4-p} \mathbf{E} \langle M \rangle_T^{p/2} \le \mathbf{E}(M_T^*)^p \le \frac{4-p}{2-p} \mathbf{E} \langle M \rangle_T^{p/2}. \tag{9.26}$$

Proof. 1) Since for each stopping time τ ($\tau < \infty$) we have $\mathbf{E} M_\tau^2 \le \mathbf{E} \langle M \rangle_\tau$ (Problem 8.6), by (9.14) we have

$$P(M_T^* \ge a^{1/p}) \le a^{-2/p} \mathbf{E}[a^{2/p} \wedge \langle M \rangle_T] + P(\langle M \rangle_T \ge a^{2/p}).$$

By this

$$\mathbf{E}(M_T^*)^p = \int_0^\infty P(M_T^* \ge a^{1/p})\, da \le \mathbf{E}\left[\int_0^{\langle M \rangle_T^{p/2}} da + \langle M \rangle_T \int_{\langle M \rangle_T^{p/2}}^\infty a^{-2/p}\, da\right]$$

$$+ \int_0^\infty P(\langle M \rangle_T \ge a^{2/p})\, da = \frac{4-p}{2-p} \mathbf{E} \langle M \rangle_T^{p/2}.$$

2) By (9.25) it suffices to establish the left-hand inequality only. To this end observe that $\mathbf{E} \langle M \rangle_\tau \le \mathbf{E}(M_\tau^*)^2$ for each Markov time τ (Problem 8.7). Hence by (9.14)

$$P(\langle M \rangle_T \ge a^{2/p}) \le a^{-2/p} \mathbf{E}[a^{2/p} \wedge (M_T^*)^2] + P(M_T^* \ge a^{1/p}).$$

Therefore

$$E\langle M\rangle_T^{p/2} = \int_0^\infty P(\langle M\rangle_T \ge a^{2/p})\,da \le E\left[\int_0^{(M_T^*)^p} da + (M_T^*)^2 \int_{(M_T^*)^p}^\infty a^{-2/p}\,da\right]$$

$$+ \int_0^\infty P(M_T^* \ge a^{1/p})\,da = \frac{4-p}{2-p} E(M_T^*)^p.$$

5. Davis' inequality.

Theorem 6. *Let* T *be a Markov time* ($T \le \infty$) *and let* M *be a local martingale* ($M \in \mathcal{M}_{loc}$) *with* $M_0 = 0$. *There exist universal constants* c *and* C *(independent of* T *and* M) *such that*

$$cE[M, M]_T^{1/2} \le EM_T^* \le CE[M, M]_T^{1/2}. \tag{9.27}$$

The proof of these inequalities is prefaced with a number of auxiliary statements formulated in seperate lemmas.

Denote $S = (S_t)_{t\ge 0}$, $V = (V_t)_{t \ge 0}$ with $S_0 = V_0 = 0$ and

$$S_t = \sup_{0<u\le t} |\Delta M_u|, \quad V_t = \sum_{0<u\le t} |\Delta M_u| I(|\Delta M_u| > 2S_{u-}).$$

Lemma 1. *The processes* $[M, M]^{1/2} - S$, $2M^* - S$ *and* $2S - V$ *are nonnegative.*

Proof. We have

$$S_t = \sup_{0<u\le t} |\Delta M_u| = (\sup_{0<u\le t} (\Delta M_u)^2)^{1/2} \le \left(\sum_{0<u\le t} (\Delta M_u)^2\right)^{1/2} \le [M, M]_t^{1/2}.$$

Further,

$$S_t = \sup_{0<u\le t} |\Delta M_u| = \sup_{0<u\le t} |M_u - M_{u-}| \le 2M_t^*.$$

Since $|\Delta M_u| \le S_u$, on the set $\{|\Delta M_u| > 2S_{u-}\}$ we have

$$|\Delta M_u| + 2S_{u-} \le 2|\Delta M_u| \le 2S_u,$$

i.e. $|\Delta M_u| \le 2\Delta S_u$. Thus

$$V_t \le 2\sum_{0<u\le t} I(|\Delta M_u| > 2S_{u-})\,\Delta S_u \le 2S_t.$$

Corollary. *The process* $V \in \mathcal{A}_{loc}^+$.

In fact, $V \le 4M^*$, while $M^* \in \mathcal{A}_{loc}^+$ (Theorem 7.7). By this the compensator $\tilde V = (\tilde V_t)_{t\ge 0}$ is defined (Theorem 6.3).

Now, define the process $Z = (Z_t)_{t \ge 0}$ by putting $Z_0 = 0$ and

$$Z_t = \sum_{0 < u \le t} \Delta M_u \, I(|\Delta M_u| > 2S_{u-}).$$

Surely $\mathrm{Var}(Z) = V$, and hence $Z \in \mathcal{A}_{loc}$. The compensator of Z is denoted by $\tilde{Z} = (\tilde{Z}_t)_{t \ge 0}$. Clearly, $Z = Z' - Z''$ with

$$Z'_t = \sum_{0 < u \le t} (\Delta M_u)^+ \, I(|\Delta M_u| > 2S_{u-}),$$

$$Z''_t = \sum_{0 < u \le t} (\Delta M_u)^- \, I(|\Delta M_u| > 2S_{u-})$$

and $\tilde{Z} = \tilde{Z}' - \tilde{Z}''$ where $\tilde{Z}'$ and $\tilde{Z}''$ are the components of the increasing processes Z' and Z''. It is evident also that $\tilde{Z}' + \tilde{Z}'' = \tilde{V}$.

An important rôle in proving the inequalities (9.27) is played by the following Davis' decomposition of a local martingale M (cf. the first decomposition (7.1)):

$$M = M' + M'' \tag{9.28}$$

with

$$M' = Z - \tilde{Z}, \quad M'' = M - M'. \tag{9.29}$$

A local martingale M' belongs to $\mathcal{M}_{loc} \cap \mathcal{A}_{loc}$ and therefore belongs to the class $\mathcal{M}^d_{loc}$ (Corollary to Theorem 7.3), hence for each $T \in \mathbf{T}$

$$[M', M']_T^{1/2} = \Big[\sum_{0 < u \le T} (\Delta M'_u)^2 \Big]^{1/2} \le \sum_{0 < u \le T} |\Delta M'_u| \le \mathrm{Var}(M')_T$$

$$\le \mathrm{Var}(Z)_T + \mathrm{Var}(\tilde{Z})_T \le V_T + \tilde{V}_T. \tag{9.30}$$

Besides

$$(M')^*_T \le \mathrm{Var}(M')_T \le V_T + \tilde{V}_T. \tag{9.31}$$

The inequality $(M'')^*_T \le M^*_T + (M'_T)^*$ and (9.31) entail

$$(M'')^*_T \le M^*_T + V_T + \tilde{V}_T. \tag{9.32}$$

Finally, by the relations

$$[M'', M'']_T = \langle M^c \rangle_T + \sum_{0 < u \le T} (\Delta M''_u)^2$$

$$\le \langle M^c \rangle_T + 2\left(\sum_{0 < u \le T} (\Delta M_u)^2 + [M', M']_T \right)$$

$$\le 2\,([M, M]_T + [M', M']_T),$$

by the inequality $(a + b)^{1/2} \le a^{1/2} + b^{1/2}$ $(a, b \ge 0)$ and by (9.30) the inequality

$$[M'', M'']_T^{1/2} \le \sqrt{2}\,([M, M]_T^{1/2} + V_T + \tilde{V}_T) \tag{9.33}$$

takes place.

Lemma 2. *For each Markov time* T

$$\mathbf{E}\,(M')^*_T \le 4\mathbf{E}\,[M, M]_T^{1/2}, \tag{9.34}$$

$$\mathbf{E}\,[M', M']_T^{1/2} \le 8\mathbf{E}\,M^*_T. \tag{9.35}$$

Proof. By (9.31)

$$\mathbf{E}\,(M')^*_T \le \mathbf{E}\,(V_T + \tilde{V}_T) = 2\mathbf{E}V_T,$$

and by Lemma 1

$$V_T \le 2S_T \le 2\,[M, M]_T^{1/2},$$

which proves (9.34).

Next, by (9.30)

$$\mathbf{E}\,[M', M']_T^{1/2} \le \mathbf{E}\,(V_T + \tilde{V}_T) = 2\mathbf{E}V_T$$

and by Lemma 1

$$V_T \le 2S_T \le 4M^*_T,$$

which proves (9.35).

Lemma 3. *For each Markov time* T

$$\mathbf{E}\,(M'')^*_T \le 9\mathbf{E}M^*_T, \tag{9.36}$$

$$\mathbf{E}\,[M'', M'']_T^{1/2} \le 5\sqrt{2}\;\mathbf{E}\,[M, M]_T^{1/2}. \tag{9.37}$$

Proof. In virtue of (9.32) and Lemma 1

$$\mathbf{E}\,(M'')^*_T \le \mathbf{E}M^*_T + 2\mathbf{E}V_T, \quad V_T \le 2S_T \le 4M^*_T,$$

which proves (9.36).

Next, by (9.33) and Lemma 1

$$\mathbf{E}\,[M'', M'']_T^{1/2} \le \sqrt{2}\,(\mathbf{E}\,[M, M]_T^{1/2} + 2\mathbf{E}V_T),$$
$$V_T \le 2S_T \le 2\,[M, M]_T^{1/2},$$

which imply (9.37).

Lemma 4. *For each finite Markov time* T

$$|\Delta M''_T| \le 4S_{T-} \quad \text{(P-a.s.)}. \tag{9.38}$$

Proof. By definition of M''

$$\Delta M''_T = \Delta M_T - \Delta M'_T$$
$$= \Delta M_T - \Delta Z_T + \Delta \tilde{Z}_T$$
$$= \Delta M_T I(|\Delta M_T| \le 2S_{T-}) + \Delta \tilde{Z}_T.$$

Therefore

$$|\Delta M''_T| \le 2S_{T-} + |\Delta \tilde{Z}_T|,$$

and it remains to prove only that $|\Delta \tilde{Z}_T| \le 2S_{T-}$.

To this end assume first that M is a uniformly integrable martingale. Then for each finite predictable time τ we have $\mathbf{E}(\Delta M_\tau | \mathcal{F}_{\tau-}) = 0$ (Problem 4.10), and hence

$$\Delta \tilde{Z}_t = \mathbf{E}(\Delta M_\tau I(|\Delta M_\tau| > 2S_{\tau-}) | \mathcal{F}_{\tau-}) = -\mathbf{E}(\Delta M_\tau I(|\Delta M_\tau| \le 2S_{\tau-}) | \mathcal{F}_{\tau-}).$$

Therefore $|\Delta \tilde{Z}_\tau| \le 2S_{\tau-}$.

Jump times τ_n, $n \ge 1$ of the predictable process $\tilde{Z}$ are predictable (Theorem 3.10). Therefore

$$\{|\Delta \tilde{Z}_T| > 2S_{T-}\} \subseteq \bigcup_{n \ge 1} \{|\Delta \tilde{Z}_{\tau_n}| > 2S_{\tau_n -}\},$$

and hence $P(|\Delta \tilde{Z}_T| > 2S_{T-}) = 0$.

In the general case $M \in \mathcal{M}_{loc}$ we denote by $(T_k)_{k \ge 1}$ a localizing sequence (for M). Then $M^{T_k} \in \mathcal{M}$, and as it has been proved $P(|\Delta \tilde{Z}_{T \wedge T_k}| > 2S_{(T \wedge T_k)-}) = 0$. Hence

$$P(|\Delta \tilde{Z}_T| > 2S_{T-}) \le P(T_k < T) \to 0, \quad k \to \infty.$$

Lemma 5. *Let* T *be a Markov time* (T ≤ ∞). *Then*

$$\mathbf{E}\,(M'')_T^* \le 3\,(5\sqrt{2} + 4)\,\mathbf{E}\,[M, M]_T^{1/2}, \tag{9.39}$$
$$\mathbf{E}\,[M'', M'']_T^{1/2} \le 51\mathbf{E}M_T^*. \tag{9.40}$$

Proof. Denote $D = 4S$. The left continuous process $D_- = (D_{t-})_{t \geq 0}$ with $D_{0-} = D_0 = 0$ is, clearly, locally bounded. By Lemma 4 we have $|\Delta M''| \leq D_-$. This implies the local boundedness of the process M'' (to prove this it suffices to take a localizing sequence $T_k = \inf(t: |M''_t| \vee D_t \geq k)$, $k \geq 1$, and to observe that $|M''_{t \wedge T_k}| \leq 2k$). Hence $M'' \in \mathfrak{M}^2_{\text{loc}}$, and therefore for each finite Markov time τ (Problems 8.6 and 8.7)

$$\mathbf{E}\,(M''_\tau)^2 \leq \mathbf{E}\,[M'', M'']_\tau \leq \mathbf{E}\,(M'')^*_\tau. \tag{9.41}$$

Besides, by Lemma 4

$$\Delta\,(M'')^* \leq D_-, \quad \Delta\,[M'', M''] \leq D^2_-. \tag{9.42}$$

Therefore by the inequality (9.23)

$$\mathsf{P}\,((M'')^*_T \geq a) \leq a^{-2}\mathbf{E}\,[a^2 \wedge ([M'', M'']_T + D^2_T)] + \mathsf{P}\,([M'', M'']_T + D^2_T \geq a^2) \tag{9.43}$$

and

$$\mathsf{P}\,([M'', M'']_T \geq a^2) \leq a^{-2}\mathbf{E}\,[a^2 \wedge ((M'')^*_T + D_T)^2] + \mathsf{P}\,((M'')^*_T + D_T \geq a). \tag{9.44}$$

Denote

$$\xi_T = [M'', M'']_T + D^2_T, \quad \eta_T = (M'')^*_T + D_T.$$

Then, according to (9.43)

$$\mathbf{E}\,(M'')^*_T = \int_0^\infty \mathsf{P}\,((M'')^*_T \geq a)\,da \leq \mathbf{E}\left[\int_0^{\xi_T^{1/2}} da + \xi_T \int_{\xi_T^{1/2}}^\infty a^{-2}\,da\right] + \int_0^\infty \mathsf{P}\,(\xi_T \geq a^2)\,da = 3\mathbf{E}\xi_T^{1/2}, \tag{9.45}$$

while, according to (9.44),

$$\mathbf{E}\,[M'', M'']_T^{1/2} = \int_0^\infty \mathsf{P}\,([M'', M'']_T \geq a^2)\,da$$

$$\leq \mathbf{E}\left[\int_0^{\eta_T} da + \eta_T \int_{\eta_T}^\infty a^{-2}\,da\right] + \int_0^\infty \mathsf{P}\,(\eta_T \geq a)\,da = 3\mathbf{E}\eta_T. \tag{9.46}$$

By (9.45) and the inequality $(a + b)^{1/2} \leq a^{1/2} + b^{1/2}$ $(a, b \geq 0)$ we get

$$\mathbf{E}\,(M'')^*_T \leq 3\mathbf{E}\,[M'', M'']^{1/2}_T + 3\mathbf{E}D_T.$$

This entails the desired inequality (9.39) by taking into consideration that in virtue of the Lemmas 1 and 3

$$D_T = 4S_T \leq 4[M, M]_T, \quad \mathbf{E}\,[M'', M'',]^{1/2}_T \leq 5\sqrt{2}\ \mathbf{E}\,[M, M]^{1/2}_T.$$

Finally, the inequality (9.40) follows from (9.46) and the Lemmas 1 and 3, because

$$D_T = 4S_T \leq 8M^*_T, \quad \mathbf{E}\,(M'')^*_T \leq 9\mathbf{E}M^*_T.$$

Proof of Theorem 6. Since

$$M^*_T \leq (M')^*_T + (M'')^*_T, \tag{9.47}$$

the right-hand inequality in (9.27) follows from the Lemmas 2 and 5 with $C = 15\sqrt{2} + 16$.

To prove the left-hand inequality in (9.27) observe that

$$[M, M] = \langle M^c\rangle_T + \sum_{0 < u \leq T} (\Delta M'_u + \Delta M''_u)^2 \leq 2\,([M', M']_T + [M'', M'']_T).$$

By the inequality $(a + b)^{1/2} \leq a^{1/2} + b^{1/2}$ $(a, b \geq 0)$ this gives

$$[M, M]^{1/2}_T \leq \sqrt{2}\ ([M', M']^{1/2}_T + [M'', M'']^{1/2}_T)$$

and the left-hand inequality in (9.27) follows from the Lemmas 2 and 5 with $c = (59\sqrt{2})^{-1}$.

6. The Burkholder-Gandy inequalities ($p > 1$) [1].

Theorem 7. *Let* T *be a Markov time* ($T \leq \infty$), M *a local martingale with* $M_0 = 0$, *and* $p > 1$. *There exist universal constants* c_p *and* C_p *(independent of* T *and* M*), such that*

$$c_p\mathbf{E}\,[M, M]^{p/2}_T \leq \mathbf{E}\,(M^*_T)^p \leq C_p\mathbf{E}\,[M, M]^{p/2}_T. \tag{9.48}$$

The proof of this theorem will be accomplished in several steps stated as lemmas.

Lemma 6. *Let* A *be an increasing locally integrable process with* $A_0 = 0$, $\tilde{A}$ *its compensator,* T *a Markov time, and* $p \geq 1$. *Then*

$$\mathbf{E}\tilde{A}^p_T \leq p^p\,\mathbf{E}A^p_T. \tag{9.49}$$

Proof. For $p = 1$ formula (9.49) holds since $\mathbf{E}\tilde{A}_T = \mathbf{E}A_T$. Let $p > 1$. By properties of the Lebesgue-Stieltjes integral (cf., by the way, Ito's formula in Ch. 2, § 3)

---

[1] The proof will be based on Ito's formula presented in Ch. 2, § 3.

$$\tilde{A}_T^p = p\int_0^T \tilde{A}_{s-}^{p-1}\, d\tilde{A}_s + \sum_{0<u\le T} [\tilde{A}_u^p - \tilde{A}_{u-}^p - p\,\tilde{A}_{u-}^{p-1}\,\Delta\tilde{A}_u]$$

$$= p\int_0^T \tilde{A}_{s-}^{p-1}\, d\tilde{A}_s^c + \sum_{0<u\le T} (\tilde{A}_u^p - \tilde{A}_{u-}^p) \qquad (9.50)$$

with

$$\tilde{A}_t^c = \tilde{A}_t - \sum_{0<u\le t} \Delta\tilde{A}_u.$$

Evidently

$$\tilde{A}_T^p \le p\int_0^T \tilde{A}_s^{p-1}\, d\tilde{A}_s.$$

Next

$$\mathbf{E}\tilde{A}_T^p \le p\mathbf{E}\int_0^T \tilde{A}_s^{p-1}\, d\tilde{A}_s = p\mathbf{E}\int_0^T \tilde{A}_s^{p-1}\, dA_s \le p\mathbf{E}\,(\tilde{A}_T^{p-1} A_T), \qquad (9.51)$$

and by Hölder's inequality (with the exponents $\frac{p}{p-1}$ and p) , this gives

$$\mathbf{E}\tilde{A}_T^p \le (\mathbf{E}\tilde{A}_T^p)^{\frac{p-1}{p}} (\mathbf{E}A_T^p)^{\frac{1}{p}}. \qquad (9.52)$$

If $\mathbf{E}\tilde{A}_T^p < \infty$, then by (9.52) we get directly the desired inequality (9.49). If $\mathbf{E}\tilde{A}_T^p = \infty$, then utilizing local boundedness of an increasing predictable process (Lemma 6.1) we get

$$\mathbf{E}\tilde{A}_{T\wedge T_k}^p \le p^p\mathbf{E}A_{T\wedge T_k}^p \le p^p\mathbf{E}A_T^p \qquad (9.53)$$

where $(T_k)_{k\ge 1}$ is a corresponding localizing sequence. By taking here the limit as $k \to \infty$, we arrive at the inequality (9.49).

Let V, S, M' and M'' be the processes defined in Subsection 5.

Corollary 1. *For* $p \ge 1$

$$\mathbf{E}\,(V_T + \tilde{V}_T)^p \le 2^{p-1}(1+p^p)\,\mathbf{E}V_T^p.$$

Corollary 2. *By* (9.32)

$$\mathbf{E}\,((M'')_T^*)^p \le 3^{p-1}\,\mathbf{E}\,[(M_T^*)^p + (1+p^p)\,V_T^p], \qquad p \ge 1.$$

Corollary 3. *By* (9.33)

$$\mathbf{E}\,[M'', M'']_T^{p/2} \le 3^{p-1}\,2^{p/2}\,\mathbf{E}\,([M, M]_T^{p/2} + (1 + p^p)\,V_T^p), \qquad p \ge 1.$$

Lemma 7. *Let* T *be a Markov time* ($T \le \infty$) *and let* $p \ge 1$. *Then a constant* C'_p *can be found such that*

$$\mathbf{E}\,((M')_T^*)^p \le C'_p\,\mathbf{E}\,[M, M]_T^{p/2}. \tag{9.54}$$

Proof. Recall that $(M')_T^* \le V_T + \tilde{V}_T$ by (9.31). Therefore (9.54) follows from Corollary 1 to Lemma 6, and from the fact that $V_T \le 2S_T \le 2\,[M, M]_T^{1/2}$ (Lemma 1).

Lemma 8. *Let* T *be a Markov time* ($T \le \infty$), *and* $p \ge 1$. *Then a constant* $c'_p > 0$ *can be found such that*

$$c'_p\mathbf{E}\,[M', M']_T^{p/2} \le \mathbf{E}\,(M_T^*)^p. \tag{9.55}$$

Proof. This assertion follows by the inequality (9.30), by the corollary to Lemma 6 and by Lemma 1 according to which $V_T \le 2S_T \le 4M_T^*$.

Lemma 9. *Let* T *be a Markov time* ($T \le \infty$) *and* $1 \le p < 2$. *Then a constant* C''_p *can be found such that*

$$\mathbf{E}\,((M'')_T^*)^p \le C''_p\mathbf{E}\,[M, M]_T^{p/2}. \tag{9.56}$$

Proof. Replace a in the inequality (9.43) by $a^{1/p}$. Then, denoting $\xi_T = [M'', M'']_T + D_T^2$, we get

$$\mathbf{E}\,((M'')_T^*)^p = \int_0^\infty P\,((M'')_T^* \ge a^{1/p})\,da$$

$$\le \mathbf{E}\left[\int_0^{\xi_T^{p/2}} da + \xi_T \int_{\xi_T^{p/2}}^\infty a^{-2/p}\,da\right] + \int_0^\infty P\,(\xi_T \ge a^{2/p})\,da = \frac{4-p}{2-p}\,\mathbf{E}\xi_T^{p/2}.$$

Next, by the inequality $(a + b)^{1/2} \le a^{1/2} + b^{1/2}$ $(a, b \ge 0)$

$$\xi_T^{p/2} \le ([M'', M'']_T^{1/2} + D_T)^p.$$

Therefore, by taking into consideration (9.33)

$$\xi_T^{p/2} \le [\sqrt{2}\,([M, M]_T^{1/2} + V_T + \tilde{V}_T) + D_T]^p.$$

By convexity of the function $f(x) = |x|^p$, $p \ge 1$, this gives

$$\xi_T^{p/2} \le 4^{p-1}\,[2^{p/2}\,([M, M]_T^{p/2} + V_T^p + \tilde{V}_T^p) + D_T^p].$$

Further, observe that by the Lemmas 1 and 6

$$\begin{aligned} 2^{p/2}\mathbf{E}\,(V_T^p + \tilde{V}_T^p) + \mathbf{E}D_T^p &\le 2^{p/2}\,(1+p^p)\,\mathbf{E}V_T^p + 4^p\mathbf{E}S_T^p \\ &\le [2^{3p/2}\,(1+p^p) + 4^p]\,\mathbf{E}S_T^p \\ &\le [2^{3p/2}\,(1+p^p) + 4^p]\,\mathbf{E}\,[M, M]_T^{p/2}. \end{aligned}$$

Consequently, the desired inequality takes place with the constant

$$C_p'' = \frac{4-p}{2-p}\,[2^{3p/2}\,(1+p^p) + 4^p]\,4^{p-1}.$$

Lemma 10. *Let* T *be a Markov time* ($T \le \infty$), $1 \le p < 2$. *Then a constant* $c_p'' > 0$ *can be found such that*

$$c_p''\mathbf{E}\,[M'', M'']_T^{p/2} \le \mathbf{E}\,(M_T^*)^p. \tag{9.57}$$

Proof. Replace a in the inequality (9.44) by $a^{1/p}$. Then, denoting $\eta_T = (M'')_T^* + D_T$ we get

$$\mathbf{E}\,[M'', M'']_T^{p/2} = \int_0^\infty P\,([M'', M'']_T \ge a^{2/p})\,da$$

$$\le \mathbf{E}\left[\int_0^{\eta_T^p} da + \eta_T^2 \int_{\eta_T^p}^\infty a^{-2/p}\,da\right] + \int_0^\infty P\,(\eta_T \ge a^{1/p})\,da = \frac{4-p}{2-p}\,\mathbf{E}\eta_T^p. \tag{9.58}$$

By (9.32) and the convexity of the function $f(x) = |x|^p$, $p \ge 1$, we have

$$\eta_T^p \le (M_T^* + V_T + \tilde{V}_T + D_T)^p \le 4^{p-1}\,((M_T^*)^p + V_T^p + \tilde{V}_T^p + D_T^p).$$

By the Lemmas 1 and 6 this gives

$$\mathbf{E}\eta_T^p \le 4^{p-1}\,\mathbf{E}\,[(M_T^*)^p + (1+p^p)\,V_T^p + 4^pS_T^p]$$

$$\le 4^{p-1}\,[1 + (1+p^p)\,4^p + 8^p]\,\mathbf{E}\,(M_T^*)^p. \tag{9.59}$$

From (9.58) and (9.59) it follows that the desired inequality (9.57) takes place with

$$c_p'' = \left\{\frac{4-p}{2-p}\,4^{p-1}\,[1 + (1+p^p)\,4^p + 8^p]\right\}^{-1}.$$

Lemma 11. *Let* T *be a Markov time* ($T \le \infty$) *and* $p \ge 2$. *Then a constant* C_p''' *can be found such that*

$$\mathbf{E}\,((M'')_T^*)^p \le C_p'''\,\mathbf{E}\,[M, M]_T^{p/2}. \tag{9.60}$$

Proof. Suppose this inequality is proved in case $(M'')_T^* \le$ const. Then it will be valid in the general case too, since by Lemma 4 the process M'' is locally bounded (the right continuous process S_- is locally bounded, hence the process $\Delta M''$ is locally bounded too, and the usual localization device can be applied).

Thus, let $(M'')_T^* \le$ const. Denote $f_p(x) = |x|^p$ and observe that for $p \ge 2$

$$f_p''(x) = p(p-1)\,f_{p-2}(x). \tag{9.61}$$

By taking this property of the function $f_p(x)$ into consideration and by using Ito's formula (cf. Ch. 2, § 3) we get

$$|M_T''|^p = f_p(M_T'') = f_p'(M_-'') \cdot M_T'' + \frac{1}{2} p(p-1)\, f_{p-2}(M_-'') \circ \langle M^c \rangle_T$$

$$+ \sum_{0 < u \le T} [f_p(M_u'') - f_p(M_{u-}'') - p f_p'(M_{u-}'')\,\Delta M_u''], \tag{9.62}$$

where $f_p'(M_-'') \cdot M_T'' = \int_0^T f_p'(M_{s-}'')\,dM_s''$ is the stochastic integral (cf. Ch. 2, § 2), while $f_{p-2}(M_-'') \circ \langle M^c \rangle_T = \int_0^T f_{p-2}(M_{s-}'')\,d\langle M^c \rangle_s$ is the Lebesgue-Stieltjes integral. Since

$$\mathbf{E} f_p'(M_-'') \cdot M_T'' = 0$$

and

$$f_p(M_u'') - f_p(M_{u-}'') - f_p'(M_{u-}'')\,\Delta M_u'' \le \frac{1}{2} f_p''((M'')_u^*)\,(\Delta M_u'')^2,$$

(9.62) entails

$$\mathbf{E}\,|M_T''|^p \le \frac{1}{2} p(p-1)\,\mathbf{E}\,(f_{p-2}((M'')^*) \circ [M'', M'']_T)$$

$$\le \frac{1}{2} p(p-1)\,\mathbf{E}\,(((M'')_T^*)^{p-2}\,[M'', M'']_T).$$

By this and Theorem 2

$$\mathbf{E}\,((M'')^*_T)^p \le \frac{1}{2}\,\frac{p^{p+1}}{(p-1)^{p-1}}\,\mathbf{E}\,(((M'')^*_T)^{p-2}\,[M'', M'']_T). \tag{9.63}$$

As $p = 2$ this inequality turns into the inequality

$$\mathbf{E}\,((M'')^*_T)^2 \le 4\mathbf{E}\,[M'', M'']_T,$$

i.e. (9.60) is valid with $C'''_2 = 4$.

As $p > 2$, using Hölder's inequality with the exponents $\frac{p}{p-2}$ and $\frac{p}{2}$, by (9.63) we get

$$\mathbf{E}\,((M'')^*_T)^p \le \frac{1}{2}\,\frac{p^{p+1}}{(p-1)^{p-1}}\,(\mathbf{E}\,((M'')^*_T)^p)^{\frac{p-2}{p}}\,(\mathbf{E}\,[M'', M'']_T^{\,p/2})^{2/p},$$

and hence

$$\mathbf{E}\,((M'')^*_T)^p \le \left(\frac{p^{p+1}}{2\,(p-1)^{p-1}}\right)^{p/2}\mathbf{E}\,[M'', M'']_T^{p/2}. \tag{9.64}$$

By the inequalities $V_T \le 2S_T \le 2\,[M, M]_T^{1/2}$ (cf. Lemma 1) and by Corollary 3 to Lemma 6

$$\mathbf{E}\,[M'', M'']_T^{p/2} \le 3^{p-1}\,2^{p/2}\,\mathbf{E}\,([M, M]_T^{p/2} + 2^p\,(1 + p^p)\,[M, M]_T^{p/2}),$$

which together with (9.64) proves (9.60) for $p > 2$.

Lemma 12. *Let* T *be a Markov time* ($T \le \infty$) *and* $p \ge 2$. *Then a constant* $c'''_p > 0$ *can be found such that*

$$c'''_p\mathbf{E}\,[M'', M'']_T^{p/2} \le \mathbf{E}\,(M^*_T)^p. \tag{9.65}$$

Proof. As in Lemma 11 it suffices to prove the inequality (9.65), assuming $(M'')^*_T \le \text{const}$. Denote $r = p/2$.

By using Ito's formula (Ch. 2, § 2) we get

$$[M'', M'']^r_T = r\,[M'', M'']^{r-1}_- \circ \langle M^c\rangle_T + \sum_{0 < u \le T} ([M'', M'']^r_u - [M'', M'']^r_{u-})$$

$$\le r\,[M'', M'']^{r-1} \circ [M'', M'']_T.$$

Observe that by Lemma 4

$$[M'', M''] \le [M'', M'']_- + D^2_-$$

with $D_t = 4S_t$. Denote $\xi_t = [M'', M'']_t + D_t^2$. Then in virtue of the inequality $A_t^r \le rA^{r-1} \circ A_t$ for $A = (A_t)_{t\ge 0} \in \mathcal{V}^+$, $r > 1$, which is deduced from Ito's formula (Ch. 2, § 3), we have

$$[M'', M'']_T^r \le r\xi_-^{r-1} \circ [M'', M'']_T = r\,[Y, Y]_T \tag{9.66}$$

where $Y = (Y_t)_{t\ge 0}$ is a local martingale with

$$Y_t = \xi_-^{\frac{r-1}{2}} \cdot M''_t.$$

By (9.66) we get (cf. Problem 8.7)

$$\mathbf{E}\,[M'', M'']_T^{p/2} \le r\mathbf{E}\,(Y_T^*)^2. \tag{9.67}$$

Next, by Ito's formula again (cf. formula (3.5) in Ch. 2)

$$\xi_T^{\frac{r-1}{2}} M''_T = \xi_-^{\frac{r-1}{2}} \cdot M''_T + M'' \circ \xi_T^{\frac{r-1}{2}} = Y_T + M'' \circ \xi_T^{\frac{r-1}{2}}.$$

Hence,

$$Y_T^* \le 2\,(M'')_T^*\,\xi_T^{\frac{r-1}{2}}.$$

This and (9.67) entail

$$\mathbf{E}\,[M'', M'']_T^{p/2} \le 2p\mathbf{E}[((M'')_T^*)^2\,\xi_T^{p/2-1}].$$

Therefore, by Hölder's inequality with the exponents p/2 and p/ (p – 2) we obtain

$$\mathbf{E}[M'', M'']_T^{p/2} \le 2p\,(\mathbf{E}\,((M'')_T^*)^p)^{2/p}\,(\mathbf{E}\xi_T^{p/2})^{\frac{p-2}{p}}. \tag{9.68}$$

Due to Lemma 1 and Corollary 2 to Lemma 6 the following estimation holds:

$$\mathbf{E}\,((M'')_T^*)^p \le l_1(p)\,\mathbf{E}\,(M_T^*)^p \tag{9.69}$$

where $l_1(p)$ is a certain constant. Further, by convexity of the function $f(x) = |x|^{p/2}$, $p \ge 2$,

$$\mathbf{E}\xi_T^{p/2} \le 2^{p/2-1}\mathbf{E}\,([M'', M'']_T^{p/2} + D_T^p).$$

Therefore by Lemma 1

$$\mathbf{E}\xi_T^{p/2} \le l_2(p)\,\mathbf{E}\,([M'', M'']_T^{p/2} + (M_T^*)^p) \tag{9.70}$$

where $l_2(p)$ is a certain constant.

Denote

$$\alpha = \mathbf{E}\,[M'', M'']_T^{p/2}, \quad \beta = \mathbf{E}\,(M_T^*)^p.$$

By (9.68) — (9.70) we get the inequality

$$\alpha \leq 1\,(p)\,\beta^{\frac{2}{p}}\,(\alpha+\beta)^{\frac{p-2}{p}} \tag{9.71}$$

where 1 (p) is a certain constant depending on p only.

As p = 2 by (9.71) we have

$$\alpha \leq 1\,(2)\,\beta,$$

which proves (9.65) with $c_2''' = (1\,(2))^{-1}$.

Let $p > 2$ and $\beta > 0$ (if $\beta = 0$, then by (9.71) we have $\alpha = 0$, and (9.65) holds indeed). Denote $\gamma = \alpha/\beta$. Then by (9.71)

$$\gamma^{\frac{2}{p}} \leq 1\,(p)\left(\frac{1+\gamma}{\gamma}\right)^{\frac{p-2}{p}},$$

and hence

$$\gamma^{\frac{p}{p-2}} \leq (1\,(p))^{\frac{p}{p-2}}\,(1+\gamma). \tag{9.72}$$

The equation

$$x^{\frac{p}{p-2}} = (1\,(p))^{\frac{p}{p-2}}\,(1+x), \qquad x \geq 0,$$

possesses a unique positive solution. Denote this solution by $(c_p''')^{-1}$. Then by (9.72) we have $\gamma \leq (c_p''')^{-1}$ which proves (9.65) for $p > 2$.

Proof of Theorem 7. The right-hand inequality in (9.48) follows from the Lemmas 7, 9 and 11 and from the evident relation

$$(M_T^*)^p \leq 2^{p-1}\,[((M')_T^*)^p + ((M'')_T^*)^p].$$

The left-hand inequality in (9.48) follows from the Lemmas 8, 10 and 12 and from the inequality (cf. the proof of Theorem 6)

$$[M, M]^{1/2} \leq \sqrt{2}\,([M', M']^{1/2} + [M'', M'']^{1/2})$$

which, in view of the convexity of the function $f(x) = |x|^p$, $p \geq 1$, gives

$$[M, M]_T^{p/2} \leq 2^{\frac{3}{2}p-1}\,([M', M']_T^{p/2} + [M'', M'']_T^{p/2}).$$

The Theorems 5 and 7 entail

Corollary. *For each continuous local martingale* M *with* $M_0 = 0$, *a Markov time* $T \leq \infty$ *and* $0 < p < \infty$

$$c_p \mathbf{E} \langle M\rangle_T^{p/2} \le \mathbf{E} (M_T^*)^p \le C_p \mathbf{E} \langle M\rangle_T^{p/2} \tag{9.73}$$

where c_p *and* C_p *are universal constants depending on* p *and independent of* M *or* T.

Problems

1. Let $M \in \mathcal{M}_{loc}^c$, $M_0 = 0$ and $\langle M\rangle_t \equiv t$. Show that for each $T \in \mathbf{T}$

$$\frac{1}{3}\mathbf{E}\sqrt{T} \le \mathbf{E}M_T^* \le 3\mathbf{E}\sqrt{T}.$$

2. Let $M^n \in \mathcal{M}_{loc}^2$, $M_0^n = 0$, $n \ge 1$, $T \in \mathbf{T}$. Show that the convergence to zero of the sequence $\langle M^n\rangle_T$, $n \ge 1$, in probability implies the convergence to zero of the sequence $((M^n)_T^*$, $n \ge 1$ in probability.

3. Let $M^n \in \mathcal{M}_{loc}^2$, $M_0^n = 0$, $n \ge 1$, $T \in \mathbf{T}$ and let the family $\sup_{0 < t \le T} (\Delta M_t^n)^2$, $n \ge 1$, be uniformly integrable. Show that

$$(M^n)_T^* \xrightarrow{P} 0 \Leftrightarrow \langle M^n\rangle_T \xrightarrow{P} 0,$$

where « $\xrightarrow{P}$ » denotes the convergence in probability.

4. Let $A^n \in \mathcal{A}_{loc}^2$, $n \ge 1$, $\tilde{A}^n$ the compensator of A^n, and $T \in \mathbf{T}$. Show that

$$\tilde{A}_T^n \xrightarrow{P} 0 \Rightarrow A_T^n \xrightarrow{P} 0$$

and

$$\tilde{A}_T^n \xrightarrow{P} 0 \Leftrightarrow A_T^n \xrightarrow{P} 0$$

as $\Delta A^n \le \text{const}$, $n \ge 1$.

5. As $M \in \mathcal{M}_{loc}$, $M_0 = 0$ show that $[M, M]^{1/2} \in \mathcal{A}_{loc}^+$.

6. Let $M, N \in \mathcal{M}_{loc}^d$, $M_0 = N_0$ and let ΔM and ΔN be indistinguishable processes. Show that the processes M and N are indistinguishable.

7. Let $M \in \mathcal{M}_{loc}^2$, $M_0 = 0$ and $T \in \mathbf{T}$. Show that

$$\mathbf{E} (M_\tau^*)^2 \le 4\mathbf{E} [M, M]_\tau = 4\mathbf{E} \langle M\rangle_\tau.$$

8. Let $(X_t)_{t \ge 0}$ be a stochastic process with trajectories from D such that there

exists $X_\infty = \lim_{t \to \infty} X_t$. For $a > 0$ put $\tau = \inf(t: |X_t| > a) \wedge T$ where T is a random variable taking on values in $[0, \infty]$. Show that

$$\{\sup_{t \le T} |X_t| > a\} \subseteq \{|X_\tau| \ge a\} \subseteq \{\sup_{t \le T} |X_t| \ge a\}.$$

CHAPTER 2

SEMIMARTINGALES. I. STOCHASTIC INTEGRAL

§ 1. Semimartingales and quasimartingales

1. We assume that a stochastic basis $(\Omega, \mathcal{F}, \mathbb{F}, P)$ with a flow of σ-algebras $\mathbb{F} = (\mathcal{F}_t)_{t \geq 0}$ is given.

Definition 1. A stochastic process $X \in \mathbb{F} \cap D$ is called a *semimartingale* if it admits the representation

$$X_t = X_0 + M_t + A_t,$$

where M is a local martingale with $M_0 = 0$ ($M \in \mathcal{M}_{loc, 0}$) and A is a process of locally bounded variation ($A \in \mathcal{V}$).

The class of semimartingales is denoted by S or S ($\mathbb{F}$), or S ($\mathbb{F}$, P). (Sometimes instead of $X \in$ S ($\mathbb{F}$) we use the expression "X is a $\mathbb{F}$-semimartingale".)

Definition 2. A semimartingale X is called *special* if it admits the representation (1.1) with A that is a process with locally integrable variation ($A \in \mathcal{A}_{loc}$).

The class of special semimartingales is denoted by **Sp** or **Sp** ($\mathbb{F}$), or **Sp** ($\mathbb{F}$, P).

If $\mathcal{K}$ is a class of processes $X = (X_t)_{t \geq 0}$ with trajectories from D, then $\mathcal{K}_0 = \{X \in \mathcal{K}: X_0 = 0\}$.

2. Theorem 1. *Let* X *be a semimartingale. The following conditions are equivalent:*

a) $X \in$ **Sp**, *i.e. there exists the decomposition* $X = X_0 + M + A$, *with* $M \in \mathcal{M}_{loc,0}$ *and* $A \in \mathcal{A}_{loc}$;

b) X *admits the decomposition* $X = X_0 + M + A$, *with* $M \in \mathcal{M}_{loc,0}$ *and* $A \in \mathcal{V} \cap \mathcal{P}$ *(then such a decomposition with a predictable* A *is unique);*

c) *in every decomposition* $X = X_0 + M + A$ ($M \in \mathcal{M}_{loc, 0}$, $A \in \mathcal{V}$) *a process* A $\in \mathcal{A}_{loc}$;

d) *a process* $(X - X_0)^* = \sup | X - X_0 | \in \mathcal{A}^+_{loc}$;

e) *a process* $(\Delta X)^* = \sup | \Delta X | \in \mathcal{A}^+_{loc}$.

Proof. a) $\Rightarrow$ b). If $A \in \mathcal{A}_{loc}$ and if $\tilde{A}$ is the compensator of A, then $A - \tilde{A} \in \mathcal{M}_{loc,0}$ (Corollary 1 to Theorem 1.6.3). Then $X = X_0 + \tilde{A} + M + (A - \tilde{A})$, with $M + (A - \tilde{A}) \in \mathcal{M}_{loc,0}$ and $\tilde{A} \in \mathcal{A}_{loc} \cap \mathcal{P}$. The decomposition $X = X_0 + M + A$ with $A \in \mathcal{V} \cap \mathcal{P}$ is unique, because if there were another decomposition $X = X_0 + M' + A'$, then the process $A - A' = M - M' \in \mathcal{M}_{loc,0} \cap \mathcal{V}(\mathbb{F}) \cap \mathcal{P}$, i.e. $A - A' = 0$ (Theorem 1.6.4).

b) $\Rightarrow$ a) follows from Lemma 1.6.1.

a) $\Rightarrow$ d). We have

$$(X - X_0)^*_t \le M^*_t + \mathrm{Var}\,(A)_t.$$

According to Theorem 1.7.7 we have $M^* \in \mathcal{A}^+_{loc}$. Consequently $(X - X_0)^* \in \mathcal{A}^+_{loc}$.

d) $\Rightarrow$ e) is obvious.

e) $\Rightarrow$ c). Consider the decomposition $X = X_0 + M + A$, $M \in \mathcal{M}_{loc,0}$ and $A \in \mathcal{V}$. Then

$$|\Delta A_s| \le |\Delta M_s| + |\Delta X_s| \le 2M^*_s + (\Delta X)^*_s.$$

By e) and Theorem 1.7.7 the process $2M^* + (\Delta X)^* \in \mathcal{A}^+_{loc}$, and hence there exists a sequence of Markov times

$$\mathbf{E}\,[\sup_{s \le \tau_n} |\Delta A_s|] < \infty, \quad \tau_n \uparrow \infty.$$

Put

$$\sigma_n = \inf\,(t\colon\ \mathrm{Var}\,(A)_t \ge n) \wedge \tau_n.$$

Then

$$\mathrm{Var}\,(A)_{\sigma_n} \le n + \sup_{s \le \tau_n} |\Delta A_s|$$

and

$$\mathbf{E}\ \mathrm{Var}\,(A)_{\sigma_n} < \infty, \quad n \ge 1.$$

Since $\sigma_n \uparrow \infty$, then $n \ge 1$.

Finally, c) $\Rightarrow$ a) is obvious. Thus

b) $\Leftrightarrow$ a) $\Rightarrow$ d) $\Rightarrow$ e) $\Rightarrow$ c) $\Rightarrow$ a).

In the following assertion an important particular case of special semimartingales is described.

Theorem 2. *If a semimartingale is such that* $|\Delta X| \le a$, $a \ge 0$, *then* $X \in \mathbf{Sp}$ *and* *in its decomposition* $X = X_0 + M + A$ *with* $A \in \mathcal{A}_{loc} \cap \mathcal{P}$ *the processes* M *and* A *are such that*

$$|\Delta M| \leq 2a, \quad |\Delta A| \leq a.$$

In particular, if X *is a continuous semimartingale, then it is special and in its decomposition* M *and* A *are continuous too.*

Proof. The property $X \in \mathbf{Sp}$ follows from Theorem 1 (Condition e)) and from the assumption $|\Delta X| \leq a$. Then by property b) in Theorem 1 we have $X = X_0 + M + A$ with $A \in \mathcal{A}_{loc} \cap \mathcal{P}$ and $M \in \mathcal{M}_{loc,\,0}$. Since ${}^{p}(\Delta M) = 0$ (Theorem 1.7.8) and $\Delta A \in \mathcal{P}$, we have ${}^{p}(\Delta X) = \Delta A$. Consequently $|\Delta A| \leq a$ and

$$|\Delta M| \leq |\Delta X| + {}^{p}(\Delta X)| \leq 2a.$$

3. In general, the semimartingale representation of type (1.1) is not unique. Let

$$X = X_0 + M_1 + A_1, \quad X = X_0 + M_2 + A_2, \quad M_i \in \mathcal{M}_{loc,\,0},$$
$$A_i \in \mathcal{V}(\mathbb{F}), \quad i = 1, 2.$$

Let us show that $M_1^c = M_2^c$, in other words, the continuous martingale component of a semimartingale is independent of a form of the decomposition (1.1).

In fact, according to Theorem 1.6.4

$$N \equiv M_1 - M_2 = A_2 - A_1 \in \mathcal{M}_{loc,\,0} \cap \mathcal{V}(\mathbb{F}) = \mathcal{M}_{loc,\,0} \cap \mathcal{A}_{loc}.$$

Let $(\tau_n)_{n \geq 1}$ be a localizing sequence such that $N^{\tau_n} \in \mathcal{M} \cap \mathcal{A}$. Then by the corollary to Theorem 1.7.3 we have $N^{\tau_n} \in \mathcal{M}^d$, $n \geq 1$, and hence $N^c = 0$. Thus $M_1^c = M_2^c$, i.e. the continuous martingale component (further on denoted by X^c) is independent of a type of the decomposition (1.1).

4. An important characteristic of a semimartingale X is presented by its *quadratic variation* [X, X] with ($\Delta X_0 = 0$)

$$[X, X]_t = \langle X^c \rangle_t + \sum_{0 \leq s \leq t} (\Delta X_s)^2 \tag{1.2}$$

(cf. Ch. 1, (8.2)).

If X and Y are two semimartingales, then by [X, Y] we denote their *quadratic covariation* defined by the formula

$$[X, Y] = \frac{1}{4}([X + Y, X + Y] - [X - Y, X - Y]), \tag{1.3}$$

or equivalently by

$$[X, Y] = \frac{1}{2}([X + Y, X + Y] - [X, X] - [Y, Y]). \tag{1.4}$$

In the following theorem a number of properties of the processes [X, X] and [X,

Y] is presented.

Theorem 3. *Let* X *and* Y *be semimartingales. Then:*

1) $[X, Y] \in \mathcal{V}$, $[X, X] \in \mathcal{V}^+$;

2) $\Delta [X, Y] = \Delta X \Delta Y$;

3) *if* $Y \in \mathcal{V}$, *then* $[X, Y]^c = 0$;

4) *if* $Y \in \mathcal{V} \cap C$, *then* $[X, Y] = 0$;

5) *if* $X \in \mathcal{M}^c_{loc}$, *then* $[X, X] = [X, X]^c = \langle X \rangle$;

6) *if* X *is a locally bounded martingale from* $\mathcal{M}_{loc, 0}$ *and if* $Y \in \mathcal{V} \cap \mathcal{P}$, *then* $[X, Y] \in \mathcal{M}_{loc, 0}$.

7) *if* $X, Y \in \mathcal{M}_{loc}$, *then* $XY - X_0Y_0 - [X, Y] \in \mathcal{M}_{loc, 0}$.

Proof. The properties 1) — 5) follow directly from the definitions of [X, X] and [X, Y]. Property 6) follows from Problem 1.6.4, since

$$[X, Y]_t = \sum_{0 \le s \le t} \Delta X_s \Delta Y_s = X \circ Y_t - X_- \circ Y_t$$

and

$$XY - X \circ Y \in \mathcal{M}_{loc}, \quad XY - X_- \circ Y \in \mathcal{M}_{loc}.$$

Property 7) is established in Corollary 2 to Theorem 2.3.1.

5. In the present subsection we discuss briefly the notion of a quasimartingale related to the notion of a semimartingale. In Theorem 4 below the relation between these two objects is described.

Let a process $X \in \mathbb{F} \cap D$. For each $n \ge 1$ and $0 \le t_1 < t_2 < \ldots < t_n$ denote

$$\operatorname{var}(X; t_1, \ldots, t_n) = \sum_{1 \le i \le n-1} | \mathbf{E} (X_{t_{i+1}} - X_{t_i} | \mathcal{F}_{t_i}) | + | X_{t_n} | \tag{1.5}$$

and

$$\operatorname{var}(X) = \sup_{n,\, t_1, \ldots,\, t_n} \mathbf{E} [\operatorname{var}(X; t_1, \ldots, t_n)]. \tag{1.6}$$

Definition 3. A stochastic process $X \in \mathbb{F} \cap D$ is called a *quasimartingale* if $\operatorname{var}(X) < \infty$. The class of quasi-martingales is denoted by **Q**, or **Q** ($\mathbb{F}$), or **Q** ($\mathbb{F}$, P).

Observe that if $X \in \mathcal{M}$, then

$$\operatorname{var}(X) = \sup_t \mathbf{E} | X_t | < \infty$$

and hence $\mathcal{M} \subseteq \mathbf{Q}$.

If a process $X \in \mathcal{A}$, then

$$\operatorname{var}(X) \le 2\mathbf{E} \int_0^\infty | dX_s | < \infty$$

and, consequently, $\mathcal{Q} \subseteq \mathbf{Q}$.

By the usual localization device we deduce from this that

$$\mathbf{Sp} \subseteq \mathbf{Q}_{loc}.$$

It is shown in the following theorem, given without proof (cf [103], Ch. 5, Theorem 5.36), that the class of special semimartingales in fact coincides with the class of local quasimartingales.

Theorem 4. *The following relation takes place:*

$$\mathbf{Sp} = \mathbf{Q}_{loc}. \tag{1.7}$$

Problems

1. Each submartingale (or supermartingale) is a special semimartingale.
2. Each special semimartingale is the difference of two local submartingales.
3. Each quasimartingale is the difference of two nonnegative submartingales.
4. Show that $\mathbf{S}_{loc} = \mathbf{S}$ and $(\mathbf{Sp})_{loc} = \mathbf{Sp}$.
5. A deterministic process $X_t(\omega) = f(t)$, $t \geq 0$, is a semimartingale where $f(t)$ is a function in the space D, if f is of finite variation over each finite interval.
6. Let $W = (W_t)_{t \geq 0}$ be a Wiener process relative to $(\mathbb{F}, P)$. Show that

$$X = \sqrt{|W|} \notin \mathbf{S}(\mathbb{F}, P).$$

7. Let $X = (X_t)_{t \geq 0}$ be a semimartingale. Show that

$$[X, X]_t = P\text{-}\lim_n \sum_{j=0}^{n-1} (X_{t^n_{j+1}} - X_{t^n_j})^2,$$

where $0 \equiv t^n_0 < t^n_1 < \ldots < t^n_n \equiv t$, $n \geq 1$, is a partition sequence of the interval $[0, t]$ such that $\max_{0 \leq j \leq n-1} (t^n_{j+1} - t^n_j) \to 0$, $n \to \infty$. (Hint: see Theorem 1.8.4.)

§ 2. Stochastic integral with respect to a local martingale and a semimartingale. Construction and properties

1. Let X be a semimartingale, defined on a stochastic basis $(\Omega, \mathcal{F}, \mathbb{F} = (\mathcal{F}_t)_{t \geq 0}, P)$. In the present subsection we intend to give a «natural» definition of the stochastic integral

$$H \cdot X = \int H_s dX_s \tag{2.1}$$

with respect to a semimartingale for a certain supply of processes H.

Let a semimartingale $X \in \mathbb{F} \cap D$ admit the representation (non-unique, in general)

$$X_t = X_0 + M_t + A_t,$$

where A is a process of locally bounded variation, while M is a local martingale. It is natural to set, by definition, that

$$H \cdot X = H \cdot M + H \circ A, \tag{2.2}$$

where $H \circ A$ is the integral with respect to a process $A \in \mathcal{V}$, while $H \cdot M$ is the integral with respect to $M \in \mathcal{M}_{loc, 0}$. For such a definition of the integral $H \cdot X$ one needs, of course, to prove that its values are independent of the type of the representation (2.2).

If a process $A \in \mathcal{V}$, i.e. it is a process of bounded variation on each interval $[0, t]$, $t \in R_+$, while H is a measurable stochastic process, then the integral $H \circ A$ is naturally understood as the Lebesgue-Stieltjes integral defined in Ch. 1, § 6 by the formula

$$(H \circ A)_t(\omega) = \begin{cases} \int_0^t H_s(\omega)\, d A_s(\omega), & \text{if } \int_0^t | H_s(\omega) |\, d \operatorname{Var}(A)_s(\omega) < \infty, \\ +\infty, & \text{if } \int_0^t | H_s(\omega) |\, d \operatorname{Var}(A)_s(\omega) = \infty \end{cases} \tag{2.3}$$

Thus, for a bounded H at least, the integral $H \circ A$ is defined «pathwise» for each $\omega \in \Omega$.

The definition of the stochastic integral $H \cdot M$ with respect to a martingale M is a much more complicated matter.

According to Ch. 1, § 7 each local martingale admits two decompositions

$$M = M^1 + M^2 \tag{2.4}$$

with $M^1 \in \mathcal{M}_{loc} \cap \mathcal{A}_{loc}$, $M^2 \in \mathcal{M}^2_{loc}$, and

$$M = M^c + M^d \tag{2.5}$$

with $M^c \in \mathcal{M}^c_{loc}$, $M^d \in \mathcal{M}^d_{loc}$.

Any of these decompositions can be used for the construction of the stochastic integral $H \cdot M$. For the construction, based on the (second) decomposition (2.5), see [103], Ch. 2. On constructing the stochastic integral below, the (first) decomposition (2.4) is utilized.

On demanding a natural additivity property of the stochastic integral, we set by definition

$$H \cdot M = H \cdot M^1 + H \cdot M^2, \tag{2.6}$$

where $H \cdot M^1$ and $H \cdot M^2$ are the stochastic integrals with respect to M^1 and M^2. Another demand met in the considered stochastic integration theory, determines the class of integrands H: for $M \in \mathcal{M}_{loc}$ it consists in requiring

$$H \cdot M \in \mathcal{M}_{loc}, \tag{2.7}$$

i.e. that the stochastic integral of H with respect to a local martingale is again a local martingale. (Appropriateness of this demand is not that clear, however it is well justified by the deep and substantial results of stochastic calculus which are based on the property (2.7) required from the stochastic integral.)

2. We start with defining $H \cdot M^1$, i.e. with defining the stochastic integral $H \cdot M$ for $M \in \mathcal{M}_{loc} \cap \mathcal{A}_{loc}$. Property (2.7) imposes certain restrictions on processes H. Taking this into account, consider first the space $L^1(\mathbb{F}, \mathrm{Var}(M))$ of predictable processes H, possessing the finite norm

$$\| H \|_1 \equiv \mathbf{E}(|H| \circ \mathrm{Var}(M)_\infty) < \infty.$$

As $H \in \mathbf{L}^1(\mathbb{F}, \mathrm{Var}(M))$ the Lebesgue-Stieltjes integral $H \circ M_t$, $t \geq 0$ is defined (cf. (2.3)). Let us show that $H \circ M = (H \circ M_t)_{t \geq 0}$ is a martingale.

By Theorem 1.7.4

$$M_t = M_0 + \sum_{s \leq t} \Delta M_s - A_t$$

with

$$A = \widetilde{\sum_s \Delta M},$$

and hence

$$H \circ M_t = \sum_{s \le t} H_s \Delta M_s - H \circ A_t.$$

Now, observe that

$$\mathbf{E} \sum_{s > 0} | H_s \Delta M_s | \le \| H \|_1 \ (< \infty). \tag{2.8}$$

Therefore, by Corollary 1 to Theorem 1.6.3 we have $H \circ A = \widetilde{\sum_s H_s \Delta M_s}$ and

$$H \circ M \in \mathcal{M}_{loc} \cap \mathcal{A}_{loc}.$$

Furthermore

$$\| H \|_1 < \infty \Rightarrow H \circ M \in \mathcal{A} \cap \mathcal{H}. \tag{2.9}$$

In fact

$$\mathrm{Var}\,(H \circ M)_\infty \le \sum_{s > 0} | H_s \Delta M_s | + | H | \circ \mathrm{Var}\,(A)_\infty,$$

$$\mathbf{E}\,(| H | \circ \mathrm{Var}\,(A)_\infty) \le \mathbf{E}\, \widetilde{\sum_{s > 0} | H_s \Delta M_s |}$$

(cf. Problem 1.6.7) and consequently, according to (2.8)

$$\mathbf{E}\, \mathrm{Var}\,(H \circ M)_\infty \le 2 \| H \|_1 < \infty, \tag{2.10}$$

and hence (2.9) takes place.

Now if H is a predictable process such that $| H | \circ \mathrm{Var}\,(M) \in \mathcal{A}^+_{loc}$, then by using a localizing sequence we establish that

$$| H | \circ \mathrm{Var}\,(M) \in \mathcal{A}^+_{loc} \Rightarrow H \circ M \in \mathcal{M}_{loc} \cap \mathcal{A}_{loc}. \tag{2.11}$$

Definition 1. For a predictable process H such that $| H | \circ \mathrm{Var}\,(M) \in \mathcal{A}^+_{loc}$, the process $H \circ M = (H \circ M_t)_{t \ge 0}$ where $H \circ M_t$ is the Lebesgue-Stieltjes integral defined by formula (2.3), is called the stochastic integral $H \cdot M$ of H with respect to a local martingale $M \in \mathcal{A}_{loc}$.

Thus, we have defined the stochastic integral $H \cdot M$ for $M \in \mathcal{M}_{loc} \cap \mathcal{A}_{loc}$ and predictable processes H with $\| H \|_1 < \infty$.

The construction given in the subsequent subsection shows how the notion of the stochastic integral with respect to $M \in \mathcal{M}_{loc} \cap \mathcal{A}_{loc}$ is extended to a larger supply of predictable processes H.

3. First, let $M \in \mathcal{M} \cap \mathcal{A}$. Then $M \in \mathcal{M}^d$ (Corollary to Theorem 1.7.3) and

$$[M, M]_t = \sum_{s \le t} (\Delta M_s)^2.$$

Introduce the space L ($\mathbb{F}$, [M, M]) of predictable processes H with the norm

$$\| H \| = \mathbf{E} (H^2 \circ [M, M]_\infty)^{1/2} < \infty.$$

Observe that $\| H \| \le \| H \|_1$. For each process H that belongs to $\mathbf{L}$ ($\mathbb{F}$, [M, M]), a sequence H^n, $n \ge 1$ can be chosen such that $H^n \in \mathbf{L}^1 (\mathbb{F}, \mathrm{Var}(M))$, $n \ge 1$ and $\| H - H_n \| \to 0$, $n \to \infty$, by setting, for instance, $H^n = H I_{\{|H| \le n\}}$. Then, by (2.9) we have $H^n \cdot M \in \mathcal{A} \cap \mathcal{M}$. Therefore by Davis inequality (Theorem 1.9.6) and by the inequality $(a + b)^{1/2} \le a^{1/2} + b^{1/2}$ $(a, b \ge 0)$ we have

$$\mathbf{E} (H^n \cdot M - H^m \cdot M)^*_\infty = \mathbf{E} ((H^n - H^m) \cdot M)^*_\infty$$
$$\le C\mathbf{E} ((H^n - H^m)^2 \circ [M, M]_\infty)^{1/2}$$
$$= C \| H^n - H^m \| \le C (\| H - H^n \| + \| H - H^m \|) \to 0, \quad n, m \to \infty.$$

This implies, in particular, that the sequence $(H^n \cdot M_\infty)$, $n \ge 1$, converges in the mean to a random variable denoted by $(H \cdot M_\infty)$ with $\mathbf{E} \,|\, (H \cdot M_\infty) \,| < \infty$. Define the random process $H \cdot M \in \mathcal{M}$, by setting $H \cdot M_\tau = \mathbf{E} ((H \cdot M_\infty) \mid \mathcal{F}_\tau)$ for each $\tau \in \mathbf{T}$. Note that $H \cdot M_\infty = (H \cdot M_\infty)$ and that by Doob's inequality (Theorem 1.9.1) $(\forall \varepsilon > 0)$

$$\mathsf{P} ((H \cdot M - H^n \cdot M)^*_\infty \ge \varepsilon)$$
$$\le \frac{1}{\varepsilon} \mathbf{E} \,|\, (H \cdot M_\infty) - H^n \cdot M_\infty \,| \to 0, \quad n \to \infty. \tag{2.12}$$

Definition 2. The process $H \cdot M \in \mathcal{M}$ obtained in this manner, independent of a type of an approximating sequence, is called the *stochastic integral* of a process $H \in \mathbf{L}$ ($\mathbb{F}$, [M, M]) with respect to a process $M \in \mathcal{M} \cap \mathcal{A}$.

It follows directly from the definition of $H \cdot M$ (Problem 3) that the equality

$$(c_1 H_1 + c_2 H_2) \cdot M = c_1 (H_1 \cdot M) + c_2 (H_2 \cdot M) \tag{2.13}$$

holds with the constants c_1, c_2 and $H_1, H_2 \in \mathbf{L}$ ($\mathbb{F}$, [M, M]), as well as the equality

$$H \cdot (M + N) = H \cdot M + H \cdot N \tag{2.14}$$

for $H \in \mathbf{L}(\mathbb{F}, [M, M]) \cap \mathbf{L}(\mathbb{F}, [N, N])$ and $M, N \in \mathcal{M} \cap \mathcal{A}$ (Problem 4).

Let us establish a number of other important properties of the process $H \cdot M$.

Theorem 1. *Let* $M \in \mathcal{M} \cap \mathcal{A}$ *and* $H \in \mathbf{L}$ ($\mathbb{F}$, [M, M]). *Then:*

1) $\Delta (H \cdot M) = H \Delta M$,

2) $H \cdot M \in \mathcal{H} \cap \mathcal{M}^d$,

3) $[H \cdot M, H \cdot M] = H^2 \circ [M, M]$,

4) $(H \cdot M)^T = H \cdot M^T \quad \forall T \in \mathbf{T}$.

Corollary . *For each* $T \in \mathbf{T}$ *we have* $M^T = I_{[\![0, T]\!]} \cdot M$.

Proof of Theorem 1. 1) Since $H \cdot M = H \circ M$ for $H \in \mathbf{L}^1(\mathbb{F}, \mathrm{Var}(M))$, the desired property holds evidently. In case $H \in \mathbf{L}(\mathbb{F}, [M, M])$ the stochastic integral $H \cdot M$ is independent of a choice of an approximating sequence H^n, $n \geq 1$. Therefore take $H^n = HI_{\{|H| \leq n\}}$ and use the fact that

$$\Delta(H^n \cdot M) = H^n \Delta M, \quad I(\tau < \infty) H^n_\tau \Delta M_\tau \to I(\tau < \infty) H_\tau \Delta M_\tau$$

for each $\tau \in \mathbf{T}$, and that

$$\sup_{t > 0} |\Delta(H \cdot M)_t - \Delta(H^n \cdot M)_t| \leq 2(H \cdot M - H^n \cdot M)^*_\infty \to 0, \quad n \to \infty,$$

in probability (cf. (2.12)). Consequently, $\Delta(H \cdot M)_\tau = H\Delta M_\tau$ for each $\tau \in \mathbf{T}(\{\tau < \infty\}, \mathrm{P}\text{-a.s.})$. Hence, the processes $\Delta(H \cdot M)$ and $H\Delta M$ are indistinguishable (Theorem 1.3.12).

2) Set $H^n = HI_{\{|H| \leq n\}}$. Since $H^n \in \mathbf{L}^1(\mathbb{F}, \mathrm{Var}(M))$, then $H^n \cdot M \in \mathcal{H} \cap \mathcal{A}$ (cf. (2.9)), and by the corollary to Theorem 1.7.3 we have $H^n \cdot M \in \mathcal{M}^d$, i.e. in the decomposition $H^n \cdot M = (H^n \cdot M)^c + (H^n \cdot M)^d$ the process $(H^n \cdot M)^c$ is negligible. Let us show now that in the decomposition $H \cdot M = (H \cdot M)^c + (H \cdot M)^d$ the process $(H \cdot M)^c$ is negligible too.

We have

$$((H \cdot M)^c)^*_\infty \leq (H \cdot M - H^n \cdot M)^*_\infty + ((H \cdot M)^d - H^n \cdot M)^*_\infty.$$

By (2.12) we have $(H \cdot M - H^n \cdot M)^*_\infty \to 0$ in probability as $n \to \infty$, while by Davis' inequality (Theorem 1.9. 6)

$$\mathbf{E}((H \cdot M)^d - H^n \cdot M)^*_\infty \leq C\mathbf{E}\left(\sum_{t > 0} (H(t)\Delta M_t - H^n(t)\Delta M_t)^2\right)^{1/2}$$

$$= C\|H - H^n\| \to 0, \quad n \to \infty.$$

Hence, $((H \cdot M)^c)^*_\infty = 0$ (P-a.s.).

This implies $H \cdot M \in \mathcal{M}^d_{\mathrm{loc}}$, and by Davis' inequality (Theorem 1.9.6)

$$\mathbf{E}\,(H\cdot M)^*_\infty \le C\mathbf{E}\left(\sum_{t>0}(H(t)\,\Delta M_t)^2\right)^{1/2} = C\ \|H\| < \infty,$$

i.e. the desired assertion takes place.

3) The required property follows from properties 1) and 2).

4) Evidently $\Delta\,(H\cdot M)^T = I_{[\![0,T]\!]}\,\Delta\,(H\cdot M)$. Therefore, by property 1)

$$\Delta\,(H\cdot M)^T = I_{[\![0,T]\!]}H\Delta M,\quad \Delta\,(H\cdot M^T) = HI_{[\![0,T]\!]}\Delta M.$$

Consequently, the process $X = (H\cdot M)^T - H\cdot M^T$ is continuous. But by property 2) we have $X \in \mathcal{M}^d$. Hence $[X, X] = 0$ and $X = 0$ by Davis' inequality (Theorem 1.9.6).

Theorem 1 is proved.

Property 4) admits the unique definition of the stochastic integral $H\cdot M$ in case of $M \in \mathcal{M}_{loc} \cap \mathcal{A}_{loc}$ and of a predictable process H such that $(H^2 \circ [M, M])^{1/2} \in \mathcal{A}^+_{loc}$, by setting

$$H\cdot M = HI_{[\![0,T_1]\!]}\cdot M + \sum_{n\ge 1} HI_{]\!]T_n, T_{n+1}]\!]}\cdot M, \tag{2.15}$$

where T_n, $n \ge 1$, is a localizing sequence and $M^{T_n} \in \mathcal{M} \cap \mathcal{A}$, $\|HI_{[\![0,T_n]\!]}\| < \infty$, $n \ge 1$. The process $H\cdot M \in \mathcal{M}^d_{loc}$ defined in such a manner is independent of choosing a localizing sequence and possesses properties 1), 3) and 4) in Theorem 1.

Definition 3. The process defined by the formula (2.15) is called the *stochastic integral* of H *with respect to a process* $M \in \mathcal{M}_{loc} \cap \mathcal{A}_{loc}$, where H is a predictable process satisfying the condition

$$(H^2 \circ [M, M])^{1/2} \in \mathcal{A}^+_{loc}.$$

4. Let us turn now to the definition of the stochastic integral $H\cdot M^2$ (cf. (2.6)) with $M^2 \in \mathcal{M}^2_{loc}$. First let M be a square integrable martingale ($M \in \mathcal{M}^2$). Denote by $\Lambda = \Lambda\,(\mathbb{F})$ the space of bounded simple predictable functions H of type

$$H = \lambda_0 I_{[\![0]\!]} + \sum_{i=0}^{n} \lambda_i I_{]\!]t_i, t_{i+1}]\!]},$$

where $0 = t_0 < t_1 < t_2 < \dots < t_n < t_{n+1} \le \infty$ and λ_i are bounded $\mathcal{F}_{t_i}$-measurable random variables.

Then the stochastic integral

$$H \cdot M_t \left(= \int_0^t H_s\, dM_s = \int_{]0,\, t]} H_s\, dM_s \right)$$

is naturally defined by the formula

$$H \cdot M_t = \sum_{i=0}^{n} \lambda_i\, [M_{t \wedge t_{i+1}} - M_{t \wedge t_i}]. \tag{2.16}$$

Clearly, $H \cdot M \in \mathfrak{M}^2$. Denote $N = M^2 - \langle M \rangle$. Then it is not difficult to find out for each $t \in \mathbf{T}$ that

$$\mathbf{E}\,(H \cdot M_\tau)^2 = \mathbf{E} \sum_{i=0}^{n} \lambda_i^2\, [M_{\tau \wedge t_{i+1}} - M_{\tau \wedge t_i}]^2 = \mathbf{E} \sum_{i=0}^{n} \lambda_i^2\, [M^2_{\tau \wedge t_{i+1}} - M^2_{\tau \wedge t_i}]$$

$$= \mathbf{E} \sum_{i=0}^{n} \lambda_i^2\, [\langle M \rangle_{\tau \wedge t_{i+1}} - \langle M \rangle_{\tau \wedge t_i}] + \mathbf{E} \sum_{i=0}^{n} \lambda_i^2\, [N_{\tau \wedge t_{i+1}} - N_{\tau \wedge t_i}]$$

$$= \mathbf{E} H^2 \circ \langle M \rangle_\tau\,. \tag{2.17}$$

Consequently, by Problem 1.4.2 we have

$$(H \cdot M)^2 - H^2 \circ \langle M \rangle \in \mathfrak{M}.$$

Since $(H \cdot M)^2 - \langle H \cdot M \rangle$ is a martingale too (cf. Ch. 1, § 8) and the process $H^2 \circ \langle M \rangle$ is predictable, then by the Doob-Meyer decomposition

$$\langle H \cdot M \rangle = H^2 \circ \langle M \rangle. \tag{2.18}$$

Denote by $\mathbf{L}^2\,(\mathbb{F}, \langle M \rangle)$ the space of predictable processes with the norm

$$\| H \|_2 = (\mathbf{E} H^2 \circ \langle M \rangle_\infty)^{1/2} < \infty.$$

The space $\Lambda\,(\mathbb{F})$ is dense in $\mathbf{L}^2\,(\mathbb{F}, \langle M \rangle)$ (cf., e.g. [188], Lemma 5.3 on p. 202). Consequently, if $H \in \mathbf{L}^2\,(\mathbb{F}, \langle M \rangle)$, then a sequence H^n, $n \geq 1$, with $H^n \in \Lambda\,(\mathbb{F})$, $n \geq 1$, can be found such that

$$\| H - H^n \|_2 \to 0, \quad n \to \infty.$$

It follows from (2.17) that

$$\mathbf{E}\,(H^n \cdot M_\infty - H^m \cdot M_\infty)^2 = \mathbf{E}\,((H^n - H^m) \cdot M_\infty)^2$$

$$= \| H^n - H^m \|_2^2 \to 0, \quad n,m \to \infty.$$

By the completeness of the Hilbert space of random variables with finite second moments, there can be found an element, denoted by $(H \cdot M)_\infty$, such that

$$\mathbf{E}\,((H^n \cdot M)_\infty - (H \cdot M)_\infty)^2 \to 0, \quad n \to \infty.$$

The value $(H \cdot M)_\infty$ here is independent of a type of an approximating sequence H^n, $n \geq 1$.

Denote the process $H \cdot M \in \mathcal{M}^2$, by setting $H \cdot M_\tau = \mathbf{E}\,((H \cdot M)_\infty \mid \mathcal{F}_\tau)$ for each $\tau \in \mathbf{T}$. By Doob's inequality (Theorem 1.9.2) we have

$$\mathbf{E}\,[(H \cdot M - H^n \cdot M)^{*}_\infty]^2$$

$$\leq 4\mathbf{E}\,[(H \cdot M)_\infty - (H^n \cdot M)_\infty]^2 \to 0, \quad n \to \infty. \tag{2.19}$$

Definition 4. The square integrable martingale $H \cdot M$ obtained in this manner, independent of an approximating sequence H^n, $n \geq 1$, is called the *stochastic integral of* $H \in \mathbf{L}^2(\mathbb{F}, \langle M \rangle)$ *with respect to a square integrable martingale* $M \in \mathcal{M}^2$.

The stochastic integral so defined possesses evidently the following properties

$$(c_1 H_1 + c_2 H_2) \cdot M = c_1 (H_1 \cdot M) + c_2 (H_2 \cdot M), \tag{2.20}$$

with the constants c_1, c_2 and $H_1, H_2 \in \mathbf{L}^2(\mathbb{F}, \langle M \rangle)$ (Problem 3);

$$H \cdot (M + N) = H \cdot M + H \cdot N \tag{2.21}$$

for $M, N \in \mathcal{M}^2$ and $H \in \mathbf{L}^2(\mathbb{F}, \langle M \rangle) \cap \mathbf{L}^2(\mathbb{F}, \langle N \rangle)$ (Problem 4).

Other important properties of $H \cdot M$ are given in the Theorems 2 and 3.

Theorem 2. *Let* $M \in \mathcal{M}^2$ *and* $H \in \mathbf{L}^2(\mathbb{F}, \langle M \rangle)$. *Then*

1) $H \cdot M \in \mathcal{H}_2$,

2) $\langle H \cdot M \rangle = H^2 \circ \langle M \rangle$,

3) $\Delta (H \cdot M) = H \Delta M$,

4) $(H \cdot M)^c = H \cdot M^c$, $(H \cdot M)^d = H \cdot M^d$,

5) $[H \cdot M, H \cdot M] = H^2 \circ [M, M]$,

6) $(H \cdot M)^T = H \cdot M^T \quad \forall T \in \mathbf{T}$.

Proof. 1) Since $H \cdot M \in \mathcal{M}^2$, the required property follows from Doob's inequality (Theorem 1.9.2).

2) This property takes place if $H \in \Lambda(\mathbb{F})$ (cf. 2.18)). Let H^n, $n \geq 1$, be an approximating sequence for H: $H^n \in \Lambda(\mathbb{F})$ $n \geq 1$, $\| H - H^n \|_2 \to 0$, $n \to \infty$. Then by the inequality

$$| a^2 - b^2 | \leq (a - b)^2 + 2\,| b |\,| a - b | \tag{2.22}$$

the following relations take place

$$\mathbf{E}\,(|\,H^2 - (H^n)^2\,| \circ \langle M\rangle_\infty)$$

$$\le \|\,H - H^n\,\|_2^2 + 2\,\|\,H\,\|_2\,\|\,H - H^n\,\|_2 \to 0, \quad n \to \infty,$$

and, by taking into consideration (2.19),

$$\mathbf{E}\sup_{t\ge 0}|\,(H\cdot M_t)^2 - (H^n\cdot M_t)^2\,|$$

$$\le \mathbf{E}\,[(H\cdot M - H^n\cdot M)_\infty^*]^2 + 2\,(\mathbf{E}[(H\cdot M)_\infty^*]^2\,\mathbf{E}[(H\cdot M - H^n\cdot M)_\infty^*]^2)^{1/2} \to 0,$$

$$n \to \infty,$$

according to which for each $\tau \in \mathbf{T}$ and $H \in L^2(\mathbb{F}, \langle M\rangle)$ we have

$$\mathbf{E}\,(H\cdot M_\tau)^2 = \mathbf{E}H^2 \circ \langle M\rangle_\tau.$$

Hence $(H\cdot M)^2 - H^2 \circ \langle M\rangle \in \mathfrak{M}$ (Problem 1.4.2), and the required property follows from the uniqueness of the Doob-Meyer decomposition and from the predictability of the process $H^2 \circ \langle M\rangle$.

3) The required property holds for $H \in \Lambda$. Let H^n, $n \ge 1$, be an approximating sequence for $H \in \mathbf{L}^2(\mathbb{F}, \langle M\rangle)$ (i.e. $H^n \in \Lambda(\mathbb{F})$, $n \ge 1$, $\|\,H - H^n\,\|_2 \to 0$, $n \to \infty$). Observe that

$$\sup_{t>0}|\,\Delta(H\cdot M)_t - H(t)\,\Delta M_t\,|$$

$$\le \sup_{t>0}|\,\Delta(H\cdot M)_t - \Delta(H^n\cdot M)_t\,| + \sup_{t>0}|\,(H(t) - H^n(t))\,\Delta M_t\,|.$$

By taking into consideration Doob's inequality (Theorem 1.9.2) and (2.19), we have

$$\mathbf{E}\sup_{t>0}|\,\Delta(H\cdot M)_t - \Delta(H^n\cdot M)_t\,|^2$$

$$\le 4\mathbf{E}\,((H\cdot M - H^n\cdot M)_\infty^*)^2$$

$$\le 16\mathbf{E}\,((H\cdot M)_\infty - (H^n\cdot M)_\infty)^2 \to 0, \quad n \to \infty. \tag{2.23}$$

Next by the inequality

$$\sup_{t>0}|\,(H(t) - H^n(t))\,\Delta M_t\,| \le \left(\sum_{t>0}(H(t) - H^n(t))^2\,(\Delta M_t)^2\right)^{1/2}$$

we have

$$\mathbf{E} \sup_{t>0} |(H(t) - H^n(t)) \Delta M_t| \le \mathbf{E}((H - H^n)^2 \circ [M, M]_\infty)^{1/2}$$

$$\le (\mathbf{E}(H - H^n)^2 \circ [M, M]_\infty)^{1/2}$$

$$= (\mathbf{E}(H - H^n)^2 \circ \langle M \rangle_\infty)^{1/2} = \| H - H^n \|_2 \to 0, \quad n \to \infty. \tag{2.24}$$

By (2.23) and (2.24) the processes $\Delta(H \cdot M)$ and $H\Delta M$ are indistinguishable.

4) Let us show now that $H \cdot M^d \in \mathcal{M}^{2,d}$. It suffices to show that $\mathbf{E}N_\tau(H \cdot M^d_\tau) = 0$ for each $N \in \mathcal{M}^{2,c}$ with $N_0 = 0$ and each $\tau \in \mathbf{T}$. If H^n, $n \ge 1$, is an approximating sequence: $H^n \in \Lambda(\mathbb{F})$, $n \ge 1$, $\| H - H^n \|_2 \to 0$, $n \to \infty$, then evidently $H^n \cdot M^d \in \mathcal{M}^{2,d}$, and hence $\mathbf{E}N_\tau(H^n \cdot M^d_\tau) = 0$. Therefore

$$|\mathbf{E}N_\tau(H \cdot M^d_\tau)| = |\mathbf{E}N_\tau((H - H^n) \cdot M^d_\tau)|$$

$$\le (\mathbf{E}N^2_\tau)^{1/2} \| H - H^n \|_2 \to 0, \quad n \to \infty. \tag{2.25}$$

By Theorem 1.5.3 the unique decomposition $H \cdot M = (H \cdot M)^c + (H \cdot M)^d$ takes place. From this and from the decomposition

$$H \cdot M = H \cdot M^c + H \cdot M^d$$

with $H \cdot M^c \in \mathcal{M}^{2,c}$ and $H \cdot M^d \in \mathcal{M}^{2,d}$ we get the desired assertion.

5) By definition (cf. Ch. 1, § 8)

$$[H \cdot M, H \cdot M] = \langle (H \cdot M)^c \rangle + \sum_s (\Delta(H \cdot M)_s)^2.$$

The desired assertion follows from this and from the properties 4), 3) and 2).

6) The required property holds for $H \in \Lambda(\mathbb{F})$. If now $H \in \mathbf{L}^2(\mathbb{F}, \langle M \rangle)$ and H^n, $n \ge 1$, is an approximating sequence: $H^n \in \Lambda(\mathbb{F})$, $n \ge 1$, $\| H - H^n \|_2 \to 0$, $n \to \infty$, then by taking into account (2.19) we have

$$\mathbf{E}[((H \cdot M)^T - (H^n \cdot M)^T)^*_\infty]^2 \le \mathbf{E}[(H \cdot M - H^n \cdot M)^*_\infty]^2 \to 0, n \to \infty, \tag{2.26}$$

and by Doob's inequality (Theorem 1.9.2), by property 2) and by the equality $\langle M^T \rangle = \langle M \rangle^T$ (Problem 1.8.16) we have

$$\mathbf{E}\,((H\cdot M^T - H^n\cdot M^T)^*_\infty)^2 \le 4\mathbf{E}\,((H - H^n)\cdot M^T_\infty)^2$$

$$= 4\mathbf{E}\,(H - H^n)^2 \circ \langle M\rangle^T_\infty$$

$$\le 4\,\| H - H^n \|_2^2 \to 0, \quad n \to \infty. \tag{2.27}$$

From (2.26) and (2.27) it follows that for each $\tau \in \mathbf{T}$ as $n \to \infty$

$$(H^n\cdot M)^T_\tau \to (H\cdot M)^T_\tau, \quad H^n\cdot M^T_\tau \to H\cdot M^T_\tau$$

in probability. Consequently (P-a.s.),

$$(H\cdot M)^T_\tau = H\cdot M^T_\tau$$

and by Theorem 1.3. 12 the processes $(H\cdot M)^T$ and $H\cdot M^T$ are indistinguishable.

Theorem 3. *Let* $M \in \mathcal{M}^2 \cap \mathcal{A}$ *and* $H \in \mathbf{L}^2(\mathbb{F}, \langle M\rangle) \cap \mathbf{L}^1(\mathbb{F}, \mathrm{Var}\,(M))$. *Then the stochastic integral* $H\cdot M$ *with respect to a square integrable martingale* M *coincides (up to* P*-indistinguishability) with the Lebesgue-Stieltjes integral* $H \circ M$, *defined by* (2.3).

Proof. Since $\mathcal{M}^2 \cap \mathcal{A} \subseteq \mathcal{M} \cap \mathcal{A}$, we have $M \in \mathcal{M}^{2,d} \cap \mathcal{M}$ by the corollary to Theorem 1.7.3, i.e. $M^c = 0$ in the decomposition $M = M^c + M^d$. Consequently, by Theorem 2 we have $H\cdot M \in \mathcal{M}^{2,d}$ and $\Delta\,(H\cdot M) = H\Delta M$. Next, by (2.9) we have $H \circ M \in \mathcal{A} \cap \mathcal{H}$, and by the corollary to Theorem 1.7.3 we have $H \circ M \in \mathcal{M}^d$ with $\Delta\,(H \circ M) = H\,\Delta M$. Therefore $X = H\cdot M - H\circ M \in \mathcal{M}^d$ with $\Delta X = 0$. Consequently, $[X, X]_\infty = 0$, and by Davis' inequality (Theorem 1.9.6) we have $\mathbf{E}X^*_\infty = 0$.

Property 6) of the stochastic integral with respect to a square integrable martingale M allows us to define the process $H\cdot M$ for a local square integrable martingale M ($M \in \mathcal{M}^2_{loc}$) and for a predictable process H such that $H^2 \circ \langle M\rangle \in \mathcal{A}^+_{loc}$, by means of the following formula (cf. (2.15))

$$H\cdot M = HI_{[\![0, T_1]\!]}\cdot M + \sum_{n\ge 1} HI_{]\!]T_n, T_{n+1}]\!]}\cdot M, \tag{2.28}$$

where T_n, $n \ge 1$, is a localizing sequence

$$M^{T_n} \in \mathcal{M}^2, \quad \mathbf{E}H^2 \circ \langle M\rangle_{T_n} < \infty, \quad n \ge 1.$$

The process $H \cdot M$ belongs to $\mathcal{M}^2_{loc}$, its value is independent of choosing a localizing sequence and it possesses the properties 2) - 6) established in Theorem 2. If in addition $M \in \mathcal{M}^2_{loc} \cap \mathcal{A}_{loc}$, then $H \cdot M$ coincides (up to P-indistinguishability) with the Lebesgue-Stieltjes integral $H \circ M$, defined by the formula (2.3).

Definition 5. The process $H \cdot M$ defined by the formula (2.28) is called the *stochastic integral* of a predictable process H with $H^2 \circ \langle M\rangle \in \mathcal{A}^+_{loc}$ with respect to M $\in \mathcal{M}^2_{loc}$.

It will be shown in the next subsection that the stochastic integral $H \cdot M$ for $M \in \mathcal{M}^2_{loc}$ can be defined for a larger supply of predictable processes H.

5. First, let $M \in \mathcal{M}^2$ and let $\mathbf{L}(\mathbb{F}, [M, M])$ be the class of predictable processes H with

$$\| H \| = \mathbf{E}\,(H^2 \circ [M, M]_\infty)^{1/2} < \infty.$$

Since

$$\| H \| \leq (\mathbf{E}H^2 \circ [M, M]_\infty)^{1/2} = (\mathbf{E}H^2 \circ \langle M\rangle_\infty)^{1/2} = \| H \|_2,$$

we have $\mathbf{L}^2(\mathbb{F}, \langle M\rangle) \subseteq \mathbf{L}(\mathbb{F}, [M, M])$ and for $H \in \mathbf{L}^2(\mathbb{F}, \langle M\rangle)$ the stochastic integral $H \cdot M$ will be understood as a square integrable martingale defined in Subsection 4 with the properties 1) - 6) of Theorem 2.

Let $H \in \mathbf{L}(\mathbb{F}, [M, M])$ and let $(H^n)_{n \geq 1}$ be a sequence with $H^n \in \mathbf{L}^2(\mathbb{F}, \langle M\rangle)$, $n \geq 1$, such that $\| H^n - H \| \to 0$, $n \to \infty$. (The sequence with $H^n = HI_{\{|H| \leq n\}}$, $n \geq 1$, serves as an example of such a sequence). Considering the sequence $(H^n \cdot M)_{n \geq 1}$ with $H^n \cdot M \in \mathcal{M}^2$, we establish that by Davis' inequality (Theorem 1.9.6)

$$\mathbf{E}\,(H^n \cdot M - H^m \cdot M)^*_\infty \leq C\mathbf{E}\,((H^n - H^m)^2 \circ [M, M]_\infty)^{1/2}$$

$$= C \| H^n - H^m \| \to 0, \quad n, m \to \infty.$$

From this it follows that the sequence $(H^n \cdot M_\infty)_{n \geq 1}$ converges in the mean to a random variable denoted by $(H \cdot M)_\infty$ with $\mathbf{E}\,| (H \cdot M)_\infty | < \infty$.

Definition 6. For $H \in \mathbf{L}(\mathbb{F}, [M, M])$ the process $H \cdot M \in \mathcal{M}$ with

$$H \cdot M_\tau = \mathbf{E}\,((H \cdot M)_\infty \mid \mathcal{F}_\tau), \ \tau \in \mathbf{T}$$

is called the *stochastic integral with respect to a square integrable martingale* M (M $\in$

$\mathcal{M}^2$).

It is easily seen that this process is independent of choosing an approximating sequence $(H_n)_{n \geq 1}$, it satisfies the relation (2.12) and possesses the properties (2.13) and (2.14).

Analogously to Theorem 1, we obtain

Theorem 4. *Let* $M \in \mathcal{M}^2$ *and* $H \in \mathbf{L}(\mathbb{F}, [M, M])$. *Then*

1) $\Delta(H \cdot M) = H\Delta M$,

2) $(H \cdot M)^c = H \cdot M^c$, $(H \cdot M)^d = H \cdot M^d$,

3) $[H \cdot M, H \cdot M] = H^2 \circ [M, M]$,

4) $H \cdot M \in \mathcal{H}$,

5) $(H \cdot M)^T = H \cdot M^T \quad \forall T \in \mathbf{T}$,

Proof. For $H \in \mathbf{L}^2(\mathbb{F}, \langle M \rangle)$ this property has been established in Theorem 2. In the general case the proof is completed as in Theorem 2.

2) As $M = M^c + M^d$ (Theorem 1.7.2), we have by (2.14) that $H \cdot M = H \cdot M^c + H \cdot M^d$. By property 1) we have $\Delta(H \cdot M^c) = 0$, i.e. $H \cdot M^c \in \mathcal{M}^c$. Hence $H \cdot M^c \in \mathcal{M}^{2, c}_{loc}$.

Let us show that $H \cdot M^d \in \mathcal{M}^d$. It suffices to show that $N(H \cdot M^d) \in \mathcal{M}$ for each bounded process N from the class $\mathcal{M}^c_{loc}$ with $N_0 = 0$ or to show that $\mathbf{E}N_\tau(H \cdot M^d_\tau) = 0$ for each $\tau \in \mathbf{T}$.

Set $H^n = HI_{\{|H| \leq n\}}$. Evidently $H^n \in \mathbf{L}^2(\mathbb{F}, \langle M \rangle)$. Therefore, by Theorem 2 (Property 4) we have $H^n \cdot M^d \in \mathcal{M}^d$, and consequently $\mathbf{E}N_\tau(H^n \cdot M^d_\tau) = 0$ for each $\tau \in \mathbf{T}$. By using this fact and taking into consideration the equality $N^*_\infty \leq c$, we get

$$|\mathbf{E}N_\tau(H \cdot M^d_\tau)| = |\mathbf{E}N_\tau((H - H^n) \cdot M^d_\tau)| \leq c\mathbf{E}|(H - H^n) \cdot M^d_\tau|$$

$$= c\mathbf{E}|\mathbf{E}((H - H^n) \cdot M^d_\infty \mid \mathcal{F}_\tau)| \leq c\mathbf{E}|(H - H^n) \cdot M^d_\infty| \to 0, \quad n \to \infty.$$

Thus $H \cdot M^c \in \mathcal{M}^c$ and $H \cdot M^d \in \mathcal{M}^d$. Therefore the desired relation takes place as the decomposition $H \cdot M = (H \cdot M)^c + (H \cdot M)^d$ is unique (Theorem 1.7.2).

3) The desired property follows from 1) and 2).

4) This property follows from 3) and Davis' inequality (Theorem 1.9.6) according to which $\mathbf{E}(H \cdot M)^*_\infty \leq C\,\|H\|$.

5) This property is obtained analogously to property 6) of Theorem 2 by using Davis' inequality instead of Doob's inequality.

Property 5) allows us to define by the formula (2.15) the process $H \cdot M \in \mathcal{M}_{loc}$, with $M \in \mathcal{M}^2_{loc}$ and a predictable process H such that $(H^2 \circ [M, M])^{1/2} \in \mathcal{A}^+_{loc}$, which preserves properties 1), 2), 3) and 5) of Theorem 4. The value $H \cdot M$ is independent of choosing a localizing sequence.

Definition 7. The process $H \cdot M$ defined by the formula (2.15) is called the *stochastic integral* of a predictable process H with $(H^2 \circ [M, M])^{1/2} \in \mathcal{A}^+_{loc}$ with respect to $M \in \mathcal{M}^2_{loc}$.

6. By utilizing the decomposition (2.6) ($M = M^1 + M^2$, $M^1 \in \mathcal{M}_{loc} \cap \mathcal{A}_{loc}$, $M^2 \in \mathcal{M}^2_{loc}$), we now present the definition of the stochastic integral $H \cdot M$ for $M \in \mathcal{M}_{loc}$ and for H from $\mathbf{L}_{loc}(\mathbb{F}, [M, M])$ that is the space of predictable processes with $(H^2 \circ [M, M])^{1/2} \in \mathcal{A}^+_{loc}$.

According to the Definitions 3 and 7 given above of the stochastic integrals $H \cdot M^1$ and $H \cdot M^2$ respectively, there exists the process $H \cdot M = H \cdot M^1 + H \cdot M^2$ that is a local martingale ($H \cdot M \in \mathcal{M}_{loc}$), provided

$$H \in \mathbf{L}_{loc}(\mathbb{F}, [M^1, M^1]) \cap \mathbf{L}_{loc}(\mathbb{F}, [M^2, M^2]).$$

It is natural to call this process $H \cdot M$ the stochastic integral of a process H with respect to a local martingale M. Of course, for the correctness of such a definition, it should be established that the value of $H \cdot M$ is independent of the form of the decomposition $M = M^1 + M^2$, $M^1 \in \mathcal{M}_{loc} \cap \mathcal{A}_{loc}$, $M^2 \in \mathcal{M}^2_{loc}$, and besides the class of predictable processes H should be described by means of characteristics of a local martingale M itself, rather then its components M^1 and M^2. The subsequent discussion is aimed at showing that the given definition is correct and the class of functions H coincides with the class $\mathbf{L}_{loc}(\mathbb{F}, [M, M])$.

An essential rôle in proving these facts are played by the Lemmas 1 - 3 given below.

Lemma 1. *Let* X *be an optional stochastic process such that* (P-a.s.) *the set* $\Gamma = \{t: X_t \neq 0\}$ *is an at most countable subset in* R_+. *Let* V (X) *and* L (X) *be increasing processes*

$$V_t(X) = \Big(\sum_{s \le t} X_s^2 \Big)^{1/2}, \quad L_t(X) = \sum_{s \le t} \frac{X_s^2}{1 + |X_s|}. \tag{2.29}$$

Then

$$V(X) \in \mathcal{A}_{loc}^{+} \Leftrightarrow L(X) \in \mathcal{A}_{loc}^{+}.$$

Proof. ($\Rightarrow$). First, let $\mathbf{E}V_\infty(X) < \infty$. Set $T_k = \inf(t\colon V_t(X) \ge k)$, $k \ge 1$, and observe that $|X_{T_k}| \le V_\infty(X)$, hence $\mathbf{E}|X_{T_k}| < \infty$. The definition of T_k implies $V_{T_k}(X) \le k$. Therefore, by taking into consideration the obvious inequality $L(X) \le V^2(X)$ we get

$$L_{T_k}(X) = L_{T_k-} + \frac{X_{T_k}^2}{1 + |X_{T_k}|} \le k^2 + |X_{T_k}|.$$

Consequently, $\mathbf{E}L_{T_k}(X) \le k^2 + \mathbf{E}|X_{T_k}| < \infty$.

Thus, for $\mathbf{E}V_\infty(X) < \infty$ the sequence $(T_k)_{k \ge 1}$ is localizing for $L(X)$.

If $V(X) \in \mathcal{A}_{loc}^{+}$, then there exists a localizing sequence $(S_k)_{k \ge 1}$ such that $\mathbf{E}V_\infty^{S_k}(X) < \infty$, $k \ge 1$. In this case the sequence $(S_k \wedge T_k)_{k \ge 1}$ is localizing for $L(X)$.

($\Leftarrow$). It suffices for this implication to obtain the inequality

$$V(X) \le \frac{1}{4} + 2L(X). \tag{2.30}$$

To this end, observe that by the mean value theorem

$$\Delta V(X) = \frac{X^2}{2\sqrt{V_-(X) + \theta X^2}}, \quad \theta \in [0, 1]. \tag{2.31}$$

Let us show that in fact $\theta \in [1/4, 1]$. Denote $a = V_-(x)$ and $b = X^2$. Then (2.31) takes the following form:

$$\sqrt{a+b} - \sqrt{a} = \frac{b}{2\sqrt{a + \theta b}}.$$

This gives

$$\theta = \frac{1}{4} + \frac{\sqrt{(a+b)\,a} - a}{2b} > \frac{1}{4}.$$

Set $T = \inf(t\colon V_t(X) \ge 1/4)$. On the set $[\![0, T[\![$ we have $V(X) \le 1/4$, while on the set $[\![T, \infty[\![$

$$\Delta V(X) \le \frac{X^2}{\sqrt{1+X^2}} \le \frac{2X^2}{1+|X|} \le 2\Delta L(X).$$

Consequently

$$V(X) = I_{[\![0,T[\![} \circ V(X) + I_{[\![T,\infty[\![} \circ V(X) \le \frac{1}{4} + 2I_{[\![T,\infty[\![} \circ L(X),$$

i.e. (2.30) takes place.

Lemma 2. *Let a process* $A \in \mathcal{A}_{loc}$, *let* $\tilde{A}$ *be its compensator and let* H *be a predictable process. Let* $f = f(x)$ *be a concave nonnegative function with* $f(0) = 0$. *Then*

$$\left\{\begin{array}{l} \left(\sum_s (H_s\Delta A_s)^2\right)^{1/2} \in \mathcal{A}^+_{loc} \\ \sum_s f(H_s\Delta A_s) \in \mathcal{A}^+_{loc} \end{array}\right\} \Rightarrow \sum_s f(H_s\Delta\tilde{A}_s) \in \mathcal{A}^+_{loc}.$$

Proof. First, let

$$\mathbf{E}\left(\sum_{s>0}(H_s\Delta A_s)^2\right)^{1/2} < \infty$$

and

$$\mathbf{E}\sum_{s>0} f(H_s\Delta A_s) < \infty.$$

Then for each Markov time τ we have $\mathbf{E}\,|H_\tau\Delta A_\tau| < \infty$ and $\mathbf{E}\,f(H_\tau\Delta A_\tau) < \infty$. Consequently, by virtue of Problem 1.3.6 and Corollary 2 to Theorem 1.6.3 we have ${}^p(H\Delta A) = H\,{}^p(\Delta A) = H\Delta\tilde{A}$. Therefore by virtue of Problem 1.3.9 and the Theorems 1.3.13 and 1.6.2 we get

$$\mathbf{E}\sum_{s>0} f(H_s\Delta\tilde{A}_s) = \mathbf{E}\sum_{s>0} I(\Delta\tilde{A}_s \ne 0)\, f(H_s\Delta\tilde{A}_s)$$

$$= \mathbf{E}\sum_{s>0} I(\Delta\tilde{A}_s \ne 0)\, f({}^p(H\Delta A_s)) \le \mathbf{E}\sum_{s>0} I(\Delta\tilde{A}_s \ne 0)^p\,(f(H\Delta A))_s$$

$$= \mathbf{E}\sum_{s>0} I(\Delta\tilde{A}_s \ne 0)\, f(H_s\Delta A_s) \le \mathbf{E}\sum_{s>0} f(H_s\Delta A_s).$$

This implies

$$\mathbf{E}\sum_{s>0} f(H_s\Delta\tilde{A}_s) < \infty.$$

In the general case, if (τ_k) and (σ_k) are localizing sequences

$$\mathbf{E}\left(\sum_{s\le\tau_k}(H_s\Delta A_s)^2\right)^{1/2} \le \infty, \quad \mathbf{E}\sum_{s\le\sigma_k} f(H_s\Delta A_s) < \infty,$$

then

$$\mathbf{E}\sum_{s\le\tau_k\wedge\sigma_k} f(H_s\Delta A_s) < \infty, \quad k\ge 1$$

and $\tau_k \wedge \sigma_k$ is a localizing sequence for

$$\sum_s f(H_s\Delta A_s).$$

Lemma 3. 1) *Let* $M \in \mathcal{M}_{loc}$ *and* $H \in \mathbf{L}_{loc}(\mathbb{F}, [M, M])$. *Then there exists the decomposition* $M = M^1 + M^2$ *with* $M^1 \in \mathcal{M}_{loc} \cap \mathcal{A}_{loc}$, $M^2 \in \mathcal{M}^2_{loc}$ *such that*

$$\mathbf{L}_{loc}(\mathbb{F}, [M, M]) = \mathbf{L}_{loc}(\mathbb{F}, [M^1, M^1]) \cap \mathbf{L}_{loc}(\mathbb{F}, [M^2, M^2]).$$

2) *A value of* $H \cdot M = H \cdot M^1 + H \cdot M^2$ *is independent of the local martingales* M^1 *and* M^2 *involved in this decomposition, i.e. if* $M = N^1 + N^2$ *is another decomposition* ($N^1 \in \mathcal{M}_{loc} \cap \mathcal{A}_{loc}$, $N^2 \in \mathcal{M}^2_{loc}$) *and* $H \in \mathbf{L}_{loc}(\mathbb{F}, [N^1, N^1]) \cap \mathbf{L}_{loc}(\mathbb{F}, [N^2, N^2])$, *then the processes* $H \cdot M^1 + H \cdot M^2$ *and* $H \cdot N^1 + H \cdot N^2$ *are indistinguishable.*

Proof. Since $M^1 \in \mathcal{M}^d_{loc}$ (Corollary to Theorem 1.7.3) and since by Theorem 1.7.2 we have $M^2 = M^{2,c} + M^{2,d}$, then

$$[M^1, M^1] + [M^2, M^2] = \langle M^{2,c}\rangle + \sum_s [(\Delta M^1_s)^2 + (\Delta M^2_s)^2].$$

Consequently, $2([M^1, M^1] + [M^2, M^2]) \ge [M, M]$ and

$$\mathbf{L}_{loc}(\mathbb{F}, [M, M]) \supseteq \mathbf{L}_{loc}(\mathbb{F}, [M^1, M^1]) \cap \mathbf{L}_{loc}(\mathbb{F}, [M^2, M^2]).$$

Let us verify the converse inclusion.

As for M^1, we can, according to the proof of Theorem 1.7.1, take $M^1 — A — \tilde{A}$ where $\tilde{A}$ is the compensator of the process

$$A = \sum_s \Delta M_s I(|\Delta M_s| \ge 1),$$

and consequently, $(\Delta M^1)^2 \le 2\,[(\Delta A)^2 + (\Delta \tilde{A})^2]$. This gives

$$\left(\sum_s H_s^2 (\Delta M_s^1)^2\right)^{1/2} \le \sqrt{2}\left[\left(\sum_s H_s^2 (\Delta A_s)^2\right)^{1/2} + \left(\sum_s H_s^2 (\Delta \tilde{A}_s)^2\right)^{1/2}\right] \quad (2.32)$$

If $H \in \mathbf{L}_{loc}(\mathbb{F}, [M, M])$, then by the inequality $(\Delta A)^2 \le (\Delta M)^2$ we get

$$\left(\sum_s H_s^2 (\Delta A_s)^2\right)^{1/2} \in \mathcal{A}^+_{loc}, \quad (2.33)$$

and consequently, by Lemma 1

$$\sum_s \frac{H_s^2 (\Delta A_s)^2}{1 + |H_s \Delta A_s|} \in \mathcal{A}^+_{loc}.$$

By Lemma 2 with $f(x) = \dfrac{x^2}{1 + |x|}$ we have

$$\sum_s \frac{H_s^2 (\Delta \tilde{A}_s)^2}{1 + |H_s \Delta \tilde{A}_s|} \in \mathcal{A}^+_{loc}$$

and hence, by Lemma 1

$$\left(\sum_s H_s^2 (\Delta \tilde{A}_s)^2\right)^{1/2} \in \mathcal{A}^+_{loc}. \quad (2.34)$$

From (2.32) - (2.34) it follows that

$$\mathbf{L}_{loc}(\mathbb{F}, [M, M]) \subseteq \mathbf{L}_{loc}(\mathbb{F}, [M^1, M^1]).$$

Next, since

$$[M^2, M^2] = \langle M^{2,c} \rangle + \sum_s (\Delta M - \Delta M^1)^2 \le 2\,([M, M] + [M^1, M^1]),$$

the inclusion

$$\mathbf{L}_{loc}(\mathbb{F}, [M, M]) \subseteq \mathbf{L}_{loc}(\mathbb{F}, [M^2, M^2])$$

takes place. Thus the spaces $\mathbf{L}_{loc}(\mathbb{F}, [M, M])$ and $\mathbf{L}_{loc}(\mathbb{F}, [M^1, M^1]) \cap \mathbf{L}_{loc}(\mathbb{F}, [M^2, M^2])$ coincide.

2) Denote

$$X = (H \cdot M^1 + H \cdot M^2) - (H \cdot N^1 + H \cdot N^2).$$

By (2.13) and (2.14) we have

$$X = H \cdot (M^1 - N^1) + H \cdot (M^2 - N^2).$$

Observe that

$$M^2 - N^2 = -(M^1 - N^1).$$

From this it follows by Problem 9 that X is a negligible process.

Definition 8. The process $H \cdot M$, defined above with $M \in \mathcal{M}_{loc}$ and $H \in \mathbf{L}_{loc}(\mathbb{F}, [M, M])$, is called the *stochastic integral with respect to a local martingale.*

In the following theorem properties of the stochastic integral $H \cdot M$ are presented.

Theorem 5. *Let* $M \in \mathcal{M}_{loc}$ *and* $H \in \mathbf{L}_{loc}(\mathbb{F}, [M, M])$. *Then the stochastic integral* $H \cdot M$ *possesses the following properties:*

1) $H \cdot M \in \mathcal{M}_{loc}$,

2) $\Delta(H \cdot M) = H\Delta M$,

3) $[H \cdot M, H \cdot M] = H^2 \circ [M, M]$,

4) $(H \cdot M)^T = H \cdot M^T \ \ \forall T \in \mathbf{T}$,

5) $(H \cdot M)^c = H \cdot M^c, \ \ (H \cdot M)^d = H \cdot M^d$.

Proof. Properties 1) - 5) follow from Theorems 1 and 4.

7. Along with the space $\mathbf{L}_{loc}(\mathbb{F}, [M, M])$ we define, by applying the notion of localization to $\mathbf{L}^1(\mathbb{F}, \mathrm{Var}(M))$ and $\mathbf{L}^2(\mathbb{F}, \langle M\rangle)$ (cf. Subsections 2 and 4), the spaces of the predictable processes $\mathbf{L}^1_{loc}(\mathbb{F}, \mathrm{Var}(M))$ and $\mathbf{L}^2_{loc}(\mathbb{F}, \langle M\rangle)$ ($H \in \mathbf{L}^1_{loc}(\mathbb{F}, \mathrm{Var}(M))$ and $H \in \mathbf{L}^2_{loc}(\mathbb{F}, \langle M\rangle)$, if $|H| \circ \mathrm{Var}(M) \in \mathcal{A}^+_{loc}$ and $H^2 \circ \langle M\rangle \in \mathcal{A}^+_{loc}$ respectively).

In the following theorem the relation between these spaces is established.

Theorem 6. 1) *If* $M \in \mathcal{M}_{loc} \cap \mathcal{A}_{loc}$, then $\mathbf{L}^1_{loc}(\mathbb{F}, \mathrm{Var}(M)) \subseteq \mathbf{L}_{loc}(\mathbb{F}, [M, M])$ *and*

$$\mathbf{L}^1_{loc}(\mathbb{F}, \mathrm{Var}(M)) = \mathbf{L}_{loc}(\mathbb{F}, [M, M]) \cap \{H \in \mathcal{P} : |H| \circ \mathrm{Var}(M) \in \mathcal{V}^+\}.$$

2) *If* $M \in \mathcal{M}^2_{loc}$, *then* $\mathbf{L}^2_{loc}(\mathbb{F}, \langle M\rangle) \subseteq \mathbf{L}_{loc}(\mathbb{F}, [M, M])$ *and*

$$\mathbf{L}^2_{loc}(\mathbb{F}, \langle M\rangle) = \mathbf{L}_{loc}(\mathbb{F}, [M, M]) \cap \{H \in \mathcal{P} : \sum (H\Delta M)^2 \in \mathcal{A}^+_{loc}\}.$$

Corollary. *The spaces* $\mathbf{L}^2_{loc}(\mathbb{F}, \langle M^c\rangle)$ and $\mathbf{L}_{loc}(\mathbb{F}, [M^c, M^c])$ *coincide:*

$$\mathbf{L}^2_{loc}(\mathbb{F}, \langle M^c\rangle) = \mathbf{L}_{loc}(\mathbb{F}, [M^c, M^c]).$$

Proof. 1) Let $(\tau_n)_{n\geq 1}$ be a localizing sequence for $|H| \circ \mathrm{Var}(M)$, i.e. $\mathbf{E}\ |H| \circ \mathrm{Var}(M)_{\tau_n} < \infty$. Then by (2.8) we have

$$\mathbf{E} \sum_{s \le \tau_n} | H_s \Delta M_s | < \infty,$$

and consequently

$$\mathbf{E}\left(\sum_{s \le \tau_n} (H_s \Delta M_s)^2 \right)^{1/2} \le \mathbf{E} \sum_{s \le \tau_n} | H_s \Delta M_s | < \infty, \ \ n \ge 1.$$

This gives

$$\mathbf{L}^1_{loc} (\mathbb{F}, \mathrm{Var}\,(M)) \subseteq \mathbf{L}_{loc}(\mathbb{F}, [M, M]).$$

Let $(\tau_n)_{n \ge 1}$ be a localizing sequence for $\left(\sum_s (H_s \Delta M_s)^2 \right)^{1/2}$ and M, i.e.

$$\mathbf{E}\left(\sum_{s \le \tau_n} (H_s \Delta M_s)^2 \right)^{1/2} < \infty$$

and $M^{\tau_n} \in \mathcal{M}$. Then for each $\tau \in \mathbf{T}$ we have

$$\mathbf{E}\ | H_{\tau \wedge \tau_n} \Delta M_{\tau \wedge \tau_n} | < \infty, \quad n \ge 1. \tag{2.35}$$

Set $\sigma_n = \inf\,(t\colon\ | H | \circ \mathrm{Var}\,(M)_t \ge n)$. Since $| H | \circ \mathrm{Var}\,(M) \in \mathcal{V}^+$, then $\sigma_n \uparrow \infty$, $n \to \infty$, and hence $\gamma_n = \tau_n \wedge \sigma_n \uparrow \infty$, $n \to \infty$.

Next,

$$| H | \circ \mathrm{Var}\,(M)_{\gamma_n} = | H | \circ \mathrm{Var}\,(M)_{\gamma_n -} + | H_{\gamma_n} |\, \Delta\, \mathrm{Var}\,(M)_{\gamma_n}$$

$$\le n + | H_{\gamma_n} |\ | \Delta M_{\gamma_n} |.$$

Due to (2.35) this gives

$$\mathbf{E}\ | H | \circ \mathrm{Var}\,(M)_{\gamma_n} \le n + \mathbf{E}\ | H_{\gamma_n} \Delta M_{\gamma_n} | < \infty, \quad n \ge 1,$$

i.e. $(\gamma_n)_{n \ge 1}$ is a localizing sequence for $| H | \circ \mathrm{Var}\,(M)$.

2) The inclusion $\mathbf{L}^2_{loc} (\mathbb{F}, \langle M \rangle) \subseteq \mathbf{L}_{loc} (\mathbb{F}, [M, M])$ takes place because for each $\tau \in \mathbf{T}$

$$\mathbf{E}\ (H^2 \circ [M, M]_\tau)^{1/2} \le (\mathbf{E} H^2 \circ [M, M]_\tau)^{1/2} = (\mathbf{E} H^2 \circ \langle M \rangle_\tau)^{1/2}$$

and hence a localizing sequence $(\tau_n)_{n \ge 1}$ for $H^2 \circ \langle M \rangle$ is, in the same time, a localizing sequence for $(H^2 \circ [M, M])^{1/2}$.

If $(\tau_n)_{n \ge 1}$ is a localizing sequence for $\sum (H \Delta M)^2$, then a localizing sequence for

$H^2 \circ \langle M\rangle$ is presented by the sequence $(\gamma_n)_{n\geq 1}$ with $\gamma_n = \tau_n \wedge \sigma_n$, $\sigma_n = \inf(t\colon H^2 \circ [M, M]_t \geq n)$, because $\sigma_n \uparrow \infty$, $n \to \infty$ and

$$\mathbf{E}H^2 \circ \langle M\rangle_{\gamma_n} = \mathbf{E}H^2 \circ [M, M]_{\gamma_n} = \mathbf{E}\,(H^2 \circ [M, M]_{\gamma_n -} + (H_{\gamma_n} \Delta M_{\gamma_n})^2)$$
$$\leq n + \mathbf{E}\,(H_{\gamma_n} \Delta M_{\gamma_n})^2 < \infty, \quad n \geq 1.$$

8. Let X be a semimartingale admitting the representation

$$X_t = X_0 + A_t + M_t \tag{2.36}$$

where $A = (A_t)_{t\geq 0} \in \mathcal{V}$ with $A_0 = 0$ and $M = (M_t)_{t\geq 0} \in \mathcal{M}_{loc}$ with $M_0 = 0$. Suppose that a predictable process H possesses the following properties:

$$|H| \circ \mathrm{Var}(A) \in \mathcal{V}^+, \tag{2.37}$$

$$(H^2 \circ [M, M])^{1/2} \in \mathcal{A}^+_{loc}. \tag{2.38}$$

Definition 9. The process

$$H \cdot X = H \circ A + H \cdot M \tag{2.39}$$

is called the *stochastic integral of* H *with respect to a semimartingale* X.

Sometimes $H \cdot X_t$ will also be denoted as

$$\int_0^t H_s\, dX_s \quad \text{or} \quad \int_{(0,\, t]} H_s\, dX_s.$$

Indeed, the given definition of the stochastic integral $H \cdot X$ is correct, provided its value $H \cdot X_t$ is independent of a type of the decomposition (2.36).

Thus let yet another decomposition

$$X_t = X_0 + B_t + N_t \tag{2.40}$$

be valid with $B = (B_t)_{t\geq 0} \in \mathcal{V}$ and $N = (N_t)_{t\geq 0} \in \mathcal{M}_{loc}$, $N_0 = 0$ and let a predictable process H be such that along with the properties (2.37) and (2.38) we have

$$|H| \circ \mathrm{Var}(B) \in \mathcal{V}^+, \tag{2.41}$$

$$(H^2 \circ [N, N])^{1/2} \in \mathcal{A}^+_{loc}. \tag{2.42}$$

We will show that the processes $(H \circ A + H \cdot M)$ and $(H \circ B + H \cdot N)$ are indistinguishable. Set

$$Y = (H \circ A + H \cdot M) - (H \circ B + H \cdot N).$$

Then evidently

$$Y = H \circ (A - B) + H \cdot (M - N). \tag{2.43}$$

On the other hand (2.36) and (2.40) imply

$$M - N = B - A, \tag{2.44}$$

i.e. $M - N \in \mathcal{M}_{loc} \cap \mathcal{V}$, and hence by Theorem 1.6.4 we have $M - N \in \mathcal{M}_{loc} \cap$

$\mathcal{A}_{loc}$. Moreover, $|H| \circ \mathrm{Var}(M - N) \le |H| \circ (\mathrm{Var}(A) + \mathrm{Var}(B))$, which together with (2.37) and (2.41) shows that

$$|H| \circ \mathrm{Var}(M - N) \in \mathcal{V}^+.$$

Therefore by Theorem 6 we have $H \in \mathbf{L}^1_{loc}(\mathbb{F}, \mathrm{Var}(M - N))$, and by the definition of the stochastic integral for such processes H the process $H \cdot (M - N)$ is indistinguishable from the process $H \circ (M - N)$, defined by the Lebesgue-Stieltjes integral. This, by (2.43) and (2.44) implies that Y is a negligible process.

9. Let us turn to the basic properties of the stochastic integrals $H \cdot X$ with respect to semimartingales.

Theorem 7. *Let* X *be a semimartingale with decomposition* (2.36) *and let* H *be a predictable process with the properties* (2.37) *and* (2.38). *Then*

1) $H \cdot X \in \mathbf{S}$,

2) $\Delta(H \cdot X) = H\Delta X$,

3) $(H \cdot X)^c = H \cdot X^c$,

4) $[H \cdot X, H \cdot X] = H^2 \circ [X, X]$,

5) $(H \cdot X)^T = H \cdot X^T \quad \forall T \in \mathbf{T}$,

6) *if* $X \in \mathcal{V}$ *and* $|H| \circ \mathrm{Var}(X) \in \mathcal{V}^+$, *then the processes* $H \cdot X$ *and* $H \circ X$ *are indistinguishable.*

Proof. Property 1) follows from the definition of $H \cdot X = H \circ A + H \cdot M$, since $H \circ A \in \mathcal{V}(\mathbb{F})$ and $H \cdot M \in \mathcal{M}_{loc}$. Property 2) takes place because by the definition of the Lebesgue-Stieltjes integral $\Delta(H \cdot A) = H\Delta A$ and by Theorem 5 $\Delta(H \cdot M) = H\Delta M$.

From the decomposition (2.36) it follows that $X^c = M^c$, and from the definition of $H \cdot X$ that $(H \cdot X)^c = (H \cdot M)^c$. Therefore property 3) follows from property 5) of Theorem 5.

By definition

$$[H \cdot X, H \cdot X] = \langle H \cdot X^c \rangle + \sum_s (\Delta(H \cdot X)_s)^2.$$

Consequently, property 4) follows from 2) and 3). Property 5) follows from property 5) of Theorem 5 and from the evident property of the Lebesgue-Stieltjes integral $(H \circ A)^T = H \circ A^T$.

Property 6) takes place because in the decomposition (2.36) one can take $A_t = X_t - X_0$ and $M = 0$,while the stochastic integral $H \cdot X$ is independent of a type of the decomposition (2.36).

10. By using the construction of the stochastic integral we establish now an important property of the quadratic characteristics of locally square integrable martingales.

Theorem 8. *Let* M, N $\in \mathcal{M}^2_{loc}$. *Then:*

1) *a* $\mathcal{P}$*-measurable function* $h = h(\omega, t)$ *can be found such that the processes* $\langle M, N\rangle$ *and* $h \circ \langle M\rangle$ *are indistinguishable* $(\langle M, N\rangle = h \circ \langle M\rangle)$;

2) $\langle N\rangle - h^2 \circ \langle M\rangle \in \mathcal{V}^+$ (*i.e.* $h^2 \circ \langle M\rangle \in \mathcal{A}^+_{loc}$);

3) *if* h' *is a* $\mathcal{P}$*-measurable function such that* $\langle M, N\rangle = h' \circ \langle M\rangle$, *then* $\langle N\rangle - (h')^2 \circ \langle M\rangle \in \mathcal{V}^+$ *and the process* $(h - h')^2 \circ \langle M\rangle$ *is negligible.*

Proof. 1) Let g be a $\mathcal{P}$-measurable function such that $g = g^2$ and $g \circ \langle M\rangle_\infty = 0$ (P-a.s.). By Doob's inequality (Theorem 1.9.1)

$$P((g \cdot M)^*_\infty \geq a) \leq \frac{1}{a^2} \mathbf{E} \langle g \cdot M\rangle_\infty = \frac{1}{a^2} \mathbf{E} g \circ \langle M\rangle_\infty = 0.$$

Hence the process $g \cdot M$ is negligible and consequently the process $(g \cdot M) N$ is negligible too. Therefore by the definition of the quadratic characteristic (cf. Ch. 1, § 8) we have that $\langle (g \cdot M), N\rangle$ is a local martingale being a predictable process at the same time. Hence $\langle (g \cdot M), N\rangle$ is negligible (Theorem 1.6.4). On the other hand, by Problem 6 we have $\langle (g \cdot M), N\rangle = g \circ \langle M, N\rangle$, and hence the process $g \circ \langle M, N\rangle$ is negligible.

On the measurable space $(\Omega \times R_+, \mathcal{P})$ define the σ-finite measures

$$Q(d\omega, dt) = P(d\omega)\, d\langle M\rangle_t(\omega), \quad R(d\omega, dt) = P(d\omega)\, d\langle M, N\rangle_t(\omega).$$

By the implication $g \circ \langle M\rangle = 0 \Rightarrow g \circ \langle M, N\rangle = 0$ for $g \in \mathcal{P}$ with $g = g^2$ the measure R is absolutely continuous with respect to the measure Q ($R \ll Q$). Denote

$$h(\omega, t) = \frac{dR}{dQ}(\omega, t).$$

Then

$$P(d\omega)\, d\langle M, N\rangle_t(\omega) = h(\omega, t)\, P(d\omega)\, d\langle M\rangle_t(\omega). \tag{2.45}$$

Let $(T_n)_{n \geq 1}$ be a localizing sequence such that $\langle M\rangle_{T_n} + \langle N\rangle_{T_n} \leq n$ (the existence of such a sequence has been established in Lemma 1.6.1). Then, taking into account Problem 1.8.9, by (2.45) we get

$$\mathbf{E} \, |h| \circ \langle M\rangle_{T_n} = \mathbf{E} \operatorname{Var}(\langle M, N\rangle)_{T_n} \leq \frac{1}{2} \mathbf{E}(\langle M\rangle_{T_n} + \langle N\rangle_{T_n}) \leq n/2,$$

i.e. $|h| \circ \langle M\rangle \in \mathcal{A}^+_{loc}$.

Denote $U = \langle M, N\rangle - h \circ \langle M\rangle$. Since $U \in \mathcal{A}_{loc} \cap \mathcal{P}$, there exists by Lemma 1.6.1 a localizing sequence $(S_n)_{n\geq 1}$ such that $\mathrm{Var}\,(U)_{S_n} \leq n$. According to Problem 2.3.4 this gives

$$\frac{1}{2} U^2_{t\wedge S_n} \leq U \circ U_{t\wedge S_n} \leq n^2, \quad n \geq 1.$$

But by (2.45) for any predictable time τ we have

$$EU \circ U_{\tau\wedge S_n} I(\tau < \infty) = 0, \ n \geq 1.$$

Consequently,

$$EU^2_{\tau\wedge S_n} I(\tau < \infty) = 0, \ n \geq 1,$$

and hence $U^2_\tau I(\tau < \infty) = 0$ (P-a.s.). By Theorem 1.3.12 this means that the process U is negligible, thus the first assertion of the theorem takes place.

2) To prove the second assertion, set $h^n = hI_{\{|h|\leq n\}}$ and consider the process $N - h^n \cdot M$, that is a local square integrable martingale.

By Problem 6 its quadratic characteristic has the following representation:

$$\langle N - h^n \cdot M\rangle = \langle N\rangle - 2h^n \circ \langle M, N\rangle + (h^n)^2 \circ \langle M\rangle.$$

Taking into consideration the equality $h^n h = (h^n)^2$ and the representation $\langle M, N\rangle = h \circ \langle M\rangle$, by this we get

$$\langle N - h^n \cdot M\rangle = \langle N\rangle - (h^n)^2 \circ \langle M\rangle.$$

Hence, $\langle N\rangle - (h^n)^2 \circ \langle M\rangle \in \mathcal{V}^+$, $n \geq 1$, and consequently $\langle N\rangle - h^2 \circ \langle M\rangle \in \mathcal{V}^+$.

3) By examining the proof of 2) we see that

$$\langle M, N\rangle = h \circ \langle M\rangle \Rightarrow \langle N\rangle - h^2 \circ \langle M\rangle \in \mathcal{V}^+.$$

Therefore the relation $\langle N\rangle - (h')^2 \circ \langle M\rangle \in \mathcal{V}^+$ takes place. Next, $\langle M, N\rangle = h \circ \langle M\rangle$ and $\langle M, N\rangle = h' \circ \langle M\rangle$ imply the negligibility of the process $(h - h') \circ \langle M\rangle$. Then the process

$$(h - h')^2 \circ \langle M\rangle = (h - h') \circ ((h - h') \circ \langle M\rangle)$$

is negligible too.

Remark. The function h in Assertion 1) will sometimes be denoted by $\dfrac{d\langle M, N\rangle}{d\langle M\rangle}$.

Corollary. *Let* $N = H \cdot M$ *with* $H \in \mathbf{L}^2_{loc}(\mathbb{F}, \langle M\rangle)$ *and let* h *be a* $\mathcal{P}$*-measurable function such that* $\langle M, N\rangle = h \circ \langle M\rangle$. *Then the processes* N *and* N'

$(N' = h \cdot M)$ *are indistinguishable.*

In fact, by Problem 6 we have $\langle M, N\rangle = H \circ \langle M\rangle$. Therefore $(h - H)^2 \circ \langle M\rangle = 0$, and by Doob's inequality (Theorem 1.9.1)

$$P((N - N')^*_\infty \geq a) \leq \frac{1}{a^2} E\langle N - N'\rangle_\infty = \frac{1}{a^2} E(h - H)^2 \circ \langle M\rangle_\infty = 0.$$

11. Let X be a semimartingale relative to a flow $\mathbb{F} = (\mathcal{F}_t)_{t \geq 0}$. Consider a natural flow $\mathcal{F}^X_+ = (\mathcal{F}^X_{t+})_{t \geq 0}$ of σ-algebras:

$$\mathcal{F}^X_{t+} = \bigcap_{\varepsilon > 0} (\sigma(X_s,\ 0 \leq s \leq t + \varepsilon) \vee \mathcal{N}),$$

where $\mathcal{N}$ is a system of sets from $\mathcal{F}$ of P-measure zero.

If $H = H(\omega, t)$ is a $\mathcal{P}(\mathbb{F}^X_+)$- measurable and $\mathbb{F}^X_+$-locally bounded function (i.e. $|H| \leq c_n$ on the set $[\![0, \tau_n]\!]$, where c_n is a constant depending on n and $(\tau_n)_{n \geq 1}$ is a sequence of Markov times with respect to $\mathbb{F}^X_+$ such that $\tau_n \uparrow \infty$), then the process $H \cdot X$ is defined, that is the stochastic integral with respect to a $\mathbb{F}$-semimartingale X.

Let us establish a number of important properties of the process $H \cdot X$.

Theorem 9. 1) *The process* $H \cdot X$ *is* $\mathbb{F}^X_+$*-adapted.*

2) *If a* $\mathbb{F}$*-semimartingale* X *is also a* $\mathbb{F}^X_+$*-semimartingale* ($X \in S(\mathbb{F}^X_+)$), *then the stochastic integral* $\overline{H \cdot X}$ *with respect to a* $\mathbb{F}^X_+$*-semimartingale* X *and the process* $H \cdot X$ *are indistinguishable.*

Proof. Without loosing generality one can assume $|H| \leq$ const. Since the σ-algebra $\mathcal{P}(\mathbb{F}^X_+)$ coincides with the σ-algebra $\mathcal{P}_6(\mathbb{F}^X_+)$, generated by sets of the type $]\!]\sigma, \tau]\!]$ and $A \times \{0\}$ with $A \in \mathcal{F}^X_{0+}$, $\sigma, \tau \in \mathbf{T}(\mathbb{F}^X_+)$ (cf. the proof of Theorem 1.2.2), one can find a sequence $(H_k)_{k \geq 1}$ of bounded ($|H_k| \leq$ const) stepwise functions (generating the σ-algebra $\mathcal{P}_6(\mathbb{F}^X_+)$) of type

$$H_k(t) = c_0 I_{[\![0]\!]}(t) + \sum_{i=1}^{k} c_i I_{]\!]\sigma_i, \tau_i]\!]}(t),$$

where c_0 is a $\mathcal{F}_0$-measurable random variable, c_i is a constant and $\sigma_i, \tau_i \in \mathbf{T}$, $i = 1, \ldots, k$, can be found such that $H_k \to H$ as $k \to \infty$ pointwise.

Evidently $H_k \cdot X$ is a $\mathbb{F}^X_+$-adapted process, $k \geq 1$. Let $T \in \mathbf{T}(\mathbb{F}^X_+)$, and $T < \infty$. Since X admits the decomposition (2.2), we have

$$\xi_k = \sup_{s \leq T} | H \cdot X_s - H_k \cdot X_s | = \sup_{s \leq T} | (H - H_k) \cdot X_s |$$

$$\leq | H - H_k | \circ \operatorname{Var}(A)_T + \sup_{s \leq T} | (H - H_k) \cdot M_s |. \tag{2.46}$$

It can be assumed by the decomposition (2.4) that in the decomposition (2.2) for X the process $M \in \mathcal{M}^2_{loc}$. Then $(H - H_k) \cdot M \in \mathcal{M}^2_{loc}(\mathbb{F})$ and $\langle (H - H_k) \cdot M \rangle = (H - H_k)^2 \circ \langle M \rangle$. Consequently, $(H - H_k)^2 \circ \langle M \rangle$ dominates the process $((H - H_k) \cdot M)^2$ (see Theorem 1.9.3). Therefore by the Lenglart-Rebolledo inequality (Theorem 1.9.3)

$$\mathsf{P}\left(\sup_{0 \leq s \leq T} | (H - H_k) \cdot M_s | \geq a\right) \leq \frac{b}{a^2} + \mathsf{P}\left((H - H_k)^2 \circ \langle M \rangle_T \geq b\right). \tag{2.47}$$

From (2.46) and (2.47) it follows that $(\varepsilon > 0)$

$$\mathsf{P}(\xi_k \geq \varepsilon)$$

$$\leq \mathsf{P}\left(| H - H_k | \circ \operatorname{Var}(A)_T \geq \frac{\varepsilon}{2} \right) + \frac{4b}{\varepsilon^2} + \mathsf{P}\left((H - H_k)^2 \circ \langle M \rangle_T \geq b\right). \tag{2.48}$$

Taking the limit $\lim_{b \to 0} \overline{\lim_{k \to \infty}}$ we see that $\xi_k \xrightarrow{P} 0$ as $k \to \infty$ and hence a certain subsequence $(\xi_{k'})_{k' \geq 1}$ converges to zero with probability one. This means the $\mathcal{F}^X_{T+}$-measurability of the random variable $H \cdot X_t$, i.e. for each $t \in R_+$ the random variable $H \cdot X_t$ is $\mathcal{F}^X_{t+}$-measurable.

2) Let $T \in \mathbf{T}(\mathbb{F}^X_+)$ and $T < \infty$. Then

$$| H \cdot X_T - \overline{H \cdot X_T} | \leq \sup_{s \leq T} | (H - H_k) \cdot X_s | + \sup_{s \leq T} | \overline{(H - H_k) \cdot X_s} |, \quad k \geq 1.$$

It is not hard to deduce from this that

$$H \cdot X_T = \overline{H \cdot X_T} \quad (\{T < \infty\};\ \text{P-a.s.})$$

for each $T \in \mathbf{T}(\mathbb{F}^X_+)$. As the process $H \cdot X$ is $\mathbb{F}^X_+$-adapted, by Theorem 1.3.12 the processes $H \cdot X$ and $\overline{H \cdot X}$ are indistinguishable.

Problems

1. Show that the stochastic integral $I = H \cdot M$ is a unique element of the space $\mathcal{H}^2$ for $H \in \mathbf{L}^2(\mathbb{F}, \langle M\rangle)$ and $M \in \mathcal{M}^2$ (up to the stochastic indistinguishability) such that for each $N \in \mathcal{H}^2$

$$\mathbf{E}(I_\infty N_\infty) = \mathbf{E}(H \circ \langle M, N\rangle_\infty) = \mathbf{E}(H \circ [M, N]_\infty).$$

2. Show that the stochastic integral $I = H \cdot M$ is a unique element of the space $\mathcal{M}^d_{loc}$ for

$$H \in \mathbf{L}_{loc}(\mathbb{F}, [M, M]) \text{ and } M \in \mathcal{M}^d_{loc}$$

(up to the stochastic indistinguishability) such that $\Delta I = H\Delta M$.

3. Let $M \in \mathcal{M}_{loc} \cap \mathcal{A}_{loc}$ or $M \in \mathcal{M}^2_{loc}$ and $H_i = \mathbf{L}_{loc}(\mathbb{F}, [M, M])$, $i = 1, 2$. Show that

$$(c_1H_1 + c_2H_2) \cdot M = c_1(H_1 \cdot M) + c_2(H_2 \cdot M).$$

4. Let $M, N \in \mathcal{M}_{loc} \cap \mathcal{A}_{loc}$ or $M, N \in \mathcal{M}^2_{loc}$ and $H \in \mathbf{L}_{loc}(\mathbb{F}, [M, M]) \cap \mathbf{L}_{loc}(\mathbb{F}, [N, N])$. Show that

$$H \cdot (M + N) = H \cdot M + H \cdot N.$$

5. Let $M \in \mathcal{M}_{loc}$, $H \in \mathbf{L}_{loc}(\mathbb{F}, [M, M])$ and let a predictable process K be such that $KH \in \mathbf{L}_{loc}(\mathbb{F}, [M, M])$. Show that

$$K \cdot (H \cdot M) = (KH) \cdot M.$$

6. Let $M, N \in \mathcal{M}^2_{loc}$ and $H \in \mathbf{L}^2_{loc}(\mathbb{F}, \langle M\rangle)$. Show that

$$\langle H \cdot M, N\rangle = H \circ \langle M, N\rangle.$$

7. Let $M \in \mathcal{M}_{loc}$, $M \geq 0$ and $T = \inf(t\colon M_{t-} = 0)$. Show that the process $MI_{[\![T, \infty[\![}$ is indistinguishable. (Hint: use the representation $I(T < \infty)\, I(M_{T-} = 0)\ \Delta M_T = I(M_- = 0) \cdot M_T$.)

8. Show that for local martingales M ($M \in \mathcal{M}_{loc}$)

$$\langle M^c \rangle + \sum_s f(\Delta M_s) \in \mathcal{A}^+_{loc}, \quad f(x) = x^2/(1 + |x|)$$

(Hint: use Lemma 1).

9. Let $M \in \mathcal{M}^2_{loc} \cap (\mathcal{M}_{loc} \cap \mathcal{A}_{loc})$, $H \in \mathbf{L}_{loc}(\mathbb{F}, [M, M])$ and let the stochastic integrals $(H \cdot M)'$ and $(H \cdot M)''$ be defined with respect to a martingale M from $\mathcal{M}_{loc} \cap \mathcal{A}_{loc}$ and $\mathcal{M}^2_{loc}$ respectively. Show that the processes $(H \cdot M)'$ and $(H \cdot M)''$ are indistinguishable. (Hint: use Theorem 3 for a bounded H .)

10. Let $M, N \in \mathcal{M}^2_{loc}$ and let H be a predictable process such that $H^2 \circ \langle M \rangle \in \mathcal{A}^+_{loc}$.

Show that one can find a sequence of simple predictable processes H^n, $n \geq 1$, such that

$$(H - H^n)^2 \circ \langle M \rangle_t \xrightarrow{P} 0, \quad t > 0$$

and for each such sequence H^n, $n \geq 1$,

$$\langle H^n \cdot M, N \rangle_t \xrightarrow{P} H \circ \langle M, N \rangle_t.$$

§ 3. Ito's formula.I

1. At the basis of the stochastic calculus for semimartingales lies the following fundamental formula of changing variables, or, the so called Ito's formula.

Theorem 1. *Let* $X = (X^1, \dots, X^d)$ *be a* d*-dimensional stochastic process with semimartingale components* (d*-dimensional semimartingale*), *and let* $f(x) = f(x_1, \dots, x_d)$ *be a twice differentiable function on* R^d. *Then* $f(X)$ *is a semimartingale and Ito's formula*

$$f(X_t) = f(X_0) + \sum_{i \le d} (D_i f(X_-)) \cdot X^i_t + \frac{1}{2} \sum_{i,j \le d} D_{ij} f(X_-) \circ \langle X^{ic}, X^{jc} \rangle_t$$

$$+ \sum_{0 < s \le t} \left[f(X_s) - f(X_{s-}) - \sum_{i \le d} D_i f(X_{s-}) \Delta X^i_s \right] \tag{3.1}$$

is valid where $\langle X^{ic}, X^{jc} \rangle$ *are the mutual quadratic characteristics of the continuous local martingales* X^{ic} *and* X^{jc}, *while*

$$D_i f(x) = \frac{\partial f(x)}{\partial x_i}, \quad D_{ij} f(x) = \frac{\partial^2 f(x)}{\partial x_i \partial x_j}.$$

The proof of this formula is given in many textbooks (cf., e.g. [83, 50, 219]) and will be omitted here. We remark only that the stochastic integral $D_i f(X_-) \cdot X^i$ with respect to a semimartingale X^i is always defined (Problem 1) and

$$\sum_s | f(X_s) - f(X_{s-}) - \sum_{i \le d} D_i f(X_{s-}) \Delta X^i_s | \in \mathcal{V}^+$$

(Problem 2).

2. A number of particular cases of this formula deserves special mentioning.

Corollary 1. *Let* X *and* Y *be semimartingales. Then their product* XY *is a semimartingale too and for each* $t > 0$

$$X_t Y_t = X_0 Y_0 + X_- \cdot Y_t + Y_- \cdot X_t + [X, Y]_t, \tag{3.2}$$

or, in a "differentional" form,

$$d(X_t Y_t) = X_{t-} dY_t + Y_{t-} dX_t + d[X, Y]_t. \tag{3.3}$$

If $X \in \mathcal{V}$ and $Y \in \mathcal{V}$, then $[X, Y] = (\Delta X) \circ Y = (\Delta Y) \circ X$ and

$$X_tY_t - X_0Y_0 = X_- \circ Y_t + Y \circ X_t = X \circ Y_t + Y_- \circ X_t. \qquad (3.4)$$

If $X \in \mathbf{S}$ and $Y \in \mathcal{V}$, then $[X, Y] = (\Delta X) \circ Y$ and

$$X_tY_t = X_0Y_0 + Y_- \cdot X_t + X \circ Y_t. \qquad (3.5)$$

If $X \in \mathbf{S}$ and $Y \in \mathcal{V} \cap \mathcal{P}$, then $[X, Y] = (\Delta X) \circ Y = (\Delta Y) \cdot X$ (cf. Problem 5) and

$$X_tY_t - X_0Y_0 = Y_- \cdot X_t + X \circ Y_t = Y \cdot X_t + X_- \circ Y_t. \qquad (3.6)$$

Corollary 2. *If* M *and* N *are local martingales* ($M, N \in \mathcal{M}_{loc}$), *then* $M_tN_t - [M, N]_t = M_0N_0 + N_- \cdot M_t + M_- \cdot N_t$. *If, in addition,* $\mathbf{E} \mid M_0N_0 \mid < \infty$, then $MN - [M, N] \in \mathcal{M}_{loc}$. *In particular, if* $M \in \mathcal{M}_{loc}$, *then*

$$M_t^2 = M_0^2 + 2M_- \cdot M_t + [M, M]_t. \qquad (3.7)$$

3. By applying Ito's formula we establish the following important property of the quadratic variation [X, X] of a semimartingale X.

Assume that $X \in \mathbf{S}(\mathbb{F})$ and that $\mathbb{F}_+^X = (\mathcal{F}_{t+}^X)_{t \geq 0}$ is the natural flow of σ-algebras

$$\mathcal{F}_{t+}^X = \bigcap_{\varepsilon > 0} \sigma((X_s, s \leq t + \varepsilon) \vee \mathcal{N}),$$

where $\mathcal{N}$ is the system of sets from $\mathcal{F}$ of P-measure zero.

Let $X \in \mathbf{S}(\mathbb{F}_+^X)$. Denote its quadratic variation and the quadratic characteristic of its continuous martingale component by $\overline{[X, X]}$ and $\langle \overline{X}^c \rangle$ respectively.

Theorem 2. 1) *The quadratic variation* [X, X] *of a semimartingale* X *is a* $\mathbb{F}_+^X$-*adapted process.*

2) *If in addition* $X \in \mathbf{S}(\mathbb{F}_+^X)$, *then its quadratic variation* $\overline{[X, X]}$ *and the quadratic variation* [X, X] *are indistinguishable.*

Corollary. *The quadratic characteristic* $\langle X^c \rangle$ *of a semimartingale* X *is a* $\mathbb{F}_+^X$-*adapted process. If, in addition,* $X \in \mathbf{S}(\mathbb{F}_+^X)$, *then* $\langle X^c \rangle$ *and* $\langle \overline{X}^c \rangle$ *are indistinguishable.*

Proof. 1) By Ito's formula (3.2)

$$X_t^2 = X_0^2 + 2X_- \cdot X_t + [X, X]_t, \tag{3.8}$$

and the desired assertion follows from Theorem 2.9.

2) By Ito's formula again

$$X_t^2 = X_0^2 + \overline{2X_- \cdot X_t} + \overline{[X, X]}_t, \tag{3.9}$$

where $\overline{2X_- \cdot X}$ is a stochastic integral with respect to a $\mathbb{F}_+^X$-semimartingale X. By Theorem 2.9 the processes $2X_- \cdot X$ and $\overline{2X_- \cdot X}$ are indistinguishable. Therefore, by (3.8) and (3.9), $[X, X]$ and $\overline{[X, X]}$ are indistinguishable.

Proof of Corollary. If $X \in S(\mathbb{F})$, then by definition (cf. § 1),

$$[X, X] = \langle X^c \rangle + \sum_s (\Delta X_s)^2. \tag{3.10}$$

Therefore $\langle M^c \rangle$ is a $\mathbb{F}_+^X$-adapted process.

If $X \in S(\mathbb{F}) \cap S(\mathbb{F}_+^X)$, then along with (3.10) we have

$$\overline{[X, X]} = \langle \overline{X}^c \rangle + \sum_s (\Delta X_s)^2.$$

Hence, $\langle X^c \rangle$ and $\langle \overline{X}^c \rangle$ are indistinguishable.

Problems

1. Let X be a semimartingale, $X_t = X_0 + A_t + M_t$ with $A_0 = M_0 = 0$, $A \in \mathcal{V}(\mathbb{F})$, $M \in \mathcal{M}_{loc}$ and $Y \in \mathbb{F} \cap D$. Show that

$$|Y_-| \circ \mathrm{Var}(A) \in \mathcal{V}^+, \quad (Y_-^2 \circ [M, M])^{1/2} \in \mathcal{A}_{loc}^+.$$

2. If $X \in D$ and

$$\sum_s (\Delta X_s)^2 \in \mathcal{V}^+,$$

then for each twice continuously differentiable function $f = f(x)$

$$\sum_s |f(X_s) - f(X_{s-}) - f'(X_{s-}) \Delta X_s| \in \mathcal{V}^+.$$

3. Let the processes X and Y be such as in Problem 1 and, besides, $Y \in \mathcal{V} \cap \mathcal{P}$. Show that

$$|\Delta Y| \circ \mathrm{Var}(A) \in \mathcal{V}^+, \quad ((\Delta Y)^2 \circ [M, M])^{1/2} \in \mathcal{A}_{loc}^+.$$

(Hint: use Lemma 1.6.1.)

4. Let $X \in \mathcal{V}$. Show that

$$X \circ X_t \geq \frac{1}{2} X_t^2, \quad X_- \circ X \leq \frac{1}{2} X_t^2.$$

5. Let $M \in \mathcal{M}_{loc}$ and $Y \in \mathcal{V}(\mathbb{F}) \cap \mathcal{P}$. Show that

$$\sum_s \Delta Y_s \Delta M_s = (\Delta Y) \cdot M.$$

(Hint: use Theorem 1.8.3 and the fact that

$$\Delta\left(\sum_s \Delta Y_s \Delta M_s - (\Delta Y) \cdot M\right) = 0.)$$

§ 4. Doléans equation. Stochastic exponential

1. Let X be a semimartingale defined on a stochastic basis $(\Omega, \mathcal{F}, \mathbb{F}, P)$. Consider the stochastic linear equation of Doléan

$$Y = 1 + Y_- \cdot X \tag{4.1}$$

that is the same as

$$Y_t = 1 + \int_0^t Y_{s-}\, dX_s. \tag{4.2}$$

This equation is often written in differential form

$$dY_t = Y_{t-}\, dX_t, \quad Y_0 = 1 \tag{4.3}$$

and called the Doléans differential equation.

Theorem 1. *The stochastic equation* (4.1) *has the unique solution* Y (*up to indistinguishability*) *within the class of semimartingales* $\mathbf{S}(\mathbb{F})$, *which is denoted by* $\mathcal{E}(X)$ *and is called the stochastic exponential; it has the following representation*

$$\mathcal{E}_t(X) = \exp\left(X_t - X_0 - \frac{1}{2}\langle X^c\rangle\right) \prod_{s \le t} (1 + \Delta X_s)\, e^{-\Delta X_s}. \tag{4.4}$$

Proof. We will show first that the product (infinite, in general)

$$V_t \equiv \prod_{s \le t} (1 + \Delta X_s) e^{-\Delta X_s}$$

absolutely converges. Let $X_t = X_0 + M_t + A_t$ with $M \in \mathcal{M}_{loc}$ and $A \in \mathcal{V}$. Since

$$\sum_s (\Delta X_s)^2 \le 2 \sum_s ((\Delta M_s)^2 + (\Delta A_s)^2) \le 2\left(\sum_s (\Delta M_s)^2 + (\mathrm{Var}(A))^2\right),$$

by Theorem 1.7.6 we have

$$\sum_{s \le t} (\Delta X_s)^2 < \infty \quad (P\text{-a.s.})$$

for each $t > 0$. As $X \in D$, there is for each $t > 0$ a finite number of times $s \le t$ at most, for which $|\Delta X_s| \ge \frac{1}{2}$.

If $|\Delta X_s| \le \frac{1}{2}$, then $|\ln(1 + \Delta X_s) - \Delta X_s| \le c\, (\Delta X_s)^2$ with a constant c, and since $\sum_{s \le t} (\Delta X_s)^2 < \infty$ (P-a.s.), the product V_t absolutely converges.

Clearly, the process $V \in \mathbb{F} \cap D$ and, in addition, $V \in \mathcal{V}(\mathbb{F})$.

Denote

$$U_t = X_t - X_0 - \frac{1}{2}\langle X^c\rangle_t. \tag{4.5}$$

Then

$$Y_t \equiv \mathcal{E}_t(X) = e^{U_t} V_t. \tag{4.6}$$

By Ito's formula (3.1), applied to the function $f(x_1, x_2) = e^{x_1}x_2$, we get

$$Y_t = 1 + Y_- \cdot U_t + e^{U_-} \circ V_t + \frac{1}{2} Y_- \circ \langle U^c\rangle_t + \sum_{s \le t} (\Delta Y_s - Y_{s-}\Delta U_s - e^{U_{s-}} \Delta V_s)$$

$$= 1 + Y_- \cdot U_t + \frac{1}{2} Y_- \circ \langle U^c\rangle + \sum_{s \le t} (\Delta Y_s - Y_{s-}\Delta U_s).$$

But

$$\Delta Y_s = e^{U_{s-} + \Delta U_s} V_{s-} (1 + \Delta X_s) e^{-\Delta X_s} - e^{U_{s-}} V_{s-} = Y_{s-}\Delta X_s$$

and, according to (4.5),

$$U_s + \frac{1}{2}\langle U^c\rangle_s = X_s - X_0.$$

Therefore

$$Y_t = 1 + Y_- \cdot X_t,$$

i.e. the stochastic exponential $Y_t \equiv \mathcal{E}_t(X)$ is a solution of the equation (4.1).

Let us show that within the class of semimartingales this solution is unique.

Let $\tilde{Y}$ be another solution of the equation (4.1). Set $\tilde{W}_t = e^{-U_t} \tilde{Y}_t$, $t \ge 0$. By applying Ito's formula we get

$$\tilde{W}_t = 1 + \int_0^t \tilde{W}_{s-}\, dA_s$$

with

$$A_t = \sum_{s \le t} ((1 + \Delta X_s) e^{-\Delta X_s} - 1).$$

As in case of the process V, it is proved that the process $A \in \mathcal{V}$.

Observe now that the process $W_t = e^{-U_t} Y_t = V_t$ and

$$W_t = 1 + \int_0^t W_{s-}\, dA_s.$$

Therefore the process $\overline{W}_t = \tilde{W}_t - W_t$ $(= \tilde{W}_t - V_t)$ satisfies the equation

$$\overline{W}_t = \int_0^t \overline{W}_{s-}\, dA_s. \tag{4.7}$$

Let us show that a solution $\overline{W}$ of this equation is such that $\overline{W}_t = 0$ for each $t \ge 0$.

To this end, denote $\sigma = \inf(t\colon \overline{W}_t \neq 0)$. Then from (4.7) it follows that on the set $\{\sigma < \infty\}$ we have $\overline{W}_\sigma = 0$. Clearly, one can find $\sigma' > \sigma$ such that

$$\int_{(\sigma, \sigma']} |\, dA_s| \le \frac{1}{2}$$

and $\{\sigma < \infty\} \subseteq \{\sigma' > \sigma\}$. Then for $t > \sigma$ we obtain by (4.7) that on the set $\{\sigma < \infty\}$

$$\overline{W}_t = \overline{W}_\sigma + \int_{(\sigma, t]} \overline{W}_{s-}\, dA_s \tag{4.8}$$

and hence

$$\sup_{t \le \sigma'} |\overline{W}_t| \le \frac{1}{2} \sup_{t \le \sigma'} |\overline{W}_t|.$$

This gives $\sup_{t \le \sigma'} |\overline{W}_t| = 0$, and since $\sigma' > \sigma$ on $\{\sigma < \infty\}$, this means that time $\sigma = \inf(t\colon \overline{W}_t \neq 0) = \infty$, i.e. $\overline{W}_t = 0$ for each $t \ge 0$.

Thus $\tilde{W}_t = e^{-U_t}\tilde{Y}_t = V_t$ and hence $\tilde{Y}_t = V_t e^{U_t} = Y_t$, $t \ge 0$.

Remark. The unique solution of the equation $Y_t = Y_0 + Y_- \cdot X_t$ is presented in the form $Y_t = Y_0 \mathcal{E}_t(X)$.

2. Consider a number of properties of the stochastic exponential $\mathcal{E}(X)$.

Theorem 2. 1) *If a process* $X = A \in \mathcal{V}$, *then*

$$\mathcal{E}(A) \in \mathcal{V}.$$

2) *If a process* $X = M \in \mathcal{M}_{loc}$, *then*

$$\mathcal{E}(M) \in \mathcal{M}_{loc}.$$

3) *Let* $X \in \mathbf{S}(\mathbb{F})$ *and* $T = \inf(t\colon \Delta X_t = -1)$. *Then* $\mathcal{E}(X) \neq 0$ *on* $[\![0, T[\![$ *and* $\mathcal{E}(X) \neq 0$ *on* $[\![T, \infty[\![$.

4) *Let* X *and* Y *be semimartingales. Then*

$$\mathcal{E}(X)\,\mathcal{E}(Y) = \mathcal{E}(X + Y + [X, Y]). \tag{4.9}$$

Proof. Assertion 1) follows from the definition of $\mathcal{E}(X)$, and assertion 2) from the definition of the stochastic integral with respect to a local martingale (§ 2) and from Problem 3.1. To prove assertion 3) it suffices to observe that $\mathcal{E}_t(X) = e^{U_t}V_t$, where $V \neq 0$ on $[\![0, T[\![$ and $V = 0$ on $[\![T, \infty[\![$ by definition. Finally, the formula (4.9) is easily derived from Ito's formula (3.3).

3. We will establish now a stochastic analogue of the well known Gronwall-Bellman inequality (cf., e.g., [188], Ch. 4, Lemma 4.13).

Theorem 3. *Let* $X \in V^+$ *and let* $Y = (Y_t)_{t \geq 0} \in D$ *possess the following property: for each* $t \geq 0$

$$0 \leq Y_t \leq a + Y_- \circ X_t$$

with a non-negative constant a.

Then for each $t \geq 0$

$$Y_t \leq a\mathcal{E}_t(X)$$

and in particular $Y_t \leq a \exp(X_t)$.

Proof. For $\varepsilon > 0$ denote

$$Y_t^\varepsilon = (a + \varepsilon)\,\mathcal{E}_t(X).$$

Define time $\tau(\varepsilon) = \inf(t\colon Y_t^\varepsilon - Y_t \leq 0)$ and observe that since $Y^\varepsilon \in D$ and $Y_0^\varepsilon - Y_0 \geq (a + \varepsilon) - a = \varepsilon$, then $\tau > 0$. As for $t < \tau(\varepsilon)$ this gives

$$Y_t \leq a + Y_- \circ X_t \leq a + Y_-^e \circ X_t = a + (a + \varepsilon)\,\mathcal{E}_-(X) \circ X_t. \qquad (4.10)$$

Use the fact that $\mathcal{E}(X)$ is a solution of Doléans equation:

$$\mathcal{E}_-(X) \circ X_t = \mathcal{E}_t(X) - 1. \qquad (4.11)$$

Then from (4.10) it follows that for each $t < \tau(\varepsilon)$

$$Y_t \leq a + (a + \varepsilon)\,[\mathcal{E}_t(X) - 1] \leq (a + \varepsilon)\,\mathcal{E}_t(X). \qquad (4.12)$$

The desired inequality follows from (4.12), provided for each $\varepsilon > 0$ the equality $\tau(\varepsilon) = \infty$ takes place.

By taking into consideration (4.12), as $\tau(\varepsilon) < \infty$ we get

$$Y_{\tau(\varepsilon)} \leq a + Y_- \circ X_{\tau(\varepsilon)} \leq a + (a + \varepsilon)\,\mathcal{E}_-(X) \circ X_{\tau(\varepsilon)}.$$

Consequently, by (4.11) and by the definition of Y_t^ε we have

$$Y_{\tau(\varepsilon)} \leq a + (a + \varepsilon)\,[\mathcal{E}_{\tau(\varepsilon)}(X) - 1] = Y_{\tau(\varepsilon)}^\varepsilon - \varepsilon < Y_{\tau(\varepsilon)}^\varepsilon,$$

which contradicts the assumption $\tau(\varepsilon) < \infty$, i.e. $\tau(\varepsilon) = \infty$.

The inequality $Y_t \leq a \exp(X_t)$ takes place because $\prod_s (1 + \Delta X_s)\, e^{-\Delta X_s} \leq 1$ as $\Delta X > 0$, and hence $\mathcal{E}_t(X) \leq \exp(X_t)$.

Problem

1. Let $X = X' + iX''$ be a complex semimartingale where X' and X'' are two real valued semimartingales, and let $Y = Y' + iY''$ be a complex valued semimartingale that is a solution of the stochastic equation

$$Y = 1 + Y_- \cdot X \qquad (4.13)$$

or equivalently

$$Y' = 1 + Y'_- \cdot X' - Y''_- \cdot X'', \quad Y'' = Y'_- \cdot X'' + Y''_- \cdot X'. \tag{4.14}$$

Show that there exists a unique solution $Y \equiv \mathcal{E}(X)$ of this equation given by the formula

$$\mathcal{E}_t(X) = e^{\left(X_t - X_0 - \frac{1}{2}[X', X']^c_t + \frac{1}{2}[X'', X'']^c_t - i\,[X', X'']^c_t\right)} \prod_{s \le t} (1 + \Delta X_s)\, e^{-\Delta X_s}. \tag{4.15}$$

In particular, if $X \in \mathcal{V}$, then

$$\mathcal{E}_t(X) = e^{(X_t - X_0)} \prod_{s \le t} (1 + \Delta X_s)\, e^{-\Delta X_s}. \tag{4.16}$$

§ 5. Multiplicative decomposition of positive semimartingales

1. Consider a special semimartingale X with the decomposition

$$X = X_0 + A + M \tag{5.1}$$

with $A \in \mathcal{V} \cap \mathcal{P}$ and $M \in \mathcal{M}_{loc}$. For each $t > 0$ assume (P-a.s.)

$$\inf_{s \le t} X_s > 0, \quad \inf_{0 < s \le t} X_{s-}^{-1} \Delta A_s > -1. \tag{5.2}$$

Under these conditions the processes $\hat{A} = (\hat{A}_t)_{t \ge 0}$ and $\hat{M} = (\hat{M}_t)_{t \ge 0}$ are defined by

$$\hat{A} = X_-^{-1} \circ A, \quad \hat{M} = (X_- + \Delta A)^{-1} \cdot M \tag{5.3}$$

such that $\hat{A} \in \mathcal{V} \cap \mathcal{P}$ and $\hat{M} \in \mathcal{M}_{loc}$.

Theorem 1. 1) *Let Conditions* (5.2) *be satisfied. Then a semimartingale* X *admits the multiplicative decomposition*

$$X = X_0 \mathcal{E}(\hat{A})\, \mathcal{E}(\hat{M}), \tag{5.4}$$

with the stochastic exponents

$$\mathcal{E}(\hat{A}) \in \mathcal{V} \cap \mathcal{P}, \qquad \mathcal{E}(\hat{M}) \in \mathcal{M}_{loc}. \tag{5.5}$$

2) *The decomposition of type* (5.4) *is unique in the sense that if*

$$X = X_0 G N, \tag{5.6}$$

with $G \in \mathcal{V} \cap \mathcal{P}$, $N \in \mathcal{M}_{loc}$ *and* $G_0 = N_0 = 1$, *then the processes* G *and* $\mathcal{E}(\hat{A})$ *are indistinguishable, as well as* N *and* $\mathcal{E}(\hat{M})$.

Proof. 1) Observe that $X_- \circ \hat{A} = A$ and (Problem 2.5)

$$X_-(1 + X_-^{-1} \Delta A) \cdot \hat{M} = X_-(1 + X_-^{-1} \Delta A)(X_- + \Delta A)^{-1} \cdot M = M.$$

Therefore

$$X_t = X_0 + X_- \circ \hat{A}_t + X_-(1 + X_-^{-1} \Delta A) \cdot M_t \tag{5.7}$$

or, equivalently,

$$X_t = X_0 + X_- \cdot (\hat{A} + \hat{M} + [\hat{A}, \hat{M}])_t. \tag{5.8}$$

By Theorem 4.1 Doléans equation (5.8) has the solution

$$X_t = X_0 \mathcal{E}_t(\hat{A} + \hat{M} + [\hat{A}, \hat{M}]). \tag{5.9}$$

In view of Assertion 4) of Theorem 4.2 this gives the desired representation (5.4). Assertions (5.5) follow also from Theorem 4.2 (Assertions 1) and 2)) and from the fact that $\hat{A} \in \mathcal{P}$.

2) Let the representation (5.6) be valid. Then by (3.5)

$$\begin{aligned} X_t &= X_0 G_t N_t \\ &= X_0 + (X_0 G) \cdot N_t + N_- \circ (X_0 G)_t \\ &= X_0 + (X_0 G) \cdot N_t + (X_0 N_-) \circ G_t \end{aligned} \tag{5.10}$$

with $(X_0G) \cdot N \in \mathcal{M}_{loc}$ and $(X_0N_-) \circ G \in \mathcal{V} \cap \mathcal{P}$. By comparing (5.10) with (5.1) and taking into account the uniqueness of the representation (5.1) with a predictable process A (Theorem 1.1), we establish indistinguishability of the processes A and $(X_0N_-) \circ G$, as well as of the local martingales M and $(X_0G) \cdot N$.

Therefore

$$G_- \circ A = G_- \circ ((X_0N_-) \circ G) = (X_0G_-N_-) \circ G = X \circ G.$$

This and (5.3) give

$$G_- \circ \hat{A}_t = G_- \circ (X_-^{-1} \circ A)_t = (X_-^{-1}G_-) \circ A_t = X_-^{-1} \circ (G_- \circ A)_t$$

$$= X_-^{-1} \circ (X_- \circ G)_t = G_t - G_0.$$

Consequently, by taking into account $G_0 = 1$ we see that G_t is a solution of Doléans equation

$$G_t = 1 + G_- \circ \hat{A}_t, \tag{5.11}$$

and hence $G = \mathcal{E}(\hat{A})$.

By (5.3), (5.11) and the indistinguishability of the processes M and $(X_0G) \cdot N$ we get (cf. also Problem 2.5)

$$N_- \cdot \hat{M} = N_- \cdot ((X_- + \Delta A)^{-1} \cdot M) = \frac{N_-}{X_- + \Delta A} \cdot M = \frac{N_-}{X_- + \Delta A} \cdot ((X_0G) \cdot N)$$

$$= \frac{N_-X_0G}{X_- + \Delta A} \cdot N = \frac{X_0N_-(G_- + \Delta G)}{X_- + \Delta A} \cdot N = \frac{N_-(X_0G_- + X_0G_-\Delta\hat{A})}{X_- + \Delta A} \cdot N$$

$$= \frac{X_0G_-N_-(1 + \Delta\hat{A})}{X_- + \Delta A} \cdot N = \frac{X_-(1 + \Delta\hat{A})}{X_-(1 + X_-^{-1}\Delta A)} \cdot N = N - N_0 = N - 1,$$

i.e. $N_t = 1 + N_- \cdot \hat{M}_t$, and hence $N = \mathcal{E}(\hat{M})$.

Problems

1. Let X be a nonnegative martingale ($X_t \geq 0$, $t \geq 0$), with the decomposition (5.1) where A is a nonincreasing process. Let $\hat{A}$ be defined by the formula (5.3) and let $\mathcal{E}(\tilde{A})$ be the stochastic exponential:

$$\mathcal{E}_t(\hat{A}) = e^{\hat{A}_t} \prod_{s \le t} (1 + \Delta\hat{A}_s)\, e^{-\Delta\hat{A}_s}.$$

Show that the process $N = (N_t)_{t \ge 0}$ with

$$N_t = X_t\,[X_0 \mathcal{E}_t(\hat{A})]^{-1}\, I\,(X_0 \mathcal{E}_t(\hat{A}) > 0)$$

is a nonnegative supermartingale.

2. Let X be a semimartingale with the decomposition (5.1) and let $X_0 = 0$ and $0 \le X \le c$. Show that as $0 < \lambda < \frac{1}{c}$ the following estimate holds:

$$\mathbf{E} e^{\lambda A_\infty} \le \frac{1}{1 - \lambda c}.$$

3. Let X be a nonnegative semimartingale with the decomposition (5.1) and let $X_0 = 0$. If $\mathbf{E}X_t^m < \infty$, then

$$\mathbf{E}A_t^m \le m^m \mathbf{E}X_t^m, \quad m \ge 1.$$

§ 6. Convergence sets and the strong law of large numbers for special martingales

1. Let $X \in \mathsf{D}$ and $X_\infty = \lim_{t \to \infty} X_t$. Denote by $\{X \to \}$ the set on which X_∞ exists and presents a finite random variable, and let $\{X \nrightarrow \} = \Omega \setminus \{X \to \}$. In this section problems are treated concerning the structure of sets $\{X \to \}$ for special semimartingales X.

If sets $\Gamma_1, \Gamma_2 \in \mathcal{F}$, then by writing $\Gamma_1 = \Gamma_2$ (P-a.s.) or $\Gamma_1 \subseteq \Gamma_2$ (P-a.s.) we mean $\mathsf{P}(\Gamma_1 \Delta \Gamma_2) = 0$ or $\mathsf{P}(\Gamma_1 \cap (\Omega \setminus \Gamma_2)) = 0$ respectively, where Δ is the symmetric difference sign between the sets.

On proving the basic results concerning convergence sets, we systematically apply

Lemma 1. *Let* $X, Y \in \mathbb{F} \cap \mathsf{D}$ *and for each* $\sigma_a = \inf(t: |X_t| \geq a)$, $a > 0$, *let* $\mathsf{P}(Y^{\sigma_a} \to) = 1$.

Then

$$\{\sup_{t \geq 0} |X_t| < \infty\} \subseteq \{Y \to \} \quad (\mathsf{P}\text{-a.s.})$$

Proof. By the assumption $\mathsf{P}(Y^{\sigma_a} \to) = 1$ we have $\mathsf{P}(Y \nrightarrow, \sigma_a = \infty) = 0$. Consequently,

$$\mathsf{P}(Y \nrightarrow, \bigcup_{a > 0} \{\sigma_a = \infty\}) = 0, \tag{6.1}$$

where $\bigcup_{a > 0}$ is taken over all integers $a > 0$.

The desired assertion follows from (6.1), by observing that

$$\{\sup_{t \geq 0} |X_t| < \infty\} = \bigcup_{a > 0} \{\sigma_a = \infty\}.$$

2. Let $X \in \mathcal{A}^+_{loc}$. According to Theorem 1.6.3 we have $X \in \mathbf{Sp}$, because X admits the representation $X = A + M$ where A is the compensator of X and $M \in \mathcal{M}_{loc}$.

Theorem 1. *Let* $X \in \mathcal{A}^+_{loc}$ *and let* A *be is its compensator. Then*:

1) $\{A_\infty < \infty\} \subseteq \{X_\infty < \infty\}$ (P-a.s);

2) *if in addition* $\mathbf{E}\Delta X_\sigma I(\sigma < \infty) < \infty$ *for each Markov time* σ, *then*

$$\{A_\infty < \infty\} = \{X_\infty < \infty\} \quad (\mathsf{P}\text{-a.s.}).$$

Proof. 1) First assume that the expression $\mathbf{E}\Delta A_\sigma I(\sigma < \infty) < \infty$ for each Markov time σ. Define the Markov times $\sigma_a = \inf(t: A_t \geq a)$, $a > 0$. Then

$$\mathbf{E}A_{\sigma_a} = \mathbf{E}(A_{\sigma_a -} + \Delta A_{\sigma_a} I(\sigma_a < \infty)) \leq a + \mathbf{E}\Delta A_{\sigma_a} I(\sigma_a < \infty) < \infty.$$

By the property of the compensator of an increasing process (Theorem 1.6.3)

$$\mathbf{E}X_{\sigma_a} = \mathbf{E}A_{\sigma_a} < \infty. \tag{6.2}$$

Therefore $\mathsf{P}(X_\infty^{\sigma_a} < \infty) = 1$. The desired assertion follows from Lemma 1, since $\{X\to\} = \{X_\infty < \infty\}$ and $\{\sup_{t\geq 0} A_t < \infty\} = \{A_\infty < \infty\}$.

Consider now the general case.

Define the increasing processes $A' = I(\Delta A > 1) \circ A$ and $A'' = I(\Delta A \leq 1) \circ A$. Evidently, A' and A'' are the compensators of the increasing processes $X' = I(\Delta A > 1) \circ X$ and $X'' = I(\Delta A \leq 1) \circ X$. Since $\Delta A'' \leq 1$, then, as has been proved,

$$\{A''_\infty < \infty\} \subseteq \{X''_\infty < \infty\} \quad (\mathsf{P}\text{-a.s.}). \tag{6.3}$$

Observe next that

$$A'_\infty = \sum_{s>0} I(\Delta A_s > 1)\,\Delta A_s, \quad X'_\infty = \sum_{s>0} I(\Delta A_s > 1)\,\Delta X_s.$$

Therefore

$$\{A'_\infty < \infty\} \subseteq \left\{\sum_{s>0} I(\Delta A_s > 1) < \infty\right\}$$

$$\subseteq \left\{\sum_{s>0} I(\Delta A_s > 1)\,\Delta X_s < \infty\right\} = \{X'_\infty < \infty\}. \tag{6.4}$$

By (6.3) and (6.4) we get

$$\{A_\infty < \infty\} = \{A'_\infty < \infty\} \cap \{A''_\infty < \infty\}$$

$$\subseteq \{X'_\infty < \infty\} \cap \{X''_\infty < \infty\} = \{X_\infty < \infty\} \quad (\mathsf{P}\text{-a.s.}).$$

2) It is established under the condition $\mathbf{E}\Delta X_\sigma I(\sigma < \infty) < \infty$ that $\{X_\infty < \infty\} \subseteq \{A_\infty < \infty\}$ (P-a.s.), analogously to the proof of the first assertion. By this and 1) we get that the desired sets coincide.

3. The following results are well known, and they are presented here without proof (cf. [96, 217] and Theorem 1.4.1).

Theorem 2. 1) *Let* X *be a submartingale and* $\sup_{t\geq 0} \mathbf{E}X_t^+ < \infty$, *with* $a^+ = a \vee 0$. *Then*

$$\mathsf{P}(X\to) = 1, \quad \mathbf{E}X_\infty^+ < \infty.$$

2) *If* X *is a nonnegative supermartingale, then*

$$\mathsf{P}(X\to) = 1, \quad \mathbf{E}X_\infty \leq \mathbf{E}X_0.$$

3) *If* X *is a martingale* ($X \in \overline{\mathcal{M}}$) *and* $\sup_{t\geq 0} \mathbf{E}|X_t| < \infty$, *then*

$$\mathsf{P}(X\to) = 1, \quad \mathbf{E}|X_\infty| < \infty.$$

4) *If* ξ *is an integrable random variable, and if* $(\mathcal{F}_n)_{n=0,\pm1,\pm2,\dots}$ *is a family of* σ*-algebras,* $\mathcal{F}_n \subseteq \mathcal{F}_{n+1}$, $\mathcal{F}_{-\infty} = \bigcap_n \mathcal{F}_n$, *and* $\mathcal{F}_\infty = \sigma\left(\bigcup_{n\geq 1} \mathcal{F}_n\right)$, *then* (P-a.s.)

$$\lim_n \mathbf{E}\,(\xi \mid \mathcal{F}_n) = \mathbf{E}\,(\xi \mid \mathcal{F}_\infty)$$

$$\lim_n \mathbf{E}\,(\xi \mid \mathcal{F}_{-n}) = \mathbf{E}\,(\xi \mid \mathcal{F}_{-\infty}).$$

Theorem 2 and Davis' inequality (Theorem 1.9.6) allow us to establish the convergence (with probability one) of a martingale the quadratic characteristic of which is dominated by the quadratic characteristic of a convergent martingale.

Theorem 3. *Let* $X, M \in \overline{\mathcal{M}}$ *and let the following conditions be satisfied:*

(α) $$\sup_{t\geq 0} \mathbf{E}\,|M_t| < \infty,$$

(β) $$[X, X] \leq [M, M].$$

Then $\mathsf{P}\,(X \rightarrow) = 1$.

Proof. By Theorem 2 and and assumption (α) we have $\mathsf{P}\,(M \rightarrow) = 1$. This implies

$$\mathsf{P}\,(\sup_{t\geq 0} |M_t| < \infty) = 1. \tag{6.5}$$

For each $a > 0$ define a Markov time $\sigma_a = \inf\,(t\colon |M_t| \geq a)$. If it were be shown that

$$\mathsf{P}\,(X^{\sigma_a} \rightarrow) = 1, \quad a > 0, \tag{6.6}$$

then by Lemma 1 we would have

$$\{\sup_{t\geq 0} |M_t| \leq \infty\} \subseteq \{X \rightarrow\} \quad (\text{P-a.s.}),$$

and the desired assertion would follow from (6.5).

To prove (6.6) we first will obtain

$$\mathbf{E} \sup_{t\leq\sigma_a} |M_t| < \infty, \quad a > 0. \tag{6.7}$$

For each $t > 0$ we have

$$\sup_{s\leq t\wedge\sigma_a} |M_s| \leq \sup_{s<t\wedge\sigma_a} |M_s| + |\Delta M_{t\wedge\sigma_a}|$$

$$\leq 2 \sup_{s<t\wedge\sigma_a} |M_s| + |M_{t\wedge\sigma_a}| \leq 2a + |M_{t\wedge\sigma_a}|. \tag{6.8}$$

Since $M \in \overline{\mathcal{M}}$, by taking into consideration Jensen's inequality we get

$$\mathbf{E}\,|M_{t\wedge\sigma_a}| = \mathbf{E}\,|\mathbf{E}\,(M_t \mid \mathcal{F}_{t\wedge\sigma_a})| \leq \mathbf{E}\,|M_t|.$$

This and (6.8) imply

$$\mathbf{E} \sup_{s\leq t\wedge\sigma_a} |M_s| \leq 2a + \sup_{s\geq 0} \mathbf{E}\,|M_s|.$$

By taking the limit $\lim_{t\to\infty}$ we arrive at the desired relation (6.7).

Condition (b) and Davis' inequality (Theorem 1.9 6) give

$$\mathbf{E}\sup_{t\le\sigma_a}|X_t|\le C\mathbf{E}\,[X,X]^{1/2}_{\sigma_a}\le C\mathbf{E}\,[M,M]^{1/2}_{\sigma_a}\le\frac{C}{c}\mathbf{E}\sup_{t\le\sigma_a}|M_t|.$$

By these inequalities and by (6.7) we have $X^{\sigma_a}\in\mathcal{H}$, and hence $X^{\sigma_a}\in\mathcal{M}$. Therefore, by Theorem 1.4.2 we have $\mathsf{P}\;(X^{\sigma_a}\to)=1$.

4. Let $X\in\mathcal{M}_{loc}$ and $X_0=0$. Then by Theorem 1.7.2 we have $X=X^c+X^d$ with $X^c\in\mathcal{M}^c_{loc}$ and $X^d\in\mathcal{M}^d_{loc}$. Besides, by Lemma 2.1 and Problem 1.9.5 the process B with

$$B_t=\sum_{0<s\le t}\frac{(\Delta X_s)^2}{1+|\Delta X_s|}\tag{6.9}$$

belongs to $\mathcal{A}^+_{loc}$. Let $\tilde{B}$ be the compensator of the process B. Denote

$$A=\langle X^c\rangle+\tilde{B}.\tag{6.10}$$

Theorem 4. *Let* $X\in\mathcal{M}_{loc}$, $X_0=0$ *and let the process* A *be defined by formula* (6.10). *Then*:

1) $\{A_\infty<\infty\}\subseteq\{X\to\}$ (P-a.s.);

2) *if for each Markov time* σ *we have* $\mathbf{E}\,|\Delta X_\sigma|\,I(\sigma<\infty)<\infty$, *then*

$$\{A_\infty<\infty\}=\{[X,X]_\infty<\infty\}=\{X\to\}\quad(\text{P-a.s.});$$

Proof. 1) Assume first that $\mathbf{E}\Delta A_\sigma I\,(\sigma<\infty)<\infty$ for each Markov time σ. For $a>0$ define a Markov time $\sigma_a=\inf\,(t\colon A_t\ge a)$. Then

$$\mathbf{E}A_{\sigma_a}=\mathbf{E}\,(A_{\sigma_a-}+\Delta A_{\sigma_a}I\,(\sigma_a<\infty))\le a+\mathbf{E}\Delta A_{\sigma_a}I\,(\sigma_a<\infty)<\infty.$$

Let us show now that

$$\mathbf{E}\,[X,X]^{1/2}_{\sigma_a}<\infty.\tag{6.11}$$

According to the formula (2.30) and Theorem 1.6.3

$$\mathbf{E}\left(\sum_{s\le\sigma_a}(\Delta X_s)^2\right)^{1/2}\le\frac{1}{4}+2\mathbf{E}B_{\sigma_a}=\frac{1}{4}+2\mathbf{E}\tilde{B}_{\sigma_a}\le\frac{1}{4}+2\mathbf{E}A_{\sigma_a}<\infty.\tag{6.12}$$

Next, by (6.10) the inequalities

$$\mathbf{E}\,\langle X^c\rangle^{1/2}_{\sigma_a}\le\mathbf{E}A^{1/2}_{\sigma_a}\le(\mathbf{E}A_{\sigma_a})^{1/2}\tag{6.13}$$

take place. Therefore the inequality (6.11) for

$$[X,X]^{1/2}_{\sigma_a}=\left(\langle X^c\rangle_{\sigma_a}+\sum_{s\le\sigma_a}(\Delta X_s)^2\right)^{1/2}$$

is deduced from (6.12), (6.13) and the obvious inequality $(a+b)^{1/2} \le a^{1/2} + b^{1/2}$ $(a, b \ge 0)$.

By (6.11) and Davis' inequality (Theorem 1.9.6) we have $\mathbf{E} \sup_{t \le \sigma_a} |X_t| < \infty$, i.e $X^{\sigma_a} \in \mathcal{M}$, and by Theorem 1.4.1 we have $\mathsf{P}(X^{\sigma_a} \rightarrow) = 1$.

The desired assertion follows from Lemma 1 because $\{A_\infty < \infty\} = \{\sup_{t \ge 0} A_t < \infty\}$.

In the general case consider the decomposition $X = X' + X''$ with $X' = I(\Delta\tilde{B} > 1) \cdot X$ and $X'' = I(\Delta\tilde{B} \le 1) \cdot X$. Clearly

$$X', X'' \in \mathcal{M}_{loc}, \; X_0' = X_0'' = 0$$

and the process B'', related to the local martingale X'', is defined by the formula

$$B_t'' = \sum_{0 < s \le t} I(\Delta\tilde{B}_s \le 1) \frac{(\Delta X_s)^2}{1 + |\Delta X_s|}.$$

Therefore by Theorem 1.6.3 the compensator $\tilde{B}''$ of the process B'' is defined in the following manner:

$$\tilde{B}'' = I(\Delta\tilde{B} \le 1) \circ \tilde{B}.$$

Hence the process $A'' = \langle X^c \rangle + \tilde{B}''$ is such that $\Delta A'' \le 1$.

As has been proved

$$\{A_\infty'' < \infty\} \subseteq \{X'' \rightarrow\} \quad (\mathsf{P}\text{-a.s.}). \tag{6.14}$$

The process B', related to the local martingale X', is evidently defined by the formula

$$B_t' = \sum_{0 < s \le t} I(\Delta\tilde{B}_s > 1) \frac{(\Delta X_s)^2}{1 + |\Delta X_s|}$$

and it has the compensator $\tilde{B}' = I(\Delta\tilde{B} > 1) \circ \tilde{B}$.

Denote $A' = \tilde{B}'$. Then the following successive inclusions of sets take place:

$$\{A_\infty < \infty\} \subseteq \{A_\infty' < \infty\} \subseteq \left\{\sum_{s>0} I(\Delta\tilde{B}_s > 1) < \infty\right\}$$

$$\subseteq \left\{\sum_{s>0} I(\Delta\tilde{B}_s >)\, \Delta X_s \rightarrow\right\}. \tag{6.15}$$

Since

$$\sum_s I(\Delta\tilde{B}_s > 1)\, \Delta X_s = I(\Delta\tilde{B} > 1) \cdot X,$$

by (6.14) and (6.15) we get

$$\{A_\infty < \infty\} = \{A'_\infty < \infty\} \cap \{A''_\infty < \infty\} \subseteq \{X' \to\} \cap \{X'' \to\} = \{X \to\} \quad (\text{P-a.s.}),$$

i.e. the desired assertion takes place.

2) For a > 0 define a Markov time $\sigma_a = \inf(t: |X_t| \ge a)$. Since

$$\sup_{t \le \sigma_a} |X_t| \le \sup_{t < \sigma_a} |X_t| + |\Delta X_{\sigma_a}| I(\sigma_a < \infty)$$

and

$$\sup_{t < \sigma_a} |X_t| \le a,$$

by the assumption $E|\Delta X_\sigma| I(\sigma < \infty) < \infty$, $\sigma \in \mathbf{T}$, we have

$$\mathbf{E} \sup_{t \le \sigma_a} |X_t| < \infty.$$

This and Davis' inequality (Theorem 1.9.6) imply $\mathbf{E}\,[X, X]_{\sigma_a}^{1/2} < \infty$. Therefore by Lemma 1

$$\{\sup_{t \ge 0} |X_t| < \infty\} \subseteq \{[X, X]_\infty < \infty\} \quad (\text{P-a.s.}).$$

Consequently, by taking into consideration $\{X \to\} \subseteq \{\sup_{t \ge 0} |X_t| < \infty\}$ we have

$$\{X \to\} \subseteq \{[X, X]_\infty < \infty\} \quad (\text{P-a.s.}).$$

Next, observe that

$$[X, X] = \langle X^c \rangle + \sum_s (\Delta X_s)^2 \ge \langle X^c \rangle + B,$$

and hence,

$$\{[X, X]_\infty < \infty\} \subseteq \{\langle X^c \rangle_\infty + B_\infty < \infty\}.$$

Since

$$\Delta(\langle X^c \rangle + B) = \Delta B = \frac{(\Delta X)^2}{1 + |\Delta X|} \le |\Delta X|,$$

by the assumption $\mathbf{E}|\Delta X_\sigma| I(\sigma < \infty) < \infty$ and by Theorem 1 we have

$$\{\langle X^c \rangle_\infty + B_\infty < \infty\} = \{\langle X^c \rangle_\infty + \tilde{B}_\infty < \infty\},$$

i.e. (cf. (6.10))

$$\{X \to\} \subseteq \{[X, X]_\infty < \infty\} \subseteq \{A_\infty < \infty\} \quad (\text{P-a.s.}),$$

and by 1) we get the desired assertion.

5. For $X \in \mathcal{M}^2_{loc}$, $X_0 = 0$, the following result can be deduced from Theorem 4.

Theorem 5. *Let* $X \in \mathcal{M}^2_{loc}$, $X_0 = 0$. *Then*

1) $\{\langle X\rangle_\infty < \infty\} \subseteq \{X \rightarrow\}$ (P-a.s.);

2) *if for each Markov time* σ *we have* $\mathbf{E}(\Delta X_\sigma)^2 I(\sigma < \infty) < \infty$, *then*

$$\{\langle X\rangle_\infty < \infty\} = \{[X, X]_\infty < \infty\} = \{X \rightarrow\} \quad \text{(P-a.s.)}.$$

Proof. 1) Observe that

$$[X, X] - (\langle X^c\rangle + B) \in \mathcal{A}^+_{loc}$$

with B defined by the formula (6.9). Since $\langle X\rangle$ is the compensator of the process $[X, X]$, and A (cf. (6.10)) the compensator of $\langle X^c\rangle + B$, then $\langle X\rangle - A$ is the compensator of the increasing process $[X, X] - (\langle X^c\rangle + B)$, and hence $\langle X\rangle \geq A$. Consequently,

$$\{\langle X\rangle_\infty < \infty\} \subseteq \{A_\infty < \infty\} \quad \text{(P-a.s.)}$$

and the desired assertion takes place.

2) Since

$$\mathbf{E}\,|\Delta X_\sigma|\, I(\sigma < \infty) \leq (\mathbf{E}(\Delta X_\sigma)^2 I(\sigma < \infty))^{1/2},$$

by Theorem 4 we have

$$\{[X, X]_\infty < \infty\} = \{X \rightarrow\} \quad \text{(P-a.s.)}.$$

Finally, by Theorem 1 we have

$$\{[X, X]_\infty < \infty\} = \{\langle X\rangle_\infty < \infty\} \quad \text{(P-a.s.)},$$

because $\langle X\rangle$ is the compensator of $[X, X]$ and

$$\Delta [X, X]_\sigma I(\sigma < \infty) = (\Delta X_\sigma)^2 I(\sigma < \infty).$$

6. Consider now the case in which a process X is a special semimartingale ($X \in \mathbf{Sp}$). By definition 2 (§ 1) the process X admits the decomposition

$$X_t = X_0 + A_t + M_t \tag{6.16}$$

with $A \in \mathcal{V} \cap \mathcal{P}$ and $M \in \mathcal{M}_{loc}$.

Clearly, $\{A \rightarrow\} \cap \{M \rightarrow\} \subseteq \{X \rightarrow\}$.

In case of nonnegative semimartingales $X \in \mathbf{Sp}$ we shall obtain more substantial results.

Meanwhile, let us establish an auxiliary fact.

Lemma 2. *Let* $X \in \mathbf{Sp}$, $X \geq 0$, $\mathbf{E}X_0 < \infty$ *and let the process* A *in the decomposition* (6.16) *possess the following property*: $A \in \mathcal{A}^+ \cap \mathcal{P}$. *Then for* M *in the decomposition* (6.16) *the inequality* $\sup_{t \geq 0} \mathbf{E}\,|M_t| < \infty$ *takes place and* M *is a supermartingale.*

Proof. If $(\tau_n)_{n \geq 1}$ is a localizing sequence for M ($M \in \mathcal{M}_{loc}$), then for each $t > 0$

$$\mathbf{E}X_{t\wedge\tau_n} = \mathbf{E}X_0 + \mathbf{E}A_{t\wedge\tau_n} \le \mathbf{E}\,(X_0 + A_\infty).$$

Therefore, by applying Fatou's lemma we get

$$\mathbf{E}X_t \le \varliminf_n \mathbf{E}X_{t\wedge\tau_n} \le \mathbf{E}\,(X_0 + A_\infty).$$

Consequently,

$$|M_t| \le X_0 + A_t + X_t \quad \text{and} \quad \sup_{t\ge 0} \mathbf{E}\,|M_t| \le 2\mathbf{E}\,(X_0 + A_\infty).$$

Let us verify now the supermartingale inequality ($t > s$, $\Gamma \in \mathcal{F}_s$)

$$\int_\Gamma M_t\,dP \le \int_\Gamma M_s\,dP. \tag{6.17}$$

To this end introduce the continuous even function $f_l = f_l(x)$ with $f_l(x) = (1 \wedge (l - x)) \vee 0$ as $x \ge 0$, where $l > 1$, and the random variable $\xi^n = f_l(M_{s\wedge\tau_n})I_\Gamma$. Evidently, for each n the equality

$$\mathbf{E}M_{t\wedge\tau_n}\xi^n = \mathbf{E}M_{s\wedge\tau_n}\xi^n. \tag{6.18}$$

takes place. Since $M_{s\wedge\tau_n}\xi^n = M_{s\wedge\tau_n}f_l(M_{s\wedge\tau_n})I_\Gamma$, we have

$$\lim_n \mathbf{E}M_{s\wedge\tau_n}\xi^n = \mathbf{E}M_s f_l(M_s)I_\Gamma. \tag{6.19}$$

This means, in particular, that there exists $\lim_n \mathbf{E}M_{t\wedge\tau_n}\xi^n$. On the other hand by the inequality $M_{t\wedge\tau_n}\xi^n \ge -(X_0 + A_\infty)\,\xi^n$ and Fatou's lemma we get

$$\lim_n \mathbf{E}M_{t\wedge\tau_n}\xi^n \ge \mathbf{E}M_t f_l(M_s)\,I_\Gamma.$$

Thus, by (6.18) and (6.19) the inequality

$$\mathbf{E}M_t f_l(M_s)\,I_\Gamma \le \mathbf{E}M_s f_l(M_s)\,I_\Gamma$$

holds which gives as $l \to \infty$ the desired inequality (6.17) by taking the limit.

Theorem 6. *Let* $X \in \mathbf{Sp}$, $X \ge 0$, $\mathbf{E}X_0 < \infty$ *and let the process* A, *involved in the decomposition* (6.16), *belong to* $\mathcal{V}^+ \cap \mathcal{P}$.

1) *If at least one of the following conditions is fullfilled:*

(α) *there exists* $\varepsilon > 0$ *such that for each* $t > 0$ *the random variable* $A_{t+\varepsilon}$ *is* $\mathcal{F}_t$-*measurable, or*

(β) *for each predictable time* σ

$$\mathbf{E}\Delta A_\sigma I\,(\sigma < \infty) < \infty,$$

then

$$\{A_\infty < \infty\} \subseteq \{X \to\ \} \quad (P\text{-a.s.}).$$

2) *If for each Markov time* σ

$$\mathbf{E}\,(\Delta X_\sigma)^+ I\,(\sigma < \infty) < \infty,$$

then $\{A_\infty < \infty\} = \{X \to\}$ (P-a.s.).

Proof. 1) If σ_a for $a > 0$ denotes any of the Markov times $\inf(t: A_{\tau+\varepsilon} \ge a)$ (under Condition (α)) or $\inf(t: A_t \ge a)$ (under Condition (β)), then

$$EA_{\sigma_a} \le c(a) \tag{6.20}$$

with a constant c(a) depending on a.

By the decomposition (6.16) we have

$$X_t^{\sigma_a} = X_0 + A_t^{\sigma_a} + M_t^{\sigma_a}.$$

Therefore, by Lemma 2 the process M^{σ_a} is a submartingale and

$$\sup_{t \ge 0} E\,|M_t^{\sigma_a}| < \infty.$$

Consequently, according to Theorem 2 we have $P(M^{\sigma_a} \to) = 1$, and hence $P(X^{\sigma_a} \to) = 1$.

Now, the desired assertion follows from Lemma 1, because $\sup_{t \ge 0} A_t = A_\infty$.

2) By Theorem 1.7.8 we have ${}^p(\Delta M) = 0$. Therefore, by the decomposition (6.16) we have $\Delta A = {}^p(\Delta X)$. If σ is a predictable time, then (cf. Theorem 1.3.13)

$$E\Delta A_\sigma I(\sigma < \infty) = E\,{}^p(\Delta X)_\sigma I(\sigma < \infty)$$

$$\le E\,{}^p((\Delta X)^+)_\sigma I(\sigma < \infty) = EE((\Delta X_\sigma)^+ \mid \mathcal{F}_{\sigma-})\, I(\sigma < \infty).$$

The set $\{\sigma < \infty\}$ belongs to $\mathcal{F}_{\sigma-}$ (Problem 1.1.6), and hence

$$E\Delta A_\sigma I(\sigma < \infty) \le E(\Delta X_\sigma)^+ I(\sigma < \infty).$$

Therefore, according to assertion 1)

$$\{A_\infty < \infty\} \subseteq \{X \to\} \quad \text{(P-a.s.)}.$$

The converse inclusion

$$\{X \to\} \subseteq \{A_\infty < \infty\} \quad \text{(P-a.s.)}$$

is established in the following manner. For $a > 0$ the Markov time $\sigma_a = \inf(t: X_t \ge a)$ is introduced. Then

$$EX_{\sigma_a} = E(X_{\sigma_a -} + \Delta X_{\sigma_a} I(\sigma_a < \infty)) \le a + E(\Delta X_{\sigma_a})^+ I(\sigma_a < \infty) < \infty.$$

If $(\tau_n)_{n \ge 1}$ is a localizing sequence for M, then

$$EA_{\sigma_a \wedge \tau_n} = EX_{\sigma_a \wedge \tau_n} - EX_0 \le EX_0 + EX_{\sigma_a \wedge \tau_n}$$

$$\le EX_0 + EI(\sigma_a > \tau_n) X_{\tau_n} + EI(\sigma_a \le \tau_n) X_{\sigma_a} \le EX_0 + a + EX_{\sigma_a},$$

and consequently,

$$\mathbf{E}A_{\sigma_a} < \infty, \text{ i.e. } \mathsf{P}\,(A^{\sigma_a} \to) = 1.$$

By Lemma 1 the desired inclusion takes place, because $\{X \to\} \subseteq \{\sup_{t \ge 0} X_t < \infty\}$ and $\{A \to\} = \{A_\infty < \infty\}$.

For the process A involved in the decomposition (6.16) Theorem 6 is proved under the assumption $A \in \mathcal{V}^+ \cap \mathcal{P}$. To relax this assumption we set $A = A^1 - A^2$ with A^1, $A^2 \in \mathcal{V}^+ \cap \mathcal{P}$.

Theorem 7. *Let* $X \in \mathbf{Sp}$, $X \ge 0$, $\mathbf{E}X_0 < \infty$.

1) *If for the process* A^1 *at least one of Conditions* (α) *or* (β) *of Theorem* 6 *is fullfilled, then*

$$\{A^1_\infty < \infty\} \subseteq \{X \to\} \cap \{A^2_\infty < \infty\} \quad (\mathsf{P}\text{-a.s.}).$$

2) *If the conditions of assertion* 1) *are fullfilled and for each Markov time* σ *we have* $\mathbf{E}\,(\Delta X_\sigma)^+ I\,(\sigma < \infty) < \infty$ *and* $\mathbf{E}\Delta A^2_\sigma I\,(\sigma < \infty) < \infty$, *then*

$$\{A^1_\infty < \infty\} = \{X \to\} \cap \{A^2_\infty < \infty\} \quad (\mathsf{P}\text{-a.s.}).$$

Proof. Denote

$$Y_t = X_0 + A^1_t + M_t$$

where M is the local martingale in the decomposition (6.16). Observe that $X = Y - A^2$. Therefore $Y \in \mathbf{Sp}$, $Y \ge 0$, $\mathbf{E}Y_0 = \mathbf{E}X_0 < \infty$. By the assumption concerning A^1, Theorem 6 implies

$$\{A^1_\infty < \infty\} \subseteq \{Y \to\} \quad (\mathsf{P}\text{-a.s.}).$$

But from the definition of Y it follows that $A^2 \le Y$. Consequently, by $A^2 \in \mathcal{V}^+$ the inclusion $\{Y \to\} \subseteq \{A^2_\infty < \infty\}$ takes place.

The desired assertion follows from this in an obvious manner.

2) By assertion 1), proved already, it suffices to show that

$$\{A^1_\infty < \infty) \supseteq \{X \to\} \cap \{A^2_\infty < \infty\}.$$

According to the equality $Y = X + A^2$ the obvious inclusion

$$\{X \to\} \cap \{A^2_\infty < \infty\} \subseteq \{Y \to\}$$

takes place. Therefore, it remains to establish the inclusion

$$\{Y \to\} \subseteq \{A^1_\infty < \infty\} \quad (\mathsf{P}\text{-a.s.}).$$

This inclusion takes place in virtue of Theorem 6, because by assumption we have

$$\mathbf{E}\,(\Delta Y_\sigma)^+ I\,(\sigma < \infty) \le \mathbf{E}\,((\Delta X_\sigma)^+ + \Delta A^2_\sigma)\, I\,(\sigma < \infty) < \infty.$$

Corollary. *If* $X \in \mathcal{M}_{loc}$ *and* $X \ge 0$, *then*

$$P(X \to) = 1.$$

Consider yet another case in which $X \in \mathbf{Sp}$, however it is not a nonnegative process.

Theorem 8. *Let* $X \in \mathbf{Sp}$, $\mathbf{E}\,|X_0| < \infty$, $|\Delta X| \le c$ *and let the process* A *involved in decomposition* (6.16) *belong to* $\mathcal{V}^+ \cap \mathcal{P}$.

Then the process M *involved in the decomposition* (6.16) *belongs to* $\mathcal{M}^2_{loc}$ *and*

$$\{A_\infty + \langle M \rangle_\infty < \infty\} = \{X \to\} \quad (P\text{-a.s.}).$$

Proof. By Theorem 1.2 we have $|\Delta M| \le 2c$. Hence $M \in \mathcal{M}^2_{loc}$ with a localizing sequence $\tau_n = \inf(t: |M_t| \ge n)$. By Theorem 5 we have $\{\langle M \rangle_\infty < \infty\} \subseteq \{M \to\}$ (P-a.s.). Consequently, by taking into consideration (6.16), the inclusion

$$\{A_\infty + \langle M \rangle_\infty < \infty\} \subseteq \{X \to\} \quad (P\text{-a.s.})$$

takes place. To prove the converse inclusion

$$\{A_\infty + \langle M \rangle_\infty < \infty\} \subseteq \{X \to\} \quad (P\text{-a.s.}) \tag{6.21}$$

introduce Markov times $\sigma_a = \inf(t: |X_t| \ge a)$, $a > 0$. Then

$$\mathbf{E}\,|X_{\sigma_a}| = \mathbf{E}\,|X_{\sigma_a -} + \Delta X_{\sigma_a} I(\sigma_a < \infty)| \le (a + c) \vee \mathbf{E}\,|X_0|.$$

If $(\tau_n)_{n \ge 1}$ is a localizing sequence introduced above, then

$$\mathbf{E}A_{\sigma_a \wedge \tau_n} = \mathbf{E}\,(X_{\sigma_a \wedge \tau_n} - X_0) \le \mathbf{E}\,|X_{\sigma_a \wedge \tau_n}| + \mathbf{E}\,|X_0|$$

$$\le \mathbf{E}\,|X_0| + \mathbf{E}I(\tau_n < \sigma_a)\,|X_{\sigma_a \wedge \tau_n}| + \mathbf{E}I(\tau_n \ge \sigma_a)\,|X_{\sigma_a \wedge \tau_n}|$$

$$\le \mathbf{E}\,|X_0| + a + (a + c) \vee \mathbf{E}\,|X_0| < \infty.$$

Therefore $\mathbf{E}A_{\sigma_a} \le \text{const}$, i.e. $P(A^{\sigma_a} \to) = 1$. By Lemma 1 we have

$$\{\sup_{t \ge 0} |X_t| < \infty\} \subseteq \{A \to\} = \{A_\infty < \infty\} \quad (P\text{-a.s.})$$

which, by the obvious inclusion $\{X \to\} \subseteq \{\sup_{t \ge 0} |X_t| < \infty\}$ yields

$$\{X \to\} \subseteq \{A_\infty < \infty\} \quad (P\text{-a.s.}). \tag{6.22}$$

This and the decomposition (6.16) imply

$$\{X \to\} \subseteq \{M \to\} \quad (P\text{-a.s.}).$$

But $|\Delta M| \le 2c$, and consequently by Theorem 5 we have $\{M \to\} = \{\langle M \rangle_\infty < \infty\}$ (P-a.s.), i.e.

$$\{X \to\} \subseteq \{\langle M \rangle_\infty < \infty\} \quad (P\text{-a.s.}). \tag{6.23}$$

The desired inclusion (6.21) takes place by (6.22) and (6.23).

7. It is said that for $X \in \mathbf{S}$ and $L \in \mathcal{V}^+ \cap \mathcal{P}$ the pair (X, L) satisfies the *strong law of large numbers* if

$$P\left(\lim_{t \to \infty} \frac{X_t}{L_t} = 0 \right) = 1. \tag{6.24}$$

Further on, the set $\left\{ \lim_{t \to \infty} \frac{X_t}{L_t} = 0 \right\}$ is defined by $\left\{ \frac{X}{L} \to 0 \right\}$.

Relate to the processes (X, L) the process $Y \in \mathbf{S}$ with

$$Y = (1 + L)^{-1} \cdot X.$$

On studying the strong law of large numbers, an essential rôle is played by the following result that is a generalization of Kronecker's lemma.

Lemma 3. *Let* $X \in \mathbf{S}$ *and* $L \in \mathcal{V}^+ \cap \mathcal{P}$. *Then*

$$\{L_\infty = \infty\} \cap \{Y \to\ \} \subseteq \left\{ \frac{X}{L} \to 0 \right\} \quad (\mathsf{P}\text{-a.s.}).$$

Proof. Since $X \in \mathbf{S}$, the decomposition $X_t = X_0 + A_t + M_t$ takes place with $A \in \mathcal{V}$ and $M \in \mathcal{M}_{loc}$, hence $Y = (1 + L)^{-1} \circ A + (1 + L)^{-1} \cdot M$.

According to Problem 2.5 we have $(1 + L) \cdot ((1 + L)^{-1} \cdot M) = M$. Therefore

$$(1 + L) \cdot Y_t = X_t - X_0. \tag{6.25}$$

Since the sets $\{L_\infty = \infty\} \cap \left\{ \frac{X}{L} \to 0 \right\}$ and $\{L_\infty = \infty\} \cap \left\{ \frac{X}{1 + L} \to 0 \right\}$ coincide, it suffices to verify the following inclusions:

$$\{L_\infty = \infty\} \cap \{Y \to\ \} \subseteq \left\{ \frac{X}{1 + L} \to 0 \right\} \quad (\mathsf{P}\text{-a.s.}). \tag{6.26}$$

To this end, observe that by Ito's formula (cf. § 3)

$$(1 + L_t) Y_t = (1 + L) \cdot Y_t + Y_- \circ L_t,$$

and consequently, by taking into consideration (6.25) we get the following relation:

$$\frac{X_t}{1 + L_t} = \frac{X_0 + Y_t}{1 + L_t} + \frac{L_t Y_t - Y_- \circ L_t}{1 + L_t}.$$

By the inclusion $\{Y \to\ \} \subseteq \{ \sup_{t \geq 0} |Y_t| < \infty\}$ on the set $\{L_\infty = \infty\} \cap \{Y \to\ \}$ we have

$$\lim_{t \to \infty} \frac{X_0 + Y_t}{1 + L_t} = 0.$$

Consequently, it suffices to show that on the same set $\{L_\infty = \infty\} \cap \{Y \to\}$ we have

$$\lim_{t \to \infty} \frac{L_t Y_t - Y_- \circ L_t}{1 + L_t} = 0. \tag{6.27}$$

To establish (6.27), observe that for each $\varepsilon > 0$ a positive random variable t_ε can be found such that $P(t_\varepsilon = \infty,\ Y \to) = 0$ and $|Y_\infty - Y_{t-}| \le \varepsilon$ on the set $\{Y \to\} \cap \{t > t_\varepsilon\}$. Then on the set $\{L_\infty = \infty\} \cap \{Y \to\}$ the following chain of inequalities take place:

$$\frac{|L_t Y_t - Y_- \circ L_t|}{1 + L_t} \le \frac{\int_0^{t \wedge t_\varepsilon} |Y_t - Y_{s-}|\, dL_s}{1 + L_t} + \frac{\int_{t \wedge t_\varepsilon}^{t} (|Y_\infty - Y_t| + |Y_\infty - Y_{s-}|)\, dL_s}{1 + L_t}$$

$$\le 2 \sup_{s \ge 0} |Y_s| \frac{L_{t \wedge t_\varepsilon}}{1 + L_t} + |Y_\infty - Y_t| + I(t_\varepsilon < t)\, \varepsilon,$$

according to which on the indicated set we have

$$\overline{\lim_{t \to \infty}} \frac{|L_t Y_t - Y_- \circ L_t|}{1 + L_t} \le \varepsilon.$$

As $\varepsilon > 0$ is arbitrary, this implies the desired relation (6.27).

8. Consider a number of examples in which the law of large numbers is verified.

Let $X \in \mathbf{S}$ and let the semimartingale decomposition

$$X_t = X_0 + A_t + M_t \tag{6.28}$$

be given with $A \in \mathcal{V}$ and $M \in \mathcal{M}_{loc}$. Denote by $\tilde{B}$ the compensator of the process B with

$$B_t = \sum_{0 < s \le t} \frac{(\Delta M_s / (1 + L_s))^2}{1 + |\Delta M_s / (1 + L_s)|}, \tag{6.29}$$

and let $\langle M^c \rangle$ be the quadratic characteristic of the continuous martingale component M^c of a local martingale M.

Theorem 9. *Let* $X \in \mathbf{S}$ *be representable in form* (6.28), $L \in \mathcal{V}^+ \cap \mathcal{P}$ *and*

$$P(L_\infty = \infty) = 1.$$

Then under condition

$$(1 + L)^{-1} \circ \mathrm{Var}(A)_\infty + (1 + L)^{-2} \circ \langle M^c \rangle_\infty + \tilde{B}_\infty < \infty \quad (P\text{-a.s.}) \tag{6.30}$$

the strong law of large numbers takes place:

$$\mathsf{P}\left(\frac{X}{L} \to 0\right) = 1.$$

Proof. By the condition $\mathsf{P}\ (L_\infty = \infty) = 1$, it suffices to show, according to Lemma 3, that

$$\mathsf{P}\ ((1 + L)^{-1} \cdot X \to\ ) = 1.$$

By Theorem 1.7.2 we have $M = M^c + M^d$ with $M^c \in \mathcal{M}^c_{loc}$ and $M^d \in \mathcal{M}^d_{loc}$.

By taking this decomposition and (6.28) into consideration we have

$$(1 + L)^{-1} \cdot X = (1 + L)^{-1} \circ A + (1 + L)^{-1} \cdot M^c + (1 + L)^{-1} \cdot M^d. \quad (6.31)$$

Therefore it suffices to show that each of the three processes involved in the expression on the right hand side of the equation (6.31) converges to a finite limit with probability one as $t \to \infty$.

From (6.30) it follows that $(1 + L)^{-1} \circ \mathrm{Var}\,(A)_\infty < \infty$ (P-a.s.), and hence $\mathsf{P}\ ((1 + L)^{-1} \circ A \to\ ) = 1$. Evidently, the process $(1 + L)^{-1} \cdot M^c$ belongs to $\mathcal{M}^c_{loc}$ and possesses the quadratic characteristic $(1 + L)^{-2} \circ \langle M^c\rangle$ (Theorem 2.2). According to (6.30) we have $(1 + L)^{-2} \circ \langle M^c\rangle_\infty < \infty$ (P-a.s.) and according to Theorem 5 we have $\mathsf{P}\ ((1 + L)^{-1} \cdot M^c \to\ ) = 1$. The process $Z = (1 + L)^{-1} \cdot M^d$ belongs to $\mathcal{M}^d_{loc}$ and by Theorem 4 we have $\mathsf{P}\ (Z \to\ ) = 1$ provided

$$\mathsf{P}\left(\widetilde{\sum_{s > 0} \frac{(\Delta Z_s)^2}{1 + |\Delta Z_s|}} < \infty\right) = 1$$

where

$$\widetilde{\sum_s \frac{(\Delta Z_s)^2}{1 + |\Delta Z_s|}}$$

is the compensator of the process

$$\sum_s \frac{(\Delta Z_s)^2}{1 + |\Delta Z_s|} = \sum_s \frac{(\Delta M_s / (1 + L_s))^2}{1 + |\Delta M_s / (1 + L_s)|},$$

i.e. $\mathsf{P}\ (Z \to\ ) = 1$ if $\mathsf{P}\ (\tilde{B}_\infty < \infty) = 1$.

The last relation follows from (6.30) as well.

Theorem 10. *Let* $X \in \mathcal{M}^2_{loc}$, $L \in \mathcal{V}^+ \cap \mathcal{P}$ *and* $P(L_\infty = \infty) = 1$.

Then under the condition

$$(1 + L)^{-2} \circ \langle X \rangle_\infty < \infty \quad (P\text{-a.s.})$$

the strong law of large numbers takes place:

$$P\left(\frac{X}{L} \to 0\right) = 1.$$

Proof. In virtue of the assumption $P(L_\infty = \infty) = 1$ it suffices by Lemma 3 to show that $P((1 + L)^{-1} \cdot X \to) = 1$. But $(1+ L)^{-1} \cdot X \in \mathcal{M}^2_{loc}$ and $\langle (1 + L)^{-1} \cdot X \rangle = (1 + L)^{-2} \circ \langle X \rangle$ (Theorem 2.2). Therefore the desired relation $P((1 + L)^{-1} \cdot X \to) = 1$ follows from Theorem 5.

Corollary 1. *If* $X \in \mathcal{M}^2_{loc}$ *and* $P(\langle X \rangle_\infty = \infty) = 1$, *then*

$$P\left(\frac{X}{\langle X \rangle} \to 0\right) = 1.$$

According to Theorem 10 it suffices that $(1 + \langle X \rangle)^{-2} \circ \langle X \rangle_\infty < \infty$ (P-a.s.), whereas by Problem 1 we have $(1 + \langle X \rangle)^{-2} \circ \langle X \rangle_\infty \le 1$.

Corollary 2. *If* $X \in \mathcal{M}^2_{loc}$ *and* $a > 0$, *then*

$$P\left(\frac{X}{a + \langle X \rangle} \to\right) = 1.$$

In fact, on the set $\{\langle X \rangle_\infty = \infty\}$ we have $\lim\limits_{t \to \infty} \dfrac{X_t}{a + \langle X \rangle_t} = 0$ by Lemma 3, while on the set $\{\langle X \rangle_\infty < \infty\}$ there exists the finite limit $\lim\limits_{t \to \infty} X_t$ (Theorem 5).

Let us establish now the strong law for large numbers for local martingales.

Theorem 11. *Let* $X \in \mathcal{M}_{loc}$, $L \in \mathcal{V}^+ \cap \mathcal{P}$ *and* $P(L_\infty = \infty) = 1$. *If for each Markov time* σ *we have* $\mathbf{E}\,|\Delta X_\sigma|\, I(\sigma < \infty) < \infty$, *then under the condition*

$$(1 + L)^{-2} \circ [X, X]_\infty < \infty \quad (P\text{-a.s.})$$

the strong law of large numbers takes place:

$$P\left(\frac{X}{L} \to 0\right) = 1.$$

Proof. In virtue of the assumption $P(L_\infty = \infty) = 1$ it suffices by Lemma 3 to show that $P((1 + L)^{-1} \cdot X \to) = 1$. The process $Z = (1 + L)^{-1} \cdot X$ belongs to $\mathcal{M}_{loc}$, and according to the assumption $\mathbf{E}\,|\Delta X_\sigma|\,I(\sigma < \infty) < \infty$, the inequality

$$\mathbf{E}\,|\Delta Z_\sigma|\,I(\sigma < \infty) \le \mathbf{E}\,\frac{|\Delta X_\sigma|}{1 + L_\sigma}\,I(\sigma < \infty) < \infty$$

takes place. Therefore, by Theorem 4

$$\{[Z, Z]_\infty < \infty\} = \{Z \to\} \quad (P\text{-a.s.}).$$

Now, it remains to observe that

$$[Z, Z] = (1 + L)^{-2} \circ [X, X].$$

This theorem implies the following result.

Theorem 12. *Let* $A \in \overset{+}{\mathcal{A}}_{loc}$, *let* $\tilde{A}$ *be the compensator of* A *and let the conditions*

$$P(\tilde{A}_\infty = \infty) = 1, \quad \mathbf{E} \sup_{t > \infty} \Delta A_t < \infty$$

be satisfied. Then

$$P\left(\lim_{t \to \infty} \frac{A_t}{\tilde{A}_t} = 1\right) = 1.$$

Proof. Since $M = A - \tilde{A} \in \mathcal{M}_{loc}$ (Theorem 1.6.3), it suffices to show that

$$P\left(\frac{M}{\tilde{A}} \to 0\right) = 1.$$

To this end, it suffices by Theorem 11 to show that $\mathbf{E}\,|\Delta M_\sigma|\,I(\sigma < \infty) < \infty$ for each Markov time σ and that

$$(1 + \tilde{A})^{-2} \circ [M, M]_\infty < \infty \quad (P\text{-a.s.}). \tag{6.32}$$

Let us show that $\mathbf{E}\,|\Delta M_\sigma|\,I(\sigma < \infty) \le 2\mathbf{E} \sup_{t > 0} \Delta A_t$. We have $|\Delta M| \le \Delta A + \Delta\tilde{A}$. Consequently, it suffices to establish that for any predictable Markov time σ ($\sigma \in \mathbf{T}_p$)

$$\mathbf{E}\Delta\tilde{A}_\sigma I(\sigma < \infty) \le \mathbf{E} \sup_{t > 0} \Delta A_t, \tag{6.33}$$

because by Theorem 1.3.6 the set $\{\Delta\tilde{A} > 0\}$ is exhausted by a sequence of predictable stopping times.

To establish (6.33), use the fact that $\Delta\tilde{A}_\sigma I(\sigma < \infty) = I_{[\![\sigma]\!]} \circ \tilde{A}_\infty$. Since $\sigma \in \mathbf{T}_p$, the process $I_{[\![\sigma]\!]}$ is predictable and by Theorem 1.6.3

$$\mathbf{E}I_{[\![\sigma]\!]} \circ \tilde{A}_\infty = \mathbf{E}I_{[\![\sigma]\!]} \circ A_\infty = \mathbf{E}\Delta A_\sigma I(\sigma < \infty).$$

This implies the inequality (6.33) in an obvious manner.

The verification of the validity of (6.32) is based on the following estimates

$$(1+\tilde{A})^{-2} \circ [M, M]_\infty \le 2 \sum_{s>0} (1+\tilde{A}_s)^{-2} ((\Delta A_s)^2 + (\Delta \tilde{A}_s)^2)$$

$$\le 2\,[\sup_{t>0} \Delta A_t (1+\tilde{A})^{-2} \circ A_\infty + (1+\tilde{A})^{-2} \Delta\tilde{A} \circ \tilde{A}_\infty].$$

By the condition $\mathbf{E}\sup_{t>0} \Delta A_t < \infty$ it suffices to show that

$$(1+\tilde{A})^{-2} \circ A_\infty < \infty, \quad (1+\tilde{A})^{-2}\Delta\tilde{A} \circ \tilde{A}_\infty < \infty \quad (\text{P-a.s.}).$$

The first inequality takes place because

$$\mathbf{E}\,(1+\tilde{A})^{-2} \circ A_\infty = \mathbf{E}\,(1+\tilde{A})^{-2} \circ \tilde{A}_\infty \le 1$$

(cf. Theorem 1.6.3 and Problem 1). The second inequality holds by Theorem 1.6.3 and Problem 1 again, because

$$\mathbf{E}\,(1+\tilde{A})^{-2}\Delta\tilde{A} \circ \tilde{A}_\infty = \mathbf{E}\,(1+\tilde{A})^{-2}\Delta\tilde{A} \circ A_\infty = \mathbf{E} \sum_{s>0} (1+\tilde{A}_s)^{-2} \Delta\tilde{A}_s \Delta A_s$$

$$\le \mathbf{E}\,(\sup_{t>0} \Delta A_t)((1+\tilde{A})^{-2} \circ \tilde{A}_\infty) \le \mathbf{E} \sup_{t>0} \Delta A_t.$$

9. Let us point out yet another application of results concerning convergence sets for local martingales.

Theorem 13. *Let* $X \in \mathcal{M}_{loc}$ *and for each Markov time* σ *let* $\mathbf{E}\,|\Delta X_\sigma|\, I(\sigma < \infty) < \infty$. *Then*

1) $\{X \to\} = \{[X, X]_\infty < \infty\}$ (P-a.s.) ;

2) $\{\varliminf_{t\to\infty} X_t = -\infty\} \cap \{\varlimsup_{t\to\infty} X_t = \infty\} = \{[X, X]_\infty = \infty\}$ (P-a.s.)

Proof. 1) The coincidence of the respective sets is established in Theorem 4.

2) It suffices to establish the inclusion

$$\{[X, X]_\infty = \infty\} \subseteq \{\varliminf_{t\to\infty} X_t = -\infty\} \cap \{\varlimsup_{t\to\infty} X_t = \infty\} \quad (\text{P-a.s.}). \qquad (6.34)$$

In fact, if (6.34) takes place, then

$$\{[X, X]_\infty < \infty\} \supseteq \{\varliminf_{t\to\infty} X_t > -\infty\} \cup \{\varlimsup_{t\to\infty} X_t < \infty\} \quad (\text{P-a.s.}),$$

and hence, by 1),

$$\{X \to\} \supseteq \{\varliminf_{t\to\infty} X_t > -\infty\} \cup \{\varlimsup_{t\to\infty} X_t < \infty\} \quad (\text{P-a.s.}).$$

Therefore, by the obvious inclusion

$$\{X \to\} \subseteq \{\varliminf_{t\to\infty} X_t > -\infty\} \cup \varlimsup_{t\to\infty} X_t < \infty\} \quad (\text{P-a.s.})$$

we get

$$\{X \to\} = \{\varliminf_{t\to\infty} X_t > -\infty\} \cup \{\varlimsup_{t\to\infty} X_t < \infty\} \quad (\text{P-a.s.}) \tag{6.35}$$

This and 1) imply the desired coincidence of sets.

The verification of the validity of (6.34) is based on the fact that the following relations

$$\bigcup_{a>0} \{\tau_a < \infty\} = \{\varliminf_{t\to\infty} X_t = -\infty\}, \tag{6.36}$$

$$\bigcup_{b>0} \{\sigma_b < \infty\} = \{\varlimsup_{t\to\infty} X_t = \infty\}, \tag{6.37}$$

take place where $\bigcup_{a>0}$ and $\bigcup_{b>0}$ are taken over all positive integers a and b, while $\tau_a = \inf(t\colon X_t \le -a)$ and $\sigma_b = \inf(t\colon X_t \ge b)$; consequently, it suffices to show that

$$\{[X, X]_\infty = \infty\} \subseteq \{\tau_a < \infty\} \quad (\text{P-a.s.}), \tag{6.38}$$

$$\{[X, X]_\infty = \infty\} \subseteq \{\sigma_b < \infty\} \quad (\text{P-a.s.}). \tag{6.39}$$

To prove (6.38), define the increasing processes V^1 and V^2 with

$$V_t^1 = (\Delta X_{\tau_a})^- I(t \ge \tau_a), \qquad V_t^2 = (\Delta X_{\tau_a})^+ I(t \ge \tau_a)$$

where, as usual, $c^+ = c \vee 0$ and $c^- = -(c \wedge 0)$. By the assumption

$$E\,|\,\Delta X_\sigma\,|\,I(\sigma < \infty) < \infty$$

we have $V^1, V^2 \in \mathcal{A}^+$. Denote by A^1 and A^2 the compensators of the processes V^1 and V^2 respectively. Then the process $N = (V^1 - A^1)$ is a local martingale (Theorem 1.6.3). Define also the random variable $Y_0 = (X_0 + a)$, the local martingale M with $M_t = (X_{t\wedge\tau_a} - X_0 + N_t)$ and a special semimartingale Y with

$$Y_t = Y_0 + A_t^1 + M_t.$$

Observe that by the definitions of Y_0, M and A^1 given above, for Y the representation

$$Y_t = X_{t\wedge\tau_a} + a + (\Delta X_{\tau_a})^- I(t \ge \tau_a) \tag{6.40}$$

takes place. This and the definition of Markov time τ_a imply that Y is a nonnegative special semimartingale. By Theorem 6

$$\{A^1_\infty < \infty\} \subseteq \{Y \to \ \} \quad (\mathsf{P}\text{-a.s.}),$$

provided $\mathbf{E}\Delta A^1_\sigma I\,(\sigma < \infty) < \infty$ for each Markov time σ. But $\mathbf{E}\Delta A^1_\sigma I\,(\sigma < \infty) \le \mathbf{E}A^1_\infty$, whereas by Theorem 1.6.3 we have $\mathbf{E}A^1_\infty = \mathbf{E}V^1_\infty$. Since $\mathbf{E}V^1_\infty \le \mathbf{E}\,|\,\Delta X_{\tau_a}\,|\,I\,(\tau_a < \infty) < \infty$ the conditions of Theorem 6 are satisfied and $\mathsf{P}\,(A^1_\infty < \infty) = 1$. Hence $\mathsf{P}\,(Y \to\ ) = 1$. This and (6.40) yield $\mathsf{P}\,(X^{\tau_a} \to\ ) = 1$. Then, in accordance with 1),

$$\mathsf{P}\,([X, X]_{\tau_a} < \infty) = 1, \tag{6.41}$$

hence

$$\mathsf{P}\,([X, X]_\infty = \infty,\ \tau_a = \infty) = 0,$$

i.e. (6.38) takes place.

The validity of (6.39) is analoguously established.

Corollary. *If* $X \in \mathcal{M}^2_{loc}$ *and for each Markov time* σ *we have* $\mathbf{E}\,(\Delta X_\sigma)^2 I\,(\sigma < \infty) < \infty$, *then*:

1) $\{X \to\ \} = \{\langle X\rangle_\infty < \infty\}$ $(\mathsf{P}\text{-a.s.})$,

2) $\{\underline{\lim}_{t\to\infty} X_t = -\infty\} \cap \{\overline{\lim}_{t\to\infty} X_t = \infty\} = \{\langle X\rangle_\infty = \infty\}$ $(\mathsf{P}\text{-a.s.})$.

For the proof it suffices to observe only that by Theorem 5 we have $\{\langle X\rangle_\infty < \infty\} = \{[X, X]_\infty < \infty\}$ $(\mathsf{P}\text{-a.s.})$.

Problems

1. Show that for $A \in \mathsf{V}^+$ the inequality $(1 + A)^{-2} \circ A_\infty \le 1$ holds.

2. If $A \in \mathcal{A}^+_{loc}$, $\tilde{A}$ is the compensator of A and $M = A - \tilde{A}$, then

$$\{\tilde{A} < \infty\} \subseteq \{M \to\ \} \quad (\mathsf{P}\text{-a.s.}).$$

3. Let θ be an unknown parameter $(\theta \in R)$ and let $M \in \mathcal{M}^2_{loc}$ have the quadratic characteristic $\langle M\rangle$. Let parameter θ be estimated by observations on the process $X \in \mathbf{Sp}$ with

$$X_t = X_0 + \theta X_- \circ \langle M\rangle_t + M_t$$

by means of the estimator $\hat{\theta}_t$, obtained by the least square method, i.e. let for a fixed t the estimator $\hat{\theta}_t$ of θ be defined by the formula

$$\hat{\theta}_t = (X_-^2 \circ \langle M \rangle_t)^{-1} (X_- \cdot X_t),$$

where $\frac{0}{0} = 0$.

Show that if $P(\langle M^c \rangle_\infty + \tilde{B}_\infty = \infty) = 1$ where $\tilde{B}$ is the compensator of the process

$$B = \sum_s (\Delta M_s)^2 \wedge 1,$$

then the sequence of estimators $(\hat{\theta}_t)_{t \geq 0}$ is strongly consistent, i.e.

$$P(\lim_{t \to \infty} \hat{\theta}_t = \theta) = 1.$$

4. Construct an example of a local square integrable martingale X with the properties:

$$P(X \to) = 1, \ P(\langle X \rangle_\infty = \infty) = 1.$$

5. Let $X \in \mathbf{Sp}$, $X \geq 0$, $\mathbf{E}X_0 < \infty$ and $X_t = X_0 + X_- \circ B_t + A_t^1 - A_t^2 + M_t$ with B, $A^1, A^2 \in \mathcal{V}^+ \cap \mathcal{P}$, $M \in \mathcal{M}_{\text{loc}, 0}$ and let for A^1 the conditions of Theorem 7 be fullfilled.

Show that

$$\{A_\infty^1 < \infty\} \cap \{B_\infty < \infty\} \subseteq \{X \to\} \cap \{A_\infty^2 < \infty\} \quad (P\text{-a.s.}).$$

(Hint: apply Theorem 7 to the process $Y = X\mathcal{E}^{-1}(B)$ where $\mathcal{E}(B)$ is the stochastic exponential, Ch. 2, § 4).

CHAPTER 3

RANDOM MEASURES AND THEIR COMPENSATORS

§ 1. Optional and predictable random measures

1. Let $(\Omega, \mathcal{F}, \mathbb{F}, P)$ be a stochastic basis with $\mathbb{F} = (\mathcal{F}_t)_{t \geq 0}$, and let $\mathcal{O}(\mathbb{F})$ and $\mathcal{P}(\mathbb{F})$ be optional and predictable σ-algebras of random sets in $\Omega \times R_+$, respectively. Also, let $(E, \mathcal{E})$ be the Lusin's space (i.e. E is a Borel subspace of a compact metric space with the Borel σ-algebra $\mathcal{E}$), and let

$$\tilde{\Omega} = \Omega \times R_+ \times E, \quad \tilde{\mathcal{F}} = \mathcal{F} \otimes B(R_+) \otimes \mathcal{E},$$

$$\tilde{\mathcal{O}} = \mathcal{O} \otimes \mathcal{E}, \quad \tilde{\mathcal{P}} = \mathcal{P} \otimes \mathcal{E}. \tag{1.1}$$

Definition 1. The family

$$\mu = \{\mu(\omega; dt, dx), \quad \omega \in \Omega\}$$

of σ-finite nonnegative measures $\mu(\omega; \cdot\,)$; $\omega \in \Omega$ on $(R_+ \times E, B(R_+) \otimes \mathcal{E})$, such that for each $A \in B(R_+) \otimes \mathcal{E}$ the variable $\mu(\,\cdot\,; A)$ is $\mathcal{F}$-measurable and

$$\mu(\omega; \{0\} \times E) = 0, \quad \forall\, \omega \in \Omega, \tag{1.3}$$

is called a *random measure* (on $R_+ \times E$).

Let $X = X(\omega, t, x)$ be a nonnegative $\tilde{\mathcal{F}}$-measurable function. For $\omega \in \Omega$ and $t \in R_+$ the Lebesgue integral

$$X * \mu_t = \int_{[\![0,\, t]\!] \times E} X(\omega, s, x)\, \mu(\omega; ds, dx) \tag{1.4}$$

can be defined.

Further on we write $X\mu$ to denote the random measure $X(\omega, t, x)\, \mu(\omega; dt, dx)$. We will also use the notation $X * \mu = (X * \mu_t)_{t \geq 0}$.

Definition 2. A random measure μ is called *optional* (*predictable*), if for any nonnegative $\tilde{\mathcal{O}}$-measurable (respectively $\tilde{\mathcal{P}}$-measurable) function X the random process $X * \mu$ is optional (respectively predictable).

2. To each random measure μ and probability measure P one can relate the Doléans measure M_μ^P with

$$M_\mu^P(d\omega, dt, dx) = P(d\omega)\, \mu(\omega; dt, dx), \tag{1.5}$$

defined on $(\tilde{\Omega}, \tilde{\mathcal{F}})$. According to this definition for every nonnegative $\tilde{\mathcal{F}}$-measurable function X (ω, t, x)

$$M_\mu^P(X) = \int_{\tilde\Omega} X(\omega, t, x)\, M_\mu^P(d\omega, dt, dx) = \mathbf{E}(X * \mu_\infty). \tag{1.6}$$

We will say that the measure M_μ^P is finite if $M_\mu^P(I_{\tilde\Omega}) < \infty$. We will say that M_μ^P is $\tilde{\mathfrak{D}}$-σ-finite (respectively $\tilde{\mathcal{P}}$-σ-finite) if there can be found a sequence of sets $(\tilde\Omega_n)_{n\ge1}$ such that $\tilde\Omega_n \uparrow \tilde\Omega$, $\tilde\Omega_n \in \tilde{\mathfrak{D}}$ (respectively $\tilde{\mathcal{P}}$) and $M_\mu^P(I_{\tilde\Omega_n}) < \infty$, $n \ge 1$.

For optional random measures μ, we introduce the classes $\tilde{\mathcal{A}}^+$, $\tilde{\mathcal{V}}^+$ and $\tilde{\mathcal{V}}^+_{\mathcal{P}}$, assuming by definition that

$$\mu \in \tilde{\mathcal{A}}^+, \text{ if } M_\mu^P(I_{\tilde\Omega}) < \infty;$$

$$\mu \in \tilde{\mathcal{V}}^+, \text{ if } M_\mu^P(I_{\tilde\Omega_n}) < \infty, \quad \tilde\Omega_n \in \tilde{\mathfrak{D}}, \quad \tilde\Omega_n \uparrow \tilde\Omega;$$

$$\mu \in \tilde{\mathcal{V}}^+_{\mathcal{P}}, \text{ if } M_\mu^P(I_{\tilde\Omega_n}) < \infty, \quad \tilde\Omega_n \in \tilde{\mathcal{P}}, \quad \tilde\Omega_n \uparrow \tilde\Omega.$$

Clearly,

$$\tilde{\mathcal{A}}^+ \subseteq \tilde{\mathcal{V}}^+_{\mathcal{P}} \subseteq \tilde{\mathcal{V}}^+.$$

3. **Definition 3.** We say that the random measures μ and η coincide P-a.s. (writing $\mu \overset{P}{=} \eta$), if for each $\tilde{\mathcal{F}}$-measurable nonnegative function X

$$M_\mu^P(X) = M_\eta^P(X). \tag{1.7}$$

Lemma 1. *Let μ and η be optional random measures such that at least one of them belongs to the class* $\tilde{\mathcal{V}}^+$. *If the equation* (1.7) *holds for any nonnegative* $\tilde{\mathfrak{D}}$-*measurable function* X, *then* $\mu \overset{P}{=} \eta$.

Proof. For definiteness, let $\mu \in \tilde{\mathcal{V}}^+$. Then there can be found a set $B \in \tilde{\mathfrak{D}}$ such that $M_\mu^P(I_B) < \infty$. Let Y = Y (ω, t) be a nonnegative bounded $\mathcal{F} \otimes B(R_+)$-measurable function with the optional projection denoted by ${}^0Y = {}^0Y(\omega, t)$. The function ${}^0YI_B$ is $\tilde{\mathfrak{D}}$-measurable. Therefore, by the assumption of the lemma

$$M_\mu^P({}^0YI_B) = M_\eta^P({}^0YI_B). \tag{1.8}$$

Define the optional increasing process $A = (A_t)_{t\ge0}$ with $A_t = I_B * \mu_t$ and note that $\mathbf{E}(Y \circ A_\infty) = M_\mu^P(YI_B) < \infty$. On the other hand $\mathbf{E}({}^0Y \circ A_\infty) = M_\mu^P({}^0YI_B)$. Therefore

(Problem 1.6.9)

$$M_\mu^P\,(YI_B) = \mathbf{E}\,(Y \circ A_\infty) = \mathbf{E}\,({}^0Y \circ A_\infty) = M_\mu^P\,({}^0YI_B).$$

By (1.8) we have $M_\eta^P\,({}^0YI_B) < \infty$, which allows us to establish analogously that $M_\eta^P\,(YI_B) = M_\eta^P\,({}^0YI_B)$. Consequently, by (1.8) we have

$$M_\mu^P\,(YI_B) = M_\eta^P\,(YI_B),$$

i.e. (1.7) holds with $X = YI_B$.

The general case is reduced to the considered one by standard arguments of the theorem on monotone classes ([81], Ch. 4, Theorem 18).

4. In the sequel we will need a definition of random measures on R_+. Such measures are defined analogously to random measures on $R_+ \times E$. To avoid repetitions we may regard a random measure $p = p\,(\omega; dt)$ on R_+ as a certain random measure on $R_+ \times E$ with E consisting of one point : $E = \{e\}$, i.e.

$$p\,(\omega; dt) = \mu\,(\omega; dt, \{e\}).$$

In this case Doléans measure $P\,(d\omega)\,\mu\,(\omega; dt, \{e\}) = P\,(d\omega) \times p\,(\omega; dt)$ will be denoted by $M_p^P\,(d\omega, dt)$. For a nonnegative and $\mathcal{F} \otimes B\,(R_+)$-measurable function $X = X\,(\omega, t)$ with $\omega \in \Omega$ and $t \in R_+$ the Lebesgue integral

$$X * p_t = \int_{[\![0,\, t]\!]} X\,(\omega, s)\, p\,(\omega; ds).$$

can be defined. A random measure p is called optional (predictable) if for any nonnegative optional (predictable) function $X = X\,(\omega, t)$ the stochastic process $X * p$ is optional (predictable). A measure p is finite if $M_p^P\,(\Omega \times R_+) < \infty$ and $\mathcal{D}$-σ-finite ($\mathcal{P}$-σ-finite), if there exists a sequence of sets $A_n \in \mathcal{D}$, $n \geq 1$ ($A_n \in \mathcal{P}$, $n \geq 1$) such that $A_n \uparrow \Omega \times R_+$, $M_p^P\,(A_n) < \infty$, $n \geq 1$.

Problem

1. Show that the predictable random measures μ and η coincide P-a.s. if at least one of them belongs to the class $\tilde{\mathcal{V}}_{\mathcal{P}}^+$ and the equality (1.7) holds for any nonnegative $\tilde{\mathcal{P}}$-measurable function X.

§ 2. Compensators of random measures. Conditional mathematical expectation with respect to the σ-algebra $\tilde{\mathcal{P}}$

1. Let a process $A \in \mathcal{A}^+_{loc}$. According to Theorem 1.6.3 there exists a unique predictable process $\tilde{A} \in \mathcal{A}^+_{loc} \cap \mathcal{P}$, called the dual predictable projection of A or the compensator, such that $A - \tilde{A} \in \mathcal{M}_{loc}$, or equivalently

$$EH \circ A_\infty = EH \circ \tilde{A}_\infty, \quad \forall H \geq 0, \quad H \in \mathcal{P}. \tag{2.1}$$

The definition of the notion of the compensator for a random measure is based on an analogous equality.

Definition 1. A predictable random measure ν is called the *dual predictable projection or compensator of a random measure* $\mu \in \mathcal{V}^+_{\mathcal{P}}$ if it is such that for any nonnegative $\tilde{\mathcal{P}}$-measurable function $X = X(\omega, t, x)$ we have

$$M^P_\mu(X) = M^P_\nu(X). \tag{2.2}$$

Theorem 1. *Each random measure* $\mu \in \tilde{\mathcal{V}}^+_{\mathcal{P}}$ *possesses the unique* (P-a.s.) *compensator* ν.

Proof. Let us first establish the uniqueness of the compensator ν. Let ν' be yet another predictable random measure such that $M^P_\mu(X) = M^P_{\nu'}(X)$ for each nonnegative $\tilde{\mathcal{P}}$-measurable function X. Then $M^P_\nu(X) = M^P_{\nu'}(X)$, and hence, (Problem 1.1), ν = ν'.

To prove the existence of the compensator, assume first that $\mu \in \tilde{\mathcal{A}}^+$. In this case $M^P_\mu(I_{\tilde{\Omega}}) < \infty$, and a finite measure $m = m(d\omega, dt)$ can be defined on the measurable space $(\Omega \times R, \mathcal{P})$ by setting

$$m(I_G) = M^P_\mu(I_G I_E) \tag{2.3}$$

for any set $G \in \mathcal{P}$.

Since E is a Lusin space, there exists a regular conditional probability B (ω, t; dx), possessing the following property (cf. [217]): for each $(\omega, t) \in \Omega \times R_+$ B (ω, t; ·) is a probability measure on $(E, \mathcal{E})$, while for each $\Gamma \in \mathcal{E}$ B (·; Γ) is a $\mathcal{P}$-measurable nonnegative function, and if X is a nonnegative $\tilde{\mathcal{P}}$-measurable function, then

$$\overline{X}(\omega, t) = \int_E X(\omega, t, x)\, B(\omega, t; dx)$$

is a nonnegative $\mathcal{P}$-measurable function and the equality

$$M_\mu^P(X) = m(\overline{X}) = \int_{\Omega \times P_+} \overline{X}(\omega, t)\, m(d\omega, dt) \tag{2.4}$$

holds.

The measure m possesses the following properties:

1) $m(\Omega \times \{0\}) = 0$,

2) $m(Y) = 0$ for any nonnegative $\mathcal{P}$-measurable negligible function $Y = Y(\omega, t)$.

In fact, property 1) takes place by the assumption $\mu(\omega; \{0\} \times E) = 0$ and the equality (2.3) with $G = \Omega \times \{0\}$, and property 2) by (2.3), because $m(Y) = M_\mu^P(YI_E) = \mathbf{E}(Y * \mu_\infty) = 0$.

Properties 1), 2) and [81], Ch. 4, Theorem 4, imply that there exists a unique predictable increasing process $\tilde{A} = (\tilde{A}_t)_{t \geq 0}$ such that for any nonnegative $\mathcal{P}$-measurable function $Y = Y(\omega, t)$

$$m(Y) = \mathbf{E}(Y \circ \tilde{A}_\infty). \tag{2.5}$$

Define a (nonnegative) measure ν by setting

$$\nu(\omega; dt, dx) = B(\omega, t; dx)\, d\tilde{A}_t(\omega) \tag{2.6}$$

Obviously, this measure (called the dual predictable projection or compensator) is predictable. Let us show that it satisfies the equality (2.2).

Let X be a nonnegative $\tilde{\mathcal{P}}$-measurable function. Then, by taking into consideration (2.6) and (2.4), we have

$$M_\nu^P(X) = \int_{\tilde{\Omega}} X(\omega, t, x)\, P(d\omega)\, B(\omega, t; dx)\, d\tilde{A}_t(\omega) = \mathbf{E}\overline{X} \circ \tilde{A}_\infty = m(\overline{X}) = M_\mu^P(X),$$

i.e. (2.2) holds.

Let us establish now the existence of the compensator ν in case $\mu \in \tilde{\mathcal{V}}^+_{\mathcal{P}}$.

Let $\tilde{\Omega}_n \in \tilde{\mathcal{P}}$, $\tilde{\Omega}_n \uparrow \tilde{\Omega}$ and $M_\mu^P(I_{\tilde{\Omega}_n}) < \infty$, $n \geq 1$. Set $\mu_n = I_{\tilde{\Omega}_n}\mu$, $n \geq 1$. Evidently, $\mu_n \in \mathcal{Q}^+$, $n \geq 1$. Therefore, as was proved, for each $n \geq 1$ the measure μ_n has the compensator ν_n. Define a predictable random measure ν by the equality

$$\nu = I_{\tilde{\Omega}_1} \nu_1 + \sum_{n \ge 1} I_{\tilde{\Omega}_{n+1} \setminus \tilde{\Omega}_n} \nu_{n+1}.$$

Then, assuming $\tilde{\Omega}_0 = \varnothing$, we have

$$M^P_\mu (X) = \lim_n M^P_\mu (XI_{\tilde{\Omega}_n}) = \lim_n \sum_{k=1}^n M^P_\mu (XI_{\tilde{\Omega}_k \setminus \tilde{\Omega}_{k-1}}) = \lim_n \sum_{k=1}^n M^P_{\mu_k} (XI_{\tilde{\Omega}_k \setminus \tilde{\Omega}_{k-1}})$$

$$= \lim_n \sum_{k=1}^n M^P_{\nu_k} (XI_{\tilde{\Omega}_k \setminus \tilde{\Omega}_{k-1}}) = \lim_n M^P_\nu (XI_{\tilde{\Omega}_n}) = M^P_\nu (X),$$

i.e. (2.2) holds.

Remark. Since $\mu (\omega; \{0\} \times E) = 0$, then by (2.2) with

$$X (\omega, t, x) = I_\Omega (\omega) I_{\{0\}} (t) I_E (x)$$

we get

$$\nu (\omega; \{0\} \times E) = 0 \ \text{P-a.s.}$$

2. Definition 2. Let $X = X (\omega, t, x)$ be a nonnegative $\tilde{\mathcal{F}}$-measurable function, $\mu \in \tilde{\mathcal{V}}^+_{\mathcal{P}}$ and $X\mu \in \tilde{\mathcal{V}}^+_{\mathcal{P}}$. A nonnegative $\tilde{\mathcal{P}}$-measurable function $M^P_\mu (X \mid \tilde{\mathcal{P}})$ is called the *conditional mathematical expectation of* X with respect to the measure M^P_μ and the σ-algebra $\tilde{\mathcal{P}}$ if for any bounded nonnegative $\tilde{\mathcal{P}}$-measurable function $Z = Z (\omega, t, x)$

$$M^P_\mu (ZX) = M^P_\mu (ZM^P_\mu (X \mid \tilde{\mathcal{P}})). \tag{2.7}$$

Theorem 2. *Let the measures* M^P_μ, $M^P_{X\mu}$ *be* $\tilde{\mathcal{P}}$-σ-*finite. Then there exists the conditional mathematical expectation* $M^P_\mu (X \mid \tilde{\mathcal{P}})$ *that is unique* M^P_μ-a.s.

Proof. Denote by $M^P_\mu \mid \tilde{\mathcal{P}}$ and $M^P_{X\mu} \mid \tilde{\mathcal{P}}$ the restrictions on the σ-algebra $\tilde{\mathcal{P}}$ of the measures M^P_μ and $M^P_{X\mu}$ respectively. By the assumptions concerning μ and X_μ

$$M^P_{X\mu} \mid \tilde{\mathcal{P}} \ll M^P_\mu \mid \tilde{\mathcal{P}}.$$

Denote by $W = W (\omega, t, x)$ the respective nonnegative version of the Radon-Nikodym derivative

$$\frac{(dM^P_{X\mu} \mid \tilde{\mathcal{P}})}{(dM^P_\mu \mid \tilde{\mathcal{P}})}.$$

Then for any $\tilde{\mathcal{P}}$-measurable bounded nonnegative function $Z = Z (\omega, t, x)$ we have

$M^P_{X\mu}(Z) = M^P_\mu(ZW)$. On the other hand $M^P_{X\mu}(Z) = M^P_\mu(XZ)$. Therefore

$$M^P_\mu(ZX) = M^P_\mu(ZW).$$

Thus, as for the desired conditional mathematical expectation $M^P_\mu(X \mid \tilde{\mathcal{P}})$, we can take the function $W = W(\omega, t, x)$, that is the corresponding Radon-Nikodym derivative. The uniqueness (M^P_μ-a.s.) of the function $M^P_\mu(X \mid \tilde{\mathcal{P}})$ follows from (2.7) in the usual manner as the considered function $Z = Z(\omega, t, x)$ is arbitrary.

3. Let $p = (\omega; dt)$ be a random $\mathcal{P}$-σ-finite measure on R_+. Then, analogously to Theorem 1, we can establish the existence of a unique (P-a.s.) random predictable measure $q = q(\omega; dt)$ on R_+ such that

$$M^P_p(X) = M^P_q(X)$$

for each nonnegative and $\mathcal{P}$-measurable function $X = X(\omega, t)$. Further on, a random measure q is called the compensator of a random measure p.

Let $X = X(\omega, t)$ be a nonnegative and $\mathcal{F} \otimes B(R_+)$-measurable function, and let the random measures $p(\omega; dt)$ and $X(\omega, t)\, p(\omega; dt)$ be $\mathcal{P}$-σ-finite. Then, analogously to Theorem 2, there exists the conditional mathematical expectation $M^P_p(X \mid \mathcal{P})$ that is unique M^P_p-a.s.

Problems

1. Let μ be an optional random measure

$$\tilde{\Omega}_{mn} \in \tilde{\mathcal{P}},\ I_{\tilde{\Omega}_{mn}}\mu \in \mathcal{A}^+,\ m, n \geq 1, \text{ and } \tilde{\Omega}_{mn} \uparrow \tilde{\Omega}_n \uparrow \tilde{\Omega}.$$

Show that μ possesses the unique compensator , defined by formula (2.2).

2. Let $\mu \in \tilde{\mathcal{V}}^+_{\mathcal{P}}$ and let ν be the compensator of a measure μ. Show that if μ is a predictable random measure, then $\mu \overset{P}{=} \nu$.

3. Let $\mu \in \mathcal{V}^+_{\mathcal{P}}$, let ν be the compensator of a measure μ and let W be a $\tilde{\mathcal{P}}$-measurable nonnegative function such that $W\mu \in \mathcal{V}^+_{\mathcal{P}}$. Show that $W\nu$ is the compensator of a measure $W\mu$.

4. Let $\mu \in \tilde{\mathcal{V}}^{+}_{\mathcal{P}}$, let ν be the compensator of a measure μ and W a $\tilde{\mathcal{P}}$-measurable function such that the process $|W| * \nu \in \mathcal{A}^{+}_{loc}$. Show that the process $W * \mu - W * \nu$ is a local martingale.

5. Let $\mu \in \tilde{\mathcal{V}}^{+}_{\mathcal{P}}$, let ν be the compensator of a measure μ, let τ be a predictable Markov time and u a $\tilde{\mathcal{P}}$-measurable function such that

$$\mathbf{E} I(\tau < \infty) \int_E |u(\omega, \tau, x)| \, \nu(\omega; \{\tau\}, dx) < \infty.$$

Show that on the set $\{\tau < \infty\}$ the equality

$$\int_E u(\omega, \tau, x) \, \nu(\omega; \{\tau\}, dx) = \mathbf{E}\left(\int_E u(\omega, \tau, x) \, \mu(\omega; \{\tau\}, dx) \mid \mathcal{F}_{\tau-} \right) \tag{2.8}$$

takes place (P-a.s.) (Hint: apply Problem 4 with $W = uI_{[\![\tau]\!]}$.)

6. Show that assertion (2.8) of Problem 5 remains valid if instead of

$$\mathbf{E} I(\tau < \infty) \int_E |u(\omega, \tau, x)| \, \nu(\omega; \{\tau\}, dx) < \infty$$

the nonnegativity of a function $u = u(\omega, t, x)$ is required.

7. Let $\mu \in \tilde{\mathcal{V}}^{+}_{\mathcal{P}}$, and let X be a $\tilde{\mathcal{F}}$-measurable function such that $|X| \mu \in \tilde{\mathcal{V}}^{+}_{\mathcal{P}}$. Show that in this case the equality (2.7) defines $M^P_\mu(X \mid \tilde{\mathcal{P}})$ uniquely.

8. Let $\mu \in \tilde{\mathcal{V}}^{+}_{\mathcal{P}}$, and let X be a $\tilde{\mathcal{F}}$-measurable and Y a $\tilde{\mathcal{P}}$-measurable function respectively, such that $|XY| \mu \in \tilde{\mathcal{V}}^{+}_{\mathcal{P}}$ and $|X| \mu \in \tilde{\mathcal{V}}^{+}_{\mathcal{P}}$. Show that M^P_μ-a.s.

$$M^P_\mu(XY \mid \tilde{\mathcal{P}}) = Y M^P_\mu(X \mid \tilde{\mathcal{P}}).$$

9. Let $\mu \in \tilde{\mathcal{V}}^{+}_{\mathcal{P}}$, and let Y be a $\tilde{\mathcal{P}}$-measurable nonnegative function. Show that M^P_μ-a.s.

$$Y = M^P_\mu(Y \mid \tilde{\mathcal{P}}).$$

10. Let $\mu \in \tilde{\mathcal{V}}^+_{\mathcal{P}}$, let ν be the compensator of a measure μ and let X be a nonnegative $\tilde{\mathcal{F}}$-measurable function with $X\mu \in \tilde{\mathcal{V}}^+_{\mathcal{P}}$. Show that $M^P_\mu(X \mid \tilde{\mathcal{P}})\,\nu$ is the compensator of the measure $X\mu$.

11. Let $f = f(x)$ be a concave function, let $\mu \in \tilde{\mathcal{V}}^+_{\mathcal{P}}$ and let X be a $\tilde{\mathcal{F}}$-measurable function such that $|X|\mu \in \tilde{\mathcal{V}}^+_{\mathcal{P}}$ and $|f(X)|\mu \in \tilde{\mathcal{V}}^+_{\mathcal{P}}$. Show that for the conditional mathematical expectation $M^P_\mu(\cdot \mid \tilde{\mathcal{P}})$ Jensen's inequality

$$M^P_\mu(f(X) \mid \tilde{\mathcal{P}}) \geq f(M^P_\mu(X \mid \tilde{\mathcal{P}})) \quad (M^P_\mu\text{-a.s.})$$

takes place.

12. Let $\mu \in \tilde{\mathcal{V}}^+_{\mathcal{P}}$, and let the processes

$$X = X(\omega, t) \text{ and } X^* = X^*(\omega, t) = \sup_{s \leq t} |X(\omega, s)|$$

be $\mathcal{F} \otimes B(R_+)$-measurable functions. Show that the conditional mathematical expectation $M^P_\mu(X \mid \tilde{\mathcal{P}})$ is defined provided $\mathbf{E}X^*(\omega, t) < \infty$ for each $t \in R_+$.

§ 3. Integer-valued random measures

1. Definition. A random measure μ is called *integer-valued* if:

(1) $\mu(\omega; \{t\} \times E) \le 1$ for every $\omega \in \Omega$ and $t \in R_+$;

(2) for each $\Gamma \in \mathcal{B}(R_+) \otimes \mathcal{E}$ the random variable $\mu(\cdot\,; \Gamma)$ takes its values in the set $\{0, 1, ..., \infty\}$;

(3) μ is optional;

(4) μ is $\tilde{\mathcal{P}}$-σ-finite ($\mu \in \tilde{\mathcal{V}}^+_{\mathcal{P}}$).

Denote

$$D = \{(\omega, t): \mu(\omega; \{t\} \times E) = 1\}.$$

Let $(\omega, t) \in D$. By Condition (2) there exists a unique point $\beta_t(\omega) \in E$ such that

$$\mu(\omega; \{t\}, dx) = \varepsilon_{\beta_t(\omega)}(dx), \tag{3.1}$$

where $\varepsilon_a(dx)$ is the Dirac measure at point a, i.e. $\varepsilon_a(\Gamma) = I_\Gamma(a)$. As $(\omega, t) \notin D$, set $\beta_t(\omega) = \delta$, where δ is a certain point outside the set E.

This definition of the process $\beta = (\beta_t(\omega))_{t \ge 0}$ allows us to express the integer-valued random measure μ symbolically in the following manner

$$\mu(\omega; dt, dx) = \sum_{s > 0} I_D(\omega, s)\, \varepsilon_{(s, \beta_s(\omega))}(dt, dx), \tag{3.2}$$

where $\varepsilon_{(s, \beta)}(dt, dx)$ is the Dirac measure.

By Condition (3) the set $D \in \mathcal{D}$ and by (4) the section $D_\omega = \{t: (\omega, t) \in D\}$ is at most countable for every $\omega \in \Omega$. Therefore, according to [81], Ch.4, Theorem 33, D coincides with the set $\bigcup_{k \ge 1} [\![T_k]\!]$ up to an P-negligible set where $(T_k)_{k \ge 1}$ are Markov times with disjoint graphs, besides $\mu(\omega; \{T_k\} \otimes E) = 1$ (P-a.s.), $k \ge 1$.

Consequently, the relation (3.2) can be interpreted in the following sense

$$\mu(\omega; dt, dx) = \sum_{k \ge 1} I(T_k < \infty)\, \varepsilon_{(T_k, \beta_{T_k})}(dt, dx). \tag{3.3}$$

Let us show that for integer-valued measures the process $\beta = (\beta_t(\omega))_{t \ge 0}$ is optional and for each Markov time T the variable $\beta_T I(T < \infty)$ is $\mathcal{F}_T$-measurable.

Since $\{(\omega, t): \beta_t(\omega) = \delta\} = (\Omega \times R_+) \setminus D \in \mathcal{D}$, it suffices to show that for each $C \in \mathcal{E}$

$$\{(\omega, t): \beta_t(\omega) \in C\} \in \mathcal{D}. \tag{3.4}$$

First, let $\mu \in \tilde{\mathcal{A}}^+$. For $C \in \mathcal{E}$ define an increasing process $A = (A_t)_{t \geq 0}$ with $A_t = I_C * \mu_t$. This process is optional, and hence $\{(\omega, t): \Delta A_t = 1\} \in \mathcal{O}$. Then the desired relation (3.4) takes place because

$$\{(\omega, t): \beta_t(\omega) \in C\} = \{(\omega, t): \Delta A_t = 1\} \in \mathcal{O}.$$

Now, let $\mu \in \tilde{\mathcal{V}}^+$. Then there exists a set $\tilde{\Omega}_n \in \tilde{\mathcal{O}}$, $n \geq 1$, such that $I_{\tilde{\Omega}_n}\mu \in \tilde{\mathcal{A}}^+$, $n \geq 1$, and $\tilde{\Omega}_n \uparrow \tilde{\Omega}$. Define the increasing processes $A^n = (A^n_t)_{t \geq 0}$ with $A^t_n = I_C I_{\tilde{\Omega}_n} * \mu_t$.

Then

$$\{(\omega, t): \beta_t(\omega) \in C\} = \bigcup_{n \geq 1} \{(\omega, t): \Delta A^n_t = 1\} \in \mathcal{O}.$$

Next, the process $I_{[\![T]\!]} I_C I_{\tilde{\Omega}_n} * \mu$ is adapted and right continuous. Therefore (Problem 1.1.1) this process is progressively measurable and (Problem 1.2.2) the set $\{I_{[\![T]\!]} I_C I_{\tilde{\Omega}_n} * \mu_T > 0\} \in \mathcal{F}_T$, $n \geq 1$. Consequently,

$$\{\omega: \beta_T(\omega) \in C, T < \infty\} = \bigcup_{n \geq 1} \{I_{[\![T]\!]} I_C I_{\Omega_n} * \mu_T > 0, T < \infty\} \in \mathcal{F}_T.$$

Remark. Observe that in the course of proving the optionality of the process β the assumption $\mu \in \tilde{\mathcal{V}}^+$ is utilized (and not $\mu \in \tilde{\mathcal{V}}^+_{\mathcal{P}}$).

2. Example. Let $X = (X_t)_{t \geq 0}$ be an optional process with trajectories in D and $\Delta X_t = X_t - X_{t-}$. Since trajectories X have an at most countable number of jumps over each finite interval $[0, t]$, one can define the integer-valued random measure μ, called the measure of jumps of a process X, by setting

$$\mu(\omega; dt, dx) = \sum_{s > 0} I(\Delta X_s(\omega) \neq 0)\, \varepsilon_{(s, \Delta X_s(\omega))}(dt, dx). \tag{3.5}$$

In this case $\beta_t(\omega) = \Delta X_t(\omega)$ and $E = R \setminus \{0\}$, $\delta = 0$.

Let us show that there exists the compensator of the measure of jumps μ of a process X. To prove this, it suffices to apply the result of Problem 2.1. To this end, set

$$\tilde{\Omega}_n = \Omega \times R_+ \times \left\{x: |x| > \frac{1}{n}\right\},$$

$$\tilde{\Omega}_{mn} = \left\{(\omega, t, x): \omega \in \Omega, t \in [0, T_{mn}], x \in \left\{x: |x| > \frac{1}{n}\right\}\right\}$$

with

$$T_{mn} = \inf\left(t: \mu\left(\omega; [0, t] \times \left\{ |x| > \frac{1}{n} \right\}\right) \geq m\right).$$

Then

$$\tilde{\Omega}_{mn} \in \tilde{\mathcal{P}},\ I_{\tilde{\Omega}_{mn}} * \mu_\infty \leq m \text{ and } \tilde{\Omega}_{mn} \uparrow \tilde{\Omega}_n \uparrow \tilde{\Omega}.$$

3. According to Theorem 2.1 there exists the compensator ν of an integer-valued random measure $\mu \in \tilde{\mathcal{V}}^+_{\mathcal{P}}$. Denote

$$a_t(\omega) = \nu(\omega; \{t\} \times E). \tag{3.6}$$

Further on the following property of the compensator ν plays an important rôle.

Lemma 1. *There exists a version of the compensator of an integer-valued random measure* $\mu \in \tilde{\mathcal{V}}^+_{\mathcal{P}}$ *such that for each* $(\omega, t) \in \Omega \times R_+$

$$a_t(\omega) \leq 1. \tag{3.7}$$

Proof. Let ν' be a certain version of the compensator of μ with $a'_t(\omega) = \nu'(\omega; \{t\} \times E)$. Set $\nu = I(a' \leq 1)\nu'$. Clearly, ν is a predictable random measure with $a = I(a' \leq 1)\, a' \leq 1$.

We will show that ν is the compensator of a measure μ. According to (2.2) it suffices for this to show that

$$I(a' > 1) * \nu'_\infty = 0 \quad (\text{P-a.s.}). \tag{3.8}$$

First, suppose $\mu \in \tilde{\mathcal{A}}_+$. Then by (2.2)

$$\mathbf{E} I(a' > 1) * \nu'_\infty = \mathbf{E} I(a' > 1) * \mu_\infty < \infty. \tag{3.9}$$

Denote

$$\alpha \equiv I(a' > 1) * \nu'_\infty - I(a' > 1) * \mu_\infty = \sum_{s>0} I(a'_s > 1)(a'_s - \mu(\omega; \{s\} \times E))$$

and observe that $\alpha \geq 0$. On the other hand $\mathbf{E}\alpha = 0$ by (3.9). Consequently, $\alpha = 0$ (P-a.s.), and this is admissible if only

$$\sum_{s>0} I(a'_s > 1) = 0 \quad (\text{P-a.s.}).$$

This yields (3.8) in an obvious manner.

Now, let $\mu \in \tilde{\mathcal{V}}^+_{\mathcal{P}}$. Then, it is established analogously that (P-a.s.)

$$I(a' > 1)\, I_{\Omega_n} * \nu_\infty = 0, \quad n \geq 1$$

with $\tilde{\Omega}_n \in \tilde{\mathcal{P}}$, $I_{\tilde{\Omega}_n} \mu \in \tilde{\mathcal{A}}^+$ and $\tilde{\Omega}_n \uparrow \tilde{\Omega}$.

Consequently, (P-a.s.)

$$I(a' > 1) * \nu'_\infty = I(a' > 1) I_{\tilde{\Omega}_1} * \nu'_\infty + \sum_{n \geq 1} I(a' > 1) I_{\tilde{\Omega}_{n+1} \setminus \tilde{\Omega}_n} * \nu'_\infty = 0.$$

4. Let μ be an integer-valued random measure ($\mu \in \tilde{\mathcal{V}}^+_{\mathcal{P}}$), and ν its compensator with $a \leq 1$.

Define on $(R_+, B(R_0))$ an integer-valued random measure $p = p(\omega; dt)$, by setting

$$p(\omega; \Gamma) = \sum_{s \in \Gamma} I(a_s(\omega) > 0)(1 - \mu(\omega; \{s\} \times E)) \tag{3.10}$$

for $\Gamma \in B(R_+)$. To the random measure p relate Doléans measure (cf. (1.5) § 1)

$$M^P_p(d\omega, dt) = P(d\omega)\, p(\omega; dt).$$

It is not difficult to see that the measure M^P_p is $\mathcal{P}$-σ-finite, and analogously to Theorem 2.1 it can be shown that there exists a compensator $q = q(\omega; dt)$ of p, i.e. a unique (P-a.s.) predictable random measure on $(R_+, B(R_+))$, defined by the formula ($\Gamma \in B(R_+)$)

$$q(\omega; \Gamma) = \sum_{s \in \Gamma} I(a_s(\omega) > 0)(1 - a_s(\omega)), \tag{3.11}$$

such that

$$M^P_p(Y) = M^P_q(Y) \tag{3.12}$$

where Y is a nonnegative $\mathcal{P}$-measurable function and $M^P_q(d\omega, dt) = P(d\omega)\, q(\omega; dt)$.

For nonnegative $\mathcal{F} \otimes B(R_+)$ measurable functions $X = X(\omega, t)$, the conditional mathematical expectation $M^P_p(X \mid \mathcal{P})$ is defined by the relation (in case in which M^P_{Xp} is a $\mathcal{P}$-σ-finite measure)

$$M^P_p(Z M^P_p(X \mid \mathcal{P})) = M^P_p(ZX) \tag{3.13}$$

for each bounded nonnegative $\mathcal{P}$-measurable function Z. In case in which $M^P_{|X|p}$ is a $\mathcal{P}$-σ-finite measure, the conditional mathematical expectation $M^P_p(X \mid \mathcal{P})$ is defined for $\mathcal{F} \in B(R_+)$-measurable functions X (cf. Problem 2.7).

Now we give a result, which establishes the relation between $M_p^P(X \mid \mathcal{P})$ and $M_\mu^P(X \mid \tilde{\mathcal{P}})$ in case $X = \Delta M$ and $M \in \mathcal{M}_{loc}$. We will use the following notation: if

$$\int_E | u(\omega, t, x) | \, \nu(\omega; \{t\}, dx) < \infty,$$

then set

$$\hat{u}(\omega, t) = \int_E u(\omega, t, x) \, \nu(\omega; \{t\}, dx).$$

Theorem 1. *Let* μ *be an integer-valued random measure* ($\mu \in \tilde{\mathcal{V}}^+_{\mathcal{P}}$), *and* ν *its compensator* ($a_t = \nu(\{t\} \times E)$, $a \leq 1$). *For* $M \in \mathcal{M}_{loc}$ *the conditional mathematical expectations* $M_p^P(\Delta M \mid \mathcal{P})$ *and* $M_\mu^P(\Delta M \mid \tilde{\mathcal{P}})$ *are defined. We have*

$$M_p^P(\Delta M \mid \mathcal{P}) = -\frac{I(0 < a < 1)}{1 - a} \widehat{M_\mu^P(\Delta M \mid \tilde{\mathcal{P}})} \quad (M_p^P\text{-a.s.})$$

where $\frac{0}{0} = 0$, *and*

$$M_p^P(\Delta M \mid \mathcal{P}) = -\frac{I(0 < a < 1)}{1 - a} \widehat{M_\mu^P(\Delta M \mid \tilde{\mathcal{P}})} \quad (\{0 < a < 1\}; \ M_\mu^P\text{-a.s.}).$$

Proof. It suffices to show for the existence of $M_p^P(\Delta M \mid \mathcal{P})$ and $M_\mu^P(\Delta M \mid \tilde{\mathcal{P}})$ that $M^P_{|\Delta M| p}$ and $M^P_{|\Delta M| \mu}$ are $\mathcal{P}$- and $\tilde{\mathcal{P}}$-σ-finite measures. To this end define an increasing process $A = (A_t)_{t \geq 0}$ with

$$A_t = \sum_{0 < s \leq t} | \Delta M_s | \, I(| \Delta M_s | > 1). \tag{3.14}$$

Then

$$| \Delta M | \, p \leq p + (\Delta A) \, p, \quad | \Delta M | \, \mu \leq \mu + (\Delta A) \, \mu,$$

and consequently it suffices to show that $M^P_{(\Delta A) p}$ and $M^P_{(\Delta A) \mu}$ are $\mathcal{P}$- and $\tilde{\mathcal{P}}$-σ-finite measures respectively.

Assume $M \in \mathcal{M}$, passing to localizing sequences if necessary. It has been established in the course of proving Theorem 1.7.1 that $A \in \mathcal{A}^+_{loc}$. If $(T_n)_{n \geq 1}$ is a localizing sequence for A ($EA_{T_n} < \infty$, $n \geq 1$), then $[\![0, T_n]\!] \in \mathcal{P}$, $n \geq 1$, and

$$\mathbf{E}(\Delta A) * p_{T_n} \le \mathbf{E}A_{T_n} < \infty, \quad \mathbf{E}(\Delta A) * \mu_{T_n} \le \mathbf{E}A_{T_n} < \infty, \quad n \ge 1.$$

Let us show that there exists the function

$$\widehat{M_\mu^P(\Delta M \mid \tilde{\mathcal{P}})}(t) = \int_E M_\mu^P(\Delta M \mid \tilde{\mathcal{P}})(t, x)\, \nu(\{t\}, dx)$$

which appears on the right hand side of the equation, defining $M_p^P(\Delta M \mid \mathcal{P})$. It suffices to show that for each Markov time T

$$I(T < \infty) \int_E | M_\mu^P(\Delta M \mid \tilde{\mathcal{P}})(T, x) | \, \nu(\{T\}, dx) < \infty \quad (P\text{-a.s.}). \tag{3.15}$$

Since $|\Delta M| \le 1 + \Delta A$ (cf. (3.14)), by Jensen's inequality (Problem 2.11) we have

$$| M_\mu^P(\Delta M \mid \tilde{\mathcal{P}}) | \le M_\mu^P(|\Delta M| \mid \tilde{\mathcal{P}}) \le 1 + M_\mu^P(\Delta A \mid \tilde{\mathcal{P}}),$$

and hence (3.15) holds provided

$$I(T < \infty)\, M_\mu^P(\Delta A \mid \tilde{\mathcal{P}}) * \nu_T < \infty \quad (P\text{-a.s.}). \tag{3.16}$$

Let $(T_n)_{n \ge 1}$ be the localizing sequence for A introduced above. Then

$$\mathbf{E}M_\mu^P(\Delta A \mid \tilde{\mathcal{P}}) * \nu_{T \wedge T_n} = \mathbf{E}M^P(\Delta A \mid \tilde{\mathcal{P}}) * \mu_{T \wedge T_n}$$

$$= \mathbf{E}(\Delta A) * \mu_{T \wedge T_n} \le \mathbf{E}A_{T_n} < \infty, \quad n \ge 1,$$

i.e.

$$M_\mu^P(\Delta A \mid \tilde{\mathcal{P}}) * \nu_{T \wedge T_n} < \infty, \quad (P\text{-a.s.}), \ n \ge 1.$$

Consequently, (3.16) holds.

Let us now establish the first representation for $M_p^P(\Delta M \mid \mathcal{P})$. If the desired representation takes place for $M \in \mathcal{H}$ (i.e. $M \in \mathcal{M}$ and $\mathbf{E}M_\infty^* < \infty$), then it takes place in the general case too. In fact, by Theorem 1.7.7 the process $M^* \in \mathcal{A}_{loc}^+$. If $(\tau_n)_{n \ge 1}$ is a localizing sequence for M*, then $M^{\tau_n} \in \mathcal{H}$, $n \ge 1$, and hence by the assumption stipulated above

$$M_p^P(\Delta M^{\tau_n} \mid \mathcal{P}) = \frac{I(0 < a < 1)}{1 - a} \widehat{M_\mu^P(\Delta M^{\tau_n} \mid \tilde{\mathcal{P}})}.$$

Since $\Delta M^{\tau_n} = I_{[\![0, \tau_n]\!]} \Delta M$ and $[\![0, \tau_n]\!] \in \mathcal{P}$, we get

$$I_{[\![0, \tau_n]\!]} M_p^P(\Delta M \mid \mathcal{P}) = \frac{I(0 < a < 1)}{1 - a} I_{[\![0, \tau_n]\!]} \widehat{M_\mu^P(\Delta M \mid \tilde{\mathcal{P}})}.$$

Thus, let $M \in \mathcal{H}$ and let Z be an $\mathcal{P}$-measurable bounded function such that

$$M_p^P(|Z\Delta M|) < \infty, \quad M_\mu^P(|Z\Delta M|) < \infty, \quad \mathbf{E}\sum_{s>0} |Z_s| I(0 < a_s < 1) < \infty.$$

In view of Problem 1 we have

$$\sum_s |Z_s\Delta M_s| I(0 < a_s < 1)(1 - \mu(\{s\}\times E)) \in \mathcal{A}^+,$$

$$\sum_s |Z_s\Delta M_s| \mu(\{s\}\times E) \in \mathcal{A}^+,$$

and hence

$$\sum_s Z_s\Delta M_s I(0 < a_s < 1) \in \mathcal{A}.$$

For each Markov time τ we get by Theorem 1.6.3 that

$$\mathbf{E}\sum_{0<s\le\tau} Z_s\Delta M_s I(0 < a_s < 1) = \mathbf{E}\sum_{0<s\le\tau} Z_s\,{}^P(\Delta M_s) I(0 < a_s < 1) = 0,$$

where we have used the fact that by Theorem 1.7.8 the predictable projection ${}^P(\Delta M)$ of jumps of a local martingale M is negligible. Consequently, the process

$$\sum_s Z_s\Delta M_s I(0 < a_s < 1)$$

is a uniformly integrable martingale and

$$\mathbf{E}\sum_{s>0} Z_s\Delta M_s I(0 < a_s < 1) = 0.$$

In view of this equation and Problem 1

$$\begin{aligned}
&\mathbf{E}\sum_{s>0} Z_s\Delta M_s I(a_s > 0)(1 - \mu(\{s\}\times E)) \\
&= \mathbf{E}\sum_{s>0} Z_s\Delta M_s I(0 < a_s < 1)(1 - \mu(\{s\}\times E)) \\
&= -\mathbf{E}\sum_{s>0} Z_s\Delta M_s I(0 < a_s < 1)\,\mu(\{s\}\times E)
\end{aligned}$$

$$= - EZ\Delta MI(0 < a < 1) * \mu_\infty$$

$$= - EZM^P_\mu(\Delta M \mid \tilde{\mathcal{P}})\, I(0 < a < 1) * \mu_\infty$$

$$= - EZM^P_\mu(\Delta M \mid \tilde{\mathcal{P}})\, I(0 < a < 1) * \nu_\infty$$

$$= - E\sum_{s>0} Z_s I(0 < a_s < 1) \int_E M^P_\mu(\Delta M \mid \tilde{\mathcal{P}})(s, x)\, \nu(\{s\}, dx)$$

$$= - E\sum_s Z_s I(0 < a_s < 1) \frac{\widehat{M^P_\mu(\Delta M \mid \tilde{\mathcal{P}})}(s)}{1 - a_s}(1 - a_s).$$

On the other hand,

$$E\sum_{s>0} Z_s \Delta M_s I(0 < a_s < 1)(1 - \mu(\{s\} \times E))$$

$$= E\sum_{s>0} Z_s M^P_p(\Delta M \mid \mathcal{P})(s)\, I(0 < a_s < 1)(1 - a_s),$$

which gives the first representation for $M^P_p(\Delta M \mid \mathcal{P})$ as Z is arbitrary.

The second equality will be deduced from the first one. To this end, observe that

$$EI(0 < a)(1 - a) \left| M^P_p(\Delta M \mid \mathcal{P}) + \frac{I(0 < a < 1)}{1 - a} M^P_\mu(\Delta M \mid \tilde{\mathcal{P}}) \right| * \mu_\infty$$

$$= EI(0 < a)(1 - a) \left| M^P_p(\Delta M \mid \mathcal{P}) + \frac{I(0 < a < 1)}{1 - a} \widehat{M^P_\mu(\Delta M \mid \tilde{\mathcal{P}})} \right| * \nu_\infty$$

$$= E\sum_{s>0} I(0 < a_s)(1 - a_s) \left| M^P_p(\Delta M \mid \mathcal{P})(s) \right.$$

$$\left. + \frac{I(0 < a_s < 1)}{1 - a_s} M^P_\mu(\Delta M \mid \tilde{\mathcal{P}})(s) \right| a_s$$

$$\le E \left| M^P_p(\Delta M \mid \mathcal{P}) + \frac{I(0 < a < 1)}{1 - a} \widehat{M^P_\mu(\Delta M \mid \tilde{\mathcal{P}})} \right| * p_\infty = 0.$$

Hence, M^P_μ-a.s.

$$I(0 < a)(1 - a)\left[M^P_p(\Delta M \mid \mathcal{P}) + \frac{I(0 < a < 1)}{1 - a} \widehat{M^P_\mu(\Delta M \mid \tilde{\mathcal{P}})} \right] = 0.$$

Consequently, $(\{0 < a < 1\};\ M^P_{\mu\text{-a.s.}})$,

$$M_p^P(\Delta M \mid \mathcal{P}) = -\frac{I(0<a<1)}{1-a} \widehat{M_\mu^P(\Delta M \mid \tilde{\mathcal{P}})}.$$

Problems

1. Let an integer-valued random measure $\mu \in \tilde{\mathcal{V}}^+_{\mathcal{P}}$, let ν be its compensator with $a \leq 1$ and

$$\alpha_t \equiv I(a_t = 1)(1 - \mu(\{t\}\times E)) = I(a_t = 1)(\nu(\{t\}\times E) - \mu(\{t\}\times E)).$$

Show that the process $\alpha = (\alpha_t)_{t\geq 0}$ is negligible.

2. Let an integer-valued random measure $\mu \in \tilde{\mathcal{V}}^+_{\mathcal{P}}$, let ν be its compensator with $a \leq 1$ and $M \in \mathcal{M}_{loc}$. Show that

$$I(a=1)\widehat{M_\mu^P(\Delta M \mid \tilde{\mathcal{P}})} = 0 \quad (M_p^P\text{-a.s.}).$$

§ 4. Multivariate point processes

1. Definition. By a *multivariate point process* on $(R_+ \times E, B(R_+) \otimes \mathcal{E})$ is usually understood the sequence

$$(T_n, X_n)_{n \geq 1}, \tag{4.1}$$

where T_n are Markov times such that $T_1 > 0$,

$$T_n < T_{n+1} \text{ on } \{T_n < \infty\} \text{ and } T_{n+1} = T_n \text{ on } \{T_n = \infty\}, \tag{4.2}$$

while X_n are random elements possessing the following property

$$X_n \in E \text{ on } \{T_n < \infty\} \text{ and } X_n = \delta \text{ on } \{T_n = \infty\}, \tag{4.3}$$

where δ is a certain "fictive" point outside E ($\delta \notin E$), and $\{X_n \in C\} \in \mathcal{F}_{T_n}$, $\forall C \in \mathcal{E}$, $n \geq 1$.

A multivariate point process is completely described by the integer-valued random measure

$$\mu(\omega; dt, dx) = \sum_{n \geq 1} I(T_n < \infty)\, \varepsilon_{(T_n, X_n)}(dt, dx). \tag{4.4}$$

Lemma 1. *The measure* μ *possesses the following properties*:

(1) $\mu(\omega; \{0\} \times E) = 0$;

(2) $\mu(\omega; [\![T_\infty, \infty[\![\times E) = 0$ *with* $T_\infty = \lim_n T_n$;

(3) $\mu \in \tilde{\mathcal{V}}^+_{\mathcal{P}}$.

Proof. The first two properties follow from the definition of the measure μ and from the fact that $T_1 > 0$. To prove property (3) note that T_∞ is a predictable Markov time and $[\![0, T_n]\!] \cup [\![T_\infty, \infty]\!] \in \mathcal{P}$. Therefore

$$\tilde{\Omega}_n = \{(\omega, t, x): \omega \in \Omega, t \in [\![0, T_n]\!] \cup [\![T_n, \infty[\![, x \in E\} \in \tilde{\mathcal{P}},\ n \geq 1,$$

and $\tilde{\Omega}_n \uparrow \tilde{\Omega}$. Besides, by property (2)

$$I_{\tilde{\Omega}_n} * \mu_\infty = I_{[\![0, T_n]\!] \cup [\![T_\infty, \infty[\![} * \mu_\infty = \sum_{k=1}^{n} I(T_k < \infty) \leq n,$$

and hence, $EI_{\tilde{\Omega}_n} * \mu_\infty \leq n$, $n \geq 1$, i.e. $\mu \in \tilde{\mathcal{V}}^+_{\mathcal{P}}$.

2. According to Theorem 2.1 every random measure $\mu \in \tilde{\mathcal{V}}^+_{\mathcal{P}}$ has a compensator. Consequently, the multivariate point process in question has a compensator ν too.

Lemma 2. *The random measure* μ, *related to a multivariate point process, possesses a version of the compensator* ν *such that*

1) $\nu(\omega; \{0\} \times E) = 0$,

2) $\nu(\omega; \{t\} \times E) \le 1$,

3) $\nu(\omega; [\![T_\infty, \infty[\![\times E) = 0$.

Proof. Let ν' be a certain version of the compensator of μ. It can be assumed by Lemma 3.1 that $\nu'(\omega; \{t\} \times E) \le 1$. Set

$$\nu = I_{]\!]0, T_\infty[\![}\,\nu'.$$

The random measure ν possesses the properties 1) — 3). It is predictable because T_∞ is a predictable Markov time and $[\![0, T_\infty[\![\in \mathcal{P}$. By the remark to Theorem 2.1 we have $\nu'(\omega; \{0\} \times E) = 0$ (P-a.s.), and according to assertion (2) of Lemma 1 the measure $\nu'(\omega; [\![T_\infty, \infty[\![\times E) = 0$ (P-a.s.). Therefore ν is the compensator of a measure μ with the desired properties 1) — 3).

3. Consider the "natural" flow of σ-algebras $\mathcal{G} = (\mathcal{G}_t)_{t \ge 0}$, related to a measure μ, with

$$\mathcal{G}_t = \sigma\{\mu(\omega; (0, s] \times B),\ s \le t,\ B \in \mathcal{E}\}. \tag{4.5}$$

For the sake of completeness let us mention (without the proof) a number of results concerning the case in which

$$\mathbb{F} = (\mathcal{F}_0 \vee \mathcal{G}_t)_{t \ge 0}, \tag{4.6}$$

and $\mathbb{F}$ satisfies the usual conditions.

Denote by $G_n(\omega; dt, dx)$ the regular conditional distribution of (T_{t+1}, X_{n+1}) given the σ-algebra $\mathcal{F}_{T_n}$, and let

$$H_n(\omega; dt) = G_n(\omega; dt, E \cup \{\delta\}).$$

Proposition 1 ([100]). *Under Condition* (4.6) *the compensator* ν *of an integer-valued random measure* μ, *defined by* (4.4), *is given by the formula* ($T_0 = 0$)

$$\nu(\omega; dt, dx) = \sum_{n \ge 0} I(T_n < t \le T_{n+1}) \frac{G_n(\omega; dt, dx)}{H_n(\omega; [t, \infty))} \tag{4.7}$$

and it satisfies the properties 1) — 3) of Lemma 2.

Suppose two measures P and $\tilde{P}$ are defined on the measurable space $(\Omega, \mathcal{F})$. Denote by P_0 and $\tilde{P}_0$ their restrictions to the σ-algebra $\mathcal{F}_0$, and let ν and $\tilde{\nu}$ be the compensators of a measure μ with respect to P and $\tilde{\mathsf{P}}$ respectively.

Proposition 2 ([100]). *Under Condition* (4.6)

$$\mathsf{P}_0 = \tilde{\mathsf{P}}_0, \quad \nu \overset{\mathsf{P} \text{ or } \tilde{\mathsf{P}}}{=} \tilde{\nu} \Rightarrow \mathsf{P} = \tilde{\mathsf{P}}. \tag{4.8}$$

This proposition admits the following generalization.

Proposition 3 ([100]). *Let σ be a Markov time and let P_σ and $\tilde{\mathsf{P}}_\sigma$ be the restrictions of measures P and $\tilde{\mathsf{P}}$ to the σ-algebra $\mathcal{F}_\sigma$.*

Under Condition (4.6)

$$\mathsf{P}_0 = \tilde{\mathsf{P}}_0, \quad I_{[\![0,\sigma]\!]}\nu \overset{\mathsf{P} \text{ or } \tilde{\mathsf{P}}}{=} I_{[\![0,\sigma]\!]}\tilde{\nu} \Rightarrow \mathsf{P}_\sigma = \tilde{\mathsf{P}}_\sigma. \tag{4.9}$$

4. Let a set E consist of a single point, say $E = \{1\}$. In this case a multivariate point process $(T_n, X_n)_{n \geq 1}$ with $X_n = 1$ on $(T_n < \infty)$ and $X_n = 0$ on $\{T_n = \infty\}$ is simply called a *point process*.

In this case set

$$N_t = \mu(\omega; [0, t] \times \{1\}), \quad t \geq 0.$$

The stochastic process $N = (N_t)_{t \geq 0}$ is usually called a *counting process*. (The process N is often called a *point process* too.) The trajectories of this process are stepwise functions starting at zero and having jumps of size + 1 at instants T_n, $n \geq 1$.

The compensator ν of the considered measure μ determines a predictable increasing process $A = (A_t)_{t \geq 0}$ with $A_t = \nu(\omega; [0, t] \times \{1\})$, that is the compensator of the increasing process N (in the sense defined in Ch. 1, § 6).

Problems

1. Let $N = (N_t)_{t \geq 0}$ be a counting process having the compensator $A = (A_t)_{t \geq 0}$ with $A_t \equiv t$. Show that the random variable N_1 has the Poisson distribution with the parameter 1.

2. Let $N = (N_t)_{t > 0}$ be a counting process with the deterministic compensator $A = (A_t)_{t \geq 0}$ (independent of ω). Show that N is a process with independent increments.

3. Let τ be a predictable Markov time, N a counting process, A its compensator. Show that

$$\Delta A_\tau = \mathsf{P}(\Delta N_\tau = 1 \mid \mathcal{F}_{\tau-}) \quad (\{\tau < \infty\}; \mathsf{P}\text{-a.s.}).$$

4. Show that for a $\mathcal{P}$-measurable function f such that $|f| \circ A \in \mathcal{A}^+_{loc}$ where A is the compensator of a counting process N, the process $f \circ N - f \circ A$ is a local martingale.

5. Show that the process $N - A$ (N is a counting process, A its compensator) is a locally square integrable martingale with the quadratic characteristics

$$\langle N - A\rangle = (1 - \Delta A) \circ A.$$

6. Show that if $N = (N_t)_{t\geq 0}$ is a point process and $A = (A_t)_{t\geq 0}$ its compensator, then for each Markov time τ we have $\mathsf{P}(N_\tau < \infty) = 1 \Leftrightarrow \mathsf{P}(A_\tau < \infty) = 1$.

§ 5. Stochastic integral with respect to a martingale measure $\mu - \nu$

1. Let μ be an integer-valued random measure, $\mu \in \tilde{\mathcal{V}}^+_{\mathcal{P}}$, ν its compensator and $U = U(\omega, t, x)$ a $\tilde{\mathcal{P}}$-measurable function such that $|U| * \nu \in \mathcal{A}^+_{loc}$. Then by (2.2) we have $|U| * \mu \in \mathcal{A}^+_{loc}$. Therefore, the integrals

$$U * (\mu - \nu)_t \equiv U * \mu_t - U * \nu_t, \quad t \geq 0, \tag{5.1}$$

are defined, and according to Problem 2.4 the process $U * (\mu - \nu) = (U * (\mu - \nu)_t)_{t \geq 0}$ is a local martingale. This fact allows one to call $\mu - \nu$ a *martingale measure.*

In this section we intend to give the definition for a large supply of functions U of the stochastic integral $U * (\mu - \nu)$ with respect to a martingale measure $\mu - \nu$, that is a process of the class $\mathcal{M}^d_{loc}$.

2. Let μ be an integer-valued random measure and ν its compensator with $a_t(\omega) = \nu(\omega; \{t\} \times E) \leq 1$. Besides, let p be an integer-valued random measure on $(R_+, B(R_+))$ with

$$p(\omega; \Gamma) = \sum_{s \in \Gamma} I(a_s(\omega) > 0)(1 - \mu(\omega; \{s\} \times E)), \quad \Gamma \in B(R_+),$$

which has the compensator q with

$$q(\omega; \Gamma) = \sum_{s \in \Gamma} I(a_s(\omega) > 0)(1 - a_s(\omega))$$

(cf. Subsection 3.4).

Let $U = U(\omega, t, x)$ be a $\tilde{\mathcal{P}}$-measurable function such that for each Markov time T

$$I(T < \infty) \int_E |U(\omega, T, x)| \nu(\omega; \{T\}, dx) < \infty \quad (\text{P-a.s.}). \tag{5.2}$$

In accordance with § 3, for such a function U we define

$$\hat{U}(\omega, t) = \int_E U(\omega, t, x) \nu(\omega; \{t\}, dx). \tag{5.3}$$

Define the increasing processes $G(U)$ and $G^i(U)$, $i = 1, 2$, by setting

$$G(U) = \frac{(U - \hat{U})^2}{1 + |U - \hat{U}|} * \nu + \frac{\hat{U}^2}{1 + |\hat{U}|} * q, \tag{5.4}$$

$$G^i(U) = |U - \hat{U}|^i * \nu + |\hat{U}|^i * q. \tag{5.5}$$

We say that a $\tilde{\mathcal{P}}$-measurable function $U \in \mathcal{G}_{loc}$ if $G(U) \in \overset{+}{\mathcal{A}}_{loc}$, and $U \in \overset{i}{\mathcal{G}}_{loc}$ if $G^i(U) \in \overset{+}{\mathcal{A}}_{loc}$, $i = 1, 2$.

Evidently,

$$\overset{i}{\mathcal{G}}_{loc} \subseteq \mathcal{G}_{loc}, \quad i = 1, 2. \tag{5.6}$$

If $U \in \overset{i}{\mathcal{G}}_{loc}$ and c is a constant, then $cU \in \overset{i}{\mathcal{G}}_{loc}$; if $U_1, U_2 \in \overset{i}{\mathcal{G}}_{loc}$, then $U_1 + U_2 \in \overset{i}{\mathcal{G}}_{loc}$.

Lemma 1. 1) *If* $U \in \mathcal{G}_{loc}$ *and* c *is a constant, then* $cU \in \mathcal{G}_{loc}$.

2) *If* $U_1, U_2 \in \mathcal{G}_{loc}$, *then* $U_1 + U_2 \in \mathcal{G}_{loc}$.

3) *If* $U \in \mathcal{G}_{loc}$, *then functions* $U_1 \in \overset{1}{\mathcal{G}}_{loc}$, *and* $U_2 \in \overset{2}{\mathcal{G}}_{loc}$ *can be found such that* $U = U_1 + U_2$.

4) *If* $U \in \overset{2}{\mathcal{G}}_{loc}$, *then*

$$G_t^2(U) = U^2 * \nu_t - \sum_{0 < s \le t} \hat{U}^2(s). \tag{5.7}$$

Proof. 1) Denote

$$f(x) = \frac{x^2}{1 + |x|}.$$

Then $f(cx) \le (|c| \vee c^2) f(x)$ and the desired assertion holds indeed.

2) The function $f = f(x)$ is concave. Therefore, by 1) and by Jensen's inequality

$$f(x + y) = f\left(2\frac{x + y}{2}\right) \le 4f\left(\frac{x + y}{2}\right) \le 2\,[f(x) + (y)].$$

Therefore, $G(U_1 + U_2) \le 2\,[G, (U_1) + G(U_2)]$.

3) We will get this assertion by setting

$$U_1 = (U - \hat{U})\, I\left(|U - \hat{U}| > \frac{1}{2}\right) + \hat{U} I\left(|\hat{U}| > \frac{1}{2}\right),$$

$$U_2 = (U - \hat{U})\, I\left(|U - \hat{U}| \le \frac{1}{2}\right) + \hat{U} I\left(|\hat{U}| \le \frac{1}{2}\right).$$

Evidently, $U_1 + U_2 = U$. Observe further that

$$|U_1| * \nu \in \overset{+}{\mathcal{A}}_{loc}, \quad U_2^2 * \nu \in \overset{+}{\mathcal{A}}_{loc},$$

since

$$|U_1| * \nu \le |U - \hat{U}|\, I\left(|U - \hat{U}| > \frac{1}{2}\right) * \nu + \sum_s |\hat{U}(s)|\, I\left(|\hat{U}(s)| > \frac{1}{2}\right)$$

$$\le 3\left[\frac{(U - \hat{U})^2}{1 + |U - \hat{U}|} * \nu + 2\sum_s \frac{\hat{U}^2(s)}{1 + |\hat{U}(s)|}(1 - a_s)\, I\left(a_s \le \frac{1}{2}\right)\right]$$

$$\le 6G(U)$$

and

$$U_2^2 * \nu \le 2\left[(U - \hat{U})^2 I\left(|U - \hat{U}| \le \frac{1}{2}\right) * \nu + \sum_s \hat{U}^2(s)\, I\left(|\hat{U}(s)| \le \frac{1}{2}\right)\right]$$

$$\le 3\left[\frac{(U - \hat{U})^2}{1 + |U - \hat{U}|} * \nu + 2\sum_s \frac{\hat{U}^2(s)}{1 + |\hat{U}(s)|}(1 - a_s)\, I\left(a_s \le \frac{1}{2}\right)\right]$$

$$\le 6G(U).$$

The desired assertion follows from the easily verified inequalities:

$$G^1(U_1) \le 2|U_1| * \nu, \quad G^2(U_2) \le U_2^2 * \nu.$$

4) Since $G^2(U) \in \mathcal{A}^+_{loc}$, then $G^2_t(U) < \infty$ (P-a.s.), $t > 0$. Therefore, by (5.5)

$$\Delta G^2_t(U) = \int_E U^2(\omega, t, x)\, \nu(\omega; \{t\}, dx) - \hat{U}^2(\omega, t), \tag{5.8}$$

and the continuous component $(G^2(U))^c = I(a = 0) \circ G^2(U)$ is defined by the formula

$$(G^2(U))^c = U^2 I(a = 0) * \nu. \tag{5.9}$$

By (5.8) and (5.9) we get the desired equation (5.7).

3. Define the process $X = (X_t)_{t \ge 0}$ for $U \in \mathcal{G}_{loc}$ with $X_0 = 0$ and

$$X_t = \int_E U(\omega, t, x)\, \mu(\omega; \{t\}, dx) - \hat{U}(\omega, t). \tag{5.10}$$

By utilizing Theorem 1.8.3 we define the *stochastic integral* $U * (\mu - \nu)$ of a function $U \in \mathcal{G}_{loc}$ with respect to a martingale measure $\mu - \nu$ (analogously to the principles lying at the basis of defining the stochastic integral with respect to a local martingale; cf. Ch. 2, Subsection 2.2) as a unique process $M \in \mathcal{M}^d_{loc,\,0}$ with $\Delta M = X$. According to Theorem 1.8.3, it suffices for the existence of such processes that

(a) $^{P}(X) = 0$,

(b) $V(X) \in \mathcal{A}^{+}_{loc}$

with

$$V_t(X) = \left(\sum_{0 < s \le t} X_s^2 \right)^{1/2}.$$

For the process X defined by (5.10), property (a) takes place in virtue of the definition of the predictable projection (Theorem 1.3.13) and Problem 2.5 (the symbol ω is omitted to simplify the display):

$$^{P}\left(\int_E U(t, x)\, \mu(\{t\}, dx) \right) = {}^{P}\left(\int_E U^{+}(t, x)\, \mu(\{t\}, dx) \right) - {}^{P}\left(\int_E U^{-}(t, x)\, \mu(\{t\}, dx) \right)$$

$$= \int_E U^{+}(t, x)\, \nu(\{t\}, dx) - \int_E U^{-}(t, x)\, \nu(\{t\}, dx) = \hat{U}(t)$$

and

$$^{P}(\hat{U}(t)) = \hat{U}(t).$$

Let us now establish property (b).

By Lemma 2.2.1 it suffices to show that $L(X) \in \mathcal{A}^{+}_{loc}$ with

$$L_t(X) = \sum_{0 < s \le t} \frac{X_s^2}{1 + |X_s|}.$$

Since μ is an integer-valued random measure, there exists (§ 3) an optional process β such that (cf. (3.2))

$$\int_E U(t, x)\, \mu(\{t\}, dx) = I(\beta_t \in E)\, U(t, \beta_t). \tag{5.11}$$

Then

$$X_t = I(\beta_t \in E)\, U(t, \beta_t) - \hat{U}(t),$$

and consequently

$$L_t(X) = \sum_{0 < s \le t} \frac{(I(\beta_s \in E)\, U(s, \beta_s) - \hat{U}(s))^2}{1 + |I(\beta_s \in E)\, U(s, \beta_s) - \hat{U}(s)|}$$

$$= \sum_{0 < s \le t} I(\beta_s \in E) \frac{(U(s, \beta_s) - \hat{U}(s))^2}{1 + |U(s, \beta_s) - \hat{U}(s)|} + \sum_{0 < s \le t} I(\beta_s \notin E) \frac{\hat{U}^2(s)}{1 + |\hat{U}(s)|}.$$

Observe now that

$$I(\beta_s \in E) = \mu(\{s\} \times E)$$

and

$$I(\beta_s \notin E)\, I(\hat{U}(s) \neq 0) = I(\beta_s \notin E)\, I(a_s > 0)\, I(\hat{U}(s) \neq 0) = I(\hat{U}(s) \neq 0)\, p(\{s\}).$$

Therefore

$$L_t(X) = \frac{(U - \hat{U})^2}{1 + |U - \hat{U}|} * \mu_t + \frac{\hat{U}^2}{1 + |\hat{U}|} * p_t.$$

Let $(T_k)_{k \geq 1}$ be a localizing sequence for G (U). Then by definition of the compensators ν and q

$$EL_{T_k}(X) = E\left(\frac{(U - U)^2}{1 + |U - \hat{U}|} * \nu_{T_k} + \frac{\hat{U}^2}{1 + |\hat{U}|} * q_{T_k}\right) = EG_{T_k}(U) < \infty, \ k \geq 1,$$

i.e. $L(X) \in \mathcal{A}^+_{loc}$ and property (b) takes place.

4. Thus, the stochastic integral $U * (\mu - \nu)$ of a function $U \in \mathcal{G}_{loc}$ is defined with respect to a martingale measure $\mu - \nu$. For clarity, we use for $(I_A U) * (\mu - \nu)$, $A \in \mathcal{E}$ the notation

$$\int_0^t \int_A U d(\mu - \nu).$$

Let us establish now its properties.

Theorem 1. *Let* $U \in \mathcal{G}_{loc}$. *Then the stochastic integral* $U * (\mu - \nu)$ *possesses the following properties*:

1) $U * (\mu - \nu) \in \mathcal{M}^d_{loc,\,0}$;

2) $\Delta(U * (\mu - \nu))_t = \int_E U(t, x)\, \mu(\{t\}, dx) - \hat{U}(t)$;

3) $[U * (\mu - \nu),\ U * (\mu - \nu)] = (U - \hat{U})^2 * \mu + \hat{U}^2 * p$;

4) $(U * (\mu - \nu))^T = UI_{[\![0,T]\!]} * (\mu - \nu) \quad \forall T \in \mathbf{T}$;

5) $U \in \mathcal{G}^1_{loc} \Leftrightarrow \mathrm{Var}(U * (\mu - \nu)) \in \mathcal{A}^+_{loc}$

and in this case

$$U * (\mu - \nu) = (U - \hat{U}) * \mu - \hat{U} * p - (U - \hat{U}) * \nu + \hat{U} * q;$$

6) $U \in \mathcal{G}^2_{loc} \Leftrightarrow U * (\mu - \nu) \in \mathcal{M}^{2,d}_{loc}$

and in this case

$$\langle U * (\mu - \nu)\rangle = G^2 (U).$$

Proof. Properties 1) and 2) follow from the definition of $U * (\mu - \nu)$. Property 3) takes place by the representation

$$(\Delta (U * (\mu - \nu))_t)^2 = \int_E (U (t, x) - \hat{U} (t))^2 \mu (\{t\}, dx) + \hat{U}^2 (t) p (\{t\}),$$

which is obtained by using (5.11) and property 1).

Property 4) holds since each of the processes $(U * (\mu - \nu))^T$ and $UI_{[\![0, T]\!]} * (\mu - \nu)$ belongs to the class $\mathcal{M}^d_{loc, 0}$ and

$$\Delta (U * (\mu - \nu))^T = \Delta (UI_{[\![0, T]\!]} * (\mu - \nu)),$$

i.e.

$$N = (U * (\mu - \nu))^T - UI_{[\![0, T]\!]} * (\mu - \nu) \in \mathcal{M}^d_{loc, 0}$$

with $\Delta N = 0$, and hence by Davis' inequality (Theorem 1.9.6) the process N is negligible.

Let us establish now property 5).

($\Rightarrow$). Since $\mathcal{G}^1_{loc} \subseteq \mathcal{G}_{loc}$, the stochastic integral $U * (\mu - \nu)$ is defined. Therefore the process $G^1 (U) \in \mathcal{A}^+_{loc}$ is the compensator of an increasing process

$$| U - \hat{U} | * \mu + | \hat{U} | * p.$$

Therefore

$$| U - \hat{U} | * \mu + | \hat{U} | * p - G^1 (U) \in \mathcal{M}^d_{loc, 0}.$$

This circumstance allows us to define the process $M' \in \mathcal{M}^d_{loc, 0}$ with

$$M' = [(U - \hat{U}) * \mu - \hat{U} * p] - [(U - \hat{U}) * \nu - \hat{U} * q].$$

It is simply checked that $\Delta M' = \Delta (U * (\mu - \nu))$. Therefore, by taking into consideration that M' and $U * (\mu - \nu)$ belong to the class $\mathcal{M}^d_{loc, 0}$, we see that the processes M' and $U * (\mu - \nu)$ are P-indistinguishable. Hence,

$$\text{Var} (U * (\mu - \nu)) = \text{Var} (M').$$

But

$$\text{Var} (M') \leq | U - \hat{U} | * \mu + | \hat{U} | * p + G^1 (U),$$

and if $(T_k)_{k \geq 1}$ is a localizing sequence for $G^1(U)$, then

$$\mathbf{E}\,\mathrm{Var}\,(U * (\mu - \nu))_{T_k} \leq \mathbf{E}\,[\,|U - \hat{U}| * \mu_{T_k} + |\hat{U}| * p_{T_k} + G^1_{T_k}(U)]$$

$$= 2\mathbf{E}G^1_{T_k}(U) < \infty, \quad k \geq 1,$$

i.e. $\mathrm{Var}\,(U * (\mu - \nu)) \in \mathcal{A}^+_{loc}$ and

$$U * (\mu - \nu) = (U - \hat{U}) * \mu - \hat{U} * p - (U - \hat{U}) * \nu + \hat{U} * q.$$

($\Leftarrow$). Let $U \in \mathcal{G}_{loc}$ and $U * (\mu - \nu)$ be a stochastic integral. Then

$$\sum_{s > 0} |\Delta (U * (\mu - \nu))_s| \leq \mathrm{Var}\,(U * (\mu - \nu)) \in \mathcal{A}^+_{loc}.$$

Observe that

$$\Delta (U * (\mu - \nu))_t = I(\beta_t \in E)\, U(t, \beta_t) - \hat{U}(t)$$
$$= I(\beta_t \in E)\,(U(t, \beta_t) - \hat{U}(t)) - I(\beta_t \notin E)\,\hat{U}(t)$$

and consequently

$$|\Delta (U * (\mu - \nu))_t| = I(\beta_t \in E)\;|U(t, \beta_t) - \hat{U}(t)| + I(\beta_t \notin E)\;|\hat{U}(t)|.$$

This gives

$$\sum_s |\Delta (U * (\mu - \nu))_s| = |U - \hat{U}| * \mu + |\hat{U}| * p.$$

If $(T_k)_{k \geq 1}$ is a localizing sequence for $\mathrm{Var}\,(U * (\mu - \nu))$, then by the definition of the compensators ν and q we get

$$\mathbf{E}G^1_{T_k}(U) = \mathbf{E}\,(|U - \hat{U}| * \nu_{T_k} + |\hat{U}| * q_{T_k})$$

$$= \mathbf{E}\,(|U - \hat{U}| * \mu_{T_k} + |\hat{U}| * p_{T_k})$$

$$= \mathbf{E} \sum_{0 < s \leq T_k} |\Delta (U * (\mu - \nu))_s| \leq \mathbf{E}\,\mathrm{Var}\,(U * (\mu - \nu))_{T_k} < \infty, \quad k \geq 1.$$

i.e. $G^1(U) \in \mathcal{A}^+_{loc}$.

Finally, we prove property 6).

($\Rightarrow$). Since $\mathcal{G}^2_{loc} \subseteq \mathcal{G}_{loc}$, then $U * (\mu - \nu) \in \mathcal{M}^d_{loc,\,0}$ and by property 3)

$$[U * (\mu - \nu),\; U * (\mu - \nu)] = (U - \hat{U})^2 * \mu + \hat{U}^2 * p. \tag{5.12}$$

Let $(T_k)_{k\geq 1}$ be a localizing sequence for $G^2(U)$. Then

$$\mathbf{E}\,[U*(\mu-\nu), U*(\mu-\nu)]_{T_k} = \mathbf{E}\,((U-\hat{U})^2 * \nu_{T_k} + \hat{U}^2 * q_{T_k})$$

$$= \mathbf{E}G^2_{T_k}(U) < \infty, \quad k \geq 1.$$

By the Burkholder-Gandy inequality (Theorem 1.9.7) with $p = 2$, this yields

$$\mathbf{E}\,((U*(\mu-\nu))^*_{T_k})^2 \leq C_2\mathbf{E}\,[U*(\mu-\nu), U*(\mu-\nu)]_{T_k} < \infty, \quad k \geq 1,$$

i.e.

$$U*(\mu-\nu) \in \mathcal{M}^{2,d}_{loc,0}.$$

($\Leftarrow$). Let $U \in \mathcal{G}_{loc}$ and $U*(\mu-\nu) \in \mathcal{M}^{2,d}_{loc,0}$. From (5.12) and the definition of $G^2(U)$ it follows that $\langle U*(\mu-\nu)\rangle = G^2(U)$. Hence $G^2(U) \in \mathcal{A}^+_{loc}$, as a localizing sequence $(T_k)_{k\geq 1}$ for $U*(\mu-\nu)$ localizes $\langle U*(\mu-\nu)\rangle$ too.

5. Let $M \in \mathcal{M}_{loc}$. The conditional mathematical expectations $M^P_\mu(\Delta M \mid \tilde{\mathcal{P}})$ and $M^P_p(\Delta M) \mid \mathcal{P})$ are defined in Theorem 3.1. Let us give some of their properties.

Theorem 2. *Let* μ *be an integer-valued random measure* ($\mu \in \tilde{\mathcal{V}}^+_{\mathcal{P}}$), ν *its compensator* ($a_t = \nu(\{t\}\times E)$) *and* $M \in \mathcal{M}_{loc}$. *Then* $\left(f(x) = \dfrac{x^2}{1+|x|}\right)$:

1) $f(M^P_\mu\, \Delta M \mid \tilde{\mathcal{P}})) * \nu \in \mathcal{A}^+_{loc}$,

$$\sum_s I(a_s > 0)\,(f(M^P_p(\Delta M \mid \mathcal{P})(s))\,(1-a_s) \in \mathcal{A}^+_{loc}$$

2) $U = M^P_\mu(\Delta M \mid \tilde{\mathcal{P}}) - M^P_p(\Delta M \mid \mathcal{P})\; I(0 < a < 1) \in \mathcal{G}_{loc}$;

3) *if* $|\Delta M| \leq c$ *and if* $N = U*(\mu-\nu)$ *with a function* U *defined in* 2), *then* $|\Delta N| \leq 4c$;

4) *if* $M = U'*(\mu-\nu)$, $U' \in \mathcal{G}_{loc}$ *and if* $N = U*(\mu-\nu)$ *with* U *defined in* 2), *then the processes* M *and* N *are indistinguishable.*

Proof. 1) By Problem 1.9.5 we have

$$\left(\sum_s (\Delta M_s)^2\right)^{1/2} \in \mathcal{A}^+_{loc}.$$

Therefore, by Lemma 2.2.1 we have

$$\sum_s f(\Delta M_s) \in \mathcal{A}^+_{loc},$$

and consequently $f(\Delta M) * \mu \in \mathcal{A}^+_{loc}$. This means that $f(\Delta M)\,\mu \in \tilde{\mathcal{V}}^+_{\mathcal{P}}$ and the conditional mathematical expectation $M^P_\mu(f(\Delta M) \mid \tilde{\mathcal{P}})$ is defined. Hence, by Jensen's inequality (Problem 2.11)

$$M^P_\mu(f(\Delta M) \mid \tilde{\mathcal{P}}) \ge f(M^P_\mu(\Delta M \mid \tilde{\mathcal{P}})) \quad M^P_\mu\text{-a.s.}$$

If $(\tau_n)_{n \ge 1}$ is a localizing sequence for $f(\Delta M) * \mu$, then

$$\mathrm{E} f(M^P_\mu(\Delta M \mid \tilde{\mathcal{P}})) * \nu_{\tau_n} = \mathrm{E} f(M^P_\mu(\Delta M \mid \tilde{\mathcal{P}})) * \mu_{\tau_n}$$

$$\le \mathrm{E} M^P_\mu(f(\Delta M) \mid \tilde{\mathcal{P}}) * \mu_{\tau_n} = \mathrm{E} f(\Delta M) * \mu_{\tau_n} < \infty, \quad n \ge 1.$$

The second relation is proved analogously by using the fact that

$$\sum_s f(\Delta M_s) \in \mathcal{A}^+_{loc} \Rightarrow \sum_s f(\Delta M_s)\, I(0 < a_s < 1)\,(1 - \mu(\{s\} \times E)) \in \mathcal{A}^+_{loc}.$$

2) The definition of U yields

$$\hat{U} = \widehat{M^P_\mu(\Delta M \mid \tilde{\mathcal{P}})} - M^P_p(\Delta M \mid \mathcal{P})\, a\, I(a < 1).$$

Let us show that

$$\sum_s f(\hat{U}_s)(1 - a_s) \in \mathcal{A}^+_{loc}. \tag{5.13}$$

In the course of proving Lemma 1 the following properties of the function f (x) are mentioned:

$$f(x + y) \le 2\,[f(x) + f(y)], \quad f(cx) \le (|c| \vee c^2)\, f(x). \tag{5.14}$$

Therefore

$$f(\hat{U}) \le 2\,[f(\widehat{M^P_\mu(\Delta M \mid \tilde{\mathcal{P}})}) + a\, f(M^P_p(\Delta M \mid \mathcal{P}))].$$

By this inequality and by assertion 1) of the theorem it suffices for the validity of (5.13) to show that

$$\sum_s f(\widehat{M^P_\mu(\Delta M \mid \tilde{\mathcal{P}})}(s))(1 - a_s) \in \mathcal{A}^+_{loc}. \tag{5.15}$$

But by Jensen's inequality and by (5.14) (on the set $\{a_s > 0\}$)

$$f\left(\widehat{M^P_\mu(\Delta M \mid \tilde{\mathcal{P}})}(s)\right) = f\left(\int_E M^P_\mu(\Delta M \mid \tilde{\mathcal{P}})(s, x)\, \nu(\{s\}, dx)\right)$$
$$\le \int_E f(a_s M^P_\mu(\Delta M \mid \tilde{\mathcal{P}})(s, x)) \frac{\nu(\{s\}, dx)}{a_s}$$
$$\le \int_E f(M^P_\mu(\Delta M \mid \tilde{\mathcal{P}})(s, x))\, \nu(\{s\}, dx).$$

Hence,

$$\sum_s f\left(\widehat{M^P_\mu(\Delta M \mid \tilde{\mathcal{P}})}(s)\right) \le \sum_s \int_E f(M^P_\mu(\Delta M \mid \tilde{\mathcal{P}})(s, x))\, \nu(\{s\}, dx)$$
$$\le f(M^P_\mu(\Delta M \mid \tilde{\mathcal{P}})) * \nu, \tag{5.16}$$

and (5.15) holds by virtue of Assertion 1) of the theorem.

Next,

$$U - \hat{U} = M^P_\mu(\Delta M \mid \tilde{\mathcal{P}}) - \widehat{M^P_\mu(\Delta M \mid \tilde{\mathcal{P}})} - M^P_p(\Delta M \mid \mathcal{P})(1 - a)\, I(0 < a).$$

By concavity of the function $f(x)$ the inequality $f(x + y + z) \le \frac{1}{3}(f(x) + f(y) + f(z))$ holds. Therefore

$$f(U - \hat{U}) \le \frac{1}{3}\left(f(M^P_\mu(\Delta M \mid \tilde{\mathcal{P}})) + f\left(\widehat{M^P_\mu(\Delta M \mid \tilde{\mathcal{P}})}\right) + f(M^P_p(\Delta M \mid \mathcal{P})(1 - a)\, I(0 < a)\right);$$

By Assertion 1) of Theorem 2 we have

$$f(M^P_\mu(\Delta M \mid \tilde{\mathcal{P}})) * \nu \in \mathcal{A}^+_{loc}.$$

According to (5.16) $f\left(\widehat{M^P_\mu(\Delta M \mid \mathcal{P})}\right) * \nu$ belongs to $\mathcal{A}^+_{loc}$ as well. By Assertion 1) of Theorem 2 we have $f(M^P_p(\Delta M \mid \mathcal{P})(1 - a)\, I(0 < a)) * \nu \in \mathcal{A}^+_{loc}$, since

$$f(M^P_p(\Delta M \mid \mathcal{P})(1 - a)\, I(0 < a)) * \nu$$
$$\le \sum_s I(0 < a_s)\, f(M^P_p(\Delta M \mid \mathcal{P})(s))\, (1 - a_s).$$

Thus

$$f(U - \hat{U}) * \nu \in \mathcal{A}^+_{loc}. \tag{5.17}$$

The desired assertion follows from (5.13) and (5.17).

3) The definition of U and Jensen's inequality for the conditional mathematical expectations $M^P_\mu(\cdot \mid \tilde{\mathcal{P}})$ and $M^P_p(\cdot \mid \mathcal{P})$ yield $|U| \le 2c$. Hence,

$$|\Delta N_t| \le 2c\,(\mu(\{t\} \times E) + a_t) \le 4c.$$

4) Since the processes M and N belong to $\mathcal{M}^d_{loc,\,0}$, then according to Problem 1.9.6 the processes M and N are indistinguishable if the processes ΔM and ΔN are indistinguishable. According to Theorem 1.3.12 it suffices for this to show that for each Markov time T on the set $\{T < \infty\}$

$$\Delta M_T = \Delta N_T.$$

Let $\beta = (\beta_t)_{t \ge 0}$ be a stochastic process involved in the definition of a measure μ (cf. § 3). Then for each Markov time T on the set $\{T < \infty\}$

$$\Delta M_T = U'(T, \beta_T) - \hat{U}'(T), \quad \Delta N_T = U(T, \beta_T) - \hat{U}(T).$$

Therefore it suffices to show that

$$U - \hat{U} = U' - \hat{U}' \quad (M^P_\mu\text{-a.s.}). \tag{5.18}$$

By Problem 7 we have $U' - \hat{U}' = M^P_\mu(\Delta M \mid \tilde{\mathcal{P}})$ $(M^P_\mu\text{-a.s.})$, and by the definition of U we have

$$\hat{U} = M^P_\mu(\Delta M \mid \mathcal{P}) + a\, M^P_p(\Delta M \mid \mathcal{P})\, I(0 < a < 1);$$

hence

$$U - \hat{U} = M^P_\mu(\Delta M \mid \tilde{\mathcal{P}}) - M^P_\mu(\Delta M \mid \tilde{\mathcal{P}}) - (1 - a)\, M^P_p(\Delta M \mid \mathcal{P})\, I(0 < a < 1).$$

Therefore, by Theorem 3.1 we have $(M^P_\mu\text{-a.s.})$

$$U - \hat{U} = M^P_\mu(\Delta M \mid \tilde{\mathcal{P}}) - \widehat{M^P_p(\Delta M \mid \tilde{\mathcal{P}})\, I(a = 1)},$$

and consequently, $(M^P_\mu\text{-a.s.})$,

$$U - \hat{U} = U' - \hat{U}' - \widehat{M^P_\mu(\Delta M \mid \tilde{\mathcal{P}})\, I(a = 1)}. \tag{5.19}$$

By integrating both parts of (5.19) with respect to the measure $\nu(\{.\}, dx)$, we get the equality

$$\hat{U}(1 - a) = \hat{U}'(1 - a) - \widehat{M^P_\mu(\Delta M \mid \tilde{\mathcal{P}})\, I(a = 1)} \quad (M^P_\mu\text{-a.s.}),$$

in virtue of which $\widehat{M^P_\mu(\Delta M \mid \tilde{\mathcal{P}})\, I(a = 1)} = 0$ $(M^P_\mu\text{-a.s.})$, and according to (5.19) the desired equality (5.18) holds.

Problems

1. Let $\delta \notin E$, $E_\delta = E \cup \{\delta\}$ and $\mathcal{E}_\delta = \sigma(\mathcal{E} \cup \{\delta\})$, $\tilde{\mathcal{P}}_\delta = \mathcal{P} \otimes \mathcal{E}_\delta$. Define on $(R_+ \times E_\delta, B_+(R_+) \otimes \mathcal{E}_\delta)$ an integer-valued random measure

$$\bar{\mu} = I(x \in E)\,\mu + I(x = \delta)\,p$$

where μ is an integer-valued random measure and p is the measure defined in (3.10).

Let $\bar{\nu} = I(x \in E)\,\nu + I(x = \delta)\,q$ be the compensator of the measure $\bar{\mu}$. Let $U \in \mathcal{G}_{loc}$ and $U_\delta = I(x \in E)(U - \hat{U}) - I(x = \delta)\,\hat{U}$.

Show that:

1) there exists the stochastic integral $U_\delta * (\bar{\mu} - \bar{\nu})$;

2) the processes $U * (\mu - \nu)$ and $U_\delta * (\bar{\mu} - \bar{\nu})$ are indistinguishable;

3) $\displaystyle\int_{E_\delta} U_\delta(t, x)\,\bar{\nu}(\{t\}, dx) = 0.$

2. Let $M \in \mathcal{M}_{loc}$, let a measure

$$\mu(dt, dx) = \sum_{s > 0} I(\Delta M_s \neq 0)\,\varepsilon_{(s, \Delta M_s)}(dt, dx), \qquad E = R \setminus \{0\} \tag{5.20}$$

(cf. the example in Subsection 3.2) and let ν be its compensator.

Show that:

1) $$\int_E x\nu(\{t\}, dx) = 0; \tag{5.21}$$

2) $$\frac{x^2}{1 + |x|} * \nu \in \mathcal{A}^+_{loc}, \quad (x^2 \wedge |x|) * \nu \in \mathcal{A}^+_{loc}; \tag{5.22}$$

3) the processes M^d and $x * (\mu - \nu)$ are indistinguishable;

4) if $M \in \mathcal{M}^{2,d}_{loc,0}$, then $\langle M^d \rangle = x^2 * \nu$.

3. Let μ be an integer-valued random measure and ν its compensator, let U be a $\tilde{\mathcal{P}}$-measurable function such that $|U| * \nu \in \mathcal{A}^+_{loc}$. Show that

$$U \in \mathcal{G}^1_{loc},$$

and the stochastic integral $U * (\mu - \nu)$ and the process $U * \mu - U * \nu$ are industinguishable.

4. Let μ be an integer-valued random measure, ν its compensator, $U \in \mathcal{G}_{loc}$ and $M = U * (\mu - \nu)$. Show that for $f \in L_{loc}(\mathbb{F}, [M, M])$ (f is a $\mathcal{P}$-measurable function with $(f^2 \circ [M, M])^{1/2} \in \mathcal{A}^+_{loc}$ (cf. Ch. 2, § 2)), $f U \in \mathcal{G}_{loc}$ and the processes $f \cdot M$ and $f U * (\mu - \nu)$ are indistinguishable.

5. Let $M \in \mathcal{M}^d_{loc}$, μ an integer-valued random measure, ν its compensator and U a $\tilde{\mathcal{P}}$-measurable function such that for each Markov time τ

$$I(\tau < \infty) \int_E |U(\tau, x)| \, \nu(\{t\}, dx) < \infty \qquad (P\text{-a.s.})$$

and

$$I(\tau < \infty) \left[\Delta M_\tau - \int_E U(\tau, x) (\mu - \nu)(\{\tau\}, dx) \right] = 0.$$

Show that $U \in \mathcal{G}_{loc}$ and the processes M and $U * (\mu - \nu)$ are indistinguishable.

6) Let $U_1, U_2 \in \mathcal{G}_{loc}$. Show that

$$U_1 + U_2 \in \mathcal{G}_{loc}$$

and

$$U_1 * (\mu - \nu) + U_2 * (\mu - \nu) = (U_1 + U_2) * (\mu - \nu).$$

7. Let $U \in \mathcal{G}_{loc}$ and $M = U * (\mu - \nu)$. Show that

$$M^P_\mu(\Delta M \mid \tilde{\mathcal{P}}) = U - \hat{U}. \quad M^P_\mu\text{-a.s.}$$

8. Let μ be an integer-valued random measure, ν its compensator and $U^1, U^2 \in \mathcal{G}_{loc}$, $U^1 = U^2$ ($\{a < 1\}$; M^P_μ-a.s.). Show that the stochastic processes $U^1 * (\mu - \nu)$ and $U^2 * (\mu - \nu)$ are indistinguishable.

§ 6. Ito's formula. II

1. Let $M \in \mathfrak{M}_{loc}$ and let f (x) be a twice differentiable function. By Ito's formula (Ch. 2, § 3)

$$f(M_t) = f(M_0) + f'(M_-) \cdot M_t + \frac{1}{2} f''(M_-) \circ \langle M^c \rangle_t$$

$$+ \sum_{0 < s \le t} [f(M_s) - f(M_{s-}) - f'(M_{s-}) \Delta M_s], \tag{6.1}$$

where $\langle M^c \rangle$ is the quadratic characteristic of the continuous component M^c of a local martingale M.

Suppose that the component $M^d = M - M^c$ admits the representation

$$M^d = U * (\mu - \nu), \tag{6.2}$$

where μ is a certain integer-valued random measure, ν its compensator and a function $U \in \mathcal{G}_{loc}$, i.e.

$$\frac{(U - \hat{U})^2}{1 + |U - \hat{U}|} * \nu + \frac{\hat{U}^2}{1 + |\hat{U}|} * q \in \mathcal{A}^+_{loc}.$$

Theorem 1. *Let* $\hat{U} = 0$. *Then the formula* (6.1) *can be represented in the following form*:

$$f(M_t) = f(M_0) + f'(M_-) \cdot M^c_t + f'(M_-) U * (\mu - \nu) + \frac{1}{2} f''(M_-) \circ \langle M^c \rangle_t$$

$$+ [f(M_- + U) - f(M_-) - f'(M_-) U] * \mu_t. \tag{6.3}$$

Proof. It suffices only to show that the following pairs of processes are indistinguishable:

1) $f'(M_-) \cdot M^d$ and $f'(M_-) U * (\mu - \nu)$,

2) $\sum_s [f(M_s) - f(M_{s-}) - f'(M_{s-}) \Delta M_s]$ and $[f(M_- + U) - f(M_-) - f'(M_-) U] * \mu$.

Since

$$M^d = U * (\mu - \nu),$$

then

$$f'(M_-) \Delta M^d = f'(M_-) \Delta (U * (\mu - \nu)).$$

It is easily deduced from this that (in accordance with the constructions in § 5) there exists the stochastic integral $f'(M_-) U * (\mu - \nu)$ and that property 1) holds (cf.

Problem 5.4).

To establish 2) observe that $\Delta M_s = I(\beta_s \in E)\, U(s, \beta_s)$, where β is the process defined in § 3. Therefore

$$f(M_s) - f(M_{s-}) - f'(M_{s-})\,\Delta M_s$$

$$= f(M_{s-} + I(\beta_s \in E)\,U(s, \beta_s)) - f(M_{s-}) - f'(M_{s-})\,I(\beta_s \in E)\,U(s, \beta_s)$$

$$= \int_E [f(M_{s-} + U(s, x)) - f(M_{s-}) - f'(M_{s-})\,U(s, x)]\,\mu(\{s\}, dx).$$

Corollary 1. *In accordance with Problem* 5.2, *the function* $U(\omega, s, x) = x$ *serves as an example of a function* $U = U(\omega, s, x)$, *satisfying the conditions of the theorem.*

Corollary 2. *Let the representation* (6.2) *hold. Then, according to the notations and assertions of Problem* 5.1,

$$M^d = U_\delta * (\bar{\mu} - \bar{\nu}) \quad \textit{and} \quad \int_{E_\delta} U_\delta(t, x)\,\bar{\nu}(\{t\}, dx) = 0.$$

Therefore,

$$f(M_t) = f(M_0) + f'(M_-) \cdot M_t^c + f'(M_-)\,U_\delta * (\bar{\mu} - \bar{\nu})_t$$

$$+ \frac{1}{2} f''(M_-) \circ \langle M^c \rangle_t + [f(M_- + U_\delta) - f(M_-) - f'(M_-)\,U_\delta] * \bar{\mu}_t, \quad (6.4)$$

besides,

$$f'(M_-)\,U_\delta * (\bar{\mu} - \bar{\nu}) = f'(M_-)\,U * (\mu - \nu). \quad (6.5)$$

Problems

1. Show that as $\hat{U} = 0$ and

$$|f(M_- + U) - f(M_-) - f'(M_-)\,U| * \nu \in \mathcal{A}^+_{loc}$$

the following formula holds:

$$f(M_t) = f(M_0) + f'(M_-) \cdot M_t^c + [f(M_- + U) - f(M_-)] * (\mu - \nu)_t$$

$$+ \frac{1}{2} f''(M_-) \circ \langle M^c \rangle_t + [f(M_- + U) - f(M_-) - f'(M_-)\,U] * \nu_t \quad (6.6)$$

2. Show that if

$$|f(M_- + U_\delta) - f(M_-) - f'(M_-)U_\delta| * \bar{\nu} \in \mathcal{A}^+_{loc},$$

then

$$f(M_t) = f(M_0) + f'(M_-) \cdot M^c_t + [f(M_- + U_\delta) - f(M_-)] * (\bar{\mu} - \bar{\nu})_t$$

$$+ \frac{1}{2} f''(M_-) \circ \langle M^c \rangle_t + [f(M_- + U_\delta) - f(M_-) - f'(M_-)U_\delta] * \bar{\nu}_t. \quad (6.7)$$

CHAPTER 4

SEMIMARTINGALES. II. CANONICAL REPRESENTATION

§ 1. Canonical representation. Triplet of predictable characteristics of a semimartingale

1. Let $X = (X_t, \mathcal{F}_t)_{t \geq 0}$ be a semimartingale. Denote by $X^a = (X_t^a, \mathcal{F}_t)_{t \geq 0}$ the process

$$X_t^a = \sum_{0 < s \leq t} \Delta X_s I(|\Delta X_s| > a), \quad a > 0. \tag{1.1}$$

Since $X \in D$, then $X^a \in \mathcal{V}$ is a semimartingale. Therefore the process $Y^{(a)} = X - X^a$ is a semimartingale too and $|\Delta Y^{(a)}| \leq a$. By Theorem 2.1.2 the semimartingale $Y^{(a)}$ is special, and by Theorem 2.1.1 there exists the unique representation

$$Y_t^{(a)} = X_0 + B_t^a + M_t^a \tag{1.2}$$

with $M^a \in \mathcal{M}_{loc, 0}$ and with a predictable process B^a of bounded variation ($B^a \in \mathcal{V} \cap \mathcal{P}$).

Hence for each $a > 0$ the representation

$$X_t = X_0 + B_t^a + M_t^a + X_t^a \tag{1.3}$$

holds, called the *canonical representation* of a semimartingale X.

A local martingale M^a admits the unique decomposition

$$M^a = M^{ac} + M^{ad} \tag{1.4}$$

with $M^{ac} \in \mathcal{M}_{loc}^c$ and $M^{ad} \in \mathcal{M}_{loc}^d$ (Theorem 1.7.2).

By (1.3) and (1.4) we get

$$X_t = X_0 + B_t^a + M_t^{ac} + M_t^{ad} + X_t^a. \tag{1.5}$$

Let $\mu = \mu\,(dt, dx)$ be a jump measure of a semimartingale X:

$$\mu\,((0, t] \times \Gamma) = \sum_{0 < s \leq t} I(\Delta X_s \in \Gamma), \quad \Gamma \in B\,(R_0),$$

where $R_0 = R \setminus \{0\}$, and let $\nu = \nu\,(dt, dx)$ be its compensator (cf. Example in Ch. 3, § 3). Then, evidently

$$X_t^a = \int_0^t \int_{|x|>a} x \, d\mu. \tag{1.6}$$

By (1.5) and (1.6)

$$\Delta B_t^a + \Delta M_t^a = \int_{|x|\le a} x\mu(\{t\}, dx).$$

Since B^a is a $\mathcal{P}$-measurable function, ${}^p(\Delta M^{ad}) = 0$ (Theorem 1.7.8) and

$${}^p\left(\int_{|x|\le a} x\mu(\{\cdot\}, dx)\right) = \int_{|x|\le a} x\nu(\{\cdot\}, dx)$$

(see Ch. 1, Subsection 3.8, the definition of the predictable projection, and Problem 3.2.5), this gives

$$\Delta B^a = \int_{|x|\le a} x\nu(\{\cdot\}, dx) \tag{1.7}$$

and

$$\Delta M^{ad} = \int_{|x|\le a} x(\mu - \nu)(\{\cdot\}, dx).$$

Therefore, a local martingale M^{ad} admits the representation (Problem 3.5.5)

$$M_t^{ad} = \int_0^t \int_{|x|\le a} x \, d(\mu - \nu). \tag{1.8}$$

In accordance with Ch. 2, Subsection 3.1, the continuous martingale component X^c of a semimartingale X is independent of a type of a semimartingale representation (in particular, of a choice of $a > 0$). Hence $M^{ac} = X^c$, and consequently, the canonical representation (1.3) takes the following form:

$$X_t = X_0 + B_t^a + X_t^c + \int_0^t \int_{|x|\le a} x \, d(\mu - \nu) + \int_0^t \int_{|x|>a} x \, d\mu. \tag{1.9}$$

For definitness, take $a = 1$. Then, denoting $B = B^1$, we may represent each semimartingale X as

$$X_t = X_0 + B_t + X_t^c + \int_0^t \int_{|x|\le 1} x \, d(\mu - \nu) + \int_0^t \int_{|x|>1} x \, d\mu, \tag{1.10}$$

where B is a predictable process of locally bounded variation, $B_0 = 0$; X^c the continuous martingale component of a semimartingale X, μ the jump measure of X and ν its compensator.

Since $X^c \in \mathcal{M}^{2,c}_{loc}$, the quadratic characteristic $C = \langle X^c \rangle$ is defined and it is a continuous process.

The collection

$$(B, C, \nu) \tag{1.11}$$

is called the *triplet of predictable characteristics* of a semimartingale X. This collection will be denoted by T or T (X).

Observe that C and ν are "inner" characteristics, independent of a type of a semimartingale representation. As for the characteristic B^a, its value depends on a (cf. (1.7)).

Observe also the properties of ν, following from the definition of the compensator of an integer-valued random measure:

$$\nu(\{0\} \times R_0) = 0, \quad \nu(\{t\} \times R_0) \le 1 \tag{1.12}$$

with $R_0 = R \setminus \{0\}$.

For brevity, we will further on use the following notations:

$$a_t = \nu(\{t\} \times R_0), \quad \nu^c(dt, dx) = I(a_t = 0)\, \nu(dt, dx). \tag{1.13}$$

2. We give a number of properties of the triplet T = (B, C, ν) of a left quasi-continuous semimartingale X.

Theorem 1. *The following conditions are equivalent:*

(a) *a semimartingale* X *is a left quasi-continuous process;*

(b) *the compensator* ν *of the jump measure of* X *possesses the following property:*

$$\sum_{s>0} \nu(\{s\} \times R_0) = 0 \quad (\text{P-a.s.}).$$

Proof. (a) $\Rightarrow$ (b). Consider the stochastic process $A \in \mathcal{V}^+ \cap \mathcal{P}$ with

$$A_t = \sum_{0<s\le t} \nu(\{s\} \times \{|x| > \varepsilon\}), \ \varepsilon > o.$$

By Theorem 1.3.6 the thin set $\{\Delta A \ne 0\}$ is exhausted by a sequence of predictable Markov times. Let $\tau \in \{\Delta A \ne 0\}$. Then

$$\Delta A_\tau I(\tau < \infty) = \nu(\{t\} \times \{|x| > \varepsilon\})\, I(\tau < \infty).$$

But

$$\nu(\{t\} \times \{|x| > \varepsilon\})\, I(\tau < \infty) = P(|\Delta X_\tau| > \varepsilon \mid \mathcal{F}_{\tau-})\, I(\tau < \infty)$$

(Problem 3.2.5) and $(\tau < \infty) \in \mathcal{F}_{\tau-}$ (Problem 1.1.6). Consequently,

$$\mathbf{E}\Delta A_\tau I(\tau < \infty) = \mathbf{E} I(|\Delta X_\tau| > \varepsilon) I(\tau < \infty).$$

Since X is a local quasi-continuous process and τ a predictable Markov time, by Theorem 1.3.9 we have $\Delta X_\tau = 0$ (P-a.s.), i.e. $\Delta A_\tau I(\tau < \infty) = 0$ (P-a.s.), and the desired assertion follows from this as ε is arbitrary.

(b) $\Rightarrow$ (a). By Problem 3.2.5

$$\mathsf{P}(\Delta X_\tau \neq 0 \mid \mathcal{F}_{\tau-}) I(\tau < \infty) = \nu(\{\tau\} \times R_0) I(\tau < \infty) = 0 \quad (\text{P-a.s.})$$

for each predictable Markov time τ, i.e. $\mathsf{P}(\Delta X_\tau \neq 0,\ \tau < \infty) = 0$, and hence X is a left quasi-continuous process (Theorem 1.3.9).

Corollary. *Since*

$$\Delta B_t = \int_{|x| \le 1} x\nu(\{t\}, dx)$$

(cf. (1.7)), *the process* B, *related to a left quasi-continuous semimartingale* X, *has continuous trajectories.*

3. The canonical representation takes place for a vector-valued semimartingale $X = (X^1, \ldots, X^k)$ too:

$$X_t = X_0 + B_t + X_t^c + \int_0^t \int_{|x| \le 1} x\, d(\mu - \nu) + \int_0^t \int_{|x| > 1} x\, d\mu$$

where $B = (B^1, \ldots, B^k)$, $X^c = (X^{1c}, \ldots, X^{kc})$, $B^i \in \mathcal{V} \cap \mathcal{P}$, $X^{ic} \in \mathcal{M}^c_{loc}$, $i = 1, \ldots, k$, $\mu(dt, dx) = \mu(dt, dx^1 \ldots dx^k)$ is the jump measure of a process X, $\nu(dt, dx) = \nu(dt, dx^1 \ldots dx^k)$ its compensator, and

$$|x| = \left(\sum_{i=1}^k (x^i)^2 \right)^{1/2}.$$

The triplet $T = (B, C, \nu)$ of predictable characteristics corresponding to this canonical representation consists of the vector $B = (B^1, \ldots, B^k)$, of the matrix C with elements $C_{ij} = \langle X^{ic}, X^{jc} \rangle$ and of the predictable measure $\nu(dt, dx) = \nu(dt, dx^1 \ldots dx^k)$ on $(R_+ \times R^k \setminus \{0\}, \mathcal{B}(R_+) \otimes \mathcal{B}(R^k \setminus \{0\}))$.

Problems

1. Show that

$$(x^2 \wedge 1) * \nu \in \mathcal{A}^+_{loc}, \quad I(|x| > c) * \nu \in \mathcal{A}^+_{loc}, \quad c > 0.$$

2. Show that

$$\sum_s \left| \int_{|x|\le 1} x\nu(\{s\}, dx) \right| \in \mathcal{A}^+_{loc}.$$

3. Let T be a predictable time and $A \in B(R_0)$ ($B(R_0)$ is the Borel σ-algebra on R_0). Show that on the set $\{T < \infty\}$

$$\nu(\{T\} \times A) = P(\Delta X_T \in A \mid \mathcal{F}_{T-}) \quad (P\text{-a.s.}).$$

4. Let $X \in S \cap \mathcal{V}$. Show that the process X^c is negligible, $C_\infty = 0$, and $(|x| \wedge 1) * \nu \in \mathcal{A}^+_{loc}$.

5. Show that as $a < b$ we have

$$B^a_t - B^b_t = \int_0^t \int_{a<|x|\le b} x\, d\nu.$$

6. Show that as $X \in \mathbf{Sp}$ (in particular, as $X \in \mathcal{M}_{loc}$) we have

$$(x^2 \wedge |x|) * \nu \in \mathcal{A}^+_{loc}.$$

7. Let $X \in \mathbf{Sp}$ and in the decomposition $X = X_0 + A + M$ let A be a predictable process belonging to $\mathcal{V}$ and $M \in \mathcal{M}^2_{loc}$. Show that

$$x^2 * \nu \in \mathcal{A}^+_{loc}.$$

8. Let $X = \mathbf{Sp}$ and let the processes A and M be such as in Problem 7. Show that

$$\langle M\rangle = B - [A, A] - 2X_- \circ A$$

where B is a predictable process belonging to $\mathcal{V}$, involved in the decomposition $X^2 = X_0^2 + B + N$ with $N \in \mathcal{M}_{loc}$.

9. Show that

$$X \in \mathbf{Sp} \Leftrightarrow [X, X]^{1/2} \in \mathcal{A}^+_{loc}.$$

10. Show that as $X \in \mathcal{M}_{loc}$ the component B of the triplet $T = (B, C, \nu)$, corresponding to $a = 1$, is given by the formula

$$B_t = -\int_0^t \int_{|x|>1} x\, d\nu.$$

11. Let X be a semimartingale having the canonical representation (1.10) and

$$M_t^a = X_t^c + \int_0^t \int_{|x| \le a} x \, d(\mu - \nu), \quad a > 0.$$

Show that

$$\langle M^a \rangle_t = \langle X^c \rangle_t + \int_0^t \int_{|x| \le a} x^2 \, d\nu - \sum_{0 < s \le t} \left(\int_{|x| \le a} x\nu(\{s\}, dx) \right)^2. \quad (1.14)$$

12. Show that a process X defined on a stochastic basis $(\Omega, \mathcal{F}, \mathbb{F}, \mathsf{P})$, is a semimartingale having the triplet of predictable characteristics (B, C, ν) and the distribution P_0 on $\mathcal{F}_0^X = \sigma(X_0)$ if and only if the measure P is a solution of the following martingale problem (relative to P_0 and (B, C, ν)):

I: $\mathsf{P} \mid \mathcal{F}_0^X = \mathsf{P}_0$;

II: the following processes are local martingales with respect to the measure P (μ is the jump measure of a process X):

a) $M_t = X_t - X_0 - B_t - \sum_{s \le t} \Delta X_s I(|\Delta X_s| > 1)$,

b) $M_t^2 - C_t - \int_0^t \int_{|x| \le 1} x^2 \, d\nu + \sum_{s \le t} \left(\int_{|x| \le 1} x\nu(\{s\}, dx) \right)^2$,

c) $\int_0^t \int_{R \setminus \{0\}} g(x) \, d\mu - \int_0^t \int_{R \setminus \{0\}} g(x) \, d\nu$

with a function $g = g(x)$ of a class G, which consists of bounded Borel functions equal to zero in a vicinity of zero and possesses the property that if m and m' are two measures on R with $m(0) = m'(0)$ and $m(|x| > \varepsilon) < \infty$, $m'(|x| > \varepsilon) < \infty \; \forall \varepsilon > 0$, then the equality

$$\int_R f(x) \, m(dx) = \int_R f(x) \, m'(dx)$$

for each function $f \in G$ implies $m = m'$.

§ 2. Stochastic exponential constructed by the triplet of a semimartingale

1. Let X be a semimartingale having the triplet of predictable characteristics $T(X) = (B, C, \nu)$. An essential rôle in studying properties of a semimartingale X, characterizing it and proving limit theorems for it, is played by the stochastic exponential $\mathcal{E}(G)$ introduced below, which is constructed by means of a complex-valued function $G = (G_t)_{t \geq 0}$, called the cumulant, with $G_t = G_t(\lambda)$ where

$$G_t(\lambda) = i\lambda B_t - \frac{\lambda^2}{2} C_t + \int_0^t \int_{R_0} (e^{i\lambda x} - 1 - i\lambda x I(|x| \leq 1))\, d\nu, \qquad (2.1)$$

$i = \sqrt{-1}$, $\lambda \in R$ and the integral with respect to $d\nu$ is defined and finite (P-a.s.) in virtue of Problem 1.1 and the inequality

$$|e^{i\lambda x} - 1 - i\lambda x I(|x| \leq 1)| \leq 2I(|x| > 1) + \frac{\lambda^2 x^2}{2} I(|x| \leq 1).$$

Observe that

a) the real and imaginary parts of the process G belong to $\mathcal{V}$ and

$$\Delta G_t = \int_{R_0} (e^{i\lambda x} - 1)\, \nu(\{t\}, dx); \qquad (2.2)$$

b) On the set $\{T < \infty\}$ with a predictable time time T (P-a.s.)

$$E(e^{i\lambda \Delta X_T} - 1 \mid \mathcal{F}_{T-}) = \int_{R_0} (e^{i\lambda x} - 1)\, \nu(\{T\}, dx) \qquad (2.3)$$

or, equivalently,

$$E(e^{i\lambda \Delta X_T} - 1 \mid \mathcal{F}_{T-}) = \Delta G_T(\lambda) \quad (\{T < \infty\};\ \text{P-a.s.}) \qquad (2.4)$$

(statement (2.3) follows by Problem 1.3);

c) the process $\Delta G = (\Delta G_t)_{t \geq 0}$ possesses the following properties: $|\Delta G| \leq 1$;

$\sum_{s \leq t} |\Delta G_s| < \infty$ (P-a.s.), $t > 0$, and $\prod_{s \leq t} |1 + \Delta G_s| > 0$ ($\{t < S(\lambda)\}$; P-a.s.) with

$$S(\lambda) = \inf(t\colon \Delta G_t = -1).$$

2. Consider Doléans equation

$$Y_t = 1 + \int_0^t Y_{s-}\, dG_s(\lambda). \qquad (2.5)$$

Since the (complex-valued) process G is a process of locally bounded variation, a unique solution of (2.5) - the stochastic exponential $\mathcal{E}$ (G) - is given by the formula

$$\mathcal{E}_t (G) = e^{G_t} \prod_{0 < s \le t} (1 + \Delta G_s)\, e^{-\Delta G_s} \tag{2.6}$$

(cf. (2.4.15)) with $G_t = G_t(\lambda)$ defined by the formula (2.1). It is easily seen that the stochastic exponential $\mathcal{E}$ (G) admits the representation

$$\mathcal{E}_t (G) = e^{G_t^c} \prod_{0 < s \le t} (1 + \Delta G_s) \tag{2.7}$$

where G^c is the continuous part of G, i.e.

$$G_t^c = i\lambda B_t^c - \frac{\lambda^2}{2} C_t + \int_0^t \int_{R_0} (e^{i\lambda x} - 1 - i\lambda x I(|x| \le 1))\, d\nu^c \tag{2.8}$$

with

$$B_t^c = B_t - \sum_{0 < s \le t} \Delta B_s, \quad d\nu^c = I(a = 0)\, d\nu$$

and

$$a_t = \nu(\{t\} \times R_0).$$

As the solution of the equation (2.5) is unique, we have $\mathcal{E}_t (G) = 0$ on the set $\{t \ge T(\lambda)\}$, with

$$T(\lambda) = \inf(t: |\mathcal{E}_t (G)| = 0). \tag{2.9}$$

Lemma 1. *The function* $|\mathcal{E}(G)|$ *possesses the following properties*:

a) $|\mathcal{E}_t (G)| \le 1$,

b) $|\mathcal{E}_t (G)| \le |\mathcal{E}_s (G)|, \ t \ge s$.

Proof. By (2.7) we have

$$|\mathcal{E}_t (G)| = \exp\left(-\frac{\lambda^2}{2} C_t - \int_0^t \int_{R_0} (1 - \cos \lambda x)\, d\nu^c \right) \prod_{0 < s \le t} |1 + \Delta G_s|, \tag{2.10}$$

and by (2.4) we have $|1 + \Delta G_s| \le 1$. Both properties a) and b) follow from this representation in an obvious manner.

Lemma 2. *Let*

$$S(\lambda) = \inf(t: \Delta G_t = -1). \tag{2.11}$$

Then $T(\lambda) = S(\lambda)$.

Proof. From (2.6) it follows that $\mathcal{E}_{S(\lambda)} (G) = 0$. Hence, $S(\lambda) \ge T(\lambda)$ on $\{T(\lambda) < \infty\}$. But (2.10) gives $|\mathcal{E}_{T(\lambda)} (G)| > 0$ on the set $\{S(\lambda) > T(\lambda)\} \cap \{T(\lambda) < \infty\}$.

This contradicts the definition of T (λ) and hence S (λ) = T (λ) on {T (λ) < ∞}. If T (λ) = ∞, then obviously S (λ) = ∞ too.

Corollary. *The time* T (λ) *is predictable* (Problem 1.3.11), *and by* (2.11) *and* (2.4) *on the set* {T (λ) < ∞}

$$\mathbf{E}\,(e^{i\lambda\Delta X_{T(\lambda)}} \mid \mathcal{F}_{T(\lambda)-}) = 0, \tag{2.12}$$

and hence ({T (λ) < ∞}; P-a.s.),

$$\mathrm{E}\,(e^{i\lambda X_{T(\lambda)}} \mid \mathcal{F}_{T(\lambda)-}) = 0. \tag{2.13}$$

Lemma 3. *For every fixed* t > 0 *the function* $\mathcal{E}_t$ (G (λ)) *is continuous in* λ *at point* 0, *i.e.*

$$\lim_{\lambda\to 0} \mathcal{E}_t\,(G(\lambda)) = 1. \tag{2.14}$$

Proof. It suffices by (2.7) to show that (P-a.s.)

$$\lim_{\lambda\to 0} G_t^c(\lambda) = 0, \tag{2.15}$$

$$\lim_{\lambda\to 0} \prod_{0<s\le t} (1 + \Delta G_s(\lambda)) = 1. \tag{2.16}$$

The relation (2.15) holds, since λB_t^c, $\lambda^2 C_t$ and $v(\lambda, x) = e^{i\lambda x} - 1 - i\lambda x I(|x| \le 1)$ are continuous functions of λ:

$$|v(\lambda, x)| \le \sqrt{2(1-\cos\lambda x)}\, I(|x|>1) + \frac{\lambda^2 x^2}{2} I(|x|\le 1),$$

and this allows us, in accordance with Problem 1.1, to apply Lebesgue's dominated convergence theorem which implies

$$\lim_{\lambda\to 0} \int_0^t \int_{R_0} v(\lambda, x)\, dv^c = 0.$$

To prove (2.16) use the fact that

$$\begin{aligned}\phi_t(\lambda) &= \prod_{0<s\le t} (1+\Delta G_s(\lambda)) = \prod_{s_i\le t} (1+\Delta G_{s_i}(\lambda)) \\ &= \prod_{s_i\le t} \left(1 + \Delta G_{s_i}(\lambda)\, I\left(|\Delta G_{s_i}(\lambda)| \le \frac{1}{2}\right)\right) \times \\ &\quad \prod_{s_i\le t} \left(1 + \Delta G_{s_i}(\lambda)\, I\left(|\Delta G_{s_i}(\lambda)| > \frac{1}{2}\right)\right) = \phi_t^1(\lambda)\, \phi_t^2(\lambda)\end{aligned}$$

with $\{s_i,\ i \ge 1\} = \{s: a_s > 0\}$ and $a_s = \nu(\{s\} \times R_0)$.

Since $G(\lambda) \in \mathcal{V}$, then $\phi_t^2(\lambda)$ consists (P-a.s.) of only a finite number of multipliers. Therefore, $\lim_{\lambda \to 0} \phi_t^2(\lambda) = 1$ by virtue of $\lim_{\lambda \to 0} \Delta G_{s_i}(\lambda) = 0$, which is implied by the representation (2.2) for $\Delta G_{s_i}(\lambda)$ and by Lebesgue's dominated convergence theorem.

Let us show that

$$\lim_{\lambda \to 0} \phi_t^1(\lambda) = 1. \tag{2.17}$$

By definition

$$\ln \phi_t^1(\lambda) = \sum_{s_i \le t} \ln\left(1 + \Delta G_{s_i}(\lambda)\, I\left(|\Delta G_{s_i}(\lambda)| \le \frac{1}{2}\right)\right).$$

By the inequality $|\ln(1+a) \le 2|a|$ for $|a| \le \frac{1}{2}$, it suffices for proving (2.17) to show that

$$\lim_{\lambda \to 0} \sum_{s_i \le t} |\Delta G_{s_i}(\lambda)| = 0.$$

By (2.2) and Problems 1.1 and 1.2 we get

$$\begin{aligned}
&\sum_{s_i \le t} |\Delta G_{s_i}(\lambda)| \\
&\le |\lambda| \sum_{s_i \le t} |\Delta B_{s_i}| + \sum_{s_i \le t} \left| \int_{R_0} (e^{i\lambda x} - 1 - i\lambda x I(|x| \le 1))\, \nu(\{s_i\}, dx) \right| \\
&\le |\lambda| \sum_{s_i \le t} |\Delta B_{s_i}| + \int_0^t \int_{R_0} \left[2I(|x| > 1) + \frac{\lambda^2 x^2}{2} I(|x| \le 1) \right] d\nu \to 0, \\
&\qquad\qquad \lambda \to 0
\end{aligned}$$

This proves the lemma.

3. Suppose X is a nondecreasing semimartingale with $X \ge 0$. In this case the triplet $T = (B, C, \nu)$ possesses the following properties:

1) $\nu(R_+ \times \{x < 0\}) = 0$ (Problem 1);

2) $xI(0 < x \le 1) * \nu \in \mathcal{A}_{loc}^+$ (Problem 3);

3) $C_\infty = 0$ (Problem 2);

4) The process $A^c = (A_t^c)_{t \ge 0}$ with

$$A_t^c = B_t^c - \int_0^t \int_{0 < x \le 1} x \, d\nu^c \tag{2.18}$$

is nondecreasing (Problem 2).

Besides (Problem 2)

$$X_t = X_0 + A_t^c + \int_0^t \int_{x > 0} x \, d\mu. \tag{2.19}$$

On studying such martingales it is often useful to consider the function

$$G_t = - \lambda A_t^c + \int_0^t \int_{x > 0} (e^{-\lambda x} - 1) \, d\nu \tag{2.20}$$

with $\lambda \ge 0$, instead of the function (2.1).

The stochastic exponential $\mathcal{E}_t(G)$ corresponding to this function is given by the formulas (2.6) and (2.7) with

$$G_t^c = - \lambda A_t^c + \int_0^t \int_{x > 0} (e^{-\lambda x} - 1) \, d\nu^c, \tag{2.21}$$

where A^c is defined in (2.18) and

$$\Delta G_t = \int_{x > 0} (e^{-\lambda x} - 1) \, \nu(\{t\}, dx). \tag{2.22}$$

It is also representable in the following manner

$$\mathcal{E}_t(G) = e^{G_t^c} \prod_{0 < s \le t} \left(1 + \int_{x > 0} (e^{-\lambda x} - 1) \, \nu(\{s\}, dx) \right). \tag{2.23}$$

Besides, $0 < \mathcal{E}_t(G) \le 1$, $\mathcal{E}_t(G) \le \mathcal{E}_s(G)$ for $t \ge s$ and $\mathcal{E}_t(G(\lambda))$ is a continuous function of λ.

Problems

1. Show that for a nondecreasing semimartingale X a version of the compensator ν can be chosen such that $\nu(R_+ \times \{x < 0\}) = 0$.

2.Let X be a nondecreasing nonnegative semimartingale. Show that the process A^c with

$$A_t^c = B_t^c - \int_0^t \int_{0 < x \le 1} x \, d\nu^c$$

is nondecreasing, $C_\infty = 0$ and the representation

$$X_t = X_0 + A_t^c + \int_0^t \int_{x>0} x \, d\mu$$

holds.

3. Let X be such as in Problem 2. Show that $xI(0 < x \leq 1) * \nu \in \mathcal{A}^+_{loc}$.

4. Let $\mathcal{E}(G)$ be the stochastic exponential where G is the cumulant defined by the representation (2.1). Show that in case $|\mathcal{E}_t(G)| > 0$, $t > 0$,

$$\mathcal{E}_t^{-1}(G) = \exp\left(-\int_0^t \frac{dG_s}{1+\Delta G_s}\right) \prod_{0<s\leq t} \left(1 - \frac{\Delta G_s}{1+\Delta G_s}\right) e^{\frac{\Delta G_s}{1+\Delta G_s}} = \mathcal{E}_t(\tilde{G})$$

with

$$\tilde{G}_t = -\int_0^t \frac{dG_s}{1+\Delta G_s}.$$

§ 3. Martingale characterization of semimartingales by means of stochastic exponentials

1. Consider a semimartingale X with the triplet $T(X) = (B, C, \nu)$, with the corresponding cumulant function

$$G_t(\lambda) = i\lambda B_t - \frac{\lambda^2}{2} C_t + \int_0^t \int_{R_0} (e^{i\lambda x} - 1 - i\lambda x I(|x| \le 1))\, d\nu \tag{3.1}$$

and with the stochastic exponential $\mathcal{E}(G(\lambda)) = (\mathcal{E}_t(G(\lambda)))_{t \ge 0}$.

Denote

$$T(\lambda) = \inf(t: |\mathcal{E}_t(G(\lambda))| = 0) \tag{3.2}$$

and

$$Z_t(\lambda) = \exp(i\lambda(X_t - X_0))\, \mathcal{E}_t^{-1}(G(\lambda))\, I(|\mathcal{E}_t(G(\lambda))| > 0). \tag{3.3}$$

Theorem 1. *For every* $\lambda \in R$ *there exists a sequence* $(T_k(\lambda))_{k \ge 1}$ *of Markov times such that* $T_k(\lambda) \uparrow T(\lambda)$, $T_k(\lambda) < T(\lambda)$, $k \ge 1$, *and for every* $k \ge 1$ *the process* $Z^k(\lambda) = (Z_{t \wedge T_k(\lambda)}(\lambda))_{t \ge 0}$ *is a uniformly bounded martingale.*

Proof. In all considerations below we will fix the parameter λ and write T, T_k, G, ... instead of $T(\lambda)$, $T_k(\lambda)$, $G(\lambda)$, ...

Set

$$T_k' = \inf\left(t: |\mathcal{E}_t(G)| \le \frac{1}{k}\right), \quad k \ge 1.$$

Clearly, $T_k' \le T$, $T_k' \uparrow T$, and T_k' is a predictable Markov time (Problem 1.3.11). Based on Theorem 1.3.4 one can construct a sequence $(T_k)_{k \ge 1}$ of Markov times such that $T_k < T_k'$ and $T_k \uparrow T$. Therefore

$$|\mathcal{E}_{t \wedge T_k}(G)| \ge |\mathcal{E}_{T_k}(G)| \ge \frac{1}{k}$$

and by (3.3) we get the estimate $\sup_{t \le T_k} |Z_t| \le k$.

Next,

$$Z_{t \wedge T_k} = \exp(i\lambda(X_{t \wedge T_k} - X_0))\, \mathcal{E}_{t \wedge T_k}^{-1}(G), \tag{3.4}$$

where by Problem 2.4

$$\mathcal{E}^{-1}_{t\wedge T_k}(G) = \exp\left(-\int_0^{t\wedge T_k} \frac{dG_s}{1+\Delta G_s}\right) \prod_{0<s\le t\wedge T_k}\left(1-\frac{\Delta G_s}{1+\Delta G_s}\right) e^{\frac{\Delta G_s}{1+\Delta G_s}}.$$

By this representation and by (2.6) it follows that $\mathcal{E}^{-1}_{t\wedge T_k}(G)$ is the solution of Doléans equation

$$\mathcal{E}^{-1}_{t\wedge T_k}(G) = 1 - \int_0^{t\wedge T_k} \mathcal{E}^{-1}_{s-}(G)\frac{dG_s}{1+\Delta G_s}. \tag{3.5}$$

Denote

$$L_t = \exp(i\lambda(X_t - X_0)). \tag{3.6}$$

Then, taking into consideration the representation

$$X_t = X_0 + B_t + \int_0^t \int_{|x|>1} x\, d\mu + X^c_t + \int_0^t \int_{|x|\le 1} x\, d(\mu-\nu) \tag{3.7}$$

and the formula (3.1), by Ito's formula (Ch. 2, § 3) we get

$$L_t = 1 + \int_0^t L_{s-}\, dG_s + i\lambda\int_0^t L_{s-}\, dX^c_s + \int_0^t\int_{R_0} L_{s-}(e^{i\lambda x} - 1)\, d(\mu-\nu). \tag{3.8}$$

By (3.5) and (3.8) it is seen, applying Ito's formula (Ch. 2, § 3), that as

$$Z_{t\wedge T_k} = L_{t\wedge T_k}\,\mathcal{E}^{-1}_{t\wedge T_k}(G)$$

we have

$$Z_{t\wedge T_k} = 1 + i\lambda\int_0^{t\wedge T_k} \frac{Z_{s-}}{1+\Delta G_s} dX^c_s + \int_0^{t\wedge T_k}\int_{R_0} Z_{s-}\frac{e^{i\lambda x}-1}{1+\Delta G_s}\, d(\mu-\nu) \tag{3.9}$$

This and the estimate $\sup_{t\le T_k} |Z_t| \le k$ imply that for each $k \ge 1$ $(Z_{t\wedge T_k})_{t\ge 0}$ is a uniformly bounded martingale (for each $\lambda \in R$).

Remark. From the formula (3.8) it follows that the process $e^{i\lambda X} - e^{i\lambda X_-} \circ G(\lambda) \in \mathcal{M}_{loc}$.

2. Let X be a nondecreasing semimartingale with $X_0 \ge 0$ and

$$Z_t(\lambda) = e^{-\lambda(X_t - X_0)}\, \mathcal{E}_t^{-1}(G(\lambda)) \tag{3.10}$$

with G (λ) defined by the formula (2.20). Then the process Z (λ) is a local martingale for each $\lambda \in R$.

The proof of this statement is based on the representation

$$Z_t = 1 + \int_0^t \int_{x>0} Z_{s-} \frac{e^{-\lambda x} - 1}{1 + \Delta G_s}\, d(\mu - \nu), \tag{3.11}$$

that is obtained with the help of Ito's formula (cf. (3.9)).

3. Thus, if X is a semimartingale then, according to Theorem 1, the stopped process $Z^k(\lambda)$ is a uniformly bounded martingale.

The reverse assertion holds too, and this makes evident the usefulness of introducing the stochastic exponential for the characterizing of semimartingales.

To show this we assume now that an initial probability space $(\Omega, \mathcal{F}, P)$ presents the space $(D, \mathcal{D}, Q)$ where D is Skorohod's space of right-continuous functions $X = (X_t)_{t \geq 0}$ having left-hand limits, $\mathcal{D} = \sigma(X_t,\ t \geq 0)$ and

$$\mathcal{D}_t^Q = \bigcap_{\varepsilon > 0} (\sigma\{X_s,\ 0 \leq s \leq t + \varepsilon\} \vee \mathcal{N}_Q)$$

with the collection of sets $\mathcal{N}_Q$ from $\mathcal{D}$, having Q-measure zero. Let

$$\mathbb{D} = (\mathcal{D}_t^Q)_{t \geq 0}.$$

The set of objects (B, C, ν) is also asssumed to be given, where

I. $B = (B_t(X))_{t \geq 0} \in \mathcal{V} \cap \mathcal{P}(\mathbb{D})$,

II. $C = (C_t(X))_{t \geq 0} \in \mathcal{V}^+ \cap C$,

III. $\nu = \nu(X;\ dt, dx)$ is a (random) measure on $(R_+ \times R_0,\ B(R_+) \otimes B(R_0))$ with the following properties (Q-a.s.):

(a) $\int_0^t \int_{R_0} (c \wedge x^2)\, d\nu < \infty,\ t > 0,\ c > 0;$

(b) $\nu(X; \{0\} \times R_0) = 0,\ \nu(X; \{t\} \times R_0) \leq 1;$

(c) $\nu(X; \{t\} \times \{|x| \leq 1\}) = \Delta B_t(X);$

(d) for each $\mathcal{P}(\mathbb{D}) \otimes B(R_0)$-measurable function f (X, t, x) with $|f| \leq c \wedge x^2$ the process

$$\left(\int_0^t \int_{R_0} f\, d\nu \right)_{t \geq 0} \in \mathcal{V} \cap \mathcal{P}(\mathbb{D}).$$

Relate to the set (B, C, ν) the cumulant $G = (G_t(\lambda, X))_{t \geq 0}$ with

$$G_t(\lambda, X) = i\lambda B_t(X) - \frac{\lambda^2}{2} C_t(X) + \int_0^t \int_{R_0} (e^{i\lambda x} - 1 - i\lambda x I(|x| \leq 1))\, d\nu(X) \tag{3.12}$$

(cf. (2.1)) and the stochastic exponential $\mathcal{E}(G) = (\mathcal{E}_t(G(\lambda, X)))_{t \geq 0}$ with

$$\mathcal{E}_t(G(\lambda, X)) = \exp(G_t(\lambda, X)) \prod_{0 < s \leq t} (1 + \Delta G_s(\lambda, X))\, e^{-\Delta G_s(\lambda, X)}. \tag{3.13}$$

For a process X we set also

$$Z_t(\lambda, X) = \exp(i\lambda(X_t - X_0))\, \mathcal{E}_t^{-1}(G(\lambda, X))\, I(|\mathcal{E}_t(G(\lambda, X))| > 0 \quad (13.4)\ (3.14)$$

$$T(\lambda, X) = \inf(t: |\mathcal{E}_t(G(\lambda, X))| = 0). \tag{3.15}$$

It is deduced from the proof of Theorem 1 that there exist Markov times $(T_k(\lambda, X))_{k \geq 1}$ such that $T_k(\lambda, X) < T(\lambda, X)$,

$$|Z_{t \wedge T_k(\lambda, X)}(\lambda, X)| \leq 1/k, \ k \geq 1, \ T_k(\lambda, X) \uparrow T(\lambda, X).$$

Theorem 2. 1) *For each* $\lambda \in R$ *and* $k \geq 1$ *let the process*

$$Z^k(\lambda, X) = (Z_{t \wedge T_k(\lambda, X)}(\lambda, X))_{t \geq 0}$$

be a local martingale. Then the process $X = (X_t)_{t \geq 0}$ *is a semimartingale.*

2) *If* $(\tilde{B}, \tilde{C}, \tilde{\nu})$ *is the triplet of the semimartingale* $X = (X_t)_{t \geq 0}$ *involved in* 1) *and if at least one of the following conditions holds:*

(a) $$Q\left(\int_{R_0} (e^{i\lambda x} - 1)\, \nu(X; \{t\}, dx) \neq -1, \ t > 0 \right) = 1$$

or

(b) $$\tilde{\nu} \overset{Q}{=} \nu$$

holds, then

$$(\tilde{B}, \tilde{C}, \tilde{\nu}) \overset{Q}{=} (B, C, \nu). \tag{3.16}$$

Proof. Let us first specify the expression

$$\tilde{\nu} \overset{Q}{=} \nu$$

that means Q-indistinguishability for each $\Gamma \in B(R_0)$ of the processes $(\tilde{\nu}((0, t] \times \Gamma))_{t \geq 0}$ and $(\nu((0, t] \times \Gamma))_{t \geq 0}$.

1) We will show that the process X is a semimartingale. To this end observe that the process $Y(X) = (Y_t(X))_{t \geq 0}$ with

$$Y_t(X) = X_0 + \sum_{0 < s \leq t} \Delta X_s I(|\Delta X_s| > 1)$$

belongs to the class $\mathcal{V}$, and hence Y (X) is a semimartingale. Therefore, it suffices to show that the process $X' = X - Y(X)$ is a semimartingale.

To prove this, let us consider a sequence of numbers $(\lambda_k)_{k \geq 1}$ with $0 < |\lambda_k| \leq \frac{\pi}{8}$, $|\lambda_k| \downarrow 0$, $k \to \infty$. For each $k \geq 1$ set $T_k(X) = T_k(\lambda_k, X)$. By the definition of $T_k(\lambda, X)$ it follows that $T_k(X) < T(\lambda_k, X)$. Moreover, by Theorem 1.3.4 one can define the Markov times $(T_k(\lambda, X))_{k \geq 1}$ in such a way that

$$Q\left(T(\lambda_k, X) - T_k(X) > \frac{1}{k} \right) \leq \frac{1}{k}. \tag{3.17}$$

Set also

$$\sigma_k = \inf\left(t: |X'_t| \geq \frac{\pi}{8|\lambda_k|} \right) \wedge T_k(X). \tag{3.18}$$

Then $\sigma_k < T(\lambda_k, X)$ and by (3.14) we have

$$\exp(i\lambda_k(X_{t \wedge \sigma_k} - X_0)) = Z_{t \wedge \sigma_k}(\lambda_k, X)\, \mathcal{E}_{t \wedge \sigma_k}(G(\lambda_k, X)).$$

Consequently,

$$\exp(i\lambda_k X'_{t \wedge \sigma_k}) = Z_{t \wedge \sigma_k}(\lambda_k, X)\, \mathcal{E}_{t \wedge \sigma_k}(G(\lambda_k, X)) \exp(-i\lambda_k Y_{t \wedge \sigma_k}(X)). \tag{3.19}$$

The right-hand side of (3.19) consists of the product of semimartingales, and hence by Ito's formula (Ch. 2, § 3) the process $(\exp(i\lambda_k X'_{t \wedge \sigma_k}))_{t \geq 0}$ and, respectively, the processes $(\sin \lambda_k X'_{t \wedge \sigma_k})_{t \geq 0}$ and $(\cos \lambda_k X'_{t \wedge \sigma_k})_{t \geq 0}$ are semimartingales for each $k \geq 1$.

Observe now that $|\Delta X'| \leq 1$. Therefore, by the definition of σ_k we have

$$\sup_{t \leq \sigma_k} |X'_t| \leq \frac{\pi}{8|\lambda_k|} + 1,$$

and since $|\lambda_k| \leq \frac{\pi}{8}$, $k \geq 1$, we have

$$\sup_{t \le \sigma_k} |\lambda_k X'_t| \le \frac{\pi}{8} + |\lambda_k| \le \frac{\pi}{4}.$$

From this and the fact that $(\sin \lambda_k X'_{t \wedge \sigma_k})_{t \ge 0}$ is a semimartingale, we deduce, by applying Ito's formula, that the process $(\lambda_k X'_{t \wedge \sigma_k})_{t \ge 0}$ is a semimartingale too.

Let us introduce Markov times $\gamma_k = \max(\sigma_1, \ldots, \sigma_k)$ and show that the process $(X'_{t \wedge \gamma_k})_{t \ge 0}$ is a semimartingale.

Since $\gamma_1 = \sigma_1$, then $(X_{t \wedge \gamma_1})_{t \ge 0}$ is a semimartingale. Next, as $k \ge 2$ we have

$$X'_{t \wedge \gamma_k} = X'_{t \wedge \gamma_{k-1}} + \int_0^t I(\gamma_{k-1} < s \le \sigma_k)\, dX'_{s \wedge \sigma_k},$$

and this gives by induction the desired assertion.

Since $|\Delta X'| \le 1$, by Theorem 2.1.2 a semimartingale $(X'_{t \wedge \gamma_k})_{t \ge 0}$ is special, and it admits the decomposition

$$X'_{t \wedge \gamma_k} = A_t^{(k)} + M_t^{(k)} \tag{3.20}$$

with $A^{(k)} \in \mathcal{V} \cap \mathcal{P}$ and $M^{(k)} \in \mathcal{M}_{loc}$.

As the decomposition (3.20) (with a predictable process $A^{(k)}$) is unique, we have

$$A^{(k+1)}_{t \wedge \gamma_k} = A^{(k)}_{t \wedge \gamma_k}, \quad M^{(k+1)}_{t \wedge \gamma_k} = M^{(k)}_{t \wedge \gamma_k} \quad (Q\text{-a.s.}). \tag{3.21}$$

By showing

$$Q(\lim_k \gamma_k = \infty) = 1, \tag{3.22}$$

one can deduce from (3.21) that $X'_t = A_t + M_t$ where

$$A = (A_t)_{t \ge 0} \in \mathcal{V} \cap \mathcal{P}, \; M = (M_t)_{t \ge 0} \in \mathcal{M}_{loc}$$

with

$$A_t = A^{(1)}_{t \wedge \gamma_1} + \sum_{k \ge 1} [A^{(k+1)}_{t \wedge \gamma_{k+1}} - A^{(k)}_{t \wedge \gamma_k}],$$

$$M_t = M^{(1)}_{t \wedge \gamma_1} + \sum_{k \ge 1} [M^{(k+1)}_{t \wedge \gamma_{k+1}} - M^{(k)}_{t \wedge \gamma_k}].$$

Thus to prove that X' is a semimartingale, it remains to establish the property

(3.22) or the equivalent relation

$$Q\,(\lim_k \gamma_k \le a) = 0 \quad \forall a > 0. \tag{3.23}$$

Since for each $j \ge 1$ we have $\lim_k \gamma_k \ge \sigma_j$, then

$$Q\,(\lim_k \gamma_k \le a) \le Q\,(\sigma_j \le a), \;\; j \ge 1 \quad \forall a > 0,$$

and to verify (3.23) it suffices to show that

$$\lim_j Q\,(\sigma_j \le a) = 0 \quad \forall a > 0. \tag{3.24}$$

From (3.18) it follows that

$$Q\,(\sigma_j \le a) = Q\left(\left(\sup_{t \le a} |X_t'| \ge \frac{\pi}{8\,|\lambda_j|}\right) \cup (T_j\,(X) \le a)\right)$$

$$\le Q\left(\left(\sup_{t \le a} |X_t'| \ge \frac{\pi}{8\,|\lambda_j|}\right)\right) + Q\,(T_j\,(X) \le a). \tag{3.25}$$

Since $|\lambda_j| \downarrow 0$ and a number a is fixed, we have

$$\lim_j Q\left(\sup_{t \le a} |X_t'| \ge \frac{\pi}{8\,|\lambda_j|}\right) = 0. \tag{3.26}$$

Next, by taking into consideration (3.17) and Lemma 2.3 which can be applied here in view of Conditions (a) – (d) imposed on the measure ν, we have

$$Q\,(T_j\,(X) \le a) \le Q\left(T\,(\lambda_j, X) \le a + \frac{1}{j}\right) + Q\left(T\,(\lambda_j, X) - T_j\,(X) > \frac{1}{j}\right)$$

$$\le Q\,(T\,(\lambda_j, X) \le a + 1) + \frac{1}{j}$$

$$= Q\,(|\,\mathcal{E}_{a+1}\,(G\,(\lambda_j, X))\,| = 0) + \frac{1}{j} \to 0, \;\; j \to \infty. \tag{3.27}$$

By (3.25) – (3.27) we get (3.24).

Thus it is proved that the process X is a semimartingale.

2) Let Condition (a) be fullfilled, i.e. Q-a.s. $\Delta G_t\,(\lambda, X) \ne -1$, $t > 0$. Then by Lemma 2.2 we have $T\,(\lambda, X) = \infty$ (Q-a.s.) for each $\lambda \in R$ and the multiplicative decomposition

$$\exp\,(i\lambda\,(X_t - X_0)) = Z_t\,(\lambda, X)\,\mathcal{E}_t\,(G\,(\lambda, X)) \tag{3.28}$$

takes place. By means of the triplet $(\tilde{B}, \tilde{C}, \tilde{\nu})$ of a semimartingale X we define the cumulant $\tilde{G}\,(\lambda, X)$ (analogously to the formula (3.12) in which (B, C, ν) is replaced by $(\tilde{B}, \tilde{C}, \tilde{\nu})$) and the process

$$\tilde{Z}_t(\lambda, X) = \exp(i\lambda(X_t - X_0))\, \mathcal{E}_t^{-1}(\tilde{G}(\lambda, X))\, I(|\mathcal{E}_t(\tilde{G}(\lambda, X))| > 0) \quad (3.29)$$

and we set

$$\tilde{T}(\lambda, X) = \inf(t: |\mathcal{E}_t(\tilde{G}(\lambda, X))| = 0). \quad (3.30)$$

Let $(T_k(\lambda, X))_{k \geq 1}$ be an announcing sequence for a predictable time $T(\lambda, X)$ such that

$$|\mathcal{E}_t(G(\lambda, X))| \geq \frac{1}{k} \text{ on } [\![0, T_k(\lambda, X)]\!]$$

(cf. the proof of Theorem 1). If $Y_t = e^{i\lambda(X_t - X_0)}$, then $(T_k = T_k(\lambda, X))$

$$Y^{T_k} = Z^{T_k} \mathcal{E}^{T_k}(G).$$

By Ito's formula (Ch. 2, § 3) the process

$$Y^{T_k} - Z_- \circ \mathcal{E}^{T_k}(G) = Y^{T_k} - Y_- \circ G^{T_k}$$

is a local martingale.

But according to Remark to Theorem 1 the process $Y^{T_k} - Y_- \circ \tilde{G}^{T_k}$ is a local martingale too. As the decomposition of special semimartingales with a predictable process of locally bounded variation is unique, we therefore have

$$Y_- \circ G^{T_k} = Y_- \circ \tilde{G}^{T_k}.$$

This gives

$$Y_-^{-1} \circ (Y_- \circ G^{T_k}) = Y_-^{-1} \circ (Y_-^{-1} \circ \tilde{G}^{T_k}),$$

hence $G^{T_k} = \tilde{G}^{T_k}$, i.e. $G = \tilde{G}$ on $[\![0, T(\lambda, X)[\![$.

Thus $T(\lambda, X) \leq \tilde{T}(\lambda, X)$ (Q-a.s.), and since $T(\lambda, X) = \infty$ as Condition (a) holds, we have $G = \tilde{G}$, $\lambda \in R$.

By this we will deduce the equality $(B, C, \nu) \overset{Q}{=} (\tilde{B}, \tilde{C}, \tilde{\nu})$.

To this end define for $t \in R$ and α, $\lambda \in R$ the function

$$G_t^\alpha(\lambda, X) = G_t(\lambda, X) - \frac{1}{2}\int_{-1}^{1} G_t(\lambda + \alpha s, X)\, ds. \quad (3.31)$$

Define analogously the function $\tilde{G}_t^\alpha(\lambda, X)$.

Since $G = \tilde{G}$ and $\lambda \in R$, then $G^\alpha = \tilde{G}^\alpha$ (in the sense of Q-indistinguishability). Next

$$G_t^\alpha(\lambda, X) = \int_R e^{i\lambda x}\, \eta_\alpha(X; [0, t], dx) \tag{3.32}$$

where, as is easily verified,

$$\eta_\alpha(X; [0, t], dx) = \frac{\alpha^2}{6} C_t(X)\, \varepsilon_0(dx) + \left(1 - \frac{\sin \alpha x}{\alpha x}\right) \nu(X; [0, t], dx) \tag{3.33}$$

(here ε_0 is Dirac's measure with mass at point $\{0\}$).

Analogously the function $\tilde{G}_t^\alpha(\lambda, X)$ is related to the measure

$$\tilde{\eta}_\alpha(X; [0, t], dx) = \frac{\alpha^2}{6} \tilde{C}_t(X)\, \varepsilon_0(dx) + \left(1 - \frac{\sin \alpha x}{\alpha x}\right) \tilde{\nu}(X; [0, t], dx). \tag{3.34}$$

By (3.32) and (3.33) it is seen that

$$\frac{6G_t^\alpha(0, X)}{\alpha^2} = C_t(X) + \int_0^t \int_{R_0} \left(1 - \frac{\sin \alpha x}{\alpha x}\right) \frac{6}{\alpha^2}\, d\nu. \tag{3.35}$$

Observe that as $\delta \in (0, 1]$ we have

$$\int_0^t \int_{R_0} \left(1 - \frac{\sin \alpha x}{\alpha x}\right) \frac{6}{\alpha^2}\, d\nu \le \frac{12}{\alpha^2} \int_0^t \int_{|x| > \delta} d\nu + \int_0^t \int_{|x| \le \delta} x^2\, d\nu,$$

and the right-hand side converges to zero as the limit $\lim_{\delta \to 0} \lim_{\alpha \to \infty}$ is taken. By this and (3.35) it is seen that

$$C_t(X) = \lim_{\alpha \to \infty} \frac{6}{\alpha^2} G_t^\alpha(0, X)$$

and, analogously,

$$\tilde{C}_t(X) = \lim_{\alpha \to \infty} \frac{6}{\alpha^2} \tilde{G}_t^\alpha(0, X).$$

Thus, from the Q-indistinguishability of the processes G^α and $\tilde{G}^\alpha$ it follows that the processes C and $\tilde{C}$ are Q-indistinguishable too. This in turn implies Q-indistinguishability of the processes

$$\left(\int_0^t \int_{R_0} e^{i\lambda x}\left(1 - \frac{\sin \alpha x}{\alpha x}\right) d\nu\right)_{t \ge 0} \quad \text{and} \quad \left(\int_0^t \int_{R_0} e^{i\lambda x}\left(1 - \frac{\sin \alpha x}{\alpha x}\right) d\tilde{\nu}\right)_{t \ge 0}$$

for each $\alpha, \lambda \in R$. From this it is deduced by standard arguments that the processes

$$\left(\int_0^t \int_{R_0} g(x)\, d\nu\right)_{t \geq 0} \quad \text{and} \quad \left(\int_0^t \int_{R_0} g(x)\, d\tilde{\nu}\right)_{t \geq 0}$$

are Q-indistinguishable for each B (R_0)-measurable functions $g = g(x)$ with $|g(x)| \leq c \wedge x^2$ where c is a positive constant.

Hence $\nu = \tilde{\nu}$. If now we set $g(x) = e^{i\lambda x} - 1 - i\lambda x$ $(|x| \leq 1)$, then by the representation for G and $\tilde{G}$ we get Q-indistinguishability of the processes B and $\tilde{B}$.

Thus, $(B, C, \nu) \overset{Q}{=} (\tilde{B}, \tilde{C}, \tilde{\nu})$.

Let us obtain now this equality under Condition (b): $\nu \overset{Q}{=} \tilde{\nu}$.

Let σ_k, $k \geq 1$, be stopping times defined in the course of proving the first part of the theorem and let $\tilde{\sigma}_k$, $k \geq 1$, be stopping times defined analogously with $(\tilde{B}, \tilde{C}, \tilde{\nu})$ instead of (B, C, ν). Set $\bar{\sigma}_k = \sigma_k \wedge \tilde{\sigma}_k$. Applying the method of proving under Condition (a) the second part of the theorem, it is easily seen that the processes $G^{\bar{\sigma}_k}(\lambda_k, X)$ and $\tilde{G}^{\bar{\sigma}_k}(\lambda_k, X)$ are Q-indistinguishable for each $k \geq 1$, where λ_k, $k \geq 1$, is a sequence of numbers defined in the course of proving the first part of the theorem. This and the equality $\nu \overset{Q}{=} \tilde{\nu}$ imply the negligibility of the processes

$$i\lambda_k\left(B^{\bar{\sigma}_k} - \tilde{B}^{\bar{\sigma}_k}\right) - \frac{\lambda_k^2}{2}\left(C^{\bar{\sigma}_k} - \tilde{C}^{\bar{\sigma}_k}\right), \quad k \geq 1.$$

This in turn implies the negligibility of the processes $B^{\bar{\sigma}_k} - \tilde{B}^{\bar{\sigma}_k}$ and $C^{\bar{\sigma}_k} - \tilde{C}^{\bar{\sigma}_k}$, $k \geq 1$. Set $\bar{\gamma}_k = \max(\bar{\sigma}_1, \ldots, \bar{\sigma}_k)$ and denote

$$\xi_t = \sup_{s \leq t}\left(|B_s - \tilde{B}_s| + |C_s - \tilde{C}_s|\right).$$

Then

$$Q(\xi_{\bar{\gamma}_k} > 0) = Q\left(\bigcup_{j=1}^{k} \{\xi_{\bar{\sigma}_j} > 0\}\right) \leq \sum_{j=1}^{k} Q(\xi_{\bar{\sigma}_j} > 0) = 0, \quad k \geq 1,$$

i.e. the processes $B^{\bar{\gamma}_k} - \tilde{B}^{\bar{\gamma}_k}$ and $C^{\bar{\gamma}_k} - \tilde{C}^{\bar{\gamma}_k}$, $k \geq 1$, are negligible. Since for each $a > 0$ we have $\lim_k Q(\sigma_k \leq a) = 0$ and $\lim_k Q(\bar{\sigma}_k \leq a) = 0$,then $\lim_k \bar{\gamma}_k = \infty$.

This gives

$$Q(\xi_\infty > 0) = Q\big(\bigcup_{k\geq 1}\{\xi_{\bar{\gamma}_k} > 0\}\big) \leq \sum_{k\geq 1} Q(\xi_{\bar{\gamma}_k} > 0) = 0.$$

The theorem is proved.

Corollary. *Let* $L(X) = (L_t(X))_{t\geq 0}$ *be a* $\mathbb{D}$*-adapted process with continuous trajectories, let* $C = (C_t(X))_{t\geq 0}$ *be a* $\mathbb{D}$*-adapted process too with trajectories in* $V^+ \cap C$ *and let* Q *be a certain probability measure on* $(D, \mathcal{D}^Q)$. *Let for each* $\lambda \in R$ *the process* $Z(\lambda, X) = Z_t(\lambda, X))_{t\geq 0}$ *with*

$$Z_t(\lambda, X) = \exp\left(i\lambda L_t(X) + \frac{\lambda^2}{2} C_t(X)\right)$$

be a local martingale (with respect to Q*). Then the process* L (X) *is a local martingale with the quadratic characteristic* $\langle L(X)\rangle = C(X)$.

4. It has been pointed out in Subsection 3.2 that if X is a semimartingale with decreasing trajectories, then it admits the representation (2.19).

Besides, the process Z (λ), defined by the formula (3.10), is a local martingale. Analogously to Theorem 2, we give a "martingale" characterization of semimartingales with nondecreasing trajectories.

The space $(D, \mathcal{D}, Q)$ is supposed to be given as well as the following objects:

I. $B = (B_t(X))_{t\geq 0} \in \mathcal{V} \cap \mathcal{P}(\mathbb{D})$;

II. $\nu = \nu(X; dt, dx)$, which is a predictable (random) measure on $(R_+ \times R_0, \mathcal{B}(R_+) \otimes \mathcal{B}(R_0))$ with the properties (Q-a.s.):

(a) $$\int_0^t \int_{R_0} (c \wedge |x|)\, d\nu < \infty, \quad t > 0, \quad c > 0;$$

(b) $\nu(X; \{0\} \times R_0) = 0, \quad \nu(X; \{t\} \times R_0) \leq 1, \quad \nu(X; R_+ \times \{x < 0\}) = 0$;

(c) $\nu(X; \{t\} \times \{0 < x \leq 1\}) = \Delta B_t(X)$;

(d) $$B_t^c(X) - B_s^c(X) + \int_s^t \int_{0<x\leq 1} x\, d\nu \geq 0.$$

Theorem 3. *Let* $X = (X_t)_{t\geq 0}$ *be a* $\mathbb{D}$*-adapted process,* $X_0 \geq 0$, $X_t - X_s \geq 0$, $t \geq s$. *Let Conditions* **I** *and* **II** *be fullfilled and let the process* Z (λ), *defined by* (3.10), *be a local martingale.*

Then X *is a semimartingale with the triplet of predictable characteristics* (B, 0, ν).

Proof. By assumption $X \in \mathcal{V}^+$, and hence X is a semimartingale.

Let its triplet be the collection $(\tilde{B}, 0, \tilde{\nu})$ (concerning the equality $C \overset{Q}{=} 0$, see Problem 2.2). Let us show that $(B, \nu) \overset{Q}{=} (\tilde{B}, \tilde{\nu})$.

By (2.19) we have

$$X_t = X_0 + \tilde{A}^c_t(X) + \int_0^t \int_{x>0} x \, d\mu \tag{3.36}$$

with

$$\tilde{A}^c_t = \tilde{B}^c_t - \int_0^t \int_{0<x\le 1} x \, d\tilde{\nu}^c.$$

For $\lambda \ge 0$ set

$$\tilde{G}_t(\lambda, X) = -\lambda \tilde{A}^c_t(X) + \int_0^t \int_{x>0} (e^{-\lambda x} - 1) \, d\tilde{\nu}$$

(cf. (2.20)).

Then, in accordance with a characterization of type (3.10)

$$e^{-\lambda(X_t - X_0)} = \tilde{Z}_t(\lambda, X) \, \mathcal{E}_t(\tilde{G}(\lambda, X)) \tag{3.37}$$

where $\tilde{Z}(\lambda, X)$ is a local martingale. Analogously

$$e^{-\lambda(X_t - X_0)} = Z_t(\lambda, X) \, \mathcal{E}_t(G(\lambda, X)) \tag{3.38}$$

where $Z(\lambda, X)$ is a local martingale and

$$G_t(\lambda, X) = -\lambda \tilde{A}^c_t(X) + \int_0^t \int_{x>0} (e^{-\lambda x} - 1) \, d\nu.$$

Since the processes $\mathcal{E}(\tilde{G}(\lambda, X))$ and $\mathcal{E}(G(\lambda, X))$ are predictable, the uniqueness of the multiplicative decomposition of positive semimartingales (Ch. 2, § 5) entails the Q-indistinguishability of $\mathcal{E}(\tilde{G}(\lambda, X))$ and $\mathcal{E}(G(\lambda, X))$ and hence of $\tilde{G}(\lambda, X)$ and $G(\lambda, X)$ too.

For $t > 0$, $\lambda \in R$, $X \in D$ and $\alpha \in [-1, 1]$ set

$$G^a_t(\lambda, X) = G_t(\lambda, X) - \frac{1}{2}\int_{-1}^{1} G_t(\lambda + \alpha s, X) \, ds,$$

and define $\tilde{G}^\alpha_t(\lambda, X)$ analogously. It is not hard to see that

$$G^\alpha_t(\lambda, X) = \int_{R_+} e^{-\lambda x} \, \eta_\alpha(X; [0, t], dx)$$

with

$$\eta_\alpha (X; [0, t], dx) = \left(1 - \frac{\operatorname{sh} \alpha x}{\alpha x}\right) \nu (X; [0, t], dx).$$

In an analogous manner the function $\tilde{G}_t^\alpha (\lambda, X)$ is related to the measure

$$\tilde{\eta}_\alpha (X; [0, t], dx) = \left(1 - \frac{\operatorname{sh} \alpha x}{\alpha x}\right) \tilde{\nu} (X; [0, t], dx).$$

As α is arbitrary and $G^\alpha (\lambda, X)$ and $\tilde{G}^\alpha (\lambda, X)$ are Q-indistinguishable, it is not hard to deduce from this that $\nu \overset{Q}{=} \tilde{\nu}$. Therefore $A^c (X)$ and $\tilde{A}^c (X)$ are Q-indistinguishable and hence $B (X)$ and $\tilde{B} (X)$ are Q-indistinguishable too.

Corollary. *Let* $g = g(x)$ *be a continuous nonnegative bounded function with*

$$g (x) = g (x) I (|x| > \varepsilon)$$

for a certain $\varepsilon > 0$. *Denote*

$$Y_t (X) = \sum_{0 < s \le t} g (\Delta X_s), \quad G_t^g (\lambda, X) = \int_0^t \int_{R_0} (e^{-\lambda g (x)} - 1) \, d\nu.$$

Suppose Q *is a certain probability measure on* $(D, \mathcal{D}^Q)$ *such that for each function* $g = g (x)$ *with the indicated properties and for each* $\lambda \in R_+$ *the process* $Z (X) = (Z_t (X))_{t \ge 0}$ *with*

$$Z_t (X) = e^{-\lambda Y_t (X)} \mathcal{E}_t^{-1} (G^g (\lambda, X))$$

is a local martingale.

Then $\nu = \nu (X; dt, dx)$ *is the compensator of the jump measure of a process* X.

Problems

1. Establish the validity of the formulas (3.8), (3.9) and (3.11).
2. Prove Corollary to Theorem 3.

§ 4. Charactarization of semimartingales with conditionally independent increments

1. Let $(\Omega, \mathcal{F}, \mathbb{F}, \mathsf{P})$ be a stochastic basis and a σ-algebra $\mathcal{G} \subseteq \mathcal{F}_0$.

Definition 1. A stochastic process $X = (X_t, \mathcal{F}_t)_{t \geq 0}$ taking values in D is called a *process with $\mathcal{G}$-conditionally independent increments* if for all $\mathcal{G}$-measurable random variables σ and τ $(0 \leq \sigma \leq \tau < \infty)$ and each $\lambda \in R$ (P-a.s.)

$$\mathrm{E}\,(e^{i\lambda (X_\tau - X_\sigma)} \mid \mathcal{F}_\sigma) = \mathrm{E}\,(e^{i\lambda (X_\tau - X_\sigma)} \mid \mathcal{G}). \tag{4.1}$$

Observe that it suffices to verify the equation (4.1) only for deterministic times σ and τ. In fact, if the equation (4.1) were valid for deterministic σ and τ, then obviously it would be valid for $\mathcal{G}$-measurable times σ and τ too which take on a finite number of values. In the general case σ and τ are approximated by the sequences $(\sigma_k)_{k \geq 1}$ and $(\tau_k)_{k \geq 1}$ where $\sigma_k \downarrow \sigma$, $\tau_k \downarrow \tau$, $\sigma_k \leq \tau_k$ and σ_k and τ_k are $\mathcal{G}$-measurable random variables taking on a finite number of values. Then (P-a.s.)

$$\mathrm{E}\,(e^{i\lambda (X_{\tau_k} - X_{\sigma_k})} \mid \mathcal{F}_{\sigma_k}) = \mathrm{E}\,(e^{i\lambda (X_{\tau_k} - X_{\sigma_k})} \mid \mathcal{G}). \tag{4.2}$$

By taking here the limit as $k \to \infty$ we arrive at the desired equality (4.1), since $X_{\tau_k} \to X_\tau$, $X_{\sigma_k} \to X_\sigma$ and $\mathcal{F}_\sigma = \bigcap_{k \geq 1} \mathcal{F}_{\sigma_k}$.

Remark. If $\mathcal{G} = \{\varnothing, \Omega\}$ is a trivial σ-algebra and if for each $s \leq t$ and $\lambda \in R$ the equality (P-a.s.)

$$\mathrm{E}\,(e^{i\lambda (X_t - X_s)} \mid \mathcal{F}_s) = \mathrm{E} e^{i\lambda (X_t - X_s)} \tag{4.3}$$

holds, then X is a process with independent increments.

Definition 2. Let X be a semimartingale. It is said that the triplet $T = (B, C, \nu)$ of a semimartingale X is *$\mathcal{G}$-measurable*, if for each $t > 0$ and each measurable function

$$g = g(x) \text{ with } |g(x)| \leq x^2 \wedge 1$$

the random variables B_t, C_t and

$$\nu([0, t]\, g) = \int_0^t \int_{R_0} g(x)\, d\nu$$

are $\mathcal{G}$-measurable (up to P-indistinguishability).

2. Let X be a semimartingale with the triplet $T = (B, C, \nu)$, and let the function $G(\lambda) = (G_t(\lambda))_{t \geq 0}$ be defined by the formula (3.1).

For $s < t$ set

$$\mathcal{E}_s^t(G(\lambda)) = e^{G_t(\lambda) - G_s(\lambda)} \prod_{s<u\le t} (1 + \Delta G_u(\lambda))\, e^{-\Delta G_u(\lambda)} \tag{4.4}$$

(cf. (2.6)).

Theorem 1. *The following conditions are equivalent:*

(α) *a semimartingale* X *(with the triplet* T = (B, C, ν)) *is a process with conditionally independent increments;*

(β) *the triplet* T = (B, C, ν) *is* $\mathcal{G}$*-measurable.*

Besides, as s < t (P-a.s.)

$$\mathbf{E}(e^{i\lambda(X_t - X_s)} \mid \mathcal{F}_s) = \mathcal{E}_s^t(G(\lambda)). \tag{4.5}$$

Corollary. *A semimartingale* X *(with the triplet* T = (B, C, ν)) *is a process with independent increments if and only if its triplet is deterministic.*

Proof of Theorem 1. (α) $\Rightarrow$ (β). Let $0 \le \sigma < \tau < \infty$ where σ and τ are $\mathcal{G}$-measurable random variables. Denote

$$L_\sigma^\tau = e^{i\lambda(X_\tau - X_\sigma)}, \quad \phi_\sigma^\tau = \mathbf{E}(L_\sigma^\tau \mid \mathcal{F}_\sigma). \tag{4.6}$$

As s = 0, we write L (τ) and ϕ (τ) instead of L_0^τ and ϕ_0^τ. The variables ϕ_s^t are $\mathcal{G}$-measurable and, in particular,

$$\phi(t) = \mathbf{E}(e^{i\lambda(X_t - X_0)} \mid \mathcal{G}). \tag{4.7}$$

According to Problem 1 there exists a version of the process $\phi = (\phi(t))_{t \ge 0}$ with trajectories in D, and this particular version will be considered below.

Let us show that the process $|\phi| = (|\phi(t)|)_{t\ge 0}$ is nonincreasing.

By (4.6) we have that $|\phi_s^t| \le 1$, and by (4.1) and (4.6)

$$\begin{aligned}\phi(\tau) &= \mathbf{E}(L(\tau) \mid \mathcal{G}) = \mathbf{E}(L(\sigma) L_\sigma^\tau \mid \mathcal{G}) = \mathbf{E}(L(\sigma)\,\mathbf{E}(L_\sigma^\tau \mid \mathcal{F}_\sigma) \mid \mathcal{G}) \\ &= \mathbf{E}(L(\sigma)\,\phi_\sigma^\tau \mid \mathcal{G}) = \mathbf{E}(L(\sigma) \mid \mathcal{G})\,\phi_\sigma^\tau = \phi(\sigma)\,\phi_\sigma^\tau.\end{aligned} \tag{4.8}$$

This gives $|\phi(t)| = |\phi(s)|\,|\phi_s^t| \le |\phi(s)|$.

Define the $\mathcal{G}$-measurable random variables

$$\hat{T}_k(\lambda) = \left(\inf\left(t: |\phi(t)| \le \frac{1}{k}\right) - \frac{1}{k}\right) \vee 0, \quad k \ge 1,$$

$$\hat{T}(\lambda) = \lim_k \hat{T}_k(\lambda).$$

Obviously,

$$\hat{T}(\lambda) = \inf(t: |\phi(t)| = 0), \quad |\phi(t \wedge \hat{T}_k(\lambda))| \ge \frac{1}{k}.$$

This fact, along with (4.1), allows us to verify that the process $\hat{Z}(\lambda) = (\hat{Z}_t(\lambda))_{t \geq 0}$ with

$$\hat{Z}_t(\lambda) = L(t)\, \phi^{-1}(t)\, I(|\phi(t)| > 0) \tag{4.9}$$

possesses the following properties:

$$\hat{Z}^{\hat{T}_k(\lambda)}(\lambda) \in \mathcal{M}_{loc}, \quad k \geq 1.$$

In fact, let $s < t$, $\sigma = s \wedge \hat{T}_k(\lambda)$ and $\tau = t \wedge \hat{T}_k(\lambda)$. Then, by taking into consideration (4.46) and (4.8), we get

$$\hat{Z}_\tau(\lambda) = L(\tau)\, \phi^{-1}(\tau) = L(\sigma)\, L_\sigma^\tau\, \phi^{-1}(\sigma)\, (\phi_\sigma^\tau)^{-1} = \hat{Z}_\sigma(\lambda)\, L_\sigma^\tau\, (\phi_\sigma^\tau)^{-1}.$$

This gives the martingale equality

$$\mathbf{E}(\hat{Z}_\tau(\lambda) \mid \mathcal{F}_\sigma) = \hat{Z}_\sigma(\lambda), \tag{4.10}$$

since $(\phi_\sigma^\tau)^{-1}$ is $\mathcal{G}$-measurable and

$$\mathbf{E}(L_\sigma^\tau (\phi_\sigma^\tau)^{-1} \mid \mathcal{F}_\sigma) = \mathbf{E}(L_\sigma^\tau \mid \mathcal{F}_\sigma)(\phi_\sigma^\tau)^{-1} = \phi_\sigma^\tau (\phi_\sigma^\tau)^{-1} = 1.$$

Notice the additional property of the process $\tilde{Z}(\lambda)$, namely

$$\sup_{t \leq \hat{T}_k(\lambda)} |\hat{Z}_k(\lambda)| \leq k, \tag{4.11}$$

which holds in virtue of

$$|\phi(t \wedge \hat{T}_k(\lambda))| \geq \frac{1}{k}, \quad |L(t)| = 1.$$

Hence

$$\phi(t \wedge \hat{T}_k(\lambda)) = L(t \wedge \hat{T}_k(\lambda))\, \hat{Z}^{-1}_{t \wedge \hat{T}_k(\lambda)}, \quad k \geq 1, \tag{4.12}$$

and, consequently, it is not hard to verify by applying Ito's formula (Ch. 2, § 3) that the processes $(\phi(t \wedge \hat{T}_k(\lambda)))_{t \geq 0}$ are semimartingales. The random variables $\phi(t \wedge \hat{T}_k(\lambda))$ are $\mathcal{G}$-measurable, $\mathcal{G} \subseteq \mathcal{F}_0$, for each $t \geq 0$. Therefore, according to the Problems 2 and 4, we have that $(\phi(t \wedge \hat{T}_k(\lambda)))_{t \geq 0} \in \mathcal{V}$ and that it is a predictable process.

Let $(T_k)(\lambda))_{k \geq 1}$ be stopping times, defined in the course of proving Theorem 3.1. Set

$$\bar{T}_k(\lambda) = \hat{T}_k(\lambda) \wedge T_k(\lambda), \quad k \geq 1.$$

By Theorem 3.1

$$L(t \wedge \bar{T}_k(\lambda)) = \mathcal{E}_{t \wedge \bar{T}_k(\lambda)}(G(\lambda))\, Z_{t \wedge \bar{T}_k(\lambda)} \tag{4.13}$$

with $(Z_{t\wedge \bar{T}_k(\lambda)})_{t\geq 0} \in \mathcal{M}_{loc}$, $k \geq 1$, and by (4.12)

$$L(t\wedge \bar{T}_k(\lambda)) = \phi(t\wedge \bar{T}_k(\lambda))\, \hat{Z}_{t\wedge \bar{T}_k(\lambda)}. \tag{4.14}$$

As the multiplicative decomposition for a semimartingale is unique (Theorem 2.5.1) from (4.13) and (4.14) it follows that the processes $(\phi(t\wedge \bar{T}_k(\lambda)))_{t\geq 0}$ and $(\mathcal{E}_{t\wedge \bar{T}_K(\lambda)}(G(\lambda)))_{t\geq 0}$ are indistinguishable for each $\lambda \in R$ and $k \geq 1$. Since $\hat{T}_k(\lambda) \uparrow \hat{T}(\lambda)$ and $T_k(\lambda) \neq T(\lambda)$ with $T(\lambda) = \inf\,(t: |\,\mathcal{E}_t(G(\lambda))\,| = 0)$, the indicated above indistinguishability of processes entails $\hat{T}(\lambda) = T(\lambda)$ (P-a.s.), and the processes ϕ and $\mathcal{E}(G(\lambda))$ are indistinguishable for any $\lambda \in R$.

Define the $\mathcal{G}$-measurable random variables $S_k(\lambda)$, $k \geq 1$ with

$$S_0(\lambda) = 0,\ \ S_1(\lambda) = T(\lambda),\quad S_{k+1}(\lambda) = \inf\,(t > S_k(\lambda): |\,\phi^t_{t\wedge S_k(\lambda)}\,| = 0).$$

It is established, analogously to the indistinguishability of the processes ϕ and $\mathcal{E}(G(\lambda))$, that the processes

$$(\mathcal{E}^t_{t\wedge S_k(\lambda)}(G(\lambda)))_{t\geq 0} \text{ and } (\phi^\tau_{t\wedge S_k(\lambda)})_{t\geq 0},\quad k \geq 1$$

are indistinguishable.

Let us show now that $\lim_k S_k(\lambda) = \infty$ (P-a.s.). Denote $S_\infty(\lambda) = \lim_k S_k(1)$ and set $P(S_\infty(\lambda) < \infty) > 0$. Analogously to Lemma 2.2, it can be established that

$$S_k(\lambda) = \inf\,(t > S_{k-1}(\lambda): \Delta G_t(\lambda) = -1),\ \ k \geq 2.$$

This in turn gives

$$P\Big(I(S_\infty(\lambda) < \infty) \sum_{s\leq S_\infty(\lambda)} I(\Delta G_s(\lambda) = -1) = \infty\Big) > 0,$$

which contradicts the following property of the cumulant:

$$P\left(\sum_{s\leq t} |\,\Delta G_s(\lambda)\,| < \infty\right) = 1,\ \ t > 0.$$

Hence $P(S_\infty(\lambda) < \infty) = 0$.

From this it follows that the random variables $G_t(\lambda)$ are $\mathcal{G}$-measurable for every $t \geq 0$. To prove this, use the fact that by (2.5)

$$G_{t\wedge S_k(\lambda)}(\lambda) = G_{t\wedge S_{k-1}(\lambda)} + \int_{t\wedge S_{k-1}(\lambda)}^{t\wedge S_k(\lambda)} \frac{dY_u}{Y_{u-}},\ \ t \leq S_k(\lambda),\ \ k \geq 1$$

with $Y_u = \mathcal{E}^u_{u\wedge S_{k-1}(\lambda)}(G(\lambda))$.

For each $t \geq 0$ and $\alpha \in R$ set

$$G_t(\lambda, \alpha) = G_t(\lambda) - \frac{1}{2}\int_{-1}^{1} G_t(\lambda + \alpha s)\, ds.$$

The random variables $G_t(\lambda, \alpha)$ are $\mathcal{G}$-measurable and

$$G_t(\lambda, \alpha) = \int_R e^{i\lambda x}\, \eta_{\alpha, t}(dx)$$

with (ε_0 is Dirac's measure with mass at point $\{0\}$)

$$\eta_{\alpha, t}(dx) = \frac{\alpha^2}{6} C_t \varepsilon_0(dx) + \left(1 - \frac{\sin \alpha x}{\alpha x}\right) \nu([0, t], dx).$$

Hence,

$$\frac{6G_t(0, \alpha)}{\alpha^2} = C_t + \int_0^t \int_{0 < |x| \leq \delta} \left(1 - \frac{\sin \alpha x}{\alpha x}\right)\frac{6}{\alpha^2} d\nu + \int_0^\tau \int_{|x| > \delta} \left(1 - \frac{\sin \alpha x}{\alpha x}\right)\frac{6}{a^2} d\nu$$

and, as in the course of proving Theorem 3.3, we have

$$C_t \leq \lim_{\alpha \to \infty} \frac{6G_t(0, \alpha)}{\alpha^2} = C_t + \int_0^t \int_{0 < |x| \leq \delta} x^2\, d\nu \to C_t,\ \delta \to 0.$$

Thus, the random variables C_t are $\mathcal{G}$-measurable. Hence the variables

$$\int_0^t \int_{R_0} g(x)\, d\nu$$

are $\mathcal{G}$-measurable too for each measurable function $g = g(x)$ with $|g(x)| \leq x^2 \wedge 1$. This implies $\mathcal{G}$-measurability of

$$\int_0^t \int_{R_0} (e^{i\lambda x} - 1 - i\lambda x\,(|x| \leq 1))\, d\nu,$$

and consequently, by (3.1) we get the $\mathcal{G}$-measurability of the variables B_t.

(β) $\Rightarrow$ (α). In view of the $\mathcal{G}$-measurability of the triplet $T = (B, C, \nu)$, the stochastic exponential $\mathcal{E}(g)$ is $\mathcal{G}$-measurable. By Theorem 3.1 we have

$$\mathbf{E}\,(Z_{t \wedge T_k(\lambda)}(\lambda) \mid \mathcal{F}_0) = 1,\ k \geq 1$$

(the Markov times $T_k(\lambda)$ and the process $Z(\lambda)$ are defined in Theorem 3.1). Therefore (P-a.s.)

$$\mathbf{E}\,(e^{i\lambda (X_{t\wedge T_k(\lambda)} - X_0)} \mid \mathcal{F}_0) = \mathcal{E}_{t\wedge T_k(\lambda)}\,(G\,(\lambda)),\ \ k \geq 1.$$

On the set $\{t < T\,(\lambda)\}$ this gives (P-a.s.)

$$\mathbf{E}\,(e^{i\lambda (X_t - X_0)} \mid \mathcal{F}_0) = \mathcal{E}_t\,(G\,(\lambda)).$$

Analogously, as $s < t$ we have (P-a.s.)

$$\mathbf{E}\,(e^{i\lambda (X_{t\wedge T_k(\lambda)} - X_{s\wedge T_k(\lambda)})} \mid \mathcal{F}_s) = \mathcal{E}_{s\wedge T_k(\lambda)}^{t\wedge T_k(\lambda)}\,(G\,(\lambda)).$$

Consequently, (P-a.s.),

$$I\,(t < T\,(\lambda))\,\mathbf{E}\,(e^{i\lambda (X_t - X_s)} \mid \mathcal{F}_s) = I\,(t < T\,(\lambda))\,\mathcal{E}_s^t\,(G\,(\lambda))$$

and

$$I\,(S_k\,(\lambda) \leq t < S_{k+1}\,(\lambda))\,\mathbf{E}\,(e^{i\lambda (X_t - X_s)} \mid \mathcal{F}_s)$$

$$= I\,(S_k\,(\lambda) \leq t < S_{k+1}\,(\lambda))\,\mathcal{E}_s^t\,(G\,(\lambda)),\ \ k \geq 1.$$

This implies, in particular, the formula (4.5), in view of which

$$\mathbf{E}\,(e^{i\lambda (X_t - X_s)} \mid \mathcal{F}_s)$$

is a $\mathcal{G}$-measurable random variable, and consequently (P-a.s.)

$$\mathbf{E}\,(e^{i\lambda (X_t - X_s)} \mid \mathcal{G}) = \mathbf{E}\,(e^{i\lambda (X_t - X_s)} \mid \mathcal{F}_s).$$

Corollary 2. *From* (4.5) *and* (4.1) *it follows that* $\mathcal{E}_s^t\,(G\,(\lambda))$ *is the conditional* (*under Condition* $\mathcal{G}$) *characteristic function of a random variable* $X_t - X_s$. *If* $\mathcal{G} = \{\varnothing, \Omega\}$ *is a trivial* σ*-algebra, then* $\mathcal{E}_s^t(G\,(\lambda))$ *is the characteristic function of* $X_t - X_s$.

Remark. If X is a nonnegative nondecreasing semimartingale the triplet of which is $\mathcal{G}$-measurable, $\mathcal{G} \subseteq \mathcal{F}_0$, then for $\lambda \geq 0,\ t > 0$ (P-a.s.)

$$\mathbf{E}\,(e^{-\lambda (X_t - X_0)} \mid \mathcal{G}) = \mathcal{E}_t\,(G\,(\lambda)) \tag{4.15}$$

with $G_t\,(\lambda)$ defined by the formula (2.20).

Problems

1. Let $Y = (Y_t)_{t \geq 0}$ be a bounded process with trajectories in the space D and and let $\mathcal{G}$ be a certain sub-σ-algebra $\mathcal{F}$. Show that there exists a modification of the process $(\mathbf{E}\,(Y_t \mid \mathcal{G}))_{t \geq 0}$ with trajectories in D.

2. Let a process $X = (X_t, \mathcal{F}_t)_{t \geq 0}$ with trajectories in D be such that X_t is $\mathcal{G}$-measurable, $\mathcal{G} \subseteq \mathcal{F}_0$ for each $t \geq 0$. Then X is a semimartingale if and only if $X \in \mathcal{V}$.

3. Prove that a process with independent increments is a semimartingale if and only if for each $\lambda \in R$ the function $f_\lambda(t) = \mathbf{E}\, e^{i\lambda X_t}$ is a function of locally bounded variation.

4. Let $\mathbb{F} = (\mathcal{F}_t)_{t \geq 0}$ be a flow of σ-algebras and $\mathcal{G} \subseteq \mathcal{F}_0$. Show that a stochastic process $X = (X_t)_{t \geq 0} \in D$ is $\mathcal{P}(\mathbb{F})$-predictable if it possesses the property that for every $t \geq 0$ a random variable X_t is $\mathcal{G}$-measurable.

5. Prove the formula (4.15).

6. Let $M \in \overset{c}{\mathcal{M}}_{loc}$, $M_0 = 0$ and let $\langle M \rangle$ be its quadratic characteristic. Show that in case of a deterministic function $\langle M \rangle$ the process M is a Gaussian martingale ($M \in \overline{\mathcal{M}}^2$).

§ 5. Semimartingales and change of probability measures. Transformation of triplets

1. Let $(\Omega, \mathcal{F}, \mathbb{F}, Q)$ be a stochastic basis, and $Q = (P + \tilde{P}) / 2$, P and $\tilde{P}$ probability measures on $(\Omega, \mathcal{F})$. By this definition of Q the measures P and $\tilde{P}$ are absolutely continuous with respect to Q ($P \ll Q$, $\tilde{P} \ll Q$). Let $\mathfrak{z} = (\mathfrak{z}_t)_{t \geq 0}$ be the optional projection of dP / dQ relative to $(\mathbb{F}, Q)$, which can be chosen in such a way that $\mathfrak{z} \in \mathcal{M}(\mathbb{F}, Q)$ (cf. Theorem 1.3.13 and Problem 1.3.10). The process $\mathfrak{z}$ will be called the *density process of the measure* P *with respect to* Q. The density process $\tilde{\mathfrak{z}} = (\tilde{\mathfrak{z}}_t)_{t \geq 0}$ of the measure $\tilde{P}$ with respect to Q is defined analogously. From the definition of Q it follows also that for each $t \in R_+$ ($\mathbf{E}_Q$ is the mathematical expectation with respect to the measure Q)

$$Q(\mathfrak{z}_t = 0, \ \tilde{\mathfrak{z}}_t = 0) \leq \frac{1}{2}\,(P(\tilde{\mathfrak{z}}_t = 0) + \tilde{P}(\mathfrak{z}_t = 0))$$

$$= \frac{1}{2}\left(\int_{\{\mathfrak{z}_t = 0\}} \mathbf{E}_Q\left(\frac{dP}{dQ} \mid \mathcal{F}_t\right) dQ + \int_{\{\tilde{\mathfrak{z}}_t = 0\}} \mathbf{E}_Q\left(\frac{d\tilde{P}}{dQ} \mid \mathcal{F}_t\right) dQ \right) = 0. \tag{5.1}$$

The restrictions of measures P and $\tilde{P}$ to σ-algebras $\mathcal{F}_\tau$, $\tau \in \mathbf{T}$, are denoted by P_τ and $\tilde{P}_\tau$.

Definition 1. We say that a measure $\tilde{P}$ is *locally absolutely continuous with respect to* P (and write $\tilde{P} \overset{loc}{\ll} P$), if $\tilde{P}_t \ll P_t$ for each $t \in R_+$.

In case $\tilde{P} \overset{loc}{\ll} P$ the property (5.1) of the processes $\mathfrak{z}$ and $\tilde{\mathfrak{z}}$ allow us to define Q-a.s. the process $Z = (Z_t)_{t \geq 0}$ with

$$Z_t = \tilde{\mathfrak{z}}_t / \mathfrak{z}_t$$

where $Z_t = d\tilde{P}_t / dP_t$. A process, defined this way, is called the *local density process* of a measure $\tilde{P}$ with respect to P, and it presents a nonnegative local martingale relative to $(\mathbb{F}, P)$, possessing the following property: for each $\tau \in \mathbf{T}$ ($\tau \in \mathbf{T}_p$)

$$I(\tau < \infty) Z_\tau = I(\tau < \infty) \frac{d\tilde{P}_\tau}{dP_\tau} \left(I(\tau < \infty) Z_{\tau-} = I(\tau < \infty) \frac{d\tilde{P}_{\tau-}}{dP_{\tau-}} \right)$$

where $\tilde{P}_{\tau-}$ and $P_{\tau-}$ are restrictions of the measures $\tilde{P}$ and P to $\mathcal{F}_{\tau-}$, and

$$\frac{d\tilde{P}_\tau}{dP_\tau} \left(\frac{d\tilde{P}_{\tau-}}{dP_{\tau-}} \right)$$

are the derivatives of the absolutely continuous part of $\tilde{P}_\tau (\tilde{P}_{\tau-})$ with respect to $P_\tau (P_{\tau-})$. Observe also that in case $\tilde{P} \ll P$ the "density process" $Z \in \mathcal{M}(\mathbb{F}, P)$.

Lemma 1. *Let* $\tilde{P} \overset{loc}{\ll} P$ *and let* Z *be the local density process of a measure* $\tilde{P}$ *with respect to* P.

Then:

1) $\tilde{P}(\inf_{t \le T} Z_t = 0,\ T < \infty) = 0 \quad \forall\, T \in \mathbf{T}$,

2) $\tilde{P}(\lim_n T_n = \infty) = 1$ *with* $T_n = \inf\left(t: Z_t < \frac{1}{n} \right)$.

Proof. 1) Let

$$S = \inf(t: Z_{t-} = 0 \text{ or } Z_t = 0).$$

Then

$$\{\inf_{t \le T} Z_t = 0,\ T < \infty\} = \{S \le T < \infty\}.$$

Therefore, by taking into consideration the properties of the process Z, we get

$$\tilde{P}(\inf_{t \le T} Z_t = 0,\ T < \infty) = \tilde{P}(S \le T < \infty) \le \tilde{P}(S < \infty) = \int_{\{S < \infty\}} Z_S dP$$

$$= \int_{\{S < \infty\} \cap \{(Z_{S-} = 0) \cup (Z_S = 0)\}} Z_S dP = \int_{\{S < \infty\} \cap \{Z_{S-} = 0\}} Z_S dP = 0$$

where the last equality holds in virtue of Problem 2.2.7.

2) Denote $T = \lim_n T_n$. Clearly $Z_T = 0$ on $\{T < \infty\}$ by Problem 2.2.7. Consequently, $\inf_{t \le T} Z_t = 0$ on $\{T < \infty\}$. But, as is proved above, $\tilde{P}(\inf_{t \le T} Z_t = 0,\ T < \infty) = 0$. This gives $\tilde{P}(T < \infty) = 0$, i. e. $\tilde{P}(T = \infty) = 1$.

Lemma 2. *Let* $V = (V_t)_{t \ge 0}$ *be a* $\mathcal{P}(\mathbb{F})$*-measurable stochastic process with* *right-continuous nondecreasing trajectories (tending, perhaps, to* $+\infty$ *within finite time)*, $V_0 = 0$.

If $\tilde{P} \overset{loc}{\ll} P$ *and Z is the local density process of a measure* $\tilde{P}$ *with respect to* P, *then the following implication takes place:*

$$\tilde{P}(V_\infty = 0) = 1 \Rightarrow P(I(Z_- > 0) \circ V_\infty = 0) = 1.$$

Proof. It suffices to establish the implication

$$\tilde{P}(V_\infty = 0) = 1 \Rightarrow P(I(Z_- > 0) \circ V_t = 0) = 1, \quad t \in R_+.$$

Set $T = \inf(t: Z_{t-} = 0 \text{ and } Z_t = 0)$. By Problems 2.2.7 and 1.1.6

$$I(Z_- > 0) \circ V_t = I(Z_- > 0) \circ V_{t \wedge T}.$$

Let us show that the process $Z_- I(V > 0)$ is P-negligible.

According to Theorem 1.3.12 it suffices to verify that

$$I(\tau < \infty) Z_{\tau-} I(V_\tau > 0) = 0 \quad (P\text{-a.s.})$$

for each $\tau \in \mathbf{T}_p$. This equality holds in virtue of the relation

$$I(\tau < \infty) Z_{\tau-} = I(\tau < \infty) \frac{d\tilde{P}_{\tau-}}{dP_{\tau-}},$$

since $V_t = 0$ ($\tilde{P}$-a.s.) and

$$\mathbf{E} I(\tau < \infty) Z_{\tau-} I(V_\tau > 0) = \tilde{\mathbf{E}} I(\tau < \infty) I(V_\tau > 0) = 0.$$

By the inequality $\Delta V \le V$ the process $Z_- I(\Delta V > 0)$ is P-negligible too. Hence, on the set $[\![0, T[\![$ the process V is P-negligible, $V_{T-} = 0$ (P-a.s.) and, consequently, $I(Z_- > 0) \circ V_{(t \wedge T)-} = 0$ (P–a.s.). Therefore by the P-negligibility of the process $Z_- I(\Delta V > 0)$ we have

$$I(Z_- > 0) \circ V_{t \wedge T} = I(Z_- > 0) \Delta V_{t \wedge T} = 0 \quad (P\text{-a.s.}).$$

2. Let $\mu = \mu(dt, dx)$ be an integer-valued random measure on $(R_+ \times E, \mathcal{B}(R_+) \times \mathcal{E})$ (cf. Ch. 3, § 3). Suppose $\mu \in \tilde{\mathcal{V}}^+_{\mathcal{P}}$ relative to P and $\tilde{P}$. The compensators of μ relative to P and $\tilde{P}$ are denoted by ν and $\tilde{\nu}$ respectively.

Let $\tilde{P} \overset{loc}{\ll} P$ and let Z be the local density process of a measure $\tilde{P}$ with respect to P. Set

$$Y(t, x) = Z_{t-}^{-1} I(Z_{t-} > 0) M^P_\mu (Z \mid \tilde{\mathcal{P}})(t, x) \tag{5.2}$$

where $M^P_\mu(\cdot \mid \tilde{\mathcal{P}})$ is the conditional mathematical expectation with respect to the σ-algebra $\tilde{\mathcal{P}}$ (cf. Ch. 3, § 2).

Theorem 1. *Let* $\tilde{\mathsf{P}} \overset{loc}{\ll} \mathsf{P}$. *Then*

1) $\tilde{\nu} \overset{\tilde{\mathsf{P}}}{=} Y\nu$,

2) $I(Z_- > 0)\,\tilde{\nu} \overset{\mathsf{P}}{=} Y\nu$

with the measure $(Y\nu)(\omega; dt, dx) = Y(\omega, t, x)\,\nu(\omega; dt, dx)$.

Proof. By the definition of the compensator (Ch. 3, § 2) it suffices to verify the validity of the equation

$$M_\mu^{\tilde{\mathsf{P}}}(X) = M_{Y\nu}^{\tilde{\mathsf{P}}}(X) \tag{5.3}$$

for each $\tilde{\mathcal{P}}$-measurable and nonnegative function $X = X(\omega, t, x)$. Clearly, it suffices to restrict the considerations to the case $M_\mu^{\tilde{\mathsf{P}}}(X) < \infty$.

Denote $A = X * \mu$. Then

$$M_\mu^{\tilde{\mathsf{P}}}(X) = \tilde{\mathbf{E}}A_\infty = \lim_{t\to\infty} \tilde{\mathbf{E}}A_t.$$

By the properties of the local density Z

$$\tilde{\mathbf{E}}A_t = \mathbf{E}Z_tA_t.$$

In view of Corollary to Theorem 1.6.1 we have $\mathbf{E}Z_tA_t = \mathbf{E}Z \circ A_t$. Therefore, this gives

$$\begin{aligned}\tilde{\mathbf{E}}A_t &= \mathbf{E}Z \circ A_t = \mathbf{E}ZX * \mu_t = \mathbf{E}M_\mu^{\mathsf{P}}(Z \mid \tilde{\mathcal{P}})\,X * \mu_t = \mathbf{E}M_\mu^{\mathsf{P}}(Z \mid \tilde{\mathcal{P}})\,X * \nu_t \\ &= \mathbf{E}I(Z_- > 0)\,Z_-Z_-^{-1}M_\mu^{\mathsf{P}}(Z \mid \tilde{\mathcal{P}})\,X * \nu_t + \mathbf{E}I(Z_- = 0)\,M_\mu^{\mathsf{P}}(Z \mid \tilde{\mathcal{P}})\,X * \nu_t.\end{aligned} \tag{5.4}$$

Next,

$$\begin{aligned}\mathbf{E}I(Z_- = 0)\,M_\mu^{\mathsf{P}}(Z \mid \tilde{\mathcal{P}})\,X * \nu_t &= \mathbf{E}I(Z_- = 0)\,M_\mu^{\mathsf{P}}(Z \mid \tilde{\mathcal{P}})\,X * \mu_t \\ &= \mathbf{E}I(Z_- = 0)\,ZX * \mu_t = 0,\end{aligned} \tag{5.5}$$

since the process $I(Z_- = 0)\,Z$ is P-negligible (see Problem 2.2.7).

Denote $B = I(Z_- > 0)\;Z_-^{-1}\,M_\mu^{\mathsf{P}}(Z \mid \tilde{\mathcal{P}})\,X * \nu$,, that is

$$B = YX * \nu. \tag{5.6}$$

Then by (5.4) and (5.5) we get

$$\tilde{\mathbf{E}}A_t = \mathbf{E}Z_- \circ B_t, \quad t \geq 0.$$

Replacing X by $XI_{[\![0, T]\!]}$ where T is a Markov time, instead of this equation we get the equation

$$\tilde{E}A_{t\wedge T} = EZ_- \circ B_{t\wedge T}. \tag{5.7}$$

Introduce the stopping times

$$T_n = \inf\left(t: Z_t < \frac{1}{n}\right).$$

Since $Z_- \geq \frac{1}{n}$ on the set $[\![0, T_n]\!]$ we have

$$EB_{t\wedge T_n} \leq nEZ_- \circ B_{t\wedge T_n} \leq n\tilde{E}A_{t\wedge T_n} \leq n\tilde{E}A_\infty < \infty.$$

Therefore in view of Corollary to Theorem 1.6.1 and properties of the process Z, we have

$$EZ_- \circ B_{t\wedge T_n} = EZ_{t\wedge T_n} B_{t\wedge T_n} = \tilde{E}B_{t\wedge T_n}.$$

Consequently, by the equation (5.7),

$$\tilde{E}A_{t\wedge T_n} = \tilde{E}B_{t\wedge T_n},\ \ n \geq 1. \tag{5.8}$$

Observe now that by Lemma 1 $\lim_n T_n = \infty$ ($\tilde{P}$-a.s.). Hence, (5.8) yields the equation

$$\tilde{E}A_t = \tilde{E}B_t,$$

in view of which (cf. (5.6))

$$M_\mu^{\tilde{P}}(X) = \lim_{t\to\infty} \tilde{E}B_t = \tilde{E}B_\infty = \tilde{E}YX * \nu = M_{Y\nu}^{\tilde{P}}(X).$$

2) Let X be a $\tilde{\mathcal{P}}$-measurable nonnegative function such that $X * \tilde{\nu} \in \mathcal{V}^+$. Denote

$$V_t = \mathrm{Var}\,(X * \tilde{\nu} - XY * \nu)_t.$$

By virtue of the first assertion proved already,

$$V_\infty = 0\ \ (\tilde{P}\text{-a.s.}), \tag{5.9}$$

and hence, according to Lemma 2, $P\,(I\,(Z_- > 0) \circ V_\infty = 0) = 1$. Therefore the desired assertion $I\,(Z_- > 0)\,\tilde{\nu} \overset{P}{=} Y\nu$ follows from the relation $I\,(Z_- > 0)\,YX * \nu = YX * \nu$, which is a consequence of (5.2).

3. Let $M \in \mathcal{M}_{loc}(\mathbb{F}, P)$, $\tilde{P} \overset{loc}{\ll} P$ and let Z be the local density process of a measure $\tilde{P}$ with respect to P. Since $Z \in \mathcal{M}_{loc}(\mathbb{F}, P)$, the process [M, Z] is defined (cf. Ch. 1, § 8). Suppose

$$[M, Z] \in \mathcal{A}_{loc}(\mathbb{F}, P). \tag{5.10}$$

Then, according to Corollary 1 to Theorem 1.6.3, the process [M, Z] has the compensator $\widetilde{[M, Z]}$ relative to $(\mathbb{F}, P)$).

Theorem 2. *Let* $M \in \mathcal{M}_{loc}(\mathbb{F}, P), \tilde{P} \overset{loc}{\ll} P$, *let* Z *be the local density process of* $\tilde{P}$ *with respect to* P, *let Condition* (5.10) *be fullfilled and let* $\widetilde{[M, Z]}$ *be the compensator of* $[M, Z]$.

Then the process

$$\tilde{M} = M - I(Z_- > 0)\, Z_-^{-1} \circ \widetilde{[M, Z]} \in \mathcal{M}_{loc}(\mathbb{F}, \tilde{P}). \tag{5.11}$$

Besides, the processes $\langle M^c \rangle$ *and* $\langle \tilde{M}^c \rangle$ *are* $\tilde{P}$*-indistinguishable, while the processes* $I(Z_- > 0) \circ \langle M^c \rangle$ *and* $I(Z_- > 0) \circ \langle \tilde{M}^c \rangle$ *are* P*-indistinguishable.*

The proof of this theorem is based on the following result.

Lemma 3. *Let* $\tilde{M} \in D \cap \mathbb{F}$, $\tilde{P} \overset{loc}{\ll} P$ *and let* Z *be the local density process of* $\tilde{P}$ *with respect to* P.

Then

$$\tilde{M}Z \in \mathcal{M}_{loc}(\mathbb{F}, P) \Rightarrow \tilde{M} \in \mathcal{M}_{loc}(\mathbb{F}, \tilde{P}).$$

If there exists a localizing sequence $(\tau_n)_{n \geq 1}$ *for* $\tilde{M}$ *such that* $P(\lim_n \tau_n = \infty) = 1$, *then the reverse implication*

$$\tilde{M}Z \in \mathcal{M}_{loc}(\mathbb{F}, P) \Leftarrow \tilde{M} \in \mathcal{M}_{loc}(\mathbb{F}, \tilde{P})$$

takes place.

Proof. Let $(\tau_n)_{n \geq 1}$ be a localizing sequence for $\tilde{M}Z$ and $\tau_\infty = \lim_n \tau_n$. By definition $P(\tau_\infty = \infty) = 1$. Observe that

$$\tilde{P}(\tau_\infty = \infty) = 1, \tag{5.12}$$

since by properties of the process Z

$$\tilde{P}(\tau_\infty < \infty) = \int_{\{\tau_\infty < \infty\}} Z_{\tau_\infty}\, dP = 0.$$

Denote $\sigma_n = \tau_n \wedge n$. Clearly, $(\sigma_n)_{n \geq 1}$ is a localizing sequence for $\tilde{M}Z$ as well and by (5.12) ($\sigma_\infty = \lim_n \sigma_n$)

$$\tilde{P}(\sigma_\infty = \infty) = 1. \tag{5.13}$$

As $(\sigma_n)_{n \geq 1}$ is a localizing sequence for $\tilde{M}$ the desired assertion holds, provided it is established that for each $n \geq 1$

$$(\tilde{M}Z)^{\sigma_n} \in \mathcal{M}(\mathbb{F}, P) \Rightarrow \tilde{M}^{\sigma_n} \in \mathcal{M}(\mathbb{F}, \tilde{P}).$$

Observe that $(\sigma_n)_{n \geq 1}$ is a localizing sequence for Z too. Therefore it suffices to prove only that

$$\tilde{M}Z \in \mathcal{M}(\mathbb{F}, P) \Rightarrow \tilde{M} \in \mathcal{M}(\mathbb{F}, P), \tag{5.14}$$

under the assumption $Z \in \mathcal{M}(\mathbb{F}, P)$ (i.e. $\tilde{P} \overset{loc}{\ll} P$).

Let T be an arbitrary Markov time. Then, according to Problem 1.4.2,

$$\mathbf{E}\,(\tilde{M}Z)_T = \mathbf{E}\,(\tilde{M}Z)_0. \tag{5.15}$$

On the other hand

$$\mathbf{E}\,|\,(\tilde{M}Z)_T\,| = \mathbf{E}\,|\,\tilde{M}_T\,|\,Z_T = \tilde{\mathbf{E}}\,|\,\tilde{M}_T\,|\,.$$

Therefore, $\tilde{\mathbf{E}}\,|\,\tilde{M}_T\,| < \infty$ and $\tilde{\mathbf{E}}\tilde{M}_T = \mathbf{E}\,(\tilde{M}Z)_T$. Analogously

$$\tilde{\mathbf{E}}\tilde{M}_0 = \mathbf{E}\,(\tilde{M}Z)_0\,.$$

Consequently,

$$\tilde{\mathbf{E}}\tilde{M}_T = \tilde{\mathbf{E}}\tilde{M}_0\,,$$

in view of (5.15). Hence $\tilde{M} \in \mathcal{M}(\mathbb{F}, P)$ (Problem 1.4.2), and the implication (5.14) takes place.

The reverse implication is verified analogously.

Corollary 1. *Let*

$$T_n = \inf\left(t\colon Z_t < \frac{1}{n}\right) \wedge n, \quad n \geq 1.$$

Then

$$(\tilde{M}Z)^{T_n} \in \mathcal{M}_{loc}(\mathbb{F}, P),\ n \geq 1 \Rightarrow \tilde{M} \in \mathcal{M}_{loc}(\mathbb{F}, \tilde{P}).$$

In fact, from Lemma 3 it follows that

$$(\tilde{M}Z)^{T_n} \in \mathcal{M}_{loc}(\mathbb{F}, P) \Rightarrow \tilde{M}^{T_n} \in \mathcal{M}_{loc}(\mathbb{F}, \tilde{P}),$$

and by Lemma 1 we have $T_n \uparrow \infty$ ($\tilde{P}$-a.s.). Therefore

$$\tilde{M}^{T_n} \in \mathcal{M}_{loc}(\mathbb{F}, \tilde{P}),\ n \geq 1, \Rightarrow \tilde{M} \in \mathcal{M}_{loc}(\mathbb{F}, \tilde{P}).$$

Corollary 2. *Let* $\tilde{M} \in \mathbf{S}(\mathbb{F}, P)$, *let* h *be a* $\mathcal{P}$*-measurable locally bounded function* ($|\,h\,|\,I_{[\![0,\tau_n]\!]} \leq c_n$, $n \geq 1$, $(\tau_n)_{n \geq 1}$ *a localizing sequence) and* $h \cdot \tilde{M}$ *the stochastic integral with respect to a* $(\mathbb{F}, P)$*-semimartingale* M.

If $(\tilde{M}Z)^{T_n} \in \mathcal{M}_{loc}(\mathbb{F}, P)$, $n \geq 1$, *then* $h \cdot \tilde{M} \in \mathcal{M}_{loc}(F, \tilde{P})$.

Proof of Theorem 2. By Corollary 1 to Lemma 3 it suffices to show that $(\tilde{M}Z)^{T_n} \in \mathcal{M}_{loc}(\mathbb{F}, P)$, $n \geq 1$. Utilizing the representation (5.11) for $\tilde{M}$ and taking into consideration the fact that on the set $[\![0, T_n]\!]$ we have $\{Z_- \geq 1/n\}$, by Ito's formula (cf. (3.5) and (3.6) in Ch. 2, § 3) we get

$$(\tilde{M}Z)_{t\wedge T_n} = M_- \cdot Z_{t\wedge T_n} + Z_- \cdot M_{t\wedge T_n} - (Z_-^{-1} \circ \widetilde{[M, Z]}) \cdot Z_{t\wedge T_n}$$

$$+ [M, Z]_{t\wedge T_n} - \widetilde{[M, Z]}_{t\wedge T_n}.$$

By Corollary 1 to Theorem 1.6.3 this yields

$$(\tilde{M}Z)^{T_n} \in \mathcal{M}_{loc}(\mathbb{F}, P),\ n \geq 1.$$

Let us establish now $\tilde{P}$-indistinguishability of the processes $\langle M^c \rangle$ and $\langle \tilde{M}^c \rangle$.

According to the representation (5.11) we have $\tilde{M}^{T_n} \in \mathbf{Sp}(\mathbb{F}, P)$, $n \geq 1$. Therefore, by Ito's formula (Corollary 1 to Theorem 2.3.1) we get

$$(\tilde{M}^{T_n})^2 = 2\tilde{M}_-^{T_n} \cdot \tilde{M}^{T_n} + [\tilde{M}^{T_n}, \tilde{M}^{T_n}]$$

where the quadratic variation of a semimartingale $\tilde{M}^{T_n}$ is given by the formula

$$[\tilde{M}^{T_n}, \tilde{M}^{T_n}] = \langle M^c \rangle^{T_n} + \sum_s (\Delta \tilde{M}_s^{T_n})^2.$$

Therefore

$$(\tilde{M}^{T_n})^2 = 2\tilde{M}_-^{T_n} \cdot \tilde{M}^{T_n} + \langle M^c \rangle^{T_n} + \sum_s (\Delta \tilde{M}_s^{T_n})^2.$$

By the property $(\tilde{M}Z)^{T_n} \in \mathcal{M}_{loc}(\mathbb{F}, P)$, $n \geq 1$ proved already, and by Corollary 2 to Lemma 3, we have

$$2\tilde{M}_-^{T_n} \cdot \tilde{M}^{T_n} \in \mathcal{M}_{loc}(\mathbb{F}, \tilde{P}).$$

Therefore the process

$$Y^n = (\tilde{M}^{T_n})^2 - \sum_s (\Delta \tilde{M}_s^{T_n})^2 \in \mathbf{Sp}(\mathbb{F}, \tilde{P})$$

and

$$Y^n = N + \langle M^c \rangle^{T_n}$$

with $N = 2\tilde{M}_-^{T_n} \cdot \tilde{M}^{T_n} \in \mathcal{M}_{loc}(\mathbb{F}, \tilde{P})$. On the other hand $\tilde{M}^{T_n} \in \mathcal{M}_{loc}(\mathbb{F}, \tilde{P})$, and by Ito's formula (Corollary 2 to Theorem 2.3.1) we get (relative to $(\mathbb{F}, \tilde{P})$)

$$(\tilde{M}^{T_n})^2 = \tilde{N} + [\tilde{M}^{T_n}, \tilde{M}^{T_n}] = \tilde{N} + \langle \tilde{M}^c \rangle^{T_n} + \sum_s (\Delta \tilde{M}_s^{T_n})^2$$

where $\tilde{N}$ is the stochastic integral $2\tilde{M}^{T_n} \cdot \tilde{M}^{T_n}$ with respect to $(F, \tilde{P})$-local martingale $\tilde{M}^{T_n}$. Hence, Y^n admits yet another decomposition

$$Y^n = \tilde{N} + \langle \tilde{M}^c \rangle^{T_n}.$$

As the decomposition of special semimartingales is unique, we get $\tilde{P}$-indistinguishability of the processes $\langle M^c \rangle^{T_n}$ and $\langle \tilde{M}^c \rangle^{T_n}$, $n \geq 1$. Consequently, as $T_n \uparrow \infty$ ($\tilde{P}$-a.s.), the processes $\langle M^c \rangle$ and $\langle \tilde{M}^c \rangle$ are $\tilde{P}$-indistinguishable too.

It follows by Lemma 2 that the processes $I(Z_- > 0) \circ \langle M^c \rangle$ and $I(Z_- > 0) \circ \langle \tilde{M}^c \rangle$ are indistinguishable relative to the measure P.

4. Let a process $X \in \mathbf{S}(\mathbb{F}, P)$ have the triplet of predictable characteristics $T = (B, C, \nu)$ and the canonical representation

$$X_t = X_0 + B_t + X_t^c + \int_0^t \int_{|x| \leq 1} x \, d(\mu - \nu) + \int_0^t \int_{|x| > 1} x \, d\mu. \tag{5.16}$$

Denote

$$M_t = X_t^c + \int_0^t \int_{|x| \leq 1} x \, d(\mu - \nu) \tag{5.17}$$

and observe that $M = (M_t)_{t \geq 0} \in \mathcal{M}_{loc}(\mathbb{F}, P)$ with

$$|\Delta M| \leq 2. \tag{5.18}$$

Lemma 4. *Let* $\tilde{P} \overset{loc}{\ll} P$ *and let* Z *be a local density process of a measure* $\tilde{P}$ *with respect to* P.

Then the processes

$$\sum_s |\Delta M_s \Delta Z_s|, \quad \sum_s |\Delta X_s I(|\Delta X_s| \leq 1) \Delta Z_s| \quad \textit{and} \sum_s |\Delta B_s \Delta Z_s|$$

belong to $\mathcal{A}_{loc}^+(P)$.

Proof. By the Cauchy-Bunjakovski inequality

$$\sum_s |\Delta M_s \Delta Z_s| \le \left(\sum_s (\Delta M_s)^2 \sum_s (\Delta Z_s)^2\right)^{1/2} \le [M, M]^{1/2} [Z, Z]^{1/2}.$$

By Theorem 1.8.1 we have

$$[Z, Z]^{1/2} \in \mathcal{A}^+_{loc}(\mathsf{P}).$$

Next, $[M, M]_{\tau_n} \le n + 4$ for $\tau_n = \inf(t\colon [M, M]_t \ge n)$, $n \ge 1$. Since $[M, M] \in \mathcal{V}^+$ (Theorem 1.8.1), we have $\tau_n \uparrow \infty$, $n \to \infty$. Hence,

$$[M, M]^{1/2} [Z, Z]^{1/2} \in \mathcal{A}^+_{loc}$$

and consequently

$$\sum_s |\Delta M_s \Delta Z_s| \in \mathcal{A}^+_{loc}.$$

The analogous evaluation

$$\sum_s |\Delta Z_s \Delta X_s I(|\Delta X_s| \le 1)| \le \left(\sum_s (\Delta Z_s)^2 \sum_s (\Delta X_s)^2 I(|\Delta X_s| \le 1)\right)^{1/2}$$
$$\le [Z, Z]^{1/2} (I(|\Delta X| \le 1) \circ [X, X])^{1/2}$$

leads to

$$\sum_s |\Delta Z_s \Delta X_s I(|\Delta X_s| \le 1)| \in \mathcal{A}^+_{loc}.$$

In fact $[Z, Z]^{1/2} \in \mathcal{A}^+_{loc}$, $I(|\Delta X| \le 1) \circ [X, X]_{\tau_n} \le n + 1$ for $\tau_n = \inf(t\colon [X, X]_t \ge n)$, and $\tau_n \uparrow \infty$, $n \to \infty$, by $[X, X] \in \mathcal{V}^+$.

Since

$$\Delta B_s = \int_{|x| \le 1} x\nu(\{s\}, dx)$$

(cf. (1.7) in § 1), (5.17) entails

$$\Delta B_s = -\Delta M_s + \Delta X_s I(|\Delta X_s| \le 1). \tag{5.19}$$

Therefore

$$\sum_s |\Delta B_s \Delta Z_s| \le \sum_s |\Delta M_s \Delta Z_s| + \sum_s |\Delta Z_s \Delta X_s I(|\Delta X_s| \le 1)| \in \mathcal{A}^+_{loc}(\mathsf{P}).$$

Lemma 5. *Let* $\tilde{\mathsf{P}} \overset{loc}{\ll} \mathsf{P}$, *let* Z *be a local density process of* $\tilde{\mathsf{P}}$ *with respect to* P *and* $M \in \mathcal{M}_{loc}(\mathbb{F}, \mathsf{P})$ *with* M *defined in* (5.17).

Then the process $[M, Z] \in \mathcal{A}_{loc}(\mathsf{P})$ *and its compensator*

$$\widetilde{[M, Z]} = \langle X^c, Z^c \rangle + I(|x| \le 1)\, x M_\mu^P (\Delta Z \mid \tilde{\mathcal{P}}) * \nu \tag{5.20}$$

with $\tilde{\mathcal{P}} = \mathcal{P} \times B(R_0)$. *Besides*

$$I(|x| \le 1)\, x M_\mu^P (\Delta Z) \mid \tilde{\mathcal{P}}) * \nu \in \mathcal{A}_{loc}(P).$$

Proof. By Lemma 4 we have $\sum_s \Delta M_s \Delta Z_s \in \mathcal{A}_{loc}(P)$. Next $\langle X^c, Z^c \rangle \in \mathcal{V} \cap \mathcal{P}$. Therefore, by Lemma 1.6.1 we have $\langle X^c, Z^c \rangle \in \mathcal{A}_{loc}(P)$. Consequently $[M, Z] \in \mathcal{A}_{loc}(P)$, since

$$[M, Z] = \langle X^c, Z^c \rangle + \sum_s \Delta M_s \Delta Z_s. \tag{5.21}$$

By Corollary to Theorem 1.6.3 the process $[M, Z]$ has the compensator $\widetilde{[M, Z]}$. Besides, by (5.21)

$$\widetilde{[M, Z]} = \langle X^c, Z^c \rangle + \widetilde{\sum_s \Delta M_s \Delta Z_s} \tag{5.22}$$

where $\widetilde{\sum_s \Delta M_s \Delta Z_s}$ is the compensator of the process $\sum_s \Delta M_s \Delta Z_s$ (relative to $(\mathbb{F}, P)$).

By (5.19) and Lemma 4

$$\sum_s \Delta M_s \Delta Z_s = \sum_s \Delta Z_s \Delta X_s I(|\Delta X_s| \le 1) - \sum_s \Delta B_s \Delta Z_s.$$

By Theorem 2.1.3 and Lemma 4 we have

$$\sum_s \Delta B_s \Delta Z_s \in \mathcal{M}_{loc}(\mathbb{F}, P) \cap \mathcal{A}_{loc}.$$

Therefore

$$\widetilde{\sum_s \Delta B_s \Delta Z_s} = 0$$

(Theorem 1.6.4), and consequently

$$\widetilde{\sum_s \Delta M_s \Delta Z_s} = \widetilde{\sum_s \Delta X_s I(|\Delta X_s| \le 1)\, \Delta Z_s} = \widetilde{(I(|x| \le 1)\, x\, \Delta Z) * \mu}. \tag{5.23}$$

By Lemma 4 we have

$$(I(|x| \le 1)\, x\, \Delta Z) * \mu \in \mathcal{A}_{loc}(P).$$

We will assume $(I(|x| \le 1)\, x\, \Delta Z) * \mu \in \mathcal{A}(P)$, making use of localizing sequences if necessary. Let t be an arbitrary Markov time. Then, taking into consideration the definition of the conditional mathematical expectation $M_\mu^P(\cdot \mid \tilde{\mathcal{P}})$ (cf. Ch. 3, § 2), we get

$$\mathbf{E}I(|x|\le 1)\,x\,\Delta Z * \mu_\tau = \mathbf{E}I(|x|\le 1)\,x M_\mu^{P}(\Delta Z\,|\,\tilde{\mathcal{P}}) * \mu_\tau$$

$$= \mathbf{E}I(|x|\le 1)\,x M_\mu^{P}(\Delta Z\,|\,\tilde{\mathcal{P}}) * \nu_\tau.$$

Consequently, according to Problem 1.4.2

$$I(|x|\le 1)\,x\,\Delta Z * \mu - I(|x|\le 1)\,x M_\mu^{P}(\Delta Z\,|\,\tilde{\mathcal{P}}) * \nu \in \mathcal{M}\,(\mathbb{F}, P).$$

But then by Theorem 1.6.3 $I(|x|\le 1)\,x M_\mu^{P}(\Delta Z\,|\,\tilde{\mathcal{P}}) * \nu$ is the compensator of the process $I(|x|\le 1)\,x\,\Delta Z * \mu$, and the desired representation (5.20) takes place by (5.22) and (5.23).

If $(\tau_n)_{n\ge 1}$ is a localizing sequence for $I(|x|\le 1)\,x\,\Delta Z * \mu$ (i.e. $\mathbf{E}I(|x|\le 1)\,|x\,\Delta Z| * \mu_{\tau_n} < \infty$, $n \ge 1$), then by Jensen's inequality for conditional mathematical expectations $M_\mu^{P}(\,\cdot\,|\,\tilde{\mathcal{P}})$ (Problem 3.2.11)

$$\mathbf{E}I(|x|\le 1)\,|x M_\mu^{P}(\Delta Z\,|\,\tilde{\mathcal{P}})| * \nu_{\tau_n} \le \mathbf{E}I(|x|\le 1)\,|x|\,M_\mu^{P}(|\Delta Z|\,|\,\tilde{\mathcal{P}}) * \nu_{\tau_n}$$

$$= \mathbf{E}I(|x|\le 1)\,|x|\,M_\mu^{P}(|\Delta Z|\,|\,\tilde{\mathcal{P}}) * \mu_{\tau_n} = \mathbf{E}I(|x|\le 1)\,|x\,\Delta Z| * \mu_{\tau_n} < \infty,\ n\ge 1.$$

Hence, $I(|x|\le 1)\,x M_\mu^{P}(\Delta Z\,|\,\tilde{\mathcal{P}}) * \nu \in \mathcal{A}_{loc}(P)$ with a localizing sequence $(\tau_n)_{n\ge 1}$.

Lemma 6. *Let* $\tilde{P} \overset{loc}{\ll} P$, *let* Z *be the local density process of a measure* $\tilde{P}$ *with respect to* P *and*

$$Y = I(Z_- > 0)\,Z_-^{-1} M_\mu^{P}(Z\,|\,\tilde{\mathcal{P}}), \tag{5.24}$$

$$\beta = I(Z_- > 0)\,Z_-^{-1}\,\frac{d\langle X^c, Z^c\rangle}{d\langle X^c\rangle}. \tag{5.25}$$

Then

$$I(|x|\le 1)\,x\,(Y-1) * \nu \in \mathcal{A}_{loc}^{+}(\tilde{P}),\quad \beta^2 \circ C \in \mathcal{A}_{loc}^{+}(\tilde{P}).$$

Proof. By Lemma 1 we have

$$\inf_{s\le t} Z_s > 0\ \ (\tilde{P}\text{-a.s.}),\ \ t\ge 0.$$

Therefore, by Lemma 5 ($\tilde{P}$-a.s.)

$$I(|x|\le 1)\,|x\,(Y-1)| * \nu_t = I(|x|\le 1)\,|x\,(Z_-^{-1} M_\mu^{P}(Z\,|\,\tilde{\mathcal{P}}) - 1)| * \nu_t$$

$$= I(|x|\le 1)\,|x Z_-^{-1} M_\mu^{P}(\Delta Z\,|\,\tilde{\mathcal{P}})| * \nu_t$$

$$\leq (\inf_{s \leq t} Z_s)^{-1} (I(|x| \leq 1)\,|\,x M_\mu^{P} (\Delta Z \,|\, \tilde{\mathcal{P}})\,| * \nu_t) < \infty, \quad t > 0.$$

Since $(I(|x| \leq 1)\,|\,x(Y - 1)\,|) * \nu$ is a predictable process, the first assertion holds by Lemma 1.6.1.

The second assertion is proved analogously, since by taking into consideration Theorem 2.2.8 we have ($\tilde{P}$-a.s.)

$$\beta^2 \circ C_t = Z_-^{-2} \left(\frac{d \langle X^c, Z^c \rangle}{d \langle X^c \rangle} \right)^2 \circ \langle X^c \rangle_t \leq (\inf_{s \leq t} Z_s)^{-2} \langle Z^c \rangle_t < \infty, \quad t > 0.$$

5. Theorem 3. *Let a semimartingale* $X \in S(\mathbb{F}, P)$, *let* $T = (B, C, \nu)$ *be the triplet of its predictable characteristics and* $\tilde{P} \overset{loc}{\ll} P$. *Then* $X \in S(\mathbb{F}, \tilde{P})$ *and its triplet* $\tilde{T} = (\tilde{B}, \tilde{C}, \tilde{\nu})$ *of predictable characteristics (with respect to the measure* $\tilde{P}$) *is determined by the triplet* $T = (B, C, \nu)$ *according to the following formulas*

$$\tilde{B} = B + \beta \circ C + I(|x| \leq 1)\, x\, (Y - 1) * \nu, \tag{5.26}$$

$$\tilde{C} = C, \tag{5.27}$$

$$\tilde{\nu} = Y\nu \tag{5.28}$$

with Y *and* β *defined in* (5.24) *and* (5.25).

Remark. Denote

$$B' = B + \beta \circ C + I(|x| \leq 1)\, x\, (Y - 1) * \nu,$$
$$\nu' = Y\nu.$$

By Lemma 6 the process $B' \in \mathcal{V} \cap \mathcal{P}$. The equalities (5.26) and (5.27) are understood in the sense that the pairs of processes B' and $\tilde{B}$, as well as C and $\tilde{C}$ are $\tilde{P}$-indistinguishable. The equality (5.28) means that $U * \tilde{\nu} \overset{\tilde{P}}{=} U * \nu'$ for each nonnegative $\tilde{\mathcal{P}} = \mathcal{P} \otimes B(R_0)$-measurable function $U = U(\omega, t, x)$.

Let us remark also that by Lemma 2 the processes $I(Z_- > 0) \circ B'$ and $I(Z_- > 0) \circ \tilde{B}$ are P-indistinguishable, as well as $I(Z_- > 0) \circ C$ and $I(Z_- > 0) \circ \tilde{C}$. By Theorem 1 we have $I(Z_- > 0)\, U * \tilde{\nu} = U * \nu'$ (P-a.s.) for each nonnegative $\tilde{\mathcal{P}}$-measurable function U.

Proof of Theorem 3. A semimartingale $X \in S(\mathbb{F}, P)$ with the triplet $T = (B, C, \nu)$ admits the canonical representation (5.16). By taking into account notation (5.17) we get

$$X_t = X_0 + B_t + M_t + \sum_{0<s\le t} \Delta X_s I(|\Delta X_s| > 1)$$

with $M \in \mathcal{M}_{loc}(\mathbb{F}, P)$.

By Lemma 5 the process $[M, Z]$ belongs to $\mathcal{A}_{loc}(\tilde{P})$ and it has the compensator $\widetilde{[M, Z]}$ defined by the formula (5.20). Since $\inf_{s\le t} Z_s > 0$ ($\tilde{P}$-a.s.), $t \ge 0$, according to Lemma 1, by (5.20) and the definitions of the functions Y and β (cf. Lemma 6) we get ($\tilde{P}$-a.s.)

$$I(Z_- > 0) Z_-^{-1} \circ (\widetilde{[M, Z]})_t = \beta \circ C_t + I(|x| \le 1)\, x\, (Y - 1) * \nu_t$$
$$= B'_t - B_t, \quad t \ge 0.$$

Therefore by Theorem 2

$$\tilde{M} = M + B - B' \in \mathcal{M}_{loc}(\mathbb{F}, \tilde{P}) \tag{5.29}$$

and the processes $\langle \tilde{M}^c \rangle$ and C are $\tilde{P}$-indistinguishable.

Hence, a process X has the following decomposition with respect to the measure $\tilde{P}$:

$$X_t = X_0 + B'_t + \tilde{M}_t + \sum_{0<s\le t} \Delta X_s I(|\Delta X_s| > 1). \tag{5.30}$$

By virtue of Lemma 6 and the assumption $\tilde{P} \overset{loc}{\ll} P$ the process $B' \in \mathcal{A}_{loc}(\tilde{P})$. This, (5.29) and (5.30) entail $X \in \mathbf{S}(\mathbb{F}, \tilde{P})$.

Let $\tilde{T} = (\tilde{B}, \tilde{C}, \tilde{\nu})$ be the triplet of predictable characteristics of this semimartingale. Since B' is a predictable process, from (5.29) and (5.30) it follows that B' and $\tilde{B}$ are $\tilde{P}$-indistinguishable processes.

Next, $\tilde{C} = \langle \tilde{M}^c \rangle$ and hence $\tilde{C}$ and C are $\tilde{P}$-indistinguishable as well. Finally, by Theorem 1 we have

$$\tilde{\nu} \overset{\tilde{P}}{=} \nu'.$$

6. It has been established by Theorem 3 that a locally absolutely continuous change of measures preserves the semimartingale property of a stochastic process. In this subsection yet another transformation of a probability measure will be considered which preserves the semimartingale property as well.

Let $(\Omega, \mathcal{F}, \mathbb{F}, P)$ be a stochastic basis, $X \in D \cap \mathbb{F}$ and P the convex combination of probability measures P' and P", i.e.

$$P = \alpha' P' + \alpha'' P'', \quad \alpha', \alpha'' > 0, \quad \alpha' + \alpha'' = 1. \tag{5.31}$$

Since $P' \ll P$, $P'' \ll P$, there exist the local density processes Z' and Z'' of the measures P' and P'' with respect to P. In the present case Z' and $Z'' \in \mathfrak{M}\ (\mathbb{F}, P)$.

Suppose $X \in S\ (\mathbb{F}, P') \cap S\ (\mathbb{F}, P'')$. We denote the triplets of predictable characteristics of a process X by $T' = (B', C', \nu')$ and $T'' = (B'', C'', \nu'')$ respectively.

The following theorem tells us that the process X is a semimartingale with respect to $(\mathbb{F}, P)$.

Theorem 4. *Let a probability measure* P *be the convex combination* (5.31) *of the probability measures* P' *and* P'', *and* Z' *and* Z'' *the density processes of the measures* P' *and* P'' *with respect to* P.

Let $X \in S\ (\mathbb{F}, P') \cap S\ (\mathbb{F}, P'')$ *and the let triplets* T' *and* T'' *of predictable characteristics possess the following properties*: B' *and* $B'' \in \mathcal{V}\ (P)$, C' *and* $C'' \in \mathcal{V}^+\ (P)$, $M^P_{\nu'}$ *and* $M^P_{\nu''}$ *are finite measures on* $(\Omega \times R_+ \times R_0, \ \mathcal{P} \otimes B\ (R_0))$.

Then $X \in S\ (\mathbb{F}, P)$ *and the triplet* $T = (B, C, \nu)$ *of its predictable characteristics is given by the formulas*

$$B = \alpha' \overset{'}{Z}_- \circ B' + \alpha'' \overset{''}{Z}_- \circ B'',$$

$$C = \alpha' \overset{'}{Z}_- \circ C' + \alpha'' \overset{''}{Z}_- \circ C'',$$

$$\nu = \alpha' \overset{'}{Z}_- \nu' + \alpha'' \overset{''}{Z}_- \nu''. \tag{5.32}$$

The proof of this theorem relies on a number of auxiliary facts.

Lemma 7. *The process* $\alpha' Z' + \alpha'' Z''$ *is* P*-indistinguishable from a process identically equal to one.*

Proof. Let $\tau \in \mathbf{T}$ and $\xi = \text{sign}\ (1 - \alpha' Z' - \alpha'' Z'')$. Then (E' and E'' are the mathematical expectations with respect to measures P' and P'')

$$\mathbf{E}\ |\ 1 - \alpha' \overset{'}{Z}_\tau - \alpha'' \overset{''}{Z}_\tau\ | = \mathbf{E}\xi_\tau\ (1 - \alpha' \overset{'}{Z}_\tau - \alpha'' \overset{''}{Z}_\tau) = \mathbf{E}\xi_\tau - \alpha' \mathbf{E}' \xi_\tau - \alpha'' \mathbf{E}'' \xi_\tau = 0,$$

and the desired assertion follows from Theorem 1.3.12.

Lemma 8. *Let a process* $W = (W_t)_{t \geq 0}$ *be* P'*-negligible (respectively* P''*-negligible).*

Then the process WZ' *(respectively* WZ''*) is* P*-negligible.*

Proof. As $\tau \in \mathbf{T}$ we have

$$\mathbf{E} I\ (W_\tau \neq 0)\ \overset{'}{Z}_\tau = \mathbf{E}' I\ (W_\tau \neq 0) = 0.$$

Therefore we see, in view of Theorem 1.3.12, that the process $I\ (W \neq 0)\ Z'$ is P-negligible. Now the desired assertion follows in an obvious manner.

Lemma 9. *Let the processes* W *and* W' *be* P'*-indistinguishable and the* *processes* W *and* W'' P''-*indistinguishable. Then the processes* W and $\alpha'Z'W' + \alpha''Z''W''$ *are* P-*indistinguishable.*

Proof. Due to the property $1 \overset{P}{=} \alpha'Z' + \alpha''Z''$ we have

$$W - \alpha'Z'W' - \alpha''Z''W'' = \alpha'Z'(W - W') + \alpha''Z''(W - W'').$$

Therefore the desired assertion follows by Lemma 8.

Proof of Theorem 4. Since the process $X \in S(\mathbb{F}, P') \cap S(\mathbb{F}, P'')$, it admits the decompositions (cf. § 1)

$$X_t = X_0 + B'_t + M'_t + \sum_{0<s\le t} \Delta X_s I(|\Delta X_s| > 1)$$

and

$$X_t = X_0 + B''_t + M''_t + \sum_{0<s\le t} \Delta X_s I(|\Delta X_s| > 1) \tag{5.33}$$

where $M' \in \mathcal{M}_{loc}(\mathbb{F}, P')$ and $M'' \in \mathcal{M}_{loc}(\mathbb{F}, P'')$ while B' and B'' are the components of the triplets T' and T''. Then by Lemmas 7 and 9 the processes

$$(X_t - X_0 - \sum_{0<s\le t} \Delta X_s I(|\Delta X_s| > 1))_{t\ge 0}$$

and

$$\alpha'Z'B' + \alpha''Z''B'' + \alpha'Z'M' + \alpha''Z''M''$$

are P-indistinguishable, i.e. relative to the measure P the process X has the decomposition

$$X_t = X_0 + \alpha'Z'_tB'_t + \alpha''Z''_tB''_t + \alpha'Z'_tM'_t + \alpha''Z''_tM''_t$$

$$+ \sum_{0<s\le t} \Delta X_s I(|\Delta X_s| \ge 1). \tag{5.34}$$

Since B', B'' $\in \mathcal{V}(P)$, then by Ito's formula (cf. (3.5) Ch. 2, § 3) we get

$$\alpha'Z'_tB'_t + \alpha''Z''_tB''_t = \alpha Z'_- \circ B'_t + \alpha''Z''_- \circ B''_t + \alpha'B' \cdot Z'_t + \alpha''B'' \cdot Z''_t \tag{5.35}$$

with

$$N = \alpha'B' \cdot Z' + \alpha''B'' \cdot Z'' \in \mathcal{M}_{loc}(\mathbb{F}, P) \tag{5.36}$$

and

$$B = \alpha'Z'_- \circ B' + \alpha''Z''_- \circ B'' \in \mathcal{V}(P) \cap \mathcal{P}, \tag{5.37}$$

i.e.

$$\alpha'Z'B' + \alpha''Z''B'' = N + B \in \mathbf{Sp}(\mathbb{F}, P). \tag{5.38}$$

From (5.33) it follows that M', M'' $\in$ D with respect to the measure P as B', B'' $\in \mathcal{V}(P)$. Consequently, $\tau_n = \inf(t: |M'_t| \ge n) \to \infty$ (P-a.s.), $n \to \infty$.

Therefore $(M')^{\tau'_n} \in \mathcal{M}\ (\mathbb{F}, P')$, and by Lemma 3 we have

$$Z'M' \in \mathcal{M}_{loc}\ (\mathbb{F}, P).$$

Analogously $Z''M'' \in \mathcal{M}_{loc}\ (\mathbb{F}, P)$. Then

$$M = N + \alpha' Z'M' + \alpha'' Z''M'' \in \mathcal{M}_{loc}\ (\mathbb{F}, P), \tag{5.39}$$

and in virtue of (5.31), (5.35) and (5.36) the process X admits the following decomposition with respect to the measure P:

$$X_t = X_0 + B_t + M_t + \sum_{0 < s \le t} \Delta X_s I\,(\,|\,\Delta X_s\,| > 1), \tag{5.40}$$

in view of which $X \in S\ (\mathbb{F}, P)$ and B is the first component of the triplet $T = (B, C, \nu)$ of this semimartingale (cf. (5.37)).

Let us calculate now the components C and ν of the triplet T. By Theorem 3 we have $C \overset{P'}{=} C'$ and $C \overset{P''}{=} C''$. Therefore by Lemma 9

$$C \overset{P}{=} \alpha' Z'C' + \alpha'' Z''C''.$$

From this it follows by Ito's formula (cf. (3.5) Ch. 2, § 3) and by taking into account $C', C'' \in \mathcal{V}^+(P)$ that

$$C = \alpha' Z'_- \circ C' + \alpha'' Z''_- \circ C'' + \alpha' C' \cdot Z' + \alpha'' C'' \cdot Z''.$$

By Theorem 1.6.4 the process $\alpha' C' \cdot Z' + \alpha'' C'' \cdot Z''$ is P-negligible since it belongs to the class $\mathcal{M}_{loc}\ (\mathbb{F}, P) \cap \mathcal{V} \cap \mathcal{P}$. Hence, the desired assertion takes place for C too.

Set

$$Y' = I\,(Z'_- > 0)\ (Z'_-)^{-1}\ M^P_\mu\ (Z' \mid \tilde{\mathcal{P}})$$

and

$$Y'' = I\,(Z''_- > 0)\ (Z''_-)^{-1}\ M^P_\mu\ (Z'' \mid \tilde{\mathcal{P}}).$$

By Theorem 1

$$Z'_-\nu' \overset{P}{=} Z'_- Y'\nu, \quad Z''_-\nu'' \overset{P}{=} Z''_- Y''\nu. \tag{5.41}$$

Let U be a $\tilde{\mathcal{P}}$-measurable nonnegative function such that $EU * \nu_\infty < \infty$. Due to Lemma 7 and the fact that the processes $I\,(Z'_- = 0)\,Z'$ and $I\,(Z''_- = 0)\,Z''$ are P-negligible, (Problem 2.2.7), we get

$$\begin{aligned}
\mathrm{E}U * \nu_\infty &= \mathrm{E}U * \mu_\infty = \mathrm{E}U\,(\alpha' Z' + \alpha'' Z'') * \mu_\infty \\
&= \mathrm{E}U\,[\alpha' Z'_- I\,(Z'_- > 0)\,(Z'_-)^{-1} Z' + \alpha'' Z''_- I\,(Z''_- > 0)\,(Z''_-)^{-1} Z''] * \mu_\infty \\
&= \mathrm{E}U\,[\alpha' Z'_- Y' + \alpha'' Z''_- Y''] * \mu_\infty \\
&= \mathrm{E}U\,[\alpha' Z'_- Y' + \alpha'' Z''_- Y''] * \nu_\infty .
\end{aligned}$$

By this and (5.41) we get

$$\mathrm{E}U * \nu_\infty = \mathrm{E}\,[U\alpha' Z'_- * \nu'_\infty + U\alpha'' Z''_- * \nu''_\infty].$$

As a function U is arbitrary the desired representation for ν follows by this equality.

Problem

1. Let the conditions of Theorem 1 be fullfilled, $a_t = \nu\,(\{t\} \times E)$ and $\tilde{a}_t = \tilde{\nu}\,(\{t\} \times E)$. Show that

$$I\,(a = 1) \overset{\tilde{P}}{=} I\,(\tilde{a} = 1) \text{ and } I\,(Z_- > 0,\ a = 1) \overset{P}{=} (Z_- > 0,\ \tilde{a} = 1).$$

§ 6. Semimartingales and reduction of a flow of σ-algebras

1. Let $(\Omega, \mathcal{F}, \mathbb{F} = (\mathcal{F}_t)_{t \geq 0}, \mathsf{P})$ be a stochastic basis, and $X = (X_t)_{t \geq 0}$ a semimartingale ($X \in \mathbf{S}(\mathbb{F}, \mathsf{P})$) with the triplet of predictable characteristics $T = (B, C, \nu)$ and the canonical representation

$$X_t = X_0 + B_t + X_t^c + \int_0^t \int_{|x| \leq 1} x \, d(\mu - \nu) + \int_0^t \int_{|x| > 1} x \, d\mu. \tag{6.1}$$

Consider the natural flow of σ-algebras $\mathcal{F}_+^X = (\mathcal{F}_{t+}^X)_{t \geq 0}$, associated with a process X,

$$\mathcal{F}_{t+}^X = \bigcap_{\varepsilon > 0} (\sigma\{X_s,\ 0 \leq s \leq t + \varepsilon\} \vee \mathcal{N})$$

where $\mathcal{N}$ is a system of sets from $\mathcal{F}$, having P-measure zero (cf. Ch. 2, § 2, Subsection 11). Let also a flow of σ-algebras $\mathsf{G} = (\mathcal{G}_t)_{t \geq 0}$ be given, satisfying the "usual" conditions (a) and (b) of Definition 1 in Ch. 1, § 1, and let

$$\mathcal{F}_{t+}^X \subseteq \mathcal{G}_t \subseteq \mathcal{F}_t, \quad t \geq 0. \tag{6.2}$$

Theorem 1. *Under the assumption* (6.2) *we have*

$$X \in \mathbf{S}(\mathbb{F}, \mathsf{P}) \rightarrow X \in \mathbf{S}(\mathsf{G}, \mathsf{P}). \tag{6.3}$$

Besides,

$$C \overset{\mathsf{P}}{=} \overline{C} \tag{6.4}$$

where $\overline{C}$ *is the quadratic characteristic of the continuous local martingale* $\overline{X}^c$ *in the canonical representation of* X *relative to* (G, P).

2. This theorem tells us that the semimartingale property is preserved under the considered reduction of a flow of σ-algebras; we consider at first its particular cases.

Theorem 2. *Let* $X = \mathbf{S}(\mathbb{F}, \mathsf{P})$ *with*

$$X_t = B_t + M_t, \quad t \geq 0, \tag{6.5}$$

where $B = (B_t)_{t \geq 0} \in \mathcal{A} \cap \mathcal{P}(\mathbb{F})$, *and* $M = (M_t)_{t \geq 0} \in \mathcal{H}(\mathbb{F}, \mathsf{P})$, $M_0 = 0$.

Then $X \in \mathbf{S}(\mathsf{G}, \mathsf{P})$. (*Recall that according to Definition* 3 *in Ch.* 1, § 5 $\mathcal{H}(\mathbb{F}, \mathsf{P})$ *is the space of local martingales* M *with* $\mathbf{E} \sup_{t \geq 0} |M_t| < \infty$).

To prove this theorem we need the following two lemmas in which the optional projections of Y relative to (G, P) are denoted by $\pi(Y) = (\pi_t(Y))_{t \geq 0}$.

Lemma 1. *Let* $A \in \mathcal{A}^+ \cap \mathbb{F}$. *Then* $\pi(A)$ *presents a submartingale of the class* $(\mathcal{D})$ *and a semimartingale relative to* (G, P).

Proof. By Problem 1.3.10 a modification $\pi(A) \in \mathsf{D} \cap \mathsf{G}$ can be chosen, and in accordance with the definition of the optional projection (Theorem 1.3.13), as $t > s$ we have (P-a.s.)

$$\mathbf{E}(\pi_t(A) \mid \mathcal{G}_s) = \mathbf{E}(\mathbf{E}(A_t \mid \mathcal{G}_t) \mid \mathcal{G}_s) = \mathbf{E}(A_t \mid \mathcal{G}_s) \geq \mathbf{E}(A_s \mid \mathcal{G}_s) = \pi_s(A).$$

Hence, $\pi(A)$ is a submartingale. Next, for each stopping time τ (relative to G) we have $\pi_t(A) \leq \mathbf{E}(A_\infty \mid \mathcal{G}_\tau)$. Therefore, by Definition 4 (Ch. 1, § 4) the family $\{\pi_t(A), \tau \in \mathbf{T}(G)\}$ is uniformly integrable, i.e. $\pi(A)$ belongs to class $(\mathcal{D})$. Hence, by the Doob-Meyer theorem (Theorem 1.6.5), the submartingale $\pi(A)$ is a semimartingale.

Lemma 2. *Let* $M = (M_t)_{t \geq 0} \in \mathcal{H}(\mathbb{F}, P)$ *with* $M_0 = 0$. *Then* $\pi(M) \in (G, P)$.

Proof. Due to Problem 1.3.10 one can choose $\pi(M) \in D \cap G$. Next, by the definition of the optional projection as $t > s$ we have (P-a.s.)

$$\mathbf{E}(\pi_t(M) \mid \mathcal{G}_s) = \mathbf{E}(\mathbf{E}(M_t \mid \mathcal{G}_t) \mid \mathcal{G}_s) = \mathbf{E}(M_t \mid \mathcal{G}_s)$$
$$= \mathbf{E}(\mathbf{E}(M_t \mid \mathcal{F}_s) \mid \mathcal{G}_s) = \mathbf{E}(M_s \mid \mathcal{G}_s) = \pi_s(M),$$

i.e. $\pi(M)$ is a martingale relative to (G, P). Finally,

$$\pi_t(M) = \mathbf{E}(\mathbf{E}(M_\infty \mid \mathcal{F}_t) \mid \mathcal{G}_t) = \mathbf{E}(M_\infty \mid \mathcal{G}_t) \quad \text{(P-a.s.)}.$$

Hence $\pi(M)$ is a uniformly integrable martingale (Theorem 1.4.1).

Remark. The assertion of the lemma remains true (with $\mathcal{M}_{loc}(G)$ replaced by $\mathcal{M}(G)$) for $M \in \mathcal{M}_{loc}(\mathbb{F}, P)$ if there exists a localizing sequence $(\tau_n)_{n \geq 1}$ such that $M^{\tau_n} \in \mathcal{H}(\mathbb{F}, P)$, $\tau^n \in \mathbf{T}(G)$, $n \geq 1$.

Proof of Theorem 2. Let $B^i = (B^i_t)_{t \geq 0} \in \mathcal{A}^+ \cap \mathcal{P}(\mathbb{F})$, $i = 1, 2$ be processes such that $B_t = B^1_t - B^2_t$ and $\mathrm{Var}(B) = B^1 + B^2$. Then

$$X = B^1 - B^2 + M. \tag{6.6}$$

Since $\mathcal{F}^X_{t+} \subseteq \mathcal{G}_t$, $t \geq 0$, we have $\pi(X) = X$. Consequently,

$$X = \pi(B^1) - \pi(B^2) + \pi(M), \tag{6.7}$$

and the desired assertion follows by Lemmas 1 and 2.

Theorem 3. *Let* $X \in \mathbf{S}(\mathbb{F}, P)$ *admit the representation* $X_t = B_t + M_t$ *with* $B \in \mathcal{V} \cap \mathcal{P}(\mathbb{F})$, $\mathrm{Var}(B)_\infty < \infty$ (P-a.s.) *and* $M \in \mathcal{M}^2(\mathbb{F}, P)$, $M_0 = 0$. *Then*

$$X \in \mathbf{S}(G, P).$$

Proof. By the assumptions of the theorem,

$$P\left(\int_0^\infty |dB_s| < \infty\right) = 1.$$

Introduce the random variables

$$\alpha = \frac{1}{1 + \int_0^\infty |dB_s|}, \quad Z_\infty = \frac{\alpha}{E\alpha} \tag{6.8}$$

and on $(\Omega, \mathcal{F})$ define a new probability measure $\tilde{P}$ with

$$d\tilde{P} = Z_\infty \, dP. \tag{6.9}$$

Let $Z = (Z_t)_{t \geq 0}$ be the density process of $\tilde{P}$ with respect to P. Since $0 < Z_\infty \leq \frac{1}{E\alpha}$, the measures $\tilde{P}$ and P are equivalent, $\tilde{P} \sim P$. By the definition of the process Z ($Z_T = E(Z_\infty \mid \mathcal{F}_T)$ for each Markov time T) it follows that $\sup_{t \geq 0} Z_t \leq \frac{1}{E\alpha}$. Therefore, as the process Z is nonnegative, we have

$$Z \in \mathcal{M}^2(\mathbb{F}, P).$$

Consider the process $\tilde{M} = (\tilde{M}_t)_{t \geq 0}$ with

$$\tilde{M}_t = M_t - Z_-^{-1} \circ \langle Z, M \rangle_t. \tag{6.10}$$

Since M, $Z \in \mathcal{M}^2(\mathbb{F}, P)$, the process $\langle Z, M\rangle$ is the compensator of the process [Z, M] relative to $(\mathbb{F}, P)$ (Theorem 1.8.1). Then by Theorem 5.2 we have $\tilde{M} \in \mathcal{M}_{loc}(F, \tilde{P})$ and $X \in \mathbf{Sp}(F, \tilde{P})$, besides

$$X_t = \tilde{B}_t + \tilde{M}_t \tag{6.11}$$

with

$$\tilde{B}_t = B_t + Z_-^{-1} \circ \langle Z, M\rangle_t = B_t + L_t. \tag{6.12}$$

Let us show that

$$\tilde{E}\,\mathrm{Var}(\tilde{B})_\infty < \infty, \quad \tilde{E} \sup_{t \geq 0} |M_t| < \infty. \tag{6.13}$$

Aiming at this, we remark that by the definition of Z_∞ we have

$$\tilde{E}\,\mathrm{Var}(B)_\infty = EZ_\infty \mathrm{Var}(B)_\infty = \frac{1}{E\alpha} E \frac{\int_0^\infty |dB_s|}{1 + \int_0^\infty |dB_s|} \leq \frac{1}{E\alpha} < \infty. \tag{6.14}$$

Next, by Problems 1.8.9 and 1.6.3

$$\tilde{\mathbf{E}}\,\mathrm{Var}\,(L)_\infty \le \frac{1}{2}\tilde{\mathbf{E}}\tilde{Z}_-^{-1} \circ (\langle Z\rangle + \langle M\rangle)_\infty = \frac{1}{2}\mathbf{E}Z_\infty (Z_-^{-1} \circ (\langle Z\rangle + \langle M\rangle)_\infty)$$

$$= \frac{1}{2}\mathbf{E}\,(\langle Z\rangle_\infty + \langle M\rangle_\infty) < \infty. \tag{6.15}$$

The first inequality in (6.13) follows now from (6.12), (6.14) and (6.15).

To establish the second inequality in (6.13), observe that

$$\tilde{\mathbf{E}} \sup_{t\ge 0} |\tilde{M}_t| \le \tilde{\mathbf{E}} \sup_{t\ge 0} |M_t| + \tilde{\mathbf{E}}\,\mathrm{Var}\,(L)_\infty, \tag{6.16}$$

and

$$\tilde{\mathbf{E}} \sup_{t\ge 0} |M_t| = \mathbf{E}Z_\infty \sup_{t\ge 0} |M_t| \le \frac{1}{\mathbf{E}\alpha}\mathbf{E} \sup_{t\ge 0}|M_t| \le \frac{3\mathbf{E}\,\langle M\rangle_\infty^{1/2}}{\mathbf{E}\alpha}, \tag{6.17}$$

where the last inequality takes place in virtue of Theorem 1.9.5. The second inequality in (6.13) follows from (6.16) and (6.17).

From (6.11) and (6.13) it follows that as $X \in \mathbf{S}\,(\mathbb{F}, \tilde{P})$ the conditions of Theorem 2 are fullfilled, and hence $X \in \mathbf{S}\,(G, \tilde{P})$. But the measure $P \ll \tilde{P}$, and consequently $X \in \mathbf{S}\,(G, P)$ by Theorem 5.3.

Theorem 4. *Let* $(T^n)_{n\ge 1}$ *be a sequence of Markov times (relative to* G*) such that* $T^n \uparrow \infty$, $n \to \infty$. *Let* $X \in \mathbf{S}\,(\mathbb{F}, P)$ *and* $X \in B + M$ *with* $B^{T^n} \in \mathcal{V}\,(\mathbb{F}) \cap \mathcal{P}\,(\mathbb{F})$, $\mathrm{Var}\,(B)_{T_n} < \infty$ (P-a.s.), $M^{T^n} \in \mathcal{M}^2\,(\mathbb{F}, P)$, $M_0 = 0$, $n \ge 1$. *Then*

$$X \in \mathbf{S}\,(G, P).$$

Proof. The process $X^n = X^{T^n} \in \mathbf{S}\,(\mathbb{F}, P)$. Since $T^n \in \mathbf{T}\,(G)$, for the process X^n Condition (6.2) is still satisfied and by Theorem 3 we have $X^n \in \mathbf{S}\,(G, P)$. Then

$$X_t^n = \overline{B}_t^n + \overline{M}_t^n, \quad t \ge 0$$

with

$$\overline{B}^n = \mathcal{V}\,(G) \cap \mathcal{P}\,(G), \ \overline{M}^n \in \mathcal{M}_{loc}\,(G, P), \ \overline{M}_0^n = 0.$$

Set

$$\overline{B}_t = \overline{B}^1_{t\wedge T^1} + \sum_{n\ge 1} (\overline{B}^{n+1}_{t\wedge T^{n+1}} - \overline{B}^n_{t\wedge T^n})$$

and

$$\overline{M}_t = \overline{M}^1_{t\wedge T^1} + \sum_{n\ge 1} (\overline{M}^{n+1}_{t\wedge T^{n+1}} - \overline{M}^n_{t\wedge T^n}).$$

Clearly $X_t = \overline{B}_t + \overline{M}_t$, since

$$X_t^n = X^{n+1}_{t\wedge T^n}, \quad \overline{B}_t^n = \overline{B}^{n+1}_{t\wedge T^n}, \quad \overline{M}_t^n = \overline{M}^{n+1}_{t\wedge T^n}$$

and $\overline{M} \in \mathcal{M}_{loc}(G, P)$, $\overline{B} \in \mathcal{V}(G)$.

Proof of Theorem 1. Let

$$X_t^1 = X_0 + \int_0^t \int_{|x|>1} x\, d\mu \left(= X_0 + \sum_{0<s\le t} \Delta X_s I(|\Delta X_s| > 1)\right).$$

Clearly, $X^1 \in \mathcal{V}(\mathbb{F})$. Therefore, $X^1 \in \mathbf{S}(\mathbb{F}_+^X, P)$ and $X^1 \in \mathbf{S}(G, P)$.

Introduce the process $Y = X - X^1$. Then $|\Delta Y| \le 1$ and $Y = B + M$ with

$$M_t = X_t^c + \int_0^t \int_{|x|\le 1} x\, d(\mu - \nu).$$

To establish the desired assertion $X \in \mathbf{S}(G, P)$ we need to show $Y \in \mathbf{S}(G, P)$. Since $\mathcal{F}^Y_{t+} \subseteq \mathcal{F}^X_{t+}$, it suffices for this, according to Theorem 4, to establish the existence of a sequence of Markov times $(T^n)_{n\ge 1}$, $T^n \in \mathbf{T}(G)$ with the properties: $T^n \uparrow \infty$, $B^{T^n} \in \mathcal{V} \cap \mathcal{P}(\mathbb{F})$ and $M^{T^n} \in \mathcal{M}^2(\mathbb{F}, P)$.

Aiming at this, we remark that the process $[Y, Y]$ is a $\mathbb{F}_+^X$-adapted process because, by the definition of the quadratic variation (cf. Ch. 2, § 1) we have

$$[Y, Y] = \langle X^c \rangle + \sum_s (\Delta X_s)^2 I(|\Delta X_s| \le 1),$$

and the process $\langle X^c\rangle$ is $\mathbb{F}_+^X$-adapted by the corollary to Theorem 2.3.2. Consequently, $[Y, Y]$ is also a G-adapted process. Set

$$T^n = \inf(t: [Y, Y]_t \ge n) \wedge n, \quad n \ge 1.$$

These times $T^n \in \mathbf{T}(\mathbb{F}_+^X)$ and hence $T^n \in \mathbf{T}(G)$.

Since $[Y, Y] \in \mathcal{V}^+$, we have $T^n \uparrow \infty$ (P-a.s.). Next, by the inequality $T^n \le n$ we have

$$\mathrm{Var}(B)_{T^n} \le \mathrm{Var}(B)_n < \infty \quad \text{(P-a.s.)},$$

i.e. $B^{T^n} \in \mathcal{V}$. To establish the property $M^{T^n} \in \mathcal{M}^2(\mathbb{F}, P)$, $n \ge 1$, it suffices to verify that

$$\mathbf{E}[M, M]_{T^n} < \infty, \quad n \ge 1 \tag{6.18}$$

(cf. Theorem 1.9.7).

To prove (6.18), observe that

$$\Delta M = \Delta Y - \Delta B.$$

Consequently,

$$[M, M]_{T^n} = \langle X^c \rangle_{T^n} + \sum_{0 < s \le T^n} (\Delta Y_s - \Delta B_s)^2$$

$$\le \langle X^c \rangle_{T^n} + 2 \sum_{0 < s \le T^n} (\Delta Y_s)^2 + 2 \sum_{0 < s \le T^n} (\Delta B_s)^2.$$

But by (1.7)

$$\mathbf{E} \sum_{0 < s \le T^n} (\Delta B_s)^2 = \mathbf{E} \sum_{0 < s \le T^n} \left(\int_{|x| \le 1} x \nu(\{s\}, dx) \right)^2 \le \mathbf{E} \sum_{0 < s \le T^n} \int_{|x| \le 1} x^2 \nu(\{s\}, dx)$$

$$\le \mathbf{E} \int_0^{T^n} \int_{|x| \le 1} x^2 \, d\nu = \mathbf{E} \int_0^{T^n} \int_{|x| \le 1} x^2 \, d\mu = \mathbf{E} \sum_{0 < s \le T^n} (\Delta Y_s)^2$$

Hence,

$$\mathbf{E}\, [M, M]_{T^n} \le \mathbf{E} \left(\langle X^c \rangle_{T^n} + 4 \sum_{0 < s \le T^n} (\Delta Y_s)^2 \right)$$

$$\le 4\mathbf{E} \left(\langle X^c \rangle_{T^n} + \sum_{0 < s \le T^n} (\Delta Y_s)^2 \right) = 4\mathbf{E}\, [Y, Y]_{T^n} \le 4\, (n + 1),$$

since by the inequality $|\Delta Y| \le 1$ we have

$$[Y, Y]_{T^n} \le n + 1.$$

Thus $Y \in \mathbf{S}(\mathbb{G}, \mathsf{P})$ and, in particular, $Y \in \mathbf{S}(\mathbb{F}^X_+, \mathsf{P})$.

Let us show that $C \overset{\mathsf{P}}{=} \overline{C}$. Since $X \in \mathbf{S}(\mathbb{F}, \mathsf{P}) \cap \mathbf{S}(\mathbb{G}, \mathsf{P}) \cap \mathbf{S}(\mathbb{F}^X_+, \mathsf{P})$, in virtue of the corollary to Theorem 2.3.2 the processes C and $\overline{C}$ are P-indistinguishable from the quadratic characteristic of the continuous component of the $\mathbb{F}^X_+$-semimartingale X. Hence, C and $\overline{C}$ are P-indistinguishable processes.

3. We will show how to calculate in one particular case the components $\overline{B}$ and $\overline{\nu}$ of the triplet $\overline{T} = (\overline{B}, \overline{C}, \overline{\nu})$ (recall that $\overline{C} \overset{\mathsf{P}}{=} C$) of predictable characteristics of a $\mathbb{G}$-semimartingale X.

Theorem 5. *Let*

$$B_t = \int_0^t \beta(s)\, db_s, \qquad \nu(dt, dx) = \gamma(t, x)\, \Lambda(dt, dx)$$

where $\beta(t)$ *and* $\gamma(t, x)$ *are* $\mathcal{P}(\mathbb{F})$*- and* $\tilde{\mathcal{P}}(\mathbb{F}) = \mathcal{P}(\mathbb{F}) \otimes \mathcal{B}(R_0)$*-measurable functions, let* $b \in \mathcal{V} \cap \mathcal{P}(\mathbb{G})$*, let* Λ *be a* $\tilde{\mathcal{P}}(\mathbb{G})$ $(= \mathcal{P}(\mathbb{G}) \otimes \mathcal{B}(R_0))$*-*$\sigma$*-finite predictable measure on* $(R_+ \times R_0)$ *and let there exist a sequence of stopping times* $(\tau_n)_{n \geq 1}$ *(relative to* $\mathbb{G}$*) such that* $\tau_n \uparrow \infty$, $n \to \infty$, *and for* $n \geq 1$

$$\mathbf{E}\, |\beta| \circ \mathrm{Var}(b)_{\tau_n} < \infty, \quad \mathbf{E}\, |\gamma| * \Lambda_{\tau_n} < \infty. \tag{6.19}$$

Then

$$\overline{B}_t = \int_0^t {}^{P}\pi_s(\beta)\, db_s, \quad \overline{\nu}(dt, dx) = M^{P}_{\Lambda}(\gamma \mid \tilde{\mathcal{P}}(\mathbb{G}))(t, x)\, \Lambda(dt, dx),$$

where ${}^{P}\pi(\beta)$ *is the predictable projection* β *relative to* $(\mathbb{G}, P)$*, and* $M^{P}_{\Lambda}(\cdot \mid \tilde{\mathcal{P}}(\mathbb{G}))$ *the conditional mathematical expectation, relative to* $\tilde{\mathcal{P}}(\mathbb{G})$.

Proof. Let Y and M be the processes defined in the course of proving Theorem 1. To prove the validity of the representation for $\overline{B}$, it suffices to establish that $\overline{M} = Y - \overline{B} = B - \overline{B} + M \in \mathcal{M}_{\mathrm{loc}}(\mathbb{G}, P)$. Set $\sigma_n = \tau_n \wedge T^n$, $n \geq 1$, where T^n is a stopping time defined in the course of proving Theorem 1. Then it suffices to show that $\overline{M}^{\sigma_n} \in \mathcal{M}(\mathbb{G}, P)$, $n \geq 1$, or that for each $\tau \in \mathbf{T}(\mathbb{G})$ (Problem 1.4.2)

$$\mathbf{E}\overline{M}^{\sigma_n}_{\tau} = 0. \tag{6.20}$$

From (6.18) it follows that $M^{T^n} \in \mathcal{M}^2(\mathbb{F}, P)$. Hence, $\mathbf{E}M^{\sigma_n}_{\tau} = \mathbf{E}M^{T_n}_{\tau \wedge \tau_n} = 0$. Besides, according to Theorem 1.6.3,

$$\mathbf{E} \int_0^{\tau \wedge \sigma_n} \beta(s)\, db_s = \mathbf{E} \int_0^{\tau \wedge \sigma_n} {}^{P}\pi_s(\beta)\, db_s.$$

Thus, (6.20) takes place and consequently the representation for $\overline{B}$ is valid.

By virtue of the second inequality in (6.19) the conditional mathematical expectation $M^{P}_{\Lambda}(\gamma \mid \tilde{\mathcal{P}}(\mathbb{G}))$ exists (Problem 3.2.7). Next, if U is a $\tilde{\mathcal{P}}(\mathbb{G})$-measurable function such that $\mathbf{E}\,|U\gamma| * \Lambda_\infty < \infty$, then

$$\mathbf{E}U * \overline{\nu}_\infty = \mathbf{E}U * \mu_\infty = \mathbf{E}U * \nu_\infty = \mathbf{E}U\gamma * \Lambda_\infty = \mathbf{E}U\, M^{P}_{\Lambda}(\gamma \mid \tilde{\mathcal{P}}(\mathbb{G})) * \Lambda_\infty,$$

and consequently the representation for $\overline{\nu}$ holds.

Problems

1. Let ν and $\bar{\nu}$ be the compensators of the jump measure of a semimartingale X relative to $(\mathbb{F}, P)$ and (G, P), $G \subseteq \mathbb{F}$ and $H \in \tilde{\mathcal{P}}(G)$. Show that

$$H * \bar{\nu} \in \mathcal{A}_{loc} \Rightarrow H * \nu \in \mathcal{A}_{loc}$$

and

$$H * \nu \in \mathcal{A} \Rightarrow H * \bar{\nu} \in \mathcal{A}.$$

2. Let $\Gamma = \{x_1, x_2, \ldots\}$ with $x_i \neq 0$, $i \geq 1$, and let the measure Λ, defined in Theorem 5, possess the following property $\Lambda(R_+ \times R_0 \setminus \Gamma) = 0$ (P-a.s.), and let the second inequality in (6.19) hold for a $\tilde{\mathcal{P}}(\mathbb{F})$-measurable function. Show that as for $M_\Lambda^P(\gamma \mid \tilde{\mathcal{P}}(G))(t, x)$ one can take a $\tilde{\mathcal{P}}(G)$-measurable function $H = H(t, x)$ such that for each $\tau \in R_+$

$$H(t, x) = \begin{pmatrix} 0, & x \in R_0 \setminus \Gamma \\ {}^P\pi_t(\gamma_i), & x = x_i,\ i \geq 1, \end{pmatrix}$$

where $\gamma_i(t) = \gamma(t, x_i)$ and ${}^P\pi$ is the predictable projection relative to (G, P).

3. Let $G \subseteq \mathbb{F}$, and let X be a continuous semimartingale relative to G and $\mathbb{F}$ with the decomposition

$$X = A + M$$

and

$$X = \bar{A} + \bar{M}$$

respectively, where $A \in \mathcal{V}^+(\mathbb{F})$ and $M \in \mathcal{M}_{loc}(\mathbb{F})$. Show that $\bar{A} \in \mathcal{V}^+(G)$, in particular that $A = 0$ implies $\bar{A} = 0$.

§ 7. Semimartingales and random change of time

1. Let $(\Omega, \mathcal{F}, \mathbb{F}, P)$ be a stochastic basis.

Definition. A stochastic process $\hat{\tau} = (\hat{\tau}_t)_{t \geq 0}$, belonging to the class V^+ (Ch. 1, § 6) and such that $\hat{\tau}_t$ is a stopping time (relative to the family $\mathbb{F}$) for each $t \geq 0$, is called a *random change of time.*

Denote

$$\hat{\tau}_\infty = \lim_{t \to \infty} \hat{\tau}_t.$$

With any process $X \in D \cap \mathbb{F}$ and any random change of time $\hat{\tau} = (\hat{\tau}_t)_{t \geq 0}$ one may associate a new process $\hat{X}$ by setting

$$\hat{X}_t(\omega) = X_{\hat{\tau}_t(\omega)}(\omega), \quad t \geq 0. \tag{7.1}$$

With the family $\mathbb{F}$ and $\hat{\tau}$ we associate also a new flow of σ-algebras $\hat{\mathbb{F}} = (\hat{\mathcal{F}}_t)_{t \geq 0}$ with

$$\hat{\mathcal{F}}_t = \mathcal{F}_{\hat{\tau}_t}. \tag{7.2}$$

It is not hard to show (Problem 1) that for the family $\hat{\mathbb{F}}$ the "usual" conditions (a) and (b) of Definition 1.1.1 are fullfilled.

Theorem 1. *The following implications take place:*

1) $X \in D \cap \mathbb{F} \Rightarrow \hat{X} \in D \cap \hat{\mathbb{F}}$,

2) $X \in \mathcal{V}(\mathbb{F}) \Rightarrow \hat{X} \in \mathcal{V}(\hat{\mathbb{F}})$,

3) $X \in \mathcal{A}(\mathbb{F}) \Rightarrow \hat{X} \in \mathcal{A}(\hat{\mathbb{F}})$,

4) $X \in \mathcal{M}(\mathbb{F}) \Rightarrow \hat{X} \in \mathcal{M}(\hat{\mathbb{F}})$,

5) $X \in \mathcal{H}^p(\mathbb{F}) \Rightarrow \hat{X} \in \mathcal{H}^p(\hat{\mathbb{F}})$, $p \geq 1$,

6) $X \in D \cap \mathcal{D}(\mathbb{F}) \Rightarrow \hat{X} \in D \cap \mathcal{D}(\hat{\mathbb{F}})$.

Proof. 1) The implication $X \in D \Rightarrow \hat{X} \in D$ follows in an obvious manner from properties of the functions $(X_t)_{t \geq 0}$ and $(\hat{\tau}_t)_{t \geq 0}$. Since $X \in D$, the random variables $X_{\hat{\tau}_t}$ are $\mathcal{F}_{\hat{\tau}_t}$-measurable (Problems 1.1.1 and 1.1.7). Thus, $\hat{X}$ is a $\hat{\mathbb{F}}$--adapted process.

1 if necessary.

Denote

$$Z = 1 + \frac{N}{2a}.$$

Then $Z \in \mathcal{H}(\mathbb{F}, P)$ and $Z \geq \frac{1}{2}$. Define the probability measure $\tilde{P}$ with $d\tilde{P} = Z_\infty dP$. We will show that $\tilde{P} \in \mathcal{L}_P$.

By construction $\tilde{P} \ll P$. By the property $Z_0 = 1$ we have $\tilde{P}_0 = P_0$. Next, by Theorem 5.3 we have $X \in S(\mathbb{F}, \tilde{P})$. From (8.27) and (8.28) it follows that $\langle Z^c, X^c \rangle = 0$ and $M^P_\mu(\Delta Z \mid \tilde{\mathcal{P}}) = 0$.

Therefore, the local absolutely continuous change of measures (see. (8.19)), leads here to the triplet transformation with $\beta = 0$ and $Y = 1$. Consequently, by Theorem 5.3 and the remark to it the triplet $\tilde{T}$ of the semimartingale $X \in S(\mathbb{F}, P)$ is P-indistinguishable from the triplet T.

Thus $\tilde{P} \in \mathcal{L}_P$. But, by definition $\mathcal{L}_P = \{P\}$. Hence $\tilde{P} = P$ and the local density process Z is P-indistinguishable from the process identically equal to one, i.e the process N is P-negligible and the processes M and M' are P-indistinguishable.

5. Let a process $X \in S(\mathbb{F}, P)$. Define the σ-algebras

$$\mathcal{F}^X_t = \sigma(\{X_s, 0 \leq s \leq t\} \vee \mathcal{N}), \quad \mathcal{F}^X_{t+} = \bigcap_{\varepsilon > 0} \mathcal{F}^X_{t+\varepsilon},$$

where $\mathcal{N}$ is a system of sets in $\mathcal{F}$ of P-measure zero.

Set

$$\mathcal{F}^X_\infty = \sigma\left(\bigcup_{t \geq 0} \mathcal{F}^X_t\right).$$

The restriction of a measure P on the σ-algebras $\mathcal{F}^X_\infty$, $\mathcal{F}^X_{0+}$ and $\mathcal{F}^X_0$ is denoted by P^X, P^X_{0+} and P^X_0 respectively.

By Theorem 6.1 we have

$$X \in S(\mathbb{F}^X_+, P^X).$$

Therefore, there arises a natural question about the integral representation for each $M \in \mathcal{M}_{loc}(\mathbb{F}^X_+, P^X)$. In the present special case one can strengthen the result of Theorem 1 to some extent.

$$X_t = X_0 + A_t + M_t$$

with $A \in \mathcal{V} \cap \mathbb{F}$ and $M \in \mathcal{M}_{loc,0}(\mathbb{F})$. Then

$$\hat{X}_t = X_0 + \hat{A}_t + \hat{M}_t$$

with

$$\hat{A}_t = A_{\hat{\tau}_t}, \quad \hat{M}_t = M_{\hat{\tau}_t}.$$

By Theorem 1 we have $\hat{A} \in \mathcal{V}(\hat{\mathbb{F}})$. Therefore $\hat{A} \in \mathbf{S}(\hat{\mathbb{F}})$, and it suffices to show only that

$$\hat{M} \in \mathbf{S}(\hat{\mathbb{F}}). \tag{7.3}$$

If $M \in \mathcal{M}(\mathbb{F})$, then by Theorem 1 we have $\hat{M} \in \mathcal{M}(\hat{\mathbb{F}})$, and hence (7.3) takes place. If $M \in \mathcal{M}_{loc}(\mathbb{F})$ and $(T_n)_{n \geq 1}$ is a localizing sequence for M, then the process $M^n = M^{T_n} \in \mathcal{M}(\mathbb{F})$, $n \geq 1$, and $\hat{M}^n \in \mathcal{M}(\hat{\mathbb{F}})$.

Define

$$\hat{T}_n = \inf(t: \hat{\tau}_t \geq T_n), \ n \geq 1.$$

By Lemma 1 we have $\hat{T}_n \in \mathbf{T}(\hat{\mathbb{F}})$, and due to the fact that $T_n \uparrow \infty$ and $\hat{\tau} = (\hat{\tau}_t)_{t \geq 0} \in \mathcal{V}^+$, we get that the sequence $(\hat{T}_n)_{n \geq 1}$ is nondecreasing with $\lim_n \hat{T}_n = \infty$.

Let us show that

$$(\hat{M})^{\hat{T}_n} \in \mathbf{S}(\hat{\mathbb{F}}), \ n \geq 1. \tag{7.4}$$

To this end observe that $\hat{\tau}_{\hat{T}_n} \in \mathbf{T}(\mathbb{F})$ (Lemma 1), $\hat{\tau}_{\hat{T}_n} \geq T_n$ and $\hat{\tau}_{\hat{T}_n -} \leq T_n$.

By definition

$$(\hat{M})_t^{\hat{T}_n} = \hat{M}_{t \wedge \hat{T}_n} = M_{\hat{\tau}_{t \wedge \hat{T}_n}}.$$

On the other hand, by definition again,

$$(\hat{M}^n)_t^{\hat{T}_n} = \hat{M}^n_{t \wedge \hat{T}_n} = M_{\hat{\tau}_{t \wedge \hat{T}_n} \wedge T_n}.$$

From this it follows that the processes $(\hat{M})^{\hat{T}_n} I_{[\![0, \hat{T}_n[\![}$ and $(\hat{M}^n)^{\hat{T}_n} I_{[\![0, T_n[\![}$ are P-indistinguishable. Consequently

$$(\hat{M})_t^{\hat{T}_n} = (M^n)_t^{\hat{T}_n} + (\hat{M}_{\hat{T}_n} - M^n_{\hat{T}_n})\, I(t \geq \hat{T}_n),$$

i.e. $\hat{M}^{\hat{T}_n}$ is the sum of processes of the class $\mathcal{M}(\hat{\mathbb{F}})$ and $\mathcal{V}(\hat{\mathbb{F}})$. Hence (7.4) takes place, and by Problem 2 we have $\hat{M} \in \mathbf{S}(\hat{\mathbb{F}})$.

Problems

1. Let $\hat{\tau} = (\hat{\tau}_t)_{t \geq 0} \in V^+$ and for each $t \in R_+$ let $\hat{\tau}_t \in \mathbf{T}(\mathbb{F})$. Set $\hat{\mathcal{F}}_t = \mathcal{F}_{\hat{\tau}_t}$, $t \geq 0$. Show that the "usual" conditions (a) and (b) (cf. Ch. 1, § 1) are fullfilled for $\hat{\mathbb{F}} = (\hat{\mathcal{F}}_t)_{t \geq 0}$ if they are fullfilled for $\mathbb{F}$.

2. Let $X \in \mathcal{M}^c_{loc}(\mathbb{F})$ and let its quadratic characteristics $\langle X \rangle$ possess the property $\langle X \rangle_\infty = \infty$ (P-a.s.). Set $\tau_t = \inf(s: \langle X \rangle_s > t)$. Show that $\hat{X}$ with $\hat{X}_t = X_{\hat{\tau}_t}$ is a Wiener process relative to $\hat{\mathbb{F}} = (\hat{\mathcal{F}}_t)_{t \geq 0}$ with $\hat{\mathcal{F}}_t = \mathcal{F}_{\hat{\tau}_t}$.

3. Let $X \in \mathbb{F}$ be a counting process with the continuous compensator $A = (A_t)_{t \geq 0}$ relative to $\mathbb{F}$ with $A_\infty = \infty$. Show that as $\hat{\tau}_t = \inf(s: A_s \geq t)$ the process $\hat{X}$ with $\hat{X}_t = X_{\hat{\tau}_t}$ is a Poisson process relative to $\hat{\mathbb{F}} = (\hat{\mathcal{F}}_t)_{t \geq 0}$ with $\hat{\mathcal{F}}_t = \mathcal{F}_{\hat{\tau}_t}$.

§ 8. Semimartingales and integral representation of martingales

1. Let $(\Omega, \mathcal{F}, \mathbb{F} = (\mathcal{F}_t)_{t \geq 0}, \mathrm{P})$ be a stochastic basis, $\mathcal{F} = \mathcal{F}_\infty$, $X \in \mathbf{S}(\mathbb{F}, \mathrm{P})$ and let $T = (B, C, \nu)$ be the triplet of its predictable characteristics.

If h is a $\mathcal{P}$-measurable function and H a $\tilde{\mathcal{P}} = \mathcal{P} \otimes B(R_0)$-measurable function such that (cf. Ch. 2, § 2 and Ch. 3, § 5)

$$h \in \mathbf{L}^2_{\mathrm{loc}}(\mathbb{F}, \langle X^c \rangle), \quad H \in \mathcal{G}_{\mathrm{loc}},$$

where $\langle X^c \rangle = C$ and $\mathcal{G}_{\mathrm{loc}}$ is related to the jump measure of X, then with a semimartingale X one may associate a local martingale M ($M \in \mathcal{M}_{\mathrm{loc}}(\mathbb{F}, \mathrm{P})$), which presents the sum of a $\mathcal{F}_0$-measurable and integrable random variable M_0 and the stochastic integrals $h \cdot X^c$ and $H * (\mu - \nu)$ with respect to the continuous component X^c and to the martingale measure $\mu - \nu$ (μ is the jump measure of X), i.e.

$$M_t = M_0 + h \cdot X^c_t + H * (\mu - \nu)_t. \tag{8.1}$$

In the present subsection the following converse problem is considered. Let $M \in \mathcal{M}_{\mathrm{loc}}(\mathbb{F}, \mathrm{P})$ and $X \in \mathbf{S}(\mathbb{F}, \mathrm{P})$. It is asked in which case the representation (8.1) for M takes place. In case when (8.1) is valid we say that a local martingale M *admits the integral representation* (with respect to the continuous martingale component X^c and the martingale measure $\mu - \nu$ of a semimartingale X).

2. In order to formulate conditions ensuring the integral representation for $M \in \mathcal{M}_{\mathrm{loc}}(\mathbb{F}, \mathrm{P})$, we define the family

$$\mathcal{L}_{\mathrm{P}} = \{\tilde{\mathrm{P}} : \tilde{\mathrm{P}} \ll \mathrm{P},\ \tilde{\mathrm{P}}_0 = \mathrm{P}_0,\ \tilde{T} \overset{\mathrm{P}}{=} T\}, \tag{8.2}$$

which presents the family of probability measures possessing the following properties: if $\tilde{\mathrm{P}} \in \mathcal{L}_{\mathrm{P}}$, then $\tilde{\mathrm{P}} \ll \mathrm{P}$, the restriction of the measure $\tilde{\mathrm{P}}$ to the σ-algebra $\mathcal{F}_0$ coincides with the restriction of P to $\mathcal{F}_0$ ($\tilde{\mathrm{P}}_0 = \mathrm{P}_0$), and the triplet $\tilde{T} = (\tilde{B}, \tilde{C}, \tilde{\nu})$ of predictable characteristics of a $(\mathbb{F}, \tilde{\mathrm{P}})$-semimartingale X ($X \in \mathbf{S}(\mathbb{F}, \tilde{\mathrm{P}})$ by Theorem 5.3) is P-indistinguishable from the triplet $T = (B, C, \nu)$.

Theorem 1. *The following conditions are equivalent:*

(α) *each local martingale* $M \in \mathcal{M}_{\mathrm{loc}}(\mathbb{F}, \mathrm{P})$ *admits the integral representation;*

(β) $\mathcal{L}_{\mathrm{P}} = \{\mathrm{P}\}$, *i.e. the set* $\mathcal{L}_{\mathrm{P}}$ *consists of one point* P *only.*

3. The prove of this theorem relies on a number of auxiliary results, formulated below as lemmas.

Lemma 1. *Let* $M \in \mathcal{M}_{\mathrm{loc}}(\mathbb{F}, \mathrm{P})$ *admit the integral representation* (8.1). *Then* *in this representation one can take*

$$h = \frac{d\langle M^c, X^c\rangle}{d\langle X^c\rangle}, \quad H = M^P_\mu(\Delta M \mid \tilde{\mathcal{P}}) - I(0 < a < 1)\, M^P_p(\Delta M \mid \mathcal{P}).$$

Proof. From (8.1) it follows that $M^c = h \cdot X^c$, and hence $\langle M^c, X^c\rangle = h \circ \langle X^c\rangle$ (Problem 2.2.6), i.e. for h the desired representation holds.

Since by Problem 3.3.1 we have

$$H * (\mu - \nu) = (H - \hat{H}\, I(a = 1)) * (\mu - \nu),$$

instead of H, involved in (8.1), we can take $h - \hat{H}\, I(a = 1)$. By Problem 3.5.7 we have $M^P_\mu(\Delta M \mid \tilde{\mathcal{P}}) = H - \hat{H}$, and hence

$$\widehat{M^P_\mu(\Delta M \mid \tilde{\mathcal{P}})} = \hat{H}(1 - a).$$

By taking into consideration the definition of $M^P_p(\Delta M \mid \mathcal{P})$ (cf. Theorem 3.3.1) this gives

$$\hat{H} = \frac{I(0 < a < 1)}{1 - a}\, \widehat{M^P_\mu(\Delta M \mid \tilde{\mathcal{P}})} + \hat{H}\, I(a = 1)$$

$$= -I(0 < a < 1)\, M^P_p(\Delta M \mid \mathcal{P}) + \hat{H}\, I(a = 1).$$

Consequently,

$$\hat{H}\, I(a < 1) = -I(0 < a < 1)\, M^P_p(\Delta M \mid \mathcal{P})$$

and

$$H - \hat{H}\, I(a = 1) = M^P_\mu(\Delta M \mid \tilde{\mathcal{P}}) + \hat{H}\, I(a < 1)$$

$$= M^P_\mu(\Delta M \mid \tilde{\mathcal{P}}) - I(0 < a < 1)\, M^P_p(\Delta M \mid \mathcal{P}).$$

Corollary. *If for* $M \in \mathcal{M}_{loc}(\mathbb{F}, P)$ *the integral representation takes place, then*

$$\langle M^c\rangle = \left(\frac{d\langle M^c, X^c\rangle}{d\langle X^c\rangle}\right)^2 \circ C, \tag{8.3}$$

$$I(\tau \le \infty)\, \Delta M_\tau = I(\tau < \infty)\,[M^P_\mu(\Delta M \mid \tilde{\mathcal{P}})(\tau, \Delta X_\tau)\, I(\Delta X_\tau \ne 0)$$
$$- I(0 < a < 1)\, M^P_p(\Delta M \mid \mathcal{P})(\tau)\, I(\Delta X_\tau = 0)] \quad \forall\, \tau \in \mathbf{T}. \tag{8.4}$$

Lemma 2. *If for each* $M \in \mathcal{M}_{loc}(\mathbb{F}, P)$ *such that*

$$\mathbf{E}\,(\langle M^c\rangle_\infty + \sum_{s\ge 0} f(\Delta M_s)) < \infty \tag{8.5}$$

with

$$f(x) = \frac{x^2}{1+|x|}$$

the integral representation takes place, then it takes place for each $M \in \mathcal{M}_{loc}(\mathbb{F}, P)$ *as well.*

Proof. According to Problem 2.2.8 for each $M \in \mathcal{M}_{loc}(\mathbb{F}, P)$ we have

$$\langle M^c\rangle + \sum_s f(\Delta M_s) \in \mathcal{A}^+_{loc}.$$

Therefore for each $M \in \mathcal{M}_{loc}(\mathbb{F}, P)$ there exists a localizing sequence $(\tau_n)_{n\ge 1}$ such that

$$\mathbf{E}\left(\langle M^c\rangle_{\tau_n} + \sum_{0<s\le\tau_n} f(\Delta M_s)\right) < \infty.$$

By assumption M^{τ_n} admits the integral representation. Besides,

$$\frac{d\langle (M^{\tau_n})^c, X^c\rangle}{d\langle X^c\rangle} = I_{[\![0,\tau_n]\!]}\frac{d\langle M^c, X^c\rangle}{d\langle X^c\rangle} = I_{[\![0,\tau_n]\!]}h,$$

$$M^P_\mu(\Delta M^{\tau_n} \mid \tilde{\mathcal{P}}) - I(0<a<1)\,M^P_p(\Delta M^{\tau_n} \mid \mathcal{P})$$

$$= I_{[\![0,\tau_n]\!]}(M^P_\mu(\Delta M \mid \tilde{\mathcal{P}}) - I(0<a<1)\,M^P_p(\Delta M \mid \mathcal{P}))$$

$$= I_{[\![0,\tau_n]\!]}H.$$

Since $h \in \mathbf{L}^2_{loc}(\mathbb{F}, \langle X^c\rangle)$ (Theorem 2.2.8) and $H \in \mathcal{G}_{loc}$ (Theorem 3.5.2), the process $M' \in \mathcal{M}_{loc}(\mathbb{F}, P)$ is defined with

$$M'_t = M_0 + h\cdot X^c_t + H*(\mu-\nu)_t.$$

Obviously, M^{τ_n} and $(M')^{\tau_n}$ are P-indistinguishable processes for every $n \ge 1$. Therefore M and M' are P-indistinguishable processes and hence M admits the integral representation.

Lemma 3. *If for any* $M \in \mathcal{M}_{loc}(\mathbb{F}, P)$, *satisfying Condition* (8.5) *and the inequality* $|\Delta M| \le \text{const}$, *the integral representation takes place, then it takes place for any* $M \in \mathcal{M}_{loc}(\mathbb{F}, P)$ *as well.*

Proof. By Lemma 2 it suffices to establish the existence of the integral representation for $M \in \mathcal{M}_{loc}(\mathbb{F}, P)$ with the property (8.5). Therefore we will assume henceforth that for M the property (8.5) holds.

For $k \geq 1$ set

$$A_t^k = \sum_{0 < s \leq t} \Delta M_s I(|\Delta M_s| > k).$$

Then

$$\mathbf{E}\,\mathrm{Var}(A^k)_\infty < \infty. \tag{8.6}$$

In fact

$$\mathrm{Var}(A^k)_\infty = \sum_{s > 0} |\Delta M_s| I(|\Delta M_s| > k) \leq \frac{k}{1+k} \sum_{s > 0} f(\Delta M_s), \tag{8.7}$$

and consequently (8.6) follows from (8.5).

Due to (8.6) there exists, by Theorem 1.6.3, the compensator $\tilde{A}_k$ of the process $A^k = (A_t^k)_{t \geq 0}$. Besides (cf. Problem 1.6.7)

$$\mathbf{E}\,\mathrm{Var}(\tilde{A}^k)_\infty \leq \mathbf{E}\,\widetilde{\mathrm{Var}(A^k)}_\infty = \mathbf{E}\,\mathrm{Var}(A^k)_\infty < \infty.$$

Consequently

$$M^k = A^k - \tilde{A}^k \in \mathcal{M}(\mathbb{F}, P) \cap \mathcal{A}(\mathbb{F}, P).$$

Consider the process $M - M^k$. Clearly $M - M^k \in \mathcal{M}_{loc}(\mathbb{F}, P)$. Besides $\Delta(M - M^k) = \Delta M\, I(|\Delta M| \leq k) + \Delta(\tilde{A}^k)$. By Corollary 2 to Theorem 1.6.3 we have $\Delta(\tilde{A}^k) = {}^p(\Delta A^k)$, while by Theorem 1.7.8 we have ${}^p(\Delta M) = 0$. Therefore

$$\Delta(\tilde{A}^k) = -{}^p(\Delta M\, I(|\Delta M| \leq k))$$

and consequently

$$\Delta(M - M^k) = \Delta M\, I(|\Delta M| \leq k) - {}^p(\Delta M\, I(|\Delta M| \leq k)). \tag{8.8}$$

By the representation (8.8) and the definition of the predictable projection (Theorem 1.3.13) we get

$$|\Delta(M - M^k)| \leq 2k. \tag{8.9}$$

Let us show that $M - M^k$ possesses the property (8.5). Since $(M - M^k)^c = M^c$, it suffices to verify that

$$\mathbf{E} \sum_{s > 0} f(\Delta(M - M^k)_s) < \infty.$$

It has been established in the course of proving Lemma 3.5.1 that

$$f(x + y) \leq 2[f(x) + f(y)].$$

Therefore by (8.8) we get

$$f(\Delta(M - M^k)) \le 2\,[f(\Delta M\, I(|\Delta M| \le k)) + f({}^P(\Delta M\, I(|\Delta M| \le k)))].$$

Further, by Problem 1.3.9

$$f({}^P(\Delta M\, I(|\Delta M| \le k))) \le {}^P f(\Delta M\, I(|\Delta M| \le k)).$$

Denote

$$B^k = \sum_s f(\Delta M_s I(|\Delta M_s| \le k)).$$

Since $f(\Delta M\, I(|\Delta M| \le k)) \le f(\Delta M)$, by (8.5) we have $\mathbf{E}B^k_\infty < \infty$, and the process B^k has the compensator $\Delta\tilde{B}^k$ (Theorem 1.6.3). Besides, by Corollary 2 to Theorem 1.6.3 we have $\Delta\tilde{B}^k = {}^P(\Delta B^k) = {}^P f(\Delta M\, I(|\Delta M| \le k))$. In view of these properties

$$\mathbf{E}\sum_{s>0} f(\Delta(M - M^k)_s \le 2\mathbf{E}\left[\sum_{s>0} f(\Delta M_s I(|\Delta M_s| \le k)) + \sum_{s>0} \Delta(\tilde{B}^k)_s\right]$$

$$\le 2\mathbf{E}\,(B^k_\infty + (\tilde{B}^k)_\infty) = 4\mathbf{E}B^k_\infty = 4\mathbf{E}\sum_{s>0} f(\Delta M_s I(|\Delta M_s| \le k))$$

$$\le 4\mathbf{E}\sum_{s>0} f(\Delta M_s) < \infty.$$

Thus $M - M^k \in \mathcal{M}_{loc}(\mathbb{F}, \mathsf{P})$ possesses the property (8.5) and satisfies the inequality $|\Delta(M - M^k)| \le$ const. By assumption for this local martingale the integral representation

$$(M - M^k)_t = M_0 + h \cdot X^c_t + H^k * (\mu - \nu)_t$$

takes place with

$$h = \frac{d\langle M^c, X^c\rangle}{d\langle X^c\rangle}$$

(since $M - M^k)^c = M^c$) and

$$H^k = M^{\mathsf{P}}_\mu(\Delta(M - M^k) \mid \tilde{\mathcal{P}}) - I(0 < a < 1)\, M^{\mathsf{P}}_p(\Delta(M - M^k) \mid \mathcal{P}),$$

by Lemma 1.

Let $M' \in \mathcal{M}_{loc}(\mathbb{F}, \mathsf{P})$ be defined by the integral representation

$$M'_t = M_0 + h \cdot X^c_t + H * (\mu - \nu)_t \tag{8.10}$$

with

$$h = \frac{d\langle M^c, X^c\rangle}{d\langle X^c\rangle}$$

and

$$H = M^P_\mu(\Delta M \mid \tilde{\mathcal{P}}) - I(0 < a < 1)\, M^P_p(\Delta M \mid \mathcal{P}).$$

Let us show that the processes M and M' are P-indistinguishable.

We have

$$M - M' = M^k + (M - M^k) - M' = M^k + H^k * (\mu - \nu) - H * (\mu - \nu)$$
$$= A^k - \tilde{A}^k - U^k * (\mu - \nu)$$

with

$$U^k = H - H^k = M^P_\mu(\Delta M^k \mid \tilde{\mathcal{P}}) - I(0 < a < 1)\, M^P_p(\Delta M^k \mid \mathcal{P}).$$

From this it follows that (cf. Theorem 3.3.1)

$$\mathrm{Var}(M - M')_\infty \le \mathrm{Var}(A^k)_\infty + \mathrm{Var}(\tilde{A}^k)_\infty + |M^P_\mu(\Delta M^k \mid \tilde{\mathcal{P}})| * (\mu + \nu)_\infty$$

$$+ \sum_{s>0} I(a_s > 0)\, |M^P_p(\Delta M^k \mid \mathcal{P})|\, [(1 - \mu(\{s\} \times R_0)) + (1 - a_s)]. \qquad (8.11)$$

Let us establish now the inequality

$$\mathbf{E}\,\mathrm{Var}(M - M')_\infty \le 10\mathbf{E}\,\mathrm{Var}(A^k)_\infty. \qquad (8.12)$$

By Problem 1.6.7

$$\mathbf{E}\,\mathrm{Var}(\tilde{A}^k)_\infty \le \mathbf{E}\,\widetilde{\mathrm{Var}(A^k)_\infty} = \mathbf{E}\,\mathrm{Var}(A^k)_\infty. \qquad (8.13)$$

This implies, in particular,

$$\mathbf{E} \sum_{s>0} |\Delta M^k_s| \le \mathbf{E}\,(\mathrm{var}(A^k)_\infty + \mathrm{Var}(\tilde{A}^k)_\infty) \le 2\mathbf{E}\,\mathrm{Var}(A^k)_\infty. \qquad (8.14)$$

Therefore, due to Jesen's inequality for $M^P_\mu(\,\cdot \mid \tilde{\mathcal{P}})$ (Problem 3.2.11), we get

$$\mathbf{E}\,|M^P_\mu(\Delta M^k \mid \tilde{\mathcal{P}})| * (\mu + \nu)_\infty = 2\mathbf{E}\; M^P_\mu(\Delta M^k \mid \tilde{\mathcal{P}})| * \mu_\infty \le 2\mathbf{E}M^P_\mu(|\Delta M^k| \mid \tilde{\mathcal{P}}) * \mu_\infty$$

$$= 2\mathbf{E}\; |\Delta M^k| * \mu_\infty \le 2\mathbf{E} \sum_{s>0} |\Delta M^k_s|$$
$$\le 4\mathbf{E}\;\mathrm{Var}(A^k)_\infty \qquad (8.15)$$

and analogously

$$\mathbf{E} \sum_{s>0} I(a_s > 0)\, |M^P_p(\Delta M^k \mid \mathcal{P})|\, [(1 - \mu(\{s\} \times R_0)) + (1 - a_s)]$$

$$= 2\mathbf{E}\, |M^P_p(\Delta M^k \mid \mathcal{P})| * p_\infty \le 2\mathbf{E}M^P_p(|\Delta M^k| \mid \mathcal{P}) * p_\infty$$

$$= 2\mathbf{E}\, |\Delta M^k| * p_\infty \le 2\mathbf{E} \sum_{s>0} |\Delta M^k_s| \le 4\mathbf{E}\,\mathrm{Var}(A^k)_\infty. \qquad (8.16)$$

The desired inequality (8.12) follows from (8.11), (8.13), (8.15) and (8.16).

To prove P-indistinguishability of the processes M and M', it suffices, by the obtained inequality (8.12), to show that

$$\lim_{k} \mathbf{E}\,\mathrm{Var}\,(A^k)_\infty = 0. \tag{8.17}$$

In virtue of (8.7) as $k \geq 1$ we have

$$\mathrm{Var}\,(A^k)_\infty \leq I\left(M^*_\infty > \frac{k}{2}\right)\sum_{s>0} |\Delta M_s|\, I\,(|\Delta M_s| > 1)$$

$$\leq \frac{1}{2} I\left(M^*_\infty > \frac{k}{2}\right)\sum_{s>0} f\,(\Delta M_s).$$

This shows that (8.17) will take place if $M \in \mathcal{H}\,(\mathbb{F}, P)$. For this in turn it suffices to show that $M^c \in \mathcal{H}\,(\mathbb{F}, P)$ and $M^d \in \mathcal{H}\,(\mathbb{F}, P)$.

Let us show that the following implication takes place:

$$\mathbf{E}\left(\langle M^c\rangle_\infty + \sum_{s>0} f\,(\Delta M_s)\right) < \infty \Rightarrow \begin{cases} M^c \in \mathcal{H}\,(\mathbb{F}, P), \\ M^d \in \mathcal{H}\,(\mathbb{F}, P). \end{cases} \tag{8.18}$$

In fact, by Doob's inequality (Theorem 1.9.2) and (8.5) we have

$$\mathbf{E}\,((M^c)^*_\infty)^2 \leq 4\mathbf{E}\,\langle M^c\rangle_\infty < \infty,$$

i.e.

$$M^c \in \mathcal{H}\,(\mathbb{F}, P).$$

Next, by the inequalities (8.5) and (2.30) in Ch. 2 we have

$$\mathbf{E}\,[M^d, M^d]^{1/2}_\infty \leq \frac{1}{4} + 2\mathbf{E} \sum_{s>0} f\,(\Delta M_s) < \infty.$$

By Davis' inequality (Theorem 1.9.6) this gives

$$\mathbf{E}\,(M^d)^*_\infty \leq C\mathbf{E}\,[M^d, M^d]^{1/2}_\infty < \infty$$

i.e.

$$M^d \in \mathcal{H}\,(\mathbb{F}, P).$$

Thus Lemma 3 is proved.

4. Proof of Theorem 1.

(α) $\Rightarrow$ (β). Let $\tilde{P} \in \mathcal{L}_P$ and $\tilde{P} \neq P$. By the definition of $\mathcal{L}_P$ we have $\tilde{P} \ll P$. Let Z be the density process of the measure $\tilde{P}$ with respect to P. Since $Z \in \mathcal{M}\,(\mathbb{F}, P)$, then Z admits the integral representation

$$Z_t = Z_0 + h \cdot X^c_t + H * (\mu - \nu)_t.$$

By the equality $\tilde{P}_0 = P_0$ the random variable $Z_0 = 1$. By Theorem 5.3 the process X belongs to $S\,(\mathbb{F}, \tilde{P})$ and possesses the triplet $\tilde{T} = (\tilde{B}, \tilde{C}, \tilde{\nu})$ with

$$\tilde{B} = B + \beta \circ C + I(|x| \le 1)\, x\,(Y - 1) * \nu,$$

$$\tilde{C} = C,$$

$$\tilde{\nu} = Y\nu, \tag{8.19}$$

where

$$Y = I(Z_- > 0)\, Z_-^{-1}\, M_\mu^{P}(Z \mid \tilde{\mathcal{P}}), \quad \beta = I(Z_- > 0)\, Z_-^{-1} \frac{d\langle X^c, Z^c\rangle}{d\langle X^c\rangle}.$$

Denote by M_C^P the measure with $M_C^P(d\omega, dt) = P(d\omega)\, dC_t(\omega)$. Since by the definition of the class $\mathfrak{T}_P$ we have $\tilde{T} \overset{P}{=} T$, then $Y = 1$ (M_ν^P-a.s.) and $\beta = 0$ (M_C^P-a.s.). Set $\tau = \inf(t\colon Z_t < 1/2)$. From the definition of Y and β it follows that on the set $[\![0, \tau]\!]$ we have

$$M_\mu^P(\Delta Z \mid \tilde{\mathcal{P}}) = 0 \quad (M_\nu^P\text{-a.s.})$$

and

$$\frac{d\langle X^c, Z^c\rangle}{d\langle X^c\rangle} = 0 \quad (M_C^P\text{-a.s.}).$$

Next (cf. Theorem 3.3.1),

$$(1 - a)\, I_{[\![0, \tau]\!]} M_p^P(\Delta Z \mid \mathcal{P}) = -\, I_{[\![0, \tau]\!]} \widehat{I(0 < a < 1)\, M_\mu^P(\Delta Z \mid \tilde{\mathcal{P}})} = 0.$$

Since by Lemma 1 in the integral representation for Z we have

$$h = \frac{d\langle X^c, Z^c\rangle}{d\langle X^c\rangle}, \quad H = M_\mu^P(\Delta Z \mid \tilde{\mathcal{P}}) - I(0 < a < 1)\, M_p^P(\Delta Z \mid \mathcal{P}),$$

then $Z^\tau = Z_0 = 1$. This gives, in particular, $P(\tau = \infty) = 1$, i.e. $Z_\infty = 1$ (P-a.s.).

Therefore $\tilde{P} = P$.

$(\beta) \Rightarrow (\alpha)$. By Lemma 3 it suffices to consider local martingales M possessing Conditions (8.5) and $|\Delta M| \le c$. Observe that in accordance with (8.18) we have $M \in \mathcal{H}(\mathbb{F}, P)$.

Set

$$h = \frac{d\langle M^c, X^c\rangle}{d\langle X^c\rangle}, \quad H = M_\mu^P(\Delta M \mid \tilde{\mathcal{P}}) - I(0 < a < 1)\, M_p^P(\Delta M \mid \mathcal{P}).$$

By Theorem 2.2.8 we have $h^2 \circ C \in \mathcal{A}_{loc}^+$ and by Theorem 3.5.2 we have $H \in \mathcal{G}_{loc}$, consequently $M' \in \mathcal{M}_{loc}(\mathbb{F}, P)$ is defined with

$$M'_t = M_0 + h \cdot X^c_t + H * (\mu - \nu)_t.$$

Let us show that by Condition (8.5) for M the local martingale $M' \in \mathcal{H}$. It suffices to show, by (8.18), that

$$E\left(\langle M' \rangle_\infty + \sum_{s>0} f(\Delta M'_s) \right) < \infty. \tag{8.20}$$

By Theorem 2.2.8 we have $\langle M'^c \rangle = h^2 \circ C_\infty \le \langle M^c \rangle_\infty$, and hence

$$E \langle M'^c \rangle_\infty < \infty. \tag{8.21}$$

Since by Problem 3.3.1 we have $H * (\mu - \nu) = (H - \hat{H} I(a = 1)) * (\mu - \nu)$, then from the definition of M' it follows that

$$\Delta M'_s = (H(s, \Delta X_s) - \hat{H}(s)) I(\Delta X_s \neq 0) - \hat{H}(s) I(a_s < 1) I(\Delta X_s = 0). \tag{8.22}$$

Therefore, by the inequality $f(x + y) \le 2[f(x) + f(y)]$ we get

$$\sum_{s>0} f(\Delta M'_s) \le 2f(H - \hat{H}) * \mu_\infty + 2f(\hat{H} I(a < 1)) * p_\infty. \tag{8.23}$$

But by the definition of H and by Problem 3.3.2 we have

$$H - \hat{H} = M^P_\mu(\Delta M \mid \tilde{\mathcal{P}}), \quad \hat{H} I(a < 1) + - I(0 < a < 1) M^P_p(\Delta M \mid \mathcal{P}). \tag{8.24}$$

In view of Jensen's inequality by this we get

$$\begin{aligned} E \sum_{s>0} f(\Delta M'_s) &\le 2Ef(M^P_\mu(\Delta M \mid \tilde{\mathcal{P}})) * \mu_\infty + 2Ef(M^P_p(\Delta M \mid \mathcal{P})) * p_\infty \\ &\le 2EM^P_\mu(f(\Delta M) \mid \tilde{\mathcal{P}}) * \mu_\infty + 2EM^P_p(f(\Delta M) \mid \mathcal{P}) * p_\infty \\ &= 2Ef(\Delta M) * \mu_\infty + 2Ef(\Delta M) * p_\infty \le 4E \sum_{s>0} f(\Delta M_s) < \infty. \end{aligned} \tag{8.25}$$

Thus, (8.20) follows from (8.21) and (8.25), and hence $M' \in \mathcal{H}(\mathbb{F}, P)$. Then $N = M - M' \in \mathcal{H}(\mathbb{F}, P)$ and the process N possesses, as it will be shown below, the following properties:

$$|\Delta N| \le |\Delta M| + |\Delta M'| \le 5c, \tag{8.26}$$

$$\langle N^c, X^c \rangle = 0, \tag{8.27}$$

$$M^P_\mu(\Delta N \mid \tilde{\mathcal{P}}) = 0 \quad (M^P_\mu\text{-a.s.}). \tag{8.28}$$

In fact, the property (8.26) takes place, since $|\Delta M| \le c$ and by Theorem 3.5.2 we have $|\Delta M'| \le 4c$. The property (8.27) holds, since $\langle M^c, X^c \rangle = \langle M'^c, X^c \rangle$. Finally, the property (8.28) follows from (8.22) and (8.24).

We assume $N^*_\infty \le a$, making use of a localizing sequence $\tau_n = \inf(t: |N_t| \ge n)$, $n \ge$

1 if necessary.

Denote

$$Z = 1 + \frac{N}{2a}.$$

Then $Z \in \mathcal{H}(\mathbb{F}, P)$ and $Z \geq \frac{1}{2}$. Define the probability measure $\tilde{P}$ with $d\tilde{P} = Z_\infty dP$. We will show that $\tilde{P} \in \mathcal{L}_P$.

By construction $\tilde{P} \ll P$. By the property $Z_0 = 1$ we have $\tilde{P}_0 = P_0$. Next, by Theorem 5.3 we have $X \in S(\mathbb{F}, \tilde{P})$. From (8.27) and (8.28) it follows that $\langle Z^c, X^c \rangle = 0$ and $M^P_\mu(\Delta Z \mid \tilde{\mathcal{P}}) = 0$.

Therefore, the local absolutely continuous change of measures (see. (8.19)), leads here to the triplet transformation with $\beta = 0$ and $Y = 1$. Consequently, by Theorem 5.3 and the remark to it the triplet $\tilde{T}$ of the semimartingale $X \in S(\mathbb{F}, P)$ is P-indistinguishable from the triplet T.

Thus $\tilde{P} \in \mathcal{L}_P$. But, by definition $\mathcal{L}_P = \{P\}$. Hence $\tilde{P} = P$ and the local density process Z is P-indistinguishable from the process identically equal to one, i.e the process N is P-negligible and the processes M and M' are P-indistinguishable.

5. Let a process $X \in S(\mathbb{F}, P)$. Define the σ-algebras

$$\mathcal{F}^X_t = \sigma(\{X_s, 0 \leq s \leq t\} \vee \mathcal{N}), \quad \mathcal{F}^X_{t+} = \bigcap_{\varepsilon > 0} \mathcal{F}^X_{t+\varepsilon},$$

where $\mathcal{N}$ is a system of sets in $\mathcal{F}$ of P-measure zero.

Set

$$\mathcal{F}^X_\infty = \sigma\left(\bigcup_{t \geq 0} \mathcal{F}^X_t\right).$$

The restriction of a measure P on the σ-algebras $\mathcal{F}^X_\infty$, $\mathcal{F}^X_{0+}$ and $\mathcal{F}^X_0$ is denoted by P^X, P^X_{0+} and P^X_0 respectively.

By Theorem 6.1 we have

$$X \in S(\mathbb{F}^X_+, P^X).$$

Therefore, there arises a natural question about the integral representation for each $M \in \mathcal{M}_{loc}(\mathbb{F}^X_+, P^X)$. In the present special case one can strengthen the result of Theorem 1 to some extent.

Denote by T^X the triplet of predictable characteristics of a $(\mathbb{F}^X_+, P^X)$-semimartingale X. If $\tilde{P}^X$ is a probability measure on $(\Omega, \mathcal{F}^X_\infty)$, absolutely continuous with respect to a measure P^X ($\tilde{P}^X \ll P^X$), then by Theorem 5.3 we have

$$X \in S(\mathbb{F}^X_+, \tilde{P}^X).$$

The triplet of predictable characteristics of this semimartingale is denoted by $\tilde{T}^X$.

Define two families of probability measures on $(\Omega, \mathcal{F}^X_\infty)$ (cf. (8.2)):

$$\mathcal{L}_{P^X_+} = \{\tilde{P}^X : \tilde{P}^X \ll P^X, \ \tilde{P}^X_{0+} = P^X_{0+}, \ \tilde{T}^X \overset{P^X}{=} T^X\},$$

$$\mathcal{L}_{P^X} = \{\tilde{P}^X : \tilde{P}^X \ll P^X, \ \tilde{P}^X_0 = P^X_0, \ \tilde{T}^X \overset{P^X}{=} T^X\},$$

where $\tilde{P}^X_{0+}$ and $\tilde{P}^X_0$ are the restrictions of the measures $\tilde{P}^X$ to $\mathcal{F}^X_{0+}$ and $\mathcal{F}^X_0$. Clearly,

$$\mathcal{L}_{P^X} \supseteq \mathcal{L}_{P^X_+}.$$

Theorem 2. *The following conditions are equivalent:*

(α) *each local martingale* $M \in \mathcal{M}_{loc}(\mathbb{F}^X_+, P^X)$ *admits the integral representation and* $\mathbb{F}^X_+ = \mathbb{F}^X$;

(β) $\mathcal{L}_{P^X} = \mathcal{L}_{P^X_+} = \{P^X\}$.

Proof. (α) $\Rightarrow$ (β). By Theorem 1 and the equality $\mathbb{F}^X_+ = \mathbb{F}^X$ we have

$$\mathcal{L}_{P^X_+} = \mathcal{L}_{P^X} = \{P\}.$$

(β) $\Rightarrow$ (α). By Theorem 1 and the condition $\mathcal{L}_{P^X_+} = \{P\}$ for each $M \in \mathcal{M}_{loc}(\mathbb{F}^X_+, P^X)$ the integral representation takes place.

Let us show now that $\mathbb{F}^X_+ = \mathbb{F}^X$. To this end we establish first that $\mathcal{F}^X_{0+} = \mathcal{F}^X_0$, and then we show that $\mathcal{F}^X_{t+} = \mathcal{F}^X_t$ for each $t > 0$.

Let $\Gamma \in \mathcal{F}^X_{0+}$. Define the random variable

$$Z_\infty = 1 + \frac{1}{2}(I_\Gamma - P^X(\Gamma \mid \mathcal{F}^X_0)).$$

Clearly,

$$\frac{1}{2} \le Z_\infty \le \frac{3}{2}, \quad \int_\Omega Z_\infty dP^X = 1$$

and for each set $A \in \mathcal{F}^X_0$ we have

$$\int_A Z_\infty dP^X = P^X(A).$$

Therefore, the probability measure $\tilde{P}^X$ on $(\Omega, \mathcal{F}^X_\infty)$ with $d\tilde{P}^X = Z_\infty dP^X$ is equivalent to the measure P^X ($\tilde{P}^X \sim P^X$), and $\tilde{P}^X_0 = P^X_0$. Let $Z = (Z_t)_{t \ge 0}$ be the density process of the measure $\tilde{P}^X$ with respect to P^X (and the family $\mathbb{F}^X_+$). Clearly, $Z_t \equiv Z_\infty$ (P^X-a.s.). This means that for the martingale Z the continuous martingale component Z^c and the jump process ΔZ are negligible. By Theorem 5.3 this gives

$$\tilde{T}^X \overset{P^X}{=} T^X.$$

Hence

$$\tilde{P}^X \in \mathcal{L}_{P^X_+}.$$

By definition

$$\mathcal{L}_{P^X} = \mathcal{L}_{P^X_+} = \{P^X\},$$

i.e. $\tilde{P}^X = P^X$, and consequently $Z_\infty = 1$. Therefore $I_\Gamma = P^X(\Gamma \mid \mathcal{F}^X_0)$, i.e. $\Gamma \in \mathcal{F}^X_0$ (recall that the σ-algebra $\mathcal{F}^X_0$ is completed by sets from $\mathcal{F}$ of zero probability P). As the set Γ is arbitrary, we have $\mathcal{F}^X_{0+} = \mathcal{F}^X_0$.

Let $t > 0$ and $\Gamma \in \mathcal{F}^X_{t+}$. Consider a process $M \in \mathcal{M}_{loc}(\mathbb{F}^X_+, P^X)$ that is the optional projection of I_Γ relative to $(\mathbb{F}^X_+, P^X)$. Clearly, $M_t = I_\Gamma$. On the other hand M admits the integral representation, in particular

$$I_\Gamma = \mathsf{P}^X(\Gamma \mid \mathcal{F}^X_0) + h \cdot X^c_t + H * (\mu - \nu)_t,$$

where X^c is the continuous $(\mathbb{F}^X_+, \mathsf{P}^X)$-martingale component of $X \in \mathcal{S}(\mathbb{F}^X_+, \mathsf{P}^X)$ and ν the compensator of the jump measure of the process X relative to $(\mathbb{F}^X_+, \mathsf{P}^X)$.

Let $s > t$. The random variables M_s are $\mathcal{F}^X_t$-measurable. Hence M_{t-} is a $\mathcal{F}^X_{t-}$-measurable random variable. Therefore, it suffices to show that ΔM_t is a $\mathcal{F}^X_t$-measurable random variable. From the integral representation it follows that

$$\Delta M_t = H(t, \Delta X_t)\, I(\Delta X_t \neq 0) - \hat{H}(t).$$

The functions $H(t, x)$ and $\hat{H}(t)$ are measurable with respect to the σ-algebras $\mathcal{P}(\mathbb{F}^X_+) \otimes \mathcal{B}(R_0)$ and $\mathcal{P}(\mathbb{F}^X_+)$. Since $\mathcal{P}(\mathbb{F}^X_+) = \mathcal{P}(\mathbb{F}^X)$ (Problem 1.2.5), for each fixed t the variables $H(t, x)$ and $\hat{H}(t)$ are $\mathcal{F}^X_{t-} \otimes \mathcal{B}(R_0)$- and $\mathcal{F}^X_{t-}$-measurable functions. Since ΔX_t is $\mathcal{F}^X_t$-measurable, this shows that ΔM_t is a $\mathcal{F}^X_t$-measurable random variable.

6. Let $X \in \mathcal{S}(\mathbb{F}, \mathsf{P})$ and Q be a probability measure on $(\Omega, \mathbb{F})$, locally absolutely continuous with respect to a measure P ($\mathsf{Q} \overset{loc}{\ll} \mathsf{P}$). By Theorem 5.3 we have $X \in \mathcal{S}(\mathbb{F}, \mathsf{Q})$. Besides, the triplets $T = (B, C, \nu)$ and $T^{\mathsf{Q}} = (B^{\mathsf{Q}}, C^{\mathsf{Q}}, \nu^{\mathsf{Q}})$ of the corresponding semimartingales are related by the following relations

$$\begin{aligned} B^{\mathsf{Q}} &= B + \beta \circ C + I(|x| \le 1)\, x\, (Y - 1) * \nu, \\ C^{\mathsf{Q}} &= C, \\ \nu^{\mathsf{Q}} &= Y\nu, \end{aligned} \tag{8.29}$$

that are understood in the sense of indistinguishability of the corresponding left-hand and right-hand parts relative to the measures Q and P on the set $[\![0, \tau[\![$, where $\tau = \inf(t\colon Z_t = 0)$ and Z is the local density process of the measure Q with respect to P. Moreover, on the set $[\![0, \tau[\![$

$$\beta = Z_-^{-1} \frac{d\langle Z^c, X^c\rangle}{d\langle X^c\rangle}, \qquad Y = Z_-^{-1} M^{\mathsf{P}}_\mu(\Delta Z \mid \tilde{\mathcal{P}}) + 1.$$

We will consider the relations (8.29) from the point of view of defining the functions β and Y by the triplets T and T^{Q}.

Suppose $\mathcal{Z}_{\mathsf{P}} = \{\mathsf{P}\}$. Then by Theorem 1 any local martingale $M \in \mathcal{M}_{loc}(\mathbb{F}, \mathsf{P})$ admits the integral representation. Since $Z \in \mathcal{M}_{loc}(\mathbb{F}, \mathsf{P})$, for Z the integral representation

$$Z_t = Z_0 + h \cdot X_t^c + H * (\mu - \nu)_t$$

takes place where

$$Z_0 = \frac{dQ_0}{dP_0}$$

(Q_0, P_0 are the restrictions of the measures Q and P to the σ-algebra $\mathcal{F}_0$),

$$h = \frac{d\langle Z^c, X^c\rangle}{d\langle X^c\rangle}, \quad H = M_\mu^P(\Delta Z \mid \tilde{\mathcal{P}}) - I(0 < a < 1)\, M_p^P(\Delta Z \mid \mathcal{P})$$

(cf. Lemma 1) with ($\{a < 1\}$; M_μ^P-a.s.)

$$M_p^P(\Delta Z \mid \mathcal{P}) = -\frac{I(0 < a < 1)}{1 - a} \widehat{M_\mu^P(\Delta Z \mid \tilde{\mathcal{P}})}.$$

Therefore, on the set $[\![0, \tau[\![$

$$h = Z_- \beta, \quad H = Z_-\left(Y - 1 + \frac{I(0 < a < 1)}{1 - a}(\hat{Y} - a)\right)$$

(cf. Problem 3.5.8). Let

$$\tau_n = \inf\left(t: Z_t < \frac{1}{n}\right), \quad n \geq 1.$$

On the set $[\![0, \tau_n[\![$ we have $Z_- \geq \frac{1}{n}$. Therefore, the process $M^n \in \mathcal{M}_{loc}(\mathbb{F}, P)$ is defined with

$$M^n = I_{[\![0, \tau_n]\!]}\beta \cdot X^c + I_{[\![0, \tau_n]\!]}\left(Y - 1 + \frac{I(0 < a < 1)}{1 - a}(\hat{Y} - a)\right) * (\mu - \nu) \tag{8.30}$$

The following theorem, which is frequently utilized in various applications, in particular in statistics of stochastic processes, describes the structure of the densities of probability measures related to two semimartingales.

Theorem 3. *Let* $X \in S(\mathbb{F}, P)$ *and* $\mathcal{L}_P = \{P\}$. *If* $Q \overset{loc}{\ll} P$ *and* Z *is the local density process of a measure* Q *with respect to* P, *then (in the sense of* P-*indistinguishability)*

$$Z_t^{\tau_n} = Z_0 \mathcal{E}(M^n)_t, \quad t \geq 0, \; n \geq 1,$$

where $\mathcal{E}(M^n)$ *is the stochastic exponential (Ch. 2, § 4),*

$$\mathcal{E}_t(M^n) = e^{M_t^n - \frac{1}{2}\langle M^{nc}\rangle_t} \prod_{0<s\le t} (1+\Delta M_s^n)\, e^{-\Delta M_s^n}. \tag{8.31}$$

Proof. From the definition of h, H and the process M^n it follows that

$$Z_t^{\tau_n} = 1 + Z_- \beta \cdot X^c_{t\wedge\tau_n} + Z_-\left(Y - 1 + \frac{I(0<a<1)}{1-a}(\hat{Y} - a)\right) * (\mu - \nu)_{t\wedge\tau_n}$$

$$= 1 + Z_-^{\tau_n} \cdot M_t^n, \tag{8.32}$$

i.e. Z^{τ_n} is defined by Doléans equation (Ch. 2, § 4) which has the unique solution (Theorem 2.4.1), and this gives the desired representation.

7. The representation for Z^{τ_n}, obtained in Theorem 3, can be used for checking the property $\mathfrak{Z}_P = \{P\}$.

Theorem 4. *Let* $X \in \mathbf{S}(\mathbb{F}, P)$ *and* $Q \overset{loc}{\ll} P$. *Then*

$$\mathfrak{Z}_P = \{P\} \Rightarrow \mathfrak{Z}_Q = \{Q\}. \tag{8.33}$$

Proof. Let $\tilde{Q} \in \mathfrak{Z}_Q$ and $\tilde{Q} \neq Q$. Since $\tilde{Q} \ll Q$ and $Q \overset{loc}{\ll} P$, then $\tilde{Q} \overset{loc}{\ll} P$. Let Z and $\tilde{Z}$ be the density processes of the measures Q and $\tilde{Q}$ with respect to P on τ_n and $\tilde{\tau}_n$ which are defined by the relations:

$$\tau_n = \inf\left(t: Z_t < \frac{1}{n}\right), \quad \tilde{\tau}_n = \inf\left(t: \tilde{Z}_t < \frac{1}{n}\right).$$

Set

$$\sigma_n = \tau_n \wedge \tilde{\tau}_n.$$

Since $\tilde{Q} \in \mathfrak{Z}_Q$, the triplets T^Q and $T^{\tilde{Q}}$, corresponding to $X \in \mathbf{S}(\mathbb{F}, Q)$ and $X \in \mathbf{S}$ (F, $\tilde{Q}$), are Q-indistinguishable. From this it is not hard to deduce, by using formulas of type (8.29), that the functions β, $\tilde{\beta}$ and Y, $\tilde{Y}$ can be chosen in such a way that

$$\beta = \tilde{\beta}\ ([\![0, \sigma_n[\![;\ M_C^P\text{-a.s.}),$$

with

$$M_C^P(d\omega, dt) = P(d\omega)\, dC_t(\omega),$$

and

$$Y = \tilde{Y} \; ([\![0, \sigma_n[\![; \; M_\mu^P\text{-a.s.}).$$

This implies the coincidence on the set $[\![0, \sigma_n]\!]$ of the local martingales M^n and $\tilde{M}^n$, involved in the representation of type (8.32). Since $Q_0 = \tilde{Q}_0$, the equality $Z_0 = \tilde{Z}_0$ holds and from Theorem 3 it follows that the processes Z^{σ_n} and $\tilde{Z}^{\sigma_n}$ are P-indistinguishable and $\tau_n = \tilde{\tau}_n$, $n \geq 1$, i.e. the processes Z and $\tilde{Z}$ are P-indistinguishable on $\bigcup_{n \geq 1} [\![0, \tau_n]\!]$.

Let $\tau = \inf(t: Z_t \wedge \tilde{Z}_t = 0)$. Since

$$[\![0, \tau[\![\; \subseteq \bigcup_{n \geq 1} [\![0, \tau_n]\!],$$

the processes Z and $\tilde{Z}$ are P-indistinguishable on $[\![0, \tau[\![$.

Let us show that $ZI_{[\![\tau, \infty[\![}$ and $\tilde{Z}I_{[\![\tau, \infty[\![}$ are P-indistinguishable processes. The set $\{t < \infty\}$ is representable P-a.s. as the sum of sets $\{Z_{\tau-} = \tilde{Z}_{\tau-} = 0\}$ and $\{Z_{\tau-} = \tilde{Z}_{\tau-} > 0\}$. By Problem 2.2.7 we have

$$ZI_{[\![\tau, \infty[\![} = \tilde{Z}I_{[\![\tau, \infty[\![} \;\; (\{Z_{\tau-} = \tilde{Z}_{\tau-} = 0\}; \; \text{P-a.s.}).$$

If

$$\omega \in \{Z_{\tau(\omega)-}(\omega) = \tilde{Z}_{\tau(\omega)-}(\omega) > 0\},$$

then for a given ω we have

$$\bigcup_{n \geq 1} [\![0, \tau_n(\omega)]\!] = [\![0, \tau(\omega)]\!], \quad Z_{\tau(\omega)}(\omega) = \tilde{Z}_{\tau(\omega)}(\omega) = 0,$$

and this gives

$$ZI_{[\![\tau, \infty[\![} = \tilde{Z}I_{[\![\tau, \infty[\![} \qquad (\{Z_{\tau-} = \tilde{Z}_{\tau-} > 0\}; \; \text{P-a.s.})$$

by Problem 1.4.11.

Thus Z and $\tilde{Z}$ are P-indistinguishable processes. This means $Q_t = \tilde{Q}_t$, $t \geq 0$, where Q_t, $\tilde{Q}_t$ are the restrictions of the measures Q, $\tilde{Q}$ to the σ-algebra $\mathcal{F}_t$. This implies $Q = \tilde{Q}$.

8. By utilizing the corollary to Lemma 1 one can construct local martingales for which which the integral representation does not exist.

Example 1. Let W^1 and W^2 be independent Wiener processes relative to $(\mathbb{F}, P)$, $X_t = W^1 \cdot W_t^2$ and $M_t = (W_t^1)^2 - t$. Obviously,

$$X = (X_t)_{t \geq 0} \in \mathcal{M}_{loc}(\mathbb{F}, P)$$

and

$$M = (M_t)_{t \geq 0} \in \mathcal{M}_{loc}(\mathbb{F}, P).$$

Besides

$$\langle X \rangle_t = \int_0^t (W_s^1)^2 ds.$$

In view of the remark to Lemma 6.2 we have

$$X \in \mathcal{M}_{loc}(\mathbb{F}_+^X, P)$$

and its quadratic characteristic coincides with $\langle X \rangle$ (Theorem 6.1). It is not hard to deduce from this that M is a $\mathbb{F}_+^X$-adapted process and consequently, by the remark to Lemma 6.2 we have

$$M \in \mathcal{M}_{loc}(\mathbb{F}_+^X, P).$$

Since $\langle M, X \rangle = 0$, if the integral representation were valid, then by (8.3) we should have $\langle M \rangle = 0$. This contradiction shows that the local martingale $M \in \mathcal{M}_{loc}(\mathbb{F}_+^X, P)$ does not admit the integral representation.

Let us consider the more sophisticated

Example 2. Let $a = (a_t)_{t \geq 0}$ be a nonnegative $B(R_+)$-measurable function such that

$$\int_0^t a_s ds < \infty, \ t > 0,$$

and for each open set Δ of a positive Lebesgue measure

$$\int_\Delta a_s^2 ds = \infty. \tag{8.34}$$

Let $\Theta = (\Theta_t)_{t \geq 0}$ and $W = (W_t)_{t \geq 0}$ be mutually independent Wiener processes, relative to $(\mathbb{F}, P)$ and $X \in \mathbf{S}(\mathbb{F}, P)$ with

$$X_t = \int_0^t a_s \Theta_s ds + W_t. \tag{8.35}$$

It will be shown that under the assumption that the σ-algebra $\mathcal{F}^X_{0+}$ is completed by sets from $\mathcal{F}$ of zero P-probability and under Condition (8.34) the processes Θ and W are $\mathbb{F}^X_+$-adapted. By the remark to Lemma 6.2 from this it follows that the processes Θ and W belong to the class $\mathcal{M}^c_{loc}(\mathbb{F}^X_+, P)$. Besides $\langle Q, W\rangle = 0$, and by the corollary to Lemma 1 we get that there is no integral representation for the process Θ.

We will prove now that the processes Θ and W are $\mathbb{F}^X_+$-adapted. By (8.35) it suffices to prove only that the process Θ is $\mathbb{F}^X_+$-adapted. Since the σ-algebra $\mathcal{F}^X_{0+}$ is completed by sets from $\mathcal{F}$ of zero P-probability, it suffices to show that for each $t \geq 0$

$$\gamma_t = \mathbf{E}\,(\Theta_t - \pi_t(\Theta))^2 = 0,$$

where $\pi(\Theta)$ is the optional projection of the process Θ relative to $\mathbb{F}^X_+$.

Aiming at this set $a^n = a\, I\,(a \leq n)$ and define the process X^n by

$$X^n_t = \int_0^t a^n_s \Theta_s\, ds + W_t. \tag{8.36}$$

Denote by $\pi^n(\Theta)$ the optional projection of Θ relative to $\mathbb{F}^{X^n}_+$, and let

$$\gamma^n_t = \mathbf{E}\,(\Theta_t - \pi^n_t(\Theta))^2.$$

In accordance with the Kalman-Busy filtration equations (cf., for instance, [188])

$$\pi^n_t(\Theta) = \int_0^t \gamma^n_s a^n_s\,(dX^n_s - a^n_s \pi^n_s(\Theta)\, ds), \tag{8.37}$$

$$\gamma^n_t = t - \int_0^t (a^n_s \gamma^n_s)^2 ds. \tag{8.38}$$

Let $\Theta^n = (\Theta^n_t)_{t \geq 0}$ be defined by the equation (cf. (8.37))

$$\Theta^n_t = \int_0^t \gamma^n_s a^n_s\,(dX_s - a^n_s \Theta^n_s ds). \tag{8.39}$$

Observe that $a^n \cdot X = a^n \cdot X^n$ (cf. the definition of a^n and the representation (8.35)). Therefore (cf. (8.37))

$$\Theta_t^n = \int_0^t \gamma_s^n a_s^n \,(dX_s^n - a_s^n \Theta_s^n ds),$$

and hence $\Theta^n = \pi^n(\Theta)$.

Since $\pi_t(\Theta)$ is the optimal estimator for Θ_t in the mean-square sense, then

$$\gamma_t \le E\,(\Theta_t - \Theta_\tau^n)^2 = E\,(\Theta_t - \pi_t^n(\Theta)) = \gamma_t^n.$$

By this inequality and (8.38) we get

$$\gamma_t^n \le t - \int_0^t (a_s^n \gamma_s)^2 ds = t - \int_{\{s \le t:\, \gamma_s > 0\}} (a_s^n \gamma_s)\, ds. \tag{8.40}$$

According to the remark to Lemma 6.2 we have $\pi(\Theta) \in \mathcal{M}_{loc}(\mathbb{F}_+^X, P)$ (and $E\pi_t^2(\Theta) < \infty$, $t > 0$). Therefore $\gamma = (\gamma_t)_{t \ge 0}$ is a continuous function of t and consequently the set $\Delta = \{s \le t: \gamma_s > 0\}$ is open. Next, observe that

$$\int_\Delta (a_s^n \gamma_s)^2\, ds \le t.$$

Hence

$$\int_\Delta (a_s \gamma_s)^2\, ds \le t.$$

Therefore, as $t > 0$ by the property (8.34) of the function $a = (a_t)_{t \ge 0}$ the equality

$$\int_\Delta ds = 0$$

takes place, i.e. $\gamma_s = 0$, $s \le t$, for each $t > 0$.

Problems

1. Let $X \in S(\mathbb{F}_+^X, P)$ and $\mathcal{F}^X = \sigma(\bigcup_{t \ge 0} \mathcal{F}_t^X)$. Show that $\mathfrak{Z}_{P^X}$ consists of a unique point if

(a) X is a process with conditionally independent increments (cf. § 4);

(b) X is a point process;

(c) X is a process with piecewise constant trajectories and a finite number of jumps over any finite time interval;

(d) X is a unique weak solution (cf. [50, 188]) of Ito's stochastic equation

$$X_t = X_0 + \int_0^t a(s, X_s)\, ds + \int_0^t b(s, X_s)\, dW_s \tag{8.41}$$

where a (s, x) and b (s, x) are $B(R_+) \otimes B(R)$-measurable functions, $b^2(s, x) > 0$ for each $(s, x) \in R_+ \times R$, and $W = (W_t)_{t \geq 0}$ is a Wiener process relative to $(\mathbb{F}^X_+, P)$.

2. Let $X \in \mathcal{M}^2_{loc}(\mathbb{F}^X, P)$, $\mathcal{F} = \sigma(\bigcup_{t \geq 0} \mathcal{F}^X_t)$ and let $\mathcal{F}^X_0$ be completed by sets from $\mathcal{F}$ of P-measure zero. If $M \in \mathcal{M}^2_{loc}(\mathbb{F}^X, P)$ and (X, M) is a Gaussian process, then

$$M_t = M_0 + h \cdot X_t,$$

where $h = h(t)$ is a $B(R_+)$-measurable function defined by the equality

$$h(t) = \frac{dE(X_t - X_0)(M_t - M_0)}{dE(X_t - X_0)^2}.$$

3. Let $X \in \mathbf{Sp}(\mathbb{F}, P)$ and $X_t = X_0 + A_t + M_t$ with $A = (A_t)_{t \geq 0} \in \mathcal{A}_{loc} \cap \mathcal{P}(\mathbb{F})$ and $M \in \mathcal{M}_{loc}(\mathbb{F}, P)$. If $Y \in \mathcal{M}^{2,d}_{loc}(\mathbb{F}^X, P)$ and (X_0, A, M, Y) present a Gaussian system, then

$$Y_t = Y_0 + \sum_{0 < s \leq t} I(\Delta\langle Y\rangle_s > 0)\, h(s)\, [\Delta X_s - \mathbf{E}(\Delta A_s \mid \mathcal{F}^X_{s-})]$$

with

$$h(s) = [\mathbf{E}(\Delta X_s - \mathbf{E}(\Delta A_s \mid \mathcal{F}^X_{s-}))^2]^{-1}\, \mathbf{E}[\Delta Y_s (\Delta X_s - \mathbf{E}(\Delta A_s \mid \mathcal{F}^X_{s-}))],$$

$s > 0$ and $\frac{0}{0} = 0$.

§ 9. Gaussian martingales and semimartingales

1. Definition. A stochastic process $X = (X_t)_{t \geq 0}$ is called *Gaussian* if all of its finite-dimensional distributions are Gaussian.

In the present section the structure of trajectories of Gaussian semimartingales is studied and their characterization is given in terms of the mean value and correlation function.

2. First of all we will dwell on a number of facts and properties of Gaussian vectors composed of (real-valued) random variables.

A random variable ξ is called *Gaussian* or *normally distributed* with parameters a and γ^2 (notation $\xi \sim N(a, \gamma^2)$), if the characteristic function $f(\lambda) = \mathbf{E}e^{i\lambda\xi}$ has the following form:

$$f(\lambda) = e^{i\lambda a - \frac{1}{2}\lambda^2\gamma^2}. \tag{9.1}$$

Here $a = \mathbf{E}\xi$ and $\gamma^2 = \mathbf{E}(\xi - a)^2$.

A random vector $\xi = (\xi_1, ..., \xi_n)$ is called *Gaussian* or *normally distributed* with parameters A and Γ if the n-dimensional characteristic function $f(\lambda) = \mathbf{E}e^{i(\lambda, \xi)}$, $\lambda = (\lambda_1, ..., \lambda_n)$ has the following form:

$$f(\lambda) = e^{i(\lambda, A) - \frac{1}{2}(\Gamma\lambda, \lambda)},$$

where $A = (a_1, ..., a_n)$, a symmetric nonnegative definite matrix $\Gamma = \| \Gamma_{ij} \|$ and the scalar products

$$(\lambda, A) = \sum_{i=1}^{n} \lambda_i a_i,$$

$$(\Gamma, \lambda) = \lambda^* \Gamma\lambda = \sum_{i,j=1}^{n} \lambda_i \lambda_j \Gamma_{ij}.$$

Here $a_i = \mathbf{E}\xi_i$ and $\Gamma_{ij} = \mathbf{E}(\xi_i - a_i)(\xi_j - a_j)$.

The fact that a random vector $\xi = (\xi_1, ..., \xi_n)$ is Gaussian is equivalent to the fact that for each vector $\lambda = (\lambda_1, ..., \lambda_n)$ a variable

$$\sum_{i=1}^{n} \lambda_i \xi_i$$

is Gaussian.

An infinite dimensional random vector $\xi = (\xi_1, \xi_2, ...)$ is called *Gaussian* if each finite dimensional subvector $(\xi_{i_1}, ..., \xi_{i_n})$ is Gaussian or, equivalently, if for each vector $\lambda = (\lambda_1, \lambda_2, ...)$ with a finite number non-zero components a random variable

$$\sum_{i=1}^{\infty} \lambda_i \xi_i$$

is Gaussian.

Random vectors $\xi = (\xi_1, \xi_2, ...)$ and $\eta = (\eta_1, \eta_2, ...)$ form a *Gaussian system* (ξ, η) if the random vector $(\xi_1, \eta_1, \xi_2, \eta_2, ...)$ is Gaussian. Independent Gaussian vectors form a Gaussian system (Problem 1). If (ξ, η) is a Gaussian system, then both of the random vectors ξ and η are Gaussian.

If (ξ, η) is a Gaussian system, $\mathcal{F}^{\eta}$ the σ-algebra generated by vector η and $\hat{\xi} = (\hat{\xi}_1, \hat{\xi}_2, ...)$ with $\hat{\xi}_i = \mathbf{E}(\xi_i \mid \mathcal{F}^{\eta})$, $i \geq 1$, then $(\hat{\xi}, \xi, \eta)$ is a Gaussian system (Problem 3).

3. The following property of Gaussian vectors is called the "zero-one" law.

Lemma 1. *Let* $\xi = (\xi_1, \xi_2, ...)$ *be a Gaussian vector with* $\mathbf{E}\xi_i = 0$, $i \geq 1$, *and let* F *be a linear subspace of the space* $R^{\infty} = R \times R \times ...$, *belonging to the Borel* σ-*algebra* $B(R^{\infty})$.

Then the alternative

$$P(\xi \in F) = 0 \text{ or } P(\xi \in F) = 1$$

takes place.

Proof. Let ξ' and ξ'' be independent copies of ξ, i.e. independent Gaussian vectors with the vector of mean values and the correlation matrix of the vector ξ. For $\theta \in [0, 2\pi]$ form the random vectors

$$\gamma'_{\theta} = \xi' \cos\theta + \xi'' \sin\theta, \quad \gamma''_{\theta} = \xi' \sin\theta - \xi'' \cos\theta.$$

By Problem 2 it may be simply checked that γ'_{θ} and γ''_{θ} are independent random vectors forming a Gaussian system and presenting independent copies ξ for each θ. Therefore, the probability of the event

$$A(\theta) = \{\omega: \gamma'_{\theta}(\omega) \in F, \ \gamma''_{\theta}(\omega) \notin F\}$$

is independent of θ. Let us show as $\theta_1 \neq \theta_2$, that $A(\theta_1) \cap A(\theta_2) = \varnothing$. In fact, if $\omega \in A(\theta_1) \cap A(\theta_2)$, then $\gamma'_{\theta_1}(\omega) \in F$ and $\gamma'_{\theta_2}(\omega) \in F$, and as the matrix

$$\begin{pmatrix} \cos\theta_1 & \sin\theta_1 \\ \cos\theta_2 & \sin\theta_2 \end{pmatrix}$$

is nonsingular, we get $\xi'(\omega) \in F$ and $\xi''(\omega) \in F$, i.e. then

$$\gamma''_{\theta_1}(\omega) \in F \text{ and } \gamma''_{\theta_2}(\omega) \in F,$$

and this contradicts the definition of the sets A (θ_1) and A (θ_2).

Consequently, there is only one possibility

$$A(\theta_1) \cap A(\theta_2) = \varnothing.$$

Since the probability of the event A (θ) is independent of θ and the events A(θ) do not intersect for different θ's, then P (A (θ)) = 0 evidently.

Hence P (A (0)) = 0, i.e.

$$0 = \mathsf{P}(\xi' \in F,\ \xi'' \notin F) = \mathsf{P}(\xi \in F)\,\mathsf{P}(\xi \notin F) = \mathsf{P}(\xi \in F)\,(1 - \mathsf{P}(\xi \in F))$$

and, consequently, the "zero-one" law takes place.

4. Let B (R^∞) be a Borel σ-algebra in R^∞. A function N (x), $x \in R^\infty$, measurable with respect to B (R^∞) and taking values in $\overline{R}_+ = R_+ \cup \{\infty\}$ is called a *pseudonorm* if the set $N^{-1}(R_+)$ is a linear subspace in R^∞, in which N (x) induces a seminorm ([302]).

The functions

$$N(x) = \left(\sum_{i=1}^{\infty} |x_i|^p \right)^{1/p}, \quad p \geq 1, \qquad N(x) = \sup_{i \geq 1} |x_i|.$$

serve as examples of pseudonorms.

Let $\xi = (\xi_1, \xi_2, \ldots)$ be a Gaussian vector with $\mathbf{E}\xi_i = 0$, $i \geq 1$, and let N (x) be a pseudonorm. Since by the definition of the pseudonorm the set $\{x: N(x) < \infty\}$ is a linear subspace of R^∞ and belongs to B (R^∞), we have by Lemma 1

$$\mathsf{P}(N(\xi) < \infty) = 1 \text{ or } \mathsf{P}(N(\xi) < \infty) = 0,$$
$$\mathsf{P}(N(\xi) = \infty) = 1 \text{ or } \mathsf{P}(N(\xi) = \infty) = 0.$$

We present yet another property of a pseudonorm for a Gaussian vector.

Lemma 2. *Let* $\xi = (\xi_1, \xi_2, \ldots)$ *be a Gaussian vector with* $\mathbf{E}\xi_i = 0$, $i \geq 1$ *and let* N (x) *be a pseudonorm. If* P (N (x) < ∞) > 0; *then there exists a number* $\varepsilon > 0$ *such that*

$$\mathbf{E} \exp(\varepsilon N^2(\xi)) < \infty.$$

Proof. Let ξ' and ξ'' be independent copies of ξ. Then

$$\frac{\xi' + \xi''}{\sqrt{2}} \text{ and } \frac{\xi' - \xi''}{\sqrt{2}}$$

are independent copies of ξ too. Therefore for all nonnegative numbers s and t

$$P(N(\xi) \le s)\, P(N(\xi) > t) = P\left(N\left(\frac{\xi' + \xi''}{\sqrt{2}}\right) > t\right) P\left(N\left(\frac{\xi' - \xi''}{\sqrt{2}}\right) \le s\right)$$

$$= P\left(N\left(\frac{\xi' + \xi''}{\sqrt{2}}\right) > t,\, N\left(\frac{\xi' - \xi''}{\sqrt{2}}\right) \le s\right). \quad (9.2)$$

For $x,\ y \in R^\infty$ we have

$$\frac{x+y}{\sqrt{2}} = \frac{x-y}{\sqrt{2}} + \sqrt{2}\ y, \quad \frac{x+y}{\sqrt{2}} = \frac{y-x}{\sqrt{2}} + \sqrt{2}\ x.$$

In view of the triangle inequality for $N(x)$ this gives

$$N\left(\frac{x+y}{\sqrt{2}}\right) \le N\left(\frac{x-y}{\sqrt{2}}\right) + \sqrt{2}\ (N(x) \wedge N(y)).$$

Consequently, as

$$N\left(\frac{x-y}{\sqrt{2}}\right) \le s$$

and

$$N\left(\frac{x+y}{\sqrt{2}}\right) > t\,,$$

the inequalities

$$N(x) > \frac{t-s}{\sqrt{2}} \text{ and } N(y) > \frac{t-s}{\sqrt{2}}$$

hold. From this and (9.2) it follows that

$$P(N(\xi) \le s)\, P(N(\xi) > t) \le P\left(N(\xi') > \frac{t-s}{\sqrt{2}},\ N(\xi'') > \frac{t-s}{\sqrt{2}}\right)$$

$$= P^2\left(N(\xi) > \frac{t-s}{\sqrt{2}}\right). \quad (9.3)$$

By the assumption $P(N(\xi) < \infty) > 0$ and the zero-one law for $F = \{x: N(x) < \infty\}$ we have $P(N(\xi) < \infty) = 1$. Therefore, one can find a number $s > 0$ such that $P(N(\xi) \le s) = q > \frac{1}{2}$. For $n = 0, 1, \ldots$ define

$$t_n = (\sqrt{2} + 1)\ (2^{(n+1)/2} - 1)\, s$$

and note that $t_0 = s$ and $t_n - s = t_{n-1}\sqrt{2}$. Set

$$c_n = q^{-1} \mathsf{P}\,(N(\xi) > t_n), \quad n = 0, 1, \ldots$$

Clearly $c_0 = \dfrac{1-q}{q} < 1$. Besides, using the inequality (9.3) with t_n instead of t, we get the recurrent relation for c_n, $n \geq 1$,

$$c_n \leq (c_{n-1})^2.$$

Iterating it we arrive at

$$\mathsf{P}\,(N(\xi) > t_n) \leq q \left(\frac{1-q}{q}\right)^{2^n}.$$

Let us estimate now $\mathbf{E} \exp(\varepsilon N^2(\xi))$. We have

$$\begin{aligned}
\mathbf{E} \exp(\varepsilon N^2(\xi)) &= \int_0^\infty \exp(\varepsilon u^2)\, d\mathsf{P}\,(N(\xi) \leq u) \\
&= \int_0^s \exp(\varepsilon u^2)\, d\mathsf{P}\,(N(\xi) \leq u) + \sum_{n=0}^\infty \int_{t_n}^{t_{n+1}} \exp(\varepsilon u^2)\, d\mathsf{P}\,(N(\xi) \leq u) \\
&\leq \exp(\varepsilon s^2)\, \mathsf{P}\,(N(\xi) \leq s) + \sum_{n=0}^\infty \exp(\varepsilon t_{n+1}^2)\, \mathsf{P}\,(N(\xi) > t_n) \\
&\leq q \left\{ \exp(\varepsilon s^2) + \sum_{n=0}^\infty \exp\left(2^n \left[\ln \frac{1-q}{q} + 4(\sqrt{2}+1)^2 \varepsilon s^2\right]\right)\right\}. \quad (9.4)
\end{aligned}$$

Since $\ln \dfrac{1-q}{q} < 0$, the right -hand side of the inequality (9.4) is bounded for sufficiently small ε.

The lemma is proved.

By Lemma 2 and Problem 5 we get the following

Corollary. *Let* $N(x)$, $x \in R^\infty$, *denote any of the functions*

$$\left(\sum_{i=1}^\infty |x_i|^p\right)^{1/p}, \quad p \geq 1$$

and $\sup_{i \geq 1} |x_i|$.

If ξ *is a Gaussian vector and* $\mathsf{P}\,(N(\xi) < \infty) = 1$, *then*

$$\mathbf{E} N^2(\xi) < \infty.$$

5. According to the corollary to Lemma 2 for a Gaussian vector x with the property

$$\mathsf{P}\left(\sum_{i=1}^{\infty} \xi_i^2 < \infty\right) = 1$$

the inequality

$$\mathbf{E} \sum_{i=1}^{\infty} \xi_i^2 < \infty$$

holds.

In the following lemma a specification of this inequality is given.

Lemma 3. *Let* ξ *be a Gaussian vector with* $\mathbf{E}\xi_i = 0, \; i \geq 1$. *Then*

$$\mathbf{E} \sum_{i=1}^{\infty} \xi_i^2 \leq \frac{1}{\left(\mathbf{E} \exp\left(-\sum_{i=1}^{\infty} \xi_i^2\right)\right)^2} . \tag{9.5}$$

Proof. Obviously, it suffices to establish only the validity of the inequalities

$$\mathbf{E} \sum_{i=1}^{n} \xi_i^2 \leq \frac{1}{\left(\mathbf{E} \exp\left(-\sum_{i=1}^{n} \xi_i^2\right)\right)^2}, \quad n \geq 1. \tag{9.6}$$

For every n it suffices to prove inequality (9.6) in case in which the correlation matrix of the random vector $(\xi_1, ..., \xi_n)$ is nonsingular. In fact, one can always replace the vector $(\xi_1, ...\xi_n)$ by the vector $(\xi_1, ..., \xi_n) + \varepsilon\,(\alpha_1, ..., \alpha_n)$ where $(\alpha_1, ..., \alpha_n)$ is an independent of ξ Gaussian vector with $\mathbf{E}\alpha_i = 0$, $\mathbf{E}\alpha_i^2 = 1$, $i = 1, ..., n$, $\mathbf{E}\alpha_i\alpha_j = 0$, $i \neq j$, ε is a positive number, and if

$$\mathbf{E} \sum_{i=1}^{n} (\xi_i + \varepsilon\alpha_i)^2 \leq \frac{1}{\left(\mathbf{E} \exp\left(-\sum_{i=1}^{n} (\xi_i + \varepsilon\alpha_i)^2\right)\right)^2},$$

then taking the limit as $\varepsilon \to 0$ one arrives at the inequality (9.6).

Fix n and assume that the matrix $\Gamma = \| \Gamma_{ij} \|$, asociated with the random vector $(\xi_1, ..., \xi_n)$, is nonsingular. As is known a symmetric positive definite matrix Γ can be reduced to a diagonal form by using an orthogonal matrix $C = \| C_{ij} \|$:

$$C\Gamma C^* = \mathrm{diag}\,(\lambda_1, ..., \lambda_n)$$

where C^* is the transposition of C, and $\lambda_1, ..., \lambda_n$ are positive numbers. Consider a Gaussian vector $(\eta_1, ..., \eta_n)$ with

$$\eta_i = \sum_{j=1}^{n} C_{ij}\xi_j.$$

It can be simply checked, taking into consideration the equality $CC^* = I$ (I is the identity matrix), that the Gaussian random variables $\eta_1, ..., \eta_n$ are uncorrelated and hence independent (Problem 2), and the equality

$$\sum_{i=1}^{n} \xi_i^2 = \sum_{i=1}^{n} \eta_i^2$$

holds; besides $\mathbf{E}\eta_i^2 = \lambda_i$, $i = 1, ..., n$.

Therefore it suffices to show that

$$\mathbf{E} \sum_{i=1}^{n} \eta_i^2 \le \frac{1}{\left(\mathbf{E}\exp\left(-\sum_{i=1}^{n} \eta_i^2\right)\right)^2}.$$

The last inequality holds, since direct calculations show that

$$\mathbf{E}\exp(-\eta_i^2) = (1 + 2\mathbf{E}\eta_i^2)^{-1/2}$$

and, by the independence of the random variables $\eta_1, ..., \eta_n$,

$$\left(\mathbf{E}\exp\left(-\sum_{i=1}^{n} \eta_i^2\right)\right)^2 = \left(\prod_{i=1}^{n} \mathbf{E}\exp(-\eta_i^2)\right)^2 = \prod_{i=1}^{n} (1 + 2\mathbf{E}\eta_i^2)^{-1}.$$

6. Studying properties of trajectories of Gaussian processes, we make use of the following two results.

Lemma 4. *Let* $(\alpha_n)_{n \ge 1}$ *be a sequence of Gaussian random variables and* $\alpha_n \to \alpha$ *in probability as* $n \to \infty$ $(\alpha_n \xrightarrow{P} \alpha)$. *Then* α *is a Gaussian random variable and* $\lim_n \mathbf{E}\alpha_n = \mathbf{E}\alpha$ *and* $\lim_n \mathbf{E}\alpha_n^2 = \mathbf{E}\alpha^2$.

Proof. Denote

$$a_n = \mathbf{E}\alpha_n \text{ and } \gamma_n^2 = \mathbf{E}(\alpha_n - a_n)^2.$$

According to (9.1) we have

$$\mathbf{E}e^{i\lambda\alpha_n} = \exp\left(i\lambda a_n - \frac{\lambda^2}{2}\gamma_n^2 \right).$$

The convergence $\alpha_n \xrightarrow{P} \alpha$ implies the following convergence of characteristic functions:

$$\mathbf{E}e^{i\lambda\alpha_n} \to \mathbf{E}e^{i\lambda\alpha}, \ \lambda \in R.$$

Therefore for each $\lambda \in R$ there exist the limits

$$\lim_n \exp\left(-\frac{\lambda^2}{2}\gamma_n^2 \right) = |\mathbf{E}e^{i\lambda\alpha}|, \ \lim_n \exp(i\lambda a_n) = \frac{\mathbf{E}e^{i\lambda\alpha}}{|\mathbf{E}e^{i\lambda\alpha}|}.$$

This implies (cf. Problem 6) the existence of the limits $\gamma^2 = \lim_n \gamma_n^2$ and $a = \lim_n a_n$. Besides

$$\mathbf{E}e^{i\lambda\alpha} = \exp\left(i\lambda a - \frac{\lambda^2}{2}\gamma^2 \right),$$

i.e. α is a Gaussian random variable and, in addition, $\mathbf{E}\alpha_n = a_n \to a = \mathbf{E}\alpha$, and $\mathbf{E}\alpha_n^2 = a_n^2 + \gamma_n^2 \to a^2 + \gamma^2 = \mathbf{E}\alpha^2$.

Corollary. *If* $(\alpha_n)_{n \geq 1}$ *is a sequence of Gaussian random variables and* $\alpha_n \xrightarrow{P} \alpha$, *then* $\lim_n \mathbf{E}|\alpha_n| \to \mathbf{E}|\alpha|$.

Lemma 5. *Let* $(\xi_{n1}, \ldots, \xi_{nn})$, $n \geq 1$ *be a sequence of Gaussian vectors with* $\mathbf{E}\xi_{ni} = 0$, $i \leq n$, $n \geq 1$.

If

$$\lim_{c \to \infty} \overline{\lim_n} \, \mathsf{P}\left(\sum_{i=1}^n \xi_{ni}^2 \geq c \right) = 0,$$

then

$$\overline{\lim_n} \, \mathbf{E} \sum_{i=1}^n \xi_{ni}^2 < \infty.$$

Proof. By Lemma 3

$$\mathbf{E} \sum_{i=1}^n \xi_{ni}^2 \leq \frac{1}{\left(\mathbf{E}\exp\left(-\sum_{i=1}^n \xi_{ni}^2 \right) \right)^2}.$$

Choose c' so large that

$$\overline{\lim_n}\, \mathsf{P}\left(\sum_{i=1}^{n} \xi_{ni}^2 \geq c' \right) \leq \frac{1}{2}.$$

Then the preceding inequality yields the estimate

$$\mathbf{E} \sum_{i=1}^{n} \xi_{ni}^2 \leq \frac{1}{\left(e^{-c'} \mathsf{P}\left(\sum_{i=1}^{n} \xi_{ni}^2 < c' \right)\right)^2} = \frac{e^{2c'}}{\left(1 - \mathsf{P}\left(\sum_{i=1}^{n} \xi_{ni}^2 \geq c' \right)\right)^2}$$

and consequently

$$\overline{\lim_n}\, \mathbf{E} \sum_{i=1}^{n} \xi_{ni}^2 \leq 4e^{2c'}.$$

7. In the theory studied here every semimartingale (by definition) has trajectories in D. Therefore we will be interested in the sequel only in Gaussian processes $X = (X_t)_{t \geq 0}$ the trajectories of which belong to the space D, writing for brevity $X \in D \cap G$.

If a process $X \in D \cap G$, then all its finite dimensional distributions are Gaussian, and hence the random vector $(X_{r1}, X_{r2}, \ldots)$, where $(r_1, r_2, \ldots)$ is a collection of rational numbers in R_+, is a Gaussian vector. The converse statement is true too. Namely, if $X \in D$ and if for each collection $(r_1, r_2, \ldots)$ of rational numbers in R_+ the random vector $(X_{r1}, X_{r2}, \ldots)$ is Gaussian, then $X \in D \cap G$. This fact is established by utilizing Lemma 4, since any linear combination

$$\sum_{i=1}^{k} c_i X_{t_i}$$

is approximated in the sense of convergence with probability one by a sequence of Gaussian random variables

$$\sum_{i=1}^{k} c_i X_{r_i^n}, \quad n \geq 1,$$

where r_i^n, $n \geq 1$ are rational numbers such that $r_i^n \downarrow t_i$ as $n \to \infty$. Lemma 4 establishes also that for each collection $(t_1, t_2, \ldots)$ the random vector $(X_{t_1}, \Delta X_{t_1}, X_{t_2}, \Delta X_{t_2}, \ldots)$ is Gaussian.

Remark 1. Since the space D is a vector space ($X, Y \in D \Rightarrow aX + bY \in D$), the "zero-one" law continues to hold also for Gaussian processes with trajectories in D.

Namely, if X is a Gaussian process with trajectories in D and F a subspace of D, belonging to a Borel σ-algebra $\mathcal{D}$ in D, then P (X ∈ F) = 1 or P (X ∈ F) = 0.

Remark 2. The assertion of Lemma 2 also continues to hold for Gaussian processes with trajectories in D. The typical examples of pseudonorms N (X), X ∈ D are presented by

$$N(X) = \left(\int_0^T |X_s|^p \, ds \right)^{1/p}, \quad p \geq 1; \quad N(X) = \sup_{s \leq T} |X_s|;$$

$$N(X) = \operatorname{var}(X)_T; \quad N(X) = \left(\sum_{s \leq T} (\Delta X_s)^2 \right)^{1/2}$$

with $T \leq \infty$.

Denote by

$$a(t) = \mathbf{E} X_t, \quad \Gamma(s,t) = \mathbf{E}(X_s - a(s))(X_t - a(t))$$

the mean value and correlation functions of a process X ∈ D ∩ G. Lemma 4 tells us that a (t) and Γ (s, t) are right-continuous functions at each point $t \in R_+$ having left-hand limits and $(s, t) \in R_+^2$.

For a stochastic process X ∈ D the definition of the stochastic continuity takes the following form:

$$\{t > 0 : \mathrm{P}(\Delta X_t \neq 0) > 0\} = \varnothing.$$

For a stochastic process X ∈ D ∩ G the stochastic continuity, by Lemma 4, is equivalent to the continuity in the mean square sense. Therefore for stochastically continuous Gaussian processes with trajectories in D the functions a (t) and Γ (s,t) are continuous.

Now we present the conditions under which a stochastically continuous Gaussian process with trajectories in D is P-a.s. continuous.

Lemma 6. *Let a stochastically continuous process* X ∈ D ∩ G. *If for each* T > 0 *and each partition sequence* $0 \equiv t_0^n < t_1^n < \ldots < t_n^n \equiv T$, $n \geq 1$, *of the interval* [0, T] *with* $\max_{0 \leq j \leq n-1} (t_{j+1}^n - t_j^n) \to 0$ *as* $n \to \infty$ *the condition*

$$\lim_{c \to \infty} \overline{\lim_{n}} \, \mathrm{P}\left(\sum_{j=0}^{n-1} (X_{t^n} - X_{t^n})^2 \geq c \right) = 0 \tag{9.7}$$

holds, then the process X *has* P*-a.s. continuous trajectories, i.e.*

$$X \in C \cap G.$$

Proof. A stochastically continuous Gaussian process X has a continuous mean value function. Therefore, without loosing generality, one can assume $EX_t \equiv 0$. Since $X \in D$, the stochastic process X has no discontinuities of the second kind. Therefore, to prove P-a.s. continuity of its trajectories, it suffices to show that for each $\varepsilon > 0$, $T > 0$ and for the indicated partition sequence of the interval [0, T]

$$\lim_n \sum_{j=0}^{n-1} P(|X_{t^n_{j+1}} - X_{t^n_j}| > \varepsilon) = 0 \tag{9.8}$$

(cf., for instance, Theorem 1 and its proof in [46], Ch. IV, § 5).

Denote

$$b^2_{nj} = E(X_{t^n_{j+1}} - X_{t^n_j})^2.$$

The random variable $(X_{t^n_{j+1}} - X_{t^n_j})$ is Gaussian with zero mathematical expectation and variance b^2_{nj}. Therefore

$$E(X_{t^n_{j+1}} - X_{t^n_j})^4 = 3(E(X_{t^n_{j+1}} - X_{t^n_j})^2)^2 = 3b^4_{nj}.$$

Consequently, in view of Chebyshev's inequality, we get

$$\sum_{j=0}^{n-1} P(|X_{t^n_{j+1}} - X_{t^n_j}| > \varepsilon) \le \sum_{j=0}^{n-1} \frac{E(X_{t^n_{j+1}} - X_{t^n_j})^4}{\varepsilon^4}$$

$$= \frac{3}{\varepsilon^4} \sum_{j=0}^{n-1} b^4_{nj} \le \frac{3}{\varepsilon^4} \max_{0 \le j \le n-1} b^2_{nj} \, E \sum_{j=0}^{n-1} (X_{t^n_{j+1}} - X_{t^n_j})^2.$$

By (9.7) we have

$$\overline{\lim_n} \, E \sum_{j=0}^{n-1} (X_{t^n_{j+1}} - X_{t^n_j})^2 < \infty,$$

due to Lemma 5. Therefore, the desired relation (9.8) takes place provided

$$\lim_n \max_{0 \le j \le n-1} b^2_{nj} = 0. \tag{9.9}$$

Observe that

$$b^2_{nj} = \Gamma(t^n_{j+1}, t^n_{j+1}) + \Gamma(t^n_j, t^n_j) - 2\Gamma(t^n_{j+1}, t^n_j)$$

where the function $\Gamma(s, t)$ is continuous, since it is the correlation function of a stochastically continuous process X. Consequently $\Gamma(s, t)$ is a uniformly continuous function on $[0, T]^2$, and hence the desired relation (9.9) takes place.

8. We will present now the conditions under which a Gaussian semimartingale has

P-a.s. continuous trajectories.

Lemma 7. *Let* X ∈ S ∩ G. *The following conditions are equivalent:*

(a) X *is a continuous process,*

(b) X *is a stochastically continuous process.*

Proof. The implication (a) ⇒ (b) is obvious. To establish the implication (b) ⇒ (a) it suffices, by Lemma 6, to show that

$$\lim_{c\to\infty}\overline{\lim_{n}}\, \mathsf{P}\left(\sum_{j=0}^{n-1}(X_{t^n_{j+1}}-X_{t^n_j})^2\geq c\right)=0$$

for each partition sequence $0\equiv t^n_0<t^n_1<\ldots<t^t_n\equiv T,\ \ T<\infty$ with

$$\lim_{n}\max_{0\leq j\leq n-1}(t^n_{j+1}-t^n_j)\to 0,\ n\to\infty.$$

By the canonical representation for semimartingales (§ 1) the following decomposition for X takes place:

$$X_t=X_0+A_t+M_t$$

with

$$A=(A_t)_{t\geq 0}\in\mathcal{V}\text{ and }M=(M_t)_{t\geq 0}\in\mathcal{M}^2_{loc}.$$

By this decomposition

$$\sum_{j=0}^{n-1}(X_{t^n_{j+1}}-X_{t^n_j})^2\leq 2\sum_{j=0}^{n-1}(A_{t^n_{j+1}}-A_{t^n_j})^2+2\sum_{j=0}^{n-1}(M_{t^n_{j+1}}-M_{t^n_j})^2.$$

Since (P-a.s.)

$$\sum_{j=0}^{n-1}(A_{t^n_{j+1}}-A_{t^n_j})^2\leq\left(\sum_{j=0}^{n-1}|A_{t^n_{j+1}}-A_{t^n_j}|\right)^2\leq(\mathrm{Var}\,(A)_T)^2,$$

it suffices to show that

$$\lim_{c\to\infty}\overline{\lim_{n}}\, \mathsf{P}\left(\sum_{j=0}^{n-1}(M_{t^n_{j+1}}-M_{t^n_j})^2\geq c\right)=0. \qquad (9.10)$$

If $M\in\mathcal{M}^2$, then the desired relation (9.10) takes place by Chebyshev's inequality, since

$$P\left(\sum_{j=0}^{n-1}(M_{t^n_{j+1}}-M_{t^n_j})^2 \geq c\right) \leq \frac{1}{c}\sum_{j=0}^{n-1} \mathbf{E}\,(M_{t^n_{j+1}}-M_{t^n_j})^2$$

$$= \frac{1}{c}\sum_{j=0}^{n-1} \mathbf{E}\,(M^2_{t^n_{j+1}}-M^2_{t^n_j}) = \frac{\mathbf{E}M_T^2}{c}. \quad (9.11)$$

If $M \in \mathcal{M}^2_{loc}$ and $(T_k)_{k\geq 1}$ is a localizing sequence for M, then by the inequality (9.11)

$$P\left(\sum_{j=0}^{n-1}(M_{t^n_{j+1}}-M_{t^n_j})^2 \geq c\right)$$

$$\leq P\left(\sum_{j=0}^{n-1}(M_{t^n_{j+1}\wedge T_k}-M_{t^n_j\wedge T_k})^2 \geq c\right) + P\,(T_k < T)$$

$$\leq \frac{\mathbf{E}M^2_{T\wedge T_k}}{c} + P\,(T_k < T). \quad (9.12)$$

The desired relation (9.10) takes place, since the right-hand side of the inequality (9.12) converges to zero as the limit $\lim_k \lim_{c\to\infty}$ is taken.

9. In the following theorem the structure of trajectories of a Gaussian semimartingale is described.

Theorem 1. *Let a process* $X \in \mathbf{S} \cap G$ *and*

$$H = \{t > 0:\ P\,(\Delta X_t \neq 0) = 1\}. \quad (9.13)$$

Then

1) *in the decomposition*

$$X_t = X_0 + X'_t + X''_t$$

with

$$X' = I_{R_+\setminus H} \cdot X, \qquad X'' = I_H \cdot X$$

the process X' *is continuous and* $X' \in \mathbf{S} \cap G$, *the process* $X'' \in \mathbf{S} \cap G$, *besides* (X_0, X', X'') *presents a Gaussian system;*

2) *jump times of trajectories of the process* X *are deterministic, taking values in the set* H;

3) X *is a special semimartingale;*

4) *if* $\mathbb{F} = \mathbb{F}^X_+$, *then in the decomposition* $X_t = X_0 + A_t + M_t$ *of a (special) semimartingale* X *the predictable process* $A = (A_t)_{t \geq 0}$ *and the local martingale* $M = (M_t)_{t \geq 0}$ *possess the following properties:* (X_0, A, M) *present a Gaussian system,* $E\,(\mathrm{Var}\,(A)_t)^2 < \infty$ *and* $M \in \overline{\mathcal{M}}^2$.

Corollary. *Each Gaussian semimartingale is a local quasimartingale relative to* $\mathbb{F}^X_+$.

Proof of Theorem 1. 1) If $X' = I_{R_+ \setminus H} \cdot X$ with H defined in (9.13), then, according to Problem 7, $X' \in S \cap G$. Next, observe that a Gaussian random variable is equal to zero with probability zero or one. Therefore

$$\{t > 0:\ P\,(\Delta X'_t \neq 0) > 0\} = \varnothing$$

due to the fact that $\Delta X'_t = I_{R_+ \setminus H}\, \Delta X_t$ for each $t \in R_+$. Consequently, X' is a stochastically continuous process and by Lemma 7 X' is a continuous process. According to Problem 7 the process $X'' = I_H \cdot X \in S \cap G$, the processes $(X - X_0, X', X'')$ form a Gaussian system and hence (X_0, X', X'') is a Gaussian system too.

2) From the decomposition

$$X_t = X_0 + X'_t + X''_t$$

for X it follows already that $\Delta X = \Delta X''$, and from the definition of the stochastic integral with respect to a semimartingale that $\Delta X = I_H \Delta X$. This means that discontinuity times for trajectories of X take values in the set H that consists of a countable number of points $t_1, t_2, \ldots$ at most (cf. Ch. 6, the space D), and the probability of the event $\{\Delta X_{t_i} \neq 0\}$ equals to 0 or 1 due to the Gaussian distribution of the random variables ΔX_{t_i}. Consequently, discontinuities of trajectories of the process X are fixed and concentrated on the set H.

3) Due to Problem 1.9 it suffices to verify that $[X, X]^{1/2} \in \mathcal{A}^+_{loc}$. In accordance with Assertion 2) proved already

$$\sum_{s > 0} (\Delta X_s)^2 = \sum_{s \in H} (\Delta X_s)^2,$$

and consequently

$$[X, X] = \langle X^c \rangle + \sum_{s \in H} (\Delta X_s)^2.$$

Since $\langle X^c \rangle \in \mathcal{A}^+_{loc}$, it suffices to show that

$$\sum_{s \in H} (\Delta X_s)^2 \in \mathcal{A}^+_{loc}.$$

The last relation takes place, since by Remark 2 to Lemma 2 (Subsection 7) for each $t \in R_+$ and sufficiently small $\varepsilon > 0$ we have

$$\mathbf{E} \exp\left(\varepsilon \sum_{s \le t,\, s \in H} (\Delta X_s)^2\right) < \infty.$$

4) According to 3) we have $X \in \mathbf{Sp}$. Consequently, X admits the decomposition $X_t = X_0 + A_t + M_t$ with $A \in \mathcal{V} \cap \mathcal{P}$ and $M \in \mathcal{M}_{loc}$. Denote $A'' = I_H \circ A$ and $M'' = I_H \circ M$. Evidently

$$X''_t = A''_t + M''_t.$$

First, we will show that (X, A'') form a Gaussian system. Notice meanwhile that for each $t \in R_+$

$$A''_t = \sum_{\substack{s \le t,\\ s \in H}} \Delta A_s = \sum_{\substack{s \le t,\\ s \in H}} {}^p(\Delta X_s).$$

Since $\mathbb{F} = \mathbb{F}^X_+$, then (X, A'') form a Gaussian system in virtue of Problem 10. Consequently, (X, A'', M'') is a Gaussian system.

Let us consider now the continuous semimartingale $X' = I_{R_+ \setminus H} \cdot X$, admitting the decomposition

$$X'_t = A'_t + M'_t$$

with $A' = I_{R_+ \setminus H} \circ A$ and $M' = I_{R_+ \setminus H} \cdot M$, and show that (X, A') is a Gaussian system.

To this end it suffices to show that for each finite collection $t_1, \ldots, t_k$ the system $(A'_{t_1}, \ldots, A'_{t_k}, X)$ is Gaussian. It suffices for this, in turn, to show that for each $n \ge 1$ there exists a Gaussian system $(A^n_{t_1}, \ldots, A^n_{t_k}, X)$ such that $A^n_{t_i} \to A'_{t_i}$ in probability as $n \to \infty$, $i = 1, \ldots, k$ (cf. Lemma 4). Let T denote any of the numbers t_i, $i = 1, \ldots, k$. As for A^n_T take the random variable

$$A^n_T = \sum_{j=0}^{n-1} \mathbf{E}\left(X'_{t^n_{j+1}} - X'_{t^n_j} \mid \mathcal{F}_{t^n_j}\right)$$

with $0 \equiv t_0^n < t_1^n < \dots < t_n^n = T$. The random variables $A'_{t_i^n}$, $i = 1, \dots, k$, defined in this manner form, together with X, a Gaussian system (Problem 9). Therefore it remains to show that $A'^n_T \to A'_T$ in probability as $n \to \infty$. To this end choose a partition sequence $(t_i^n)_{0 \le i \le n}$, $n \ge 1$, in such manner that

$$\max_{0 \le j \le n-1} (t_{j+1}^n - t_j^n) \to 0$$

as $n \to \infty$. Observe that

$$A'_T = \sum_{j=0}^{n-1} (A'_{t_{j+1}^n} - A'_{t_j^n}),$$

and consequently it suffices to show that

$$\sum_{j=0}^{n-1} [(A'_{t_{j+1}^n} - A'_{t_j^n}) - \mathbf{E}(X'_{t_{j+1}^n} - X'_{t_j^n} \mid \mathcal{F}_{t_j^n})] \to \infty \tag{9.14}$$

in probability as $n \to \infty$. Since X' is a continuous process, the processes A' and M' involved in the decomposition for X, are continuous processes (cf. § 1). Assume first

$$\operatorname{Var}(A')_T + \sup_{t \le T} |M'_t| \le c.$$

Then

$$\mathbf{E}(X'_{t_{j+1}^n} - X'_{t_j^n} \mid \mathcal{F}_{t_j^n}) = \mathbf{E}(A'_{t_{j+1}^n} - A'_{t_j^n} \mid \mathcal{F}_{t_j^n})$$

and

$$\mathbf{E}\left(\sum_{j=0}^{n-1} [(A'_{t_{j+1}^n} - A'_{t_j^n}) - \mathbf{E}(X'_{t_{j+1}^n} - X'_{t_j^n} \mid \mathcal{F}_{t_j^n})]\right)^2$$

$$= \mathbf{E} \sum_{j=0}^{n-1} [(A'_{t_{j+1}^n} - A'_{t_j^n})^2 - (\mathbf{E}(A'_{t_{j+1}^n} - A'_{t_j^n} \mid \mathcal{F}_{t_j^n}))^2]$$

$$\le \mathbf{E} \sum_{j=0}^{n-1} (A'_{t_{j+1}^n} - A'_{t_j^n})^2 \le \mathbf{E} \max_{0 \le j \le n-1} |A'_{t_{j+1}^n} - A'_{t_j^n}| \operatorname{Var}(A')_T$$

$$\le c\mathbf{E} \max_{0 \le j \le n-1} |A'_{t_{j+1}^n} - A'_{t_j^n}| \to 0, \quad n \to \infty.$$

In the general case one needs to introduce a localizing sequence

$$T_k = \inf\,(t\colon \mathrm{Var}\,(A')_t + \sup_{s\le T} |M'_s| \ge k),\quad k \ge 1.$$

Then for $\varepsilon > 0$

$$\mathsf{P}\left(\left|\sum_{j=0}^{n-1}\left[(A'_{t^n_{j+1}} - A'_{t^n_j}) - \mathbf{E}\,(X'_{t^n_{j+1}} - X'_{t^n_j} \mid \mathcal{F}_{t^n_j})\right]\right| > \varepsilon\right)$$

$$\le \mathsf{P}\left(\left|\sum_{j=0}^{n-1}\left[(A'_{t^n_{j+1}\wedge T_k} - A'_{t^n_j\wedge T_k}) - \mathbf{E}\,(X'_{t^n_{j+1}\wedge T_k} - X'_{t^n_j\wedge T_k} \mid \mathcal{F}_{t^n_j})\right]\right| > \varepsilon\right)$$

$$+ \mathsf{P}\,(T_k < T)$$

$$\le \frac{k}{\varepsilon^2}\mathbf{E} \max_{0\le j\le n-1} |A'_{t^n_{j+1}\wedge T_k} - A'_{t^n_j\wedge T_k}| + \mathsf{P}\,(T_k < T) \to \mathsf{P}\,(T_k < T),\ n \to \infty,$$

and the desired relation (9.14) takes place, since $\mathsf{P}\,(T_k < T) \to 0$ as $k \to \infty$.

Thus, (X, A') is a Gaussian system. Then (X, A', M') and (X, A'', M'') are Gaussian systems too. Since $A \in \mathsf{D} \cap G$ and for each $t \in R_+$ we have $\mathsf{P}\,(\mathrm{Var}\,(A)_t < \infty) = 1$, then, in view of Remark 2 in Subsection 7, we have $\mathbf{E}\exp(\varepsilon\,(\mathrm{Var}\,(A)_t)^2) < \infty$ as $\varepsilon > 0$ is sufficiently small. This yields $\mathbf{E}\,(\mathrm{Var}\,(A)_t)^2 < \infty$. Finally, since M is a Gaussian local martingale, we have $M \in \overline{\mathcal{M}}^2$, because $\mathbf{E}M_t^2 < \infty$ for each $t \in R_+$ and by Doob's inequality $\mathbf{E}\,(M^*_t)^2 \le 4\mathbf{E}M_t^2$ (Theorem 1.9.2).

Proof of Corollary. If $X \in \mathbf{S}\,(\mathbb{F}) \cap G$, then according to Theorem 6.1 we have $X \in S\,(\mathbb{F}^X_+) \cap G$. Hence in the semimartingale decomposition $X_t = X_0 + A_t + M_t$ for $X \in S\,(\mathbb{F}^X_+) \cap G$ the process A is Gaussian with $\mathbf{E}\,(\mathrm{Var}\,(A)_t)^2 < \infty$, $t \in R_+$. Consequently, for each partition $0 = t^n_0 < t^n_1 < \ldots < t^n_n \equiv t$ of the interval $[0, t]$

$$\mathbf{E}\sum_{j=0}^{n-1} |\mathbf{E}\,(X_{t^n_{j+1}} - X_{t^n_j} \mid \mathcal{F}_{t^n_j})| \le \mathbf{E}\sum_{j=0}^{n-1} |A_{t^n_{j+1}} - A_{t^n_j}| \le \mathbf{E}\,\mathrm{Var}\,(A)_t < \infty.$$

10. A Gaussian semimartingale, as any Gaussian process, is completely specified by the mean-value function $a\,(t)$ and the correlation function $\Gamma\,(s, t)$. In this subsection we characterize a Gaussian semimartingale in terms of properties of $a\,(t)$ and $\Gamma\,(s, t)$.

We will need here the following notions. We say that a right-continuous function g = g (s, t) having left-hand limits, defined on R_+^2 and taking values in R, is a function of locally bounded variation (denoting $g \in V^2$) if for each $s, t \in R_+$ and partitions $0 = s_0 < s_1 < \ldots < s_k = s$ and $0 = t_0 < t_1 < \ldots < t_l = t$ we have

$$V_g^{(k)}(s, t) = |g(0, 0)| + \sum_i |g(s_{i+1}, 0) - g(s_i, 0)| + \sum_j |g(0, t_{j+1}) - g(0, t_j)|$$

$$+ \sum_{i, j} |\Delta g[(s_i, t_j), (s_{i+1}, t_{j+1})]| \le \text{const} < \infty \tag{9.15}$$

with

$$\Delta g[(s_1, t_1), (s_2, t_2)] = g(s_2, t_2) + g(s_1, t_1) - g(s_1, t_2) - g(s_2, t_1).$$

Total variation of a function $g \in V^2$ on $[0, s] \times [0, t]$ is denoted by $V_g(s, t)$ with

$$V_g(s, t) = \sup V_g^{(k)}(s, t) \tag{9.16}$$

where sup is taken over all partitions of the intervals [0, s] and [0, t]. Evidently, for a $B(R_+) \otimes B(R_+)$-measurable function f = f (s, t) such that

$$\int_{[0, T_1] \times [0, T_2]} |f(s, t)| \, dV_g(s, t) < \infty,$$

the Lebesgue-Stieltjes integral

$$\int_{[0, T_1] \times [0, T_2]} f(s, t) \, dg(s, t)$$

is defined.

A function $g \in V^2$ is called *nonnegative definite* if for each finite $B(R_+)$-measurable function f = f (t) for which

$$\int_{R_+^2} |f(s) f(t)| \, dV_g(s, t) < \infty$$

the inequality

$$\int_{R_+^2} F(s) f(t) \, dg(s, t) \ge 0$$

takes place.

A correlation function $\Gamma = \Gamma(s, t)$, being a function of locally bounded variation, ($\Gamma \in V^2$), possesses the property of nonnegative definiteness (Problem 11). For a stochastic process $X \in D$ with the mean-value function $a(t) \equiv 0$ and a correlation

V^2, and for a B (R_+)-measurable function $f = f(t)$ such that

$$\int_0^T \int_0^T f(s)\, f(t)\, d\Gamma(s, t) < \infty,$$

the "mean-square" integral

$$\int_0^T f(t)\, dX_t$$

is defined that is understood as the limit in the mean square sense of the sequence

$$\int_0^T f_n(t)\, dX_t, \quad n \geq 1$$

where $f_n = f_n(t)$, $n \geq 1$, are elementary functions of type

$$\sum_i c_i I_{]\!]t_i, t_{i+1}]\!]}(t),$$

if

$$\lim_n \int_0^T \int_0^T (f_n(s) - f(s))(f_n(t) - f(t))\, d\Gamma(s, t) = 0$$

(cf. [203], Ch. X, § 34.3). The integral defined in this manner possesses the following properties:

1) $$\mathbf{E} \int_0^T f(s)\, dX_s = 0; \tag{9.17}$$

2) $$\mathbf{E} \int_0^{T_1} f_1(s)\, dX_s \int_0^{T_2} f_2(t)\, dX_t = \int_0^{T_1} \int_0^{T_2} f_1(s)\, f_2(t)\, d\Gamma(s, t); \tag{9.18}$$

3) if $X \in \mathrm{D} \cap \mathbb{F}$, then one may choose a $\mathcal{F}_T$-measurable version of the random variable

$$\int_0^T f(s)\, dX_s;$$

4) if $X \in \mathrm{D} \cap \mathrm{G}$, then the random variable

$$\int_0^T f(s)\, dX_s$$

is Gaussian.

In the following theorem a Gaussian special semimartingale is characterized in terms of the mean-value function and correlation function.

Theorem 2. *Let* $X \in D \cap G$, *and let* $a = a(t)$ *and* $\Gamma = \Gamma(s, t)$ *be the mean-value function and correlation function of a process* X *respectively.*

The following conditions are equivalent:

(a) *The process* $X \in \mathbf{Sp}(\mathbb{F}^X_+) \cap G$,

(b) *The functions* $a \in V$, $\Gamma \in V^2$, *and there exists a nondecreasing right-continuous function* F *such that for each* $s < t$ *and each elementary function*

$$f_s = f_s(u), \quad u \in R_+$$

with $f_s(u) = 0$ *as* $u > s$ *the inequality*

$$\frac{\left| \int_s^t \int_0^s f_s(v)\, d\Gamma(v, u) \right|}{\sqrt{\int_0^s \int_0^s f_s(v) f_s(u)\, d\Gamma(v, u)}} \le F(t) - F(s) \tag{9.19}$$

takes place.

Proof. (a) $\Rightarrow$ (b). Let us show that $a \in V$. Suppose the contrary, that is, let $a \notin V$. Then there can be indicated an interval $[0, T]$ and a partition sequence $0 = t^n_0 < t^n_1 < \ldots < t^n_n \equiv T$ of this interval such that the sequence V_n, $n \ge 1$ with

$$V_n = \sum_{j=0}^{n-1} | a(t^n_{j+1}) - a(t^n_j) |$$

tends to infinity as $n \to \infty$. As $t^n_j < t \le t^n_{j+1}$ set

$$f_n(t) = \frac{1}{\sqrt{V_n}} \operatorname{sign}(a(t^n_{j+1}) - a(t^n_j))$$

and define the Gaussian random variables $f_n \cdot X_T$, $n \ge 1$. Let us show that

$$f_n \cdot X_T \xrightarrow{P} 0$$

as $n \to \infty$. By Theorem 1 X admits the decomposition

$$X_t = X_0 + A_t + M_t \tag{9.20}$$

with $A \in \mathcal{V} \cap \mathcal{P}$ and $M \in \overline{\mathcal{M}}^2$. Consequently

$$|f_n \cdot X_T| \le |f_n| \circ \mathrm{Var}\,(A)_T + \sup_{t \le T} |f_n \cdot M_t|.$$

Since

$$|f_n| \le \frac{1}{\sqrt{V_n}},$$

we have

$$|f_n| \circ \mathrm{Var}\,(A)_T \to 0, \ n \to \infty \ (\mathrm{P}\text{-a.s.}).$$

Next, by the Lenglart-Rebolledo inequality (Theorem 1.9.3) and by the inequality $f_n^2 \le \frac{1}{V_n}$

$$\mathrm{P}\left(\sup_{t \le T} |f_n \cdot M_t| \ge \varepsilon\right) \le \frac{\delta}{\varepsilon^2} + \mathrm{P}\,(f_n^2 \circ \langle M\rangle_T \ge \delta) \to 0$$

when the limit $\lim\limits_{\delta \to 0} \overline{\lim\limits_n}$ is taken.

Thus

$$f_n \cdot X_T \xrightarrow{\mathrm{P}} 0, \ n \to \infty.$$

Then by Lemma 4 we have

$$\lim_n \mathbf{E} f_n \cdot X_T = 0.$$

On the other hand

$$\mathbf{E} f_n \cdot X_T = \sum_{j=0}^{n-1} f_n\,(t_{j+1}^n)\,(a\,(t_{j+1}^n) - a\,(t_j^n)) = \sqrt{V_n} \to \infty, \ n \to \infty.$$

This contradiction tells us that the assumption $a \notin V$ is invalid.

Let us establish now the property $\Gamma \in V^2$.

By the property $a \in V$ proved already the process $X - a \in \mathbf{S}\,(\mathbb{F}_+^X) \cap G$. Therefore, on proving the property $\Gamma \in V^2$ one may assume $a\,(t) \equiv 0$ without loosing generality.

Observe that

$$\Gamma\,(s_{i+1}, 0) - \Gamma\,(s_i, 0) = \mathbf{E}\,(X_{s_{i+1}} - X_{s_i})\,X_0,$$

$$\Gamma\,(0, t_{j+1}) - \Gamma\,(0, t_j) = \mathbf{E} X_0\,(X_{t_{j+1}} - X_{t_j}),$$

$$\Delta\Gamma\,[(s_i, t_j), (s_{i+1}, t_{j+1})] = \mathbf{E}\,(X_{s_{i+1}} - X_{s_i})\,(X_{t_{j+1}} - X_{t_j}).$$

Therefore, on verifying (9.15) it suffices to show that

$$\sum_k |\,\mathbf{E}\,(X_{u_{k+1}} - X_{u_k})\,X_0\,| \le \text{const},$$
$$\sum_{k,\,l} |\,\mathbf{E}\,(A_{u_{k+1}} - A_{u_k})\,(A_{u_{l+1}} - A_{u_l})\,| \le \text{const},$$
$$\sum_{k,\,l} |\,\mathbf{E}\,(A_{u_{k+1}} - A_{u_k})\,(M_{u_{l+1}} - M_{u_l})\,| \le \text{const}, \tag{9.21}$$
$$\sum_{k,\,l} |\,\mathbf{E}\,(M_{u_{k+1}} - M_{u_k})\,(M_{u_{l+1}} - M_{u_l})\,| \le \text{const},$$

where (u_k) denotes any of the sequences (s_i) or (t_j). On verifying the inequalities (9.21) we use systematically the following equality

$$\mathbf{E}\,|\,\alpha\,| = \sqrt{\frac{2}{\pi}\,\mathbf{E}\alpha^2}\,,$$

valid for a Gaussian random variable α with $\mathbf{E}\alpha = 0$.

In the decomposition (9.20) A and M are Gaussian processes. Therefore

$$|\,\mathbf{E}\,(X_{u_{k+1}} - X_{u_k})\,X_0\,| = |\,\mathbf{E}\,(A_{u_{k+1}} - A_{u_k})\,X_0\,|$$
$$\le \left(\frac{\pi}{2}\,\mathbf{E}X_0^2\right)^{1/2} \mathbf{E}\,|\,A_{u_{k+1}} - A_{u_k}\,|,$$

and consequently the first inequality in (9.21) holds with the constant

$$\left(\frac{\pi}{2}\,\mathbf{E}X_0^2\right)^{1/2} \mathbf{E}\ \mathrm{Var}\,(A)_T.$$

The second inequality in (9.21) holds with the constant $\frac{\pi}{2}\,(\mathbf{E}\ \mathrm{Var}\,(A)_T)^2$, since

$$|\,\mathbf{E}\,(A_{u_{k+1}} - A_{u_k})\,(A_{u_{l+1}} - A_{u_l})\,| \le \frac{\pi}{2}\,\mathbf{E}\,|\,A_{u_{k+1}} - A_{u_k}\,|\,\mathbf{E}\,|\,A_{u_{l+1}} - A_{u_l}\,|.$$

The third inequality holds with the constant

$$\left(\frac{\pi}{2}\,\mathbf{E}M_T^2\right)^{1/2} \mathbf{E}\ \mathrm{Var}\,(A)_T.$$

To establish this fact define for each k the function

$$f_k\,(t) = \mathrm{sign}\ \mathbf{E}\,(A_{u_{k+1}} - A_{u_k})\,(M_{u_{l+1}} - M_{u_l})$$

as $u_l < t \le u_{l+1}$. Then

$$\sum_{k,\,l} |\mathbf{E}(A_{u_{k+1}} - A_{u_k})(M_{u_{l+1}} - M_{u_l})| = \sum_k \mathbf{E}(A_{u_{k+1}} - A_{u_k})(f_k \cdot M_T)$$

$$\le \sum_k \left(\frac{\pi}{2}\mathbf{E}(f_k \cdot M_T)^2\right)^{1/2} \mathbf{E}\,|A_{u_{k+1}} - A_{u_k}|$$

$$= \sum_k \left(\frac{\pi}{2}\mathbf{E}f_k^2 \circ \langle M\rangle_T\right)^{1/2} \mathbf{E}\,|A_{u_{k+1}} - A_{u_k}|,$$

and the desired estimate with the indicated constant follows from $f_k^2 \le 1$. The fourth inequality, evidently, takes place with the constant $\mathbf{E}M_T^2$.

Let us show that the inequality (9.19) holds.

Denote by $G^0(s, t)$ the correlation function of the process $X^0 = (X_t^0)_{t \ge 0}$ with $X_t^0 = X_t - X_0$. Obviously,

$$\Gamma^0(s, t) = \Gamma(s, t) - \Gamma(s, 0) - \Gamma(0, t) + \Gamma(0, 0),$$

and hence $d\Gamma^0(s, t) = d\Gamma(s, t)$.

Therefore, on verifying the inequality (9.19) one may assume $X_0 = 0$ without loosing generality.

Note that for a stepwise function $f_s(t)$ the "mean-square" integral

$$\int_0^s f_s(v)\, dX_v$$

is defined and the relations

$$\int_0^s\int_0^s f_s(v) f_s(u)\, d\Gamma(v, u) = \mathbf{E}\left(\int_0^s f_s(v)\, dX_v\right)^2,$$

$$\int_s^t\int_0^s f_s(v)\, d\Gamma(v, u) = \mathbf{E}(X_t - X_s)\int_0^s f_s(v)\, dX_v$$

are valid. Due to the construction of the "mean-square" integral and completeness of the family of σ-algebras $\mathbb{F}_+^X$ the random variable

$$\int_0^s f_s(v)\, dX_v$$

is $\mathcal{F}_s^X$-measurable. Consequently, by (9.20) and Assertion 4 of Theorem 1

$$\mathbf{E}\,(X_t - X_s) \int_0^s f_s(v)\, dX_v = \mathbf{E}\,(A_t - A_s) \int_0^s f_s(v)\, dX_v,$$

and hence

$$\frac{\left| \int_s^t \int_0^s f_s(v)\, d\Gamma(v, u) \right|}{\sqrt{\int_0^s \int_0^s f_s(v)\, f_s(u)\, d\Gamma(v, u)}} = \frac{\left| \mathbf{E}\,(A_t - A_s) \int_0^s f_s(v)\, dX_v \right|}{\sqrt{\mathbf{E}\left(\int_0^s f_s(v)\, dX_v \right)^2}}$$

$$\leq \mathbf{E}\,(A_t - A_s)^2 = \sqrt{\frac{\pi}{2}}\, \mathbf{E}\,|A_t - A_s| \leq \sqrt{\frac{\pi}{2}}\, [\mathbf{E}\, \mathrm{Var}\,(A)_t - \mathbf{E}\, \mathrm{Var}\,(A)_s],$$

i.e. the inequality (9.19) holds with

$$F(t) = \sqrt{\frac{\pi}{2}}\, \mathbf{E}\, \mathrm{Var}\,(A)_t.$$

(b) $\Rightarrow$ (a). Since $a \in \mathrm{V}$, it suffices to show that $X - a \in \mathbf{S}\,(\mathbb{F}_+^X)$. Therefore, it can be assumed at once that $a(t) \equiv 0$.

The inequality (9.19) may be rewritten in the following manner:

$$\frac{\left| \mathbf{E}\, \mathbf{E}\,(X_t - X_s \mid \mathcal{F}_s^X) \int_0^s f_s(v)\, dX_v \right|}{\sqrt{\mathbf{E}\left(\int_0^s f_s(v)\, dX_v \right)^2}} \leq F(t) - F(s). \tag{9.22}$$

It can be assumed without loosing generality that $X_0 = 0$ (P-a.s.).

Define the σ-algebras

$$\mathfrak{F}^X_{s,n} = \sigma\{X_{t_1^n} - X_{t_0^n}, \ldots, X_{t_n^n} - X_{t_{n-1}^n}\}$$

where $0 \equiv t_0^n < t_1^n < \ldots < t_n^n \equiv s$ is a partition sequence of the interval [0, s], such that

$$\lim_n \max_{0 \le j \le n-1} (t_{j+1}^n - t_j^n) = 0.$$

Clearly,

$$\mathfrak{F}^X_{s,n} \uparrow \mathfrak{F}^X_s \text{ as } n \to \infty.$$

According to the theorem on the normal correlation (cf., for instance, [188], Ch. 13, Theorem 13.1)

$$\mathbf{E}(X_t - X_s \mid \mathfrak{F}^X_{s,n}) = \int_0^s g_s(v)\, dX_v.$$

Hence, as the random variable $\mathbf{E}(X_t - X_s \mid \mathfrak{F}^X_{s,n})$ is Gaussian and $a(t) \equiv 0$, the following equality takes place:

$$\mathbf{E}\,|\,\mathbf{E}(X_t - X_s \mid \mathfrak{F}^X_{s,n})\,| = \sqrt{\frac{2}{\pi}\int_0^s\int_0^s g_s(v)\, g_s(u)\, d\Gamma(v, u)}.$$

Assume $f_s = g_s$. Then (9.22) entails

$$\mathbf{E}\,|\,\mathbf{E}(X_t - X_s \mid \mathfrak{F}^X_{s,n})\,| \le \sqrt{\frac{2}{\pi}}\,(F(t) - F(s)),$$

and hence

$$\mathbf{E}\,|\,\mathbf{E}(X_t - X_s \mid \mathfrak{F}^X_s)\,| \le \sqrt{\frac{2}{\pi}}\,(F(t) - F(s)). \tag{9.23}$$

Let us show that

$$\mathbf{E}\,|\,\mathbf{E}(X_t - X_s \mid \mathfrak{F}^X_{s+})\,| \le \sqrt{\frac{2}{\pi}}\,(F(t) - F(s)). \tag{9.24}$$

In fact, by (9.23) as $t > s$ and $\varepsilon > 0$ is sufficiently small we have

$$\mathbf{E}\,|\,\mathbf{E}(X_t - X_{s+\varepsilon} \mid \mathfrak{F}^X_{s+\varepsilon})\,| \le \sqrt{\frac{2}{\pi}}\,(F(t) - F(s+\varepsilon)) \le \sqrt{\frac{2}{\pi}}\,(F(t) - F(s)).$$

This gives the following inequality:

$$\mathbf{E}\,|\,\mathbf{E}(X_t - X_s \mid \mathfrak{F}^X_{s+\varepsilon})\,| \le \sqrt{\frac{2}{\pi}}\,(F(t) - F(s)) + \mathbf{E}\,|\,X_{s+\varepsilon} - X_s\,|.$$

By taking the limit as $\varepsilon \downarrow 0$ in this inequality we arrive at (9.24), since

$$X_{s+\varepsilon} - X_s \to 0$$

(the function X_t, $t \geq 0$, is right-continuous),

$$\mathbf{E}\,(X_t - X_s \mid \mathcal{F}^X_{s+\varepsilon}) \to \mathbf{E}\,(X_t - X_s \mid \mathcal{F}^X_{s+})$$

and $X_{s+\varepsilon} - X_s$ and $\mathbf{E}\,(X_t - X_s \mid \mathcal{F}^X_{s+\varepsilon})$ are Gaussian random variables (hence the corollary to Lemma 4 is applicable).

Therefore for each partition $0 = t_0 < t_1 < \ldots < t_n$ of the interval $[0, T]$

$$\sum_i \mathbf{E}\,|\,\mathbf{E}\,(X_{t_{i+1}} - X_{t_i} \mid \mathcal{F}^X_{t_i+})\,| \leq \sqrt{\frac{2}{\pi}}\,(F(T) - F(0)),$$

i.e. X is a local quasimartingale. Then by Theorem 2.1.4 we have

$$X \in \mathbf{Sp}\,(\mathbb{F}^X_+).$$

11. Example 1. Let the process $X = (X_t)_{t \geq 0} \in \mathrm{D} \cap \mathrm{G}$ be a Markov process with respect to the family $\mathbb{F}^X_+$. In order that $X \in \mathbf{Sp}\,(\mathbb{F}^X_+)$ it is necessary and sufficient that the following conditions hold: $a \in \mathrm{V}$, $\Gamma \in \mathrm{V}^2$ and for each $s < t$ there exists a nondecreasing right-continuous function $F(t)$ such that

$$\frac{|\,\Gamma(s, t) - \Gamma(s, s)\,|}{\sqrt{\Gamma(s, s)}} \leq F(t) - F(s) \tag{9.25}$$

where $\frac{0}{0} = 0$.

To prove this it suffices to show, according to Theorem 2, that (in the Markovian case) Conditions (9.19) and (9.25) are equivalent.

Assume $a(t) \equiv 0$ and $X_0 = 0$ (P-a.s.). Due to the Markovian property of the process X and the theorem on the normal correlation ([188], Ch. 13, Theorem 13.1)

$$\mathbf{E}\,(X_t - X_s \mid \mathcal{F}^X_{s+}) = \mathbf{E}\,(X_t - X_s \mid X_s) = \frac{\Gamma(s, t) - \Gamma(s, s)}{\Gamma(s, s)}\,X_s \quad \text{(P-a.s.)},$$

where $\frac{0}{0} = 0$. Observe that the inequality (9.19) has, along with (9.22), also the following representation:

$$\frac{\left| \mathbf{E}\left(\mathbf{E}\,(X_t - X_s \mid \mathfrak{F}^X_{s+}) \int_0^s f_s\,(v)\,dX_v \right) \right|}{\sqrt{\mathbf{E}\left(\int_0^s f_s\,(v)\,dX_v \right)^2}} \le F\,(t) - F\,(s).$$

Therefore, in the Markovian case the inequality (9.19) is equivalent to the inequality

$$\frac{|\,\Gamma\,(s,t) - \Gamma\,(s,s)\,| \left| \mathbf{E}\left(\frac{X_s}{\sqrt{\Gamma\,(s,s)}} \int_0^s f_s\,(v)\,dX_v \right) \right|}{\sqrt{\Gamma\,(s,s)}\sqrt{\mathbf{E}\left(\int_0^s f_s\,(v)\,dX_v \right)^2}} \le F\,(t) - F\,(s).$$

The equivalence of (9.19) and (9.25) is a consequence of the inequality

$$\frac{\left| \mathbf{E} \frac{X_s}{\sqrt{\Gamma\,(s,s)}} \int_0^s f_s\,(v)\,dX_v \right|}{\sqrt{\mathbf{E}\left(\int_0^s f_s\,(v)\,dX_v \right)^2}} \le 1,$$

which turns into an equality as $f_s\,(v) = I\,(s \ge v)$.

12. Let $a\,(t) \equiv 0$ and $F \in V^+$. We will show that, if for each $s < t$

$$\mathbf{E}\,|\,X_t - X_s\,| \le F\,(t) - F\,(s), \tag{9.26}$$

then the function $\Gamma \in V^2$ and the inequality (9.19) hold.

To verify the property $\Gamma \in V^2$ let us establish (9.15). Clearly,

$$|\Gamma(t,0)-\Gamma(s,0)| = |\mathbf{E}(X_t - X_s)X_0| \le \sqrt{\frac{\pi}{2}\mathbf{E}X_0^2}\,\mathbf{E}|X_t - X_s|$$
$$\le \sqrt{\frac{\pi}{2}\mathbf{E}X_0^2}\,(F(t)-F(s))$$

and the analogous estimate holds for $|\Gamma(0,t)-\Gamma(0,s)|$. Next

$$\Gamma(s_2,t_2)+\Gamma(s_1,t_1)-\Gamma(s_1,t_2)-\Gamma(s_2,t_1)| = \mathbf{E}(X_{t_2}-X_{s_2})(X_{t_1}-X_{s_1})|$$

$$\le \frac{\pi}{2}\mathbf{E}|X_{t_2}-X_{s_2}|\,\mathbf{E}|X_{t_1}-X_{s_1}| \le \frac{\pi}{2}(F(t_2)-F(s_2))(F(t_1)-F(s_1)).$$

The relation (9.15) follows from the obtained estimates in an obvious manner. To establish (9.19) it suffices to observe that the left-hand side of (9.16) coincides with the variable

$$c(s,t) = \frac{\left|\mathbf{E}\left(\mathbf{E}(X_t - X_s \mid \mathcal{F}^X_{s+})\int_0^s f_s(v)\,dX_v\right)\right|}{\sqrt{\mathbf{E}\left(\int_0^s f_s(v)\,dX_v\right)^2}},$$

for which, by the Cauchy-Bunjanovski inequality, the estimate

$$c(s,t) \le \sqrt{\mathbf{E}(\mathbf{E}(X_t - X_s \mid \mathcal{F}^X_{s+}))^2} \le \sqrt{\mathbf{E}(X_t - X_s)^2}$$
$$= \sqrt{\frac{\pi}{2}}\,\mathbf{E}|X_t - X_s|$$
$$\le \sqrt{\frac{\pi}{2}}\,(F(t)-F(s))$$

holds.

Consequently, under Condition (9.26) and Condition $a \in V$ the process $X \in D \cap G$ is a special semimartingale. In the situation under consideration, this process $X \in D \cap G$ may be characterized in more details.

The Gaussian process X with trajectories in D of locally bounded variation (in general, we do not assume $X_0 = 0$) will be denoted by $X - X_0 \in \mathcal{V} \cap G$.

Theorem 3. *Let* $X \in D \cap G$ *and let* a *and* Γ *be the mean-value function and the correlation function of the process* X *respectively. The following conditions are equivalent:*

(a) $X - X_0 \in \mathcal{V} \cap G$,

(b) $a \in V$ *and there exists a nondecreasing right-continuous function* F *such that for each* $s \le t$

$$(\Gamma(t, t) + \Gamma(s, s) - 2\Gamma(s, t))^{1/2} \le F(t) - F(s). \tag{9.27}$$

Proof. (a) $\Rightarrow$ (b). Since $X - X_0 \in \mathcal{V} \cap G$, then $X \in \mathbf{S}(\mathbb{F}^X_+)$, and hence by Theorem 1 we have $X \in \mathbf{Sp}(\mathbb{F}^X_+) \cap G$. Therefore by Theorem 2 we have $a \in V$. Consequently $X - a \in \mathbf{Sp}(\mathbb{F}^X_+) \cap G$, furthermore $X - a - (X - a)_0 \in \mathcal{V} \cap G$. In accordance with Remark 2 in Subsection 7 of the present section $\mathbf{E}\,\mathrm{Var}\,(X - a)_t < \infty$. Set

$$F(t) = \sqrt{\frac{\pi}{2}}\,\mathbf{E}\,\mathrm{Var}\,(X - a)_t.$$

The inequality (9.27) holds with this function F (t), since the left-hand side of the inequality (9.27) coincides with

$$(\mathbf{E}\,[(X_t - a_t) - (X_s - a_s)]^2)^{1/2} \le \sqrt{\frac{\pi}{2}}\,\mathbf{E}\,|(X_t - a_t) - (X_s - a_s)|$$

$$\le F(t) - F(s).$$

(b) $\Rightarrow$ (a). Since $a \in V$, it suffices to show that $X - a - (X_0 - a_0) \in \mathcal{V}$. It may be assumed, without loosing generality, that $a = 0$. Let us show that $\mathbf{E}\,\mathrm{Var}\,(X)_T < \infty$ for each T in R_+. By definition

$$\mathrm{Var}\,(X)_T = \sup \sum_j |X_{t_{j+1}} - X_{t_j}|$$

where sup is taken over all finite partitions of the interval [0, T]. Naturally, only a sequence of nested partitions may be considered. If n is the number of a sequence and t^n_j of partition points, then for nested sequences

$$\sum_{j=0}^{n-1} |X_{t^n_{j+1}} - X_{t^n_j}| \le \sum_{j=0}^{n} |X_{t^{n+1}_{j+1}} - X_{t^{n+1}_j}| \tag{9.28}$$

and consequently

$$\mathrm{Var}\,(X)_T = \lim_n \sum_{j=0}^{n-1} |X_{t^n_{j+1}} - X_{t^n_j}|.$$

Therefore by (9.27) and (9.28) we get

$$E\,\mathrm{Var}\,(X)_T = \lim_n \sum_{j=0}^{n-1} E\,|\,X_{t^n_{j+1}} - X_{t^n_j}\,| \le \sqrt{\frac{2}{\pi}}\,(F(T) - F(0)).$$

13. We present a device allowing in certain cases to establish that a continuous Gaussian process is not a semimartingale.

Theorem 4. *Let* $X \in D \cap G$ *and* a *and* Γ *its mean-value and correlation functions such that* $a \in V$ *and* $\Gamma \in V^2$.

If for Γ *Condition* (9.27) *fails to hold and if for each* $T \in R_+$ *and for each partition sequence* $0 \equiv t^n_0 < t^n_1 < \dots < t^n_n \equiv T$, $n \ge 1$, *of the interval* $[0, T]$ with

$$\max_{0 \le j \le n-1} (t^n_{j+1} - t^n_j) \to 0$$

as $n \to \infty$

$$\sum_{j=0}^{n-1} (X_{t^n_{j+1}} - X_{t^n_j})^2 \to 0, \; n \to \infty \tag{9.29}$$

in probability, then X *is a continuous Gaussian process that is not a semimartingale.*

Proof. The process X is continuous by Lemma 6. Assume $X \in S$. Then by Theorem 1 we have $X \in \mathbf{Sp}$, and hence X has the semimartingale decomposition

$$X_t = X_0 + A_t + M_t \tag{9.30}$$

with $A \in \mathcal{V} \cap \mathcal{P}$ and $M \in \mathcal{M}_{loc}$. Besides, A and M are continuous processes (Theorem 2.1.2). Let us show that under Condition (9.29)

$$\langle M \rangle_\infty = 0 \;\; (P\text{-a.s.}). \tag{9.31}$$

In view of (9.30) and Ito's formula (Ch. 2, § 3) for each $s < t$ we have

$$(X_t - X_s)^2 = 2\int_s^t (X_u - X_s)\,dA_u + 2\int_s^t (X_u - X_s)\,dM_s + \langle M \rangle_t - \langle M \rangle_s.$$

This gives

$$\langle M \rangle_t - \langle M \rangle_s \le (X_t - X_s)^2 + 2 \sup_{s \le u \le t} |X_u - X_s|\,[\mathrm{Var}\,(A)_t - \mathrm{Var}\,(A)_s] + 2\left|\int_s^t (X_u - X_s)\,dM_u\right|.$$

Therefore, the partition $0 \equiv t^n_0 < \dots < t^n_n \equiv T$ of the interval $[0, T]$ yields

$$\langle M\rangle_T \le \sum_{j=0}^{n-1} (X_{t^n_{j+1}} - X_{t^n_j})^2 + 2 \max_{0\le j\le n-1} \sup_{t^n_j \le u \le t^n_{j+1}} |X_u - X_{t^n_j}| \operatorname{Var}(A)_T$$

$$+ \sup_{0\le t\le T} |f_n \cdot M_t| \qquad (9.32)$$

with $f_n(u) = 2(X_u - X_{t^n_j})$, $t^n_j \le u < t^n_{j+1}$. The first term on the right-hand side of (9.32) converges to zero in probability by Condition (9.29), the second one by the continuity of the process X and the third one by the Lenglart-Rebolledo inequality (Theorem 1.9.3), in view of which

$$P\left(\sup_{t\le T} |f_n \cdot M_t| \ge \varepsilon\right) \le \frac{\delta}{\varepsilon^2} + P(f_n^2 \circ \langle M\rangle_T \ge \delta), \quad \varepsilon, \delta > 0,$$

with

$$\lim_n P(f_n^2 \circ \langle M\rangle_T \ge \delta) = 0$$

for each $\delta > 0$ due to the continuity of the process X. By this for each $T \in R_+$ we have $\langle M\rangle_T = 0$. Hence (9.31) holds.

By (9.31) and Theorem 1.9.5 we get

$$EM^*_\infty \le 3E\langle M\rangle_\infty = 0.$$

Consequently $M = 0$ and $X \in \mathcal{V} \cap G$, which is a contradiction since the correlation function Γ of the process X does not satisfy Condition (9.27) (Theorem 3).

The obtained contradiction indicates that $X \notin S$.

Let us show how Theorem 4 is applied in one concrete case.

Example 2. Let X be a Gaussian process with trajectories in D, $a(t) \equiv 0$ and

$$\Gamma(s, t) = \frac{1}{2}(t^\alpha + s^\alpha - |t - s|^\alpha), \quad 1 < \alpha < 2$$

(Problem 14). Condition (9.27) fails because as $t > s$

$$(\Gamma(t, t) + \Gamma(s, s) - 2\Gamma(s, t))^{1/2} = |t - s|^{\alpha/2}.$$

Next, as $t > s$,

$$E(X_t - X_s)^2 = |t - s|^\alpha$$

and Condition (9.29) is satisfied since

$$\sum_{j=0}^{n-1} E(X_{t^n_{j+1}} - X_{t^n_j})^2 = \max_{0\le j\le n-1} (t^n_{j+1} - t^n_j)^{\alpha-1}\, T \to 0, \ n \to \infty.$$

Thus, a Gaussian process X with this correlation function is not a semimartingale.

14. Let us consider certain properties of a process X that is a Gaussian local martingale ($X \in \mathcal{M}_{loc} \cap G$).

Theorem 5. *Let* $X \in \mathcal{M}_{loc,\,0} \cap G$. *Then the following properties take place:*

1) $X \in \mathcal{M}^2_{loc}$, $EX_t \equiv 0$;

2) X *is a process with independent increments ,* $EX_t^2 = \langle X\rangle_t$ *and*

$$\Gamma(s, t) = (EX_s^2) \wedge (EX_t^2);$$

3) *the local martingales* X^c *and* X^d, *involved in the decomposition* $X = X^c + X^d$, *form a Gaussian system, besides*

$$\int_{R_0} e^{i\lambda x} \nu(\{t\}, dx) = \exp\left\{ -\frac{\lambda^2 \Delta(EX_t^2)}{2} \right\}, \quad \lambda \in R, \quad t \in H$$

with $H = \{t > 0:\ \Delta(EX_t^2) > 0\}$;

4) *the triplet* (B, C, ν) *of predictable characteristics corresponding to the canonical representation for X has the form*

$$B_t \equiv 0, \quad C_t = (EX_t^2)^c, \quad \nu^c(R_+ \times R_0) = 0.$$

Proof. 1) Since X_t is a Gaussian random variable, we have $EX_t^2 < \infty$. Therefore $X \in \mathcal{M}^2_{loc}$ and by Doob's inequality (Theorem 1.9.2)

$$E \sup_{s \le t} X_s^2 \le 4EX_t^2 < \infty,$$

i.e., as for localizing Markov times one may take times $\tau_n = n$, $n \ge 1$. Consequently, for each $s < t$

$$E(X_t \mid \mathcal{F}_s) = X_s. \tag{9.33}$$

Due to the condition $X_0 = 0$ (P-a.s.) this gives $EX_t = 0$.

2) From (9.33) it follows also that X is a process with uncorrelated increments:

$$E(X_{t_1} - X_{s_1})(X_{t_2} - X_{s_2}) = 0, \quad [s_1\, t_1] \cap [s_2, t_2] = \varnothing.$$

Then by Problem 2 X is a process with independent increments.

According to Problem 3.5.2

$$\langle X\rangle = C + x^2 * \nu,$$

where C and ν are elements of the triplet (B, C, ν) of predictable characteristics of the process X. Since X is a process with independent increments, its triplet (B, C, ν) is deterministic (Corollary 1 to Theorem 4.1). Therefore $\langle X\rangle$ is a deterministic function.

The definition of $\langle X\rangle$ (Ch. 1, § 8) yields $Y = X^2 - \langle X\rangle \in \mathcal{M}_{\text{loc}}$. Besides, since a random variable X_t has a Gaussian distribution, we have

$$\mathbf{E}Y_t^2 \le 2\,(\mathbf{E}X_t^4 + \langle X\rangle_t^2) < \infty,$$

i.e. $Y \in \mathcal{M}^2_{\text{loc}}$. By Doob's theorem 1.9.2 we have $\mathbf{E}\sup_{s\le t} Y_s^2 \le 4\mathbf{E}Y_s^2 < \infty$, and hence the sequence $t_n = n$, $n \ge 1$, may serve as a localizing sequence for Y. Then $\mathbf{E}Y_t = 0$ and the desired equality $\mathbf{E}X_t^2 = \langle X\rangle_t$ holds.

The representation for the correlation function $\Gamma(s, t)$ follows from the equality $\mathbf{E}X_t = 0$ as the increments of the process X are uncorrelated.

3) Since ΔX_t is a Gaussian random variable, the set $\{\Delta X_t \ne 0\}$ has probability one if and only if $\mathbf{E}\,(\Delta X_t)^2 > 0$. But $\mathbf{E}\,(\Delta X_t)^2 = \Delta\,(\mathbf{E}X_t^2)$ as the increments of the process X are uncorrelated. Therefore the sets $\{t > 0\colon \mathsf{P}\,(\Delta X_t \ne 0) = 1\}$ and $\{t > 0\colon \Delta\,(\mathbf{E}X_t^2) > 0\}$ coincide and the assertion of this part of the theorem follows from Theorem 1.

4) Jump times of X are fixed and belong to the set

$$H = \{t > 0\colon \Delta\mathbf{E}\,(X_t^2) > 0\}.$$

Therefore $\nu^c(R^+ \times R_0) = 0$. By independence of the increments of the process X and by Problem 3.2.5

$$\int_{R_0} e^{i\lambda x}\,\nu(\{t\}, dx) = \mathbf{E}e^{i\lambda\Delta X_t}, \quad t \in H,$$

and consequently $\nu(\{t\}, dx)$ possesses the indicated property, equivalent to

$$\nu(\{t\}, dx) = \frac{1}{\sqrt{2\pi\,\Delta\mathbf{E}\,(X_t^2)}}\exp\left(-\frac{x^2}{2\Delta\,(\mathbf{E}X_t^2)}\right)dx$$

as $t \in H$. This gives

$$x^2 * \nu_t = \sum_{s\le t}\Delta\,(\mathbf{E}X_s^2) = \sum_{s\le t}\Delta\langle X\rangle_s,$$

i.e.

$$C_t = \langle X\rangle_t - \sum_{s\le t}\Delta\langle X\rangle_s = \langle X\rangle_t^c.$$

Finally

$$B_t = -I\,(|x| > 1)\,x * \nu_t = 0, \quad t \ge 0.$$

The following result is a generalization of Levi's theorem (cf., for instance, [188], Ch. 4, Theorem 4.1) on a martingale characterization of a Wiener process.

Theorem 6. *Let* $X \in S(\mathbb{F})$, $X_0 = 0$ *and a* σ*-algebra* $\mathcal{G} \subseteq \mathcal{F}_0$. *Let the triplet* $T = (B, C, \nu)$ *of predictable characteristics, corresponding to the canonical representation of* X, *be* $\mathcal{G}$*-measurable, possessing the following properties* (ν^c *and* a_t *are defined by the formula* (1.13)):

$$B = 0, \quad \nu^c(R_+ \times R_0) = 0,$$

$$\int_{R_0} e^{i\lambda x} \nu(\{t\}, dx) = I(a_t > 0)\, e^{-\frac{\lambda^2}{2}\Delta_t}, \quad \lambda \in R,$$

with

$$\Delta_t = \int_{R_0} x^2 \nu(\{t\}, dx).$$

Then

$$X \in \mathcal{M}^2_{loc}, \quad \langle X \rangle_t = C_t + \sum_{s \le t} \Delta_s$$

and X *is a process with* $\mathcal{G}$*-conditionally independent and* $\mathcal{G}$*-conditionally Gaussian increments:*

$$E(e^{i\lambda(X_t - X_s)} \mid \mathcal{F}_s) = E(e^{i\lambda(X_t - X_s)} \mid \mathcal{G}) = e^{-\frac{\lambda^2}{2}(\langle X \rangle_t - \langle X \rangle_s)} \qquad \text{(P-a.s.)}$$

for each $s \le t$ *and* $\lambda \in R$.

Proof. By Theorem 4.1 X is a process with $\mathcal{G}$-conditionally independent increments and

$$E(e^{i\lambda(X_t - X_s)} \mid \mathcal{F}_s) = E(e^{i\lambda(X_t - X_s)} \mid \mathcal{G}) = \mathcal{E}^t_s(G(\lambda)), \quad \lambda \in R$$

with

$$\mathcal{E}^t_s(G(\lambda)) = e^{(G_t(\lambda) - G_s(\lambda))} \prod_{s < u \le t} (1 + \Delta G_u(\lambda))\, e^{-\Delta G_u(\lambda)},$$

$$G_t(\lambda) = i\lambda B_t - \frac{\lambda^2}{2} C_t + \int_0^t \int_{R_0} (e^{i\lambda x} - 1 - i\lambda x I(|x| \le 1))\, d\nu.$$

Due to the indicated properties of the triplet $T = (B, C, \nu)$ we have

$$G_t(\lambda) = -\frac{\lambda^2}{2} C_t + \sum_{s \le t} (e^{-\frac{\lambda^2}{2}\Delta_s} - 1).$$

Consequently, for $\mathcal{E}_s^t(G(\lambda))$ the following representation takes place:

$$\mathcal{E}_s^t(G(\lambda)) = \exp\left(-\frac{\lambda^2}{2}\left[C_t - C_s + \sum_{s<u\leq t} \Delta_u\right]\right).$$

As s = 0 this gives

$$\sum_{0<u\leq t} \Delta_u < \infty \quad (\text{P-a.s.}),$$

since the stochastic exponential $\mathcal{E}(G(\lambda))$ turns into zero over the interval [0, t] only on account of $\Delta G_u(\lambda) = -1$, $u \in [0, t]$ (Lemma 2.2). Therefore by the properties of the compensator ν we get

$$x^2 * \nu_t = \sum_{0<s\leq t} \Delta_s < \infty \quad (\text{P-a.s.}).$$

Consequently $|x| I(|x| > 1) * \nu_t < \infty$ and $xI(|x| > 1) * \nu_t = 0$, $t \in R_+$.

Therefore the canonical representation for X may be presented in the following form

$$X_t = X_t^c + x * (\mu - \nu)_t$$

where X^c is the continuous martingale component of the process X, while μ is the measure of its jumps. It can be deduced from this representation that $X \in \mathcal{M}_{loc}$, and from the property $x^2 * \nu_t < \infty$ (P-a.s.), $t \in R_+$, that $X \in \mathcal{M}^2_{loc}$ (Theorem 3.5.1). Besides

$$\langle X\rangle_t = C_t + \sum_{s\leq t} \Delta_s$$

(Problem 3.5.2), i.e.

$$\mathcal{E}_s^t(G(\lambda)) = e^{-\frac{\lambda^2}{2}(\langle X\rangle_t - \langle X\rangle_s)}.$$

Corollary. *If* $\mathcal{G} = \{\varnothing, \Omega\}$ *is a trivial* σ*-algebra, then* X *is a Gaussian martingale with* $\langle X\rangle_t = \mathbf{E}X_t^2$.

Problems

1. Let ξ and η be independent Gaussian vectors. Show that (ξ, η) is a Gaussian system.

2. Let vectors ξ and η in a Gaussian system be uncorrelated, i.e.

$$\mathbf{E}\,(\xi_i - \mathbf{E}\,\xi_i)\,(\eta_j - \mathbf{E}\,\eta_j) = 0, \quad i, j \geq 1.$$

Show that ξ and η are independent vectors.

3. Let (ξ, η) be a Gaussian system, $\mathcal{F}^{\eta} = \sigma(\eta)$ a σ-algebra generated by a random vector η and $\hat{\xi}$ a vector with the components $\hat{\xi}_i = \mathbf{E}\,(\xi_i \mid \mathcal{F}^{\eta})$. Show that $(\hat{\xi}, \xi, \eta)$ is a Gaussian system.

4. Let ξ be a Gaussian vector, a a vector of its mean-values and F a linear subspace of R^{∞}. Show that as $\mathsf{P}\,(\xi \in F) = 1$ a vector $a \in F$.

5. Let ξ be a Gaussian vector with a nonzero vector of mean-values. Show that under the condition $\mathsf{P}\,(N\,(\xi) < \infty) = 1$ where $N\,(\cdot)$ is a pseudonorm, one may choose a positive number ε such that

$$\mathbf{E} \exp\,(\varepsilon\, N^2\,(\xi)) < \infty.$$

6. Let $(a_n)_{n \geq 1}$ be a sequence of numbers and for each $\lambda \in R$ let the limit $\lim_n e^{i\lambda a_n}$ exist. Show that there exists the limit

$$\lim_n a_n \;(= a), \; |\, a \,| < \infty.$$

7. Let $X \in \mathbf{S} \cap G$, let f and g be bounded $B\,(R_+)$-measurable functions, $X' = f \cdot X$ and $X'' = g \cdot X$. Show that $X' \in \mathbf{S} \cap G$, $X'' \in \mathbf{S} \cap G$ and the processes (X', X'') form a Gaussian system.

8. Let $X \in \mathbf{S}$. If for each bounded and $B\,(R_+)$-measurable function f and for each $T \in R_+$ the random variable $f \cdot X_T$ is Gaussian, then $X \in \mathbf{S} \cap G$.

9. Let $X \in \mathsf{D} \cap G$ and let ξ be a random variable such that (X, ξ) is a Gaussian system, and let $\mathcal{F}^X_t$ for $t \in R_+$ be the σ-algebra generated by the random variables $(X_s, 0 \leq s \leq t)$. Show that

$$(X, \mathbf{E}\,(\xi \mid \mathcal{F}^X_{t_i}), \quad i = 1, \ldots, k)$$

is a Gaussian system.

10 Let $X \in \mathbf{S} \cap G$ and $\mathbb{F} = \mathbb{F}^X_+$. Show that

$$(X, {}^{P}(\Delta X)_s, \; s \in \{t > 0\colon \mathsf{P}\,(\Delta X_t \neq 0) = 1\})$$

is a Gaussian system.

11. If $\Gamma = \Gamma\,(s, t)$ is a correlation function and $\Gamma \in \mathsf{V}^2$, then for each $B\,(R_+)$-measurable function $f = f\,(t)$ and each $T \in R_+$ for which

$$\int_{[0,T]^2} |f(s) f(t)| \, dV_\Gamma(s,t) < \infty,$$

the inequality

$$\int_{[0,T]^2} f(s) f(t) \, d\Gamma(s,t) \geq 0$$

holds.

12. If a deterministic function $m \in V$ and $X \in D \cap G$, then the processes $X_- \circ m$, $X \circ m$ form a Gaussian system.

13. Let $W = (W_t)_{t \geq 0}$ be a Wiener process, and $X_t = W_t - W_{t/2}$. Then $\mathbb{F}^X = \mathbb{F}^W$ and X is not a semimartingale relative to $\mathbb{F}^W$.

14. Show that the function

$$F(s,t) = \frac{1}{2}(t^\alpha + s^\alpha - |t - s|^\alpha), \quad 1 < \alpha < 2,$$

is positive definite.

15. Let $X \in \mathcal{M}^2_{loc}$ be a process with $\mathcal{G}$-conditionally independent and $\mathcal{G}$-conditionally Gaussian increments. Besides $\langle X \rangle_t = \eta A_t$ where η is a $\mathcal{G}$-measurable nonnegative random variable, $A = (A_t)_{t \geq 0}$ a deterministic function in V^+. Show that for $\eta > 0$ X admits the representation

$$X_t = \sqrt{\eta}\, Y_t$$

with a Gaussian martingale $Y = (Y_t)_{t \geq 0}$ such that $\langle Y \rangle_t = A_t$ is independent of η (the assertion remains valid for $\eta \geq 0$ if the initial probability space is sufficiently rich).

§ 10. Filtration of special semimartingales

1. Let $(X, Y) = (X_t, Y_t)_{t \geq 0}$ be a partially observable stochastic process where Y is an unobservable component, while X is a process that is observed. The problem of filtering out the process Y by means of the process X is understood as constructing for each $t \geq 0$ an "optimal" (in one or another sense) estimate of variables Y_t by observations on X_s, $s \leq t$. Based on previous results, in the present section the problem will be studied of filtering out a semimartingale Y by means of a semimartingale X.

Thus, let $(\Omega, \mathcal{F}, \mathbb{F}, P)$ be a stochastic basis and $Y \in \mathbf{Sp}(\mathbb{F}, P)$, $X \in S(\mathbb{F}, P)$. It is useful to define the process X by its canonical representation

$$X_t = X_0 + B_t + X_t^c + \int_0^t \int_{|x| \leq 1} x \, d(\mu - \nu) + \int_0^t \int_{|x| > 1} x \, d\mu. \tag{10.1}$$

Let $T = (B, C, \nu)$ be the triplet of predictable characteristics of the process X. Assume that the process Y is a special semimartingale

$$Y_t = Y_0 + A_t + M_t \tag{10.2}$$

with $A = (A_t)_{t \geq 0} \in \mathcal{A}_{loc} \cap \mathcal{P}(\mathbb{F})$ and $M = (M_t)_{t \geq 0} \in \mathcal{M}_{loc, 0}(\mathbb{F})$. Assume

$$\begin{aligned} &E|Y_0| < \infty, \quad E \operatorname{Var}(A)_\infty < \infty, \\ &M \in \mathcal{H}(\mathbb{F}), \end{aligned} \tag{10.3}$$

and denote by $\mathbb{F}_+^X = (\mathcal{F}_{t+}^X)_{t \geq 0}$ the family of σ-algebras $\mathcal{F}_{t+}^X = \bigcap_{\varepsilon > 0} \mathcal{F}_{t+\varepsilon}^X$ where

$$\mathcal{F}_t^X = \sigma\{X_s, s \leq t\} \vee \mathcal{N}$$

and $\mathcal{N}$ is the family of sets from $\mathcal{F}$ of P-measure zero.

Denote by $\pi(Y) = (\pi_t(Y))_{t \geq 0}$ the optional projection of Y with respect to $\mathbb{F}_+^X$.

Analogously to proving Theorem 6.2, it can be established that $\pi(Y) \in \mathbf{Sp}(\mathbb{F}_+^X, P)$, i.e.

$$\pi_t(Y) = \pi_0(Y) + A_t^X + M_t^X, \tag{10.4}$$

with $A^X \in \mathcal{A}_{loc} \cap \mathcal{P}(\mathbb{F}_+^X)$ and $M^X \in \mathcal{M}_{loc}(\mathbb{F}_+^X)$.

If instead of (10.3) the conditions

$$\begin{aligned} &EY_0^2 < \infty, \quad E(\operatorname{Var}(A)_\infty)^2 < \infty, \\ &M \in \mathcal{H}^2(\mathbb{F}) \end{aligned} \tag{10.5}$$

are fullfilled, then evidently $\mathbf{E}\sup_{t\geq 0} Y_t^2 < \infty$ and for each $\tau \in \mathbf{T}(\mathbb{F}_+^X)$

$$\mathbf{E}\pi_\tau^2(Y) = \mathbf{E}(\mathbf{E}(Y_\tau \mid \mathcal{F}_{\tau+}^X))^2 \leq \mathbf{E}Y_\tau^2 < \infty.$$

Therefore, in this case the process $\pi(Y)$ may be interpreted as the filtration estimate of the process Y by observations on X, optimal in the mean square sense. More specifically, $\pi_t(Y)$ is the optimal in the mean square sense estimate of Y by observations on $\bigcap_{\varepsilon>0}\{X_s,\ 0 \leq s \leq t+\varepsilon\}$. From the point of view of applications it is more natural to assume the flow of observations $\mathbb{F}^X = (\mathcal{F}_t^X)_{t\geq 0}$ is given, instead of $\mathbb{F}_+^X$.

If $\mathbb{F}_+^X = \mathbb{F}^X$, then clearly these cases become equivalent. It will be shown below that in fact under certain additional conditions the flows $\mathbb{F}^X$ and $\mathbb{F}_+^X$ coincide. Also, we will clarify the structure of the processes A^X and M^X (see (10.4)), determining the filtration estimate $\pi_t(Y)$.

2. We begin with formulating the conditions under which the procedure determining the process $\pi(Y)$ will be worked out.

The group of Conditions (A):

(A_1) $\mathbf{E}Y_0^2 < \infty$,

(A_2) $\mathbf{E}(\mathrm{Var}(A)_t)^2 < \infty,\ t \geq 0$,

(A_3) $\mathbf{E}M_t^2 < \infty,\ t \geq 0.$

By Theorem 6.1

$$X \in \mathbf{S}(\mathbb{F}_+^X, P).$$

Therefore

$$X \in \mathbf{S}(\mathbb{F}_+^X, P^X)$$

where P^X is the restriction of the measure P to the σ-algebra

$$\mathcal{F}_\infty^X = \sigma(\bigcup_{t\geq 0} \mathcal{F}_t^X).$$

The triplet of a semimartingale $X \in \mathbf{S}(\mathbb{F}_+^X, P^X)$ is denoted by $T^X = (B^X, C^X, \nu^X)$. Assume $a_t^X = \nu(\{t\} \times R_0)$. If $\tilde{P}^X$ is a probability measure on $(\Omega, \mathcal{F}_\infty^X)$ and $\tilde{P}^X \ll P^X$, then by Theorem 5.3 we have

$$X \in S(\mathbb{F}_+^X, \tilde{P}^X)$$

(the triplet of this semimartingale is denoted by $\tilde{T}^X$). Define the family of probability measures on $(\Omega, \mathcal{F}_\infty^X)$:

$$\mathfrak{L}_{P^X} = \{\tilde{P}^X : \tilde{P}^X \ll P^X,\ \tilde{P}_0^X = P_0^X,\ \tilde{T}^{X\, P^X} = T^X\}$$

(for more details see § 8, Subsection 5).

The group of Conditions (B):

(B_1) $\mathfrak{L}_{P^X} = \{P^X\}$, i.e. $\mathfrak{L}_{P^X}$ consists of a single point P^X,

(B_2) $I(|x| \le 1)\, x * (\nu - \nu^X)^c \in \mathcal{A}_{loc}$,

(B_3) there exists a $\mathcal{P}(\mathbb{F})$-measurable function $\phi = \phi(\omega, t)$ such that

$$\phi \circ C = B^c - (B^X)^c - I(|x| \le 1)\, x * (\nu - \nu^X)^c,$$

(B_4) $E\,|Y_- \phi| \circ C_t < \infty$, $t \ge 0$.

Further on the optional and predictable projections with respect to $\mathbb{F}_+^X$ are denoted by $\pi(\cdot)$ and ${}^p\pi(\cdot)$ respectively. Also, assume

$$\tilde{\mathcal{P}}(\mathbb{F}_+^X) = \mathcal{P}(\mathbb{F}_+^X) \otimes B(R_0).$$

Theorem 1. 1) *If Condition* (B_1) *is fullfilled, then* $\mathbb{F}_+^X = \mathbb{F}^X$.

2) *If Conditions* (A) *and* (B_1) *are fullfilled, then* $(Y) \in S(\mathbb{F}^X, P)$ *and the processes* A^X *and* M^X, *involved in the semimartingale decomposition* (10.4), *possess the following property:*

$$E(\mathrm{Var}(A^X)_t)^2 < \infty, \quad E(M_t^X)^2 < \infty, \quad t \ge 0.$$

Besides M^X *admits the integral representation*

$$M_t^X = h \cdot \overline{X}_t^c + H * (\mu - \nu^X)_t \tag{10.6}$$

where $\overline{X}^c$ *is the continuous martingale component of a* $(\mathbb{F}_+^X, P)$-*semimartingale* X and h *and* H *are* $\mathcal{P}(\mathbb{F}_+^X)$- *and* $\tilde{\mathcal{P}}(\mathbb{F}_+^X)$- *measurable functions respectively.*

3) *If Conditions* (A) *and* (B) *are fullfilled, then in the integral representation* (10.6)

$$h = {}^{P}\pi \left(\frac{d\langle M^c, X^c\rangle}{d\langle X^c\rangle} + Y_{-}\phi \right), \tag{10.7}$$

$$H = U + \frac{I(0 < a^X < 1)}{1 - a^X}\hat{U} \tag{10.8}$$

with

$$\hat{U}(t) = \int_{R_0} U(t, x)\, \nu^X(\{t\}, dx),$$

$$U = M^P_\mu(\Delta M \mid \tilde{\mathcal{P}}(\mathbb{F}^X)) + [M^P_\mu(Y_{-} \mid \tilde{\mathcal{P}}(\mathbb{F}^X)) - \pi_{-}(Y)] + M^P_\mu(\Delta A \mid \tilde{\mathcal{P}}(\mathbb{F}^X)) - \Delta A^X]$$

Proof. 1) The equality $\mathbb{F}^X_+ = \mathbb{F}^X$ holds by Theorem 8.2 and Condition (B_1).

2) Instead of (A) suppose the stronger conditions (10.5) are fullfilled. Then obviously $\pi(Y) \in \mathbf{Sp}(\mathbb{F}^X_+, P)$ and the decomposition (10.4) holds. Let us show that

$$\mathbf{E}(\mathrm{Var}(A^X)_\infty)^2 < \infty, \quad M^X \in \mathcal{H}^2(\mathbb{F}^X_+). \tag{10.9}$$

Let A^i, $i = 1, 2$, be increasing processes involved in the decomposition of A: $A = A^1 - A^2$, $\mathrm{Var}(A) = A^1 + A^2$. Then by the second inequality in (10.5) we have $\mathbf{E}(A^i_\infty)^2 < \infty$, $i = 1, 2$. According to Lemma 6.1 $\pi(A^i)$ are submartingales of the class $(\mathcal{D})$ and they admit the semimartingale decomposition

$$\pi(A^i) = A^{X,i} + M^{X,i} \tag{10.10}$$

with

$$A^{X,i} \in \mathcal{A}^+ \cap \mathcal{P}(\mathbb{F}^X_+)$$

and

$$M^{X,i} \in \mathcal{M}(\mathbb{F}^X_+), \quad i = 1, 2.$$

Let us show that by the second inequality in (10.5)

$$\mathbf{E}(A^{X,i}_\infty)^2 < \infty, \quad M^{X,i} \in \mathcal{H}^2(\mathbb{F}^X_+), \qquad i = 1, 2. \tag{10.11}$$

Due to Ito's formula (Corollary 1 to Theorem 2.3.1) we get

$$(\pi(A^i))^2 = 2\pi_{-}(A^i) \circ A^{X,i} + 2(\pi_{-}(A^i) + \Delta A^{X,i}) \cdot M^{X,i}$$

$$+ \sum_s (\Delta A^{X,i}_s)^2 + [M^{X,i}, M^{X,i}]. \tag{10.12}$$

Let $(\tau_n)_{n \ge 1}$ be a localizing sequence for

$$(\pi_{-}(A^i) + \Delta A^{X,i}) \cdot M^{X,i} \in \mathcal{M}_{loc}(\mathbb{F}^X_+).$$

Then (10.12) entails the inequality

$$\mathbf{E}\,[M^{X,i}, M^{X,i}]_{\tau_n} \le \mathbf{E}\,(\pi_{\tau_n}(A^i))^2 \le \mathbf{E}\,(A^i_{\tau_n})^2,\ \ n \ge 1,$$

and hence

$$\mathbf{E}\,[M^{X,i}, M^{X,i}]_\infty \le \mathbf{E}\,(A^i_\infty)^2 < \infty.$$

Therefore, by the Burkholder-Gandy inequality (Theorem 1.9.7) the second property in (10.11) holds. But then

$$\mathbf{E}\,(A^{X,i}_\infty)^2 \le 2\mathbf{E}\,[(\pi_\infty, (A^i))^2 + (M^{X,i}_\infty)^2] \le 2\mathbf{E}\,[(A^i_\infty)^2 + (M^{X,i}_\infty)^2],$$

i.e. the first property in (10.11) holds.

On proving Theorem 6.2 it has been established that

$$A^X = A^{X,1} - A^{X,2}, \quad M^X = M^{X,1} - M^{X,2} + \pi\,(M). \tag{10.13}$$

The first property in (10.9) holds by (10.11), since

$$\mathrm{Var}\,(A^X) \le A^{X,1} + A^{X,2}.$$

The second property in (10.9) holds also by (10.11), since by Lemma 6.2 we have

$$\pi\,(M) \in \mathcal{M}\,(\mathbb{F}^X_+)$$

and by Doob's inequality (Theorem 1.9.2)

$$\mathbf{E}\,\sup_{t\ge 0} \pi^2\,(M) \le 4\mathbf{E}\,(\pi_\infty\,(M))^2 \le 4\mathbf{E}M^2_\infty < \infty.$$

If exclusively Condition (A) is fullfilled, then consider semimartingales Y^n, $n \ge 1$ with $Y^n_t = I_{[\![0,n]\!]} \cdot Y_t$. For each $n \ge 1$

$$Y^n_t = Y_0 + A^n_t + M^n_t,$$

with $A^n = I_{[\![0,n]\!]} \circ A$ and $M^n = I_{[\![0,n]\!]} \cdot M$ which satisfy the conditions in (10.5). Therefore $\pi\,(Y^n)$ admits the semimartingale decomposition

$$\pi_t\,(Y^n) = \pi_0\,(Y) + A^{n,X}_t + M^{n,X}_t,\ \ n \ge 1$$

with

$$A^{n,X} \in \mathcal{A} \cap \mathcal{P}\,(\mathbb{F}^X_+),\ \mathbf{E}\,(\mathrm{Var}\,(A^{n,X})_\infty)^2 < \infty \text{ and } M^{n,X} \in \mathcal{H}^2\,(\mathbb{F}^X_+).$$

The definition of Y^n yields

$$\pi_{t\wedge n}\,(Y^{n+1}) = \pi_{t\wedge n}\,(Y^n) = \pi_t\,(Y^n).$$

Therefore the equalities

$$A^{n+1,X}_{t\wedge n} + M^{n+1,X}_{t\wedge n} = A^{n,X}_t + M^{n,X}_t$$

hold. As the semimartingale decomposition for special martingales is unique (Theorem 2.1.1), we have

$$A^{n+1,X}_{t\wedge n} \equiv A^{n,X}_t, \quad M^{n+1,X}_{t\wedge n} \equiv M^{n,X}_t,\ \ n \ge 1.$$

This fact allows one to define the processes A^X and M^X

$$A_t^X = A_t^{n,1} + \sum_{n \geq 1} (A_t^{n+1, X} - A_t^{n, X}),$$

$$M_t^X = M_t^{n,1} + \sum_{n \geq 1} (M_t^{n+1, X} - M_t^{n, X})$$

with the properties indicated in the semimartingale decomposition (10.4), such that

$$\mathbf{E}\, (\mathrm{Var}\, (A^X)_t)^2 < \infty, \quad \mathbf{E}\, (M_t^X)^2 < \infty, \quad t > 0.$$

The integral representation (10.6) for M^X takes place by Theorem 8.2 and Condition (B_1).

3) Let us establish that in the integral representation (9.6) for M^X the functions h and H are defined by the formulas (10.7) and (10.8).

Examining the proof of Assertion 2) of the theorem, we see that it suffices to proof the formulas (10.7) and (10.8) under the stronger condition (10.5) instead of Condition (A) and under the condition

$$\mathbf{E}\, |\, Y\_\phi\, |\, \circ C_\infty < \infty \tag{10.14}$$

instead of Condition (B_4).

By Lemma 8.1 (see also Theorem 3.3.1)

$$h = \frac{d \langle M^{X, c}, \overline{X}^c \rangle}{d \langle \overline{X}^c \rangle},$$

$$H = M_\mu^P (\Delta M^X \mid \tilde{\mathcal{P}} (\mathbb{F}_+^X)) + \frac{I(0 < a^X < 1)}{1 - a^X} M_\mu^P \left(\widehat{\Delta M^X \mid \tilde{\mathcal{P}} (\mathbb{F}_+^X)}\right) \tag{10.15}$$

where $\overline{X}^c$ is the continuous martingale component of a process X, belonging to $S(\mathbb{F}_+^X, P)$. To calculate the right-hand sides in the equalities (10.15) we need two auxiliary results formulated as the following lemmas .

Lemma 1. *Let Conditions* (B_2) *and* (B_3) *be fullfilled. Then*

$$\overline{X}^c = X^c + \phi \circ C.$$

Proof. A process $X \in \mathbf{S}\, (\mathbb{F}_+^X, P)$ has the canonical representation (cf. (10.1))

$$X_t = X_0 + B_t^X + \overline{X}_t^c + \int_0^t \int_{|x| \leq 1} x\, d(\mu - \nu^X) + \int_0^t \int_{|x| > 1} x\, d\mu. \tag{10.16}$$

By (10.1), (10.16), the representation for $\phi \circ C$ (see (B_3)) and the fact that

$$\Delta B_t = \int_{|x| \le 1} x\nu(\{t\}, dx), \quad \Delta B_t^X = \int_{|x| \le 1} x\nu^X(\{t\}, dx)$$

(see § 1), we get

$$\overline{X}^c - X^c - \phi \circ C = I(|x| \le 1)\, x * (\mu - \nu)$$
$$- I(|x| \le 1)\, x * (\mu - \nu^X) + I(|x| \le 1)\, x * (\mu - \nu^X). \tag{10.17}$$

Let us show that the right-hand side of the equation (10.17) determines a negligible process. To this end define the process

$$M^\varepsilon = I(|x| \le \varepsilon)\, x * (\mu - \nu)$$

with $0 < \varepsilon < 1$. Clearly,

$$M^\varepsilon \in \mathcal{M}^2_{loc}(\mathbb{F})$$

and (see Lemma 3.5.1 and Theorem 3.5.1)

$$\langle M^\varepsilon \rangle = I(|x| \le \varepsilon)\, x^2 * \nu - \sum_s \left(\int_{|x| \le \varepsilon} x\nu(\{s\}, dx) \right)^2 \le I(|x| \le \varepsilon)\, x^2 * \nu. \tag{10.18}$$

Since $\mathbf{E}(M_\tau^\varepsilon)^2 \le \mathbf{E}\langle M^\varepsilon \rangle_\tau$ for each stopping time τ relative to $\mathbb{F}_+^X$ (Problem 1.8.6), by (10.18) we have

$$\mathbf{E}(M_\tau^\varepsilon)^2 \le \mathbf{E} I(|x| \le \varepsilon)\, x^2 * \nu_\tau.$$

Therefore by the Lenglart-Rebolledo inequality (Theorem 1.9.3)

$$\mathsf{P}(\sup_{s \le t} |M_s^\varepsilon| \ge a) \le \frac{b}{a^2} + \mathsf{P}(I(|x| \le \varepsilon)\, x^2 * \nu_t \ge b) \tag{10.19}$$

for each $a > 0$, $b > 0$ and $t > 0$. For each $t > 0$ we have $I(|x| \le 1)\, x^2 * \nu_t < \infty$ (P-a.s.) (Problem 4.1). Consequently,

$$I(|x| \le \varepsilon)\, x^2 * \nu_t \to 0 \text{ (P-a.s.) as } \varepsilon \to 0.$$

In view of this, taking the limit $\lim_{b \to 0} \overline{\lim_{\varepsilon \to 0}}$ in (10.19), we get

$$\lim_{\varepsilon \to 0} \mathsf{P}(\sup_{s \le t} |M_s^\varepsilon| \ge a) = 0 \tag{10.20}$$

for each $a > 0$ and $t > 0$.

It is established analogously that

$$\lim_{\varepsilon \to 0} \mathsf{P}(\sup_{s \le t} |M_s^{X,\varepsilon}| \ge a) = 0 \tag{10.21}$$

with $M^{X,\varepsilon} = I(|x| \le \varepsilon)\, x * (\mu - \nu^X)$.

Besides, by Assumption (B_2)

$$I(|x| \le 1)\, x * (\nu - \nu^X) \in \mathcal{A}_{loc},$$

and consequently (P-a.s.),

$$\mathrm{Var}\,(I(|x| \le \varepsilon)\, x * (\nu - \nu^X))_t \to 0, \quad \varepsilon \to 0, \tag{10.22}$$

for each $t > 0$.

Observe now that by Problem 3.5.3 as $0 < \varepsilon < 1$

$$I(\varepsilon < |x| \le 1)\, x * (\mu - \nu)$$

$$= I(\varepsilon < |x| \le 1)\, x * \mu - I(\varepsilon < |x| \le 1)\, x * \nu.$$

The analogous representation takes place for $I(\varepsilon < |x| \le 1)\, x * (\mu - \nu^X)$ too. Therefore the process

$$I(\varepsilon < |x| \le 1)\, x * (\mu - \nu) - I(\varepsilon < |x| \le 1)\, x * (\mu - \nu^X) + I(\varepsilon < |x| \le 1)\, x * (\nu - \nu^X)$$

is negligible. By taking this and equality (10.17) into consideration, for each $t > 0$ we get

$$\sup_{s \le t} |\overline{X}_s^c - X_s^c - \phi \circ C|$$

$$\le \sup_{s \le t} |M_s^{\varepsilon}| + \sup_{s \le t} |M_s^{X,\varepsilon}| + \mathrm{Var}\,(I(|x| \le \varepsilon)\, x * (\nu - \nu^X))_t.$$

From this and (10.20) — (10.22) it follows that the process $\overline{X}^c - X^c - \phi \circ C$ is negligible.

Lemma 2. *Let Conditions* (A) *be fullfilled. Then*

$$M_\mu^P(\pi(Y) \mid \tilde{\mathcal{P}}(\mathbb{F}_+^X)) = M_\mu^P(Y \mid \tilde{\mathcal{P}}(\mathbb{F}_+^X)) \quad (M_\mu^P\text{-a.s.}).$$

Proof. By (A)

$$\mathbf{E} \sup_{s \le t} |Y_s| \le \mathbf{E}\,(|Y_0| + \mathrm{Var}\,(A)_t + \sup_{s \le t} |M_s|) < \infty,$$

since according to the Cauchy-Bunjakovski and Doob inequalities (Theorem 1.9.2)

$$(\mathbf{E} \sup_{s \le t} |M_s|)^2 \le \mathbf{E} \sup_{s \le t} (M_s)^2 \le 4\mathbf{E}M_t^2.$$

Analogously, using the representation (10.4) for $\pi(Y)$, Assertion 2) of the theorem proved already and the inequality $\mathbf{E}\,|\pi_0(Y)| \le \mathbf{E}\,|Y_0|$, we get

$$\mathbf{E} \sup_{s \le t} |\pi_s(Y) < \infty.$$

Therefore, in accordance with Problem 3.2.12 there exist the conditional mathematical expectations $M_\mu^P(Y \mid \tilde{\mathcal{P}}(\mathbb{F}_+^X))$ and $M_\mu^P(\pi(Y) \mid \tilde{\mathcal{P}}(\mathbb{F}_+^X))$.

Hence, it suffices to show that

$$M_\mu^P(Y^+ \mid \tilde{\mathcal{P}}(\mathbb{F}_+^X)) = M_\mu^P(\pi(Y^+) \mid \tilde{\mathcal{P}}(\mathbb{F}_+^X)) \tag{10.23}$$

and

$$M^P_\mu(Y^- \mid \tilde{\mathcal{P}}(\mathbb{F}^X_+)) = M^P_\mu(\pi(Y^-) \mid \tilde{\mathcal{P}}(\mathbb{F}^X_+)), \tag{10.24}$$

with $Y^+ = Y \vee 0$ and $Y^- = -(Y \wedge 0)$.

Let Z be a $\tilde{\mathcal{P}}(\mathbb{F}^X_+)$-measurable nonnegative function such that $Z * \mu_\infty \le \text{const}$. Denote $B = Z * \mu$. Clearly $B \in \mathcal{A}^+$. According to Problem 1.6.9

$$EY^+ Z * \mu_\infty = EY^+ \circ B_\infty = E\,\pi(Y^+) \circ B_\infty = E\,\pi(Y^+)\, Z * \mu_\infty.$$

This easily gives the equality

$$EY^+ Z * \mu_\infty = E\,\pi(Y^+)\, Z * \mu_\infty$$

for each nonnegative $\tilde{\mathcal{P}}(\mathbb{F}^X_+)$-measurable function Z. Consequently

$$E\, M^P_\mu(Y^+ \mid \tilde{\mathcal{P}}(\mathbb{F}^X_+))\, Z * \mu_\infty = E\, M^P_\mu(\pi(Y^+) \mid \tilde{\mathcal{P}}(\mathbb{F}^X_+))\, Z * \mu_\infty$$

and the equality (10.23) takes place. The equality (10.24) is proved analogously.

We will turn now to determining the right-hand sides of the equalities (10.15) that define h and H, assuming Conditions (10.5) and (10.14) hold.

Determination of h. Since by Theorem 6.1 $\langle \overline{X}^c \rangle = C$, by (10.15) we have $\langle M^{X,c}, \overline{X}^c \rangle = h \circ C$. On the other hand,

$$N = M^{X,c}\, \overline{X}^c - \langle M^{X,c}, \overline{X}^c \rangle \in \mathcal{M}_{loc}(\mathbb{F}^X_+)$$

(see Ch. 1, § 8). Therefore the process $M^{X,c}\,\overline{X}^c \in \mathbf{Sp}(\mathbb{F}^X_+, P)$ has the semimartingale decomposition

$$M^{X,c}\,\overline{X}^c = \langle M^{X,c}, \overline{X}^c \rangle + N.$$

As this decomposition is unique (Theorem 2.1.1), the process $\langle M^{X,c}\, \overline{X}^c \rangle$ may be defined as a certain process $L = (L_t)_{t \ge 0} \in \mathcal{V} \cap \mathcal{P}(\mathbb{F}^X_+)$ such that

$$M^{X,c}\,\overline{X}^c - L \in \mathcal{M}_{loc}(\mathbb{F}^X_+). \tag{10.25}$$

It has been established in the course of proving Assertion 2) of the theorem that under Condition (10.5) we have $M^X \in \mathcal{H}^2(\mathbb{F}^X_+)$ (see (10.9)). In the subsequent considerations we will assume for convenience

$$(\overline{X}^c)^*_\infty + C_\infty + \operatorname{Var}(A^X)_\infty \le \text{const}, \tag{10.26}$$

passing to localizing sequences relative to $\mathbb{F}^X_+$ if necessary (see Lemma 1.6.1).

Under the assumtion (10.26) we have $\overline{X}^c \in \mathcal{H}^2(\mathbb{F}^X_+)$. Therefore

$$E\,\mathrm{Var}\,(\langle M^{X,c}, \overline{X}^c\rangle)_\infty < \infty$$

(Problem 1.8.9). Now, to verify (10.25) it suffices to show, by Problem 1.4.2, that for each $\tau \in \mathbf{T}(\mathbb{F}^X_+)$

$$\mathbf{E}M^{X,c}_\tau \overline{X}^c_\tau = \mathbf{E}L_\tau. \tag{10.27}$$

Next, since the martingales $M^X - M^{X,c}$ and $\overline{X}^c$ are strongly orthogonal, instead of (10.27) it suffices to verify the equality

$$\mathbf{E}M^X_\tau \overline{X}^c_\tau = \mathbf{E}L_\tau. \tag{10.28}$$

By the semimartingale decomposition (10.4) for $\pi(Y)$ we have

$$\mathbf{E}M^X_\tau \overline{X}^c_\tau = \mathbf{E}\,(\pi_\tau(Y) - \pi_0(Y) - A^X_\tau)\,\overline{X}^c_\tau = \mathbf{E}\,(Y_\tau - A^X_\tau)\,\overline{X}^c_\tau.$$

Under the assumption (10.5) we have

$$\mathbf{E}\,\mathrm{Var}\,(A^X)_\infty < \infty$$

(see (10.9)). Therefore

$$\mathbf{E}A^X_\tau \overline{X}^c_\tau = \mathbf{E}\overline{X}^c \circ A^X_\tau$$

(Corollary to Theorem 1.6.1).

Thus

$$\mathbf{E}M^{X,c}_\tau \overline{X}^c_\tau = \mathbf{E}\,(Y_\tau \overline{X}^c_\tau - \overline{X}^c \circ A^X_\tau). \tag{10.29}$$

Observe now that if $(\sigma_n)_{n\geq 1}$ is a sequence of Markov times relative to $\mathbb{F}$ such that $\sigma_n \uparrow \infty$, $n \to \infty$, then under the assumptions (10.5) and (10.26) made, we get, by the Lebesgue dominated convergence theorem, that

$$\mathbf{E}Y_\tau \overline{X}^c_\tau = \lim_n \mathbf{E}Y_{\tau_n} \overline{X}^c_{\tau_n}, \quad \tau_n = \tau \wedge \sigma_n. \tag{10.30}$$

Set

$$\sigma_n = \inf\,(t\colon \sup_{s\leq t} |X^c_s| + |\phi| \circ C_t \geq n)$$

and apply the decomposition (10.2) for Y and Lemma 1. Then

$$\mathbf{E}Y_{\tau_n} \overline{X}^c_{\tau_n} = \mathbf{E}\,(Y_0 + A_{\tau_n} + M_{\tau_n})\,(X^c_{\tau_n} + \phi \circ C_{\tau_n}),$$

and by Corollary to Theorem 1.6.1 and the definition of the mutual quadratic characteristic for square integrable martingales (see Ch. 1, § 8), we get

$$\mathbf{E}Y_{\tau_n} \overline{X}^c_{\tau_n} = \mathbf{E}\,(X^c \circ A_{\tau_n} + \langle M^c, X^c\rangle_{\tau_n} + (Y_0\,\phi) \circ C_{\tau_n} + A_{\tau_n}(\phi \circ C_{\tau_n}) + M_- \phi \circ C_{\tau_n}). \tag{10.31}$$

Next, by Ito's formula (Ch. 2, § 3) we get

$$A_{\tau_n}(\phi \circ C_{\tau_n}) = A_- \phi \circ C_{\tau_n} + (\phi \circ C) \circ A_{\tau_n}. \tag{10.32}$$

From (10.31) and (10.32) it follows that

$$\mathbf{E}Y_{\tau_n}\overline{X}^c_{\tau_n} = \mathbf{E}\,[(X^c + \phi \circ C) \circ A_{\tau_n} + \langle M^c, X^c\rangle_{\tau_n} + \phi\,(Y_0 + A_- + M_-) \circ C_{\tau_n}]$$

which, in view of the representation

$$\langle M^c, X^c\rangle = \frac{d\,\langle M^c, X^c\rangle}{d\,\langle X^c\rangle} \circ C$$

(Theorem 2.2.8), yields

$$\mathbf{E}Y_{\tau_n}\overline{X}^c_{\tau_n} = \mathbf{E}\left[\overline{X}^c \circ A_{\tau_n} + \frac{d\,\langle M^c, X^c\rangle}{d\,\langle X^c\rangle} \circ C_{\tau_n} + \phi\, Y_- \circ C_{\tau_n}\right].$$

By taking the limit $\lim\limits_n$ (Lebesgue's dominated convergence theorem) we arrive at the equality

$$\mathbf{E}Y_\tau \overline{X}^c_\tau = \mathbf{E}\left[\overline{X}^c \circ A_{\tau_n} + \left(\frac{d\,\langle M^c, X^c\rangle}{d\,\langle X^c\rangle} + \phi\, Y_-\right) \circ C_\tau\right], \tag{10.33}$$

since by (10.26) we have $(\overline{X}^c)^*_\infty \le \text{const}$, by (10.14) we have $\mathbf{E}\,|\,\phi\, Y_-\,| \circ C_\infty < \infty$ and finally, by the Cauchy-Bunjakovski inequality, Theorem 2.2.8, the inequality (10.26) (according to which $C_\infty \le \text{const}$) and the condition $M \in \mathcal{H}^2\,(\mathbb{F})$ (see (10.5)) we have

$$\mathbf{E}\left|\frac{d\,\langle M^c, X^c\rangle}{d\,\langle X^c\rangle}\right| \circ C_\infty \le \left(\mathbf{E}C_\infty \left(\frac{d\,\langle M^c, X^c\rangle}{d\,\langle X^c\rangle}\right)^2 \circ C_\infty\right)^{1/2}$$

$$\le (\mathbf{E}C_\infty\,\langle M^c\rangle_\infty)^{1/2} \le \text{const}\,(\mathbf{E}\,\langle M\rangle_\infty)^{1/2} < \infty.$$

Thus, (10.29) and (10.33) entail

$$\mathbf{E}M^{X,c}_\tau \overline{X}^c_\tau = \mathbf{E}\left(\frac{d\,\langle M^c, X^c\rangle}{d\,\langle X^c\rangle} + \phi\, Y_-\right) \circ C_\tau + \mathbf{E}\,[\overline{X}^c \circ A_\tau - \overline{X}^c \circ A^X_\tau].$$

Let A^i, $i = 1, 2$, be increasing processes involved in the decomposition of $A : A = A^1 - A^2$, let $\text{Var}\,(A) = A^1 + A^2$, and let $A^{X,\,i}$, $i = 1, 2$, be increasing processes involved in the semimartingale decomposition of $\pi\,(A^i)$ (see (10.10)). From (10.10) (see also (10.11)) it follows that

$$\mathbf{E}A^i_\tau = \mathbf{E}A^{X,\,i}_\tau, \quad i = 1, 2,$$

for each $\tau \in \mathbf{T}(\mathbb{F}^X_+)$. It is not hard to deduce from this that

$$\mathbf{E}g \circ A^i_\infty = \mathbf{E}g \circ A^{X,i}_\infty, \qquad i = 1, 2, \tag{10.34}$$

for any nonnegative bounded function $g = g(\omega, t)$, measurable relative to $\mathcal{P}_6(\mathbb{F}^X_+)$, that is the σ-algebra in $\Omega \times R_+$, generated by sets of type $A \times \{0\}$ $(A \in \mathcal{F}^X_{0+})$ and of type $]\!]\sigma, \tau]\!]$ with $\sigma, \tau \in \mathbf{T}(\mathbb{F}^X_+)$. It has been established in the course of proving Theorem 1.2.2 that $\mathcal{P}_6 = \mathcal{P}$, i.e. $\mathcal{P}_6(\mathbb{F}^X_+) = \mathcal{P}(\mathbb{F}^X_+)$. Hence the equality (10.34) holds for each bounded and $\mathcal{P}(\mathbb{F}^X_+)$-measurable function g.

But then by the equality $A^X = A^{X,1} - A^{X,2}$ (see (10.13)) we have

$$\mathbf{E}\overline{X}^c \circ A_\tau = \mathbf{E}\overline{X}^c \circ A^X_\tau,$$

and hence

$$\mathbf{E}M^{X,c}_\tau \overline{X}^c_\tau = \mathbf{E}\left(\frac{d\langle M^c, X^c\rangle}{d\langle X^c\rangle} + \phi Y_-\right) \circ C_\tau.$$

Therefore, by Problem 1.6.9,

$$\mathbf{E}M^{X,c}_\tau \overline{X}^c_\tau = \mathbf{E}\,{}^p\pi\left(\frac{d\langle M^c, X^c\rangle}{d\langle X^c\rangle} + \phi Y_-\right) \circ C_\tau.$$

Consequently,

$$L = {}^p\pi\left(\frac{d\langle M^c, X^c\rangle}{d\langle X^c\rangle} + \phi Y_-\right) \circ C,$$

and the representation (10.17) for h holds.

Determination of H. By the representation (10.15) for H it suffices to calculate $M^P_\mu(\Delta M^X \mid \tilde{\mathcal{P}}(\mathbb{F}^X_+))$. (Conditional expectations $M^P_\mu(\cdot \mid \tilde{\mathcal{P}}(\mathbb{F}^X_+))$ calculated below exist by Theorem 3.3.1 and Problem 3.2.12.)

By (10.4)

$$\Delta M^X = \pi(Y) - \pi_-(Y) - \Delta A^X.$$

Therefore, taking into account Lemma 2 we get

$$M^P_\mu(\Delta M^X \mid \tilde{\mathcal{P}}(\mathbb{F}^X_+)) = M^P_\mu(\pi(Y) \mid \tilde{\mathcal{P}}(\mathbb{F}^X_+)) - \pi_-(Y) - \Delta A^X$$

$$= M^P_\mu(Y \mid \tilde{\mathcal{P}}(\mathbb{F}^X_+)) - \pi_-(Y) - \Delta A^X.$$

The representation (10.2) for Y yields

$$Y = Y_- + \Delta A + \Delta M.$$

Consequently

$$M^P_\mu(\Delta M^X \mid \tilde{\mathcal{P}}(\mathbb{F}^X_+)) = [M^P_\mu(Y_- \mid \tilde{\mathcal{P}}(\mathbb{F}^X_+)) - \pi_-(Y)]$$

$$+ [M^P_\mu(\Delta A \mid \tilde{\mathcal{P}}(\mathbb{F}^X_+)) - \Delta A^X] + M^P_\mu(\Delta M \mid \tilde{\mathcal{P}}(\mathbb{F}^X_+)),$$

and hence $M^P_\mu(\Delta M^X \mid \tilde{\mathcal{P}}(\mathbb{F}^X_+))$ coincides with the function U, involved in the definition of H by the formula (10.8).

3. In order to illustrate the possibilities of applying Theorem 1, we present the following example. (Numerous examples of applying theorems of this type are given in the authors monograph [188]).

Example. Let $W = (W_t)_{t \geq 0}$ be a Wiener process relative to $(\mathbb{F}, P)$ and $\tau = \inf(t: W_t \geq 1)$.

Set

$$Y_t = W_{t \wedge \tau}, \quad X_t = I(t \geq \tau).$$

Clearly,

$$Y \in \mathcal{M}^c_{loc}(\mathbb{F}).$$

The process X is counting and $\mathcal{P}(\mathbb{F})$-measurable, since t is a Markov time ($\tau_n = \inf\left(t: W_t \geq 1 - \frac{1}{n}\right)$, $n \geq 1$, is an announcing sequence). Therefore, the $\mathbb{F}$-compensator of the process X coincides with it and hence $X^0 = 0$.

Let us show now that Conditions (A) and (B) of Theorem 1 are fullfilled. Condition (A) is fullfilled obviously. Condition (B_1) holds by Proposition 2 in Ch. 3, § 4. Condition (B_4) holds, since $C_\infty = 0$. Conditions (B_2) and (B_3) are automatically fullfilled.

Then due to Theorem 1 we get, by taking into account Lemma 6.2 and specific features of the counting process X, that

$$\pi_t(Y) = H \circ (X - B^X)_t,$$

where B^X is the compensator of the process X relative to $(\mathbb{F}^X_+, P)$, while H is a certain $\mathcal{P}(\mathbb{F}^X_+)$-measurable function.

As is known, the random variable τ has the distribution (see, for instance, [188], Ch. 1, formula (1.42))

$$G(t) = \sqrt{\frac{2}{\pi}} \int_{1/\sqrt{t}}^{\infty} \exp\left(-\frac{x^2}{2}\right) dx.$$

Therefore, in accordance with (4.7) (Ch. 3, § 4)

$$B_t^X = \ln \frac{1}{1 - G(t \wedge \tau)}.$$

This gives $\Delta B^X = 0$, and hence $a^X = 0$. Consequently (see (10.8)),

$$H = M_\mu^P (Y \mid \tilde{\mathcal{P}} (\mathbb{F}_+^X)) - \pi_- (Y).$$

Let us show that $M_\mu^P (Y \mid \tilde{\mathcal{P}} (\mathbb{F}_+^X)) = 1$. In fact, if $\phi = \phi(\omega, t, x)$ is a $\tilde{\mathcal{P}} (\mathbb{F}_+^X)$-measurable function such that $|\phi| \le$ const, then by setting

$$\bar{\phi}(\omega, t) = \phi(\omega, t, 1)$$

we have

$$E\phi Y * \mu_\infty = E\bar{\phi} Y * X_\infty = E\bar{\phi}(\tau) Y_\tau.$$

The distribution function G (t) is such that

$$P(\tau = \infty) = 0.$$

Consequently $Y_\tau = 1$ and hence

$$E\phi Y * \mu_\infty = E\bar{\phi}(\tau) = E\bar{\phi} * X_\infty = E\phi * \mu_\infty.$$

Therefore

$$M_\mu^P (Y \mid \tilde{\mathcal{P}} (\mathbb{F}_+^X)) = 1$$

in view of the definition of $M_\mu^P (\cdot\ \tilde{\mathcal{P}})$.

Thus

$$\pi_t (Y) = \int_0^t (1 - \pi_{s-} (Y))\, d(X - B^X).$$

Problems

1. Prove that the representation (10.4) is valid.

2. Let M' and M'' $\in \mathcal{M}_{loc} (\mathbb{F}, P)$, let A' and A'' be deterministic functions from V, X_0 and Y_0 $\mathcal{F}_0$-measurable random variables and (Y_0, X_0, M', M'') a Gaussian system. Let

$$Y_t = Y_0 + Y_- \circ A'_t + M'_t,$$
$$X_t = X_0 + Y_- \circ A''_t + M''_t.$$

Show that if $A'' = g \circ \langle M'' \rangle$ and $\mathbf{E} Y_-^2 g^2 \circ \langle M'' \rangle \in \mathcal{V}^+$, then the relations

$$\pi_t(Y) = \pi_0(Y) + \pi_-(Y) \circ A'_t + q \cdot (X - \pi_-(Y) \circ A'')_t$$

hold (the generalized Kalman-Busy filter) with

$$q = \frac{d\,[\langle M', M'' \rangle + \gamma_-(1 + \Delta A') \circ A'']}{d\,[\langle M'' \rangle + \gamma_- \Delta A'' \circ A'']}, \quad \gamma_t = \mathbf{E}\,(Y_t - \pi_t(Y))^2$$

and

$$\gamma_t = \gamma_0 + \gamma_-(2 + \Delta A') \circ A'_t + \langle M' \rangle_t - q^2 \circ (\langle M'' \rangle + \gamma_- \Delta A'' \circ A'')_t.$$

§ 11. Semimartingales and helices. Ergodic theorems

1. Let $(\Omega, \mathcal{F}, \mathsf{P})$ be a probability space and $\theta = (\theta_t)_{t \in R}$ a measure preserving a group of transformations from Ω to Ω with a group operation:

$$\theta_t \theta_s = \theta_{t+s} \ (\theta_0 \omega = \omega),$$

besides let the mapping $(t, \omega) \to \theta_t \omega$ be $B(R) \otimes \mathcal{F}/\mathcal{F}$-measurable.

Let $\mathcal{F}_0$ be a certain sub-σ-algebra of $\mathcal{F}$ such that

$$\theta_t^{-1}(\mathcal{F}_0) \subset \mathcal{F}_0, \ t < 0.$$

Define σ-algebras $\mathcal{F}_t$, $t \in R$ by setting

$$\mathcal{F}_t = \theta_t^{-1}(\mathcal{F}_0).$$

Then,

$$\mathcal{F}_{s+t} = \theta_t^{-1}(\mathcal{F}_s)$$

evidently, and the family $\mathbb{F} = (\mathcal{F}_t)_{t \in R}$ is nondecreasing. Denote by $\mathcal{F}^{\mathsf{P}}$ the completion $\mathcal{F}$ with respect to the measure P; write

$$\mathcal{F}_t^{\mathsf{P}} = \sigma(\mathcal{F}_t \cup \mathcal{N})$$

where $\mathcal{N}$ is the system of sets from $\mathcal{F}^{\mathsf{P}}$ of P-measure zero, and

$$\mathbb{F}^{\mathsf{P}} = (\mathcal{F}_t^{\mathsf{P}})_{t \in R}.$$

Set

$$\mathcal{F}_{-\infty}^{\mathsf{P}} = \bigcap_{t \in R} \mathcal{F}_t^{\mathsf{P}}, \quad \mathcal{F}_{\infty}^{\mathsf{P}} = \sigma\Big(\bigcup_{t \in R} \mathcal{F}_t^{\mathsf{P}}\Big)$$

and further on assume $\mathcal{F}^{\mathsf{P}} = \mathcal{F}_{\infty}^{\mathsf{P}}$.

Denote by J the σ-algebra of almost invariant sets:

$$J = \{A \in \mathcal{F}^{\mathsf{P}} : \mathsf{P}(I_A(\omega) = I_A(\theta_t \omega)) = 1, \ \forall t \in R\}.$$

Remark. As the σ-algebra $\mathcal{F}_0$ is completed, we have an important property:

$$J \subseteq \mathcal{F}_0.$$

If $\xi = \xi(\omega)$ is a J-measurable random variable, then

$$\xi(\omega) = \xi(\theta_t \omega) \ \ \mathsf{P}\text{-a.s.}, \ t \in R.$$

We present a useful property of conditional mathematical expectations $\mathbf{E}(\cdot \mid \mathcal{F}_t^{\mathsf{P}})$.

Lemma 1. *Let* $\xi = \xi(\omega)$ *be a random variable with* $\mathbf{E}|\xi| < \infty$. *Then for each* s, $t \in R$

$$\mathbf{E}(\xi(\theta_s \omega) \mid \mathcal{F}_{t+s}^{\mathsf{P}})(\omega) = \mathbf{E}(\xi(\omega) \mid \mathcal{F}_t^{\mathsf{P}})(\theta_s \omega) \ \ \mathsf{P}\text{-a.s.}$$

Proof. If $\eta = \eta(\omega)$ is a bounded and $\mathcal{F}^{P}_{t+s}$-measurable random variable, then the random variable $\eta(\theta_{-s}\omega)$ is $\mathcal{F}^{P}_{t}$-measurable. Hence

$$\mathbf{E}\,\eta(\omega)\,\mathbf{E}(\xi(\theta_s\omega) \mid \mathcal{F}^{P}_{t+s})(\omega) = \mathbf{E}\,\eta(\omega)\,\xi(\theta_s\omega) = \mathrm{E}\,\eta(\theta_{-s}\omega)\,\xi(\omega)$$
$$= \mathbf{E}(\eta(\theta_{-s}\omega)\,\mathbf{E}(\xi(\omega) \mid \mathcal{F}^{P}_{t})(\omega))$$
$$= \mathbf{E}(\eta(\omega)\,\mathbf{E}(\xi(\omega) \mid \mathcal{F}^{P}_{t})(\theta_s\omega)),$$

and this entails the desired assertion in an obvious manner as η is arbitrary.

The given property of a conditional mathematical expectation allows us to establish an important property of the family $\mathbb{F}^{P}$.

Theorem 1. *The family* $\mathbb{F}^{P}$ *is right-continuous.*

Proof. Set

$$\mathcal{F}^{P}_{t+} = \bigcap_{\varepsilon > 0} \mathcal{F}^{P}_{t+\varepsilon}.$$

We will show that $\mathcal{F}^{P}_{t+} = \mathcal{F}^{P}_{t}$ for each $t \in R$.

For a certain $t \in R$ assume

$$\mathcal{F}^{P}_{t+} \supset \mathcal{F}^{P}_{t},$$

i.e. there exists

$$A \in \mathcal{F}^{P}_{t+}, \quad A \notin \mathcal{F}^{P}_{t} \quad \text{and} \quad \mathrm{P}(A) > 0.$$

We will show that in fact $A \in \mathcal{F}^{P}_{t}$, i.e.

$$\mathcal{F}^{P}_{t+} = \mathcal{F}^{P}_{t}$$

for given t. As the σ-algebras $\mathcal{F}^{P}_{t+}$ and $\mathcal{F}^{P}_{t}$ are complete, it suffices to show that

$$\mathbf{E}\,|\,I_A - \mathbf{E}(I_A \mid \mathcal{F}^{P}_{t})\,| = 0. \tag{11.1}$$

In view of Lemma 1 for $h > 0$ we get

$$\mathbf{E}\,|\,I_A(\omega) - \mathbf{E}(I_A(\omega) \mid \mathcal{F}^{P}_{t})(\omega)\,| = \mathbf{E}\,|\,I_A(\theta_h\omega) - \mathbf{E}(I_A(\omega) \mid \mathcal{F}^{P}_{t})(\theta_h\omega)\,|$$
$$= \mathbf{E}\,|\,I_A(\theta_h\omega) - \mathbf{E}(I_A(\theta_h\omega) \mid \mathcal{F}^{P}_{t+h})(\omega)\,|$$
$$= \mathbf{E}\,|\,(I_A(\theta_h\omega) - I_A(\omega)) - \mathbf{E}(I_A(\theta_h\omega) - I_A(\omega) \mid \mathcal{F}^{P}_{t+h})(\omega)\,|$$
$$\le 2\mathbf{E}\,|\,I_A(\theta_h\omega) - I_A(\omega)\,|$$

$$= 2\mathbf{E} \mid I_A(\theta_{h+s}\omega) - I_A(\theta_s\omega) \mid$$

$$2\mathbf{E} \int_0^1 \mid I_A(\theta_{h+s}\omega) - I_A(\theta_s\omega) \mid ds \to 0, \ h \to 0,$$

since for each ω we have

$$\lim_{h \to 0} \int_0^1 \mid I_A(\theta_{h+s}\omega) - I_A(\theta_s\omega) \mid ds = 0, \tag{11.2}$$

as a consequence of the well-known fact: a function in L_1 is continuous in L_1.

Corollary. *There exists a modification of the conditional mathematical expectation* $\mathbf{E}(\xi(\theta_s\omega) \mid \mathcal{F}_t^P)(\omega)$, $t \geq 0$, $s \in R$ *which is measurable with respect to the* σ*-algebra* $\mathcal{F}^P \otimes B_+ \otimes B_{R_+}$, as a function of (ω, s, t).

In fact, if $\pi(\xi) = (\pi_t(\xi))_{t \geq 0}$ is the optional projection (see § 3, Ch. 1) of a random variable $\xi = \xi(\omega)$ with respect to $(\mathcal{F}_t^P)_{t \geq 0}$, i.e. such a process that

$$\mathbf{E}(\xi I(\tau < \infty) \mid \mathcal{F}_\tau^P) = \pi_\tau(\xi) I(\tau < \infty)$$

for each $\mathbb{F}^P$-Markov time τ, then $\pi_t(\xi)(\omega)$ is a $\mathcal{F}^P \otimes B_{R_+}$-measurable function of (ω, t).

On the other hand

$$\mathbf{E}(\xi(\theta_s\omega) \mid \mathcal{F}_t^P)(\omega) = \pi_{t-s}(\xi)(\theta_s\omega) \ \text{P-a.s.}$$

due to Lemma 1, and hence the desired measurability of $\pi_{t-s}(\xi)(\theta_s\omega)$, that is a modification of $\mathbf{E}(\xi(\theta_s\omega) \mid \mathcal{F}_t^P)(\omega)$, takes place in view of the $B_R \otimes \mathcal{F}^P/\mathcal{F}^P$-measurability of $\theta_s\omega$.

2. Analogously to Definition 1 (Ch. 1, § 1) $(\Omega, \mathcal{F}^P, \mathbb{F}^P = (\mathcal{F}_t^P)_{t \in R}, P)$ is called a *stochastic basis*.

Further on we will need results concerning Markov times and local martingales with respect to $(\mathcal{F}_t^P)_{t \geq 0}$.

Lemma 2. *Let* $T = T(\omega)$ *be a Markov time with respect to* $(\mathcal{F}_t^P)_{t \geq 0}$, $(M_t(\omega), \mathcal{F}_t^P) \in \mathcal{M}_{loc}$ *and* $h > 0$.

Then

1) $T(\theta_h\omega)$ *is a Markov time with respect to* $(\mathcal{F}^P_{t+h})_{t\ge 0}$,

2) $(M_t(\theta_h\omega), \mathcal{F}^P_{t+h}) \in \mathcal{M}_{loc}$.

Proof. 1) For each $t \ge 0$

$$\{\omega: T(\theta_h\omega) \le t\} = \{\theta_h^{-1}\omega: T(\omega) \le t\} \in \mathcal{F}^P_{t+h}.$$

2) Let $(T_k(\omega))_{k\ge 1}$ be a localizing sequence for $(M_t(\omega), \mathcal{F}^P_t)$. Then for each $t \ge 0$ and each bounded and $\mathcal{F}^P_t$-measurable random variable $\xi = \xi(\omega)$

$$\mathbf{E}\,\xi(\omega)\,M_{T_k(\omega)}(\omega) = \mathbf{E}\,\xi(\omega)\,M_{T_k(\omega)\wedge t}(\omega),\quad k \ge 1.$$

Consequently,

$$\mathbf{E}\,\xi(\theta_h\omega)\,M_{T_k(\theta_h\omega)}(\theta_h\omega) = \mathbf{E}\,\xi(\theta_h\omega)\,M_{T_k(\theta_h\omega)\wedge t}(\theta_h\omega),\quad k \ge 1.$$

This entails the desired assertion, since $\xi(\theta_h\omega)$ presents a $\mathcal{F}^P_{t+h}$-measurable random variable and $(T_k(\theta_h\omega))_{k\ge 1}$ a localizing sequence for $(M_t(\theta_h\omega), \mathcal{F}_{t+h})_{t\ge 0}$.

3. The following definition plays a central rôle in this section.

Definition 1. A stochastic process $X = (X_t)_{t\in R}$ with $X_0 = 0$ and right-continuous trajectories defined on a stochastic basis $(\Omega, \mathcal{F}^P, \mathbb{F}^P = (\mathcal{F}^P_t)_{t\in R}, \mathsf{P})$ is called a *helix* if for each t, s, $h \in R$ and $\omega \in \Omega$ the following equality holds:

$$X_{t+h}(\omega) - X_{s+h}(\omega) = X_t(\theta_h\omega) - X_s(\theta_h\omega). \tag{11.3}$$

Obviously, a process X, being a helix, presents at the same time a process with strictly stationary increments. From the distributional point of view processes with strictly stationary increments may be considered as helices, in view of Meyer and De Sam Lazaro's result (see also Subsection 8) given here without the proof.

Lemma 3 ([358]). *Let* X *be a stochastic process, satisfying Definition* 1, *however with the equality* (11.3) *which holds only* P-a.s.

Then a helix X' *can be found such that* X *and* X' *are indistinguishable.*

A process $X = (X_t)_{t\ge 0}$ with $X_0 = 0$ and with right-continuous trajectories which satisfies the equality (11.3) for each nonnegative t, s, t +h, s + h and $\omega \in \Omega$ is also called a helix as we extend the definition to nonnegative t by the formula

$$X_{-t}(\omega) = -X_t(\theta_{-t}\omega),\quad t > 0. \tag{11.4}$$

4. We will establish now properties of semimartingale helices, i.e. of processes being semimartingales and helices simultaneously.

Theorem 2. *Let* $(M_t, \mathcal{F}^P_t)_{t\ge 0}$ *be a local semimartingale helix. Then the decomposition*

$$M = M^c + M^d \tag{11.5}$$

involves the processes $(M^c_t, \mathcal{F}^P_t)_{t \geq 0}$ and $(M^d_t, \mathcal{F}^P_t)_{t \geq 0}$ *which present local martingale helices of the classes* $\mathcal{M}^c_{loc}$ *and* $\mathcal{M}^d_{loc}$ *respectively, and the process of quadratic variation* $[M, M]$ *presents a helix.*

Proof. Let $s > 0$ and $t > 0$. Set $N_t = M_{s+t} - M_s$. Obviously, the process $(N_t, \mathcal{F}^P_{t+s})_{t \geq 0}$ is a local martingale and it admits the decomposition (Theorem 1.7.2)

$$N = N^c + N^d,$$

where the continuous and purely discontinuous martingale components are given by the formulas

$$N^c_t = M^c_{s+t} - M^c_s, \qquad N^d_t = M^d_{s+t} - M^d_s.$$

On the other hand, by the definition of a helix

$$N_t(\omega) = M_{s+t}(\omega) - M_s(\omega) = M_t(\theta_s\omega) = M^c_t(\theta_s\omega) + M^d_t(\theta_s\omega).$$

Since

$$(M^c_t(\theta_s\omega), \mathcal{F}^P_{t+s}) \in \mathcal{M}^c_{loc},$$

$$(M^d_t(\theta_s\omega) - M_0(\theta_s\omega), \mathcal{F}^P_{t+s} \in \mathcal{M}^d_{loc}$$

we have

$$M^c_{s+t}(\omega) - M^c_s(\omega) = M^c_t(\theta_s\omega),$$

$$M^d_{s+t}(\omega) - M^d_s(\omega) = M^d_t(\theta_s\omega) - M_0(\theta_s\omega)$$

as the decomposition of type (11.5) is unique (Theorem 1.7.2); hence the first assertion holds as t and s are arbitrary.

The second assertion holds, since by Theorem 1.8.4 we have

$$\text{P-}\lim_n \sum_{j=0}^{n-1} \left(M_{t^n_{j+1}} - M_{t^n_j}\right)^2 = [M, M]_t - [M, M]_s,$$

where $t > s > 0$ and $s \equiv t^n_0 < t^n_1 < \ldots < t^n_n \equiv t$, $n \geq 1$, is the dense partition of the interval $[s, t]$. Consequently, $[M, M]_t(\theta_h\omega) - [M, M]_s(\theta_h\omega)$ with $t + h \geq 0$ and $s + h \geq 0$ is the limit in probability of the sequence

$$\sum_{j=0}^{n-1} \left(M_{t^n_{j+1}}(\theta_h\omega) - M_{t^n_j}(\theta_h\omega)\right)^2, \quad n \geq 1,$$

since M is a helix.

Therefore P-a.s.

$$[M, M]_t(\theta_h\omega) — [M, M]_s (\theta_h\omega) = [M, M]_{t+h} (\omega) — [M, M]_{s+h} (\omega)$$

and hence one may assume, in view of Lemma 3, that the last equality holds for each $\omega \in \Omega$.

Theorem 3. *Let* $A = (A_t, \overset{P}{\mathcal{F}}_t)_{t \geq 0} \in \mathcal{A}_{loc}$, *and let* $\tilde{A} = (\tilde{A}_t, \overset{P}{\mathcal{F}}_t)_{t \geq 0}$ *be the compensator* A. *If* A *is a helix, then* $\tilde{A}$ *is a helix.*

Proof. Denote $M_t = A_t — \tilde{A}_t$. By the definition of the compensator $\tilde{A}$ the process $(M_t, \overset{P}{\mathcal{F}}_t)_{t \geq 0} \in \mathcal{M}_{loc}$ (Theorem 1.6.3). Hence, as s > 0, we have

$$(M_{t+s}, \overset{P}{\mathcal{F}}_{t+s})_{t \geq 0} \in \mathcal{M}_{loc}.$$

Since

$$M_{t+s}(\omega) = A_{t+s}(\omega) — \tilde{A}_{t+s}(\omega) = A_t (\theta_s\omega) — \tilde{A}_{t+s}(\omega),$$

then $(\tilde{A}_{t+s}(\omega), \overset{P}{\mathcal{F}}_{t+s})$ is the compensator of the process $(A_t(\theta_s\omega), \overset{P}{\mathcal{F}}_{t+s})$.

On the other hand $(\tilde{A}_t(\theta_s\omega), \overset{P}{\mathcal{F}}_{t+s})$ is the compensator of the same process. By this and the uniqueness of the compensator the processes $(\tilde{A}_{t+s}(\omega))$ and $(\tilde{A}_t(\theta_s\omega))$ are indistinguishable, i.e. $\tilde{A}$ is a helix.

Corollary. *Let* $(M_t, \overset{P}{\mathcal{F}}_t) \in \mathcal{M}^2_{loc,0}$ *be a helix. Then the process of square characteristic* $\langle M \rangle$ *presents a helix.*

Proof. Due to Theorem 2 [M, M] is a helix. Consequently, $\langle M \rangle$ is a helix, since $\langle M \rangle$ is the compensator of [M, M].

Theorem 4. *Let* $X = (X_t, \overset{P}{\mathcal{F}}_t) \in$ **Sp** *be a helix. Then the decomposition*

$$X = A + M$$

with $A = (A_t, \overset{P}{\mathcal{F}}_t) \in V \cap \mathcal{P}$ *and* $(M_t, \overset{P}{\mathcal{F}}_t) \in \mathcal{M}_{loc}$ *involves the processes* A *and* M *that are helices.*

Proof. As $s \geq 0$ and $t \geq 0$ we have

$$X_t(\theta_s\omega) = (X_{t+s}(\omega) — X_s(\omega),$$

on the other hand however

$$X_t(\theta_s\omega) = A_t(\theta_s\omega) + M_t(\theta_s\omega),$$

$$X_{t+s}(\omega) — X_s(\omega) = (A_{t+s}(\omega) — A_s(\omega)) + (M_{t+s}(\omega) — M_s(\omega)).$$

Due to the uniqueness of the decomposition of a type X = A + M with a predictable process A this entails the indistinguishability of the processes $(A_t(\theta_s\omega))_{t \geq 0}$ and $(A_{t+s}(\omega) — A_s(\omega))_{t \geq 0}$ (analogously, $(M_t(\theta_s\omega))_{t \geq 0}$ and $(M_{t+s}(\omega) — M_s(\omega))_{t \geq 0}$ are indistinguishable). Hence A and M are helices.

Theorem 5. *Let* $X = (X_t, \mathcal{F}^P_t) \in S$ *be a helix. Then*

1) *if* $g = g(x)$ *is a measurable function such that*

$$\sum_{s \le t} | g(\Delta X_s) | < \infty, \quad t > 0,$$

then the process $Y = (Y_t)_{t \ge 0}$ *with*

$$Y_t = \sum_{s \le t} g(\Delta X_s)$$

is a helix;

2) *the decomposition* $X = A + M$ *with* $A \in V$ *and* $M \in \mathcal{M}_{loc}$ *may involve processes* A *and* M *that are helices;*

3) *the continuous martingale component* X^c *is a helix;*

4) *the quadratic characteristic* $[X, X]$ *is a helix;*

5) *if* $T = (B, C, \nu)$ *is the triplet of predictable characteristics* X, *then* B, C *and* $g * \nu$ *are helices, where* $g = g(x)$ *is a measurable function such that* $| g | * \nu \in V^+$.

Proof. Assertion 1) is obvious. In particular the process Y is a helix as

$$g(x) = x\, I(| x | > 1).$$

2) Denote

$$X'_t = X_t - \sum_{0 < s \le t} \Delta X_s I(| \Delta X_s | > 1).$$

According to Assertion 1) the process $X' = (X'_t)_{t \ge 0}$ is a helix. Besides $(X'_t, \mathcal{F}^P_t) \in \mathbf{Sp}$ (Theorem 2.1.1). Therefore the decomposition $X' = A' + M$ involves the processes A' and M that are helices, by Theorem 3. Now it suffices to define the process A by the following formula

$$A_t = A'_t + \sum_{s \le t} \Delta X_s I(| \Delta X_s | > 1).$$

3) The continuous martingale component $X^c = M^c$, where M is a local martingale in the decomposition $X = A + M$ with the process introduced in the course of proving 2) (see Ch. 4, § 1). Since M is a helix, M^c is a helix too (Theorem 2).

4) is proved analogously to the corresponding assertion of Theorem 2 in view of Problem 2.1.7.

5) Due to the canonical decomposition (see Ch. 4, § 1)

$$X_t = B_t + X^c_t + \sum_{s \le t} \Delta X_s (| \Delta X_s | > 1) + \int_0^t \int_{| x | \le 1} x \, d(\mu - \nu)$$

where $\mu = \mu(dt, dx)$ is the jump measure of X. Note that $B = A'$ where A' is the process involved in the decomposition $X' = A' + M$, used in the course of proving 2). Hence B is a helix.

The process C is a helix, because $C = \langle X^c \rangle$, X^c is a helix due to Assertion 3) and the corollary to Theorem 3 holds. Finally, the process $g * \nu$ is a helix in virtue of Theorem 2, since $g * \nu$ is the compensator of the process

$$\left(\sum_{s \le t} g(\Delta X_s) \right)_{t \ge 0},$$

that is a helix due to Assertion 1) of the theorem.

5. We will present a construction of semimartingale helices by means of the stochastic integration. For this we will need the following

Definition 2. A random process $H = (H_t)_{t \in R}$, defined on a stochastic basis $(\Omega, \mathcal{F}^P, \mathbb{F}^P, P)$, is called *homogeneous*, if for each t, s ∈ R we have

$$H_t(\theta_s \omega) = H_{t+s}(\omega), \ \omega \in \Omega.$$

Obviously, a homogeneous process H is a strictly stationary process.

Let H be a homogenous process, let a helix $A = (A_t)_{t \ge 0} \in V$ and

$$|H| \circ \operatorname{Var}(A) \in V^+. \tag{11.6}$$

In this case the Lebesgue-Stieltjes integral $H \circ A_t$ is defined (see Ch. 2, § 2). The stochastic process $H \circ A = (H \circ A_t)_{t \ge 0}$ here is a helix (Problem 7).

Consider another situation. Let H be a homogeneous predictable process and $M = (M_t, \mathcal{F}^P_t)_{t \ge 0} \in \mathcal{M}_{loc}$ be a helix. If

$$(H \circ [M, M])^{1/2} \in \mathcal{A}^+_{loc}, \tag{11.7}$$

then the stochastic integral $H \cdot M = (H \cdot M_t)_{t \ge 0}$ is defined (Theorem 2.2.5).

Theorem 6. *Let* H *be a homogeneous predictable process,* M *a local martingale helix and let Condition* (11.7) *be satisfied. Then the process* H · M *is a local martingale helix.*

Proof. We will utilize the decomposition

$$H \cdot M = H \cdot M^c + H \cdot M^d,$$

and we will show that each of these processes are helices.

Let $t \ge 0$, $s \ge 0$ and

$$X^c_t = H \cdot M^c_{t+s}(\omega) - H \cdot M^c_s(\omega),$$

$$X^d_t = H \cdot M^d_{t+s}(\omega) - H \cdot M^d_s(\omega).$$

Denote

$$\overline{X}^c_t = H \cdot M^c_t(\theta_s\omega)$$

and

$$\overline{X}^d_t = H \cdot M^d_t(\theta_s\omega).$$

It suffices to show that the processes X^c and $\overline{X}^c$ are indistinguishable (X^d and $\overline{X}^d$ are indistinguishable). By construction $(X^c_t, \mathcal{F}^P_{t+s}) \in \mathcal{M}^c_{loc}$ and $(X^d_t, \mathcal{F}^P_{t+s}) \in \mathcal{M}^d_{loc}$.

On the other hand $(\overline{X}^c_t, \mathcal{F}^P_{t+s}) \in \mathcal{M}^c_{loc}$ and $(\overline{X}^d_t, \mathcal{F}^P_{t+s}) \in \mathcal{M}^d_{loc}$ by Lemma 2. Hence

$$X^c - \overline{X}^c \in \mathcal{M}^c_{loc}, \quad X^d - \overline{X}^d \in \mathcal{M}^d_{loc}.$$

Since $\Delta(X^d - \overline{X}^d) = 0$, we have

$$[X^d - \overline{X}^d, X^d - \overline{X}^d]_\infty = 0.$$

Therefore the process $X^d - \overline{X}^d$ is indistinguishable from zero, due to Davis' inequality (Theorem 1.9.6). To prove that the processes X^c and $\overline{X}^c$ are indistinguishable, it suffices to show that $\langle X^c - \overline{X}^c\rangle_\infty = 0$ and then to apply Davis' inequality. We have

$$\langle X^c - \overline{X}^c\rangle = \langle X^c\rangle + \langle\overline{X}^c\rangle - 2\langle X^c, \overline{X}^c\rangle.$$

By Definition 5 in Ch. 2, Subsection 2.4,

$$\langle X^c\rangle_t = H^2 \circ \langle M^c\rangle_{t+s}(\omega) - H^2 \circ \langle M^c\rangle_s(\omega),$$

$$\langle\overline{X}^c\rangle_t = H^2 \circ \langle M^c\rangle_t(\theta_s\omega).$$

Due to Theorem 2 M^c is a helix, and due to the corollary to Theorem 3 $\langle M^c\rangle$ is a helix. This entails $\langle\overline{X}^c\rangle = \langle X^c\rangle$. Therefore it suffices to show that $\langle X^c, \overline{X}^c\rangle = \langle X^c\rangle$.

Let H^n, $n \geq 1$, be a sequence of elementary predictable functions (not necessarily presenting homogeneous processes) such that (Problem 2.2.10)

$$(H, H^n) \circ \langle M^c\rangle_{t+s} \xrightarrow{P} 0.$$

Then

$$(H^n \cdot M^c_t(\theta_s\omega), \mathcal{F}^P_{t+s}) \in \mathcal{M}^c_{loc}.$$

In virtue of the Lenglart-Rebolledo inequality (Theorem 1.9.3)

$$\sup_{u \le t} |\overline{X}^c_t - H^n \cdot M^c_t(\theta_s\omega)| \xrightarrow{P} 0,$$

and besides

$$\langle X^c, H^n \cdot M^c(\theta_s\omega)\rangle_t \xrightarrow{P} \langle X^c, \overline{X}^c\rangle_t$$

(Problem 2.2.10). Observe now that

$$H^n \cdot M^c_t(\theta_s\omega) = \int_0^t H^n_u(\theta_s\omega)\, dM^c_u(\theta_s\omega) = \int_s^{t+s} H^n_{u-s}(\theta_s\omega)\, dM^c_u(\omega).$$

Consequently,

$$\langle X^c, H^n \cdot M^c(\theta_s\omega)\rangle_t = \int_s^{t+s} H_u(\omega)\, H^n_{u-s}(\theta_s\omega)\, d\langle M^c\rangle_u(\omega).$$

Therefore

$$\langle X^c, \overline{X}^c\rangle_t = \int_s^{t+s} H_u(\omega)\, H_{u-s}(\theta_s\omega)\, d\langle M^c\rangle_u(\omega)$$

$$= \int_s^{t+s} H^2_u(\omega)\, d\langle M^c\rangle_u(\omega) = \langle X^c\rangle_t.$$

The definition of the stochastic integral with respect to a semimartingale (see Ch. 2, § 2), Theorem 6 and Problem 7 imply

Theorem 7. *Let* X *be a semimartingale helix with the decomposition* $X = A + M$, *where* $A \in \mathcal{V}$, $M \in \mathcal{M}_{loc}$, *and* A *and* M *are helices. If* H *is a predictable homogeneous process such that*

$$|H| \circ \mathrm{Var}(A) \in \mathcal{V}^+, \quad (H^2 \circ [M, M])^{1/2} \in \mathcal{A}^+_{loc},$$

then $H \cdot X$ *is a semimartingale helix.*

6. Let $a = a(t)$, $t \in R$ be a measurable function such that

$$\int_{-\infty}^{\infty} a^2(t)\, dt < \infty$$

and let $X = (X_t)_{t \in R}$ be a helix that is a process with independent increments and

$$\mathbf{E}(X_t - X_s) = 0, \quad \mathbf{E}(X_t - X_s)^2 = |t - s|, \ t, \ s \in R$$

with trajectories having left-hand limits.

In this case X is a square integrable martingale and the stochastic integral $a \cdot X = (a \cdot X_t)_{t \in R}$ is defined by

$$a \cdot X_t = \int_{-\infty}^{t} a(s)\, dX_s,$$

that is a square integrable martingale with the quadratic characteristic $\langle a \cdot X\rangle = (\langle a \cdot X_t\rangle)_{t \in R}$ with

$$\langle a \cdot X\rangle_t = \int_{-\infty}^{t} a^2(s)\, ds.$$

For a measurable function $a = a(t)$ with

$$\int_0^{\infty} a^2(t)\, dt < \infty$$

the random variables

$$\xi_t = \int_{-\infty}^{t} a(t-s)\, dX_s$$

are defined at each $t \in R$, which present a strictly stationary process $\xi = (\xi_t)_{t \in R}$ with

$$\mathbf{E}\, \xi_t = 0 \text{ and } \mathbf{E}\, \xi_t^2 = \int_0^{\infty} a^2(s)\, ds.$$

7. We will present now a number of results analogous to the ergodic theorem. We begin with the following result.

Theorem 8. *Let* $X = (X_t)_{t \geq 0}$ *be a helix with trajectories in* $D\,[0, \infty)$.

If $\mathbf{E}\,|X_1| < \infty$, *then*

$$\frac{X_t}{t} \xrightarrow{P} \mathbf{E}\,(X_1 \mid J), \quad t \to \infty.$$

If $\mathbf{E} \sup_{0 \leq s \leq 1} |X_s| < \infty$, *then*

$$\mathbf{E}\left|\frac{X_t}{t} - \mathbf{E}\,(X_1 \mid J)\right| \to 0, \quad t \to \infty.$$

Proof. Let $[t]$ denote the integer part of t. Then

$$X_{[t]}(\omega) = \sum_{k=1}^{[t]} (X_k(\omega) - X_{k-1}(\omega)) = \sum_{k=1}^{[t]} X_1(\theta_{k-1}\omega).$$

Therefore due to the Birkhof-Khintchine theorem for discrete time [289], [332]

$$\frac{X_{[t]}}{[t]} \to \mathbf{E}\,(X_1 \mid J)$$

P-a.s. and in the mean. Hence

$$\frac{X_{[t]}}{t} \to \mathbf{E}\,(X_1 \mid J),\ \ t \to \infty.$$

P-a.s. and in the mean. But

$$\left| \frac{X_t - X_{[t]}}{t} \right| \le \frac{1}{t} \sup_{0 \le s \le 1} | X_{[t]+s} - X_{[t]} |.$$

This means that the desired assertion holds provided

$$\frac{1}{t} \sup_{0 \le s \le 1} | X_{[t]+s} - X_{[t]} |$$

converges to zero as $t \to \infty$ in probability and in the mean. Observe that

$$P\left(\frac{1}{t} \sup_{0 \le s \le 1} | X_{[t]+s}(\omega) - X_{[t]}(\omega) | > \varepsilon \right)$$

$$= P\left(\frac{1}{t} \sup_{0 \le s \le 1} | X_s(\theta_{[t]}\omega) | > \varepsilon \right)$$

$$= P\left(\frac{1}{t} \sup_{0 \le s \le 1} | X_s(\omega) | > \varepsilon \right) \to 0,\ \ t \to \infty$$

since $X \in D$. Next

$$\mathbf{E} \frac{1}{t} \sup_{0 \le s \le 1} | X_{[t]+s}(\omega) - X_{[t]}(\omega) | = \frac{1}{t} \mathbf{E} \sup_{0 \le s \le 1} | X_s(\theta_{[t]}\omega) |$$

$$= \frac{1}{t} \mathbf{E} \sup_{0 \le s \le 1} | X_s(\omega) | \to 0,\ \ t \to \infty$$

under the condition

$$\mathbf{E} \sup_{0 \le s \le 1} | X_s | < \infty.$$

Corollary. *If* $X \in \overline{\mathfrak{M}}$ *is a helix (and hence* $X_0 = 0$*), then*

$$\frac{X_t}{t} \xrightarrow{P} 0,\ \ t \to \infty.$$

If, besides, $\mathbf{E}\,[X, X]_t^{1/2} < \infty,\ t > 0$, *then*

$$\mathbf{E} \left| \frac{X_t}{t} \right| \to 0,\ \ t \to \infty.$$

The first assertion holds, since $J \subseteq \mathcal{F}_0$, and hence

$$\mathbf{E}\,(X_1 \mid J) = \mathbf{E}\,(\mathbf{E}\,(X_1 \mid \mathcal{F}_0) \mid J) = \mathbf{E}\,(X_0 \mid J) = 0.$$

The second assertion is based on Davis' inequality (Theorem 1.9.6), according to which

$$\mathbf{E} \sup_{s \le t} | X_s | \le C\mathbf{E}\,[X, X]_t^{1/2}.$$

Theorem 9. *Let* X *be a semimartingale helix with the decomposition* X = A + M, *where* $A \in V$, $M \in \mathfrak{M}_{loc}$ *and* A *and* M *are helices. If the conditions* $\mathbf{E}\operatorname{Var}(A)_t < \infty$ *and* $\mathbf{E}M_t^2 < \infty$ *are satisfied,, then*

$$\frac{X_t}{t} \to \mathbf{E}(A_1 \mid J)$$

as $t \to \infty$ *with probability one and in the mean.*

Proof. The convergence in the mean follows from Theorem 8, since

$$\mathbf{E} \sup_{0 \le s \le 1} |X_s| \le \mathbf{E}\operatorname{Var}(A)_1 + \mathbf{E} \sup_{0 \le s \le 1} |M_s| \le \mathbf{E}\operatorname{Var}(A_1) + 2\,(\mathbf{E}M_1^2)^{1/2}$$

in view of Doob's inequality (Theorem 1.9.2) and

$$\mathbf{E}(X_1 \mid J) = \mathbf{E}(A_1 \mid J) + \mathbf{E}(M_1 \mid J) = \mathbf{E}(A_1 \mid J) \quad (\text{P-a.s.})$$

(as $\mathbf{E}(M_1 \mid J) = 0$).

Next, observe that $\dfrac{M_t}{t} \to 0$ P-a.s. as $t \to \infty$, since

$$\int_0^\infty \frac{d\langle M\rangle_t}{(1+t)^2} < \infty \quad \text{P-a.s.}$$

(Theorem 2.6.10). The last inequality takes place due to Problem 5 which gives

$$\mathbf{E}\int_0^\infty \frac{d\langle M\rangle_t}{(1+t)^2} = \mathbf{E}\langle M\rangle_1 \int_0^t \frac{dt}{(1+t)^2} = \mathbf{E}M_1^2.$$

Therefore to prove the theorem, it suffices to show that

$$\frac{A_t}{t} \to \mathbf{E}(A_1 \mid J) \quad \text{P-a.s.}$$

Repeting the arguments used in the course of proving Theorem 8 we see that it suffices to establish that

$$\frac{1}{t} \sup_{0 \le s \le 1} |A_{[t]+s} - A_{[t]}| \to 0, \quad t \to \infty \quad \text{P-a.s.}$$

To establish this relation observe that

$$\sup_{0 \le s \le 1} |A_{[t]+s} - A_{[t]}| \le \operatorname{Var}(A)_{[t+1]} - \operatorname{Var}(A)_{[t]}.$$

Since A is a helix, Var (A) is a helix too (Problem 4). By the assumption of the theorem

$$\mathbf{E}\operatorname{Var}(A_1) < \infty.$$

Then by the Birkhof-Khintchine theorem for discrete time [289], [332]

$$\lim_{t \to \infty} \frac{\text{Var}(A)_{[t]}}{t} = \lim_{t \to \infty} \frac{\text{Var}(A)_{[t]}}{[t]} = \mathbf{E}(\text{Var}(A)_1 \mid J) \ (\text{P-a.s.})$$

and hence

$$\lim_{t \to \infty} \frac{1}{t} (\text{Var}(A)_{[t+1]} - \text{Var}(A)_{[t]} = 0. \ (\text{P-a.s.})$$

Corollary (the Birkhof-Khintchine theorem for discrete time). *Let* $\xi(\omega)$ *be an integrable random variable. Then* $\frac{1}{t}\int_0^t \xi(\theta_s\omega)\, ds$ *converges as* $t \to \infty$ to $\mathbf{E}(\xi \mid J)_t$ P-a.s. *and in the mean.*

Proof. Set

$$X_t = \int_0^t \xi(\theta_s\omega)\, ds.$$

Obviously, $X_t = (X_t)_{t \geq 0}$ is a semimartingale helix with the decomposition $X_t = A_t$. Besides

$$\mathbf{E}\, \text{Var}(A)_t = \int_0^t \mathbf{E} \mid \xi(\theta_s\omega) \mid ds = t\, \mathbf{E} \mid \xi \mid.$$

Hence, in view of the theorem, $\frac{X_t}{t}$ converges as $t \to \infty$ to

$$\mathbf{E}(\int_0^1 \xi(\theta_s\omega)\, ds \mid J)$$

P-a.s. and in the mean. In fact (P-a.s.)

$$\mathbf{E}(\int_0^1 \xi(\theta_s\omega)\, ds \mid J) = \int_0^1 \mathbf{E}(\xi(\theta_s\omega) \mid J)\, ds = \int_0^1 \mathbf{E}(\xi(\omega) \mid J)\, ds = \mathbf{E}(\xi \mid J).$$

(see Problem 6.)

8. Let $\xi = (\xi_t)_{t \in R}$ be a strictly stationary process, defined on a probability space $(\Omega, \mathcal{F}, P)$, with trajectories in $(\mathbf{X}, \mathcal{X})$ ($\mathbf{X}$ is a space of measurable functions $X = (X_t)_{t \in R}$ and $\mathcal{X}$ a σ-algebra of cylindric sets)

$$\mathcal{X}_\Delta = \{\pi^{-1}_{t_1, \dots, t_k}(A) \colon A \in B(R^k)\}$$

with $t_j \in R$ and

$$\pi^{-1}_{t_1, \dots, t_k}(A) = \{(X_{t_1}, \dots, X_{t_k}) \in A\}.$$

Let Q be a probability measure on $(\mathbf{X}, \mathfrak{X})$ such that for each $\Gamma \in \xi$

$$Q(\Gamma) = P(\xi \in \Gamma).$$

Then a stochastic process $X = (X_t)_{t \in R}$, given on a probability space $(\mathbf{X}, \mathfrak{X}, Q)$ coincides in distribution with a stochastic process $\xi = (\xi_t)_{t \in R}$, given on $(\Omega, \mathcal{F}, P)$.

One may define on $(\mathbf{X}, \mathfrak{X}, Q)$ a measure preserving group of transformations $\theta = (\theta_t)_{t \in R}$ from $\mathbf{X}$ to $\mathbf{X}$, by setting

$$X_t(\theta_s X) = X_{t+s}(X),$$

i.e. θ_s is a shift transformation of lag s. Therefore a stochastic basis

$$(\mathbf{X}, \mathfrak{X}^Q, \mathbb{F}^Q = (\mathcal{F}_t^Q)_{t \in R}, Q)$$

may be defined, which is naturally called a coordinate basis. Denote by J the σ-algebra of invariant sets generated by θ.

Define the σ-algebras

$$G_t^\xi = \sigma\{\xi_s, -\infty < s \le t\}$$

and

$$\mathcal{F}_t^\xi = \bigcap_{\varepsilon > 0} G_{t+\varepsilon}^\xi \vee \mathcal{N}^P,$$

where $\mathcal{N}^P$ is the system of sets of P-measure zero from the $\mathcal{F}^P$-completion of the σ-algebra $\mathcal{F}$ with respect to the measure P.

To a stationary process ξ relate the σ-algebra J^ξ of invariant sets (i.e. sets $A \in \mathcal{F}$ such that there exists $B \in \mathfrak{X}$ for which

$$A = \{\omega: (\xi_s(\omega))_{s \ge t} \in B\}$$

for each $t \in R$), completed by sets from $\mathcal{F}^P$ of P-measure zero.

If $\eta = (\eta_t)_{t \in R}$ is an $\mathbb{F}^\xi = (\mathcal{F}_t^\xi)_{t \in R}$-adapted process, then an $\mathbb{F}^Q$-adapted process $\bar{\eta}(X) = (\bar{\eta}_t(X))_{t \in R}$ can be found such that $\bar{\eta}(\xi) = (\bar{\eta}_t(\xi))_{t \ge 0}$ presents a modification of the process η. This implies that in case of a strictly stationary process η the equality

$$\bar{\eta}_{t+s}(X) = \bar{\eta}_t(\theta_s X) \quad (Q\text{-a.s.}),$$

takes place, while in case of a process η with strictly stationary increments

$$\bar{\eta}_{t+h}(X) - \bar{\eta}_{s+h}(X) = \bar{\eta}_t(\theta_h X) - \bar{\eta}_s(\theta_h X) \quad (Q\text{-a.s.})$$

Therefore, if the process $\bar{\eta}(X)$ has right-continuous trajectories and $\bar{\eta}_0(X) = 0$, then in view of Lemma 3 there exists a process $\eta'(X)$, which is a helix, indistinguishable from $\bar{\eta}(X)$.

In particular, if $(\eta_t)_{t \ge 0} \in S(\mathbb{F}^\xi)$ with $\eta_0 = 0$ presents a process with strictly stationary increments, then there exists a process $(\bar{\eta}_t(X))_{t \ge 0} \in S(\mathbb{F}^Q)$, which is a helix. There are analogous assertions concerning a local martingale, a local square integrable martingale and an increasing process.

If $\alpha = \alpha(X)$ is a J-measurable random variable, then $\alpha(\xi)$ is J^{ξ}-measurable.

Problems

1. Show that $J \subseteq \overset{P}{\mathfrak{F}}_t$, $t \in R$, and hence $J \subseteq \overset{P}{\mathfrak{F}}_0$.

2. Let $X = (X_t)_{t \in R}$ be a helix and $\mathbf{E}|X_t| < \infty$, $t \in R$. Show that
$$\mathbf{E}X_t = t\,\mathbf{E}X_1.$$

3. Let $X = (X_t)_{t \in R}$ be a helix and $\mathbf{E}|X_t| < \infty$, $t \in R$. Show that
$$\mathbf{E}(X_t \mid J) = t\,\mathbf{E}(X_1 \mid J).$$

4. Let $X \in V$ be a helix. Then $\mathrm{Var}(X)$ is a helix.

5. Let $X \in V$ be a helix and $\mathbf{E}\,\mathrm{Var}(X)_t < \infty$. If $H = H(t)$ is a measurable function such that
$$\int_0^t |H(s)|\,ds < \infty,$$
then the stochastic integral $H \cdot X$ is defined and
$$\mathbf{E}\,H \cdot X_t = \mathbf{E}\,X_1 \int_0^t H(s)\,ds.$$

6. Let $\xi(\omega)$ be an integrable random variable. Show that (P-a.s.)
$$\mathbf{E}(\xi(\theta_t\omega) \mid J) = \mathbf{E}(\xi(\omega) \mid J),\quad t \in R.$$

7. Let H be a homogeneous process, and $A \in V$ a helix. Show that as $|H| \circ \mathrm{Var}(A) \in V$, then the process $H \circ A$ is a helix.

§ 12. Semimartingales - stationary processes

1. In the following two theorems the properties of a semimartingale are presented, which are obtained under the assumption that it is a stationary process.

Theorem 1. *Let a strictly stationary process* $X = (X_t)_{t \geq 0}$ *be a semimartingale with the decomposition*

$$X_t = X_0 + A_t + M_t,$$

where

$$EX_0^2 < \infty, \quad M = (M_t)_{t \geq 0} \in \overline{\mathcal{M}}^2, \quad E\,\mathrm{Var}^2(A)_t < \infty, \quad t > 0.$$

Then

$$\frac{1}{\sqrt{n}} \sup_{t \leq nT} |X_t| \xrightarrow{P} 0, \quad n \to \infty, \quad \forall\, T > 0.$$

Proof. Denote

$$\xi_k = \sup |X_t|,$$

where sup is taken over $t \in ((k-1)T, kT]$. Then

$$\sup_{t \leq nT} |X_t| = \max_{1 \leq k \leq n} \xi_k.$$

Therefore it suffices to show

$$\frac{1}{\sqrt{n}} \max_{1 \leq k \leq n} \xi_k \xrightarrow{P} 0, \quad n \to \infty.$$

In turn, this relation takes place provided

$$\max_{k \geq m} \frac{\xi_k}{\sqrt{k}} \xrightarrow{P} 0, \quad m \to \infty \tag{12.1}$$

by the following succession of inequalities:

$$\frac{1}{\sqrt{n}} \max_{1 \leq k \leq n} \xi_k \leq \frac{1}{\sqrt{n}} \max_{1 \leq k \leq m-1} \xi_k + \frac{1}{\sqrt{n}} \max_{m \leq k \leq n} \xi_k$$

$$\leq \frac{1}{\sqrt{n}} \sum_{k=1}^{m-1} \xi_k + \max_{k \geq m} \frac{\xi_k}{\sqrt{k}}.$$

The relation (12.1) means that

$$\frac{\xi_n}{\sqrt{n}} \to 0 \quad \text{P-a.s.}$$

The relation (12.1) is certainly valid when for each $\varepsilon > 0$

$$\sum_{n=1}^{\infty} \mathsf{P}\left(\frac{|\xi_n|}{\sqrt{n}} > \varepsilon \right) < \infty.$$

Since x_n, $n \geq 1$, are identically distributed random variables, then

$$\sum_{n=1}^{\infty} P\left(\frac{|\xi_n|}{\sqrt{n}} > \varepsilon\right) = \sum_{n-1}^{\infty} P\left(\frac{|\xi_1|}{\sqrt{n}} > \varepsilon\right)$$

$$= \sum_{n=1}^{\infty} P(\xi_1^2 \varepsilon^{-2} > n) = \varepsilon^{-2} E\xi_1^2.$$

Consequently, the relation (12.1) holds, provided

$$E\xi_1^2 = E \sup_{t \leq T} X_t^2 < \infty.$$

But this is satisfied under the condition of the theorem in virtue of the estimate

$$\sup_{t \leq T} X_t^2 \leq 3 (X_0^2 + \mathrm{Var}^2(A)_T + \sup_{t \leq T} M_t^2)$$

and Doob's inequality

$$E \sup_{t \leq T} M_t^2 \leq 4 \, EM_T^2$$

(Theorem 1.9.2).

Theorem 2. *Let a strictly stationary process* $X = (X_t)_{t \geq 0}$ *be a semimartingale with the decomposition*

$$X_t = X_0 + A_t + M_t,$$

where $A = (A_t)_{t \geq 0} \in \mathcal{V}$ *and* $M = (M_t)_{t \geq 0} \in \overline{\mathcal{M}}$. *Let the following conditions be satisfied:*

$$E|X_0| < \infty, \quad E\int_0^t (1 + |X_{s-}|) \, d\,\mathrm{Var}(A)_s < \infty, \ t > 0.$$

Then

$$E[X, X]_t = -2E\int_0^t X_{s-} dA_s, \ t > 0.$$

If, besides, $A \in \mathcal{V} \cap \mathcal{P}$, *then*

$$M \in \overline{\mathcal{M}}^2, \quad E[A, A]_t < \infty, \ t > 0$$

and

$$EM_t^2 = -E\left(2\int_0^t X_{s-} dA_s + [A, A]_t\right).$$

Proof. Approximate the function $f(x) = x^2$ by the sequence of functions $f_N(x)$, $N \geq 1$, such that

$$f_N(x) = \int_0^x \int_0^y g_N(z)\,dzdy, \tag{12.2}$$

where $g_N(z) = g_N(-z)$ and, as $z \geq 0$,

$$g_N(z) = \begin{cases} 2, & z \leq N, \\ 2(N+1-z), & N < z \leq N+1 \\ 0, & z > N+1. \end{cases}, \tag{12.3}$$

Obviously,

$$g_N(z) \to 2,\ f_N'(z) \to 2z,\ f_N(z) \to z^2 \text{ as } N \to \infty,$$

and, besides,

$$|f_N'(z)| \leq 2|z| \wedge L(N), \quad f_N(z) \leq L(N)(1+|z|),$$

where a constant $L(N)$ depends on N only. By Ito's formula (Ch. 2, § 3) applied to $f_N(X_t)$ we get

$$f_N(X_t) = f_N(X_0) + f_N'(X-) \circ A_t + f_N'(X-) \circ M_t + \frac{1}{2} g_N(X) \circ \langle M^c \rangle_t$$

$$+ \sum_{s \leq t} [f_N(X_s) - f_N(X_{s-}) - f_N'(X_{s-})\Delta X_s] \tag{12.4}$$

with

$$f_N(X_s) - f_N(X_{s-}) - f_N'(X_{s-})\Delta X_s$$

$$= \int_0^{\Delta X_s} \int_0^y g_N(X_{s-} + z)\,dz\,dy. \tag{12.5}$$

Let $(\tau_n)_{n \geq 1}$ be a localizing sequence for $f_N'(X-) \cdot M \in \mathcal{M}_{loc}$. By (12.4) and (12.5) we get the equality

$$\mathbf{E} f_N(X_{t \wedge \tau_n}) = \mathbf{E} f_N(X_0) + \mathbf{E} f_N'(X-) \circ A_{t \wedge \tau_n}$$

$$+ \mathbf{E}\left(\frac{1}{2} g_N(X) \circ \langle M^c \rangle_{t \wedge \tau_n} + \sum_{s \leq t \wedge \tau_n} \int_0^{\Delta X_s} \int_0^y g_N(X_{s-} + z)\,dz\,dy\right). \tag{12.6}$$

We will now show that

$$\lim_n \mathbf{E} f_N(X_{t \wedge \tau_n}) = \mathbf{E} f_N(X_t). \tag{12.7}$$

To this end, we note that

$$|f_N(X_{t\wedge\tau_n}) - f_N(X_t)| \le L(N)\,|X_{t\wedge\tau_n} - X_t|$$

$$\le L(N)\,[\mathrm{Var}(A)_t - \mathrm{Var}(A)_{t\wedge\tau_n} + |M_t - M_{t\wedge\tau_n}|].$$

Since

$$\mathbf{E}\,\mathrm{Var}(A)_t = \lim_n \mathbf{E}\,\mathrm{Var}(A)_{t\wedge\tau_n},$$

to prove (12.7) it suffices to show that

$$\lim_n \mathbf{E}\,|M_t - M_{t\wedge\tau_n}| = 0.$$

The last relation holds, because $M \in \overline{\mathcal{M}}$ and hence $(M_{s\wedge\tau})_{s\ge 0}$ is a uniformly integrable martingale.

We note also that

$$\frac{1}{2} g_N(X) \circ \langle M^c\rangle_{t\wedge\tau_n} + \sum_{s\le t\wedge\tau_n} \int_0^{\Delta X_s}\int_0^y g_N(X_{s-}+z)\,dz\,dy$$

is a monotone function of τ_n and consequently

$$\lim_n \mathbf{E}\left(\frac{1}{2} g_N(X) \circ \langle M^c\rangle_{t\wedge\tau_n} + \sum_{s\le t\wedge\tau_n} \int_0^{\Delta X_s}\int_0^y g_N(X_{s-}+z)\,dz\,dy\right)$$

$$= \mathbf{E}\left(\frac{1}{2} g_N(X) \circ \langle M^c\rangle_t + \sum_{s\le t} \int_0^{\Delta X_s}\int_0^y g_N(X_{s-}+z)\,dz\,dy\right). \tag{12.8}$$

Besides, the sequence $f'_N(X-) \circ A_{t\wedge\tau_n}$, $n \ge 1$, has the integrable majorant

$$2\,|X-| \circ \mathrm{Var}(A)_t$$

and consequently

$$\lim_n \mathbf{E}f'_N(X-) \circ A_{t\wedge\tau_n} = \mathbf{E}f'_N(X-) \circ A_t. \tag{12.9}$$

Therefore, taking into consideration (12.7) - (12.9) and the equality

$$\mathbf{E}f_N(X_t) = \mathbf{E}f_N(X_0),$$

which holds as the process $X = (X_t)_{t\ge 0}$ is strictly stationary, and taking the limit $\lim_n$ in (12.6) we get

$$\mathbf{E}\left(\frac{1}{2}g_N(X)\circ\langle M^c\rangle_t+\sum_{s\le t}\int_0^{\Delta X_s}\int_0^y g_N(X_{s-}+z)\,dz\,dy\right)$$

$$= -\,\mathbf{E}f_N'(X_-)\circ A_t. \tag{12.10}$$

Since

$$\lim_N\left(\frac{1}{2}g_N(X)\circ\langle M^c\rangle_t+\sum_{s\le t}\int_0^{\Delta X_s}\int_0^y g_N(X_{s-}+z)\,dz\,dy\right)=[X,X]_t,$$

taking the limit $\lim_N$ in (12.10) we arrive at the desired representation for $\mathbf{E}\,[X, X]_t$.

If $A \in \mathcal{V} \cap \mathcal{P}$, then the representation for $\mathbf{E}M_t^2$ and the inequality $\mathbf{E}\,[A, A]_t < \infty$ are obtained in the following manner.

Denote $N = [M, A]$ and observe that

$$[X, X] = [M, M] + 2N + [A, A].$$

The process N is a local martingale (Problem 2.4.5), and if $(\sigma_k)_{k\ge1}$ is a localizing sequence, then

$$\mathbf{E}\,[X, X]_t = \lim_k \mathbf{E}\,[X, X]_{t\wedge\sigma_k}$$

$$= \lim_k \mathbf{E}\,[M, M]_{t\wedge\sigma_k} + \lim_k \mathbf{E}\,[A, A]_{t\wedge\sigma_k}$$

$$= \mathbf{E}\,[M, M]_t + \mathbf{E}\,[A, A]_t. \tag{12.11}$$

Since $\mathbf{E}\,[X, X]_t < \infty$, $t > 0$, for each $t > 0$ we have $\mathbf{E}\,[M, M]_t < \infty$ and $\mathbf{E}\,[A, A]_t < \infty$. By the Burkholder-Gandy inequality (Theorem 1.9.7 as p = 2) we get

$$\mathbf{E}M_t^2 \le C\mathbf{E}\,[M, M]_t < \infty$$

and hence

$$M \in \overline{\mathcal{M}}^2,$$

i.e.

$$\mathbf{E}M_t^2 = \mathbf{E}\,[M, M]_t.$$

This formula, (12.11) and the representation for $\mathbf{E}\,[X, X]_t$ give the formula for $\mathbf{E}M_t^2$.

The theorem is proved.

Problems

1. Let $X = (X_t, \mathcal{F}_t)$ be a semimartingale with the decomposition

$$X_t = X_0 + A_t + M_t,$$

where

$$A_t = \int_0^t \xi_s ds$$

and $\xi = (\xi_t)_{t \geq 0}$ is a strictly stationary process.

If

$$\mathbf{E}X_0^2 < \infty, \quad M \in \overline{\mathcal{M}}, \quad \mathbf{E}\,|\xi_0| < \infty, \quad \mathbf{E}\,|\xi_0 X_0| < \infty$$

and (X, ξ) is a strictly stationary process, then

$$\mathbf{E}M_t^2 = -2t\,\mathbf{E}X_0\xi_0.$$

2. Let the conditions of Problem 1 be satisfied and let G be a σ-algebra such that $G \subseteq \mathcal{F}_0$. Then (P-a.s.)

$$\mathbf{E}(M_t^2 \mid G) = -2t\,\mathbf{E}(X_0\xi_0 \mid G).$$

§ 13. Exponential inequalities for large deviation probabilities

1. Usually, it is said that an exponential convergence (in probability) takes place for a convergent (in probability) sequence of random variables X^n, $n \le 1$ to a random variable X, if for each $\varepsilon > 0$

$$P(|X^n - X| \ge \varepsilon) \le a e^{-bn} \tag{13.1}$$

with certain constants $a = a(\varepsilon)$ and $b = b(\varepsilon)$. Naturally, to satisfy an exponential inequality of type (13.1) one is bounded to impose considerable restrictions on random variables involved, among which the so-called Cramér condition. (A random variable ξ satisfies Cramér's condition if $\mathbf{E}e^{\lambda|\xi|} < \infty$). For instance, it is well known that if ξ_1, ξ_2, ... is a sequence of independent identically distributed random variables with $\mathbf{E}\xi_1 = m$, satisfying Cramér's condition, then $X^n = S_n/n$ with $S_n = \xi_1 + \dots + \xi_n$ converges to m with the exponential rate:

$$P\left(\left|\frac{S_n}{n} - m\right| \ge \varepsilon\right) \le 2e^{-n \min[h(m+\varepsilon),\, h(m-\varepsilon)]}, \tag{13.2}$$

where $h(a) = \sup_{\lambda} [a\lambda - \psi(\lambda)]$ is Cramér's transformation of the function $\psi(\lambda) = \log \mathbf{E} \exp(\lambda\xi_1)$.

In contrast with the central limit theorem which describes the deviation of S_n from nm of order of magnitude $\sqrt{n}$, in (13.2) the deviation is estimated of S_n from nm of a larger order of magnitude (then $\sqrt{n}$), namely of magnitude n. This explains the commonly used terminology according to which estimates of type (13.2) are called *inequalities of large deviation type*.

In the present section we consider exponential estimates of large deviation type for semimartingales which generalize estimates of type (13.2).

2. Let $X = (X_t)_{t \ge 0}$ be a semimartingale, $X_0 = 0$, given on a stochastic basis $(\Omega, \mathcal{F}, \mathbb{F} = (\mathcal{F}_t)_{t \ge 0}, P)$, with the triplet of predictable characteristics (B, C, ν). We will be interested in cases of validity of exponential estimates from above for the probabilities

$$P(\sup_{s \le t} X_s \ge \delta\phi(t)), \quad P\left(\sup_{s \le t} \frac{X_s}{A_s} \ge \delta\right)$$

with $\delta > 0$, a (nonrandom) function $\phi = \phi(t) \in V^+$ and $A = (A_t)_{t \le 0} \in \mathcal{V}^+ \cap \mathcal{P}$, which entail, in particular, estimates for probabilities $P(X_t \ge \delta\phi(t))$ and $P(X_t/A_t \le \delta)$.

It is assumed throughout this section that the compensator ν of the jump measure μ

satisfy the condition

$$\int_0^t \int_{|x|>1} |x| \, \nu(ds, dx) < \infty \quad (\mathsf{P}\text{-a.s.}), \quad t > 0 \tag{13.3}$$

and Cramér's condition: there exists $\lambda_0 > 0$ such that for all $\lambda \in \Lambda = \{\lambda > 0; \lambda \le \lambda_0\}$

$$\int_0^t \int_{|x|>1} e^{\lambda x} \nu(ds, dx) < \infty \quad (\mathsf{P}\text{-a.s.}), \quad t > 0. \tag{13.4}$$

(Conditions (13.3) and (13.4) are equivalent to the local integrability of the processes

$$\sum_{s \le t} |\Delta X_s| \, I(|\Delta X_s| > 1))_{t \ge 0}$$

and

$$\sum_{s \le t} e^{\lambda \Delta X_s} I(|\Delta X_s| > 1)_{t \ge 0}$$

for each $0 \le \lambda \le \lambda_0$).

Under Conditions (13.3) and (13.4) the cumulant

$$G(\lambda) = (G_t(\lambda))_{t \ge 0}, \quad 0 \le \lambda \le \lambda_0$$

is defined by

$$G_t(\lambda) = \lambda B_t + \frac{\lambda^2}{2} C_t + \int_0^t \int_{R_0} (e^{\lambda x} - 1 - \lambda x \, I(|x| \le 1)) \, \nu(ds, dx). \tag{13.5}$$

Setting

$$\tilde{B}_t = B_t + \int_0^t \int_{|x|>1} x \, \nu(ds, dx) \tag{13.6}$$

we get

$$G_t(\lambda) = \lambda \tilde{B}_t + \frac{\lambda^2}{2} C_t + \int_0^t \int_{R_0} (e^{\lambda x} - 1 - \lambda x) \, \nu(ds, dx). \tag{13.7}$$

With the function $G(\lambda)$ we associate the stochastic exponential (cf. Ch. 4, § 2)

$$\mathcal{E}_t(\lambda) = e^{G_t(\lambda)} \prod_{s \le t} (1 + \Delta G_s(\lambda)) \, e^{-\Delta G_s(\lambda)}, \tag{13.8}$$

which plays an essential rôle in constructing estimates of an exponential type. In this connection, we present a number of properties of the stochastic exponential needed below, assuming $0 \le \lambda \le \lambda_0$.

A. Since

$$\Delta G_t(\lambda) = \int_{R_0} (e^{\lambda x} - 1)\,\nu(\{t\}, dx) > 1$$

and $0 < (1+a)\,e^{-a} \le 1$ as $a > -1$, then

$$\prod_{s \le t} (1 + \Delta G_s(\lambda))\, e^{-\Delta G_s(\lambda)} \le 1.$$

Therefore

$$0 < \mathcal{E}_t(\lambda) = e^{G_t(\lambda)} \le e^{G_t^*(\lambda)}, \tag{13.9}$$

with the (increasing) process

$$G_t^*(\lambda) = \lambda \sup_{s \le t} \tilde{B}_s + \frac{\lambda^2}{2} C_t + \int_0^t \int_{R_0} (e^{\lambda x} - 1 - \lambda x)\,\nu(ds, dx). \tag{13.10}$$

B. If the initial semimartingale $X = (X_t)_{t \ge 0}$ is a locally integrable increasing process with the compensator $\tilde{X} = (\tilde{X}_t)_{t \ge 0}$, then

$$G_t(\lambda) = \lambda \tilde{X}_t + \int_0^t \int_{x > 0} (e^{\lambda x} - 1 - \lambda x)\,\nu(ds, dx) \tag{13.11}$$

(Problem 1) and the cumulant $G(\lambda) = (G_t(\lambda))_{t \ge 0}$ is an increasing process.

C. If the initial semimartingale $X = (X_t)_{t \ge 0}$ is a local martingale, then $\tilde{B}_t \equiv 0$ (see Problem 1.10) and, consequently, the cumulant $G(\lambda) = (G_t(\lambda))_{t \ge 0}$ is an increasing process.

D. The positive process $Z(\lambda) = (Z_t(\lambda))_{t \ge 0}$ with

$$Z_t(\lambda) = e^{\lambda X_t} \mathcal{E}_t^{-1}(\lambda) \tag{13.12}$$

is a nonnegative local martingale with $Z_0(\lambda) = 1$ (Problem 2), hence

$$\mathbf{E} Z_\tau(\lambda)\, I(\tau < \infty) \le 1 \tag{13.13}$$

for each Markov time τ (Problem 3).

3. The rôle played by the stochastic exponential in obtaining exponential estimates, is clarified first for a semimartingale X that is a process with independent increments and for A that is a deterministic function in the class V^+.

Theorem 1. *Let a semimartingale* X *be a process with independent increments, and for a certain* $\lambda_0 > 0$ *let Cramér's condition* (13.4) *be satisfied on the set* $\Lambda = \{\lambda: 0 < \lambda \le \lambda_0\}$, *as well as Condition* (13.3).

1) *For each* $t > 0$ *and* $\delta > 0$

1) *For each* t > 0 *and* δ > 0

$$P\left(\sup_{s \le t} X_s \ge \delta\phi(t)\right) \le \exp\left(-\phi(t)\sup_{\lambda \in \Lambda}\left[\lambda\delta - \frac{G_t^*(\lambda)}{\phi(t)}\right]\right). \tag{13.14}$$

2) *Let* $A \in V^+$, *and for* t > 0 *and* $\lambda \in \Lambda$

$$\frac{\ln \mathcal{E}_t(\lambda)}{A_t} \le H(\lambda), \tag{13.15}$$

where H = H (λ) *is a certain concave (on the set* $\{0\} \cup \Lambda$*) function which is bounded from below and such that* H (0) = 0.

Then for each t > 0 *and* δ > 0

$$P\left(\sup_{s \ge t} \frac{X_s}{A_s} \ge \delta\right) \le \exp\left(-A_t \sup_{\lambda \in \Lambda}[\lambda\delta - H(\lambda)]\right). \tag{13.16}$$

Corollary. *If for each* δ > 0

$$\sup_{\lambda \in \Lambda}[\lambda\delta - H(\lambda)] > 0$$

and

$$\lim_{t \to \infty} A_t = 0,$$

then

$$\lim_{t \to \infty} X_t / A_t = 0 \quad P\text{-a.s.}$$

Proof. 1) Define the stopping time

$$\tau = \inf\{s \le t : X_s \ge \delta\phi(t)\} \wedge t.$$

Taking then into consideration (13.13), we have

$$\mathbf{E}\, I\{\sup_{s \le t} X_s \ge \delta\phi(t)\} Z_\tau(\lambda) \le 1, \ \lambda \in \Lambda.$$

In view of (13.12), this gives

$$1 \ge \mathbf{E}\, I\{\sup_{s \le t} X_s \ge \delta\phi(t)\} e^{\lambda X_\tau} \mathcal{E}_\tau^{-1}(\lambda)$$

$$\ge \mathbf{E}\, I\{\sup_{s \le t} X_s \ge \delta\phi(t)\} e^{\lambda\delta\phi(t) - G_\tau^*(\lambda)}$$

$$\ge \mathbf{E}\, I(\sup_{s \le t} X_s \ge \delta\phi(t)\} e^{\phi(t)\left[\lambda\delta - \frac{G_t^*(\lambda)}{\phi(t)}\right]} \tag{13.17}$$

According to the corollary to Theorem 4.1 the triplet (B, C, ν) of the semimartingale X is independent of ω (up to indistinguishability), and hence $G_t^*(\lambda)$ is independent of (ω). Therefore, by (13.7) for each $\lambda \in \Lambda$ we get

$$P\left(\sup_{s\le t} X_s \ge \delta\phi(t)\right) \le e^{-\phi(t)\left[\lambda\delta - \frac{G_t^*(\lambda)}{\phi(t)}\right]}$$

which gives the desired inequality (13.14) in a natural manner.

2) Denote

$$\tau = \inf\{s \ge t: \frac{X_s}{A_s} \ge \delta\}$$

setting, as usual, $\tau = \infty$ if $X_s/A_s < \delta$ for each $s \ge t$. Then for $\lambda' \in \Lambda$ such that $\lambda'\delta$ — $H(\lambda') > 0$ we have, on the set $\{\tau < \infty\}$, that

$$Z_\tau(\lambda') = e^{\lambda' X_\tau}\,\mathcal{E}_\tau^{-1}(\lambda') \ge e^{\lambda'\delta A_\tau - \ln \mathcal{E}_\tau(\lambda')}$$

$$= \exp\left\{A_\tau\left[\lambda'\delta - \frac{\ln \mathcal{E}_\tau(\lambda')}{A_\tau}\right]\right\}.$$

Thus, in virtue of (13.13)

$$1 \ge \mathbf{E}\, Z_\tau(\lambda')\, I(\tau < \infty) = \mathbf{E}\, Z_\tau(\lambda')\, I\left(\frac{X_\tau}{A_\tau} \ge \delta\right) I(\tau < \infty)$$

$$\ge \exp\{A_t[\lambda'\delta - H(\lambda')]\}\, P\left(\frac{X_\tau}{A_\tau} \ge \delta,\ \tau < \infty\right),$$

i.e.

$$P\left(\sup_{s\ge t} \frac{X_s}{A_s} \ge \delta\right) \le e^{-A_t[\lambda'\delta - H(\lambda')]}, \tag{13.18}$$

where $\lambda' \in \Lambda$ is such that $\lambda'\delta - H(\lambda') > 0$. It is not hard to deduce from this that $\lambda'\delta - H(\lambda')$ in (13.18) can be substituted by $\sup_{\lambda\in\Lambda}[\lambda\delta - H(\lambda)]$, hence (13.16) is proved.

4. Let us turn now to the general case. Let $H = H(\lambda)$ be a concave function with $H(0) = 0$, bounded from below. Set

$$L_H(x) = \sup_{\lambda\in\Lambda}[\lambda x - H(\lambda)] \tag{13.19}$$

and

$$\lambda^x = \arg\max_{\lambda\in\Lambda}[\lambda x - H(\lambda)]. \tag{13.20}$$

Theorem 2. *Let Condition* (13.3) *be satisfied, as well as Cramér's condition* (13.4) *on the set* $\Lambda = \{\lambda > 0: \lambda \le \lambda_0\}$, *and for a given* $\delta > 0$ *let* $L_H(\delta) > 0$.

Then for each $\varepsilon \in [0, L_H(\delta)]$ *and* $t > 0$

$$P\left(\sup_{s\le t} X_s \ge \delta\phi(t),\ G_t^*(\lambda^\delta) \le [H(\lambda^\delta)+\varepsilon]\,\phi(t)\right)$$

$$\le \exp(-\phi(t)\,[L_H(\delta)-\varepsilon]). \qquad (13.21)$$

Proof. For a given $t > 0$ denote

$$A = \{\sup_{s\le t} X_s \ge \delta\phi(t)\} \cap \{G_t^*(\lambda^\delta) \le [H(\lambda^\delta)+\varepsilon]\,\phi(t)\}$$

and

$$\tau = \inf(s \le t\colon X_s \ge \delta\phi(t)\} \wedge t.$$

On the set A we have

$$Z_\tau(\lambda^\delta) = \exp(\lambda^\delta X_\tau)\,\mathcal{E}_\tau^{-1}(\lambda^\delta) \ge \exp(\lambda^\delta \delta\phi(t) - G_t^*(\lambda^\delta))$$

$$\ge \exp(\phi(t)\,[\lambda^\delta\delta - H(\lambda^\delta) - \varepsilon]) = \exp(\phi(t)\,[L_H(\delta)-\varepsilon]).$$

This and the inequality

$$\mathbf{E}\, I_A Z_\tau(\lambda^\delta) \le 1,$$

derived from (13.13), give the desired assertion (13.21).

Corollary. *By* (13.21) *we have*

$$P\left(\sup_{s\le t} X_s \ge \delta\phi(t)\right) \le \exp(-\phi(t)[L_H(\delta)-\varepsilon])$$

$$+ P(G_t^*(\lambda^\delta) \ge [H(\lambda^\delta)+\varepsilon]\,\phi(t)). \qquad (13.22)$$

Remark. Analogously to the proof of Theorem 2, it is established that

$$P(X_t \ge \delta\phi(t),\ G_t(\lambda^\delta) \le [H(\lambda^\delta)+\varepsilon]\,\phi(t)) \le \exp(-\phi(t)[L_H(\delta)-\varepsilon]) \qquad (13.23)$$

and hence

$$P(X_t \ge \delta\phi(t)) \le \exp(-\phi(t)[L_H(\delta)-\varepsilon])$$

$$+ P(G_t(\lambda^\delta) \ge [H(\lambda^\delta)+\varepsilon]\,\phi(t)). \qquad (13.24)$$

We present yet another estimate for the probability $P(X_t \ge \delta\phi(t))$.

Theorem 3. *For sufficiently small* $\lambda > 0$ *and* $q - 1 > 0$ *let*

$$\mathbf{E}\exp[(q-1)\,G_t(\lambda)] < \infty. \qquad (13.25)$$

Then for each $\delta > 0$

$$P(X_t \ge \delta\phi(t)) \le \inf_{\lambda>0,\ q>1}\left(\mathbf{E}\exp\left\{-(q-1)\,\phi(t)\left[\lambda\delta - \frac{G_t(\lambda)}{\phi(t)}\right]\right\}\right)^{1/q}. \qquad (13.26)$$

Proof. In view of Chebishev's inequality, for each $b > 0$ we have

$$P(X_t \ge \delta\phi(t)) \le e^{-b\delta\phi(t)}\,\mathbf{E}\, e^{bX_t}. \qquad (13.27)$$

Set p = q/q − 1 and λ = pb. Then

$$\mathbf{E}\, e^{bX_t} = \mathbf{E}\, e^{\frac{1}{p}\lambda X_t}\, \mathcal{E}_t^{-\frac{1}{p}}(\lambda)\, \mathcal{E}_t^{\frac{1}{p}}(\lambda) = \mathbf{E}\, Z_t^{\frac{1}{p}}(\lambda)\, \mathcal{E}_t^{\frac{q-1}{q}}(\lambda)$$

and taking into consideration (13.13) and (13.9) we get by Hölder's inequality that

$$\mathbf{E}\, e^{bX_t} \le (\mathbf{E}\, Z_t(\lambda))^{1/p}\, (\mathbf{E}\, \mathcal{E}_t^{q-1}(\lambda))^{1/q} \le (\mathbf{E}\, \mathcal{E}_t^{q-1}(\lambda))^{1/q} \le \mathbf{E}\, e^{(q-1)\, G_t(\lambda)}.$$

This and (13.27) give the inequality

$$\mathsf{P}\,(X_t \ge \delta\phi\,(t)) \le e^{-\frac{q-1}{q}\lambda\delta\phi\,(t)} \left(\mathbf{E}\, e^{(q-1)\, G_t(\lambda)}\right)^{1/q}, \tag{13.28}$$

which implies (13.26).

Remark. For sufficiently small λ > 0 and q − 1 > 0 the right hand side of (13.28) is finite. Furthermore, it takes the unit value for λ = 0 and q = 1. Therefore, the right hand side of (13.26) does not exceed the unity.

5. Let us turn to estimates for the probabilities

$$\mathsf{P}\left(\sup_{s \ge t} \frac{X_s}{A_s} \ge \delta\right).$$

Theorem 4. *Let Cramér's condition* (13.4) *be satisfied, and*

$$\frac{\ln \mathcal{E}_t(\lambda)}{A_t} \le H\,(\lambda)$$

where H = H (λ) *is a certain concave function with* H (0) = 0, *bounded from below and such that for a given* δ > 0 *we have* $L_H(\delta) > 0$.

Then for a function a = a (t) ∈ V^+

$$\mathbf{E}\, I\left(\sup_{s \ge t} \frac{X_s}{A_s} \ge \delta,\ A_t \ge a\,(t)\right) \le e^{-a\,(t)\, L_H(\delta)}. \tag{13.29}$$

Proof. Denote

$$B = \left\{\sup_{s \ge t} \frac{X_s}{A_s} \ge \delta\right\} \cap (A_t \ge a\,(t)\}$$

and

$$\tau = \inf\left(s \ge t: \frac{X_s}{A_s} \ge \delta\right).$$

In virtue of (13.13)

$$\mathbf{E}\, I_B\, I\,(\tau < \infty)\, Z_\tau\,(\lambda^\delta) \le 1. \tag{13.30}$$

We estimate $Z_\delta\,(\lambda^\delta)$ on the set B ∩ {τ < ∞}. We have

$$Z_\tau(\lambda^\delta) = 1^{\lambda^\delta X_\tau} \mathcal{E}_\tau^{-1}(\lambda^\delta) \geq e^{\lambda^\delta \delta A_\tau - \ln \mathcal{E}_\tau(\lambda^\delta)} = e^{A_\tau\left[\lambda^\delta \delta - \frac{\ln \mathcal{E}_\tau(\lambda)}{A_\tau}\right]}$$

$$\geq e^{A_\tau[\lambda^\delta \delta - H(\lambda^\delta)]} = e^{A_\tau L_H(\delta)} \geq e^{a(t) L_H(\delta)}.$$

This and (13.30) give the desired inequality (13.29) in an obvious manner.

Corollary. *Under the conditions of Theorem 4*

$$\mathsf{P}\left(\sup_{s \geq t} \frac{X_s}{A_s} \geq \delta\right) \leq e^{-a(t) L_H(\delta)} + \mathsf{P}(A_t \leq a(t)). \tag{13.31}$$

6. In the present subsection the result of Theorem 2 is applied to the case where X is a locally square integrable martingale with the quadratic characteristic $\langle X\rangle = (\langle X\rangle_t)_{t \geq 0}$ and bounded jumps ($|\Delta X| \leq c$).

In this case $\tilde{B} = 0$, hence

$$G_t^*(\lambda) = G_t(\lambda) = \frac{\lambda^2}{2} C_t + \int_0^t \int_{|x| \leq c} (e^{\lambda x} - 1 - \lambda x)\, \nu(ds, dx).$$

Set

$$\phi_c(\lambda) = \begin{cases} \dfrac{e^{\lambda c} - 1 - \lambda c}{c^2}, & c > 0, \\[2mm] \dfrac{\lambda^2}{2}, & c = 0. \end{cases} \tag{13.32}$$

Since

$$\phi_c(\lambda) \geq \frac{\lambda^2}{2}, \quad e^{\lambda x} - 1 - \lambda x \leq x^2 \phi_c(\lambda)$$

as $|x| \leq c$, then (cf., e.g., (1.14) in § 1)

$$G_t(\lambda) \leq \phi_c(\lambda) \langle X\rangle_t.$$

By specifying $H(\lambda)$ in Theorems 2 and 4 as the function $K\phi_c(\lambda)$ with $K > 0$ and $\Lambda = \{\lambda: \lambda > 0\}$, we get from (13.22) the following results.

Theorem 5. *Let* $X \in \mathcal{M}^2_{\text{loc},0}$ and $|\Delta X| \leq c$. *Then*

$$\mathsf{P}(\sup_{s \leq t} X_s \geq \delta\phi(t)) \leq \exp(-\phi(t) \sup_{\lambda > 0} [\lambda\delta - K\phi_c(\lambda)])$$

$$+ \mathsf{P}(\langle X\rangle_t \geq K\phi(t)). \tag{13.33}$$

Corollary. *If* $X \in \mathcal{M}^2_{loc,\,0}$ *and* $|\Delta X| \le c$, *then*

$$P\left(\sup_{s \le t} |X_s| \ge \delta\phi(t)\right) \le 2\left\{\exp\left(-\phi(t) \sup_{\lambda > 0} [\lambda\delta - K\phi_c(\lambda)]\right)\right.$$

$$\left. + P(\langle X\rangle_t \ge K\phi(t))\right\}. \tag{13.34}$$

Theorem 6. *Let* $X \in \mathcal{M}^2_{loc,\,0}$ and $|\Delta X| \le c$, *and there exists a function* $b = b(t) \in V^+$ *such that* (P-a.s.)

$$\langle X\rangle_t \ge b(t), \ t > 0.$$

Then

$$P\left(\sup_{s \ge t} \frac{X_s}{\langle X\rangle_s} \ge \delta\right) \le e^{-b(t)\, L\phi_c(\delta)} \tag{13.35}$$

and

$$P\left(\sup_{s \ge t} \frac{|X_s|}{\langle X\rangle_s} \ge \delta\right) \le 2e^{-b(t)\, L\phi_c(\delta)}. \tag{13.36}$$

7. We apply now Theorem 2 to the case where X is an increasing locally integrable process with bounded jumps ($\Delta X \le c$) and the compensator $\tilde{X}$.

Define the function

$$\psi_c(\lambda) = \begin{cases} \dfrac{e^{\lambda c} - 1}{c}, & c > 0, \\ \lambda, & c = 0. \end{cases} \tag{13.37}$$

It is not hard to see that

$$\psi_c(\lambda) \ge \lambda, \quad \left(e^{\lambda x} - 1\right) \le \psi_c(\lambda)\, x$$

as $0 \le x \le c$. It follows from (13.7) that

$$G_t(\lambda) \le \psi_c(\lambda)\, \tilde{X}_t.$$

Assuming $H(\lambda) = K\psi_c(\lambda)$, $K > 0$, it is easily seen that this function satisfies the conditions of Theorem 2, so that its corollary gives the following result concerning the case in question.

Theorem 7. *Let* X *be an increasing locally integrable process with* $\Delta X \le c$ *and the compensator* $\tilde{X}$.

Then

$$P(X_t \geq \delta\phi(t)) \leq e^{-\phi(t) L_K \psi_c(\delta)} + P(\tilde{X}_t \geq K\phi(t)). \tag{13.38}$$

Corollary. *If* $\tilde{X}_t \leq b\phi(t)$, $b > 0$ *for each* $t > 0$, *then*

$$P(X_t \geq \delta\phi(t)) \leq e^{-\phi(t) L_b \psi_c(\delta)}. \tag{13.39}$$

8. Based on the results obtained above, we consider a number of examples.

Example 1. Let a semimartingale X present a solution of the stochastic equation

$$X_t = -\int_0^t X_s ds + M_t, \tag{13.40}$$

where $M = (M_t)_{t \geq 0}$ is a continuous local martingale having the quadratic characteristic

$$\langle M \rangle_t = \int_0^t \alpha(s)\, ds$$

with a nonnegative predictable process $\alpha = (\alpha(t))_{t \geq 0}$ such that $\alpha(t) \leq c$, $t \geq 0$.

We will show that for each $t > 0$ and $\delta > 0$

$$P(X_t \geq \delta\phi(t)) \leq \exp\left(-\frac{\delta^2}{c}\phi^2(t)\frac{e^{2t}}{e^{2t}-1}\right). \tag{13.41}$$

To this end, denote

$$Y_t = \int_0^t e^s dM_s$$

and observe that $X_t = e^{-t} Y_t$. Hence

$$P(X_t \geq \delta\phi(t)) = P(Y_t \geq \delta\phi(t)\, e^t). \tag{13.42}$$

Since $Y = (Y_t)_{t \geq 0}$ is a continuous martingale, $\tilde{B} = 0$ and

$$G_t(\lambda) = \frac{\lambda^2}{2}\int_0^t e^{2s} d\langle M \rangle_s \leq \frac{\lambda^2 c}{2}\int_0^t e^{2s} ds = \frac{\lambda^2 c}{4}\left(e^{2t} - 1\right).$$

Next

$$\sup_{\lambda > 0}\left[1\, d - \frac{G_t(\lambda)}{f(t)}\right] \geq \phi(t)\frac{\delta^2}{c}\cdot\frac{e^t}{e^{2t}-1}$$

and by Theorem 3

$$\mathsf{P}\,(\sup_{s\le t} Y_s \ge \delta\phi(t)\,e^t) \le \inf_{q>1} \exp\left(-\frac{q-1}{q}\phi^2(t)\frac{\delta^2}{c}\cdot\frac{e^{2t}}{e^{2t}-1}\right),$$

which, together with (13.42), gives the desired estimate (13.41).

Example 2. Let a Wiener process $W = (W_t)_{t\ge 0}$ and a (strictly) stationary predictable process $\xi = (\xi_t)_{t\ge 0}$ with $\mathbf{E}\, e^{\xi_0^2} < \infty$ be given on a stochastic basis $(\Omega, \mathcal{F}, \mathbb{F}, \mathsf{P})$. Let

$$X_t = \int_0^t \xi_s dW_s.$$

We will show that

$$\mathsf{P}\,(X_t \ge \delta t) \le \sqrt{\mathbf{E}\, e^{\xi_0^2}}\; e^{-\frac{\delta\sqrt{t}}{2\sqrt{2}}}, \quad t > 0,\ \delta > 0. \tag{13.43}$$

Observe that

$$G_t(\lambda) = \frac{\lambda^2}{2}\int_0^t \xi_s^2\, ds$$

and by Jensen's inequality

$$\exp\,(G_t(\lambda)) = \exp\left(\frac{1}{t}\int_0^t \frac{t\lambda^2}{2}\xi_s^2\, ds\right) \le \frac{1}{t}\int_0^t \exp\left(\frac{t\lambda^2}{2}\xi_s^2\right) ds.$$

Hence

$$\mathbf{E}\exp G_t(\lambda) \le \mathbf{E}\exp\left(\frac{t\lambda^2}{2}\xi_0^2\right) \tag{13.44}$$

and since by assumption $\mathbf{E}\, e^{\xi_0^2} < \infty$, then

$$\mathbf{E}\, e^{G_t(\lambda)} < \infty \text{ as}$$

as $\frac{t\lambda^2}{2} \le 1$. Therefore, applying Theorem 3 as $q = 2$, we get from (13.26) that

$$\mathsf{P}\,(X_t \ge \delta t) \le e^{-\frac{\lambda\delta}{2}t}\,(\mathbf{E}\exp G_t(\lambda))^{1/2}.$$

Setting $\lambda = \sqrt{\frac{2}{t}}$ and applying the inequality (13.44), we get the estimate (13.43).

Example 3. Let $\xi = (\xi_t)_{t \in R}$ be a (strictly) stationary process, $\mathbf{E}\, \xi_0^2 < \infty$, $\mathbf{E}\, \xi_0 = 0$ and

$$X_t = \int_0^t \xi_s ds. \tag{13.45}$$

We will present (exponential) estimates for the probabilities $\mathsf{P}(X_t \geq \delta t)$, restricting ourself by the case where a process ξ admits the representation

$$\xi_t = \int_{-\infty}^{t} a(t-s)\, d\, Y_s, \tag{13.46}$$

(Section 6.6) where $a = a(t)$ is a measurable function with

$$\int_0^\infty a^2(s)\, ds < \infty, \tag{13.47}$$

while $Y = (Y_t)_{t \in R}$ is a process with independent increments having trajectories in D, such that

$$\mathbf{E}(Y_t - Y_s)^2 = |t - s|, \ \mathbf{E}(Y_t - Y_s) = 0, \ s, t \in R;$$

besides it is a helix (see Section 11.2) given on a stochastic basis $(\Omega, \mathcal{F}, (\mathcal{F}_t)_{t \in R}, \mathsf{P})$.

In virtue of this assumption the process $(Y_t)_{t \geq 0}$ is a square integrable martingale. Its triplet (B, C, ν) is deterministic and (Theorem 11.5)

$$C_t = ct, \quad \nu(dt, dx) = K(dx)\, dt,$$

where $c \geq 0$ and $K(dx)$ is a measure on R_0 with

$$\int_{R_0} x^2 K(dx) < \infty,$$

besides

$$c + \int_{R_0} x^2 K(dx) = 1.$$

We assume Cramér's condition

$$\int_{R_0} e^{\lambda x} K(dx) < \infty, \ \lambda \in R \tag{13.48}$$

which guarantees, in particular, the existence of a function $h = h(\lambda)$ with

$$h(\lambda) = \int_{R_0} \left(e^{\lambda x} - 1 - \lambda x\right) K(dx). \tag{13.49}$$

Along with (13.47) we assume

$$\int_0^\infty | a(t) | dt < \infty. \tag{13.50}$$

Set

$$\alpha(t) = \int_t^\infty a(s)\, ds \tag{13.51}$$

and assume

$$\int_0^\infty [h(\alpha(t)) + h(-\alpha(t))]\, dt < \infty, \tag{13.52}$$

and

$$\int_0^\infty \sqrt{\int_t^\infty a^2(s)\, ds}\ dt < \infty. \tag{13.53}$$

as well. By the last condition the process X admits the decomposition

$$X_t = V_t - V_0 - M_t, \tag{13.54}$$

where $V = (V_t)_{t \geq 0}$ is a strictly stationary process, and $M = (M_t)_{t \geq 0}$ a square integrable martingale presenting a helix (see Lemma 9.2.1 and Ch. 9, § 2, Example 2).

We will show that

$$M_t = \alpha(0)\, Y_t,\ \ V_0 \overset{d}{=} \int_0^\infty \alpha(s)\, d\, Y_s, \tag{13.55}$$

i.e. V_0 coincides in distribution with $\lim_{t \to \infty} \mathcal{V}_t$, where $\mathcal{V} = (\mathcal{V}_t)_{t \geq 0}$ is a square integrable martingale with

$$\mathcal{V}_t = \int_0^t \alpha(s)\, d\, Y_s.$$

According to Ch.2, § 2, the limit $\lim_{t \to \infty} \mathcal{V}_t$ exists (P-a.s.) if

$$\int_0^\infty \alpha^2(s)\, ds < \infty. \tag{13.56}$$

But $\alpha^2(s) \leq 2h(\alpha(s))$, so that (13.56) follows from (13.52). In accordance with Lemma 9.2.1, the process M presents a modification of the process

$$\int_0^\infty [\mathbf{E}(\xi_s \mid \mathcal{F}_t) - \mathbf{E}(\xi_s \mid \mathcal{F}_0)]\, ds.$$

In view of (13.46)

$$\int_0^\infty [\mathbf{E}(\xi_s \mid \mathcal{F}_t) - \mathbf{E}(\xi_s \mid \mathcal{F}_0)]\, ds$$

$$= \int_0^t \int_0^s a(s-u)\, dY_u\, ds + \int_t^\infty \int_0^t a(s-u)\, dY_u\, ds. \tag{13.57}$$

Since (13.56) takes place, we have P-a.s. (Problem 4)

$$\int_0^t \int_0^s a(s-u)\, dY_u\, ds = \int_0^t \int_u^t a(s-u)\, ds\, dY_u,$$

$$\int_t^\infty \int_0^t a(s-u)\, dY_u\, ds = \int_t^\infty \int_t^\infty a(s-u)\, ds\, dY_u. \tag{13.58}$$

Therefore the right hand side of (13.57) coincides with

$$\int_0^t \int_u^\infty a(s-u)\, ds\, dY_u.$$

Consequently, in virtue of

$$\alpha(0) = \int_u^\infty a(s-u)\, ds,$$

we get the first equality in (13.55). To prove the second one observe that due to Lemma 9.2.1

$$V_0 = \int_0^\infty \mathbf{E}(\xi_s \mid \mathcal{F}_0)\, ds = \int_0^\infty \int_{-\infty}^0 a(s-u)\, dY_u\, ds.$$

Since Y is a helix, we have (see Theorem 11.6)

$$\int_{-\infty}^0 a(s-u)\, dY_u \overset{d}{=} \int_0^\infty a(s+u)\, dY_u.$$

Therefore

$$V_0 \stackrel{d}{=} \int_0^\infty \int_0^\infty a(s+u)\, dY_u ds.$$

Next, in view of the condition

$$\int_0^\infty \alpha^2(s)\, ds < \infty$$

we have P-a.s. (Problem 4)

$$\int_0^\infty \int_0^\infty a(s+u)\, dY_u ds = \int_0^\infty \int_0^\infty a(s+u)\, ds\, dY_u = \int_0^\infty \alpha(u)\, dY_u \tag{13.59}$$

and consequently the second equality in (13.55) is verified.

We will show now that

$$\mathbf{E} \exp\left(\int_0^\infty \alpha(u)\, dY_u\right) \le \exp\left(\frac{c}{2}\int_0^\infty \alpha^2(u)\, du + \int_0^\infty h(\alpha(u))\, du\right). \tag{13.60}$$

To this end, observe that by Fatou's lemma

$$\mathbf{E} \exp\left(\int_0^\infty \alpha(u)\, dY_u\right) \le \varliminf_{t \to \infty} \mathbf{E} \exp(\mathcal{V}_t), \tag{13.61}$$

with

$$\mathcal{V}_t = \int_0^t \alpha(u)\, dY_u.$$

The process $\mathcal{V} = (\mathcal{V}_t)_{t \ge 0}$ is a square integrable martingale with independent increments. Therefore

$$\mathbf{E} \exp(\mathcal{V}_t) = \exp\left(\frac{c}{2}\int_0^t \alpha^2(u)\, du + \int_0^t \int_{R_0} \left[e^{\alpha(u)x} - 1 - \alpha(u)x\right] K(dx)\, du\right)$$

$$= \exp\left(\frac{c}{2}\int_0^t \alpha^2(u)\, du + \int_0^t h(\alpha(u))\, du\right)$$

due to Problem 2. This and (13.61) give the inequality (13.60) in an obvious manner. Analogously, it is shown that

$$E \exp\left(-\int_0^\infty \alpha(u)\, d Y_u\right) = \exp\left(\frac{c}{2}\int_0^\infty \alpha^2(u)\, du + \int_0^\infty h(-\alpha(u))\, du\right). \quad (13.62)$$

Denote

$$g(y) = e^y - 1 - y + \frac{c}{2} y^2.$$

It is deduced from (13.60), (13.62) and Chebishev's inequality that for each $\varepsilon > 0$

$$\begin{aligned} P(V_0 \geq \varepsilon t) &\leq e^{-\varepsilon t} \exp\left(\int_0^\infty g(\alpha(u))\, du\right), \\ P(-V_0 \geq \varepsilon t) &\leq e^{-\varepsilon t} \exp\left(\int_0^\infty g(-\alpha(u))\, du\right). \end{aligned} \quad (13.63)$$

Let us estimate now $P(-M_t \geq \varepsilon t)$. By (13.54) the cumulant $G_t(\lambda)$ of the square integrable martingale $(-M_t)_{t \geq 0}$ has the form

$$G_t(\lambda) = \left[\frac{\lambda^2 \alpha^2(0)}{2} + g(-\lambda\alpha(0))\right] t.$$

Therefore by Theorem 1

$$P(-M_t \geq \varepsilon) \leq e^{-t \sup_{\lambda > 0}\left[\lambda\varepsilon - \frac{\lambda^2\alpha^2(0)}{2} - g(-\lambda\alpha(0))\right]}. \quad (13.64)$$

We utilize now the fact that $X_t = V_t - V_0 - M_t$ and $V_t \overset{d}{=} V_0$. We have

$$P(X_t \geq \delta t) = P\left(V_0 \geq \frac{\delta}{3} t\right) + P\left(-V_0 \geq \frac{\delta}{3} t\right) + P\left(-M_t \geq \frac{\delta}{3} t\right).$$

From this, (13.63) and (13.64) we deduce that

$$P(X_t \geq \delta t) \leq d_1 e^{-\frac{\delta}{3} t} + e^{-d_2 t},$$

where

$$d_1 = \exp\left(\int_0^\infty g(\alpha(u))\, du\right) + \exp\left(\int_0^\infty g(-\alpha(u))\, du\right),$$

$$d_2 = \sup_{\lambda > 0}\left[\lambda\frac{\delta}{3} - \frac{\lambda^2\alpha^2(0)}{2} - g(-\lambda\alpha(0))\right].$$

Problems

1. Let X be a semimartingale with the triplet of predictable characteristics (B, C, ν) presenting a locally integrable increasing process with the compensator $\tilde{X}$. Show that its cumulant is presented by

$$G_t(\lambda) = \lambda \tilde{X}_t + \int_0^t \int_{|x|>0} (e^{\lambda x} - 1 - \lambda x)\, \nu(ds, dx).$$

2. Let X be a semimartingale with $X_0 = 0$ whose triplet of predictable characteristics (B, C, ν) determines the cumulant

$$G_t(\lambda) = \lambda B_t + \frac{\lambda^2}{2} C_t + \int_0^t \int_{R_0} (e^{\lambda x} - 1 - \lambda x\, I(|x| \le 1))\, \nu(ds, dx),$$

and the related stochastic exponential

$$\mathcal{E}_t(\lambda) = e^{G_t(\lambda)} \prod_{s \le t} (1 + \Delta G_s(\lambda))\, e^{-\Delta G_s(\lambda)}.$$

Show that the process $Z(\lambda) = (Z_t(\lambda))_{t \ge 0}$ with

$$Z_t(\lambda) = e^{\lambda X_t} \mathcal{E}_t^{-1}(\lambda)$$

is a local martingale.

3. Let $M = (M_t)_{t \ge 0}$ be a nonnegative local martingale with $M_0 = 1$ and τ a stopping time. Show that

$$E\, M_\tau \le 1.$$

If τ is a Markov time, than the present inequality remains valid with

$$M_\tau I(\tau = \infty) = \lim_{t \to \infty} M_t I(\tau = \infty).$$

4. Establish that (13.58) and (13.59) are valid.

PART II

CHAPTER 5

WEAK CONVERGENCE OF FINITE-DIMENSIONAL DISTRIBUTIONS OF SEMIMARTINGALES TO DISTRIBUTIONS OF PROCESSES WITH CONDITIONALLY INDEPENDENT INCREMENTS

§ 1. Method of stochastic exponentials. I. Convergence of conditional characteristic functions

1. Let $(\Omega, \mathcal{F}, P)$ be a probability space on which families of σ-algebras

$$\mathbb{F} = (\mathcal{F}_t)_{t \geq 0}, \quad \mathbb{F}^n = (\mathcal{F}_t^n)_{t \geq 0}, \quad n \geq 1,$$

are defined, satisfying Conditions (a) and (b) in Definition 1 (Ch. 1, § 1).

Also, let $X = (X_t, \mathcal{F}_t)$ and $X^n = (X_t^n, \mathcal{F}_t^n)$, $n \geq 1$, be semimartingales and

$$\mathcal{F}_\infty = \sigma(\bigcup_{t \geq 0} \mathcal{F}_t), \quad \mathcal{F}_\infty^n = \sigma(\bigcup_{\tau \geq 0} \mathcal{F}_t^n).$$

If S is a nonempty subset of $R_+ = [0, \infty)$, then the expression

$$X^n \xrightarrow{d_f(S)} X$$

denotes the weak convergence of a sequence of distributions of vectors $(X_{t_1}^n, \ldots, X_{t_m}^n)$, $n \geq 1$, to the distribution of a vector $(X_{t_1}, \ldots, X_{t_m})$ for each finite subset $\{t_1, \ldots, t_m\} \in S$, while the expression

$$X^n \xrightarrow{d_f(S)} X \quad (\mathcal{G}\text{-stably})$$

means that the relation

$$\lim_n \mathbf{E}\xi h(X_{t_1}^n, \ldots, X_{t_m}^n) = \mathbf{E}\xi h(X_{t_1}, \ldots, X_{t_m})$$

is valid for each bounded function $h(x_1, \ldots, x_m)$ continuous in all variables $(x_1, \ldots, x_m)$, and for each bounded $\mathcal{G}$-measurable random variable ξ, where $\mathcal{G}$ is a sub-σ-algebra of $\mathcal{F}$.

Clearly the convergence « $\xrightarrow{d_f(S)}$ » ($\mathcal{G}$-stably) implies the convergence « $\xrightarrow{d_f(S)}$ ».

If S consists of a single point $S = \{t_1\}$ or $S = R_+$, then instead of « $\xrightarrow{d_f(t_1)}$ » and « $\xrightarrow{d_f(R_+)}$ » we will write « $\xrightarrow{d}$ » and « $\xrightarrow{d_f}$ » respectively.

Thus the expression « $\xrightarrow{d}$ » means the convergence of random variables in distribution, while the expression « $\xrightarrow{d_f}$ » means the weak convergence of all finite dimensional distributions. Further on the expression « $\xrightarrow{P}$ » is also used to define the convergence in probability.

This chapter is aimed at describing the method, called the method of stochastic exponentials, which allows us to formulate in "predictable terms" the conditions for the convergence

$$X^n \xrightarrow{d_f(S)} X \quad (\mathcal{G}\text{-stably}).$$

Within this chapter "limiting" semimartingales $X = (X_t, \mathcal{F}_t)$ are assumed to be processes with $\mathcal{G}$-conditionally independent increments. This means that the σ-algebra $\mathcal{G} \subseteq \mathcal{F}_0$ and that for each $\mathcal{G}$-measurable random variable τ, $0 \le \tau < \infty$,

$$E(e^{i\lambda(X_\tau - X_0)} \mid \mathcal{F}_0) = E(e^{i\lambda(X_\tau - X_0)} \mid \mathcal{G}) = \mathcal{E}_\tau(G(\lambda))$$

$$(P\text{-a.s.}), \quad \lambda \in R, \tag{1.1}$$

where $\mathcal{E}_t(G(\lambda))$ is the stochastic exponential

$$\mathcal{E}_t(G(\lambda)) = e^{G_t(\lambda)} \prod_{0 < s \le t} (1 + \Delta G_s(\lambda))\, e^{-\Delta G_s(\lambda)} \tag{1.2}$$

(Theorem 4.41), while the function $G_t(\lambda)$ is calculated by means of the triplet $T = (B, C, \nu)$ of a semimartingale $X = (X_t, \mathcal{F}_t)$ according to the formula

$$G_t(\lambda) = i\lambda B_t - \frac{\lambda^2}{2} C_t + (e^{i\lambda x} - 1 - i\lambda x I(|x| \le 1)) * \nu_t \tag{1.3}$$

(see Ch. 4, § 2).

In dealing with the conditions for the convergence

$$X^n \xrightarrow{d_f(S)} X \ (\mathcal{G}\text{-stably}),$$

an essential rôle is played by the conditions ensuring the convergence of conditional characteristic functions:

$$\mathbf{E}(e^{i\lambda(X_t^n - X_0^n)} \mid \mathcal{F}_0^n) \xrightarrow{P} \mathbf{E}(e^{i\lambda(X_t - X_0)} \mid \mathcal{G}).$$

According to (1.1)

$$\mathbf{E}(e^{i\lambda(X_t - X_0)} \mid \mathcal{G}) = \mathcal{E}_t(G(\lambda)).$$

It is impossible, however, to calculate explicitely the characteristic function

$$\mathbf{E}(e^{i\lambda(X_t^n - X_0^n)} \mid \mathcal{F}_0^n).$$

Therefore it's rôle is taken over by the stochastic exponential $\mathcal{E}_t(G^n(\lambda))$, defined by the formula (1.2) with $G^n(\lambda)$ instead of $G(\lambda)$,where $G_t^n(\lambda)$ is calculated by means of the formula (1.3) with $T = (B, C, \nu)$ replaced by the triplet $T^n = (B^n, C^n, \nu^n)$ of a semimartingale $X^n = (X_t^n, \mathcal{F}_t^n)$.

2. The indicated rôle of the stochastic exponential $\mathcal{E}(G^n(\lambda))$ becomes apparent in the following three theorems.

Theorem 1. *Let the following conditions be fullfilled:*

1) $\mathcal{G} \subseteq \bigcap_{n \geq 1} \mathcal{F}_0^n$,

2) *for given* t *and* λ $(t \in R_+,\ \lambda \in R)$

$$\mathcal{E}_t(G^n(\lambda)) \xrightarrow{P} \mathcal{E}_t(G(\lambda)).$$

Then for these t *and* λ *as* $n \to \infty$

$$I(|\mathcal{E}_t(G(\lambda))| > 0)\, \mathbf{E}(e^{i\lambda(X_t^n - X_0^n)} \mid \mathcal{F}_0^n) \xrightarrow{P} \mathbf{E}(e^{i\lambda(X_t - X_0)} \mid \mathcal{G}).$$

For every $n \geq 1$ let $\hat{\tau}_n = (\hat{\tau}_n(t))_{t \geq 0}$ be a random change of time (Ch. 4, § 7), i.e. $\hat{\tau}_n \in V$ and for every $t \geq 0$ let $\hat{\tau}_n(t)$ be a stopping time (relative to the family $\mathbb{F}^n$). Define the family of σ-algebras $\hat{\mathbb{F}}^n = (\hat{\mathcal{F}}_t^n)_{t \geq 0}$ and the random process $(\hat{X}_t^n)_{t \geq 0}$ with

$$\hat{\mathcal{F}}_t^n = \mathcal{F}_{\hat{\tau}_n(t)}^n, \quad \hat{X}_t^n = X_{\hat{\tau}_n(t)}^n.$$

By Theorem 4.7.2 $\hat{X}^n = (\hat{X}^n_t, \hat{\mathcal{F}}^n_t)$ is a semimartingale. Concerning the sequence $\hat{X}^n = (\hat{X}^n_t, \hat{\mathcal{F}}^n_t)$, $n \geq 1$, we present the statement analogous to Theorem 1, assuming $X = (X_t, \mathcal{F}_t)$ is a semimartingale with $\mathcal{G}$-conditionally independent increments.

Theorem 2. *Let the following conditions be fullfilled:*

1) $\mathcal{G} \subseteq \bigcap_{n \geq 1} \mathcal{F}^n_0$,

2) *for given* t *and* λ ($t \in R_+$, $\lambda \in R$)

$$\mathcal{E}_{\hat{\tau}_n(t)}(G(\lambda)) \xrightarrow{P} \mathcal{E}_t(G(\lambda)).$$

Then for these t *and* λ *as* $n \to \infty$

$$I(|\mathcal{E}_t(G(\lambda))| > 0)\, \mathbf{E}(e^{i\lambda(\hat{X}^n_t - \hat{X}^n_0)} \mid \hat{\mathcal{F}}^n_0) \xrightarrow{P} \mathbf{E}(e^{i\lambda(X_t - X_0)} \mid \mathcal{G}).$$

For a fixed λ assume $P(|\mathcal{E}_t(G(\lambda))| > 0) = 1$, and for the same λ assume that Condition 2) of Theorem 1 holds. Then for this λ by Theorem 1 we get

$$\lim_n \mathbf{E}\xi e^{i\lambda(X^n_t - X^n_0)} = \mathbf{E}\xi e^{i\lambda(X_t - X_0)} \tag{1.4}$$

for each $\mathcal{G}$-measurable and bounded random variable ξ.

The relation (1.4), satisfied for each $\lambda \in R$, defines the $\mathcal{G}$-stable convergence of a sequence of distributions of random variables $(X^n_t - X^n_0)$, $n \geq 1$, to the distribution of a random variable $(X_t - X_0)$.

In applications the condition

$$\mathcal{G} \subseteq \bigcap_{n \geq 1} \mathcal{F}^n_0$$

of Theorem 1 is not always fullfilled. In this connection we give another type of conditions imposed below on flows of σ-algebras $\mathbb{F}^n$, $n \geq 1$. To clarify their sense we present the following example of defining flows $\mathbb{F}^n$, $n \geq 1$. Let $\mathbb{H} = (\mathcal{H}_t)_{t \geq 0}$ be a family of σ-algebras $\mathcal{H}_t \subseteq \mathcal{F}$, $t \geq 0$, satisfying Conditions (a) and (b) in Definition 1 (Ch. 1, § 1), $\mathcal{H}_\infty = \sigma(\bigcup_{t \geq 0} \mathcal{H}_t)$ and $(k_n)_{n \geq 1}$ a nondecreasing sequence of nonnegative numbers such that $\lim_n k_n = \infty$. Set

$$\mathcal{F}^n_t = \mathcal{H}_{tk_n}, \quad n \geq 1,$$

and denote

$$s_n = 1/\sqrt{k_n}\,.$$

Then flows of σ-algebras $\mathbb{F}^n$, $n \ge 1$, possess the following properties:

$$\mathcal{F}^n_{s_n} \subseteq \mathcal{F}^{n+1}_{s_{n+1}}, \quad n \ge 1, \qquad \mathcal{H}_\infty = \sigma(\bigcup_{n \ge 1} \mathcal{F}^n_{s_n}).$$

It is natural to call flows of σ-algebras $\mathbb{F}^n$, $n \ge 1$, with this property *nested.*

Theorem 3. *Let the following conditions be fullfilled:*

1) *the families of σ-algebras* $\mathbb{F}^n$, $n \ge 1$, *are nested, i.e. there exists a sequence of numbers* $(s_n)_{n \ge 1}$, $s_n \downarrow 0$ *as* $n \to \infty$ *such that*

$$\mathcal{F}^n_{s_n} \subseteq \mathcal{F}^{n+1}_{s_{n+1}}, \qquad \sigma(\bigcup_{n \ge 1} \mathcal{F}^n_{s_n}) = \sigma(\bigcup_{n \ge 1} \mathcal{F}^n_\infty)$$

and

$$\mathcal{G} \subseteq \sigma(\bigcup_{n \ge 1} \mathcal{F}^n_\infty);$$

2) *for given* t *and* λ ($t \in R_+$, $\lambda \in R$)

$$\mathcal{E}_t(G^n(\lambda)) \xrightarrow{P} \mathcal{E}_t(G(\lambda)), \qquad |\mathcal{E}_t(G(\lambda))| \ge 0 \quad \text{(P-a.s.)};$$

3) $X^n_{s_n} - X^n_0 \xrightarrow{P} 0, \quad \mathcal{E}_{s_n}(G^n(\lambda)) \xrightarrow{P} 1.$

Then for these t *and* λ *as* $n \to \infty$

$$\mathbf{E}(e^{i\lambda(X^n_t - X^n_0)} \mid \mathcal{F}^n_{s_n}) \xrightarrow{P} \mathbf{E}(e^{i\lambda(X_t - X_0)} \mid \mathcal{G}).$$

3. Observe that as $\hat{\tau}_n(t) \equiv t$ the assertion of Theorem 1 follows from Theorem 2. Consequently it suffices to prove Theorem 2 only. Note also that by Lemma 4.2.1 we have $\mathcal{E}_t(G) = \mathcal{E}_t(G)\, I(|\mathcal{E}_t(G)| > 0)$.

Let us prove Theorem 2 under the assumption $P(|\mathcal{E}_t(G)| > 0) = 1$. In the general case the proof is analogous.

The proof of Theorem 2 will be provided in several steps.

Step 1. Let us show that for any $\varepsilon > 0$ and every $n \ge 1$ there can be found a Markov time $\sigma_n \in \mathbf{T}(\mathbb{F}^n)$ such that

$$|\mathcal{E}^{-1}_{\hat{\tau}_n(t) \wedge \sigma_n}(G^n)| < \frac{2}{\varepsilon}. \tag{1.5}$$

To this end set (assuming $\inf\{\varnothing\} = \infty$)

$$\tilde{\sigma}_n = \inf\left(s: |\mathcal{E}^{\hat{\tau}_n(t)}_s (G^n)| \le \frac{\varepsilon}{2} \right).$$

By Lemma 4.2.1 the process $1 - |\mathcal{E}^{\hat{\tau}_n(t)}(G^n)| \in \mathcal{A}^+$. By Corollary 2 to Theorem 1.6.3 $|\mathcal{E}^{\hat{\tau}_n(t)}(G^n)|$ is a predictable process, and by Problem 1.3.11 $\tilde{\sigma}_n$ is a predictable Markov time. Consequently, by Theorem 1.3.4 there can be found a Markov time σ_n such that $\sigma_n < \tilde{\sigma}_n$ on the set $\{\tilde{\sigma}_n < \infty\}$. The function $|\mathcal{E}_s^{\hat{\tau}_n(t)}(G^n)|$ is nonincreasing in s (Lemma 4.2.1). Consequently $|\mathcal{E}_{\hat{\tau}_n(t)\wedge\sigma_n}(G^n)| > \varepsilon/2$, and this proves (1.5).

Step II. Markov times σ_n, $n \geq 1$, involved in (1.5), can be chosen in such a way that the implication

$$\mathcal{E}_{\hat{\tau}_n(t)}(G^n) \xrightarrow{P} \mathcal{E}_t(G) \Rightarrow \lim_n P(\sigma_n \leq \hat{\tau}_n(t),\ |\mathcal{E}_t(G)| > \varepsilon) = 0 \tag{1.6}$$

takes place.

Observe that for a fixed $\varepsilon > 0$ a Markov time σ_n can be chosen in such a way that

$$P\left(\tilde{\sigma}_n - \sigma_n > \frac{1}{n}\right) \leq \frac{1}{n}$$

(see Theorem 1.3.4). Next, due to the inclusion

$$\{\sigma_n \leq \hat{\tau}_n(t)\} \subseteq \left\{\tilde{\sigma}_n \leq \hat{\tau}_n(t) + \frac{1}{n}\right\} \cup \left\{\tilde{\sigma}_n - \sigma_n > \frac{1}{n}\right\}$$

and to $\tilde{\sigma}_n \leq \hat{\tau}_n(t)$ or to $\tilde{\sigma}_n = \infty$, we get

$$\begin{aligned}\{\sigma_n \leq \hat{\tau}_n(t)\} &\subseteq \{\tilde{\sigma}_n \leq \hat{\tau}_n(t)\} \cup \left\{\tilde{\sigma}_n - \sigma_n > \frac{1}{n}\right\} \\ &\subseteq \left\{|\mathcal{E}_{\hat{\tau}_n(t)}(G^n)| \leq \frac{\varepsilon}{2}\right\} \cup \left\{\tilde{\sigma}_n - \sigma_n > \frac{1}{n}\right\} \\ &\subseteq \left\{|\mathcal{E}_{\hat{\tau}_n(t)}(G^n)| \leq \frac{\varepsilon}{2}, |\mathcal{E}_t(G)| > \varepsilon\right\} \cup \left\{\tilde{\sigma}_n - \sigma_n > \frac{1}{n}\right\} \cup \{|\mathcal{E}_t(G)| \leq \varepsilon\}\end{aligned}$$

This gives

$$\begin{aligned}&\{\sigma_n \leq \hat{\tau}_n(t), |\mathcal{E}_t(G)| > \varepsilon\} \\ &\subseteq \left\{|\mathcal{E}_{\hat{\tau}_n(t)}(G^n) - \mathcal{E}_t(G)| \geq \frac{\varepsilon}{2}\right\} \cup \left\{\tilde{\sigma}_n - \sigma_n > \frac{1}{n}\right\},\end{aligned} \tag{1.7}$$

and consequently implication (1.6) holds.

Step III. Denote

$$Z_s^n = \exp[i\lambda(X^n_{\hat{\tau}_n(t)\wedge\sigma_n\wedge s} - X_0^n)]\, \mathcal{E}^{-1}_{\hat{\tau}_n(t)\wedge\sigma_n\wedge s}(G^n(\lambda)). \tag{1.8}$$

Analogously to proving Theorem 4.3.1 it can be established that $(Z_s^n, \mathcal{F}_s^n)$ is a local martingale. Besides, by (1.5)

$$|Z_s^n| \le \frac{2}{\varepsilon}, \quad s \in R_+,$$

and consequently $(Z_s^n, \mathcal{F}_s^n)$ is a uniformly bounded martingale with $(\hat{\mathcal{F}}_0^n = \mathcal{F}_0^n)$

$$E(Z^n_{\hat{\tau}_n(t)\wedge\sigma_n} \mid \hat{\mathcal{F}}_0^n) = 1 \quad (P\text{-a.s.}). \tag{1.9}$$

Step IV. Denoting

$$J_n' = E(\exp[i\lambda(X^n_{\hat{\tau}_n(t)\wedge\sigma_n} - X_0^n)] \mid \hat{\mathcal{F}}_0^n) - \mathcal{E}_t(G(\lambda))$$

we will show that

$$\lim_n E|J_n'|\, I(|\mathcal{E}_t(G)| > \varepsilon) = 0. \tag{1.10}$$

Since $\mathcal{E}_t(G)$ is a $\mathcal{F}_0$-measurable random variable, J_n' admits (by step III) the following representation:

$$J_n' = E(e^{i\lambda(X^n_{\hat{\tau}_n(t)\wedge\sigma_n} - X_0^n)} - \mathcal{E}_t(G(\lambda))\, Z^n_{\hat{\tau}_n(t)\wedge\sigma_n} \mid \hat{\mathcal{F}}_0^n).$$

In view of the representation (1.8) for Z_s^n, this and (1.5) imply

$$|J_n'| \le E(|1 - \mathcal{E}^{-1}_{\hat{\tau}_n(t)\wedge\sigma_n}(G^n(\lambda))\, \mathcal{E}_t(G(\lambda))| \mid \hat{\mathcal{F}}_0^n)$$

$$\le \frac{2}{\varepsilon} E(|\mathcal{E}_{\hat{\tau}_n(t)\wedge\sigma_n}(G^n(\lambda)) - \mathcal{E}_t(G(\lambda))| \mid \hat{\mathcal{F}}_0^n),$$

and hence to prove (1.10) it suffices to show that

$$\lim_n E|\mathcal{E}_{\hat{\tau}_n(t)\wedge\sigma_n}(G^n(\lambda)) - \mathcal{E}_t(G(\lambda))|\, I(|\mathcal{E}_t(G(\lambda))| > \varepsilon) = 0. \tag{1.11}$$

The relation (1.11) holds, since firstly

$$\overline{\lim_n} E|\mathcal{E}_{\hat{\tau}_n(t)\wedge\sigma_n}(G^n(\lambda)) - \mathcal{E}_t(G(\lambda))|\, I(|\mathcal{E}_t(G(\lambda))| > \varepsilon)\, I(\sigma_n > \hat{\tau}_n(t))$$

$$\le \overline{\lim_n} E|\mathcal{E}_{\hat{\tau}_n(t)}(G^n(\lambda)) - \mathcal{E}_t(G(\lambda))| = 0,$$

and secondly (see (1.6)),

$$\overline{\lim_n}\, \mathbf{E}\,|\, \mathcal{E}_{\hat{\tau}_n(t)\wedge\sigma_n}(G^n(\lambda)) - \mathcal{E}_t(G(\lambda))\,|\, I(|\mathcal{E}_t(G(\lambda))| > \varepsilon)\, I(\sigma_n \le \hat{\tau}_n(t))$$

$$\le 2\, \overline{\lim_n}\, \mathrm{P}(\sigma_n \le \hat{\tau}_n(t), |\mathcal{E}_t(G(\lambda))| > \varepsilon) = 0.$$

Step V. Denote

$$J''_n = \mathbf{E}\left(e^{i\lambda(\hat{X}^n_t - \hat{X}^n_0)} - e^{i\lambda(X^n_{\hat{\tau}_n(t)\wedge\sigma_n} - X^n_0)} \,|\, \hat{\mathcal{F}}^n_0\right).$$

We will show that

$$\lim_n \mathbf{E}\,|J''_n|\, I(|\mathcal{E}_t(G(\lambda))| > \varepsilon) = 0.$$

Indeed, due to the estimate

$$|J''_n| \le 2\mathrm{P}(\sigma_n \le \hat{\tau}_n(t) \,|\, \hat{\mathcal{F}}^n_0)$$

and $\hat{\mathcal{F}}^n_0$ $(= \mathcal{F}^n_0)$-measurability of the random variable $\mathcal{E}_t(G(\lambda))$ we get (see (1.6))

$$\overline{\lim_n}\, \mathbf{E}\,|J''_n|\, I(|\mathcal{E}_t(G(\lambda))| > \varepsilon) \le 2\, \overline{\lim_n}\, \mathrm{P}(\sigma_n \le \hat{\tau}_n(t), \quad |\mathcal{E}_t(G(\lambda))| > \varepsilon) = 0.$$

Step VI. Denote

$$J_n = \mathbf{E}\left(e^{i\lambda(\hat{X}^n_t - \hat{X}^n_0)} \,|\, \hat{\mathcal{F}}^n_0\right) - \mathcal{E}_t(G(\lambda)).$$

Since $J_n = J'_n + J''_n$, by (1.10) and (1.12)

$$\lim_n \mathbf{E}\,|J_n|\, I(|\mathcal{E}_t(G(\lambda))| > \varepsilon) = 0.$$

Consequently

$$\overline{\lim_n}\, \mathbf{E}\,|J_n| \le \overline{\lim_n}\, \mathbf{E}\,|J_n|\, I(0 < |\mathcal{E}_t(G(\lambda))| \le \varepsilon)$$

$$\le 2\mathrm{P}(0 < |\mathcal{E}_t(G(\lambda))| \le \varepsilon) \to 0, \ \ \varepsilon \to 0.$$

The theorem is proved.

4. The proof of Theorem 3 is based on the following auxiliary fact.

Lemma 1. *Let Condition* 1) *of Theorem* 3 *be satisfied and let* β_n *be a* $\mathcal{F}^n_\infty$-*measurable bounded random variable* $|\beta_n| \le c$, $n \ge 1$, *and* ξ *a* $\mathcal{G}$-*measurable bounded random variable* $|\xi| \le c$.

Then the following implication holds:

$$\lim_n \mathbf{E}\xi\beta_n = 0 \Leftrightarrow \lim_n \mathbf{E}\xi\, \mathbf{E}(\beta_n \,|\, \mathcal{F}^n_{s_n}) = 0.$$

Proof. The assertion of the lemma is a consequence of the relation

$$\mathbf{E}\xi\mathbf{E}(\beta_n \,|\, \mathcal{F}^n_{s_n}) = \mathbf{E}\beta_n\mathbf{E}(\xi \,|\, \mathcal{F}^n_{s_n})$$

and

$$\mathbf{E}\xi\beta_n = \mathbf{E}\beta_n\mathbf{E}(\xi \mid \mathcal{F}^n_{s_n}) + \mathbf{E}\beta_n(\xi - \mathbf{E}(\xi \mid \mathcal{F}^n_{s_n})).$$

Indeed

$$|\mathbf{E}\beta_n(\xi - \mathbf{E}(\xi \mid \mathcal{F}^n_{s_n}))| \le c\mathbf{E}\,|\xi - \mathbf{E}(\xi \mid \mathcal{F}^n_{s_n})| \to 0,$$

since by Theorem 2.6.2 (Assertion 4)) we have

$$\mathbf{E}(\xi \mid \mathcal{F}^n_{s_n}) \to \mathbf{E}(\xi \mid \sigma(\bigcup_{n\ge 1} \mathcal{F}^n_{s_n})), \quad \mathcal{G} \subseteq \sigma(\bigcup_{n\ge 1} \mathcal{F}^n_{s_n})$$

and as ξ ($|\xi| \le c$) is bounded one can apply here Lebesgue's theorem on dominated convergence.

On proving Theorem 3 we utilize in the first step the fact that (see (1.1)) (P-a.s.)

$$\mathcal{E}_t(G(\lambda)) = \mathbf{E}(e^{i\lambda(X_t - X_0)} \mid \mathcal{G}),$$

and that by Condition 1) of Theorem 3 and Assertion 4) of Theorem 2.6.2

$$\mathcal{E}_t(G(\lambda)) = \lim_n \mathbf{E}(\mathcal{E}_t(G(\lambda)) \mid \mathcal{F}^n_{s_n}).$$

Therefore it suffices to prove that

$$\mathbf{E}(e^{i\lambda(X^n_t - X^n_0)} - \mathcal{E}_t(G(\lambda)) \mid \mathcal{F}^n_{s_n}) \xrightarrow{P} 0. \tag{1.13}$$

Notice one more useful fact. Namely, the relation (1.13) holds provided

$$\lim_n \mathbf{E}\,|\mathbf{E}(e^{i\lambda(X^n_t - X^n_0)} - \mathcal{E}_t(G(\lambda)) \mid \mathcal{F}^n_{s_n})|\, I(|\mathcal{E}_t(G(\lambda)) > \varepsilon) = 0 \tag{1.14}$$

for each $\varepsilon > 0$, since

$$\mathbf{E}\,|\mathbf{E}(e^{i\lambda(X^n_t - X^n_0)} - \mathcal{E}_t(G(\lambda)) \mid \mathcal{F}^n_{s_n})|\, I(0 < |\mathcal{E}_t(G(\lambda))| \le \varepsilon)$$

$$\le 2\mathsf{P}(0 < |\mathcal{E}_t(G(\lambda))| \le \varepsilon) \to 0, \quad \varepsilon \to 0.$$

Thus to prove the theorem it suffices to verify (1.14). We will verify this in several steps analogous to those proving Theorem 2.

Step I consists in choosing a Markov time σ_n such that

$$|\mathcal{E}^{-1}_{t\wedge\sigma_n}(G^n)| < \frac{2}{\varepsilon}. \tag{1.15}$$

Time σ_n is determined as in step I of the proof of Theorem 2 with $\hat{\tau}_n(t) \equiv t$.

Step II consists in proving the implication

$$\mathcal{E}_t(G^n) \xrightarrow{P} \mathcal{E}_t(G) \Rightarrow \lim_n \mathsf{P}(\sigma_n \le t, |\mathcal{E}_t(G)| > \varepsilon) = 0, \tag{1.16}$$

which completely coincides with step II of the proof of Therorem 2 with $\hat{\tau}_n(t) \equiv t$.

Step III. The process Z^n_s, defined by the representation (1.8) with $\hat{\tau}_n(t) \equiv t$ and σ_n of steps I and II of the present proof, possesses the following properties:

$$|Z^n_s| \le \frac{2}{\varepsilon}, \quad s \in R_+, \tag{1.17}$$

$(Z^n_s, \mathcal{F}^n_s)$ is a uniformly bounded martingale and

$$\mathbf{E}(Z^n_t \mid \mathcal{F}^n_{s_n}) = Z^n_{t \wedge s_n}. \tag{1.18}$$

Let us show that

$$\lim_n \mathbf{E}\,|1 - Z^n_{t\wedge s_n}|\, I(|\mathcal{E}_t(G(\lambda))| > \varepsilon) = 0. \tag{1.19}$$

By the estimate (1.17) it suffices to show that as $n \to \infty$

$$|1 - Z^n_{t\wedge s_n}|\, I(|\mathcal{E}_t(G(\lambda))| > \varepsilon) \xrightarrow{P} 0,$$

and the latter relation in turn holds, by the definition of Z^n_s, provided

$$g^n_{t\wedge s_n \wedge \sigma_n}\, I(|\mathcal{E}_t(G(\lambda))| > \varepsilon) \xrightarrow{P} 0 \tag{1.20}$$

with

$$g^n_t = |X^n_t - X^n_0| + |\mathcal{E}_t(G^n(\lambda)) - 1|.$$

For $a > 0$ we have

$$\begin{aligned}
&P(g^n_{t\wedge s_n\wedge\sigma_n} \ge a, |\mathcal{E}_t(G(\lambda))| > \varepsilon) \\
&\le P(g^n_{s_n} \ge a, s_n \le t \wedge \sigma_n, |\mathcal{E}_t(G(\lambda))| > \varepsilon) + P(s_n > t\wedge\sigma_n, |\mathcal{E}_t(G(\lambda))| > \varepsilon) \\
&\le P(g^n_{s_n} \ge a) + P(s_n > t, \sigma_n > t, |\mathcal{E}_t(G(\lambda))| > \varepsilon) + P(\sigma_n \le t, |\mathcal{E}_t(G(\lambda))| > \varepsilon) \\
&\le P(g^n_{s_n} \ge a) + I(s_n > t) + P(\sigma_n \le t, |\mathcal{E}_t(G(\lambda))| > \varepsilon).
\end{aligned}$$

This implies the relation (1.20) in an obvious manner: by Condition 3) of the theorem we have

$$P(g^n_s \ge a) \to 0, \quad n \to \infty,$$

by Condition 1) of the theorem we have

$$s_n \downarrow 0,$$

and by (1.16) we have

$$P(\sigma_n \le t, |\mathcal{E}_t(G(\lambda))| > \varepsilon) \to 0, \ n \to \infty.$$

Step IV. Denoting

$$J_n' = \mathbf{E}(\exp[i\lambda(X^n_{t\wedge\sigma_n} - X^n_0)] - \mathcal{E}_t(G(\lambda)) \mid \mathcal{F}^n_{s_n}),$$

we will show that

$$\lim_n \mathbf{E} \mid J_n' \mid I(\mid \mathcal{E}_t(G(\lambda)) \mid > \varepsilon) = 0. \tag{1.21}$$

To this end represent the variable $\mathbf{E}(\mathcal{E}_t(G) \mid \mathcal{F}^n_{s_n})$ in the following way:

$$\mathbf{E}(\mathcal{E}_t(G) \mid \mathcal{F}^n_{s_n}) = \mathbf{E}(\mathcal{E}_t(G) \mid \mathcal{F}^n_{s_n})(1 - Z^n_{t\wedge s_n}) + \mathbf{E}(\mathbf{E}(\mathcal{E}_t(G) \mid \mathcal{F}^n_{s_n}) Z^n_t \mid \mathcal{F}^n_{s_n}).$$

Denote

$$\delta_n = \mid \mathbf{E}(\mathcal{E}_t(G) \mid \mathcal{F}^n_{s_n})(1 - Z^n_{s_n}) \mid,$$

$$\gamma_n = \mid \mathcal{E}_t(G) - \mathbf{E}(\mathcal{E}_t(G) \mid \mathcal{F}^n_{s_n}) \mid .$$

Taking into account these notations and the definition of Z^n_s, we get the following estimates for $\mid J_n' \mid$:

$$\mid J_n' \mid \le \mathbf{E}(\mid 1 - \mathbf{E}(\mathcal{E}_t(G) \mid \mathcal{F}^n_{s_n})\, \mathcal{E}^{-1}_{t\wedge\sigma_n}(G^n) \mid\mid \mathcal{F}^n_{s_n}) + \delta_n$$

$$\le \frac{2}{\varepsilon}\mathbf{E}(\mid \mathcal{E}_{t\wedge\sigma_n}(G^n) - \mathbf{E}(\mathcal{E}_t(G) \mid \mathcal{F}^n_{s_n}) \mid\mid \mathcal{F}^n_{s_n}) + \delta_n$$

$$\le \frac{2}{\varepsilon}\mathbf{E}(\mid \mathcal{E}_{t\wedge\sigma_n}(G^n) - \mathcal{E}_t(G) \mid\mid \mathcal{F}^n_{s_n}) + \frac{2}{\varepsilon}\gamma_n + \delta_n.$$

Observe that by (1.19) and the estimate $\delta_n \le \mid 1 - Z^n_{s_n} \mid$ we have

$$\lim_n \mathbf{E}\delta_n I(\mid \mathcal{E}_t(G) \mid > \varepsilon) = 0.$$

By Theorem 2.6.2 (Assertion 4)) we have $\gamma_n \to 0$ (P-a.s.), $\mid \gamma_n \mid \le 2$, and hence

$$\lim_n \mathbf{E}\gamma_n = 0.$$

Therefore the desired relation (1.21) will take place if

$$\lim_n \mathbf{E}I(\mid \mathcal{E}_t(G) \mid > \varepsilon)\, \mathbf{E}(\mid \mathcal{E}_{t\wedge\sigma_n}(G^n) - \mathcal{E}_t(G) \mid\mid \mathcal{F}^n_{s_n}) = 0$$

or, taking into consideration Lemma 1, if

$$\lim_n \mathbf{E}I(\mid \mathcal{E}_t(G) \mid > \varepsilon) \mid \mathcal{E}_{t\wedge\sigma_n}(G^n) - \mathcal{E}_t(G) \mid = 0.$$

The last relation holds since

$$I(|\mathcal{E}_t(G)| > \varepsilon)\,|\mathcal{E}_{t\wedge\sigma_n}(G^n) - \mathcal{E}_t(G)|$$

$$\leq 2I(|\mathcal{E}_t(G)| > \varepsilon,\ \sigma_n \leq t) + |\mathcal{E}_t(G^n) - \mathcal{E}_t(G)| \xrightarrow{P} 0$$

by (1.16) and Condition 2) of the theorem.

Step V. Denote

$$J''_n = \mathbf{E}\,(e^{i\lambda(X^n_t - X^n_0)} - e^{i\lambda(X^n_{t\wedge\sigma_n} - X^n_0)} \mid \mathcal{F}^n_{s_n}).$$

We will show that

$$\lim_n \mathbf{E}\,|J''_n|\,I(|\mathcal{E}_t(G)| > \varepsilon) = 0. \tag{1.22}$$

Indeed, by the estimate

$$|J''_n| \leq 2P(\sigma_n \leq t \mid \mathcal{F}^n_{s_n})$$

and Lemma 1 it suffices to show that

$$\lim_n \mathbf{E}I(\sigma_n \leq t, |\mathcal{E}_t(G)| > \varepsilon) = 0.$$

But this fact is already established in Step II (see (1.16)).

Step VI. Observe that

$$\mathbf{E}\,(e^{i\lambda(X^n_t - X^n_0)} - \mathcal{E}_t(G(\lambda)) \mid \mathcal{F}^n_{s_n}) = J'_n + J''_n,$$

and that the desired relation (1.14) follows from (1.21) and (1.22).

5. As is well known in the probability theory, dealing with nonnegative random variables, it is often convenient to resort to the Laplace transform rather then to the Fourier transform. In this regard it is useful to present analogous to Theorems 1- 3 statements for nondecreasing semimartingales.

In accordance with Ch. 4, § 2 a nondecreasing semimartingale $X = (X_t, \mathcal{F}_t)$ with $X_0 \geq 0$ and the triplet (B, C, ν) ($C = 0$ here) admits the representation

$$X_t = X_0 + A^c_t + \int_0^t \int_{x>0} x\, d\mu$$

with

$$A^c_t = B^c_t - \int_0^t \int_{0<x\leq 1} x\, d\nu^c.$$

Denoting

$$G_t(\lambda) = -\lambda A_t^c + \int_0^t \int_{x>0} (e^{-\lambda x} - 1)\, d\nu,\ \lambda \in R_+,$$

we get (cf. (2.20) — (2.23), Ch. 4, § 2)

$$\mathcal{E}_t(G(\lambda)) = e^{G_t^c(\lambda)} \prod_{0<s\le t} \left(1 + \int_{x>0} (e^{-\lambda x} - 1)\, \nu(\{s\}, dx)\right). \tag{1.23}$$

The following statement is analogous to Theorem 1.

Theorem 4. *Let* $X = (X_t, \mathcal{F}_t)$ *and* $X^n = (X_t^n, \mathcal{F}_t^n)$ *be nondecreasing semimartingales with the stochastical exponentials* $\mathcal{E}(G(\lambda))$ *and* $\mathcal{E}(G^n(\lambda))$, *defined by the formula* (1.23), *and* X *a semimartingale with* $\mathcal{G}$*-conditionally independent increments.*

If the following conditions are fullfilled:

1) $\mathcal{G} \subseteq \bigcap_{n\ge 1} \mathcal{F}_0^n$,

2) *for given* t *and* λ ($t \in R_+$, $\lambda \in R_+$)

$$\mathcal{E}_t(G^n(\lambda)) \xrightarrow{P} \mathcal{E}_t(G(\lambda)),$$

then for these t *and* λ *as* $n \to \infty$

$$E(e^{-\lambda(X_t^n - X_0^n)} \mid \mathcal{F}_0^n) \xrightarrow{P} E(e^{-\lambda(X_t - X_0)} \mid \mathcal{G}).$$

The statements analogous to Theorems 2 and 3 are similarly formulated. The proof of Theorem 4 and the analogies to Theorems 2 and 3 is carried out as that of Theorems 1 - 3.

6. Let us show how Theorem 2 works in the particular, but frequently encountered case of "schemes of series".

On a probability space $(\Omega, \mathcal{F}, P)$ let sequences (series) of random variables $(\xi_{nk})_{k\ge 0}$, $n \ge 1$ be defined. For every $n \ge 1$ a random variable ξ_{nk} is measurable relative to a σ-algebra $\mathcal{H}_k^n$, where a sequence $(\mathcal{H}_k^n)_{k\ge 0}$ possesses the following properties: $\mathcal{H}_0^n = \{\emptyset, \Omega\}$, $\mathcal{H}_k^n \subseteq \mathcal{H}_{k+1}^n$, $k \ge 1$. (In the simplest case of $\xi_{n0} = 0$ $\mathcal{H}_k^n$ can be replaced by the σ-algebra $\sigma\{\xi_{n0}, \ldots, \xi_{nk}\}$).

Assuming further for simplicity $\xi_{n0} \equiv 0$, denote ([a] is the integer part of a)

$$X_t^n = \sum_{k=0}^{[nt]} \xi_{nk}, \quad \mathcal{F}_t^n = \mathcal{H}_{[nt]}^n, \quad t \geq 0. \tag{1.24}$$

The process $(X_t^n)_{t \geq 0}$ has trajectories in V, and consequently $X^n = (X_t^n, \mathcal{F}_t^n)$ is a semimartingale. Its triplet $T^n = (B^n, C^n, \nu^n)$ is calculated by the formulas ($\mathcal{H}_{-1}^n = \{\varnothing, \Omega\}$)

$$B_t^n = \sum_{k=0}^{[nt]} \mathbf{E}\,(\xi_{nk} I\,(\,|\,\xi_{nk}\,| \leq 1) \mid \mathcal{H}_{k-1}^n), \quad C_t^n \equiv 0, \tag{1.25}$$

$$\nu^n\,((0, t] \times \Gamma) = \sum_{k=0}^{[nt]} \nu_{nk}\,(\Gamma), \quad \Gamma \in B\,(R_0),$$

where ν_{nk} is the random measure on $(R_0, B\,(R_0))$ such that (P-a.s.)

$$\nu_{nk}\,(\Gamma) = \mathrm{P}\,(\xi_{nk} \in \Gamma \mid \mathcal{H}_{k-1}^n).$$

Denote

$$g_{nk}\,(\lambda) = \int_{R_0} (e^{i\lambda x} - 1)\,\nu_{nk}\,(dx).$$

Then (P-a.s.)

$$g_{nk}\,(\lambda) = \mathbf{E}\,(e^{i\lambda\xi_{nk}} - 1 \mid \mathcal{H}_{k-1}^n)$$

and $G_t^n\,(\lambda)$ is given by the formula (cf. (1.3))

$$G_t^n\,(\lambda) = i\lambda B_t^n + \sum_{k=0}^{[nt]} \int_{R_0} (e^{i\lambda x} - 1 - i\lambda x I\,(\,|\,x\,| \leq 1))\,\nu_{nk}\,(dx).$$

In accordance with (1.2) the stochastic exponential

$$\mathcal{E}_t\,(G^n\,(\lambda)) = \prod_{k=0}^{[nt]} (1 + g_{nk}\,(\lambda)) = \prod_{k=0}^{[nt]} \mathbf{E}\,(e^{i\lambda\xi_{nk}} \mid \mathcal{H}_{k-1}^n).$$

Suppose now that a sequence $(\gamma_n)_{n \geq 1}$ of random variables is given such that for every $n \geq 1$ γ_n takes values in the set $\{0, 1, 2, \ldots\}$ and it is a Markov time relative to $(\mathcal{H}_k^n)_{k \geq 0}$, i.e. $\{\gamma_n \leq m\} \in \mathcal{H}_m^n$, $m \geq 0$. Denote $\tau_n = \gamma_n/n$. This random variable is a Markov time relative to $\mathbb{F}^n = (\mathcal{F}_t^n)_{t \geq 0}$ with $\mathcal{F}_t^n = \mathcal{H}_{[nt]}^n$, since

$$\{\tau_n \leq t\} = \{\gamma_n \leq nt\} = \{\gamma_n \leq [nt]\} \in \mathcal{H}_{[nt]}^n = \mathcal{F}_t^n.$$

Besides

$$\mathcal{E}_{\tau_n}(G^n(\lambda)) = \prod_{k=0}^{\gamma_n} \mathbf{E}(e^{i\lambda\xi_{nk}} \mid \mathcal{H}^n_{k-1}), \quad X^n_{\tau_n} = \sum_{k=0}^{\gamma_n} \xi_{nk}.$$

From Theorem 2 it follows that the convergence

$$\prod_{k=0}^{\gamma_n} \mathbf{E}(e^{i\lambda\xi_{nk}} \mid \mathcal{H}^n_{k-1}) \xrightarrow{P} e^{-\lambda^2/2}, \quad n \to \infty, \tag{1.26}$$

implies the convergence of the characteristic functions

$$\mathbf{E}e^{i\lambda X^n_{\tau_n}} \to e^{-\lambda^2/2}, \quad n \to \infty. \tag{1.27}$$

Thus the validity of the condition (1.26) for all $\lambda \in R$ implies the convergence in distribution of the sequence

$$\left(\sum_{k=0}^{\gamma_n} \xi_{nk}\right)_{n \geq 1}$$

to a Gaussian random variable with the parameters (0, 1).

If $\gamma_n \equiv n$ (Theorem 1 is directly applicable to this version), $\mathcal{H}^n_k = \sigma\{\xi_{n0}, \ldots, \xi_{nk}\}$ and random variables $\xi_{n0}, \xi_{n1}, \ldots$ are mutually independent, then Condition (1.26) turns into Condition (1.27). Thus the statement given in Theorem 1 is meaningful especially in case of dependent variables.

Problems

1. Prove Theorem 4.

2. Prove the validity of the representation (1.24) for the triplet T^n of the semimartingale $X^n = (X^n_t, \mathcal{F}^n_t)$, defined by (1.25).

§ 2. Method of stochastic exponentials. II. Weak convergence of finite dimensional distributions

1. It is supposed in the present section that $X^n = (X_t^n, \mathcal{F}_t^n)$, $n \geq 1$, and $X = (X_t, \mathcal{F}_t)$ are semimartingales with the triplets $T^n = (B^n, C^n, \nu^n)$, $n \geq 1$ and $T = (B, C, \nu)$ and the stochastic exponentials $\mathcal{E}(G^n)$, $n \geq 1$ and $\mathcal{E}(G)$ respectively. Here X is a semimartingale with $\mathcal{G}$-conditionally independent increments.

The method of stochastic exponentials for determining conditions for the weak convergence of finite dimensional distributions (in "predictable" terms) is developed in the following three theorems.

Theorem 1. *Let the following conditions be fullfilled:*

1) $\mathcal{G} \subseteq \bigcap_{n \geq 1} \mathcal{F}_0^n$;

2) *for each* $\lambda \in R$ *and points* t_j *of the set* $S = \{t_1, ..., t_m\}$ *with* $0 = t_0 < t_1 < ... < t_m$

$$\mathcal{E}_{t_j}(G^n(\lambda)) \xrightarrow{P} \mathcal{E}_{t_j}(G(\lambda)), \ n \to \infty;$$

3) $|\mathcal{E}_{t_m}(G(\lambda))| > 0$ (P-a.s.), $\lambda \in R$;

4) $X_0^n \xrightarrow{d} X_0$ ($\mathcal{G}$-*stably*).

Then

$$X^n \xrightarrow{d_f(S)} X \ (\mathcal{G}\text{-stably}).$$

For every $n \geq 1$ let $\hat{\tau}_n = (\hat{\tau}_n(t))_{t \geq 0}$ be a random change of time (Ch. 4, § 7). Define the family of σ-algebras $\hat{\mathbb{F}}^n = (\hat{\mathcal{F}}_t^n)_{t \geq 0}$ and a stochastic process $(\hat{X}_t^n)_{t \geq 0}$ with

$$\hat{\mathcal{F}}_t^n = \mathcal{F}^n_{\hat{\tau}_n(t)}, \quad \hat{X}_t^n = X^n_{\hat{\tau}_n(t)}.$$

Concerning the semimartingales $\hat{X}^n = (\hat{X}_t^n, \hat{\mathcal{F}}_t^n)$, $n \geq 1$, the following theorem is valid.

Theorem 2. *Let the following conditions be fullfilled:*

1) $\mathcal{G} \subseteq \bigcap_{n \geq 1} \mathcal{F}_0^n$;

2) *for every* $\lambda \in R$ *and points* t_j *of the set* $S = \{t_j, ..., t_m\}$ *with* $0 = t_0 < t_1 < ... < t_m$

$$\mathcal{E}_{\hat{\tau}_n(t_j)}(G^n(\lambda)) \xrightarrow{P} \mathcal{E}_{t_j}(G(\lambda)), \ n \to \infty;$$

3) $|\mathcal{E}_{t_m}(G(\lambda))| > 0$ (P-a.s.), $\lambda \in R$;

$$4) \ X_0^n \xrightarrow{d} X_0 \quad (\mathcal{G}\text{-stably}).$$

Then

$$\hat{X}^n \xrightarrow{d_f(S)} X \quad (\mathcal{G}\text{-stably}).$$

Theorem 3. *Let the following conditions be fullfilled:*

1) *families of* σ*-algebras* $\mathbb{F}^n$, $n \geq 1$, *are nested: i.e. there exists a sequence of numbers* $(s_n)_{n \geq 1}$, $s_n \downarrow 0$, *as* $n \to \infty$ *such that*

$$\mathcal{F}^n_{s_n} \subseteq \mathcal{F}^{n+1}_{s_{n+1}}, \quad \sigma\left(\bigcup_{n \geq 1} \mathcal{F}^n_{s_n}\right) = \sigma\left(\bigcup_{n \geq 1} \mathcal{F}^n_{\infty}\right) \ \textit{and} \ \mathcal{G} \subseteq \sigma\left(\bigcup_{n \geq 1} \mathcal{F}^n_{\infty}\right);$$

2) *for every* $\lambda \in R$ *and points* t_j *of the set* $S = \{t_1, ..., t_m\}$ *with* $0 = t_0 < t_1 < ... < t_m$

$$\mathcal{E}_{t_j}(G^n(\lambda)) \xrightarrow{P} \mathcal{E}_{t_j}(G(\lambda)), \ n \to \infty;$$

3) $|\mathcal{E}_{t_m}(G(\lambda))| > 0$ (P-a.s.), $\lambda \in R$;

4) $X_0^n \xrightarrow{d} X_0$ ($\mathcal{G}$-*stably*);

5) $X^n_{s_n} - X^n_0 \xrightarrow{P} 0$, $\mathcal{E}_{s_n}(G^n(\lambda)) \xrightarrow{P} 1$, $\lambda \in R$.

Then

$$X^n \xrightarrow{d_f(S)} X \quad (\mathcal{G}\text{-stably}).$$

2. To prove Theorems 1-3 we need an auxiliary fact.

Let S be a set involved in the statement of Theorems 1-3, H_S a class of functions $h = h(t)$ of type

$$h(t) = c_j \ \text{for} \ t_j < t \leq t_{j+1}, \ h(0) = 0, \ h(t) = 0,$$
$$t > t_m, \ t_j \in S, \ j = 1, ..., m, \ m \geq 1,$$

and $h^n(t) = h(\hat{\tau}_n(t))$, where $\hat{\tau}_n = (\hat{\tau}_n(t))_{t \geq 0}$ is a random change of time.

Define the semimartingales $X^{n, h^n} = (X_t^{n, h^n}, \mathcal{F}^n_t)$ and $X^h = (X^h_t, \mathcal{F}_t)$ with

$$X_t^{n, h^n} = c_0 X_0^n + h^n \cdot X_t^n, \quad X_t^h = c_0 X_0 + h \cdot X_t \tag{2.1}$$

and denote by $\mathcal{E}(G^{n,h^n}(\lambda))$ and $\mathcal{E}(G^h(\lambda))$ the stochastic exponentials related to these semimartingales.

Lemma 1. *For any function* $h \in H_S$ *the semimartingale* X^h *is a process with* $\mathcal{G}$-*conditionally independent increments. If Condition* 3) *of Theorem* 2 *is fullfilled, then*

$$|\mathcal{E}_{t_m}(G^h(\lambda))| > 0 \quad (P\text{-a.s.}), \quad \lambda \in R.$$

If in addition Condition 2) *of Theorem* 2 *is fullfilled, then* $(h^n(t) = h(\hat{\tau}_n(t)))$

$$\mathcal{E}_{\hat{\tau}_n(t_m)}(G^{n,h^n}(\lambda)) \xrightarrow{P} \mathcal{E}_{t_m}(G^h(\lambda)), \quad n \to \infty, \quad \lambda \in R.$$

Proof. Denote by $T^h = (B^h, C^h, \nu^h)$ the triplet of the semimartingale X^h. It is not hard to show (Problem 1) that the components of T^h are determined in the following way:

$$B^h = h \circ B + hx\,(I_{(|hx| \le 1)} - I_{(|x| \le 1)}) * \nu, \quad C^h = h^2 \circ C,$$

$$g(x)\,\nu^h(dt, dx) = g(xh(t))\,\nu(dt, dx) \tag{2.2}$$

for each $B(R_0)$-measurable function $g = g(x)$ with $|g(x)| \le \text{const}\,(x^2 \wedge 1)$. According to the representation (2.2) the triplet T^h is $\mathcal{G}$-measurable and hence by Theorem 4.4.1 the semimartingale X^h is a process with $\mathcal{G}$-conditionally independent increments.

Denote

$$\mathcal{E}_{t_{j-1}}^{t_j}(G(\lambda)) = \exp(G_{t_j}(\lambda) - G_{t_{j-1}}(\lambda)) \prod_{t_{j-1} < u \le t_j} (1 + \Delta G_u(\lambda))\, e^{-\Delta G_u(\lambda)} \tag{2.3}$$

and observe that

$$\mathcal{E}_{t_m}(G(\lambda)) = \prod_{j=1}^{m} \mathcal{E}_{t_{j-1}}^{t_j}(G(\lambda)). \tag{2.4}$$

Due to Condition 3) of Theorem 2 and Lemma 4.2.1 $|\mathcal{E}_t(G(\lambda))|$ is a nonincreasing function of t, and in view of this we have

$$|\mathcal{E}_{t_j}(G(\lambda))| > 0 \;(P\text{-a.s.}), \; j = 1, \ldots, m, \; \lambda \in R.$$

Therefore

$$\mathcal{E}_{t_{j-1}}^{t_j}(G(\lambda)) = \mathcal{E}_{t_j}(G(\lambda)) / \mathcal{E}_{t_{j-1}}(G(\lambda)) \quad (P\text{-a.s.}), \tag{2.5}$$

and hence (P-a.s.)

$$|\mathcal{E}_{t_{j-1}}^{t_j}(G(\lambda))| > 0 \tag{2.6}$$

for each $\lambda \in R$, $j = 1, \ldots, m$.

Let $h \in H_S$. Denote $\lambda_j = \lambda h(t_j)$. Then by (2.2) and (2.3) we get

$$\mathcal{E}_{t_m}(G^h(\lambda)) = \prod_{j=1}^{m} \mathcal{E}_{t_{j-1}}^{t_j}(G^h(\lambda)) = \prod_{j=1}^{m} \mathcal{E}_{t_{j-1}}^{t_j}(G(\lambda_j)), \tag{2.7}$$

and hence in accordance with (2.6) we have

$$|\mathcal{E}_{t_m}(G^h(\lambda))| > 0 \text{ (P-a.s.)}$$

for each $h \in H_S$ and $\lambda \in R$.

To prove the last assertion of the lemma, observe that the triplet $T^{n,h^n} = (B^{n,h^n}, C^{n,h^n}, \nu^{n,h^n})$ of the semimartingale X^{n,h^n} is determined in the following manner (Problem 1):

$$B^{n,h^n} = h^n \circ B^n + h^n x\,(I_{(|h^n x| \le 1)} - I_{(|x| \le 1)}) * \nu^n, \quad C^{n,h^n} = (h^n)^2 \circ C^n,$$

$$g(x)\,\nu^{n,h^n}(dt, dx) = g(xh^n(t))\,\nu^n(dt, dx), \quad |g(x)| \le \text{const}\,(x^2 \wedge 1).$$

The variable $\mathcal{E}_{\hat{\tau}_n(t_m)}(G^{n,h^n}(\lambda))$ is representable in the form

$$\mathcal{E}_{\hat{\tau}_n(t_m)}(G^{n,h^n}(\lambda)) = \prod_{j=1}^{m} \mathcal{E}_{\hat{\tau}_n(t_{j-1})}^{\hat{\tau}_n(t_j)}(G^{n,h^n}(\lambda))$$

(cf. (2.7)) with $\mathcal{E}_{\hat{\tau}_n(t_{j-1})}^{\hat{\tau}_n(t_j)}(G^{n,h^n}(\lambda))$ given by the formula (2.3) in which $G(\lambda)$ is replaced by $G^{n,h^n}(\lambda)$ and t_j, t_{j-1} by $\hat{\tau}_n(t_j)$, $\hat{\tau}_n(t_{j-1})$. Therefore it suffices to show that for $j = 1, \ldots, m$

$$\mathcal{E}_{\hat{\tau}_n(t_{j-1})}^{\hat{\tau}_n(t_j)}(G^{n,h^n}(\lambda)) \xrightarrow{P} \mathcal{E}_{t_{j-1}}^{t_j}(G^h(\lambda)), \quad n \to \infty, \tag{2.8}$$

for each $\lambda \in R$.

Observe now that as $t_{j-1} < t \le t_j$ we have $h(t) = c_j$ and as $\hat{\tau}_n(t_{j-1}) < t \le \hat{\tau}_n(t_j)$ we have $h^n(t) = c_j$. By this observation and by the representation of the triplets T^h and T^{n,h^n} the desired relation (2.8) holds, provided for each $\lambda \in R$

$$\mathcal{E}_{\hat{\tau}_n(t_{j-1})}^{\hat{\tau}_n(t_j)}(G^n(\lambda)) \xrightarrow{P} \mathcal{E}_{t_{j-1}}^{t_j}(G(\lambda)), \quad n \to \infty, \quad j = 1, \ldots, m.$$

On the set

$$\{\mathcal{E}_{\hat{\tau}_n(t_{j-1})}(G^n(\lambda))\,|\geq \varepsilon\},\ \ \varepsilon > 0$$

the function

$$\mathcal{E}_{\hat{\tau}_n(t_{j-1})}^{\hat{\tau}_n(t_j)}(G^n(\lambda))$$

is defined by the formula (2.5) in which $G(\lambda)$ is replaced by $G^n(\lambda)$ and t_j, t_{j-1} by $\hat{\tau}_n(t_j)$, $\hat{\tau}_n(t_{j-1})$. Therefore by Condition 2) of Theorem 2 for every $a > 0$ we get

$$\lim_n P\left(|\mathcal{E}_{\hat{\tau}_n(t_{j-1})}^{\hat{\tau}_n(t_j)}(G^n(\lambda)) - \mathcal{E}_{t_{j-1}}^{t_j}(G(\lambda))| \geq a,\ |\mathcal{E}_{\hat{\tau}_n(t_{j-1})}(G^n(\lambda))| \geq \varepsilon\right) = 0$$

for each $j = 1, \ldots, m$ and each $\lambda \in R$.

Hence

$$\overline{\lim_n}\, P\left(|\mathcal{E}_{\hat{\tau}_n(t_{j-1})}^{\hat{\tau}_n(t_j)}(G^n(\lambda)) - \mathcal{E}_{t_{j-1}}^{t_j}(G(\lambda))| \geq a\right)$$

$$\leq \overline{\lim_n}\, P\left(|\mathcal{E}_{\hat{\tau}_n(t_{j-1})}(G^n(\lambda))| < \varepsilon\right) \leq \overline{\lim_n}\, P\left(|\mathcal{E}_{\hat{\tau}_n(t_{j-1})}(G^n(\lambda))| < \varepsilon,\right.$$

and

$$|\mathcal{E}_{t_{j-1}}(G(\lambda))| \geq 2\varepsilon) + P\left(0 < |\mathcal{E}_{t_{j-1}}(G(\lambda))| < 2\varepsilon\right)$$

$$\leq \overline{\lim_n}\, P\left(|\mathcal{E}_{\hat{\tau}_n(t_{j-1})}(G^n(\lambda)) - \mathcal{E}_{t_{j-1}}(G(\lambda))| \geq \varepsilon\right) + P\left(0 < |\mathcal{E}_{t_{j-1}}(G(\lambda))| < 2\varepsilon\right)$$

$$= P\left(0 < |\mathcal{E}_{t_{j-1}}(G(\lambda))| \leq 2\varepsilon\right) \to 0,\ \ \varepsilon \to 0,$$

due to Condition 3) of Theorem 2.

3. Since Theorem 1 is a consequence of Theorem 2 as $\hat{\tau}_n(t) \equiv t$, only Theorems 2 and 3 will be proved.

Proof of Theorem 2. First assume $m = 1$, i.e. S consists of a single point $S = \{t_1\}$. We need to prove that for each $\lambda \in R$

$$\lim_n \mathbf{E}\xi \exp(i\lambda X^n_{\hat{\tau}_n(t_1)}) = \mathbf{E}\xi \exp(i\lambda X_{t_1}) \tag{2.9}$$

where ξ is an arbitrary $\mathcal{G}$-bounded measurable random variable ($|\xi| \leq c$).

Since $X = (X_t, \mathcal{F}_t)$ is a semimartingale with $\mathcal{G}$-conditionally independent increments ($\mathcal{G} \subseteq \mathcal{F}_0$), then (see (1.1) in § 1)

$$\begin{aligned}
\mathbf{E}\xi \exp(i\lambda X_{t_1}) &= \mathbf{E}\xi \exp(i\lambda X_0) \exp(i\lambda (X_{t_1} - X_0)) \\
&= \mathbf{E}\xi \exp(i\lambda X_0)\, \mathbf{E}(\exp(i\lambda(X_{t_1} - X_0)) \mid \mathcal{F}_0) \\
&= \mathbf{E}\xi \exp(i\lambda X_0)\, \mathbf{E}(\exp(i\lambda(X_{t_1} - X_0)) \mid \mathcal{G}) \\
&= \mathbf{E}\xi \exp(i\lambda X_0)\, \mathcal{E}_{t_1}(G(\lambda)).
\end{aligned} \tag{2.10}$$

On the other hand by Condition 1) of the theorem

$$\mathbf{E}\xi \exp(i\lambda X^n_{\hat{\tau}_n(t_1)}) = \mathbf{E}\xi e^{i\lambda X^n_0}\, \mathbf{E}(\exp(i\lambda\, X^n_{\hat{\tau}_n(t_1)} - X^n_0)) \mid \mathcal{F}^n_0). \tag{2.11}$$

Denote

$$J_n(\lambda)) = \mathbf{E}(\exp(i\lambda(X^n_{\hat{\tau}_n(t_1)} - X^n_0)) \mid \mathcal{F}^n_0).$$

Then (2.10) and (2.11) imply

$$\begin{aligned}
&|\mathbf{E}(\xi[\exp(i\lambda X^n_{\hat{\tau}_n(t_1)}) - \exp(i\lambda X_{t_1})])| \\
&= |\mathbf{E}(\xi[\exp(i\lambda X^n_0)\, J_n(\lambda) - \exp(i\lambda X_0)\, \mathcal{E}_{t_1}(G(\lambda))])| \\
&\le \mathbf{E}\,|\xi \exp(i\lambda X^n_0)(J_n(\lambda) - \mathcal{E}_{t_1}(G(\lambda)))| + |\mathbf{E}(\xi \mathcal{E}_{t_1}(G(\lambda))[\exp(i\lambda X^n_0) - \exp(i\lambda X_0)]|.
\end{aligned}$$

This gives the desired relation (2.9), since

$$\mathbf{E}\,|\xi \exp(i\lambda X^n_0)(J_n(\lambda) - \mathcal{E}_{t_1}(G(\lambda)))| \le c\mathbf{E}\,|J_n(\lambda) - \mathcal{E}_{t_1}(G(\lambda))|,$$

$$J_n(\lambda) \xrightarrow{\mathsf{P}} \mathcal{E}_{t_1}(G(\lambda))$$

by Theorem 1.2, besides

$$\mathbf{E}\xi \mathcal{E}_{t_1}(G(\lambda))[\exp(i\lambda X^n_0) - \exp(i\lambda X_0)] \to 0$$

due to Condition 4), since $\mathcal{E}_{t_1}(G(\lambda))$ is a $\mathcal{G}$-measurable random variable, $|\xi| \le c$ and $|\mathcal{E}_{t_1}(G(\lambda))| \le 1$ (Lemma 4.2.1).

Thus, for $m = 1$ the theorem is proved.

The case of $m > 1$ is easily reduced to the case already considered. In fact, we need to prove that for each $\lambda_j \in R$, $j = 1, \ldots, m$,

$$\lim_n \mathbf{E}\xi \exp\left(i \sum_{j=1}^{m} \lambda_j X^n_{\hat{\tau}_n(t_j)}\right) = \mathbf{E}\xi \exp\left(i \sum_{j=1}^{m} \lambda_j X_{t_j}\right). \tag{2.12}$$

Set

$$c_j = \sum_{k=j}^{m} \lambda_k, \quad j = 1,\dots, m, \quad c_0 = c_1.$$

Then

$$\sum_{j=1}^{m} \lambda_j X^n_{\hat{\tau}_n(t_j)} = c_0 X^n_0 + \sum_{j=1}^{m} c_j \left(X^n_{\hat{\tau}_n(t_j)} - X^n_{\hat{\tau}_n(t_{j-1})}\right) = c_0 X^n_0 + h^n \cdot X^n_{t_m}$$
$$= X^{n,\, h^n}_{t_m}$$

where $h^n \in H^n_S$ with $(\hat{\tau}_n(t_j)) = c_j$. Obviously, the analogous representation holds for $\sum_{j=1}^{m} \lambda_j X_{t_j}$ too:

$$\sum_{j=1}^{m} \lambda_j X_{t_j} = c_0 X_0 + h \cdot X_{t_m} = X^h_{t_m}$$

where $h \in H_S$ with $h(t_j) = c_j$.

Therefore (2.12) is represented in the following form:

$$\lim_n \mathbf{E}\xi \exp (iX^{n,\, h^n}_{t_m}) = \mathbf{E}\xi \exp (iX^h_{t_m}), \tag{2.13}$$

i.e. it suffices to prove Theorem 1 for the set S, consisting of a single point $\{t_m\}$, but this time for the semimartingales $X^{n,\, h^n}$, $n \geq 1$, and X^h. By Lemma 1 X^h is a process with $\mathcal{G}$-conditionally independent increments. Consequently, in accordance with the assertion already proved concerning the case of $m = 1$, it suffices to verify that Condition 1) of the theorem and the following conditions hold:

2') $\mathcal{E}_{t_m}(G^{n,\, h^n}(\lambda)) \xrightarrow{P} \mathcal{E}_{t_m}(G^h(\lambda))$, $n \to \infty$, $\lambda \in R$,

3') $|\mathcal{E}_{t_m}(G^h(\lambda))| > 0$ (P-a.s.), $\lambda \in R$,

4') $X^{n,\, h^n}_0 \xrightarrow{d} X^h_0$ ($\mathcal{G}$-stably).

Conditions 2') and 3') are verified in Lemma 1. Condition 4') follows from Condition 4) of Theorem 1, since $X^{n,\, h^n}_0 = X^n_0$ and $X^h_0 = X_0$.

The theorem is proved.

Proof of Theorem 3. First, assume $m = 1$. We need to prove that for each $\lambda \in R$

$$\lim_n \mathbf{E}\xi \exp(i\lambda X^n_{t_1}) = \mathbf{E}\xi \exp(i\lambda X_{t_1}) \tag{2.14}$$

where ξ is a bounded ($|\xi| \le c$) and $\mathcal{G}$-measurable random variable. Denote $\xi_n = \mathbf{E}(\xi \mid \mathcal{F}^n_{s_n})$. By Theorem 2.6.2 (Assertion 4)) $\xi_n \to \xi$ (P-a.s.). Consequently

$$|\mathbf{E}(\xi - \xi_n)\exp(i\lambda X^n_{t_1})| \le \mathbf{E}|\xi - \xi_n| \to 0,\ n \to \infty,$$

and (2.14) takes place provided

$$\lim_n \mathbf{E}\xi \exp(i\lambda X^n_{t_1}) = \mathbf{E}\xi \exp(i\lambda X_{t_1}). \tag{2.15}$$

By (2.10) we have

$$\mathbf{E}\xi \exp(i\lambda X_{t_1}) = \mathbf{E}\xi \exp(i\lambda X_0)\, \mathcal{E}_{t_1}(G(\lambda)).$$

Analogously to (2.11) with $\hat{\tau}_n(t_1) = t_1$ we establish that

$$\mathbf{E}\xi_n \exp(i\lambda X^n_{t_1}) = \mathbf{E}\xi_n \exp(i\lambda X^n_0)\, \mathbf{E}(\exp(i\lambda(X^n_{t_1} - X^n_0)) \mid \mathcal{F}^n_{s_n}).$$

Therefore, denoting

$$J_n(\lambda) = \mathbf{E}(\exp(i\lambda(X^n_{t_1} - X^n_0) \mid \mathcal{F}^n_{s_n}),$$

we have the estimate

$$|\mathbf{E}[\xi_n \exp(i\lambda X^n_{t_1}) - \xi \exp(i\lambda X_{t_1})]|$$

$$\le |\mathbf{E}[\xi_n \exp(i\lambda X^n_0) J_n(\lambda) - \xi \exp(i\lambda X_0)\, \mathcal{E}_{t_1}(G(\lambda))]|$$
$$\le \mathbf{E}|(\xi - \xi_n)\exp(i\lambda X^n_0) J_n(\lambda)| + \mathbf{E}|\xi \exp(i\lambda X^n_0)(J_n(\lambda) - \mathcal{E}_{t_1}(G(\lambda)))|$$

$$+ |\mathbf{E}\xi \mathcal{E}_{t_1}(G(\lambda))[\exp(i\lambda X^n_0) - \exp(i\lambda X_0)]|$$
$$\le \mathbf{E}|\xi - \xi_n| + \mathbf{E}|J_n(\lambda) - \mathcal{E}_{t_1}(G(\lambda))|$$

$$+ |\mathbf{E}\xi \mathcal{E}_{t_1}(G(\lambda))[\exp(i\lambda X^n_0) - \exp(i\lambda X_0)]|.$$

Due to the construction of variables ξ_n we have $\lim_n \mathbf{E}|\xi - \xi_n| = 0$. By Theorem 1.3, the conditions of which are satisfied in virtue of assumptions 1) - 3), 5) of Theorem 3, we have

$$\lim_n \mathbf{E}|J_n(\lambda) - \mathcal{E}_{t_1}(G(\lambda))| = 0.$$

Next, by Condition 4) of Theorem 3

$$\lim_n |\mathbf{E}(\xi \mathcal{E}_{t_1}(G(\lambda))[\exp(i\lambda X_0^n) - \exp(i\lambda X_0)])| = 0.$$

Hence (2.15) holds.

The proof of the theorem in case m > 1 is analogous to the proof of Theorem 2 for this case as $\hat{\tau}_n(t_j) \equiv t_j$, $j = 1, \ldots, m$.

4. Let $X^n = (X_t^n, \mathcal{F}_t^n)$, $n \geq 1$, and $X = (X_t, \mathcal{F}_t)$ be nondecreasing semimartingales having the stochastic exponentials $(G^n(\lambda))$, $n \geq 1$, and $\mathcal{E}(G(\lambda))$, defined by the formula (1.23) (§1), and let X be a semimartingale with $\mathcal{G}$-conditionally independent increments. Theorem 1.4 allows us to prove the following result.

Theorem 4. *Let the following conditions be fullfilled:*

1) $\mathcal{G} \subseteq \bigcap_{n \geq 1} \mathcal{F}_0^n$;

2) *for all* $\lambda > 0$ *and points* t_j *of the set* $S = \{t_1, \ldots, t_n\}$ *with* $0 = t_0 < t_1 < \ldots < t_m$

$$\mathcal{E}_{t_j}(G^n(\lambda)) \xrightarrow{P} \mathcal{E}_{t_j}(G(\lambda)), \ n \to \infty;$$

3) $X_0^n \xrightarrow{d} X_0$ ($\mathcal{G}$-*stably*).

Then

$$X^n \xrightarrow{d_f(S)} X \ (\mathcal{G}\text{-}stably).$$

The proof of this theorem is analogous to the proof of Theorem 2 as $\hat{\tau}_n(t_j) \equiv t_j$, $j = 1, \ldots, m$.

Problems

1. Let $X = (X_t, \mathcal{F}_t)$ be a semimartingale with the triplet $T = (B, C, \nu)$ and $h = h(t)$ a $\mathcal{B}(R_+)$-measurable and bounded function. Show that the semimartingale $X^h = (X_t^h, \mathcal{F}_t)$ with

$$X_t^h = h \cdot X_t$$

has the triplet $T^h = (B^h, C^h, \nu^h)$ with

$$B^h = h \circ B + hx\,(I_{(|hx| \leq 1)} - I_{(|x| \leq 1)} * \nu, \quad C^h = h^2 \circ C,$$

$$g(x)\,\nu^h(dt, dx) = g(xh(t))\,\nu(dt, dx)$$

for each $B(R_0)$-measurable function $g = g(x)$ with $|g(x)| \le \text{const}(x^2 \wedge 1)$.

2. Let $X^n \xrightarrow{d_f(S)} X$ ($\mathcal{G}$-stably) and for every $t \in S$ let $X_t^n - Y_t^n \xrightarrow{P} 0$, $n \to \infty$, where $Y^n = (Y_t^n)_{t \ge 0}$, $n \ge 1$, is a certain sequence of stochastic processes. Show that

$$Y^n \xrightarrow{d_f(S)} X \quad (\mathcal{G}\text{-stably}).$$

3. Prove Theorem 4.

§ 3. Weak convergence of finite dimensional distributions of point processes and semimartingales to distributions of point processes

1. In the present section the method of stochastic exponentials exposed in §§ 1 and 2 is applied to the study of conditions for the weak convergence of finite dimensional distributions of semimartingales $X^n = (X^n_t, \mathcal{F}^n_t)$, presenting a point (counting) process (see Ch. 3, § 4). We start with the case in which X^n is a point (counting) process for each $n \geq 1$. This is explained by the simplicity of conditions for the weak convergence of finite dimensional distributions, as well as by the fact that on treating just this example (see [134]) it became clear that the method of stochastic exponentials is applicable to a much more general "semimartingale" scheme.

2. Thus let $X^n = (X^n_t, \mathcal{F}^n_t)$, $n \geq 1$, and $X = (X_t, \mathcal{F}_t)$ be semimartingales presenting point (counting) processes. Denote by A^n, $n \geq 1$, and A the compensators of the indicated increasing processes X^n, $n \geq 1$, and X respectively (see Ch. 3, § 4).

Since X is a semimartingale, we have

$$X_t < \infty \text{ (P-a.s.)}, \quad t \in R_+.$$

Consequently, in accordance with Problem 3..4.6 we have

$$A_t < \infty \text{ (P-a.s.)}, \quad t \in R_+.$$

Let $T = (B, C, \nu)$ be the triplet of predictable characteristics. We will show that

$$B = A, \qquad C = 0, \tag{3.1}$$

and the compensator ν of the jump measure μ of the process X, obviously, possesses the following properties:

$$\nu((0, t] \times \{1\}) = A_t, \qquad \int_0^t \int_{R_0} f(x)\, d\nu = f(1)\, A_t \tag{3.2}$$

($f = f(x)$ is a $B(R_0)$-measurable function with $|f(x)| \leq x^2 \wedge 1$). Indeed, the properties (3.1) and (3.2) hold, since the canonic representation for X is of form

$$X_t = A_t + \int_0^t \int_{\{1\}} x\, d(\mu - \nu). \tag{3.3}$$

From (3.2) it follows that

$$V^c \equiv B^c - I(0 < x \leq 1) * \nu^c = 0.$$

Consequently (see Subsection 1.5),

$$G_t(\lambda) = \int_0^t \int_{x>0} (e^{-\lambda x} - 1)\, d\nu = (e^{-\lambda} - 1)\, A_t,\ 1 \in R_+,$$

and by (1.23) the stochastic exponential $\mathcal{E}(G(\lambda))$, related to a counting process X, is given by the formula

$$\mathcal{E}_t(G(\lambda)) = e^{-aA_t^c} \prod_{0<s\le t} (1 - a\,\Delta A_s) = e^{-aA_t} \prod_{0<s\le t} (1 - a\,\Delta A_s)\, e^{a\Delta A_s} \qquad (3.4)$$

with $a = 1 - e^{-\lambda}$, $\lambda \in R_+$.

Observe that $\Delta A \le 1$ (see the definition of the compensator A and Lemma 3.4.2). Consequently, as $\lambda > 0$ the inequality $a\Delta A < 1$ holds and

$$\prod_{0<s\le t} (1 - a\Delta A_s)\, e^{a\Delta A_s} = \exp\left(\sum_{0<s\le t} [\ln(1 - a\Delta A_s) + a\Delta A_s] \right)$$

$$= \exp\left(\sum_{k=2}^{\infty} \frac{(-a)^k}{k} \sum_{0<s\le t} (\Delta A_s)^k \right).$$

Thus the stochastic exponential $\mathcal{E}(G(\lambda))$, related to a counting process X admits the following representation:

$$\mathcal{E}_t(G(\lambda)) = \exp\left(-aA_t + \sum_{k=2}^{\infty} \frac{(-a)^k}{k} \sum_{0<s\le t} (\Delta A_s)^k \right). \qquad (3.5)$$

Clearly, the stochastic exponential $\mathcal{E}(G^n(\lambda))$, related to a counting process X^n is given by the representation (3.5) in which A is replaced by A^n.

Theorem 1. *Let semimartingales* $X^n = (X_t^n, \mathcal{F}_t^n)$, $n \ge 1$, *and* $X = (X_t, \mathcal{F}_t)$ *be counting processes having the compensators* A^n, $n \ge 1$, *and* A *respectively,* X *a process with* $\mathcal{G}$*-conditionally independent increments* ($\mathcal{G} \subseteq \mathcal{F}_0$) *and* S *a nonempty subset of* R_+.

Let the following conditions be fullfilled:

(o) $\mathcal{G} \subseteq \bigcap_{n\ge 1} \mathcal{F}_0^n$,

(a) $A_t^n \xrightarrow{P} A_t$, $t \in S$,

(a_k) $\sum_{0<s\le t} (\Delta A_s^n)^k \xrightarrow{P} \sum_{0<s\le t} (\Delta A_s)^k$, $t \in S$, $k \ge 2$.

Then

$$X^n \xrightarrow{d_f(S)} X \quad (\mathcal{G}\text{-}stably).$$

Corollary 1. *If* S *is a subset dense in* R_+ *and Conditions* (o) *and* (a) *of Theorem* 1 *are fullfilled as well as the condition*

(a₂)
$$\sum_{0<s\le t} (\Delta A_s^n)^2 \xrightarrow{P} \sum_{0<s\le t} (\Delta A_s)^2, \quad t \in S,$$

then

$$X^n \xrightarrow{d_f(S)} X \quad (\mathcal{G}\text{-}stably).$$

Corollary 2. *If the compensator* A *has continuous trajectories* ($A \in \mathcal{V}^+ \cap C$), *and Conditions* (o) *and* (a) *of Theorem 1 are fullfilled, as well as the condition*

(e)
$$\sup_{0<s\le t} \Delta A_s^n \xrightarrow{P} 0, \quad t \in S,$$

then

$$X^n \xrightarrow{d_f(S)} X \quad (\mathcal{G}\text{-}stably).$$

Corollary 3. *If* $A \in \mathcal{V}^+ \cap C$, S *is a subset dense in* R_+ *and Conditions* (o) *and* (a) *of Theorem* 1 *are fullfilled, then*

$$X^n \xrightarrow{d_f(S)} X \quad (\mathcal{G}\text{-}stably).$$

Proof of Theorem 1. In accordance with Theorem 2.4 it suffices to verify only that

$$\mathcal{E}_t(G^n(\lambda)) \xrightarrow{P} \mathcal{E}_t(G(\lambda)), \quad \lambda > 0, \quad t \in S, \tag{3.6}$$

since Condition 1) of Theorem 2.4 holds by Condition (o) of Theorem 1, and Condition 3) of Theorem 2.4 holds as well because $X_0^n \equiv 0$ and $X_0 = 0$.

By Condition (a) we have $A_t^n \xrightarrow{P} A_t$, $t \in S$. Therefore, in view of the representation (3.5) for stochastic exponentials, we need for the validity of (3.6) to establish that as $t \in S$

$$\sum_{k=2}^{\infty} \frac{(-a)^k}{k} \sum_{0<s\le t} (\Delta A_s^n)^k \xrightarrow{P} \sum_{k=2}^{\infty} \frac{(-a)^k}{k} \sum_{0<s\le t} (\Delta A_s)^k. \tag{3.7}$$

To prove the validity of (3.7) denote ($k \ge 2$)

$$V_t^n(k) = \sum_{0<s\le t} (\Delta A_s^n)^k, \quad V_t(k) = \sum_{0<s\le t} (\Delta A_s)^k \tag{3.8}$$

and observe that since $\Delta A \le 1$, $\Delta A^n \le 1$, $n \ge 1$, then

$$V_t(k) \le A_t, \quad V_t^n(k) \le A_t^n, \ n \ge 1.$$

In view of these inequalities we get

$$J_t^n = \left| \sum_{k=2}^{\infty} \frac{(-a)^k}{k} V_t^n(k) - \sum_{k=2}^{\infty} \frac{(-a)^k}{k} V_t(k) \right|$$

$$\le \sum_{k=2}^{L} \frac{a^k}{k} | V_t^n(k) - V_t(k) | + (A_t^n + A_t) \sum_{k=L+1}^{\infty} \frac{a^k}{k}$$

$$\le \sum_{k=2}^{L} | V_t^n(k) - V_t(k) | + | A_t^n - A_t | \sum_{k=L+1}^{\infty} \frac{a^k}{k} + 2A_t \sum_{k=L+1}^{\infty} \frac{a^k}{k}. \tag{3.9}$$

By Conditions (a) and (a_k), $k \ge 2$,

$$\sum_{k=2}^{L} | V_t^n(k) - V_t(k) | + | A_t^n - A_t | \sum_{k=L+1}^{\infty} \frac{a^k}{k} \xrightarrow{P} 0, \ t \in S.$$

Consequently, for $\varepsilon > 0$

$$\overline{\lim_n} P(J_t^n \ge \varepsilon) \le P\left(2A_t \sum_{k=L+1}^{\infty} \frac{a^k}{k} \ge \frac{\varepsilon}{2} \right) \to 0, \ L \to \infty,$$

since as $a \in (0, 1)$

$$\sum_{k=L+1}^{\infty} \frac{a^k}{k} \to 0, \ L \to \infty.$$

Thus the desired relation (3.7) holds.

Corollary 1 follows from the following results of independent interest.

Lemma 1. *Let* A^n, $n \ge 1$, *and* A *be functions (deterministic) in* V^+ *and let* Δ *be a subset, dense in* R_+. *If for each* $t \in \Delta$ *as* $n \to \infty$

$$A_t^n \to A_t, \quad \sum_{0<s\le t} (\Delta A_s^n)^2 \to \sum_{0<s\le t} (\Delta A_s)^2,$$

then for each $t \in \Delta$ *as* $n \to \infty$

$$\sum_{0<s\le t} (\Delta A_s^n)^k \to \sum_{0<s\le t} (\Delta A_s)^k, \ k \ge 3.$$

Lemma 2. *Let* A^n, $n \ge 1$ *and* A *be stochastic processes with trajectories in* V^+, *defined on one and the same probability space, and let* Δ *be a subset, dense in* R_+.

If for each $t \in \Delta$ *as* $n \to \infty$

$$A_t^n \xrightarrow{P} A_t, \quad \sum_{0<s\le t} (\Delta A_s^n)^2 \xrightarrow{P} \sum_{0<s\le t} (\Delta A_s)^2,$$

then for all $t \in \Delta$ *as* $n \to \infty$

$$\sum_{0<s\le t} (\Delta A_s^n)^k \xrightarrow{P} \sum_{0<s\le t} (\Delta A_s)^k, \quad k \ge 3.$$

Proof of Lemma 1. As $k \ge 2$ and $t > s$ we get by Ito's formula (Ch. 2, § 3) that

$$(A_t - A_s)^k = k \int_{(s,\, t]} (A_{u-} - A_s)^{k-1} dA_u$$

$$+ \sum_{s<u\le t} [(A_u - A_s)^k - (A_{u-} - A_s)^k - k\,(A_{u-} - A_s)^{k-1} \Delta A_u].$$

Due to Newton's binomial formula, this gives

$$(A_t - A_s)^k = \sum_{s<u\le t} (\Delta A_u)^k$$

$$+ \sum_{j=1}^{k-1} \binom{k}{j} \int_{(s,\, t]} (A_{u-} - A_s)^{k-j} (\Delta A_u)^{j-1} dA_u, \tag{3.10}$$

where $\binom{k}{j}$ is the number of combinations out of k objects j at a time.

Let $t \in \Delta$ and let (t_j^r), $t_l^r \in \Delta$, $l = 0, 1, \ldots, r$, $r \ge 1$, be a partition sequence of the interval $[0, t]$, $0 = t_0^r < t_1^r < \ldots < t_r^r = t$ such that $\max_{1\le j\le r} (t_j^r - t_{j-1}^r) \to 0$, $r \to \infty$. We assume that the sequence t_l^r, $l = 0, 1, \ldots, r$, contains points

$$T_k^r = \inf (t > T_{k-1}^r : A_t - A_{T_{k-1}^r} \ge 1/r), \quad k \ge 1,$$

with $T_0^r = 0$, under the condition that $T_k^r < t$. Let us show that

$$\lim_r \sum_{l=0}^{r-1} (A_{t_{l+1}^r} - A_{t_l^r})^k = \sum_{0<s\le t} (\Delta A_s)^k. \tag{3.11}$$

By (3.10)

$$\sum_{l=0}^{r-1} (A_{t_{l+1}^r} - A_{t_l^r})^k = \sum_{0<s\le t} (\Delta A_s)^k + J_r^k(A) \tag{3.12}$$

with

$$J_r^k(A) = \sum_{l=0}^{r-1}\sum_{j=1}^{k-1}\binom{k}{j}\int_{(t_l^r, t_{l+1}^r]} (A_{u-} - A_{t_l^r})^{k-j}(\Delta A_u)^{j-1} dA_u$$

$$\leq \sum_{j=1}^{k-1}\binom{k}{j} A_t^{k-2}\sum_{l=0}^{r-1}\int_{(t_l^r, t_{l+1}^r]} (A_{u-} - A_{t_l^r})\, dA_u. \tag{3.13}$$

From this estimate and from (3.12) it follows that the relation (3.11) takes place provided

$$\lim_{r\to\infty}\sum_{l=0}^{r-1}\int_{(t_l^r, t_{l+1}^r]} (A_{u-} - A_{t_l^r})\, dA_u = 0. \tag{3.14}$$

To prove (3.14), observe that the definition of t_l^r implies the inequality

$$\int_{(t_l^r, t_{l+1}^r]} (A_{u-} - A_{t_l^r})\, dA_u \leq 1/r\,(A_{t_{l+1}^r} - A_{t_l^r}).$$

Thus as r is fixed we have

$$\sum_{l=0}^{r-1}\int_{(t_l^r, t_{l+1}^r]} (A_{u-} - A_{t_l^r})\, dA_u \leq 1/r\, A_t. \tag{3.15}$$

Consequently, (3.14) follows from (3.15).

Let us turn now to the proof of the desired assertion.

By (3.12) we have for any $k \geq 3$ the following estimate:

$$I_n^k = \left|\sum_{0<s\leq t}(\Delta A_s^n)^k - \sum_{0<s\leq t}(\Delta A_s)^k\right|$$

$$\leq \sum_{l=0}^{r-1}\left|(A^n_{t_{l+1}^r} - A^n_{t_l^r})^k - (A_{t_{l+1}^r} - A_{t_l^r})^k\right| + J_r^k(A) + J_r^k(A^n),$$

which entails

$$\overline{\lim_n}\, I_n^k \leq J_r^k(A) + \overline{\lim_n}\, J_r^k(A^n).$$

By the inequality (3.13) and the relation (3.14) we have

$$\lim_{r\to\infty} J_r^k(A) = 0.$$

Therefore it suffices to show that

$$\lim_{r\to\infty} \overline{\lim_{n}}\, J_r^k(A^n) = 0,\ k \ge 3. \tag{3.16}$$

To this end observe that as $k \ge 3$ (3.13) implies the inequality

$$J_r^k(A) \le J_r^2(A) \sum_{j=1}^{k-1} \binom{k}{j} (A_t)^{k-2}.$$

Consequently

$$J_r^k(A^n) \le J_r^2(A^n) \sum_{j=1}^{k-1} \binom{k}{j} (A_t^n)^{k-2},$$

and hence (3.16) takes place provided

$$\lim_{r\to\infty} \overline{\lim_{n}}\, J_r^2(A^n) = 0.$$

The last relation holds in virtue of the assumptions of Lemma 1 and the representation (see (3.12))

$$J_r^2(A^n) = \sum_{l=0}^{r-1} (A^n_{t^r_{l+1}} - A^n_{t^r_l})^2 - \sum_{0<s\le t} (\Delta A_s^n)^2.$$

Proof of Lemma 2. Let $t \in \Delta$, $\varepsilon > 0$ and

$$\xi_t^n = | A_t^n - A_t | + \left| \sum_{0<s\le t} (\Delta A_s^n)^2 - \sum_{0<s\le t} (\Delta A_s)^2 \right|,$$

$$a^n = \mathsf{P}\left(\left| \sum_{0<s\le t} (\Delta A_s^n)^k - \sum_{0<s\le t} (\Delta A_s)^k \right| \ge \varepsilon \right),\ k \ge 3.$$

Let (n') be a subsequence of (n) such that $\overline{\lim_n}\, a^n = \lim_{n'} a^{n'}$. Since $\xi_t^n \xrightarrow{\mathsf{P}} 0$, we have $\xi_t^{n'} \xrightarrow{\mathsf{P}} 0$. Let (n'') be a subsequence of (n') such that $\xi_t^{n''} \to 0$ (P-a.s.), $n'' \to \infty$. Then by Lemma 1 we have $\lim_{n''} a^{n''} = 0$. But $\lim_{n''} a^{n''} = \lim_{n'} a^{n'}$. Hence $\overline{\lim_n}\, a^n = 0$.

Proof of Corollary 2. Under the assumption $A \in \mathcal{V}^+ \cap C$, Condition (a_k) of Theorem 1 takes the form

$$\sum_{0<s\le t} (\Delta A_s^n)^k \xrightarrow{\mathsf{P}} 0,\ t \in S,\ k \ge 2. \tag{3.17}$$

By the inequality $\Delta A^n \le 1$ we have the estimate

$$\sum_{0<s\le t} (\Delta A_s^n)^k \le \sum_{0<s\le t} (\Delta A_s^n)^2 \le \sup_{0<s\le t} \Delta A_s^n A_t^n.$$

Therefore (3.17) follows from Condition (a) of Theorem 1 and Condition (e).

Proof of Corollary 3. If $A \in \mathcal{V}^+ \cap C$, then S is a subset dense in R_+ and Condition (a) of Theorem 1 holds, then by Problem 2 we have

$$\sup_{s \le t} | A^n_s - A_s | \xrightarrow{P} 0, \quad t \in R_+,$$

and consequently

$$\sup_{0 < s \le t} \Delta A^n_s = \sup_{0 < s \le t} | \Delta A^n_s - \Delta A_s | \le 2 \sup_{s \le t} | A^n_s - A_s | \xrightarrow{P} 0, \quad t \in R_+,$$

i.e. Condition (e) of Corollary 2 is fullfilled and hence the desired assertion holds.

3. In the present subsection we assume that a semimartingale $X = (X_t, \mathcal{F}_t)$ presents a counting process, while $X^n = (X^n_t, \mathcal{F}^n_t)$, $n \ge 1$, are semimartingales (not necessarily presenting counting processes).

Theorem 2. *Let* $X^n = (X^n_t, \mathcal{F}^n_t)$, $n \ge 1$, *be semimartingales with the triplets* $T^n = (B^n, C^n, \nu^n)$, $n \ge 1$, *let a semimartingale* $X = (X_t, \mathcal{F}_t)$ *present a counting process with the compensator* A *and* $\mathcal{G}$*-conditionally independent increments* ($\mathcal{G} \subseteq \mathcal{F}_0$), *and let* S *be a nonempty subset of* R_+, $0 < \delta < \frac{1}{2}$.

Let the following conditions be fullfilled:

(o) $\mathcal{G} \subseteq \bigcap_{n \ge 1} \mathcal{F}^n_0$;

(a) $I(|x - 1| \le \delta) * \nu^n_t \xrightarrow{P} A_t$, $t \in S$;

(a_k) $\sum_{0 < s \le t} (\nu^n(\{s\} \times \{|x - 1| \le \delta\}))^k \xrightarrow{P} \sum_{0 < s \le t} (\Delta A_s)^k$, $t \in S$;

(b) $B^n_t - xI(\delta < |x| \le 1) * \nu^n_t \xrightarrow{P} 0$, $t \in S$;

(c) $C^n_t + x^2 I(|x| \le \delta) * \nu^n_t - \sum_{0 < s \le t} \left(\int_{|x| \le \delta} x \nu^n(\{s\}, dx) \right)^2 \xrightarrow{P} 0$, $t \in S$;

(d) $\nu^n((0, t] \times \Delta_\delta) \xrightarrow{P} 0$, $t \in S$,

$$\Delta_\delta = \{x: R_0 \setminus (\{|x| \le \delta\} \cup \{|x - 1| \le \delta\})\};$$

(f) $X^n_0 \xrightarrow{P} 0$.

Then

$$X^n \xrightarrow{d_f(S)} X \quad (\mathcal{G}\textit{-stably}).$$

Remark. 1. If S presents a subset, dense in R_+, then Condition (a_k) can be replaced by Condition (a_2), i.e. it suffices to assume that Condition (a_k) is satisfied only for $k = 2$.

2. If $A \in \mathcal{V}^+ \cap C$, then Condition (a_k) can be replaced by Condition

(e) $\sup_{0 < s \le t} \nu^n(\{s\} \times \{|x - 1| \le \delta\}) \xrightarrow{P} 0, \quad t \in S.$

3. If $A \in \mathcal{V}^+ \cap C$ and S is a subset dense in R_+, then the assertion of Theorem 2 is true without assuming Condition (a_k).

Proof of Theorem 2. Let $\mu^n = \mu^n(dt, dx)$ be the jump measure of a semimartingale X^n. Define the semimartingale $Y^{n,\delta} = (Y_t^{n,\delta}, \mathcal{F}_t^n)$ with

$$Y_t^{n,\delta} = I(|x - 1| \le \delta) * \mu_t^n. \tag{3.18}$$

Obviously, $Y^{n,\delta}$ is a counting process. Besides, its compensator A^n is given by the formula

$$A_t^n = I(|x - 1| \le \delta) * \nu_t^n. \tag{3.19}$$

Observe that the Conditions (a) and (a_k) of the present theorem coincide with Conditions (a) and (a_k) of Theorem 1 concerning the compensator A^n. Therefore by Theorem 1

$$Y^{n,\delta} \xrightarrow{d_f(S)} X \quad (\mathcal{G}\text{-stably}).$$

Due to Problem 2.2 the desired assertion takes place provided

$$X_t^n - Y_t^{n,\delta} \xrightarrow{P} 0, \quad t \in S, \tag{3.20}$$

and consequently the proof is completed by verifying the validity of the implications

$$\text{(b), (c), (d), (f)} \Rightarrow (3.20). \tag{3.21}$$

To prove (3.21) we utilize the canonical representation for X^n:

$$X_t^n = X_0^n + B_t^n + X_t^{nc} + \int_0^t \int_{|x| > 1} x \, d\mu^n + \int_0^t \int_{|x| \le 1} x \, d(\mu^n - \nu^n) \tag{3.22}$$

(see Ch. 4, § 1).

Denote

$$M_t^{n,\delta} = X_t^{nc} + \int_0^t \int_{|x| \le \delta} x \, d(\mu^n - \nu^\nu) \tag{3.23}$$

and observe that $M^{n,\delta} = (M_t^{n,\delta}, \mathcal{F}_t^n)$ is a local square integrable martingale with the quadratic characteristic

$$\langle M^{n,\delta}\rangle_t = C_t^n + x^2 I(|x| \le \delta) * \nu_t^n - \sum_{0 < s \le t} \left(\int_{|x| \le \delta} x\nu^n(\{s\}, dx) \right)^2 \quad (3.24)$$

(see Theorem 3.5.1 (Assertion 6) and Lemma 3.5.1).

From (3.22), (3.18) and (3.23) it follows that

$$|X_t^n - Y_t^{n,\delta}| \le |X_0^n| + |B_t^n - xI(\delta < |x| \le 1) * \nu_t^n|$$

$$+ |M_t^{n,\delta}| + I(x \in \Delta_\delta)|x| * \mu_t^n, \quad (3.25)$$

with

$$\Delta_\delta = \{x: R \setminus (\{|x| \le \delta\} \cup \{|x - 1| \le \delta\})\}.$$

By Conditions (b) and (f) and the estimate (3.25) the desired implication (3.21) takes place, provided

$$(c) \Rightarrow M_t^{n,\delta} \xrightarrow{P} 0, \ t \in S, \quad (3.26)$$

$$(d) \Rightarrow I(x \in \Delta_\delta)|x| * \mu_t^n \xrightarrow{P} 0, \ t \in S \quad (3.27)$$

is established.

In accordance with the representation (3.24) Condition (c) is equivalent to $\langle M^{n,\delta}\rangle_t \xrightarrow{P} 0$, $t \in S$. Therefore the implication (3.26) holds (Problem 1.9.2). To prove (3.27), observe that for $0 < a < \delta$

$$\{I(x \in \Delta_\delta)|x| * \mu_t^n \ge a\} \subseteq \{I(x \in \Delta_\delta) * \mu_t^n \ge 1\},$$

and hence (3.27) takes place, provided

$$(d) \Rightarrow I(x \in \Delta_\delta) * \mu_t^n \xrightarrow{P} 0, \ t \in S \quad (3.28)$$

is shown.

Observe now that

$$I(x \in \Delta_\delta) * \mu^n \in \mathcal{A}_{loc}^+$$

and the process $I(x \in \Delta_\delta) * \nu^n$ is its compensator (Problem 1), and hence Condition (d) means that

$$I(x \in \Delta_\delta) * \nu_t^n \xrightarrow{P} 0, \ t \in S.$$

Consequently, in virtue of Problem 1.9.4 (3.28) holds.

The remarks to Theorem 2 are valid by the Corollaries 1-3 to Theorem 1.

4. Example. Let $\pi = (\pi_t, \tilde{\mathcal{F}}_t)$ be a counting process with the compensator $A^{\pi}_t \equiv t$, i.e. π is a Poisson process with the unite parameter (Problem 3.4.1). For every $n \geq 1$ let $\Theta^n = (\Theta^n_t, \tilde{\mathcal{F}}_t)$ be a Markov process with two states (0, 1), with right-continuous trajectories and with the transition matrix

$$n\begin{pmatrix} -\lambda_1 & \lambda_1 \\ \lambda_2 & -\lambda_2 \end{pmatrix},$$

where $\lambda_1, \lambda_2 > 0$.

Consider a counting process $X^n = (X^n_t, \mathcal{F}^n_t)$ with $\mathcal{F}^n_t \equiv \tilde{\mathcal{F}}_t$ and $X^n_t = \Theta^n \circ \pi_t$. Obviously, the compensator of this process is presented by the process

$$A^n_t = \int_0^t \Theta^n_s \, ds, \quad t \geq 0.$$

Let us show that

$$X^n \xrightarrow{d_f} X$$

where $X = (X_t)_{t \geq 0}$ is a Poisson process with the parameter $\frac{\lambda_1}{\lambda_1 + \lambda_2}$, i.e. let us show that all finite dimensional distributions of the sequence of counting processes X^n, $n \geq 1$, converge weakly to a Poisson process with the parameter $\frac{\lambda_1}{\lambda_1 + \lambda_2}$.

Since a Poisson process with the parameter $\frac{\lambda_1}{\lambda_1 + \lambda_2}$ has the compensator $\frac{\lambda_1 t}{\lambda_1 + \lambda_2}$ (Problem 3.4.1), it suffices to show according to Theorem 1 that

$$\int_0^t \Theta^n_s \, ds \xrightarrow{P} \frac{\lambda_1 t}{\lambda_1 + \lambda_2}, \quad t \in R_+. \tag{3.29}$$

To prove this let us use the fact that the process $M^n = (M^n_t, \mathcal{F}^n_t)$ with

$$M^n_t = \Theta^n_t - \Theta^n_0 - n \int_0^t [\lambda_1 - (\lambda_1 + \lambda_2)\, \Theta^n_s] \, ds \tag{3.30}$$

is a square integrable martingale (Lemma 9.2 in [188]). From (3.30) it follows that

$$\int_0^t \Theta_s^n \, ds = \frac{\lambda_1 t}{\lambda_1 + \lambda_2} - \frac{\Theta_t^n - \Theta_0^n}{n(\lambda_1 + \lambda_2)} + \frac{M_t^n}{n(\lambda_1 + \lambda_2)},$$

and hence the desired relation (3.29) takes place, provided

$$\frac{M_t^n}{n} \xrightarrow{P} 0, \quad t \in R_+. \tag{3.31}$$

To establish (3.31) we need the representation of the quadratic characteristic $\langle M^n \rangle$ and its estimate. Due to (3.30), Θ^n is a special semimartingale with the semimartingale representation

$$\Theta_t^n = \Theta_0^n + \tilde{A}_t^n + M_t^n \tag{3.32}$$

where

$$\tilde{A}_t^n = n \int_0^t [\lambda_1 - (\lambda_1 + \lambda_2)\, \Theta_s^n]\, ds.$$

Since $(\Theta^n)^2 = \Theta^n$, then $(\Theta^n)^2$ is a special semimartingale with the same semimartingale decomposition (3.32). Then, according to Problem 4.1.8,

$$\langle M^n \rangle_t = \tilde{A}_t^n - [\tilde{A}^n, \tilde{A}^n]_t - 2\Theta_-^n \circ \tilde{A}_t^n = (1 - 2\Theta_-^n) \circ \tilde{A}_t^n$$

$$= n \int_0^t [\lambda_1 (1 - \Theta_s^n) + \lambda_2 \Theta_s^n]\, ds$$

$$\le nt\,(\lambda_1 + \lambda_2).$$

Therefore, by Doob's inequality (Theorem 1.9.1)

$$P\left(\frac{|M_t^n|}{n} \ge a \right) \le \frac{1}{n^2 a^2} \mathbf{E} \langle M^n \rangle_t \le \frac{t(\lambda_1 + \lambda_2)}{n a^2} \to 0, \quad n \to \infty,$$

for each $t \in R_+$.

Problems

1. Let $X = (X_t, \mathcal{F}_t)$ be a semimartingale with the triplet $T = (B, C, \nu)$, μ the jump measure of X and Y a counting process with $Y_t = I_\Gamma * \mu_t$, where $\Gamma \in B(R_0)$ and $\Gamma \cap \{\, |x| \le \varepsilon\} = \varnothing$ for a certain $\varepsilon > 0$. Show that $A_t = I_\Gamma * \nu_t$ is the compensator of the counting process $Y = (Y_t, \mathcal{F}_t)$.

2. Let $A^n \in \mathcal{V}^+$, $n \geq 1$, $A \in \mathcal{V}^+ \cap C$, and for each t from a set dense in R_+ let $A_t^n \to A_t$ (P-a.s.) (respectively $A_t^n \xrightarrow{P} A_t$). Show that for each $t \in R_+$

$$\sup_{s \leq t} | A_s^n - A_s | \to 0 \quad \text{(P-a.s.)}$$

(respectively $\sup_{s \leq t} | A_s^n - A_s | \xrightarrow{P} 0$).

§ 4. Weak convergence of finite dimensional distributions of semimartingales to distributions of a left quasi-continuous semimartingale with conditionally independent increments

1. In the present section the limiting process will be presented by a semimartingale $X = (X_t, \mathcal{F}_t)$ with $\mathcal{G}$-conditionally independent increments ($\mathcal{G} \subseteq \mathcal{F}_0$) and with the compensator $\nu = \nu\,(dt, dx)$ of the jump measure which possesses the following property:

$$\sum_{s>0} \nu\,(\{s\} \times R_0) = 0. \tag{4.1}$$

By Theorem 4.1.1 the process X is left quasi-continuous and by the corollary to this theorem $\Delta B = 0$. Thus the triplet $T = (B, C, \nu)$ of the semimartingale X possesses the following properties: $B = B^c$, $\nu = \nu^c$, and the corresponding to this triplet stochastic exponential $\mathcal{E}\,(G\,(\lambda))$ is given by the formula

$$\mathcal{E}_t\,(G\,(\lambda)) = \exp\,(G_t\,(\lambda)) \tag{4.2}$$

with

$$G_t\,(\lambda) = i\lambda B_t - \frac{\lambda^2}{2} C_t + \int_0^t \int_{R_0} (e^{i\lambda x} - 1 - i\lambda x I\,(\,|\,x\,| \le 1))\,d\nu. \tag{4.3}$$

Let $X^n = (X_t^n, \mathcal{F}_t^n)$, $n \ge 1$, be semimartingales with the triplets $T^n = (B^n, C^n, \nu^n)$, $n \ge 1$, and the corresponding stochastic exponentials $\mathcal{E}\,(G^n\,(\lambda))$ given by

$$\mathcal{E}_t\,(G^n\,(\lambda)) = \exp\,(G_t^n\,(\lambda)) \prod_{0<s\le t} (1 + \Delta G_s^n\,(\lambda))\, e^{-\Delta G_s^n\,(\lambda)}, \tag{4.4}$$

where

$$G_t^n\,(\lambda) = i\lambda B_t^n - \frac{\lambda^2}{2} C_t^n + \int_0^t \int_{R_0} (e^{i\lambda x} - 1 - i\lambda x I\,(\,|\,x\,| \le 1))\,d\nu^n. \tag{4.5}$$

For $\delta \in (0, 1]$, $n \ge 1$, define the local square integrable martingale

$$M^{n,\delta} = (M_t^{n,\delta}, \mathcal{F}_t^n)$$

with

$$M_t^{n,\delta} = X_t^{nc} + xI\,(\,|\,x\,| \le \delta) * (\mu^n - \nu^n)_t, \tag{4.6}$$

where $X^{nc} = (X_t^{nc}, \mathcal{F}_t^n)$ is the continuous martingale component of a semimartingale X^n and $\mu^n = \mu^n(dt, dx)$ the jump measure of X^n. The quadratic characteristic $\langle M^{n,\delta}\rangle$ is defined by the formula

$$\langle M^{n,\delta}\rangle_t = C_t^n + x^2 I(|x| \le \delta) * v_t^n - \sum_{0 < s \le t} \left(\int_{|x| \le \delta} x v^n(\{s\}, dx) \right)^2. \quad (4.7)$$

Theorem 1. *Let* S *be a nonempty subset of* R_+ *and let the following conditions be fullfilled:*

(o) $\mathcal{G} \subseteq \bigcap_{n \ge 1} \mathcal{F}_0^n$;

(a) *for each bounded function* $g = g(x)$, *continuous on* $R_0 \setminus (\{-1\} \cup \{1\})$ *and such that* $g(x) = I(|x| > \varepsilon)\, g(x)$ *for a certain* $\varepsilon > 0$, *we have*

$$g * v_t^n \xrightarrow{P} g * v_t, \quad t \in S;$$

(b) $B_t^n \xrightarrow{P} B_t, \quad t \in S$;

(c) $\lim_{\delta \to 0} \overline{\lim_n} P(|\langle M^{n,\delta}\rangle_t - C_t| \ge a) = 0, \quad a > 0, \quad \delta \in (0, 1], \quad t \in S$;

(d) $\sup_{s \le t} v^n(\{s\} \times \{|x| > \delta\}) \xrightarrow{P} 0, \quad \delta \in (0, 1], \quad t \in S$;

(e) $X_0^n \xrightarrow{d} X_0$ ($\mathcal{G}$-*stably*).

Then

$$X^n \xrightarrow{d_f(S)} X \quad (\mathcal{G}\text{-stably}).$$

Remark. If S is a subset dense in R_+, then (a) $\Rightarrow$ (d).

2. The proof of Theorem 1 consists in verifying that Conditions (o), (a), (b), (c), (d) and (e) guarantee the validity of Conditions 1- 4 of Theorem 2.1. Clearly, (o) $\Rightarrow$ 1), (e) $\Rightarrow$ 4). Condition 3 is fullfilled since $\mathcal{E}_t(G(\lambda))$ is given by the formula (4.2) and consequently $|\mathcal{E}_t(G(\lambda))| > 0$ for each $t > 0$. Therefore, on proving Theorem 1, the basic step consists in establishing the implication

$$\text{(a), (b), (c), (d)} \Rightarrow 2). \quad (4.8)$$

The validity of the implication (4.8) is verified by a number of statements which are formulated as lemmas .

Lemma 1. *Let*

$$\hat{x}_s^{n,\delta} = \int_{|x| \le \delta} x\nu^n(\{s\}, dx), \quad \Delta_t^{n,\delta} = \sum_{0 < s \le t} \hat{x}_s^{n,\delta}, \tag{4.9}$$

$$G_t^{n,\delta}(\lambda) = i\lambda(B_t^n - \Delta_t^{n,\delta}) - \frac{\lambda^2}{2} C_t^n$$
$$+ \int_0^t \int_{R_0} [e^{i\lambda(x - \hat{x}_s^{n,\delta})} - 1 - i\lambda(xI(|x| \le 1) - \hat{x}_s^{n,\delta})] \, d\nu^n$$
$$+ \sum_{0 < s \le t} (e^{i\lambda\hat{x}_s^{n,\delta}} - 1 + i\lambda\hat{x}_s^{n,\delta})(1 - a_s^n)$$

with

$$a_s^n = \nu^n(\{s\} \times R_0).$$

Then

$$\mathcal{E}_t(G^n(\lambda)) = \mathcal{E}_t(G^{n,\delta}(\lambda)) \, e^{i\lambda\Delta_t^{n,\delta}}. \tag{4.10}$$

Proof. Denoting

$$V_t^{n,\delta}(\lambda) = e^{-i\lambda\Delta_t^{n,\delta}} \mathcal{E}_t(G^n(\lambda))$$

we will show that

$$V_t^{n,\delta} \equiv \mathcal{E}_t(G^{n,\delta}(\lambda)).$$

By Ito's formula (Ch. 2, § 3) we get

$$V_t^{n,\delta}(\lambda) = 1 + V_-^{n,\delta}(\lambda) \circ \tilde{G}_t^{n,\delta}(\lambda) \tag{4.11}$$

with

$$\tilde{G}_t^{n,\delta}(\lambda) = G_t^n(\lambda) + b_t^{n,\delta}(\lambda) + \sum_{0 < s \le t} \Delta G_s^n(\lambda) \, \Delta b_s^{n,\delta}(\lambda), \tag{4.12}$$

$$b_t^{n,\delta}(\lambda) = \sum_{0 < s \le t} (e^{-i\lambda\hat{x}_s^{n,\delta}} - 1), \tag{4.13}$$

where we have used the fact that $\mathcal{E}(G^n(\lambda))$ satisfies Doléans equation (Ch. 2, § 4)

$$\mathcal{E}_t(G^n(\lambda)) = 1 + \mathcal{E}_-(G^n(\lambda)) \circ G_t^n(\lambda)$$

and

$$e^{-i\lambda\Delta_t^{n,\delta}} = 1 + e^{-i\lambda\Delta_-^{n,\delta}} \circ b_t^{n,\delta}.$$

By (4.12) and (4.13), along with (4.9) and the equality

$$\Delta B_s^n = \int_{|x| \le 1} x \nu^n (\{s\}, dx),$$

it follows that

$$\tilde{G}_t^{n,\delta}(\lambda) = G_t^{nc}(\lambda) + b_t^{n,\delta}(\lambda) + \sum_{0 < s \le t} (1 + \Delta b_s^{n,\delta}(\lambda))\, \Delta G_s^n(\lambda)$$

$$= i\lambda B_t^{nc} - \frac{\lambda^2}{2} C_t^n + \int_0^t \int_{R_0} (e^{i\lambda x} - 1 - i\lambda x I(|x| \le 1))\, d\nu^{nc}$$

$$+ \sum_{0 < s \le t} (e^{-i\lambda \hat{x}_s^{n,\delta}} - 1) + \sum_{0 < s \le t} \int_{R_0} e^{-i\lambda \hat{x}_s^{n,\delta}} (e^{i\lambda x} - 1)\, \nu^n(\{s\}, dx)$$

$$= i\lambda (B_t^n - \Delta_t^{n,\delta}) - \frac{\lambda^2}{2} C_t^n + \int_0^t \int_{R_0} (e^{i\lambda x} - 1 - i\lambda x I(|x| \le 1))\, d\nu^{nc}$$

$$+ \sum_{0 < s \le t} (e^{-i\lambda \hat{x}_s^{n,\delta}} - 1 + i\lambda \hat{x}_s^{n,\delta})$$

$$+ \sum_{0 < s \le t} \int_{R_0} [e^{-i\lambda \hat{x}_s^{n,\delta}} (e^{i\lambda x} - 1) - i\lambda x I(|x| \le 1)]\, \nu^n(\{s\}, dx)$$

$$= i\lambda (B_t^n - \Delta_t^{n,\delta}) - \frac{\lambda^2}{2} C_t^n + \int_0^t \int_{R_0} (e^{i\lambda x} - 1 - i\lambda x I(|x| \le 1))\, d\nu^{nc}$$

$$+ \sum_{0 < s \le t} (e^{-i\lambda \hat{x}_s^{n,\delta}} - 1 + i\lambda \hat{x}_s^{n,\delta})(1 - a_s)$$

$$+ \sum_{0 < s \le t} \int_{R_0} [e^{i\lambda (x - \hat{x}_s^{n,\delta})} - 1 - i\lambda (x I(|x| \le 1) - \hat{x}_s^{n,\delta})]\, \nu^n(\{s\}, dx).$$

It is not hard to deduce from this representation for $\tilde{G}_t^{n,\delta}(\lambda)$ that

$$\tilde{G}_t^{n,\delta}(\lambda) \equiv G_t^{n,\delta}(\lambda)$$

(see (4.9)). Therefore by (4.11)

$$V_t^{n,\delta}(\lambda) = 1 + V_-^{n,\delta}(\lambda) \circ G_t^{n,\delta}(\lambda),$$

i.e. $V_t^{n,\delta}(\lambda)$ is a solution of Doléans equation. As the solution of Doléans equation is unique (Theorem 2.4.1), $V_t^{n,\delta}(\lambda) \equiv \mathcal{E}_t(G^{n,\delta}(\lambda))$.

Lemma 2. *Let condition* (d) *be fullfilled. Then*

$$\sup_{s \le t} |\hat{x}_s^{n,\delta}| \xrightarrow{P} 0, \quad \delta \in (0, 1], \quad t \in S. \tag{4.14}$$

Proof. Let $0 < \varepsilon < \delta$. Then

$$|\hat{x}_s^{n,\delta}| = \left| \int_{|x| \le \delta} x v^n(\{s\}, dx) \right| \le \varepsilon + \delta v^n(\{s\} \times \{|x| > \varepsilon\}).$$

Consequently, for each $a > 0$ and $0 < \varepsilon \le \dfrac{\delta \wedge a}{2}$

$$P\left(\sup_{s \le t} |\hat{x}_s^{n,\delta}| \ge a\right) \le P\left(\sup_{s \le t} v^n(\{s\} \times \{|x| > \varepsilon\}) > \delta^{-1}\left(a - \frac{\delta \wedge a}{2}\right)\right).$$

This implies the desired assertion, in view of Condition (d).

Lemma 3. *The following inequality holds* $(0 < \delta \le 1)$:

$$\sum_{0 < s \le t} |\Delta G_s^{n,\delta}(\lambda)| \le \frac{\lambda^2}{2} \langle M^{n,\delta} \rangle_t + 4I(|x| > \delta) * v_t^n. \tag{4.15}$$

Proof. Let us show that

$$|\Delta G_s^{n,\delta}(\lambda)| \le \frac{\lambda^2}{2} \Delta \langle M^{n,\delta} \rangle_s + 4v^n(\{s\} \times \{|x| > \delta\})$$

or, in view of (4.7), (4.9) and the notation

$$a_s^{n,\delta} = v^n(\{s\} \times \{|x| \le \delta\}), \tag{4.16}$$

that

$$|\Delta G_s^{n,\delta}(\lambda)| \le \frac{\lambda^2}{2} \int_{|x| \le \delta} x^2 v^n(\{s\}, dx) - \frac{\lambda^2}{2} (\hat{x}_s^{n,\delta}) + 4(a_s^n - a_s^{n,\delta}). \tag{4.17}$$

By (4.9) we get

$$\Delta G_s^{n,\delta}(\lambda) = \int_{R_0} (e^{i\lambda(x - \hat{x}_s^{n,\delta})} - 1) v^n(\{s\}, dx) + (e^{-i\lambda \hat{x}_s^{n,\delta}} - 1)(1 - a_s^n)$$

$$= \int_{|x| \le \delta} [e^{i\lambda(x - \hat{x}_s^{n,\delta})} - 1 - i\lambda(x - \hat{x}_s^{n,\delta})] v^n(\{s\}, dx)$$

$$+ \int_{|x|>\delta} (e^{i\lambda(x - \hat{x}_s^{n,\delta})} - 1)\, \nu^n(\{s\}, dx) + (e^{-i\lambda \hat{x}_s^{n,\delta}} - 1)(1 - a_s^n)$$

$$+ i\lambda \hat{x}_s^{n,\delta}(1 - a_s^{n,\delta})$$

$$= \int_{|x|\le\delta} [e^{i\lambda(x - \hat{x}_s^{n,\delta})} - 1 - i\lambda(x - \hat{x}_s^{n,\delta})]\, \nu^n(\{s\}, dx)$$

$$+ (e^{-i\lambda \hat{x}_s^{n,\delta}} - 1 + i\lambda \hat{x}_s^{n,\delta})(1 - a_s^{n,\delta})$$

$$+ \int_{|x|>\delta} (e^{i\lambda(x - \hat{x}_s^{n,\delta})} - 1)\, \nu^n(\{s\}, dx) + (e^{-i\lambda \hat{x}_s^{n,\delta}} - 1)(a_s^{n,\delta} - a_s^n).$$

Therefore

$$|\Delta G_s^{n,\delta}(\lambda)| \le \frac{\lambda^2}{2} \int_{|x|\le\delta} (x - \hat{x}_s^{n,\delta})^2\, \nu^n(\{s\}, dx)$$

$$+ \frac{\lambda^2}{2} (\hat{x}_s^{n,\delta})^2 (1 - a_s^{n,\delta}) + 2\nu^n(\{s\} \times \{|x| > \delta\})$$

$$+ 2[\nu^n(\{s\} \times R_0) - \nu^n(\{s\} \times \{|x| \le \delta\})].$$

This gives the desired inequality (4.17) in an obvious manner, due to (4.16) and the relation

$$\int_{|x|\le\delta} (x - \hat{x}_s^{n,\delta})^2\, \nu^n(\{s\}, dx) + (\hat{x}_s^{n,\delta})^2 (1 - a_s^{n,\delta})$$

$$= \int_{|x|\le\delta} x^2 \nu^n(\{s\}, dx) - (\hat{x}_s^{n,\delta})^2. \tag{4.18}$$

Lemma 4. *Let Condition* (d) *be fullfilled. Then*

$$\sup_{s \le t} |\Delta G_s^{n,\delta}(\lambda)| \xrightarrow{P} 0, \quad \delta \in (0, 1], \quad \lambda \in R, \quad t \in S.$$

Proof. For $0 < \varepsilon < \delta$ the inequality (4.17) gives

$$\sup_{s \le t} | \Delta G_s^{n,\delta}(\lambda) |$$
$$\le \frac{\lambda^2}{2} [\varepsilon^2 + \delta^2 \sup_{s \le t} \nu^n(\{s\} \times \{ |x| > \varepsilon\})] + 4 \sup_{s \le t} \nu^n(\{s\} \times \{ |x| > \delta\})$$
$$\le \frac{\lambda^2}{2} \left[\varepsilon^2 + \left(\frac{8}{\lambda^2} + \delta^2 \right) \sup_{s \le t} \nu^n(\{s\} \times \{ |x| > \varepsilon\}) \right].$$

Therefore, for each $a > 0$ and $0 < \varepsilon^2 \le \frac{a}{\lambda^2} \wedge \delta^2$

$$\mathsf{P}\left(\sup_{s \le t} | \Delta G_s^{n,\delta}(\lambda) | \ge a \right) \le \mathsf{P}\left(\sup_{s \le t} \nu^n(\{s\} \times \{ |x| > \varepsilon\}) \ge \left(\frac{8}{\lambda^2} + \delta^2 \right)^{-1} \frac{a}{\lambda^2} \right),$$

which proves the desired assertion by Condition (d).

Lemma 5. *Let Conditions* (a), (c) *and* (d) *be fullfilled. Then*

$$\lim_{\delta \to 0} \overline{\lim_n} \, \mathsf{P}\left(\sup_{s \le t} \left| \prod_{0 < u \le s} (1 + \Delta G_u^{n,\delta}(\lambda)) \, e^{-\Delta G_u^{n,\delta}(\lambda)} - 1 \right| \ge a \right) = 0,$$

$a > 0, \quad \lambda \in R, \quad t \in S.$

Proof. Define the set

$$\Gamma_t^{n,\delta} = \left\{ \sup_{s \le t} | \Delta G_s^{n,\delta}(\lambda) | \ge \frac{1}{2} \right\}$$

and denote

$$J_s^{n,\delta} = \prod_{0 < u \le s} (1 + \Delta G_u^{n,\delta}(\lambda)) \, e^{-\Delta G_u^{n,\delta}(\lambda)} - 1.$$

Then ($a > 0$)

$$\mathsf{P}\left(\sup_{s \le t} | J_s^{n,\delta} | \ge a \right) \le \mathsf{P}\left(\sup_{s \le t} | J_s^{n,\delta} | \ge a, \Omega \setminus \Gamma_t^{n,\delta} \right) + \mathsf{P}(\Gamma_t^{n,\delta}). \quad (4.19)$$

For $t \in S$ and $\delta \in (0, 1]$ we have

$$\lim_n \mathsf{P}(\Gamma_t^{n,\delta}) = 0 \quad (4.20)$$

by Lemma 4, while on the set $\Omega \setminus \Gamma_t^{n,\delta}$ we have

$$\sup_{s\le t} | J_s^{n,\delta} | \le \sup_{s\le t} \left| \exp\left(\sum_{0<u\le s} [\ln(1+\Delta G_s^{n,\delta}(\lambda)) - \Delta G_s^{n,\delta}(\lambda)] \right) - 1 \right|$$

$$= \sup_{s\le t} \left| \exp\left(\sum_{0<u\le s} \sum_{k=2}^{\infty} \frac{(-1)^{k+1}}{k} (\Delta G_u^{n,\delta}(\lambda))^k \right) - 1 \right|$$

$$\le \exp\left(\frac{1}{2} \sum_{0<u\le t} \sum_{k=2}^{\infty} | \Delta G_u^{n,\delta}(\lambda) |^k \right) - 1$$

$$= \exp\left(\frac{1}{2} \sum_{0<u\le t} \frac{| \Delta G_u^{n,\delta}(\lambda) |^2}{1 - | \Delta G_u^{n,\delta}(\lambda) |} \right) - 1$$

$$\le \exp\left(\sup_{s\le t} | \Delta G_s^{n,\delta}(\lambda) | \sum_{0<u\le t} | \Delta G_u^{n,\delta}(\lambda) | \right) - 1.$$

On the set $\Omega \setminus \Gamma_t^{n,\delta}$ this inequality and Lemma 3 entail

$$\sup_{s\le t} | J_s^{n,\delta} | \le \exp\left(\sup_{s\le t} | \Delta G_s^{n,\delta}(\lambda) | \left[\frac{\lambda^2}{2} \langle M^{n,\delta} \rangle_t + 4I(|x|>\delta) * \nu_t^n \right] \right) - 1. \tag{4.21}$$

Hence, by (4.19) and (4.20) the desired assertion takes place, provided

$$\lim_{\delta\to 0} \overline{\lim_n} \, P(\sup_{s\le t} | \Delta G_s^{n,\delta}(\lambda) | \langle M^{n,\delta} \rangle_t \ge a) = 0, \tag{4.22}$$

$$\lim_n P(\sup_{s\le t} | \Delta G_s^{n,\delta}(\lambda) | I(|x|>\delta) * \nu_t^n \ge a) = 0,$$

$$a > 0, \; t \in S. \tag{4.23}$$

The relation (4.22) takes place, since by Lemma 4

$$\sup_{s\le t} | \Delta G_s^{n,\delta}(\lambda) | \xrightarrow{P} 0, \; n \to \infty$$

and

$$P(\langle M^{n,\delta} \rangle_t > L) \le P\left(\langle M^{n,\delta} \rangle_t > L, \; C_t \le \frac{L}{2} \right) + P\left(C_t > \frac{L}{2} \right)$$

$$\le P\left(| \langle M^{n,\delta} \rangle_t - C_t | \ge \frac{L}{2} \right) + P\left(C_t > \frac{L}{2} \right) \to 0$$

as $t \in S$ and the limit $\lim_{L \to \infty} \overline{\lim}_{\delta \to 0} \overline{\lim}_{n}$ is taken. (By Condition (c)

$$\overline{\lim}_{\delta \to 0} \overline{\lim}_{n} P\left(|\langle M^{n,\delta}\rangle_t - C_t| \geq \frac{L}{2} \right) = 0$$

for each $L > 0$).

To prove (4.23) it suffices to show that

$$\lim_{L \to \infty} \overline{\lim}_{n} P(|x| > \delta) * \nu_t^n > L) = 0, \tag{4.24}$$

$\delta \in (0, 1]$, $t \in S$, since by Lemma 4 we have

$$\sup_{s \leq t} |\Delta G_s^{n,\delta}(\lambda)| \xrightarrow{P} 0, \quad n \to \infty,$$

for $\delta \in (0, 1]$ and $t \in S$. Aiming at the proof of (4.24), consider a continuous bounded nonnegative function $g = g(x)$ with the properties

$$g(x) = g(x) I\left(|x| > \frac{\delta}{2} \right), \quad g(x) I(|x| > \delta) = I(|x| > \delta).$$

Then in accordance with the inequality

$$I(|x| > \delta) * \nu_t^n \leq g * \nu_t^n$$

we have

$$P(|x| > \delta) * \nu_t^n > L) \leq P(g * \nu_t^n > L)$$

$$\leq P\left(g * \nu_t^n > L,\ g * \nu_t \leq \frac{L}{2} \right) + P\left(g * \nu_t > \frac{L}{2} \right)$$

$$\leq P\left(|g * \nu_t^n - g * \nu_t| \geq \frac{L}{2} \right) + P\left(g * \nu_t > \frac{L}{2} \right) \to 0$$

as $t \in S$ and the limit $\lim_{L \to \infty} \overline{\lim}_{n}$ is taken. (By Condition (a)

$$P\left(|g * \nu_t^n - g * \nu_t| \geq \frac{L}{2} \right) \to 0$$

as $n \to \infty$, $t \in S$ for each $L > 0$.)

Lemma 6. *Let a function* $h = h(x)$, $x \in R_0$ *be continuous on* $R_0 \setminus (\{-1\} \cup \{1\})$ *and such that*

$$|h(x)| \leq c_1 (c_2 \wedge x^2)$$

with positive constants c_1 *and* c_2.

Then under Condition (a)

$$\lim_{\delta \to 0} \overline{\lim_{n}} \ P(|hI(|x|>\delta) * \nu^n_t - hI(|x|>\delta) * \nu_t| \geq a) = 0,$$

$a > 0,\ t \in S.$

Proof. Obviously, there exists a function $h_\delta = h_\delta(x)$ such that Condition (a) of the theorem is satisfied and

$$h_\delta(x) = h_\delta(x) I\left(|x| > \frac{\delta}{2} \right), \qquad h_\delta(x) I(|x|>\delta) = h(x) I(|x|>\delta),$$

$$|h_\delta(x)| \leq |h(x)|.$$

Then

$$|h(x) I(|x|>\delta) - h_\delta(x)| = |h_\delta(x)| I(|x| \leq \delta)$$

and one can choose a nonnegative function $\tilde{h}_\delta = \tilde{h}_\delta(x)$, $x \in R_0$, continuous on $R_0 \setminus (\{-1\} \cup \{1\})$ and such that

$$\tilde{h}_\delta(x) I(|x| > 2\delta) = 0, \qquad \tilde{h}_\delta(x) I(|x| \leq \delta) = |h_\delta(x)| I(|x| \leq \delta),$$

$$\tilde{h}_\delta(x) \leq |h(x)| I(|x| \leq 2\delta).$$

In view of these properties of functions h_δ and $\tilde{h}_\delta$ we get

$$\begin{aligned}
J^{n,\delta} &= |hI(|x|>\delta) * \nu^n_t - hI(|x|>\delta) * \nu_t| \\
&\leq |h_\delta * \nu^n_t - h_\delta * \nu_t| + |hI(|x|>\delta) - h_\delta| * \nu^n_t + |hI(|x|>\delta) - h_\delta| * \nu_t \\
&= |h_\delta * \nu^n_t - h_\delta * \nu_t| + |h_\delta| I(|x| \leq \delta) * \nu^n_t + |h_\delta| I(|x| \leq \delta) * \nu_t \\
&\leq |h_\delta * \nu^n_t - h_\delta * \nu_t| + \tilde{h}_\delta * \nu^n_t + \tilde{h}_\delta * \nu_t \\
&\leq |h_\delta * \nu^n_t - h_\delta * \nu_t| + |\tilde{h}_\delta * \nu^n_t - \tilde{h}_\delta * \nu_t| + 2|h| I(|x| \leq 2\delta) * \nu_t \\
&\leq |h_\delta * \nu^n_t - h_\delta * \nu_t| + |\tilde{h}_\delta * \nu^n_t - \tilde{h}_\delta * \nu_t| + 2c_1 (c_2 \wedge x^2) I(|x| \leq 2\delta) * \nu_t.
\end{aligned}$$

By Condition (a) for each $\delta \in (0, 1]$ and $t \in S$

$$|h_\delta * \nu^n_t - h_\delta * \nu_t| + |\tilde{h}_\delta * \nu^n_t - \tilde{h}_\delta * \nu_t| \xrightarrow{P} 0, \quad n \to \infty.$$

Therefore for $a > 0$

$$\overline{\lim_{n}} \ P(J^{n,\delta} \geq a) \leq P\left(2c_1 (c_2 \wedge x^2) I(|x| \leq 2\delta) * \nu \geq \frac{a}{2} \right) \to 0, \quad \delta \to 0. \tag{4.25}$$

3. Proof of implication (4.8). Let us show that Conditions (a), (b), (c) and (d) imply

$$\mathcal{E}_t(G^n(\lambda)) \xrightarrow{P} \mathcal{E}_t(G(\lambda)), \quad n \to \infty, \quad \lambda \in R, \quad t \in S.$$

By Lemma 1 we have

$$\mathcal{E}_t(G^n(\lambda)) = e^{i\lambda\Delta_t^{n,\delta}} \mathcal{E}_t(G^{n,\delta}(\lambda)).$$

By Lemma 5

$$\lim_{\delta \to 0} \overline{\lim_n} \, P\left(\left| \prod_{0 < s \le t} (1 + \Delta G_s^{n,\delta}(\lambda))\, e^{-\Delta G_s^{n,\delta}(\lambda)} - 1 \right| \ge a \right) = 0,$$

$$a > 0, \quad \lambda \in R, \quad t \in S.$$

Therefore the desired relation takes place, provided

$$\lim_{\delta \to 0} \overline{\lim_n} \, P(|G_t^{n,\delta}(\lambda) + i\lambda\Delta_t^{n,\delta} - G_t(\lambda)| \ge a) = 0, \tag{4.26}$$

$$a > 0, \quad \lambda \in R, \quad t \in S.$$

To verify (4.26) denote

$$R_t^{n,\delta}(\lambda) = -\frac{\lambda^2}{2} C_t^n + \int_0^t \int_{|x| \le \delta} [e^{i\lambda(x - \hat{x}_s^{n,\delta})} - 1 - i\lambda(x - \hat{x}_s^{n,\delta})]\, dv^n$$

$$+ \sum_{0 < s \le t} [e^{-i\lambda\hat{x}_s^{n,\delta}} - 1 + i\lambda\hat{x}_s^{n,\delta}](1 - a_s^{n,\delta}), \tag{4.27}$$

$$Q_t^{n,\delta}(\lambda) = \int_0^t \int_{|x| > \delta} [e^{i\lambda(x - \hat{x}_s^{n,\delta})} - 1 - i\lambda(x\, I(|x| \le 1) - \hat{x}_s^{n,\delta})]\, dv^n$$

$$+ \sum_{0 < s \le t} (e^{-i\lambda\hat{x}_s^{n,\delta}} - 1 + i\lambda\hat{x}_s^{n,\delta})(a_s^{n,\delta} - a_s^n) \tag{4.28}$$

and observe that

$$G_t^{n,\delta}(\lambda) + i\lambda\Delta_t^n = i\lambda B_t^n + R_t^{n,\delta}(\lambda) + Q_t^{n,\delta}(\lambda). \tag{4.29}$$

An analogous representation takes place for $G_t(\lambda)$ too:

$$G_t(\lambda) = i\lambda B_t + R_t^\delta(\lambda) + Q_t^\delta(\lambda)$$

with

$$R_t^{\delta}(\lambda) = -\frac{\lambda^2}{2} C_t + \int_0^t \int_{|x| \le \delta} (e^{i\lambda x} - 1 - i\lambda x)\, d\nu,$$

$$Q_t^{\delta}(\lambda) = \int_0^t \int_{|x| \ge \delta} (e^{i\lambda x} - 1 - i\lambda x\, I(|x| \le 1))\, d\nu. \tag{4.30}$$

By Condition (b) we have $B_t^n \xrightarrow{P} B_t$, $n \to \infty$, $t \in S$. Therefore the desired relation (4.26) takes place, provided

$$\lim_{\delta \to 0} \overline{\lim_{n}}\, P(|R_t^{n,\delta}(\lambda) - R_t^{\delta}(\lambda)| \ge a) = 0, \tag{4.31}$$

$$\lim_{\delta \to 0} \overline{\lim_{n}}\, P(|Q_t^{n,\delta}(\lambda) - Q_t^{\delta}(\lambda)| \ge a) = 0, \tag{4.32}$$

$$a > 0,\ \lambda \in R,\ t \in S.$$

The proof of (4.31) is based on the estimate deduced from the representation (4.7) for $\langle M^{n,\delta}\rangle$ and the equality (4.18) which gives

$$\langle M^{n,\delta}\rangle_t = C_t^n + \int_0^t \int_{|x| \le \delta} (x - \hat{x}_s^{n,\delta})^2\, d\nu^n + \sum_{0 < s \le t} (\hat{x}_s^{n,\delta})^2 (1 - a_s^{n,\delta}).$$

Indeed

$$\begin{aligned}
&\left| R_t^{n,\delta}(\lambda) + \frac{\lambda^2}{2} \langle M^{n,\delta}\rangle_t \right| \\
&\le \frac{|\lambda^3|}{6} \int_0^t \int_{|x| \le \delta} |x - \hat{x}_s^{n,\delta}|^3\, d\nu^n + \frac{|\lambda^3|}{6} \sum_{0 < s \le t} |\hat{x}_s^{n,\delta}|^3 (1 - a_s^{n,\delta}) \\
&\le \frac{|\lambda^3|\delta}{3} \left(\int_0^t \int_{|x| \le \delta} (x - \hat{x}_s^{n,\delta})^2\, d\nu^n + \sum_{0 < s \le t} (\hat{x}_s^{n,\delta})^2 (1 - a_s^{n,\delta}) \right) \\
&\le \frac{|\lambda^3|\delta}{3} \langle M^{n,\delta}\rangle_t.
\end{aligned} \tag{4.33}$$

Consequently

$$
\begin{aligned}
&| R_t^{n,\delta}(\lambda) - R_t^{\delta}(\lambda) | \\
&\leq \left| R_t^{n,\delta}(\lambda) + \frac{\lambda^2}{2} \langle M^{n,\delta} \rangle_t \right| + \frac{\lambda^2}{2} \, | \langle M^{n,\delta} \rangle_t - C_t | + \left| \frac{\lambda^2}{2} C_t + R_t^{\delta}(\lambda) \right| \\
&\leq \frac{|\lambda^3| \delta}{3} \langle M^{n,\delta} \rangle + \frac{\lambda^2}{2} \, | \langle M^{n,\delta} \rangle_t - C_t | + \frac{\lambda^2}{2} \int_0^t \int_{|x| \leq \delta} x^2 \, d\nu \\
&\leq \left(\frac{\lambda^2}{2} + \frac{|\lambda^3| \delta}{3} \right) | \langle M^{n,\delta} \rangle - C_t | + \frac{|\lambda^3| \delta}{3} C_t + \frac{\lambda^2}{2} \int_0^t \int_{0 < |x| \leq \delta} x^2 \, d\nu,
\end{aligned}
$$

which gives the desired relation (4.31) by Condition (c) and

$$
\lim_{\delta \to 0} \left(\frac{|\lambda^3| \delta}{3} C_t + \frac{\lambda^2}{2} \int_0^t \int_{|x| \leq \delta} x^2 \, d\nu \right) = 0.
$$

Let us establish now (4.32). Let

$$
\tilde{Q}_t^{n,\delta}(\lambda) = \int_0^t \int_{|x| > \delta} (e^{i\lambda x} - 1 - i\lambda x \, I(|x| \leq 1)) \, d\nu^n. \tag{4.34}
$$

Then

$$
\begin{aligned}
| Q_t^n(\lambda) - \tilde{Q}_t^{n,\delta}(\lambda) | &\leq \int_0^t \int_{|x| > \delta} | e^{i\lambda (x - \hat{x}_s^{n,\delta})} - e^{i\lambda x} | \, d\nu^n \\
&+ \sum_{0 < s \leq t} | e^{-i\lambda \hat{x}_s^{n,\delta}} - 1 | \, (a_s^n - a_s^{n,\delta}) \\
&\leq 2 |\lambda| \sup_{s \leq t} | \hat{x}_s^{n,\delta} | \, \nu^n((0, t] \times \{ |x| > \delta \}).
\end{aligned} \tag{4.35}
$$

This implies

$$
| Q_t^{n,\delta}(\lambda) - \tilde{Q}_t^{n,\delta}(\lambda) | \xrightarrow{P} 0, \quad n \to \infty,
$$

$\delta \in (0, 1]$, $\lambda \in R$, $t \in S$, since by Lemma 2 we have

$$
\sup_{s \leq t} | \hat{x}_s^{n,\delta} | \xrightarrow{P} 0, \quad n \to \infty, \quad \delta \in (0, 1], \quad t \in S,
$$

and

$$
\lim_{L \to \infty} \overline{\lim_{n}} \, P(\nu^n((0, t] \times \{ |x| > \delta \}) > L) = 0,
$$

$$
\delta \in (0, 1], \quad t \in S
$$

as has been established in the course of proving Lemma 5 (see (4.24)).

Thus to complete the proof (4.32) it suffices to show only that

$$\lim_{\delta \to 0} \overline{\lim_{n}} \mathsf{P}(|\tilde{Q}_t^{n,\delta}(\lambda) - Q_t^{\delta}(\lambda)| \geq a) = 0,$$

$$a > 0, \ \lambda \in R, \ t \in S.$$

But this relation holds by Lemma 5.

The remark to the theorem is a simple consequence of Problem 3.2.

4. Let $\hat{\tau}_n = (\hat{\tau}_n(t))_{t \geq 0}$ be a random change of time, $n \geq 1$. Define a semimartingale $\hat{X}^n = (\hat{X}_t^n, \hat{\mathcal{F}}_t^n)$, with

$$\hat{X}_t^n = X_{\hat{\tau}_n(t)}^n, \qquad \hat{\mathcal{F}}_t^n = \mathcal{F}_{\hat{\tau}_n(t)}^n.$$

A generalization of Theorem 1 on the case of semimartingales $\hat{X}^n$, $n \geq 1$, is presented by

Theorem 2. *Let* S *be a nonempty subset of* R_+ *and let the following conditions be fullfilled:*

(o) $\mathcal{G} \subseteq \bigcap_{n \geq 1} \mathcal{F}_0^n$;

(a) *for each bounded function* $g = g(x)$, $x \in R_0$, *continuous on* $R_0 \setminus (\{-1\} \cup \{1\})$ *and possessing the property* $g(x) = I(|x| > \varepsilon)\, g(x)$ *for a certain* $\varepsilon > 0$, *the relation*

$$g * \nu_{\hat{\tau}_n(t)}^n \xrightarrow{\mathsf{P}} g * \nu_t, \ t \in S$$

holds;

(b) $B_{\hat{\tau}_n(t)}^n \xrightarrow{\mathsf{P}} B_t$, $t \in S$;

(c) $\lim_{\delta \to 0} \overline{\lim_{n}} \mathsf{P}(|\langle M^{n,\delta}\rangle_{\hat{\tau}_n(t)} - C_t| \geq a) = 0$, $a > 0$, $t \in S$;

(d) $\sup_{s \leq \hat{\tau}_n(t)} \nu^n(\{s\} \times \{|x| > \delta\}) \xrightarrow{\mathsf{P}} 0$, $\delta \in (0, 1]$, $t \in S$;

(e) $X_0^n \xrightarrow{d} X_0$ ($\mathcal{G}$*-stably*).

Then

$$\hat{X}^n \xrightarrow{d_f(S)} X \quad (\mathcal{G}\text{-}stably).$$

Remark. If S is a subset dense in R_+, then (a) $\Rightarrow$ (d).

The proof of this theorem, based on Theorem 2.2, is analogous to the proof of Theorem 1.

5. Consider yet another version of conditions for the convergence $X^n \xrightarrow{d_f(S)} X$ ($\mathcal{G}$-stably) in case in which a family of σ-algebras $\mathbb{F}^n$, $n \geq 1$, is nested.

Theorem 3. *Let* S *be a nonempty subset of* R_+ *and let the following conditions be fullfilled:*

(o) *a family of* σ*-algebras* $\mathbb{F}^n$, $n \geq 1$, *is nested: there exists a sequence of numbers* $(s_n)_{n \geq 1}$, $s_n \downarrow 0$, $n \to \infty$, *such that*

$$\mathcal{F}^n_{s_n} \subseteq \mathcal{F}^{n+1}_{s_{n+1}}, \quad \sigma\left(\bigcup_{n \geq 1} \mathcal{F}^n_{s_n}\right) = \sigma\left(\bigcup_{n \geq 1} \mathcal{F}^n_{\infty}\right)$$

and

$$\mathcal{G} \subseteq \sigma\left(\bigcup_{n \geq 1} \mathcal{F}^n_{\infty}\right);$$

(a) *for each bounded function* $g = g(x)$, $x \in R_0$, *continuous on* $R_0 \setminus (\{-1\} \cup \{1\}$ *and such that* $g(x) = I(|x| > \varepsilon)\, g(x)$ *for a certain* $\varepsilon > 0$, *the relation*

$$g * \nu^n_t \xrightarrow{P} g * \nu_t, \quad t \in S$$

takes place;

(b) $B^n_t \xrightarrow{P} B_t$, $t \in S$;

(c) $\lim_{\delta \to 0} \overline{\lim_n}\, P(|\langle M^{n,\delta}\rangle_t - C_t| \geq a) = 0$, $a > 0$, $t \in S$;

(d) $\sup_{s \leq t} \nu^n(\{s\} \times \{|x| > \delta\}) \xrightarrow{P} 0$, $\delta \in (0, 1]$, $t \in S$;

(e) $X^n_0 \xrightarrow{d} X_0$ ($\mathcal{G}$*-stably*);

(f) $B^n_{s_n} \xrightarrow{P} 0$, $\lim_{\delta \to 0} \overline{\lim_n}\, P(\langle M^{n,\delta}\rangle_{s_n} \geq a) = 0$,

$$g * \nu^n_{s_n} \xrightarrow{P} 0$$

for each nonnegative function $g = g(x)$ *involved in Condition* (a).

Then

$$X^n \xrightarrow{d_f(S)} X \quad (\mathcal{G}\textit{-stably}).$$

Remarks. 1. If S is a subset dense in R_+, then (a) $\Rightarrow$ (d).

2. If S is a subset dense in the interval [0, a] for a certain a > 0, then Condition (f) is a consequence of Conditions (a), (c) and

(b') $\sup_{t \le a} | B_t^n - B_t | \xrightarrow{P} 0.$

The proof of Theorem 3 is based on Theorem 2.3 and consists in verifying that Conditions (o), (a), (b), (c), (d), (e) and (f) of the present theorem guarantee the validity of Conditions 1) - 5) of Theorem 2.3.

Clearly, (o) ⇒ 1), (e) ⇒ 4), (a), (b), (c) and (d) ⇒ 2) (this fact is established in the course of proving Theorem 1) and, by the representation (4.2) for $\mathcal{E}_t (G(\lambda))$, Condintion 3) is satisfied.

Therefore it remains to show that

$$(f) \Rightarrow 5),$$

i.e. that under Condition (f) we have

$$X_{s_n}^n - X_0^n \xrightarrow{P} 0$$

and

$$\mathcal{E}_{s_n} (G^n (\lambda)) \xrightarrow{P} 1.$$

To this end, we utilize the canonical representation for

$$X_t^n = X_0^n + B_t^n + X_t^{nc} + \int_0^t \int_{|x|>1} x \, d\mu^n + \int_0^t \int_{|x| \le 1} x \, d(\mu^n - \nu^n)$$

where $X^{nc} = (X_t^{nc}, \mathcal{F}_t^n)$ is the continuous martingale component of a semimartingale X^n and $\mu^n = \mu^n (dt, dx)$ is the jump measure of X^n, and also we take into consideration the representation (4.6) for the local square integrable martingale $M^{n, d}$, which give the following estimate for $| X_{s_n}^n - X_0^n |$:

$$| X_{s_n}^n - X_0^n | \le | B_{s_n}^n | + \sup_{s \le s_n} | M_s^{n, \delta} | + \int_0^{s_n} \int_{|x|>\delta} |x| \, d\mu^n + \int_0^{s_n} \int_{\delta < |x| \le 1} |x| \, d\nu^n.$$

With the help of this estimate we will show that

$$(f) \Rightarrow X_{s_n}^n - X_0^n \xrightarrow{P} 0, \quad n \to \infty.$$

Indeed

1) (f) ⇒ $B_{s_n}^n \xrightarrow{P} 0$;

2) according to the Lenglart-Rebolledo inequality and by Problem 1.8.6 we have

$$P\left(\sup_{s \le s_n} |M_s^{n,\delta}| \ge a\right) \le \frac{b}{a^2} + P\left(\langle M^{n,\delta}\rangle_{s_n} \ge b\right),\ b > 0,$$

and consequently

$$(f) \Rightarrow \lim_{\delta \to 0} \overline{\lim_n}\ P\left(\sup_{s \le s_n} |M_s^{n,\delta}| \ge a\right) \le \frac{b}{a^2} \to 0,\ b \to 0,\ a > 0;$$

3) by the inequality

$$\int_0^{s_n} \int_{\delta < |x| \le 1} |x|\, d\nu^n \le g * \nu_{s_n}^n$$

with a function $g = g(x)$ which satisfies Condition (f) and has the property

$$g(x) \ge |x|\, I(\delta < |x| \le 1),$$

we get

$$(f) \Rightarrow \int_0^{s_n} \int_{\delta < |x| \le 1} |x|\, d\nu^n \xrightarrow{P} 0;$$

4) by the inclusion

$$\left\{ \int_0^{s_n} \int_{|x| > \delta} |x|\, d\mu^n \ge a \right\} \subseteq \left\{ \int_0^{s_n} \int_{|x| > \delta} d\mu^n \ge 1 \right\}$$

and by the following consequence of the Lenglart-Rebolledo inequality

$$P\left(\int_0^{s_n} \int_{|x| > \delta} d\mu^n \ge 1 \right) \le b + P\left(\int_0^{s_n} \int_{|x| > \delta} d\nu^n \ge b \right),\ b > 0,$$

as well as by the inequality

$$\int_0^{s_n} \int_{|x| > \delta} d\nu^n \le g * \nu_{s_n}^n$$

with a function $g = g(x)$ as in Condition (f), such that $g(x) \ge I(|x| > \delta)$, we get

$$(f) \Rightarrow \overline{\lim_n}\ P\left(\int_0^{s_n} \int_{|x| > \delta} |x|\, d\mu^n \ge a \right) \le b + \overline{\lim_n}\ P\left(g * \nu_{s_n}^n \ge b\right) = b \to 0$$

as $b \to 0$.

Let us establish now the implication

$$(f) \Rightarrow \mathcal{E}_{s_n}(G^n(\lambda)) \xrightarrow{P} 1,\ n \to \infty,\ \lambda \in R. \tag{4.36}$$

By Lemma 1

$$\mathcal{E}_{s_n}(G^n(\lambda)) = e^{i\lambda \Delta^{n,\delta}_{s_n} + G^{n,\delta}_{s_n}(\lambda)} \prod_{0<s\le s_n} (1+\Delta G^{n,\delta}_s(\lambda))\, e^{-i\lambda\,\Delta G^{n,\delta}_s(\lambda)}.$$

Therefore it suffices to show that

$$\lim_{\delta\to 0} \overline{\lim_n}\, \mathsf{P}\left(\left| \prod_{0<s\le s_n} (1+\Delta G^{n,\delta}_s(\lambda))\, e^{-\Delta G^{n,\delta}_s(\lambda)} - 1 \right| \ge a \right) = 0,$$

$$a > 0,\quad \lambda \in R \tag{4.37}$$

and

$$(f) \Rightarrow \lim_{\delta\to 0} \overline{\lim_n}\, \mathsf{P}\left(\left| i\lambda \Delta^{n,\delta}_{s_n} + G^{n,\delta}_{s_n}(\lambda)\right| \ge a\right) = 0,$$

$$a > 0,\quad \lambda \in R. \tag{4.38}$$

The relation (4.37) follows from the corresponding uniform convergence established in Lemma 5 and from the fact that for a sufficiently large n we have $s_n \le t \in S$. To establish (4.38), we utilize the fact that

$$\begin{aligned} |i\lambda \Delta^{n,\delta}_{s_n} + G^{n,\delta}_{s_n}(\lambda)| &\le |\lambda B^n_{s_n}| + |R^{n,\delta}_{s_n}(\lambda)| + |Q^{n,\delta}_{s_n}(\lambda)| \\ &\le |\lambda|\,|B^n_{s_n}| + \left| R^{n,\delta}_{s_n}(\lambda) + \frac{\lambda^2}{2}\langle M^{n,\delta}\rangle_{s_n} \right| + \frac{\lambda^2}{2}\langle M^{n,\delta}\rangle_{s_n} \\ &\quad + |Q^{n,\delta}_{s_n}(\lambda) - \tilde{Q}^{n,\delta}_{s_n}(\lambda)| + |\tilde{Q}^{n,\delta}_{s_n}(\lambda)|. \end{aligned} \tag{4.39}$$

By (f) we have

$$|B^n_{s_n}| \xrightarrow{\mathsf{P}} 0. \tag{4.40}$$

By (4.33) we have

$$\left| R^{n,\delta}_{s_n}(\lambda) + \frac{\lambda^2}{2}\langle M^{n,\delta}\rangle_{s_n} \right| \le \frac{|\lambda^3|\,\delta}{3}\langle M^{n,\delta}\rangle_{s_n}. \tag{4.41}$$

By (4.34) we have

$$|\tilde{Q}^{n,\delta}_{s_n}(\lambda)| \le (2+|\lambda|)\, I(|x|>\delta) * \nu^n_{s_n}$$

and hence

$$|\tilde{Q}^{n,\delta}_{s_n}(\lambda)| \le (2+|\lambda|)\, g * \nu^n_{s_n}, \tag{4.42}$$

where $g = g(x)$ is a function satisfying Condition (f) and such that $g(x) \ge I(|x|>\delta)$.

By (4.35) we have

$$|Q^{n,\delta}_{s_n}(\lambda) - \tilde{Q}^{n,\delta}_{s_n}(\lambda)| \le 2|\lambda| \sup_{s \le s_n} |\hat{x}^{n,\delta}_s| \, \nu^n((0, s_n] \times \{|x| > \delta\})$$

and consequently (for the notation $\hat{x}^{n,\delta}_s$ see Lemma 1),

$$|Q^{n,\delta}_{s_n}(\lambda) - \tilde{Q}^{n,\delta}_{s_n}(\lambda)| \le 2|\lambda| \delta g * \nu^n_{s_n}, \tag{4.43}$$

where $g = g(x)$ is a function satisfying Condition (f) and such that $g(x) = I(|x| > \delta)$ $g(x)$.

By (4.39) - (4.43) we get the desired implication (4.38).

Remark 1 has been established in the course of proving Theorem 1.

Let us show now the validity of Remark 2. We assume $n \ge n_0$ and $s_{n_0} \le a$. Then

$$|B^n_{s_n}| \le \sup_{s \le a} |B^n_s - B_s| + \sup_{s \le s_n} |B_s| \xrightarrow{P} 0$$

by Condition (b') and the fact that $B \in \mathcal{V}$. Next, let $\varepsilon_k \in (0, a] \cap S$, $k \ge 1$ and $\varepsilon_k \downarrow 0$ as $k \to \infty$. Then by setting $n \ge n_k$ with $s_{n_k} \le \varepsilon_k$, we get

$$\langle M^{n,\delta}\rangle_{s_n} \le \langle M^{n,\delta}\rangle_{\varepsilon_k} \le |\langle M^{n,\delta}\rangle_{\varepsilon_k} - C_{\varepsilon_k}| + C_{\varepsilon_k}.$$

By Condition (c) and the fact that $C \in \mathcal{V}^+$, this gives (as $b > 0$)

$$\lim_{\delta \to 0} \overline{\lim_{n}} \, P(\langle M^{n,\delta}\rangle_{s_n} \ge b)$$

$$\le \lim_{\delta \to 0} \overline{\lim_{n}} \, P\left(|\langle M^{n,\delta}\rangle_{\varepsilon_k} - C_{\varepsilon_k}| \ge \frac{b}{2}\right) + P\left(C_{\varepsilon_k} \ge \frac{b}{2}\right)$$

$$= P\left(C_{\varepsilon_k} \ge \frac{b}{2}\right) \to 0, \quad k \to \infty.$$

Finally, for a nonnegative function $g = g(x)$ satisfying Condition (a) (as $n \ge n_k$)

$$g * \nu^n_{s_n} \le g * \nu^n_{\varepsilon_k} \le |g * \nu^n_{\varepsilon_k} - g * \nu_{\varepsilon_k}| + g * \nu_{\varepsilon_k} \xrightarrow{P} 0$$

as the limit $\lim_k \overline{\lim_n}$ is taken, by Condition (a) and the fact that $g * \nu \in \mathcal{V}^+$.

Problems

1. Let $X^n = (X^n_t, \mathcal{F}^n_t)$, $n \ge 1$, be semimartingales with the triplets $T^n = (B^n, C^n, \nu^n)$, $n \ge 1$. Let the following conditions be fullfilled:

(a) for each continuous on $R_0 \setminus (\{-1\} \cup \{1\})$ and bounded function $g = g(x)$, $x \in R_0$, such that $g(x) = I(|x| > \varepsilon)\, g(x)$ for a certain $\varepsilon > 0$ the relation

$$g * \nu_t^n \xrightarrow{P} 0$$

takes place;

(b) $B_t^n \xrightarrow{P} 0$;

(c) $\lim_{\delta \to 0} \overline{\lim_n} \mathsf{P} (\langle M^{n,\delta} \rangle_t \geq a) = 0, \ a > 0$;

(e) $X_0^n \xrightarrow{d} 0$.

Show that for a given t we have $X_t^n \xrightarrow{P} 0$.

2. Prove Theorem 2.

§ 5. The central limit theorem. I. "Classical" version

1. In this section the "limiting" process $X = (X_t, \mathcal{F}_t)$ will be presented by a semimartingale with $X_0 = 0$ and $\mathcal{G}$-conditionally independent increments ($\mathcal{G} \subseteq \mathcal{F}_0$), and with the triplet $T = (B, C, \nu)$ which satisfies the conditions

$$B = 0, \quad \nu^c (R_+ \times R_0) = 0,$$

$$\int_{R_0} e^{i\lambda x} \nu(\{t\}, dx) = I(a_t > 0) \exp\left(-\frac{\lambda^2}{2} \int_{R_0} x^2 \nu(\{t\}, dx) \right), \quad \lambda \in R.$$

By Theorem 4.9.6, X is a locally square integrable martingale with $\mathcal{G}$-conditionally Gaussian increments, with the quadratic characteristic

$$\langle X \rangle_t = C_t + \sum_{0 < s \le t} \int_{R_0} x^2 \nu(\{s\}, dx) \tag{5.1}$$

and with the stochastic exponential

$$\mathcal{E}_t(G(\lambda)) = \exp\left(-\frac{\lambda^2}{2} \langle X \rangle_t \right). \tag{5.2}$$

If $\mathcal{G} = \{\emptyset, \Omega\}$ is a trivial σ-algebra, then X is a Gaussian locally square integrable martingale such that

$$E e^{i\lambda X_t} = e^{-\frac{\lambda^2}{2} \langle X \rangle_t}.$$

Let $X^n = (X^n_t, \mathcal{F}^n_t)$, $n \ge 1$, be semimartingales with the triplets $T^n = (B^n, C^n, \nu^n)$, $n \ge 1$ and $M^{n,\delta} = (M^{n,\delta}_t, \mathcal{F}^n_t)$, $n \ge 1$, a locally square integrable martingale with

$$M^{n,\delta}_t = X^{nc}_t + \int_0^t \int_{|x| \le \delta} x \, d(\mu^n - \nu^n), \quad \delta \in (0, 1], \tag{5.3}$$

and

$$\langle M^{n,\delta} \rangle_t = C^n_t + \int_0^t \int_{|x| \le \delta} x^2 \, d\nu^n - \sum_{0 < s \le t} (\hat{x}^{n,\delta}_s)^2 \tag{5.4}$$

where

$$\hat{x}^{n,\delta}_s = \int_{|x| \le \delta} x \nu^n (\{s\}, dx). \tag{5.5}$$

Theorem 1. *Let S be a nonempty subset of* R_+ *and let the following conditions be satisfied:*

(o) $\mathcal{G} \subseteq \bigcap_{n \geq 1} \mathcal{F}_0^n$;

(a) $\nu^n((0, t] \times \{ |x| > \delta\}) \xrightarrow{P} 0, \ \delta \in (0, 1], \ t \in S$;

(b) $B_t^n \xrightarrow{P} 0, \ t \in S$;

(c) $\langle M^{n, \delta}\rangle_t \xrightarrow{P} \langle X\rangle_t, \ \delta \in (0, 1], \ t \in S$;

(e) $X_0^n \xrightarrow{P} 0$.

Then

$$X^n \xrightarrow{d_f(S)} X \quad (\mathcal{G}\textit{-stably}).$$

Remark. Condition (a) of Theorem 1 is equivalent to the "classical" condition of uniform asymptotic negligibility of jumps:

$$\text{(a*)} \qquad \sup_{s \leq t} |\Delta X_s^n| \xrightarrow{P} 0, \ t \in S$$

(see Lemma 1 below). Just the presence of this condition is the reason for giving to the present section the subtitle: "classical" version.

2. Proof of Theorem 1. Assume first the condition $\nu^c(R_+ \times R_0) = 0$ is replaced by the stronger condition $\nu(R_+ \times R_0) = 0$. In this case $\langle X\rangle = C$ and the assertion of Theorem 1 follows from Theorem 4.1.

To prove Theorem 1 in the general situation (i.e. under the condition $\nu^c(R_+ \times R_0) = 0$) we utilize Theorem 2.1.

Thus let us show that Conditions (o), (a), (b), (c) and (e) of Theorem 1 imply Conditions 1) - 4) of Theorem 2.1.

Obviously (o) $\Rightarrow$ 1) and (e) $\Rightarrow$ 4). Condition 3) of Theorem 2.1 is fullfilled in view of the representation (5.2) for a stochastic exponential.

Therefore it remains to verify that the implication

$$\text{(a), (b), (c)} \Rightarrow \mathcal{E}_t(G^n(\lambda)) \xrightarrow{P} e^{-\frac{\lambda^2}{2}\langle X\rangle_t}, \ \lambda \in R, \ t \in S, \tag{5.6}$$

takes place, where $\mathcal{E}(G^n(\lambda))$ is the stochastic exponential related to the triplet $T^n = (B^n, C^n, \nu^n)$.

The proof of the relation ($\lambda \in R, \ t \in S$)

$$\mathcal{E}_t(G^n(\lambda)) \xrightarrow{P} e^{-\frac{\lambda^2}{2}\langle X\rangle_t}, \quad n \to \infty,$$

is analogous to the proof of the relation $\mathcal{E}_t(G^n(\lambda)) \xrightarrow{P} \mathcal{E}_t(G(\lambda))$, $n \to \infty$, in Theorem 4.1. Therefore, we utilize the decomposition

$$\mathcal{E}_t(G^n(\lambda)) = \mathcal{E}_t(G_t^{n,\delta}(\lambda))\, e^{i\lambda \Delta_t^{n,\delta}} \tag{5.7}$$

in which, according to Lemma 4.1,

$$\Delta_t^{n,\delta} = \sum_{0<s\le t} \hat{x}_s^{n,\delta}, \quad \hat{x}_s^{n,\delta} = \int_{|x|\le\delta} x\nu^n(\{s\}, dx)$$

and

$$G_t^{n,\delta}(\lambda) = i\lambda(B_t^n - \Delta_t^{n,\delta}) - \frac{\lambda^2}{2}C_t^n$$

$$+ \int_0^t \int_{R_0} [e^{i\lambda(x - \hat{x}_s^{n,\delta})} - 1 - i\lambda(xI(|x|\le 1) - \hat{x}_s^{n,\delta})]\, d\nu^n$$

$$+ \sum_{0<s\le t} (e^{-i\lambda \hat{x}_s^{n,\delta}} - 1 + i\lambda \hat{x}_s^{n,\delta})(1 - a_s^n).$$

Taking into consideration (5.7) and the definition of the stochastic exponential, we have

$$\mathcal{E}_t(G^n(\lambda)) =$$

$$\exp(G_t^{n,\delta}(\lambda) + i\lambda\Delta_t^{n,\delta}) \prod_{0<s\le t} (1 + \Delta G_s^{n,\delta}(\lambda))\, e^{-\Delta G_s^{n,\delta}(\lambda)}. \tag{5.8}$$

Therefore it suffices to show that ($\lambda \in R$, $t \in S$)

$$\lim_{\delta\to 0} \overline{\lim_n} P\left(|G_t^{n,\delta}(\lambda) + i\lambda\Delta_t^{n,\delta} + \frac{\lambda^2}{2}\langle X_t\rangle| \ge a \right) = 0, \quad a > 0,$$

$$\prod_{0<s\le t} (1 + \Delta G_s^{n,\delta}(\lambda))\, e^{-\Delta G_s^{n,\delta}(\lambda)} \xrightarrow{P} 1, \quad \delta \in (0, 1]. \tag{5.9}$$

Observe first that by the inequality

$$\sup_{s\le t} \nu^n(\{s\} \times \{|x| > \delta\}) \le \nu^n((0, t] \times \{|x| > \delta\})$$

Condition (a) of the present theorem implies Condition (d) of Theorem 4.1. Consequently, by Lemma 4.4 we have

$$\sup_{s \le t} |\Delta G_s^{n,\delta}(\lambda)| \xrightarrow{P} 0, \quad \delta \in (0, 1], \quad \lambda \in R, \quad t \in S. \tag{5.10}$$

Define the set

$$\Gamma_t^{n,\delta} = \left\{ \sup_{s \le t} |\Delta G_s^{n,\delta}(\lambda)| \ge \frac{1}{2} \right\}$$

and denote

$$J_s^{n,\delta} = \prod_{0 < u \le s} (1 + \Delta G_u^{n,\delta}(\lambda))\, e^{-\Delta G_u^{n,\delta}(\lambda)} - 1.$$

By (5.10)

$$\lim_n P(\Gamma_t^{n,\delta}) = 0, \tag{5.11}$$

and on the set $\Omega \setminus \Gamma_t^{n,\delta}$ the following estimate for $J_s^{n,\delta}$ takes place

$$\sup_{s \le t} |J_s^{n,\delta}| \le \exp\left(\sup_{u \le t} |\Delta G_u^{n,\delta}(\lambda)| \sum_{0 < u \le t} |\Delta G_u^{n,\delta}(\lambda)| \right) - 1, \tag{5.12}$$

which has been established in the course of proving Lemma 4.5.

Since by Lemma 4.3

$$\sum_{0 < u \le t} |\Delta G_u^{n,\delta}(\lambda)| \le \frac{\lambda^2}{2} \langle M^{n,\delta} \rangle_t + 4\, I(|x| > \delta) * \nu_t^n,$$

from (5.10) and (5.12) is deduced, by Conditions (a) and (c) of Theorem 1, that ($\lambda \in R$, $t \in S$, $\delta \in (0, 1]$)

$$\lim_n P\left(\sup_{s \le t} |J_s^{n,\delta}| \ge a,\ \Omega \setminus \Gamma_t^{n,\delta} \right) = 0, \quad a > 0.$$

This and (5.11) entail the second relation in (5.9).

To establish the first relation in (5.9), let us utilize the decomposition (4.29) of the previous section:

$$G_t^{n,\delta}(\lambda) + i\lambda\, \Delta_t^{n,\delta} = i\lambda\, B_t^n + R_t^{n,\delta}(\lambda) + Q_t^{n,\delta}(\lambda)$$

where $R_t^{n,\delta}(\lambda)$ and $Q_t^{n,\delta}(\lambda)$ are defined by the formulas (4.27) and (4.28). By Condition (b) we have $B_t^n \xrightarrow{P} 0$, $t \in S$. By (4.28) $|Q_t^{n,\delta}(\lambda)|$ admits the estimate

$$|Q_t^{n,\delta}(\lambda)| \le (4 + 3|\lambda|)\, \nu^n((0, t] \times \{|x| > \delta\}),$$

and hence by Condition (a) of Theorem 1 we have $Q_t^{n,\delta}(\lambda) \xrightarrow{P} 0$. Next, due to (4.33)

$$\left| R_t^{n,\delta}(\lambda) + \frac{\lambda^2}{2} \langle M^{n,\delta} \rangle_t \right| \le \frac{|\lambda^3|}{3}\, \delta\, \langle M^{n,\delta} \rangle_t,$$

and hence

$$\left| R_t^{n,\delta}(\lambda) + \frac{\lambda^2}{2}\langle X\rangle_t \right|$$

$$\leq \frac{|\lambda^3|}{3}\,\delta\,|\langle M^{n,\delta}\rangle_t - \langle X\rangle_t| + \frac{|\lambda^3|}{3}\,\delta\,\langle X\rangle_t + \frac{\lambda^2}{2}\,|\langle M^{n,\delta}\rangle_t - \langle X\rangle_t|.$$

This implies the first relation in (5.9) in an obvious manner.

3. The generalization of Theorem 1 on the case of semimartingales $\hat{X}^n = (\hat{X}_t^n, \hat{\mathcal{F}}_t^n)$, $n \geq 1$, with $\hat{X}_t^n = X^n_{\hat{\tau}_n(t)}$, $\hat{\mathcal{F}}_t^n = \mathcal{F}^n_{\hat{\tau}_n(t)}$, where $\hat{\tau}_n = (\hat{\tau}_n(t))_{t \geq 0}$ is a random change of time, $n \geq 1$ is given in

Theorem 2. *Let* S *be a nonempty subset of* R_+, *and let the following conditions be fullfilled:*

(o) $\mathcal{G} \subseteq \bigcap_{n \geq 1} \mathcal{F}_0^n$;

($\hat{a}$) $\nu^n((0, \hat{\tau}_n(t)] \times \{|x| > \delta\}) \xrightarrow{P} 0, \ \ \delta \in (0, 1], \ \ t \in S$;

($\hat{b}$) $B^{(n)}_{\hat{\tau}_n(t)} \xrightarrow{P} 0, \ \ t \in S$;

($\hat{c}$) $\langle M^{n,\delta}\rangle_{\hat{\tau}_n(t)} \xrightarrow{P} \langle X\rangle_t, \ \ t \in S$;

($\hat{e}$) $\hat{X}_0^n \xrightarrow{P} 0$.

Then

$$\hat{X}^n \xrightarrow{d_f(S)} X \quad (\mathcal{G}\text{-stably}).$$

The proof of this theorem is left to the reader.

4. Consider a version of conditions for the convergence

$$X^n \xrightarrow{d_f(S)} X,$$

when the family of σ-algebras $\mathbb{F}^n$, $n \geq 1$, is nested (see § 1).

Theorem 3. *Let* S *be a nonempty subset of* R_+, *and let the following conditions be fullfilled:*

(o) *the family of σ-algebras* $\mathbb{F}^n$, $n \geq 1$, *is nested, i.e. there exists a sequence of numbers* $(s_n)_{n \geq 1}$, $s_n \downarrow 0$, $n \to \infty$, *such that*

$$\mathcal{F}^{n}_{s_n} \subseteq \mathcal{F}^{n+1}_{s_{n+1}},\ n \geq 1, \sigma\big(\bigcup_{n\geq 1} \mathcal{F}^{n}_{s_n}\big) = \sigma\big(\bigcup_{n\geq 1} \mathcal{F}^{n}_{\infty}\big)$$

and

$$\mathcal{G} \subseteq \sigma\big(\bigcup_{n \geq 1} \mathcal{F}^{n}_{\infty}\big);$$

(a) $\nu^n((0, t] \times \{|x| > \delta\}) \xrightarrow{P} 0,\ \delta \in (0, 1],\ t \in S;$

(b) $B^n_t \xrightarrow{P} 0,\ t \in S;$

(c) $\langle M^{n,\delta}\rangle_t \xrightarrow{P} \langle X\rangle_t,\ \delta \in (0, 1],\ t \in S;$

(e) $X^n_0 \xrightarrow{P} 0;$

(f) $B^n_{s_n} \xrightarrow{P} 0,\ \ \lim_{\delta \to 0} \overline{\lim_{n}}\ P(\langle M^{n,\delta}\rangle_{s_n} \geq a) = 0,\ a > 0.$

Then

$$X^n \xrightarrow{d_f(S)} X\ \ (\mathcal{G}\text{-}stably).$$

Remarks. 1. Condition (a) implies the condition

$$\nu^n((0, s_n] \times \{|x| > \delta\}) \xrightarrow{P} 0,\ \delta \in (0, 1].$$

2. If S is a subset dense in the interval [0, a] for a certain a > 0, then Condition (f) is a consequence of Condition (c) and

(b') $$\sup_{t \leq a} |B^n_t| \xrightarrow{P} 0.$$

The proof of Theorem 3 is analogous to the proof of Theorem 4.3

5. Let us establish now the relation between Condition (a) for semimartingales $X^n = (X^n_t, \mathcal{F}^n_t)$,, $n \geq 1$, and the "classical" condition of uniform asymptotic negligibility of jumps ΔX^n, $n \geq 1$.

Lemma 1. *The the condition*

(a) $$\nu^n((0, t] \times \{|x| > \delta\}) \xrightarrow{P} 0,\ \delta \in (0, 1],\ t \in S,$$

is equivalent to the condition

(a*) $$\sup_{s \leq t} |\Delta X^n_s| \xrightarrow{P} 0,\ t \in S.$$

Remark. If every semimartingale X^n is given on "its own" probability space $(\Omega^n, \mathcal{F}^n, P^n)$, then the obvious reformulation of Lemma 1 is presented by the following assertion: the condition

(a) $$\lim_n P^n(\nu^n((0,t]\times\{|x|>\delta\})\ge\varepsilon)=0,\ \ \varepsilon>0,\ \ \delta\in(0,1],\ \ t\in S$$

is equivalent to the condition

(a*) $$\lim_n P^n(\sup_{s\le t}|\Delta X^n_s|\ge\varepsilon)=0,\ \ \varepsilon>0,\ \ t\in S.$$

Proof of Lemma 1. (a) ⇒ (a*). We start with the proof of the implication

$$(a)\Rightarrow\sum_{0<s\le t}|\Delta X^n_s|^j I(|\Delta X^n_s|>\delta)\xrightarrow{P}0,\ \ \delta\in(0,1],\ \ j=1,2,\ \ t\in S. \quad (5.13)$$

We have

$$\left\{\sum_{0<s\le t}|\Delta X^n_s|^j I(|\Delta X^n_s|>\delta)\ge a\right\}\subseteq\left\{\sum_{0<s\le t}I(|\Delta X^n_s|>\delta)\ge 1\right\},$$

and consequently it suffices to show that

$$(a)\Rightarrow I(|x|>\delta)*\mu^n_t\to 0,\ \ \delta\in(0,1],\ \ t\in S, \quad (5.14)$$

where $\mu^n=\mu^n(dt,dx)$ is the jump measure of X^n.

Since $I(|x|>\delta)*\nu^n$ is the compensator of the process $I(|x|>\delta)*\mu^n$ and since by Condition (a) we have

$$I(|x|>\delta)*\nu^n_t\xrightarrow{P}0,\ \ \delta\in(0,1],\ \ t\in S,$$

then

$$I(|x|>\delta)*\mu^n_t\xrightarrow{P}0$$

by Problem 1.9.4.

Now, the desired assertion (a) ⇒ (a*) is implied by (5.13) and the inequality

$$\sup_{s\le t}|\Delta X^n_s|\le\delta+\sum_{0<s\le t}|\Delta X^n_s|I(|\Delta X^n_s|>\delta),$$

according to which for $b\ge 2\delta$, $\delta\in(0,1]$ and $t\in S$ we have

$$\overline{\lim_n}\, P(\sup_{s\le t}|\Delta X^n_s|\ge b)\le\lim_n P\left\{\sum_{0<s\le t}|\Delta X^n_s|I(|\Delta X^n_s|>\delta)\ge\delta\right\}=0.$$

(a*) ⇒ (a). We utilize the fact that

$$\{\sup_{s\le t}|\Delta X^n_s|>\delta\}=\{I(|x|>\delta)*\mu^n_t\ge 1\}$$

$$=\{I(|x|>\delta)*\mu^n_t\ge b\}$$

with $b\in(0,1]$. Then

$$(a^*)\Rightarrow I(|x|>\delta)*\mu^n_t\xrightarrow{P}0,\ \ \delta\in(0,1],\ \ t\in S,$$

and the desired assertion follows from this by Problem 1.9.4.

The Lemma is proved.

6. Denote

$$V_t^n = C_t^n + \sum_{0<s\le t} (\Delta X_s^n - \hat{x}_s^{n,1})^2 \tag{5.15}$$

where $\hat{x}_s^{n,1}$ is given by the formula (5.5) with $\delta = 1$, and consider the local square integrable martingale $M^{n,\delta}$, defined by (5.3). Clearly

$$[M^{n,\delta}, M^{n,\delta}]_t = C_t^n + \sum_{0<s\le t} (\Delta X_s^n I(|\Delta X_s^n| \le \delta) - \hat{x}_s^{n,\delta})^2. \tag{5.16}$$

Lemma 2. *Under Condition* (a) *(or, equivalently, Condition (a*)) the following relations take place*

$$\sup_{s\le t} |V_s^n - [M^{n,1}, M^{n,1}]_s| \xrightarrow{P} 0, \quad t \in S, \tag{5.17}$$

$$\sup_{s\le t} |[M^{n,1}, M^{n,1}]_s - [M^{n,\delta}, M^{n,\delta}]_s| \xrightarrow{P} 0, \quad \delta \in (0,1], \quad t \in S, \tag{5.18}$$

$$\sup_{s\le t} |\langle M^{n,\delta}\rangle_s - \langle M^{n,\varepsilon}\rangle_s| \xrightarrow{P} 0, \quad \delta \in (0,1], \ \varepsilon \in (0,\delta), \ t \in S. \tag{5.19}$$

Proof. The relation (5.17) holds due to the implication (5.13) and estimate

$$\sup_{s\le t} |V_s^n - [M^{n,1}, M^{n,1}]_s| \le \sum_{0<s\le t} I(|\Delta X_s^n| > 1)\,[(\Delta X_s^n)^2 + 2|\Delta X_s^n|],$$

and the relation (5.18) holds due to the same implication (5.13) and estimate

$$\sup_{s\le t} |[M^{n,1}, M^{n,1}]_s - [M^{n,\delta}, M^{n,\delta}]_s|$$

$$\le 4\left(\sum_{0<s\le t} I(|\Delta X_s^n| > \delta)\,|\Delta X_s^n| + \sum_{0<s\le t} \nu^n(\{s\}\times\{|x|>\delta\})\right)$$

(observe that

$$\sum_{0<s\le t} \nu^n(\{s\}\times\{|x|>\delta\}) \le \nu^n((0,t]\times\{|x|>\delta\}) \xrightarrow{P} 0$$

by Condition (a)).

In accordance with the expression for $\langle M^{n,\delta}\rangle$ (see (5.4)), for $0 < \varepsilon < \delta \le 1$ we get

$$\sup_{s \le t} |\langle M^{n,\delta}\rangle_s - \langle M^{n,\varepsilon}\rangle_s|$$

$$\le \delta^2 \nu^n((0,t] \times \{|x| > \varepsilon\}) + (\delta + \varepsilon)\,\delta \sum_{0 < s \le t} \nu^n(\{s\} \times \{|x| > \varepsilon\})$$

$$\le 3\nu^n((0,t] \times \{|x| > \varepsilon\}) \xrightarrow{P} 0$$

by Condition (a), and this proves (5.19).

Lemma 3. *Let Condition* (a) *be fullfilled (or, equivalently, Condition* (a*)), *as well as one of the following conditions:*

(c) $$\langle M^{n,\delta}\rangle_t \xrightarrow{P} \langle X\rangle_t, \quad \delta \in (0,1], \quad t \in S,$$

or

(c*) $$V_t^n \xrightarrow{P} \langle X\rangle_t, \quad t \in S.$$

Then

$$\sup_{s \le t} |V_s^n - \langle M^{n,\delta}\rangle_s| \xrightarrow{P} 0, \quad \delta \in (0,1], \quad t \in S. \tag{5.20}$$

Corollary. *Under the assumptions of Lemma 3 Conditions (c) and (c*) are equivalent.*

Proof. We will show first that under any of Conditions (c) or (c*) the following relations take place:

$$\lim_{b \to \infty} \overline{\lim_{\delta \to 0}}\, \overline{\lim_{n}}\, P(\langle M^{n,\delta}\rangle_t \ge b) = 0, \quad t \in S, \tag{5.21}$$

$$\lim_{b \to \infty} \overline{\lim_{\delta \to 0}}\, \overline{\lim_{n}}\, P([M^{n,\delta}, M^{n,\delta}]_t \ge b) = 0, \quad t \in S. \tag{5.22}$$

In fact, if Condition (c) is fullfilled, then by Condition (c)

$$P(\langle M^{n,\delta}\rangle_t \ge b) \le P\left(\langle M^{n,\delta}\rangle_t \ge b, \langle X\rangle_t < \frac{b}{2}\right) + P\left(\langle X\rangle_t \ge \frac{b}{2}\right)$$

$$\le P\left(|\langle M^{n,\delta}\rangle_t - \langle X\rangle_t| \ge \frac{b}{2}\right) + P\left(\langle X_t\rangle \ge \frac{b}{2}\right) \to 0$$

as the required limit $\lim_{b \to \infty} \overline{\lim_{\delta \to 0}}\, \overline{\lim_{n}}$ is taken, and hence (5.21) takes place. Next, by Theorem 1.8.1 $\langle M^{n,\delta}\rangle$ is the compensator of the process $[M^{n,\delta}, M^{n,\delta}]$. Therefore by the Lenglart-Rebolledo inequality (Theorem 1.9.3)

$$P([M^{n,\delta}, M^{n,\delta}]_t \geq b) \leq \frac{1}{\sqrt{b}} + P(\langle M^{n,\delta}\rangle_t \geq \sqrt{b}),$$

and consequently (5.22) follows from the relation (5.21) which is already proved (under Condition (c)).

If Condition (c*) is fullfilled, then from (5.17) and (5.18) it follows that

$$[M^{n,\delta}, M^{n,\delta}]_t \xrightarrow{P} \langle X_t\rangle, \quad \delta \in (0, 1], \quad t \in S.$$

Therefore the proof of (5.22) under Condition (c*) is analogous to the proof of (5.21) under Condition (c). Next, observe that

$$\Delta [M^{n,\delta}, M^{n,\delta}] \leq 2\delta^2$$

and according to the Lenglart-Rebolledo inequality (Theorem 1.9.3)

$$P(\langle M^{n,\delta}\rangle_t \geq b) \leq \frac{\sqrt{b} + 2\delta^2}{b} + P([M^{n,\delta}, M^{n,\delta}]_t \geq \sqrt{b}).$$

Consequently, under Condition (c*) (5.21) follows from (5.22).

Let us establish now the relation (5.20). To this end, let us utilize the inequality

$$\sup_{s \leq t} | V_s^n - \langle M^{n,\delta}\rangle_s | \leq \sup_{s \leq t} | V_s^n - [M^{n,1}, M^{n,1}]_s |$$

$$+ \sup_{s \leq t} | [M^{n,1}, M^{n,1}]_s - [M^{n,\varepsilon}, M^{n,\varepsilon}]_s |$$

$$+ \sup_{s \leq t} | [M^{n,\varepsilon}, M^{n,\varepsilon}]_s - \langle M^{n,\varepsilon}\rangle_s | + \sup_{s \leq t} | \langle M^{n,\varepsilon}\rangle_s - \langle M^{n,\delta}\rangle_s |$$

with $\varepsilon \in (0, \delta)$. By Lemma 2 and this inequality the desired relation (5.20) takes place, provided

$$\lim_{\varepsilon \to 0} \overline{\lim_n} P(\sup_{s \leq t} | [M^{n,\varepsilon}, M^{n,\varepsilon}]_s - \langle M^{n,\varepsilon}\rangle_s | \geq b) = 0, \quad b > 0, \quad t \in S. \tag{5.23}$$

Denote

$$Y_t^{n,\varepsilon} = [M^{n,\varepsilon}, M^{n,\varepsilon}]_t - \langle M^{n,\varepsilon}\rangle_t. \tag{5.24}$$

According to Problem 1.8.11 the process $Y^{n,\varepsilon} = (Y_t^{n,\varepsilon}, \mathcal{F}_t^n)$ belongs to the class $\mathcal{M}_{loc}^d$, while the definition of $Y^{n,\varepsilon}$ yields

$$| \Delta Y^{n,\varepsilon} | \leq \Delta [M^{n,\varepsilon}, M^{n,\varepsilon}] + \Delta \langle M^{n,\varepsilon}\rangle \leq 5\varepsilon^2.$$

From this it follows that the process $Y^{n,\varepsilon}$ belongs to the class $\mathcal{M}_{loc}^{2,d}$ (Problem 1.5.5), besides its quadratic characteristic possesses the following properties:

$$\Delta\,[Y^{n,\varepsilon}, Y^{n,\varepsilon}] \le 25\varepsilon^4,$$
$$[Y^{n,\varepsilon}, Y^{n,\varepsilon}] \le 5\varepsilon^2\,([M^{n,\varepsilon}, M^{n,\varepsilon}] + \langle M^{n,\varepsilon}\rangle). \tag{5.25}$$

Therefore under Conditions (c) or (c*) by (5.21) and (5.22) we get

$$\lim_{\varepsilon\to 0}\ \overline{\lim_n}\ \mathsf{P}\,([Y^{n,\varepsilon}, Y^{n,\varepsilon}]_t \ge b) = 0. \tag{5.26}$$

For each stopping time $\tau \in \mathbf{T}\,(\mathbb{F}^n)$ (Problem 1.8.6)

$$\mathbf{E}\,(Y_\tau^{n,\varepsilon})^2 \le \mathbf{E}\,[Y^{n,\varepsilon}, Y^{n,\varepsilon}]_\tau.$$

Therefore by the Lenglart-Rebolledo inequality (Theorem 1.9.3) and by the estimate (5.25) we get

$$\mathsf{P}\,(\sup_{s\le t} |Y_s^{n,\varepsilon}| \ge b) \le \frac{a + 25\varepsilon^4}{b^2} + \mathsf{P}\,([Y^{n,\varepsilon}, Y^{n,\varepsilon}]_t \ge a),\quad a > 0,\quad b > 0.$$

As $t \in S$ this gives

$$\overline{\lim_{\varepsilon\to 0}}\ \overline{\lim_n}\ \mathsf{P}\,(\sup_{s\le t} |Y_s^{n,\varepsilon}| \ge b) \le \frac{a}{b^2} \to 0,\quad a\to 0,$$

i.e. (see (5.24)) the desired relation (5.23) holds.

The lemma is proved.

7. Obviously, Condition (a) of Theorem 1 is fullfilled under any of the conditions

$$(L_1)\qquad |x|\, I(|x| > \delta) * \nu_t^n \xrightarrow{\mathsf{P}} 0,\quad \delta\in(0,1],\ t\in S, \tag{5.27}$$

$$(L_2)\qquad x^2\, I(|x| > \delta) * \nu_t^n \xrightarrow{\mathsf{P}} 0,\quad \delta\in(0,1],\ t\in S \tag{5.28}$$

(It is natural to call Condition (L_2) the functional Lindberg condition.) Clearly,

$$(L_2) \Rightarrow (L_1). \tag{5.29}$$

Under Condition (a) in Lemma 3 the equivalence of Conditions (c) and (c*) has been established. Let us give now other conditions equivalent to Condition (c) under assumptions (L_1) and (L_2).

Lemma 4. *Let processes* $X^n = (X_t^n, \mathcal{F}_t^n)$ *belong to the class* $\mathcal{M}_{loc}$, $n > 1$. *Under assumption* (L_1) *(see 5.27)) let any of the following conditions be fullfilled:*

$$(c)\qquad \langle M^{n,\delta}\rangle_t \xrightarrow{\mathsf{P}} \langle X\rangle_t,\quad \delta\in(0,1],\ t\in S,$$

or

$$(c_1)\qquad [X^n, X^n]_t \xrightarrow{\mathsf{P}} \langle X\rangle_t,\quad t\in S.$$

Then

$$\sup_{s \le t} |\langle M^{n,\delta}\rangle_s - [X^n, X^n]_s| \xrightarrow{P} 0, \ \ \delta \in (0, 1], \ t \in S. \tag{5.30}$$

Corollary. *Under assumption* (L_1) *Conditions* (c) and (c_1) *are equivalent.*

Lemma 5. *Let processes* $X^n = (X^n_t, \mathcal{F}^n_t)$ *belong to the class* $\mathcal{M}^2_{loc}$, $n \ge 1$. *Under assumption* (L_2) *(see* (5.28)*) let any of the following conditions be fullfilled:*

(c_1)
$$[X^n, X^n]_t \xrightarrow{P} \langle X\rangle_t, \ t \in S,$$

or

(c_2)
$$\langle X^n\rangle_t \xrightarrow{P} \langle X\rangle_t, \ t \in S.$$

Then

$$\sup_{s \le t} |[X^n, X^n]_t - \langle X^n\rangle_t| \xrightarrow{P} 0, \ t \in S. \tag{5.31}$$

Corollary. *Under assumption* (L_2) *Conditions* (c), (c_1) *and* (c_2) *are equivalent.*

Proof of Lemma 4. Let us show first that

$$\sup_{s \le t} |V^n_s - [X^n, X^n]_s| \xrightarrow{P} 0, \ t \in S \tag{5.32}$$

with V^n_t defined by the formula (5.15).

We have

$$\sup_{s \le t} |V^n_s - [X^n, X^n]_s| \le \sum_{0 < s \le t} |\hat{x}^{n,1}_s (2\Delta X^n_s - \hat{x}^{n,1}_s)|$$

$$\le \sum_{0 < s \le t} |\hat{x}^{n,1}_s| (3 + I(|\Delta X^n_s| > 1) |\Delta X^n_s|). \tag{5.33}$$

According to Problem 3.5.2

$$\int_{R_0} x\nu^n(\{s\}, dx) = 0, \ s \in R_+.$$

Consequently,

$$\hat{x}^{n,1}_s = -\int_{|x|>1} x\nu^n(\{s\}, dx)$$

and hence, in accordance with (5.33), we have

$$\sup_{s \le t} |V^n_s - [X^n, X^n]_s|$$

$$\le 3|x| I(|x| > 1) * \nu^n_t + \sum_{0 < s \le t} I(|\Delta X^n_s| > 1) |\Delta X^n_s|. \tag{5.34}$$

Since $(L_1) \Rightarrow$ (a), the desired relation (5.32) follows from (5.34) by (L_1) and the implication (5.13).

Thus (L_1), $(c_1) \Rightarrow (c^*)$ with Condition (c^*) defined in Lemma 3. Therefore, under assumption (L_1) and any of the Conditions (c) or (c_1) the assumptions of Lemma 3 are fullfilled and hence the relation (5.20) takes place.

The desired relation (5.30) follows from (5.20) and (5.32) in an obvious manner.

Proof of Lemma 5. Let $\mu^n = \mu^n(dt, dx)$ be the jump measure of a process X^n. Let us show that under any of Conditions (c_1) and (c_2)

$$\lim_{\varepsilon \to 0} \overline{\lim_n}\ P(\varepsilon^2 x^2 I(|x| \le \varepsilon) * \nu_t^n \ge b) = 0,\ b > 0,\ t \in S, \tag{5.35}$$

$$\lim_{\varepsilon \to 0} \overline{\lim_n}\ P(\varepsilon^2 x^2 I(|x| \le \varepsilon) * \mu_t^n \ge b) = 0,\ b > 0,\ t \in S. \tag{5.36}$$

Let Condition (c_2) be fullfilled. Since

$$\langle X^n \rangle_t = C_t^n + x^2 * \nu_t^n$$

according to Problem 3.5.2, the inequality

$$\varepsilon^2 x^2 I(|x| \le \varepsilon) * \nu_t^n \le \varepsilon^2 \langle X^n \rangle_t$$

takes place and (5.35) follows, as the limit $\lim_{\varepsilon \to 0} \overline{\lim_n}$ is taken, from the following row of inequalities:

$$P(\varepsilon^2 x^2 I(|x| \le \varepsilon) * \nu_t^n \ge b) \le P\left(\langle X^n \rangle_t \ge \frac{b}{\varepsilon^2} \right)$$

$$\le P\left(\langle X \rangle_t \ge \frac{b}{\varepsilon^2},\ \langle X \rangle_t < \frac{b}{2\varepsilon^2} \right) + P\left(\langle X \rangle_t \ge \frac{b}{2\varepsilon^2} \right)$$

$$\le P\left(|\langle X^n \rangle_t - \langle X \rangle_t| \ge \frac{b}{2\varepsilon^2} \right) + P\left(\langle X \rangle_t \ge \frac{b}{2\varepsilon^2} \right).$$

Observe now that $\varepsilon^2 x^2 I(|x| \le \varepsilon) * \nu^n$ presents the compensator of the process $\varepsilon^2 x^2 I(|x| \le \varepsilon) * \mu^n$. Then by the Lenglart-Rebolledo inequality (Theorem 1.9.3) we get

$$P(\varepsilon^2 x^2 I(|x| \le \varepsilon) * \mu_t^n \ge b) \le \frac{a}{b} + P(\varepsilon^2 x^2 I(|x| \le \varepsilon) * \nu_t^n \ge a).$$

Consequently, by (5.35)

$$\overline{\lim_{\varepsilon \to 0}}\ \overline{\lim_n}\ P(\varepsilon^2 x^2 I(|x| \le \varepsilon) * \mu_t^n \ge b) \le \frac{a}{b} \to 0,\ a \to 0,$$

i.e. (5.36) holds.

If Condition (c_1) holds, then apply the inequality

$$\varepsilon^2 x^2 I(|x| \le \varepsilon) * \mu_t^n \le \varepsilon^2 \sum_{0<s\le t} (\Delta X_s^n)^2 I(|\Delta X_s^n| \le \varepsilon) \le \varepsilon^2 [X^n, X^n]_t,$$

in effect of which (5.36) is established under Condition (c_1) precisely in the same manner as (5.35) under Condition (c_2). Now, to establish (5.35), observe that jumps of the process $\varepsilon^2 x^2 I(|x| \le \varepsilon) * \mu^n$ do not exceed ε^4 and by the Lenglart-Rebolledo inequality (Theorem 1.9.3)

$$P(\varepsilon^2 x^2 I(|x| \le \varepsilon) * \nu_t^n \ge b) \le \frac{a + \varepsilon^4}{b} + P(\varepsilon^2 x^2 I(|x| \le \varepsilon) * \mu_t^n \ge a),$$
$$a > 0,\ b > 0.$$

Consequently, by (5.36)

$$\overline{\lim_{\varepsilon \to 0}}\ \overline{\lim_{n}}\ P(\varepsilon^2 x^2 I(|x| \le \varepsilon) * \nu_t^n \ge b) \le \frac{a}{b} \to 0,\ a \to 0,$$

i.e. (5.35) holds.

To prove the desired assertion (5.31) of the lemma apply the inequality

$$\sup_{s \le t} |[X^n, X^n]_s - \langle X^n \rangle_s|$$

$$\le \sum_{0<s\le t} (\Delta X_s^n)^2 I(|\Delta X_s^n| > \varepsilon) + x^2 I(|x| > \varepsilon) * \nu_t^n + \sup_{s \le t} |Y_s^{n,\varepsilon}| \qquad (5.37)$$

with

$$Y_t^{n,\varepsilon} = \sum_{0<s\le t} (\Delta X_s^n)^2 I(|\Delta X_s^n| \le \varepsilon) - x^2 I(|x| \le \varepsilon) * \nu_t^n.$$

Since (L_2) $\Rightarrow$ (a), the first term on the right-hand side of the inequality (5.37) converges in probability to zero as $n \to \infty$ for each $\varepsilon > 0$ by the implication (5.13). The second term on the right-hand side of the inequality (5.37) converges in probability to zero as $n \to \infty$ for each $\varepsilon > 0$ by Condition (L_2).

Therefore, to prove (5.31) it remains to verify that

$$\lim_{\varepsilon \to 0}\ \overline{\lim_{n}}\ P(\sup_{s \le t} |Y_s^{n,\varepsilon}| \ge b) = 0,\quad b > 0,\quad t \in S. \qquad (5.38)$$

Observe that the process $Y^{n,\varepsilon} = (Y_t^{n,\varepsilon}, \mathcal{F}_t)$ belongs to the class $\mathcal{M}_{loc}^d$ (Theorems 1.6.3, 1.6.4 and 1.7.3). Besides, by the estimate $|\Delta Y^{n,\varepsilon}| \le 2\varepsilon^2$ the process $Y^{n,\varepsilon}$ belongs to the class $\mathcal{M}_{loc}^{2,d}$ (Problem 1.5.5). Utilize now the fact that the process $Y^{n,\varepsilon}$ has the following representation

$$Y_t^{n,\varepsilon} = x^2 I(|x| \le \varepsilon) * (\mu^n - \nu^n)_t,$$

and its quadratic variation possesses the following properties:

$$\Delta [Y^{n,\varepsilon}, Y^{n,\varepsilon}] \le 4\varepsilon^4, \tag{5.39}$$

$$[Y^{n,\varepsilon}, Y^{n,\varepsilon}] \le 2 (\varepsilon^2 x^2 I(|x| \le \varepsilon) * \mu^n + \varepsilon^2 x^2 I(|x| \le \varepsilon) * \nu^n). \tag{5.40}$$

From the inequality (5.40) and the relations (5.35) and (5.36) it follows under Conditions (c_1) or (c_2) that

$$\lim_{\varepsilon \to 0} \overline{\lim_n}\, \mathsf{P}([Y^{n,\varepsilon}, Y^{n,\varepsilon}]_t \ge b) = 0, \quad b > 0, \quad t \in S. \tag{5.41}$$

For each stopping time $\tau \in \mathbf{T}(\mathbb{F}^n)$

$$\mathbf{E}(Y_\tau^{n,\varepsilon})^2 \le \mathbf{E}[Y^{n,\varepsilon}, Y^{n,\varepsilon}]_\tau$$

in virtue of Problem 1.8.6. Therefore by the Lenglart-Rebolledo inequality (Theorem 1.9.3) and by taking into account (5.39) we get

$$\mathsf{P}(\sup_{s \le t} |Y_s^{n,\varepsilon}| \le b) \le \frac{a + 4\varepsilon^4}{b^2} + \mathsf{P}([Y^{n,\varepsilon}, Y^{n,\varepsilon}]_t \ge a), \quad a > 0, \quad b > 0.$$

This gives

$$\lim_{\varepsilon \to 0} \overline{\lim_n}\, \mathsf{P}(\sup_{s \le t} |Y_s^{n,\varepsilon}| \ge b) \le \frac{a}{b^2} \to 0, \quad a > 0.$$

The lemma is proved.

8. Let us dwell on an interesting case in which

$$(c_1) \Rightarrow (L_1).$$

Lemma 6. *Let processes* $X^n = (X_t^n, \mathcal{F}_t^n)$ *belong to the class* $\mathcal{M}_{loc}$, $n \ge 1$, *let* $\langle X \rangle = C$, *let* S *be a subset dense in* R_+ *and let the following condition be fullfilled:*

(ρ) *the family of random variables* $(\sup_{s \le t} |\Delta X_s^n|)_{n \ge 1}$ *is uniformly integrable for each* $t \in R_+$.

Then

$$[X^n, X^n]_t \xrightarrow{\mathsf{P}} C_t, \quad t \in S \Rightarrow |x| I(|x| > \delta) * \nu_t^n \xrightarrow{\mathsf{P}} 0, \quad \delta \in (0, 1] \quad t \in R_+.$$

Proof. Since $C = (C_t)_{t \ge 0}$ is a continuous function in V^+ and S a subset dense in R_+, then by Problem 5.3.2

$$\sup_{s \le t} |[X^n, X^n]_s - C_s| \xrightarrow{\mathsf{P}} 0, \quad t \in R_+.$$

This and the estimate

$$\sup_{s \le t} \Delta [X^n, X^n]_s \le 2 \sup_{s \le t} |[X^n, X^n]_s - C_s|$$

entail

$$\sup_{s \le t} |\Delta X_s^n| \xrightarrow{P} 0, \quad t \in R_+, \tag{5.42}$$

i.e. for each $t \in R_+$ Condition (a*) of Lemma 1 is fullfilled. But then by Lemma 1 the condition

(a)
$$I(|x| > \delta) * \nu_t^n \xrightarrow{P} 0, \quad \delta \in (0, 1], \quad t \in R_+$$

is fullfilled and consequently, in accordance with the implication (5.13),

$$\sum_{0 < s \le t} |\Delta X_s^n| I(|\Delta X_s^n| > \delta) \xrightarrow{P} 0, \quad \delta \in (0, 1], \quad t \in R_+. \tag{5.43}$$

Observe now that

$$\sum_s |\Delta X_s^n| I(|\Delta X_s^n| > \delta) = |x| I(|x| > \delta) * \mu^n$$

where μ^n is the jump measure of the process X^n. By Condition (ρ) the process $|x| I(|x| > \delta) * \mu^n \in \mathcal{A}_{loc}^+$ (with the localizing sequence $\tau_k^n = \inf(t: |x| I(|x| > \delta) * \mu_t^n \ge k) \wedge k, \ k \ge 1$). Consequently $|x| I(|x| > \delta) * \nu^n$ presents the compensator of the process $|x| I(|x| > \delta) * \mu^n$ and by the Lenglart-Rebolledo inequality (Theorem 1.9.3)

$$P(|x| I(|x| > \delta) * \nu_t^n \ge b) \le \frac{a + E \sup_{s \le t} |\Delta X_s^n|}{b} + P(|x| I(|x| > \delta) * \mu_t^n \ge a).$$

From this and Condition (ρ) it follows that

$$\lim_n E \sup_{s \le t} |\Delta X_s^n| = 0. \tag{5.44}$$

By this and (5.43) and (5.44) we get

$$\overline{\lim_n}\ P(|x| I(|x| > \delta) * \nu_t^n \ge b) \le \frac{a}{b} \to 0, \quad a \to 0, \quad \delta \in (0, 1], \quad t \in S.$$

The lemma is proved.

9. Lemmas 1 - 6 allow us to establish a number of useful corollaries to Theorems 1 - 3 in case in which processes $X^n = (X_t^n, \mathcal{F}_t^n)$, $n \ge 1$, are local martingales.

To this end introduce the following conditions:

(o) $\mathcal{G} \subseteq \bigcap_{n \ge 1} \mathcal{F}_0^n$;

(L_1) $|x| I(|x| > \delta) * \nu_t^n \xrightarrow{P} 0, \quad \delta \in (0, 1], \quad t \in S$;

(L_2) $x^2 I(|x| > \delta) * \nu_t^n \xrightarrow{P} 0, \quad \delta \in (0, 1], \quad t \in S$;

(c_1) $[X^n, X^n]_t \xrightarrow{P} \langle X\rangle_t$, $t \in S$;

(c_2) $\langle X^n\rangle_t \xrightarrow{P} \langle X\rangle_t$, $t \in S$;

(ρ) the family of the random variables

$$(\sup_{s\le t} |\Delta X^n_s|)_{n\ge 1}$$

is uniformly integrable for each $t \in R_+$;

(f_1) $[X^n, X^n]_{s_n} \xrightarrow{P} 0$ where $(s_n)_{n\ge 1}$ is a sequence of numbers with $s_n \downarrow 0$;

(f_2) $\langle X^n\rangle_{s_n} \xrightarrow{P} 0$ where $(s_n)_{n\ge 1}$ is a sequence of numbers with $s_n \downarrow 0$.

Theorem 4. *Let any of the following sets of conditions be fullfilled:*

I: $X^n = (X^n_t, \mathcal{F}^n_t) \in \mathcal{M}_{loc,\,0}$, $n \ge 1$, S *is a nonempty subset of* R_+, (o), (L_1), (c_1);

II: $X^n = (X^n_t, \mathcal{F}^n_t) \in \mathcal{M}^2_{loc,\,0}$, $n \ge 1$, S *is a nonempty subset of* R_+, (o), (L_2), (c_1) *or* (c_2);

III: $X^n = (X^n_t, \mathcal{F}^n_t) \in \mathcal{M}_{loc,\,0}$, $n \ge 1$, S *is a nonempty subset of* R_+, $\langle X\rangle = C$, (o), (c_1), (ρ).

Then

$$X^n \xrightarrow{d_f(S)} X \quad (\mathcal{G}\text{-}stably).$$

Proof. It suffices to verify the validity of Conditions (o), (a), (b), (c) and (e) of Theorem 1.

Condition (o) is contained in I - III, and Condition (e) is fullfilled under I - III since $X^n_0 \equiv 0$. Condition (a) is fulfilled since under the conditions of set I (L_1) $\Rightarrow$ (a), of set II (L_2) $\Rightarrow$ (a), of set III (ρ), (c_1) $\Rightarrow$ (L_1) (Lemma 6) and (L_1) $\Rightarrow$ (a).

By the canonical representation for a local martingale we have

$$B^n_t = -\, xI(|x| > 1) * \nu^n_t$$

(Problem 4.1.10). Consequently

$$|B^n_t| \le |x| I(|x| > 1) * \nu^n_t, \tag{5.45}$$

and hence Condition (b) holds by Condition (L_1), which is satisfied under I, II, III ((L_2) $\Rightarrow$ (L_1); (o), (c_1) $\Rightarrow$ (L_1) by Lemma 6).

Condition (c) in set I is fullfilled by (c_1) and (5.30), in set II by implication (c_1),

$(\rho) \Rightarrow (L_1)$ (Lemma 6) and (5.30).

10. Theorem 5. *Let families of σ-algebras* $\mathbb{F}^n$, $n \geq 1$, *be nested, i.e. there exists a sequence of numbers* $(s_n)_{n \geq 1}$, $s_n \downarrow 0$, *such that*

$$\mathcal{F}^n_{s_n} \subseteq \mathcal{F}^{n+1}_{s_{n+1}}, \quad n \geq 1, \quad \sigma\left(\bigcup_{n \geq 1} \mathcal{F}^n_{s_n}\right) = \sigma\left(\bigcup_{n \geq 1} \mathcal{F}^n_{\infty}\right), \quad \mathcal{G} \subseteq \sigma\left(\bigcup_{n \geq 1} \mathcal{F}^n_{\infty}\right)$$

and let any of the following sets of conditions be fullfilled:

I: $X^n = (X^n_t, \mathcal{F}^n_t) \in \mathcal{M}_{\text{loc}, 0}$, $n \geq 1$, S *is a nonempty subset of* R_+, (L_1), (c_1), (f_1);

II: $X^n = (X^n_t, \mathcal{F}^n_t) \in \mathcal{M}^2_{\text{loc}, 0}$, $n \geq 1$, S *is a nonempty subset of* R_+, (L_2), (c_1) *or* (c_2), (f_1) *or* (f_2);

III: $X^n = (X^n_t, \mathcal{F}^n_t) \in \mathcal{M}_{\text{loc}, 0}$, $n \geq 1$, S *is a nonempty subset of* R_+, (c_1), (ρ), $\langle X \rangle = C$.

Then

$$X^n \xrightarrow{d_f(S)} X \quad (\mathcal{G}\text{-}stably).$$

Remark. If S contains a subset dense in the interval [0, a] for a certain a > 0, then

$$(c_1) \Rightarrow (f_1), \quad (c_2) \Rightarrow (f_2).$$

Proof. It suffuces to verify Conditions (a), (b), (c), (e) and (f) of Theorem 3.

Conditions (a), (b), (c) and (e) were verified in the course of proving Theorem 4. Let us verify Condition (f). To this end, observe that Condition (L_1) is fullfilled in I - III (in II by $(L_2) \Rightarrow (L_1)$, in III by (ρ), $(c_1) \Rightarrow (L_1)$ (Lemma 6)). Therefore, by inequality (5.45) for a sufficiently large n, such that $s_n \leq t \in S$ we have

$$|B^n_{s_n}| \leq |x| I(|x| > 1) * \nu^n_t \xrightarrow{P} 0$$

by (L_1). Condition

$$\lim_{\delta \to 0} \overline{\lim_{n}} \, P(\langle M^{n,\delta} \rangle_{s_n} \geq a) = 0, \quad a > 0$$

follows from (f_1) and (f_2) by (5.30) and (5.31). We remark only that in case III (f_1) is a consequence of (c_1) and of the fact that S is a dense subset of R_+.

The remark to the theorem is proved analogously to Remark 2 to Theorem 4.3.

11. We will give now the assertion analogous to Theorem 2 in case of local martingales $X^n = (X^n_t, \mathcal{F}^n_t)$, $n \geq 1$.

If $\hat{\tau}_n = (\hat{\tau}_n(t))_{n \geq 1}$ is a random change of time, i.e. $\hat{\tau}_n \in V^+$, $n \geq 1$, and $\hat{\tau}_n(t) \in T(\mathbb{F}^n)$ as t is fixed, then $\hat{X}^n = (\hat{X}^n_t, \hat{\mathcal{F}}^n_t)$ with $\hat{X}^n_t = X^n_{\hat{\tau}_n(t)}$ and $\hat{\mathcal{F}}^n_t = \mathcal{F}^n_{\hat{\tau}_n(t)}$ can fail, in general, to be a local martingale (see Ch. 4, § 7). However, there is an assertion completely analogous to Theorem 4. To formulate it we introduce the following conditions:

(o) $\mathcal{G} \subseteq \bigcap_{n \geq 1} \mathcal{F}^n_0$;

($\hat{L}_1$) $|x| I(|x| > \delta) * \nu^n_{\hat{\tau}_n(t)} \xrightarrow{P} 0, \ \delta \in (0, 1], \ t \in S$;

($\hat{L}_2$) $x^2 I(|x| > \delta) * \nu^n_{\hat{\tau}_n(t)} \xrightarrow{P} 0, \ \delta \in (0, 1], \ t \in S$;

($\hat{c}_1$) $[X^n, X^n]_{\hat{\tau}_n(t)} \xrightarrow{P} \langle X \rangle_t, \ t \in S$;

($\hat{c}_2$) $\langle X^n \rangle_{\hat{\tau}_n(t)} \xrightarrow{P} \langle X \rangle_t, \ t \in S$;

($\hat{\rho}$) the family of random variables $(\sup_{s \leq \hat{\tau}_n(t)} |\Delta X^n_s|)_{n \geq 1}$ is uniformly integrable for each $t \in R_+$.

Theorem 6. *Let any of the following three sets of conditions be fullfilled:*

I: $X^n = (X^n_t, \mathcal{F}^n_t) \in \mathcal{M}_{loc,0}$, $n \geq 1$, S *is a nonempty subset of* R_+, (o), ($\hat{L}_1$), ($\hat{c}_1$);

II: $X^n = (X^n_t, \mathcal{F}^n_t) \in \mathcal{M}^2_{loc,0}$, $n \geq 1$, S *is a nonempty subset of* R_+, (o), ($\hat{L}_2$), ($\hat{c}_1$) *or* ($\hat{c}_2$);

III: $X^n = (X^n_t, \mathcal{F}^n_t) \in \mathcal{M}_{loc,0}$, $n \geq 1$, S *is a nonempty subset of* R_+, $\langle X \rangle = C$, (o), ($\hat{c}_1$), ($\hat{\rho}$).

Then

$$\hat{X}^n \xrightarrow{d_f(S)} X \quad (\mathcal{G}\text{-}stably).$$

The proof of this theorem is analogous to the proof of Theorem 4, it is based on Problem 3 and left to the reader.

12. Example 1 [301]. Let $Y = (Y_t)_{t \geq 0}$ be a point process (relative to $(\mathbb{H}, P)$ where $\mathbb{H} = (\mathcal{H}_t)_{t \geq 0}$ is a family of σ-algebras satisfying Conditions (a) and (b), (Ch. 1, § 1) with the compensator $A = (A_t)_{t \geq 0}$,

$$A_t = \int_0^t (1 + Y_s)\, ds. \tag{5.46}$$

Denote $M = Y - A$. The process M belongs to $\mathcal{M}^2_{loc}(\mathbb{H})$ and has the quadratic characteristic

$$\langle M \rangle = A \tag{5.47}$$

(Problem 3.4.5).

Let $X^n = (X^n_t, \mathcal{F}^n_t)$ with

$$X^n_t = \frac{M_{nt}}{(EA_n)^{1/2}}, \quad \mathcal{F}^n_t = \mathcal{H}_{nt}, \quad \mathcal{G} = \mathcal{H}_\infty = \sigma\Big(\bigcup_{t \geq 0} \mathcal{H}_t\Big). \tag{5.48}$$

Also, let X_1 be a random variable with the characteristic function

$$Ee^{i\lambda X_1} = E\exp\left(-\frac{\lambda^2}{2}\eta^2\right), \quad \lambda \in R$$

with

$$\eta^2 = 1 + \int_0^\infty e^{-s} dM_s. \tag{5.49}$$

Let us show that

$$X^n_1 \xrightarrow{d} X_1 \quad (\mathcal{G}\text{-stably}). \tag{5.50}$$

First of all let us establish the existence and finiteness of the integral in (5.49) which determines η^2.

From (5.46) and the equality $EY_t = EA_t$, $t \in R_+$, (see Ch. 3, § 4) it follows that

$$EA_t = \int_0^t (1 + EA_s)\, ds,$$

and hence

$$EA_t = e^t - 1. \tag{5.51}$$

From (5.46) and the definition of M it follows that the process $Y = (Y_t)_{t \geq 0}$ is defined by the linear equation

$$Y_t = \int_0^t (1 + Y_s)\, ds + M_t.$$

This gives

$$Y_t = (e^t - 1) + e^t \int_0^t e^{-s} dM_s. \tag{5.52}$$

The validity of this representation for Y_t can be verified by means of Ito's formula (Ch. 2, § 3).

Consider the process $N = (N_t)_{t \geq 0}$ with

$$N_t = \int_0^t e^{-s} dM_s.$$

Since $N \in \mathcal{M}^2_{loc}$ (H), it suffices for the existence and finiteness of

$$N_\infty = \int_0^\infty e^{-s} dM_s$$

that $\langle N \rangle_\infty < \infty$ (P-a.s.) (Theorem 2.6.5). By Theorem 2.2.2 we have

$$\langle N \rangle_t = \int_0^t e^{-2s} d\langle M \rangle_s.$$

Consequently, by (5.46), (5.47) and (5.51)

$$\begin{aligned} \mathbf{E}\, \langle N \rangle_\infty &= \mathbf{E} \int_0^\infty e^{-2s} dA_s = \mathbf{E} \int_0^\infty e^{-2s} (1 + Y_s)\, ds \\ &= \int_0^\infty e^{-2s} (1 + \mathbf{E} A_s)\, ds = \int_0^\infty e^{-s} ds = 1. \end{aligned}$$

Thus the integral

$$N_\infty = \int_0^\infty e^{-s} dM_s$$

in (5.49) is defined and it is finite.

To prove Assertion (5.50), observe that the process $X^n = (X_t^n, \mathcal{F}_t^n)$ belongs to the class $\mathcal{M}^2_{loc,0}$. Therefore it suffices to verify the conditions of Theorem 5 (II). (Observe that the nested property of the family of σ-algebras $\mathbb{F}^n$, $n \geq 1$, is fullfilled by construction, and that the rôle of the limiting process can be played by the process $X = (X_t, \mathcal{F}_t)$ where $\mathcal{G} \subseteq \mathcal{F}_0$, $X_t = \sqrt{\eta^2}\, W_t$, and $W = (W_t, \mathcal{F}_t)$ is a Wiener process, independent of h^2). Obviously, it suffices to verify Conditions (c_2) and (f_2) only, i.e. it suffices to verify that

$$\langle X^n \rangle_1 \xrightarrow{P} \eta^2, \tag{5.53}$$

$$\langle X^n \rangle_{s_n} \xrightarrow{P} 0 \tag{5.54}$$

with $s_n = 1/\sqrt{n}$, $n \geq 1$.

By the definition of X_t^n (see (5.48) and (5.47)) we have

$$\langle X^n \rangle_1 = \frac{\langle M \rangle_n}{EA_n} = \frac{A_n}{EA_n} = \frac{Y_n}{EA_n} - \frac{M_n}{EA_n}.$$

From (5.52) and (5.41) it follows that

$$\frac{Y_n}{EA_n} = \frac{e^n - 1 + e^n \int_0^n e^{-s} dM_s}{e^n - 1} = 1 + \frac{e^n}{e^n - 1} N_n \to \eta^2$$

(P-a.s.). Therefore (5.53) takes place, provided

$$\frac{M_n}{EA_n} \to 0, \ n \to \infty \ \text{(P-a.s.)}$$

or

$$\frac{M_t}{EA_t} \to 0, \ t \to \infty \ \text{(P-a.s.)}.$$

The last relation will be established by applying Lemma 2.6.3 and Theorem 2.6.5, according to which it suffices to verify that

$$\int_0^\infty \frac{d\langle M \rangle_s}{(1 + EA_s)^2} < \infty \ \text{(P-a.s.)}.$$

In fact, by (5.47), (5.46) and (5.51)

$$\mathbf{E}\int_0^\infty \frac{d\langle M\rangle_s}{(1+\mathbf{E}A_s)^2} = \mathbf{E}\int_0^\infty e^{-2s}dA_s = \mathbf{E}\int_0^\infty e^{-2s}(1+Y_s)\,ds$$

$$= \int_0^\infty e^{-2s}(1+\mathbf{E}A_s)\,ds = \int_0^\infty e^{-s}ds = 1.$$

Hence

$$\frac{M_t}{\mathbf{E}A_t} \to 0 \ \ (\text{P-a.s.}).$$

Finally, the property (5.54) is satisfied, since from (5.48), (5.47) and (5.51) it follows that

$$\mathbf{E}\langle X^n\rangle_{1/\sqrt{n}} = \frac{\mathbf{E}\langle M\rangle_{\sqrt{n}}}{\mathbf{E}A_n} = \frac{\mathbf{E}A_{\sqrt{n}}}{\mathbf{E}A_n} = \frac{e^{\sqrt{n}}-1}{e^n-1} \to 0, \quad n\to\infty.$$

13. We will formulate seperately Theorems 1 - 6 in the particular case of the "scheme of series" . In accordance with Subsection 1.6 the sequence $(\xi_{n,k}, \mathcal{H}^n_k)_{k\geq 0}$, $n \geq 1$, is assumed to be given, where ξ_{nk} is a $\mathcal{H}^n_k$-measurable random variable, $\xi_{n0} \equiv 0$, $\mathcal{H}^n_0 = \{\varnothing, \Omega\}$, $\mathcal{H}^n_k \subseteq \mathcal{H}^n_{k+1}$, $k \geq 0$, $n \geq 1$.

Let $N(0, \sigma^2)$ be a Gaussian random variable with the parameters $(0, \sigma^2)$ and let

$$S_n = \sum_{k=1}^n \xi_{nk}, \quad n \geq 1.$$

Theorem 7 (Corollary to Theorem 1). *Let the following conditions be fullfilled:*

(a) $$\sum_{k=1}^n \mathrm{P}(|\xi_{nk}| > \delta) \mid \mathcal{H}^n_{k-1}) \xrightarrow{\mathrm{P}} 0, \quad \delta \in (0, 1],$$

(b) $$\sum_{k=1}^n \mathbf{E}(\xi_{nk} I(|\xi_{nk}| \leq 1) \mid \mathcal{H}^n_{k-1}) \xrightarrow{\mathrm{P}} 0,$$

(c) $$\sum_{k=1}^n (\mathbf{E}(\xi^2_{nk} I(|\xi_{nk}| \leq \delta) \mid \mathcal{H}^n_{k-1}) - (\mathbf{E}(\xi_{nk} I(|\xi_{nk}| \leq \delta) \mid \mathcal{H}^n_{k-1})^2) \xrightarrow{\mathrm{P}} \sigma^2.$$

Then

$$S_n \xrightarrow{d} N(0, \sigma^2).$$

Theorem 8 (Corollary to Theorem 4). 1) *Let* $(\xi_{nk}, \mathcal{H}^n_k)_{k \geq 1}$ *be martingale differences,* $n \geq 1$, (*i.e.* $\mathbf{E}\,|\,\xi_{nk}\,| < \infty$, $\mathbf{E}\,(\xi_{nk}\,|\,\mathcal{H}^n_{k-1}) = 0$ (P-a.s.)) *and let the following conditions be fullfilled:*

$$(l_1)\ \sum_{k=1}^{n} \mathbf{E}\,|\,\xi_{nk}\,|\,I\,(\,|\,\xi_{nk}\,| > \delta)\,|\,\mathcal{H}^n_{k-1}) \xrightarrow{P} 0,\ \ \delta \in (0, 1],$$

$$(c_1)\ \sum_{k=1}^{n} \xi^2_{nk} \xrightarrow{P} \sigma^2.$$

Then

$$S_n \xrightarrow{d} N\ (0, \sigma^2).$$

2) *Let* $(\xi_{nk}, \mathcal{H}^n_k)_{k \geq 1}$ *be martingale differences,* $n \geq 1$, $\mathbf{E}\xi^2_{nk} < \infty$, $k \geq 1$, $n \geq 1$, *and let any of the following sets of conditions be fullfilled:*

$$\text{I:}\ \ (l_2)\ \sum_{k=1}^{n} \mathbf{E}\,(\xi^2_{nk} I\,(\,|\,\xi_{nk}\,| > \delta)\,|\,\mathcal{H}^n_{k-1}) \xrightarrow{P} 0,\ \ \delta \in (0, 1],$$

$$(c_1)\ \sum_{k=1}^{n} \xi^2_{nk} \xrightarrow{P} \sigma^2,$$

or

$$\text{II}\ \ (l_2)\ \sum_{k=0}^{n} \mathbf{E}\,(\xi^2_{nk} I\,(\,|\,\xi_{nk}\,| > \delta)\,|\,\mathcal{H}^n_{k-1}) \xrightarrow{P} 0,\ \ \delta \in (0, 1],$$

$$(c_2)\ \sum_{k=1}^{n} \mathbf{E}\,(\xi^2_{nk}\,|\,\mathcal{H}^n_{k-1}) \xrightarrow{P} \sigma^2.$$

Then

$$S_n \xrightarrow{\delta} N\ (0, \sigma^2).$$

Let $(\gamma_n)_{n \geq 1}$ be random variables taking values in the set $\{0, 1, 2, ...\}$ and such that γ_n is a Markov time relative to $(\mathcal{H}^n_k)_{k \geq 0}$. Denote

$$\hat{S}_n = \sum_{k=1}^{\gamma_n} \xi_{nk}.$$

Theorem 9 (Corollary to Theorem 6). *Let* $(\xi_{nk}, \mathcal{H}^n_k)_{k \geq 1}$ *be martingale differences,* $n \geq 1$, *and let the following conditions be fullfilled:*

$$(\hat{1}_1) \sum_{k=1}^{\gamma_n} \mathbf{E}(|\xi_{nk}| I(|\xi_{nk}| > \delta) \mid \mathcal{H}^n_{k-1}) \xrightarrow{P} 0, \ \delta \in (0, 1],$$

$$(\hat{c}_1) \sum_{k=1}^{\gamma_n} \xi^2_{nk} \xrightarrow{P} \sigma^2.$$

Then

$$\hat{S}_n \xrightarrow{d} N(0, \sigma^2).$$

2) *Let* $(\xi_{nk}, \mathcal{H}^n_k)_{k \geq 1}$ *be martingale differences,* $n \geq 1$, $\mathbf{E}\xi^2_{nk} < \infty$ *and let any set of the following conditions be fullfilled:*

$$\text{I:} \ (\hat{1}_2) \sum_{k=1}^{\gamma_n} \mathbf{E}(\xi^2_{nk} I(|\xi_{nk}| > \delta) \mid \mathcal{H}^n_{k-1}) \xrightarrow{P} 0, \ \delta \in (0, 1],$$

$$(\hat{c}_1) \sum_{k=1}^{\gamma_n} \xi^2_{nk} \xrightarrow{P} \sigma^2$$

or

$$\text{II:} \ (\hat{1}_2) \sum_{k=1}^{\gamma_n} \mathbf{E}(\xi^2_{nk} I(|\xi_{nk}| > \delta) \mid \mathcal{H}^n_{k-1}) \xrightarrow{P} 0, \ \delta \in (0, 1],$$

$$(\hat{c}_2) \sum_{k=1}^{\gamma_n} \mathbf{E}(\xi^2_{nk} \mid \mathcal{H}^n_{k-1}) \xrightarrow{P} \sigma^2.$$

Then

$$\hat{S}_n \xrightarrow{d} N(0, \sigma^2).$$

Assume now that a family of σ-algebras $(\mathcal{H}^n_k)_{k \geq 1}$, $n \geq 1$, is nested:

$$\mathcal{H}^n_k \subseteq \mathcal{H}^{n+1}_k, \ k \leq n, \ n \geq 1. \tag{5.55}$$

Theorem 10 (Corollary to Theorem 5). *Let the nested condition* (5.55) *be fullfilled, let* η^2 *be a* $\mathcal{G}$*-measurable random variable where*

$$\mathcal{G} \subseteq \sigma(\bigcup_{n \ge 1} \mathcal{H}^n_\infty), \quad S_n = \sum_{k=1}^{n} \xi_{nk}, \ n \ge 1,$$

and let Z *be a random variable with the characteristic function*

$$\mathbf{E}e^{i\lambda Z} = \mathbf{E}\exp\left(-\frac{\lambda^2}{2}\eta^2\right), \quad \lambda \in R.$$

1) *Let* $(\xi_{nk}, \mathcal{H}^n_k)_{k \ge 1}$ *be martingale differences,* $n \ge 1$, *and let the following conditions be fullfilled:*

$$(l_1) \quad \sum_{k=1}^{n} \mathbf{E}(|\xi_{nk}| I(|\xi_{nk}| > \delta) \mid \mathcal{H}^n_{k-1}) \xrightarrow{P} 0, \quad \delta \in (0, 1],$$

$$(c_1) \quad \sum_{k=1}^{n} \xi^2_{nk} \xrightarrow{P} \eta^2,$$

$$(f_1) \quad \sum_{k=1}^{[nc_n]} \xi^2_{nk} \xrightarrow{P} 0$$

for a certain sequence $(c_n)_{n \ge 1}$ *with* $c_n \downarrow 0$, $nc_n \to \infty$, $n \to \infty$.

Then

$$S_n \xrightarrow{d} Z \quad (\mathcal{G}\text{-}stably).$$

2) *Let* $(\xi_{nk}, \mathcal{H}^n_k)_{k \ge 1}$ *be martingale differences,* $n \ge 1$, $\mathbf{E}\xi^2_{nk} < \infty$, $k \ge 1$, $n \ge 1$, *and*

$$(l_2) \quad \sum_{k=1}^{n} \mathbf{E}(\xi^2_{nk} I(|\xi_{nk}| > \delta) \mid \mathcal{H}^n_{k-1}) \xrightarrow{P} 0, \quad \delta \in (0, 1],$$

$$(c_2) \quad \sum_{k=1}^{n} \mathbf{E}(\xi^2_{nk} \mid \mathcal{H}^n_{k-1}) \xrightarrow{P} \eta^2,$$

$$(f_2) \quad \sum_{k=1}^{[nc_n]} \mathbf{E}(\xi^2_{nk} \mid \mathcal{H}^n_{k-1}) \xrightarrow{P} 0$$

for a certain sequence $(c_n)_{n \ge 1}$ *with* $c_n \downarrow 0$, $nc_n \to \infty$, $n \to \infty$.

If any of the following sets of conditions is fullfilled:

I: (l_2), (c_1), (f_1);

II: (l_2), (c_2), (f_1);
III: (l_2), (c_1), (f_2);
IV: (l_2), (c_2), (f_2),

then

$$S_n \xrightarrow{d} Z \quad (\mathcal{G}\text{-}stably).$$

14. Example 2 [321]. Let the nested condition (5.55) be fullfilled, let

$$\mathcal{G} \subseteq \Big(\bigcup_{n \ge 1} \mathcal{H}^n_\infty \Big)$$

and let η^2 be a $\mathcal{G}$-measurable random variable.

Also, let the following conditions be satisfied:

$$\max_{1 \le k \le n} |\xi_{nk}| \xrightarrow{P} 0, \tag{5.56}$$

$$\sum_{k=1}^{n} \xi^2_{nk} \xrightarrow{P} \eta^2, \tag{5.57}$$

$$\sup_n \mathbf{E} \max_{1 \le k \le n} |\xi_{nk}|^{1+\varepsilon} < \infty, \quad \varepsilon > \infty, \tag{5.58}$$

$$\mathbf{E} (\xi_{nk} \mid \mathcal{H}^n_{k-1}) = 0 \ \ (\mathsf{P}\text{-a.s.}). \tag{5.59}$$

Then

$$S_n \xrightarrow{d} Z \quad (\mathcal{G}\text{-stably})$$

where

$$S_n = \sum_{k=1}^{n} \xi_{nk},$$

and Z is a random variable with the characteristic function

$$\mathbf{E} \exp\left(-\frac{\lambda^2}{2} \eta^2 \right), \quad \lambda \in R.$$

To prove this let us verify Conditions (l_1) and (f_1) in Assertion 1) of Theorem 10.

By Lemma 1 Condition (5.56) is equivalent to the condition

$$\sum_{k=1}^{n} \mathsf{P}(|\xi_{nk}| > \delta \mid \mathcal{H}^n_{k-1}) \xrightarrow{P} 0, \quad \delta \in (0, 1].$$

By the implication (5.13) this gives

$$\sum_{k=1}^{n} |\xi_{nk}| I(|\xi_{nk}| > \delta) \xrightarrow{P} 0, \quad \delta \in (0, 1]. \tag{5.60}$$

Next, by the Lenglart-Rebolledo inequality (Theorem 1.9.3) we have

$$P\left(\sum_{k=1}^{n} \mathbf{E}(|\xi_{nk}| I(|\xi_{nk}| > \delta) \mid \mathcal{H}^{n}_{k-1}) \ge a \right)$$

$$\le \frac{b + \mathbf{E} \max\limits_{1 \le k \le n} |\xi_{nk}|}{a} + P\left(\sum_{k=1}^{n} |\xi_{nk}| I(|\xi_{nk}| > \delta) \ge b \right), \tag{5.61}$$

$a > 0,\ b > 0$. By (5.56) and (5.58) we have

$$\lim_{n} \mathbf{E} \max_{1 \le k \le n} |\xi_{nk}| = 0.$$

Therefore

$$\overline{\lim_{n}}\, P\left(\sum_{k=1}^{n} \mathbf{E}(|\xi_{nk}| I(|\xi_{nk}| > \delta) \mid \mathcal{H}^{n}_{k-1}) \ge a \right) \le \frac{b}{a} \to 0,\ b \to 0,$$

i.e. Condition (l_1) is satisfied.

Let us show now that (f_1) is a consequence of (5.56) and (5.58). Let

$$d_n = \frac{1}{(n\, \mathbf{E} \max\limits_{1 \le k \le n} |\xi_{nk}|) \vee (n^{1/2})}, \quad n \ge 1.$$

Obviously, $d_n \to 0,\ n \to \infty$. Besides, by (5.56) and (5.57) we have

$$\mathbf{E} \max_{1 \le k \le n} |\xi_{nk}| \to 0,\ n \to \infty,$$

and hence

$$n d_n = \frac{1}{\mathbf{E} \max\limits_{1 \le k \le n} |\xi_{nk}|} \wedge n^{1/2} \to \infty, \quad n \to \infty.$$

Finally,

$$n d_n \mathbf{E} \max_{1 \le k \le n} |\xi_{nk}| \le 1.$$

Denote $c_n = \min\limits_{1 \le j \le n} d_j$. Clearly $c_n \downarrow 0,\ n \to \infty$ and

$$n c_n \mathbf{E} \max_{1 \le k \le n} |\xi_{nk}| \le 1. \tag{5.62}$$

Let us show that $nc_n \to \infty,\ n \to \infty$. The definition of c_n implies $c_n = d_{n'(n)}$ with $n'(n) \in \{0, 1, 2, \ldots, n\}$. Since $d_n \to 0,\ n \to \infty$, then $n'(n) \to \infty,\ n \to \infty$.

Consequently

$$n c_n = n d_{n'(n)} \ge n'(n)\, d_{n'(n)} \to \infty, \quad n \to \infty.$$

Since

$$\sum_{k=1}^{[nc_n]} \xi_{nk}^2 \leq \max_{1 \leq k \leq n} |\xi_{nk}| \sum_{k=1}^{[nc_n]} |\xi_{nk}|,$$

to verify Condition (f_1) it suffices, in accordance with (5.56), to show that

$$\lim_{c \to \infty} \overline{\lim_{n}} \, \mathsf{P}\left(\sum_{k=1}^{[nc_n]} |\xi_{nk}| \geq c \right) = 0. \qquad (5.63)$$

But by Chebyshev's inequality and (5.62) we have

$$\mathsf{P}\left(\sum_{k=1}^{[nc_n]} |\xi_{nk}| \geq c \right) \leq \frac{1}{c} nc_n \mathbf{E} \max_{1 \leq k \leq n} |\xi_{nk}| \leq \frac{1}{c},$$

and consequently (5.63) is fullfilled.

Example 3 (Donsker's theorem, [16]). Let $(\xi_k)_{k \geq 1}$ be a sequence of independent and identically distributed random variables with

$\mathbf{E}\xi_1 = 0$ and $\mathbf{E}\xi_1^2 = \sigma^2$. Introduce

$$S_n = \frac{1}{\sqrt{n}} \sum_{k=1}^{n} \xi_k$$

and denote by Z a Gaussian random variable with the parameter $(0, \sigma^2)$.

Donsker's theorem asserts that

$$S_n \xrightarrow{d} Z, \quad n \to \infty.$$

To prove this weak convergence one may apply Theorem 8.

If $\mathcal{H}_k = \sigma\{\xi_1, ..., \xi_k\}$, $k \geq 1$, then

$$\left(\frac{1}{\sqrt{n}} \xi_k, \mathcal{H}_k \right)$$

are square integrable martingale differences and, besides, Conditions (l_2) and (c_2) of Theorem 8 hold. Indeed,

$$(l_2) \sum_{k=1}^{n} \mathbf{E}\left(\frac{1}{n} \xi_k^2 I\left(\frac{1}{\sqrt{n}} |\xi_k| > \delta \right) | \mathcal{H}_{k-1} \right)$$

$$= \mathbf{E}\, \xi_1^2 I(|\xi_1| > \delta\sqrt{n}) \to 0, \quad n \to \infty,$$

$$(c_1)\ \frac{1}{n}\sum_{k=1}^{n}\xi_k^2 \to \sigma^2\ (\text{P-a.s.}),\ n\to\infty$$

due to the Birkhoff-Khintchine theorem ([289], [332]).

15. We will give a multivariate version of Theorem 1. In this subsection all vectors taking part in algebraic operations are considered as column vectors, and the sign * means the transposition of vectors and matrices.

Let $X_t = (X_t^1, \ldots, X_t^k)$ and let $X = (X_t, \mathcal{F}_t)$ be a local square integrable vector-valued martingale with $\mathcal{G}$-conditionally Gaussian increments, with the quadratic characteristic

$$\langle X_t\rangle = C_t + \int_0^t \int_{R^k\setminus\{0\}} xx^*\,d\nu$$

where $C_t = \langle X^c\rangle_t$, and with the stochastic exponential

$$\mathcal{E}_t\,(G\,(\lambda)) = \exp\left(-\frac{1}{2}\lambda^*\,\langle X\rangle_t\,\lambda\right)$$

Let $X^n = (X_t^n, \mathcal{F}_t^n)$, $n \geq 1$, be vector-valued semimartingales, $X_t^n = (X_t^{n1}, \ldots, X_t^{nk})$, with the triplets of predictable characteristics $T^n = (B^n, C^n, \nu^n)$ (see Ch. 4, Subsection 1.3) and let $M^{n,\delta} = (M_t^{n,\delta}, \mathcal{F}_t^n)$, $n \geq 1$, be locally square integrable martingales with ($\delta \in (0, 1]$)

$$M_t^{n,\delta} = X_t^{nc} + \int_0^t \int_{|x|\leq\delta} x\,d\,(\mu^n - \nu^n) \qquad \left(|x| = \left(\sum_{j=1}^{k}(x^j)^2\right)^{1/2}\right)$$

and with the quadratic characteristic

$$\langle M^{n,\delta}\rangle_t = C_t^n + \int_0^t \int_{|x|\leq\delta} xx^*\,d\nu^n - \sum_{0<s\leq t} \hat{x}_s^{n,\delta}\,(\hat{x}_s^{n,\delta})^*$$

where

$$\hat{x}_s^{n,\delta} = \int_{|x|\leq\delta} x\nu^n\,(\{s\}, dx).$$

Theorem 11. *Let S be a nonempty subset of R_+ and let the following conditions be fullfilled:*

(o) $\mathcal{G} \subseteq \bigcap_{n \ge 1} \mathcal{F}_0^n$;

(a) $\nu^n((0, t] \times \{|x| > \delta\}) \xrightarrow{P} 0, \quad \delta \in (0, 1], \quad t \in S$;

(b) $B_t^n \xrightarrow{P} 0, \quad t \in S$;

(c) $\langle M^{n, \delta} \rangle_t \xrightarrow{P} \langle X \rangle_t, \quad \delta \in (0, 1], \quad t \in S$;

(e) $X_0^n \xrightarrow{P} 0$.

Then

$$X^n \xrightarrow{d_f(S)} X \quad (\mathcal{G}\text{-}stably).$$

By utilizing vector and matrix notations, the proof of this theorem does not differ from the proof of Theorem 1 and is left to the reader.

Problems

1. Proof Theorem 2.
2. Proof Theorem 3.

3. Let $\hat{\tau}_n = (\hat{\tau}_n(t))_{t \ge 0}$ be a random change of time in Theorem 2. Show that

1) the condition

$(\hat{a})$
$$\nu^n((0, \hat{\tau}_n(t)] \times \{|x| > \delta\}) \xrightarrow{P} 0, \quad \delta \in (0, 1], \quad t \in S,$$

is equivalent to the condition

$(\hat{a}*)$
$$\sup_{s \le \hat{\tau}_n(t)} |\Delta X_s^n| \xrightarrow{P} 0, \quad t \in S;$$

2) if $X^n = (X_t^n, \mathcal{F}_t^n) \in \mathcal{M}_{loc, 0}$, $n \ge 1$, and if the condition

$(\hat{L}_1)$
$$|x| I(|x| > \delta) * \nu^n_{\hat{\tau}_n(t)} \xrightarrow{P} 0, \quad \delta \in (0, 1], \quad t \in S$$

is fullfilled, as well as any of the conditions

$(\hat{c})$
$$\langle M^{n, \delta} \rangle_{\hat{\tau}_n(t)} \xrightarrow{P} \langle X \rangle_t, \quad t \in S,$$

or

$(\hat{c}_1)$
$$[X^n, X^n]_{\hat{\tau}_n(t)} \xrightarrow{P} \langle X \rangle_t, \quad t \in S,$$

then

$$\sup_{s \le \hat{\tau}_n(t)} |\langle M^{n,\delta}\rangle_s - [X^n, X^n]_s| \xrightarrow{P} 0, \quad \delta \in (0, 1], \quad t \in S;$$

3) if $X^n = (X_t^n, \mathcal{F}_t^n) \in \mathcal{M}^2_{loc,0}$, $n \ge 1$, and if the condition

$$(\hat{L}_2) \qquad x^2 I(|x| > \delta) * \nu^n_{\hat{\tau}_n(t)} \xrightarrow{P} 0, \quad \delta \in (0, 1], \quad t \in S$$

is fullfilled, as well as any of the conditions

$$(\hat{c}_1) \qquad [X^n, X^n]_{\hat{\tau}_n(t)} \xrightarrow{P} \langle X\rangle_t, \quad t \in S,$$

or

$$(\hat{c}_2) \qquad \langle X^n\rangle_{\hat{\tau}_n(t)} \xrightarrow{P} \langle X\rangle_t, \quad t \in S,$$

then

$$\sup_{s \le \hat{\tau}_n(t)} |[X^n, X^n]_s - \langle X^n\rangle_s| \xrightarrow{P} 0, \quad t \in S;$$

4) if $X^n = (X_t^n, \mathcal{F}_t^n) \in \mathcal{M}_{loc,0}$, $[X^n, X^n] \xrightarrow{P} C_t$ for each t belonging to a subset dense in R_+ and if the following condition is fullfilled:

$(\hat{\rho})$ the familly of random variables $(\sup_{s \le \hat{\tau}_n(t)} |\Delta X_s^n|)$, $n \ge 1$, is uniformly integrable

for each $t \in R_+$, then

$$|x| I(|x| > \delta) * \nu^n_{\hat{\tau}_n(t)} \xrightarrow{P} 0, \quad \delta \in (0, 1], \quad t \in R_+.$$

4. Prove Theorem 6.

5. Prove Theorem 11.

6. Formulate and prove the multivariate versions of Theorems 2 - 10.

§ 6. The central limit theorem. II. "Nonclassical" version

1. In all theorems of the preceding section the "classical" condition is presented of uniform asymptotical negligibility of jumps

$$\text{(a*)} \qquad \sup_{s \le t} |\Delta X^n_s| \xrightarrow{P} 0, \quad n \to \infty, \tag{6.1}$$

of prelimiting semimartingales X^n, $n \ge 1$, which is equivalent to the condition (Lemma 5.1)

$$\text{(a)} \qquad I(|x| > \delta) * \nu^n_t \xrightarrow{P} 0, \quad \delta \in (0, 1], \ n \to \infty. \tag{6.2}$$

It is wellknown, however, that for the validity of the central limit theorem Condition (a*) is by no means necessary.

In the present section the question will be considered on the validity of the central limit theorem in absence of the "classical" Condition (a*), more precisely in absence of the condition

$$\sup_{s \in \{s \le t:\, a^n_s > 0\}} |\Delta X^n_s| \xrightarrow{P} 0, \quad n \to \infty \tag{6.3}$$

with

$$a^n_s = \nu^n(\{s\} \times R_0).$$

2. Denote

$$\Phi(x) = \frac{1}{\sqrt{2\pi}} \int_{-\infty}^{x} e^{-y^2/2} dy, \tag{6.4}$$

$$F^n_s(x) = I(a^n_s > 0)\, \mathsf{P}(\Delta X^n_s \le x \mid \mathcal{F}^n_{s-}) \tag{6.5}$$

where $\mathsf{P}(\Delta X^n_s \le x \mid \mathcal{F}^n_{s-})$ is a regular conditional probability. On the set $\{a^n_s > 0\}$ (see Problem 3.2.5)

$$\mathsf{P}(\Delta X^n_s \le x \mid \mathcal{F}^n_{s-}) = \nu^n(\{s\} \times (-\infty, x] \cap R_0) + I(x \ge 0)(1 - a^n_s), \tag{6.6}$$

since

$$1 - a^n_s = 1 - \nu^n(\{s\} \times R_0)$$

$$= 1 - \mathsf{P}(\Delta X^n_s \ne 0 \mid \mathcal{F}^n_{s-})$$

$$= \mathsf{P}(\Delta X^n_s = 0 \mid \mathcal{F}^n_{s-}).$$

We consider the following three cases:

(0) the function $F_s^n(x)$ has, in general, no first moment for each $s \in \{s > 0: a_s^n > 0\}$ with probability one;

$$(1)\quad \int_{-\infty}^{\infty} |x|\, dF_s^n(x) < \infty, \quad \int_{-\infty}^{\infty} x\, dF_s^n(x) = 0, \quad s \in \{s > 0: a_s^n > 0\} \text{ (P-a.s.)};$$

$$(2)\quad \int_{-\infty}^{\infty} x^2\, dF_s^n(x) < \infty, \quad \int_{-\infty}^{\infty} x\, dF_s^n(x) = 0, \quad s \in \{s > 0: a_s^n > 0\} \text{ (P-a.s.)}.$$

With each of these cases we associate the functions $g_j = g_j(x)$, $j = 0, 1, 2$:

$$\begin{aligned} g_0(x) &= x\, I(|x| \le 1), \\ g_1(x) &= x\, I(|x| \le 1) + I(|x| > 1)\, \operatorname{sign} x, \\ g_2(x) &= x, \end{aligned} \tag{6.7}$$

and we define the stochastic processes $b^{nj} = (b_t^{nj})_{t \ge 0}$ and $\Delta^{nj} = (\Delta_t^{nj})_{t \ge 0}$, $j = 0, 1, 2$, in the following manner:

$$b^{n1} = 0, \quad b^{n2} = 0, \tag{6.8}$$

$$\int_{-1}^{1} F_t^n(x + b_t^{n0})\, dx = 1, \tag{6.9}$$

$$\int_0^{\infty} g_j(x) \left[\left(1 - \Phi\left(\frac{x}{\sqrt{\Delta_t^{nj}}} \right) \right) + \Phi\left(- \frac{x}{\sqrt{\Delta_t^{nj}}} \right) \right] dx$$
$$= \int_0^{\infty} g_j(x)\, [(1 - F_t^n(x + b_t^{nj})) + F_t^n(-x + b_t^{nj})]\, dx. \tag{6.10}$$

These processes b^{n0} and Δ^{nj}, $j = 0, 1, 2$, satisfying in addition the following relations

$$b_t^{n0} = I(a_t^n > 0)\, b_t^{n0}, \quad \Delta_t^{nj} = I(a_t^n > 0)\, \Delta_t^{nj}, \quad j = 0, 1, 2, \tag{6.11}$$

exist due to the following properties (Problem 4):

1) the function

$$v(b) = \int_{-1}^{1} F_s^n(x + b)\, dx$$

takes values in the set [0, 2], it is continuous,

$$\lim_{b \to -\infty} v(b) = 0 \text{ and } \lim_{b \to \infty} v(b) = 2;$$

2) $\Delta_t^{n2} = \int_{-\infty}^{\infty} x^2 \, dF_t^n(x) = \int_{-\infty}^{\infty} x^2 v^n(\{t\}, dx);$

3) the function

$$v_1(\Delta) = \int_0^{\infty} g_1(x) \{(1 - \Phi(x/\sqrt{\Delta})) + \Phi(-x/\sqrt{\Delta})\} \, dx, \quad \Delta > 0,$$

is continuous, takes values in the set $(0, \infty)$, while

$$\int_0^{\infty} g_1(x) \, [(1 - F_t^n(x)) + F_t^n(-x)] \, dx$$

$$= \int_{-\infty}^{\infty} \left[\frac{x^2}{2} I(|x| \le 1) + \left(|x| - \frac{1}{2} \right) I(|x| > 1) \right] dF_t^n(x);$$

4) the function

$$v_0(\Delta) = \int_0^{\infty} g_0(x) \, [(1 - \Phi(x/\sqrt{\Delta})) + \Phi(-x/\sqrt{\Delta})] \, dx, \quad 0 < \Delta \le \infty,$$

is continuous and it takes values in the set $(0, \frac{1}{2}]$, while

$$0 < \int_0^{1} x \, [(1 - F_t(x + b_t^{n0})) + F_t(-x + b_t^{n0})] \, dx \le \frac{1}{2}.$$

Further on the process b^{n0} will be denoted by b^n.

In case of a locally square integrable martingale $X^n = (X_t^n, \mathcal{F}_t^n)$ with the quadratic characteristic $\langle X^n \rangle$ we have

$$\Delta_t^{n2} = \Delta \langle X^n \rangle_t \tag{6.12}$$

(see Problem 3.5.2).

3. In this section (as well as in § 5) we assume that $X = (X_t, \mathcal{F}_t)$ is a locally square integrable martingale with $X_0 = 0$ and with $\mathcal{G}$-conditionally independent ($\mathcal{G} \subseteq \mathcal{F}_0$) and $\mathcal{G}$-conditionally Gaussian increments. From this assumption follows, in particular, that

$$\mathcal{E}_t(G(\lambda)) = \exp\left(-\frac{\lambda^2}{2}\langle X\rangle_t\right) \tag{6.13}$$

where $\langle X\rangle$ is the square characteristic of $X = (X_t, \mathcal{F}_t)$.

Theorem 1. *Let* $X^n = (X_t^n, \mathcal{F}_t^n)$, $n \geq 1$, *be locally square integrable martingales with* $X_0^n = 0$, *with the triplets* $T^n = (B^n, C^n, \nu^n)$ *and with the quadratic characteristics* $\langle X^n\rangle$, $n \geq 1$, *let* S *be a nonempty subset of* R_+, *and let the following conditions be fullfilled:*

(o) $\mathcal{G} \subseteq \bigcap_{n\geq 1} \mathcal{F}_0^n$;

(L_2^c) $x^2 I(|x| > \delta) * \nu_t^{nc} \xrightarrow{P} 0$, $\delta \in (0, 1]$, $t \in S$;

(L_2) $$\sum_{s\in\{s\leq t:\, a_s^n > 0\}} \int_{|x|>\delta} |x| \left| F_s^n(x) - \Phi\left(\frac{x}{\sqrt{\Delta\langle X_s^n\rangle}}\right)\right| dx \xrightarrow{P} 0,$$

$\delta \in (0, 1]$, $t \in S$;

(c_2) $\langle X^n\rangle_t \xrightarrow{P} \langle X\rangle_t$, $t \in S$.

Then

$$X^n \xrightarrow{d_f(S)} X \quad (\mathcal{G}\text{-stably}).$$

Theorem 2. *Let* $X^n = (X_t^n, \mathcal{F}_t^n)$, $n \geq 1$, *be local martingales with* $X_0^n = 0$, *and with the triplets* $T^n = (B^n, C^n, \nu^n)$, $n \geq 1$, *let* S *be a nonempty subset of* R_+ *and let the following conditions be fullfilled:*

(o) $\mathcal{G} \subseteq \bigcap_{n\geq 1} \mathcal{F}_0^n$;

(L_1^c) $|x| I(|x| > \delta) * \nu_t^{nc} \xrightarrow{P} 0$, $\delta \in (0, 1]$, $t \in S$;

(Λ_1) $$\sum_{s\in\{s\leq t:\, a_s^n > 0\}} \int_{|x|>\delta} \left| F_s^n(x) - \Phi\left(\frac{x}{\sqrt{\Delta_s^{n1}}}\right)\right| dx \xrightarrow{P} 0, \quad \delta \in (0, 1],$$

$t \in S$;

(c_1^Δ) $C_t^n + x^2 I(|x| \le \delta) * \nu_t^{nc} + \sum_{0 < s \le t} \Delta_s^{n1} \xrightarrow{P} \langle X \rangle_t$, $\delta \in (0, 1]$, $t \in S$.

Then

$$X^n \xrightarrow{d_f(S)} X \quad (\mathcal{G}\text{-}stably).$$

Theorem 3. *Let* $X^n = (X_t^n, \mathcal{F}_t^n)$, $n \ge 1$, *be semimartingales with the triplets* $T^n = (B^n, C^n, \nu^n)$, $n \ge 1$, *let* S *be a nonempty subset of* R_+, *and let the following conditions be fullfilled:*

(o) $\mathcal{G} \subseteq \bigcap_{n > 1} \mathcal{F}_0^n$;

(a^c) $I(|x| > \delta) * \nu_t^{nc} \xrightarrow{P} 0$, $\delta \in (0, 1]$, $t \in S$;

(b) $B_t^{nc} + \sum_{0 < s \le t} b_s^n \xrightarrow{P} 0$, $t \in S$,

$$\lim_{c \to \infty} \overline{\lim_{n}}\, P\left(\sum_{0 < \le t} |b_s^n| \ge c \right) = 0, \quad t \in S;$$

$$(\Lambda 0) \quad \sum_{s \in \{s \le t:\, a_s^n > 0\}} \left(\int_{\delta < |x| < 1} \left| F_s^n(x + b_s^n) - \Phi\left(\frac{x}{\sqrt{\Delta_s^{n0}}} \right) \right| dx \right.$$

$$\left. + \left(\sup_{|x| \ge 1} \left| F_s^n(x + b_s^n) - \Phi\left(\frac{x}{\sqrt{\Delta_s^{n0}}} \right) \right| \right) \right) \xrightarrow{P} 0,$$

$$\delta \in (0, 1], \quad t \in S;$$

(c_0^Δ) $C_t^n + x^2 I(|x| \le \delta) * \nu_t^{nc} + \sum_{0 < s \le t} \Delta_s^{n0} \xrightarrow{P} \langle X \rangle_t$, $\delta \in (0, 1]$, $t \in S$.

Then

$$X^n \xrightarrow{d_f(S)} X \quad (\mathcal{G}\text{-}stably).$$

4. The proof of Theorems 1 - 3 is based on Theorem 2.1 and it utilizes auxiliary facts formulated here as lemmas.

Denote

$$\Psi_t^{nj} = F_t^n (x + b_t^{nj}) - \Phi (x/\sqrt{\Delta_t^{nj}}\,), \quad j = 0, 1, 2. \tag{6.14}$$

Lemma 1. *Let for* j = 0, 1, 2 *Condition* (Λ_j) *be fullfilled as well as the condition*

$(\tilde{\Lambda}_j)$
$$\lim_{c \to \infty} \overline{\lim_{n}} \; \mathsf{P} \Big(\sum_{0 < s \le t} \Delta_s^{nj} \ge c \Big) = 0,$$

and in the cases j = 1 *and* j = 2 *let, in addition,* $F_t^n(x)$ *satisfy Assumptions* (1) *and* (2) *respectively .*

Then for j = 0, 1, 2 *respectively*

$$\sum_{s \in \{s \le t:\, a_s^n > 0\}} \Big| \int_{-\infty}^{\infty} e^{i\lambda x} d\Psi_s^{nj}(x) \Big| \xrightarrow{\mathsf{P}} 0, \quad \lambda \in R.$$

Proof. The case j = 0. We will get certain estimates for

$$I_1^n(t) = \int_{|x| \le 1} e^{i\lambda x} d\Psi_t^{n0}(x), \qquad I_2^n(t) = \int_{|x| > 1} e^{i\lambda x} d\Psi_t^{n0}(x).$$

The integration by parts gives

$$I_1^n(t) = \Psi_t^{n0}(1)\, e^{i\lambda} - \Psi_t^{n0}(-1)\, e^{-i\lambda} - i\lambda \int_{|x| \le 1} e^{i\lambda x}\, \Psi^{n0}(x)\, dx. \tag{6.15}$$

Observe now that the following two equalities hold:

$$\int_{|x| \le 1} \Psi_t^{n0}(x)\, dx = 0, \qquad \int_{|x| \le 1} x\Psi_t^{n0}(x)\, dx = 0. \tag{6.16}$$

The first relation in (6.16) holds by (6.9) and the obvious equality

$$\int_{|x| \le 1} \Phi\left(\frac{x}{\sqrt{\Delta}}\right) dx = 1, \quad \Delta > 0, \tag{6.17}$$

and the second by (6.10), since

$$\int_{|x| \le 1} x\Psi_t^{n0}(x)\, dx = \int_0^1 x\, ([(1 - F_t^n(x + b_t^n)) + F_t^n(-x + b_t^n)]$$

$$- [1 - \Phi(-x/\sqrt{\Delta_t^{n0}}\,) + \Phi(-x/\sqrt{\Delta_t^{n0}}\,)])\, dx.$$

In view of (6.16) by (6.15) we get the following estimate for $| I_1^n(t) |$:

$$| I_1^n(t) | \le 2 \sup_{|x| \ge 1} | \Psi_t^{n0}(x) | + | \lambda | \left| \int_{|x| \le 1} (e^{i\lambda x} - 1 - i\lambda x)\, \Psi_t^{n0}(x)\, dx \right|. \quad (6.18)$$

Next, for $\delta \in (0, 1]$

$$\left| \int_{|x| \le 1} (e^{i\lambda x} - 1 - i\lambda x)\, \Psi_t^{n0}(x)\, dx \right|$$

$$\le (2 + |\lambda|) \int_{\delta < |x| \le 1} | \Psi_t^{n0}(x) |\, dx + \frac{\lambda^2 \delta}{2} \int_{|x| \le \delta} |x|\, | \Psi_t^{n0}(x) |\, dx. \quad (6.19)$$

Let us estimate the second term on the right-hand side of this inequality (6.19). In view of (6.10) we have

$$\int_{|x| \le \delta} |x|\, | \Psi_t^{n0}(x) |\, dx \le \int_{|x| \le 1} |x|\, | \Psi_t^{n0}(x) |\, dx$$

$$= \int_0^\infty g_0(x) \Big(| (1 - F_t^n(x + b_t^n)) - (1 - \Phi(x/\sqrt{\Delta_t^{n0}})) |$$

$$+ | F_t^n(-x + b_t^n) - \Phi(-x/\sqrt{\Delta_t^{n0}}) | \Big)\, dx$$

$$\le \int_0^\infty g_0(x) \big[(1 - F_t^n(x + b_t)) + F_t^n(-x + b_t^n)$$

$$+ (1 - \Phi(x/\sqrt{\Delta_t^{n0}})) + \Phi(-x/\sqrt{\Delta_t^{n0}}) \big]\, dx$$

$$= 2 \int_0^\infty g_0(x) \big[(1 - \Phi(x/\sqrt{\Delta_t^{n0}})) + \Phi(-x/\sqrt{\Delta_t^{n0}}) \big]\, dx \le \Delta_t^{n0}. \quad (6.20)$$

This, (6.18) and (6.19) give the following estimate for $| I_1^n(t) |$, which we will use in the sequel:

$$| I_1^n(t) | \le 2 \sup_{|x| \ge 1} | \Psi_t^{n0}(x) | + | \lambda |) \int_{\delta < |x| \le 1} | \Psi_t^{n0}(x) | \, dx$$

$$+ \frac{1}{2} | \lambda^3 | \delta \Delta_t^{n0}. \tag{6.21}$$

Let us estimate now $| I_2^n(t) |$. For $\varepsilon > 0$ denote

$$h_1^{n\varepsilon}(t) = \int_{|x| > 1} e^{i\lambda x - \varepsilon x^2} d\Psi_t^{n0}(x),$$

$$h_2^{n\varepsilon}(t) = \int_{|x| > 1} e^{i\lambda x}(1 - e^{-\varepsilon x^2}) \, d\Psi_t^{n0}(x)$$

and observe that

$$I_2^n(t) = h_1^{n\varepsilon}(t) + h_2^{n\varepsilon}(t).$$

Integrating by parts we get

$$h_1^{n\varepsilon}(t) = - e^{i\lambda - \varepsilon} \Psi_t^{n0}(1) - \int_1^\infty \Psi_t^{n0}(x) (i\lambda - 2\varepsilon x) e^{i\lambda x - \varepsilon x^2} dx$$
$$+ e^{-i\lambda - \varepsilon} \Psi_t^{n0}(-1) - \int_{-\infty}^{-1} \Psi_t^{n0}(x) (i\lambda - 2\varepsilon x) e^{i\lambda x - \varepsilon x^2} dx.$$

It is not hard to deduce from this that the inequality

$$| h_1^{n\varepsilon}(t) | \le c(\varepsilon) \sup_{|x| \ge 1} | \Psi_t^{n0}(x) | \tag{6.22}$$

holds with the constant

$$c(\varepsilon) = 2 + \frac{|\lambda|}{\sqrt{2\varepsilon}} \int_{-\infty}^{\infty} e^{-y^2/2} dy + \int_{-\infty}^{\infty} | y | e^{-y^2/2} dy.$$

Further

$$| h_2^{n\varepsilon}(t) | \le \int_{|x| > 1} (1 - e^{-\varepsilon x^2}) \, | d\Psi_t^{n0}(x) |$$

$$\le - \int_1^\infty (1 - e^{-\varepsilon x^2}) \, d\,[(1 - F_t^n(x + b_t^n)) + (1 - \Phi(x/\sqrt{\Delta_t^{n0}}))$$

$$+ F_t^n(-x + b_t^n) + \Phi(-x/\sqrt{\Delta_t^{n0}})].$$

Integrating by parts the integral on the right-hand side of the last inequality, we arrive at the estimate

$$| h_2^{n\varepsilon}(t) | \le (1 - e^{-\varepsilon}) [(1 - F_t^n(1 + b_t^n)) + (1 - \Phi(1/\sqrt{\Delta_t^{n0}}))$$

$$+ F_t^n(-1 + b_t^n) + \Phi(-1/\sqrt{\Delta_t^{n0}})] + 2\varepsilon \int_1^\infty [(1 - F_t^n(x + b_t^n))$$

$$+ (1 - \Phi(x/\sqrt{\Delta_t^{n0}})) + F_t^n(-x + b_t^n) + \Phi(-x/\sqrt{\Delta_t^{n0}})] x e^{-\varepsilon x^2} dx,$$

in view of which

$$| h_2^{n\varepsilon}(t) | \le 4 \sup_{|x| \ge 1} | \Psi_t^{n0}(x) | + 2(1 - e^{-\varepsilon})(1 - \Phi(1/\sqrt{\Delta_t^{n0}}) + \Phi(-1/\sqrt{\Delta_t^{n0}}))$$

$$+ 4\varepsilon \int_0^\infty x (1 - \Phi(x/\sqrt{\Delta_t^{n0}}) + \Phi(-x/\sqrt{\Delta_t^{n0}})) dx.$$

Observe that

$$1 - \Phi(1/\sqrt{\Delta_t^{n0}}) + \Phi(-1/\sqrt{\Delta_t^{n0}}) = \int_{|x| > 1} d\Phi(x/\sqrt{\Delta_t^{n0}}) \le \Delta_t^{n0},$$

$$2 \int_0^\infty x (1 - \Phi(x/\sqrt{\Delta_t^{n0}}) + \Phi(-x/\sqrt{\Delta_t^{n0}})) dx = \Delta_t^{n0},$$

and consequently

$$h_2^{n\varepsilon}(t) \le 4 \sup_{|x| \ge 1} | \Psi_t^{n0}(x) | + 2(1 - e^{-\varepsilon} + \varepsilon) \Delta_t^{n0}. \tag{6.23}$$

From (6.22) and (6.23) it follows that

$$| l_2^n(t) | \le (4 + c(\varepsilon)) \sup_{|x| \ge 1} | \Psi_t^{n0}(x) | + 2(1 - e^{-\varepsilon} + \varepsilon) \Delta_t^{n0}. \tag{6.24}$$

Since

$$\left| \int_{-\infty}^{\infty} e^{i\lambda x} d\Psi_t^{n0}(x) \right| \le | l_1^n(t) | + | l_2^n(t) |,$$

the estimates (6.21) and (6.24) entail

$$\sum_{s \in \{s \le t:\, a_s^n > 0\}} \left| \int_{-\infty}^{\infty} e^{i\lambda x}\, d\Psi_s^{n0}(x) \right|$$

$$\le \sum_{s \in \{s \le t:\, a_s^n > 0\}} \left\{ |\lambda|(2+|\lambda|) \int_{\delta < |x| \le 1} |\Psi_s^{n0}(x)|\, dx + (6 + c(\varepsilon)) \sup_{|x| \ge 1} |\Psi_s^{n0}(x)| \right.$$

$$\left. + \left(\frac{|\lambda^3|}{2}\delta + 2\left(1 - e^{-\varepsilon} + \varepsilon\right) \right) \sum_{0 < s \le t} \Delta_s^{n0} \right. = I_1(n, \varepsilon) + I_2(n, \varepsilon, \delta).$$

By Condition (Λ_0) (see Theorem 3) we have $I_1(n, \varepsilon) \xrightarrow{P} 0$. Consequently, to prove the lemma in case $j = 0$ it remains to show that

$$\lim_{\varepsilon, \delta \to 0} \overline{\lim_n}\, \mathsf{P}(I_2(n, \varepsilon, \delta) \ge a) = 0, \quad a > 0. \tag{6.25}$$

But

$$\lim_{\varepsilon, \delta \to 0} \overline{\lim_n}\, \mathsf{P}\left(I_2(n, \varepsilon, \delta) \ge a, \quad \sum_{0 < s \le t} \Delta_s^{n0} < c \right) = 0, \quad a > 0, \quad c > 0,$$

and by Condition $(\tilde{\Lambda}_0)$ of the lemma

$$\lim_{c \to \infty} \overline{\lim_n}\, \mathsf{P}\left(\sum_{0 < s \le t} \Delta_s^{n0} \ge c \right) = 0, \quad t \in S.$$

The case $j = 1$. Denote

$$f(x) = \frac{1}{2} x^2 I(|x| \le 1) + \left(|x| - \frac{1}{2} \right) I(|x| > 1)$$

and observe that $g_1(x) = f'(x)$. Moreover, it is not hard to deduce from (6.10) that

$$\int_{-\infty}^{\infty} f(x)\, dF_t^n(x) = \int_{-\infty}^{\infty} f(x)\, d\Phi\left(x/\sqrt{\Delta_t^{n1}}\right). \tag{6.26}$$

Since

$$\int_{-\infty}^{\infty} x\, d\Psi_s^{n1}(x) = 0$$

by Assumption (1) in Subsection 2, from (6.26) it follows that

$$\int_{-\infty}^{\infty} e^{i\lambda x} d\Psi_s^{n1}(x) = \int_{-\infty}^{\infty} \left(e^{i\lambda x} - 1 - i\lambda x + \lambda^2 f(x)\right) d\Psi_s^{n1}(x). \tag{6.27}$$

Integrating by parts the right-hand side of (6.27), we get

$$\int_{-\infty}^{\infty} e^{i\lambda x}\, d\Psi_s^{n1}(x) = i\lambda \int_{-\infty}^{\infty} (e^{i\lambda x} - 1 - i\lambda g_1(x))\, \Psi_s^{n1}(x)\, dx. \tag{6.28}$$

Denoting

$$I_1^n(s) = \int_{|x| \le \delta} (e^{i\lambda x} - 1 - i\lambda g_1(x))\, \Psi_s^{n1}(x)\, dx,$$

$$I_2^n(s) = \int_{|x| > \delta} (e^{i\lambda x} - 1 - i\lambda g_1(x))\, \Psi_s^{n1}(x)\, dx,$$

we will estimate $|I_1^n(s)|$ and $|I_2^n(s)|$.

We have (as $\delta \in (0, 1]$)

$$|I_1^n(s)| \le \int_{|x| \le \delta} |e^{i\lambda x} - 1 - i\lambda x|\, |\Psi_s^{n1}(x)|\, dx$$

$$\le \frac{\lambda^2}{2} \int_{|x| \le \delta} |x|\, |\Psi_s^{n1}(x)|\, dx$$

$$= \frac{\lambda^2}{2}\delta \int_0^{\delta} x \{ |(1 - F_s^n(x)) - (1 - \Phi(x/\sqrt{\Delta_s^{n1}}))|$$

$$+ |F_s^n(-x) - \Phi(-x/\sqrt{\Delta_s^{n1}})| \}\, dx$$

$$\le \frac{\lambda^2}{2}\delta \int_0^{\infty} g_1(x) \{ (1 - F_s^n(x)) + (1 - \Phi(x/\sqrt{\Delta_s^{n1}}))$$

$$+ F_s^n(-x) + \Phi(-x/\sqrt{\Delta_s^{n1}}) \}\, dx.$$

In view of (6.10), this gives

$$|I_1^n(s)| \le \lambda^2\delta \int_0^\infty g_1(x)\{1 - \Phi(x/\sqrt{\Delta_s^{n1}}) + \Phi(-x/\sqrt{\Delta_s^{n1}})\}\,dx$$

$$\le \lambda^2\delta \int_0^\infty x\{1 - \Phi(x/\sqrt{\Delta_s^{n1}}) + \Phi(-x/\sqrt{\Delta_s^{n1}})\}\,dx$$

$$= \frac{\lambda^2}{2}\Delta_s^{n1}. \tag{6.29}$$

Next,

$$|I_2^n(s)| \le (2 + |\lambda|) \int_{|x|>\delta} |\Psi_s^{n1}(x)|\,dx$$

and consequently

$$\left| \int_{-\infty}^{\infty} e^{i\lambda x}\,d\Psi_s^{n1}(x) \right| \le |\lambda|(2 + |\lambda|) \int_{|x|>\delta} |\Psi_s^{n1}(x)|\,dx + \frac{|\lambda^3|\delta}{2}\Delta_s^{n1}.$$

Hence

$$\sum_{s \in \{s \le t:\, a_s^n > 0\}} \left| \int_{-\infty}^{\infty} e^{i\lambda x}\Psi_s^{n1}(x)\,dx \right|$$

$$\le \sum_{s \in \{s \le t:\, a_s^n > 0\}} |\lambda|(2 + |\lambda|) \int_{|x|>\delta} |\Psi_s^{n1}(x)|\,dx + \frac{|\lambda^3|\delta}{2}\sum_{0 < s \le t} \Delta_s^{n1},$$

and the proof in this case is accomplished in the same manner as in the case $j = 0$.

The case $j = 2$. Analogously to the case $j = 1$, it is established that

$$\int_{-\infty}^{\infty} e^{i\lambda x}\,d\Psi_s^{n2}(x) = i\lambda \int_{-\infty}^{\infty} (e^{i\lambda x} - 1 - i\lambda x)\,d\Psi_s^{n2}(x).$$

Next, for

$$I_1^n(s) = \int_{|x|\le\delta} (e^{i\lambda x} - 1 - i\lambda x)\,\Psi_s^{n2}(x)\,dx$$

and

$$I_2^n(s) = \int_{|x|>\delta} (e^{i\lambda x} - 1 - i\lambda x)\,\Psi_s^{n2}(x)\,dx$$

the inequalities

$$|I_1^n(s)| \le \frac{\lambda^2\delta}{2}\Delta_s^{n2}, \quad |I_2^n(s)| \le 2|\lambda| \int_{|x|>\delta} |x|\,|\Psi_s^{n2}(x)|\,dx$$

are established in virtue of which

$$\sum_{s\in\{s\le t:\, a_s^n>0\}} \left| \int_{-\infty}^{\infty} e^{i\lambda x}\, d\Psi_s^{n2}(x) \right|$$

$$\le 2\lambda^2 \sum_{s\in\{s\le t:\, a_s^n>0\}} \int_{|x|>\delta} |x|\,|\Psi_s^{n2}(x)|\,dx + \frac{|\lambda^3|\delta}{2} \sum_{0<s\le t} \Delta_s^{n2}.$$

The proof is accomplished in the same manner as in the case j = 0.

Lemma 2. *Let the conditions of Lemma* 1 *be fullfilled. Then (setting* $\frac{0}{0} = 0$), *for* $t \in S$, $j = 0, 1, 2$

$$\prod_{s\in\{s\le t:\, a_s^n>0\}} \int_{-\infty}^{\infty} e^{i\lambda x}\, dF_s^n(x + b_s^{nj}) \Big/ \prod_{s\in\{s\le t:\, a_s^n>0\}} \int_{-\infty}^{\infty} e^{i\lambda x}\, d\Phi\left(x/\sqrt{\Delta_s^{nj}}\right) \xrightarrow{P} 1.$$

Proof. On the set

$$\left\{ \sum_{s\in\{s\le t:\, a_s^n>0\}} \Delta_s^{nj} < c \right\}$$

we have

$$\prod_{s\in\{s\le t:\, a_s^n>0\}} \int_{-\infty}^{\infty} e^{i\lambda x}\, d\Phi\left(x/\sqrt{\Delta_s^{nj}}\right) = \exp\left(-\frac{\lambda^2}{2} \sum_{0<s\le t} \Delta_s^{nj} \right) \ge \exp\left(-\frac{\lambda^2}{2} c \right).$$

Therefore, by the assumption

$$\lim_{c\to\infty} \overline{\lim_{n}}\, P\left(\sum_{0<s\le t} \Delta_s^{nj} \ge c \right) = 0$$

it suffices to show that

$$\lim_n \mathrm{P}\Big(\Big|\prod_{s\in\{s\le t:\, a^n_s>0\}}\int_{-\infty}^{\infty} e^{i\lambda x}\, dF^n_s(x+b^{nj}_s)$$

$$-\prod_{s\in\{s\le t:\, a^n_s>0\}}\int_{-\infty}^{\infty} e^{i\lambda x}\, d\Phi\{x/\sqrt{\Delta^{nj}_s}\,)\Big|\ge a,\ \sum_{0<s\le t}\Delta^{nj}_s<c\Big)=0, \quad (6.30)$$

$$a>0,\quad c>0.$$

But according to Problem 1

$$\Big|\prod_{s\in\{s\le t:\, a^n_s>0\}}\int_{-\infty}^{\infty} e^{i\lambda x}\, dF^n_s(x+b^{nj}_s)-\prod_{s\in\{s\le t:\, a^n_s>0\}}\int_{-\infty}^{\infty} e^{i\lambda x}\, d\Phi(x/\sqrt{\Delta^{nj}_s}\,)\Big|$$

$$\le\sum_{s\in\{s\le t:\, a^n_s>0\}}\Big|\int_{-\infty}^{\infty} e^{i\lambda x}\, d\Psi^{nj}_s(x)\Big|,$$

where $\Psi^{nj}_s(x)$ is defined by the formula (6.14), so that the relation (6.30) holds by Lemma 1.

Lemma 3. *Let* $X^n=(X^n_t, \mathcal{F}^n_t)$ *be a semimartingale with the triplet* $T^n=(B^n, C^n, \nu^n)$, *and* b^{nj} *and* Δ^{nj} *processes defined by* (6.8) - (6.11). *Then on the set*

$$\Big\{\sum_{0<s\le t}(|b^{nj}_s|+\Delta^{nj}_s)<c\Big\}$$

the stochastic exponential $\mathcal{E}(G^n(\lambda))$, *related to the triplet* T^n, *admits the representation*

$$\mathcal{E}_t(G^n(\lambda))=\Big[\,i\lambda B^{nc}_t+i\lambda\sum_{0<s\le t}b^{nj}_s-\frac{\lambda^2}{2}C^n_t-\frac{\lambda^2}{2}\sum_{0<s\le t}\Delta^{nj}_s$$
$$+(e^{i\lambda x}-1-i\lambda x I(|x|\le 1))*\nu^{nc}_t\,\Big]$$
$$\times\prod_{s\in\{s\le t:\, a^n_s>0\}}\int_{-\infty}^{\infty} e^{i\lambda x}\, dF^n_s(x+b^{nj}_s)\Big/\prod_{s\in\{s\le t:\, a^n_s>0\}}\int_{-\infty}^{\infty} e^{i\lambda x}\, d\Phi(x/\sqrt{\Delta^{nj}_s}\,). \quad (6.31)$$

Proof. By the representation (2.7) (Ch. 4, § 2) we have

$$\mathcal{E}_t(G^n(\lambda)) = \exp\left[i\lambda B_t^{nc} - \frac{\lambda^2}{2} C_t^n + (e^{i\lambda x} - 1 - i\lambda x I(|x| \le 1)) * \nu_t^{nc}\right]$$

$$\times \prod_{0 < s \le t} \left(1 + \int_{R_0} (e^{i\lambda x} - 1)\, \nu^n(\{s\}, dx)\right)$$

and by the definition of $F_s^n(x)$ we have on the set $\{a_s^n > 0\}$

$$1 + \int_{R_0} (e^{i\lambda x} - 1)\, \nu^n(\{s\}, dx) = \int_{-\infty}^{\infty} e^{i\lambda x}\, dF_s^n(x).$$

The representation for $\mathcal{E}_t(G^n(\lambda))$ is obtained from this in an obvious manner.

5. Proof of Theorem 1. We will show that Conditions 1) - 4) of Theorem 2.1 are fullfilled. Conditions 1), 3) and 4) are obviously fullfilled. Let us verify the validity of Condition 2).

By Condition (c_2) and the fact that $b^{n2} = 0$ (see (6.10)) and $\Delta^{n2} = \Delta\langle X\rangle$ (see (6.12)) we have

$$\lim_{c\to\infty} \overline{\lim_n}\, P\left(\sum_{0 < s \le t} (|b_s^{n2}| + \Delta_s^{n2} \ge c)\right) = 0, \quad t \in S.$$

Therefore it suffices to show that

$$\lim_n P(|\mathcal{E}_t(G^n(\lambda)) - \mathcal{E}_t(G(\lambda))| \ge a, \sum_{0 < s \le t} (|b_s^{n2}| + \Delta_s^{n2}) < c) = 0, \qquad (6.32)$$

$$a > 0, \quad c > 0, \quad t \in S;$$

to this end we may use the representation (6.31) for $\mathcal{E}_t(G^n(\lambda))$.

In view of Lemma 2 the proof is reduced to establishing that ($b^{n2} = 0$)

$$i\lambda B_t^{nc} - \frac{\lambda^2}{2}\left(C_t^n + \sum_{0 < s \le t} \Delta\langle X^n\rangle_s\right) + (e^{i\lambda x} - 1 - i\lambda\, x I(|x| \le 1)$$

$$* \nu_t^{nc} \xrightarrow{P} -\frac{\lambda}{2} \langle X\rangle_t, \quad t \in S. \qquad (6.33)$$

According to Problem 4.1.10 we have

$$B_t^{nc} = -xI(|x| > 1) * \nu_t^{nc},$$

while by Problem 3.5.2 we have

$$\langle X^n\rangle_t = C_t^n + x^2 * \nu_t^n = C_t^n + \sum_{0 < s \le t} \Delta\langle X^n\rangle_s + x^2 * \nu_t^{nc}.$$

In view of these relations and Conditions (L_c^2) and (c_2) the relation (6.33) takes place, provided

$$\left(e^{i\lambda x} - 1 - i\lambda x I(|x| \le 1) + \frac{\lambda^2 x^2}{2}\right) * \nu_t^{nc} \xrightarrow{P} 0, \tag{6.34}$$

The proof of (6.34) is based on the inequality ($\delta \in (0, 1]$)

$$\left|\left(e^{i\lambda x} - 1 - i\lambda x I(|x| \le 1) + \frac{\lambda^2 x^2}{2}\right) * \nu_t^{nc}\right|$$

$$\le \left(2 + |\lambda| + \frac{\lambda^2}{2} x^2\right) I(|x| > \delta) * \nu_t^{nc} + \frac{|\lambda^3| \delta}{6} \langle X^n \rangle_t.$$

By Condition (L_2^c) we have

$$\left(2 + |\lambda| + \frac{\lambda^2}{2} x^2\right) I(|x| > \delta) * \nu_t^{nc} \xrightarrow{P} 0, \quad t \in S.$$

Consequently it remains to show that

$$\lim_{\delta \to 0} \overline{\lim_n} \, \mathsf{P}\left(\frac{|\lambda^3| \delta}{6} \langle X^n \rangle_t \ge a\right) = 0, \quad a > 0, \quad t \in S.$$

The last relation holds since on the set $\{\langle X^n \rangle_t < c\}$ we have

$$\frac{|\lambda^3| \delta}{6} \langle X^n \rangle_t \to 0, \quad \delta \to 0,$$

while by Condition (c_2) we have

$$\lim_{c \to \infty} \overline{\lim_n} \, \mathsf{P}(\langle X^n \rangle_t \ge c) = 0, \quad t \in S.$$

6. Proof of Theorem 2. Just as in the course of proving Theorem 1, it suffices to show that

$$i\lambda B_t^{nc} - \frac{\lambda^2}{2}\left(C_t^n + \sum_{0 < s \le t} \Delta_s^{n1}\right) + (e^{i\lambda x} - 1 - i\lambda x I(|x| \le 1)) * \nu_t^{nc} \xrightarrow{P} \frac{\lambda^2}{2} \langle X \rangle_t,$$

$$t \in S. \tag{6.35}$$

By the representation

$$B_t^{nc} = -xI(|x|>1) * v_t^{nc}$$

(Problem 4.1.10) and Conditions (L_1^c) and (c_1^Δ) the relation (6.35) takes place, provided

$$(e^{i\lambda x} - 1 - i\lambda x I(|x|\le 1) + \frac{\lambda^2 x^2}{2} I(|x|\le \delta)) * v_t^{nc} \xrightarrow{P} 0,$$
$$\delta \in (0,1], \quad t \in S.$$

The last relation takes place by the inequality

$$\left| \left(e^{i\lambda x} - 1 - i\lambda x I(|x|\le 1) + \frac{\lambda^2 x^2}{2} I(|x|\le\delta) \right) * v_t^{nc} \right|$$

$$\le 2|\lambda||x|I(|x|>\delta) * v_t^{nc} + \frac{|\lambda^3|\delta}{6} x^2 I(|x|\le\delta) * v_t^{nc}$$

$$\le 2|\lambda||x|I(|x|>\delta) * v_t^{nc} + \frac{|\lambda^3|\delta}{6} V_t^{n1}$$

with

$$V_t^{n1} = C_t^n + x^2 I(|x|\le\delta) * v_t^{nc} + \sum_{0<s\le t} \Delta_s^{n1}. \tag{6.36}$$

The proof is accomplished as in Theorem 1 by utilizing Conditions (L_1^c) and (c_1^Δ).

7. Proof of Theorem 3. Just as in the course of proving Theorems 1 and 2, it suffices to show that

$$i\lambda\left(B_t^{nc} + \sum_{0<s\le t} b_s^{n0} \right) - \frac{\lambda^2}{2}\left(C_t^n + \sum_{0<s\le t} \Delta_s^{n0} \right)$$

$$+ (e^{i\lambda x} - 1 - i\lambda x I(|x|\le 1)) * v_t^{nc} \xrightarrow{P} -\frac{\lambda^2}{2}\langle X\rangle_t, \quad t \in S. \tag{6.37}$$

By Conditions (b) and (c_0^Δ) the relation (6.37) takes place, provided

$$(e^{i\lambda x} - 1 - i\lambda x I(|x|\le 1) + \frac{\lambda^2 x^2}{2} I(|x|\le\delta)) * v_t^{nc} \xrightarrow{P} 0,$$
$$\delta \in (0,1], \quad t \in S.$$

Observe that

$$\left| \left(e^{i\lambda x} - 1 - i\lambda x I(|x| \le 1) + \frac{\lambda^2 x^2}{2} I(|x| \le \delta) \right) \right| * v_t^{nc}$$

$$\le (2 + |\lambda|) I(|x| > \delta) * v_t^{nc} + \frac{|\lambda^3| \delta}{6} x^2 I(|x| \le \delta) * v_t^{n0}$$

$$\le (2 + |\lambda|) I(|x| > \delta) * v_t^{nc} + \frac{|\lambda^3| \delta}{6} V_t^{n0}$$

with

$$V_t^{n0} = C_t^n + x^2 I(|x| \le \delta) * v_t^{nc} + \sum_{0 < s \le t} \Delta_s^{n0}.$$

This accomplishes the proof as in Theorems 1 and 2, by utilizing Conditions (a^c) and (c_0^Δ).

8. Remarks. 1) Under Conditions (a*) (see 6.1)) and (L_2^c), (c_2) (Theorem 1) Condition (Λ_2) (Theorem 1) is equivalent to Condition (L_2) (Theorem 5.4) (see Problem 2).

2) Under Conditions (a*) (see (6.1)) and (L_1^c) (Theorem 2), Conditions $((c_1^\Delta)$, $(\Lambda_1))$ (Theorem 2) are equivalent to Conditions $((c_1, (L_1))$ (Theorem 5.4) (see Problem 3).

These remarks show the relationship between the "classical" and "nonclassical" conditions under Conditions (a*), (L_2^c) and (L_1^c).

9. Let

$$(\xi_{nk}, \mathcal{H}_k^n)_{k \ge 1}, \ n \ge 1,$$

be sequences introduced in Subsection 5.13. Denote by

$$F_k^n(x) = \mathsf{P}(\xi_{nk} \le x \mid \mathcal{H}_{k-1}^n)$$

the regular conditional probability and define random variables b_k^n, Δ_k^{nj}, $j = 0, 1, 2$, by the following relations

$$\int_{-1}^{1} F_k^n(x + b_k^n)\, dx = 1$$

and

$$\int_0^\infty g_j(x)\,[\,(1-\Phi(x/\sqrt{\Delta_k^{nj}}\,))+\Phi(-x/\sqrt{\Delta_k^{nj}}\,)\,]\,dx$$

$$=\int_0^\infty g_j(x)\,[\,(1-F_k^n(x+b_k^{nj}))+F_k^n(-x+b_k^{nj})\,]\,dx$$

with $g_j(x)$, $j = 0, 1, 2$, given by the formulas (6.7), $b_k^{n0} = b_k^n$, $b_k^{n1} = b_k^{n2} = 0$ (it is assumed here that in each of the cases $j = 1, 2$

$$\int_{-\infty}^{\infty} |x|^j\, dF_k^n(x) < \infty, \quad \int_{-\infty}^{\infty} x\, dF_k^n(x) = 0 \text{ (P-a.s.)}).$$

Observe that

$$\Delta_k^{n2} = \int_{-\infty}^{\infty} x^2\, dF_k^n(x).$$

Denote

$$S_n = \sum_{k=1}^{n} \xi_{nk},$$

and let $N(0, \sigma^2)$ be a Gaussian random variable with the parameters $(0, \sigma^2)$.

Theorem 4 (Corollary to Theorem 1). *Let*

$$\int_{-\infty}^{\infty} x^2\, dF_k^n(x) < \infty, \quad \int_{-\infty}^{\infty} x\, dF_k^n(x) = 0 \quad \text{(P-a.s.)}, \; k \geq 1, \; n \geq 1,$$

and let the following conditions be fullfilled:

$$\sum_{k=1}^{n} \int_{|x|>\delta} |x|\,|F_k^n(x) - \Phi(x/\sqrt{\Delta_k^{n2}}\,)|\,dx \xrightarrow{P} 0, \quad \delta \in (0, 1],$$

$$\sum_{k=1}^{n} \Delta_k^{n2} \xrightarrow{P} \sigma^2.$$

Then

$$S_n \xrightarrow{d} N(0, \sigma^2).$$

Theorem 5 (Corollary to Theorem 2). *Let*

$$\int_{-\infty}^{\infty} |x|\, dF_k^n(x) < \infty, \quad \int_{-\infty}^{\infty} x\, dF_k^n(x) = 0 \ \ (\text{P-a.s.}),\ k \geq 1,\ n \geq 1,$$

and let the following conditions be fullfilled:

$$\sum_{k=1}^{n} \int_{|x|>\delta} |F_k^n(x) - \Phi(x/\sqrt{\Delta_k^{n1}})|\, dx \xrightarrow{P} 0, \quad \delta \in (0, 1],$$

$$\sum_{k=1}^{n} \Delta_k^{n1} \xrightarrow{P} \sigma^2.$$

Then

$$S_n \xrightarrow{d} N(0, \sigma^2).$$

Theorem 6 (Corollary to Theorem 3). *Let the following conditions be fullfilled:*

$$\sum_{k=1}^{n} b_k^n \xrightarrow{P} 0,$$

$$\sum_{k=1}^{n} \left(\int_{\delta<|x|\leq 1} \left| F_k^n(x + b_k^n) - \Phi(x/\sqrt{\Delta_k^{n0}}) \right| dx \right.$$

$$\left. + \sup_{|x|\geq 1} |F_k^n(x + b_k^n) - \Phi(x/\sqrt{\Delta_k^{n0}})| \right) \xrightarrow{P} 0, \quad \delta \in (0, 1],$$

$$\sum_{k=0}^{n} \Delta_k^{n0} \xrightarrow{P} \sigma^2.$$

Then

$$S_n \xrightarrow{\delta} N(0, \sigma^2).$$

Problems

1. Let p_s and q_s be complex variables, $s \in I$ where I is a countable subset of R_+, $|p_s| \leq 1$, $|q_s| \leq 1$.

Show that

$$\left| \prod_{s \in I} p_s - \prod_{s \in I} q_s \right| \leq \sum_{s \in I} |p_s - q_s|.$$

(Hint: apply Ito's formula.)

2. Show that under Condition (a*) (see (6.1)) and (L_2^c), (c_2) (Theorem 1)

$$(\Lambda_2) \text{ (Theorem 1)} \Leftrightarrow (L_2) \text{ (Theorem 5.4).}$$

3. Show that under Condition (a*) (see (6.1)) and (L_1^c) (Theorem 2)

$$((c_1^{\Delta}), (\Lambda_1)) \text{ (Theorem 2)} \Leftrightarrow ((c_1), (L_1)) \text{ (Theorem 5.4).}$$

4. Prove the existence of processes b^{n0}, Δ^{nj}, $j = 0, 1, 2$, with properties (6.8) - (6.11).

§ 7. Evaluation of a convergence rate for marginal distributions in the central limit theorem

1. Let $X^n = (X_t^n, \mathcal{F}_t^n)$, $n \geq 1$, be semimartingales with the triplets $T^n = (B^n, C^n, \nu^n)$, $X_0^n = 0$, $n \geq 1$. Assume that for a given $t > 0$ the following conditions are fullfilled: as $n \to \infty$

(a) $\nu^n((0, t] \times \{|x| > \delta\}) \xrightarrow{P} 0$, $\delta \in (0, 1]$,

(b) $B_t^n \xrightarrow{P} 0$,

(c) $\langle M^{n, \delta} \rangle_t \xrightarrow{P} \sigma^2$, $\delta \in (0, 1]$,

where $\langle M^{n, \delta} \rangle$ is defined by the formula (5.4) (§ 5) and σ is a positive constant.

Denote

$$\Phi(x) = \frac{1}{\sqrt{2\pi}} \int_{-\infty}^{x} e^{-\frac{y^2}{2}} dy. \tag{7.1}$$

Theorem 5.1 implies

$$P(X_t^n \leq x) \to \Phi\left(\frac{x}{\sigma}\right), \quad n \to \infty, \quad x \in R. \tag{7.2}$$

(As for $X = (X_s, \mathcal{F}_s)$ in Theorem 5.1, it suffices to take $X_s = \xi I(s \geq t)$ where ξ is a Gaussian random variable with parameters $(0, \sigma^2)$, and $\mathcal{F}_s = \sigma\{X_u, 0 \leq u \leq s\}$.)

Since $\Phi(x)$ is a continuous function, (7.2) and Polya's theorem imply

$$\sup_{x \in R} \left| P(X_t^n \leq x) - \Phi\left(\frac{x}{\sigma}\right) \right| \to 0, \quad n \to \infty.$$

The aim of the present section consists in giving an upper bound for

$$r^n = \sup_{x \in R} \left| P(X_t^n \leq x) - \Phi\left(\frac{x}{\sigma}\right) \right|, \quad n \geq 1.$$

2. The method of evaluating variables r^n is based on the wellknown Esseen inequality [240]: if $F(x)$ is a distribution function with the characteristic function

$$\phi(\lambda) = \int_{-\infty}^{\infty} e^{i\lambda x} dF(x), \quad \lambda \in R,$$

then for every $z \geq 0$

$$\sup_{x \in R} \left| F(x) - \Phi\left(\frac{x}{\sigma}\right) \right| \le \frac{2}{\pi} \int_0^{z^{-1}} \frac{\left| \phi(\lambda) - e^{-\lambda^2\sigma^2/2} \right|}{\lambda} \, d\lambda + \frac{24z}{\pi\sqrt{2\pi\sigma^2}} \qquad (7.3)$$

The main difficulty consists in getting a "good" upper bound for

$$\left| E e^{i\lambda X_t^n} - e^{-\frac{\lambda^2}{2}\sigma^2} \right|,$$

which allows the evaluation of r^n. The final step in evaluating r^n will be based on the inequality given in Lemma 1 below. In the course of proving this lemma, the properties are utilized of Lévy's distance $\mathcal{L}$ (F, G) between functions F = F (x) and G = G (x). It is defined in the following manner ([314], Ch. 4, § 19.2):

$$\mathcal{L}(F, G) = \sup_{x \in R} \min (h: G(x - h) - h \le F(x) \le G(x + h) + h). \qquad (7.4)$$

Let us give at once a number of properties of this distance $\mathcal{L}$ (F, G), necessary in the sequel:

(1) $\mathcal{L}(F, G) \le \sup_{x \in R} |F(x) - G(x)|$;

(2) if g (x) is the density (relative to Lebesgue's measure) of a distribution function G (x), then

$$\sup_{x \in R} |F(x) - G(x)| \le (1 + \sup_{x \in R} g(x))\, \mathcal{L}(F, G);$$

(3) if α and β are random variables on one and the same probability space, with distribution functions F and G, then

$$\mathcal{L}(F, G) \le d + P(|\alpha - \beta| > d), \quad d > 0,$$

$$\mathcal{L}(F, G) \le (c + 1)\, c^{-c/(c+1)} (E|\alpha - \beta|^c)^{1/(1+c)}, \quad c \ge 1.$$

Property (1) is easily derived from the definition (7.4) of $\mathcal{L}$ (F, G). In fact, if $a = \mathcal{L}(F, G)$, $b = \sup_{x \in R} |F(x) - G(x)|$ and the inequality a > b holds, then a point y (or y') can be found such that G (y — b) — b > F (y) (or F (y') > G (y' + b) + b). By this G (y) — F (y) > b (or F (y') — G (y') > b), which contradicts to the definition of b.

Property (2) is also derived directly from (7.4), since for $a = \mathcal{L}(F, G)$ we get by (7.4) that

$$F(x) - G(x) \le a + G(x + a) - G(x),$$

$$F(x) - G(x) \ge -a + G(x - a) - G(x).$$

Property (3) follows from the inequality (as $1 = d + P(|\alpha - \beta| > d)$):

$$G(x-1) = P(\alpha \le x-1) \le P(\alpha \le x-1,\ \beta > x) + P(\beta \le x)$$

$$\le P(|\alpha-\beta| > 1) + F(x) \le P(|\alpha-\beta| > d) + F(x)$$

$$\le 1 + F(x),\ x \in R$$

and the following inequality, verified analogously:

$$F(x) \le G(x+1) + 1, \quad x \in R,$$

in view of which $1 \ge \mathfrak{L}(F, G)$. The second property in (3) is based on the inequality

$$\min_d\, [d + P(|\alpha-\beta| > d)] \le \min_d \left[d + \frac{E|\alpha-\beta|^c}{d^c} \right], \tag{7.5}$$

the right-hand side of which coincides with the right-hand side of the estimate for $\mathfrak{L}(F, G)$ in (3).

From Properties (1) - (3) of $\mathfrak{L}(F, G)$ the following result can be deduced.

Lemma 1. *Let* F, G *and* Φ_σ *be distribution functions where*

$$\Phi_\sigma(x) = \Phi\left(\frac{x}{\sigma}\right), \quad \sigma > 0,$$

and let α *and* β *be random variables on one and the same probability space with the distribution functions* F *and* G *respectively.*

Then

$$\sup_{x \in R} \left| F(x) - \Phi\left(\frac{x}{\sigma}\right) \right|$$

$$\le \left(1 + \frac{1}{\sqrt{2\pi\sigma^2}}\right)\left[\sup_{x \in R}\left| G(x) - \Phi\left(\frac{x}{\sigma}\right)\right| + 2\,(E(\alpha-\beta)^2)^{1/3}\right]. \tag{7.6}$$

Proof. By Property (2)

$$\sup_{x \in R}\left| F(x) - \Phi\left(\frac{x}{\sigma}\right)\right| \le \left(1 + \frac{1}{\sqrt{2\pi s^2}}\right)\mathfrak{L}(F, \Phi_\sigma).$$

Next, by the triangle inequality

$$\mathfrak{L}(F, \Phi_\sigma) \le \mathfrak{L}(F, G) + \mathfrak{L}(G, \Phi_\sigma)$$

and in view of Properties (1) and (3) (as c = 2)

$$\mathfrak{L}(G, \Phi_\sigma) \le \sup_{x \in R}\left| G(x) - \Phi\left(\frac{x}{\sigma}\right)\right|, \quad \mathfrak{L}(F, G) \le 2\,(E(\alpha-\beta)^2)^{1/3}.$$

The desired relation follows from this in an obvious manner.

3. Let us turn now to evaluating

$$\left| \mathbf{E} e^{i\lambda X_t} - e^{-\frac{\lambda^2}{2}\sigma^2} \right|$$

from above, where $X = (X_t, \mathcal{F}_t)$ is a semimartingale. First, let us consider particular cases.

Lemma 2. *Let* $X = (X_s, \mathcal{F}_s)$ *be a locally square integrable martingale with* $X_0 = 0$ *and the quadratic characteristic* $\langle X \rangle$, *and for a fixed* $t > 0$ *let the inequality*

$$\langle X \rangle_t \le \sigma^2 \tag{7.7}$$

hold. Then for $\lambda \in R$ *and such value of* t *we have*

$$\left| \mathbf{E} e^{i\lambda X_t} - e^{-\frac{\lambda^2}{2}\sigma^2} \right| \le c_1 |\lambda|^3 \mathbf{E} \sum_{0 < s \le t} |\Delta X_s|^3 + \frac{\lambda^2}{2} \mathbf{E} \left| \langle X \rangle_t - \sigma^2 \right| \tag{7.8}$$

with $c_1 = 1/2^{3/2} + 1/6$.

Proof. By the inequality

$$\left| \mathbf{E} e^{i\lambda X_t} - e^{-\frac{\lambda^2}{2}\sigma^2} \right|$$

$$\le \left| \mathbf{E} e^{i\lambda X_t + \frac{\lambda^2}{2}(\langle X \rangle_t - \sigma^2)} - e^{-\frac{\lambda^2}{2}\sigma^2} \right| + \left| \mathbf{E} \left(e^{i\lambda X_t} - e^{i\lambda X_t + \frac{\lambda^2}{2}(\langle X \rangle_t - \sigma^2)} \right) \right|$$

and the obvious estimate

$$\left| \mathbf{E} \left(e^{i\lambda X_t} - e^{i\lambda X_t + \frac{\lambda^2}{2}(\langle X \rangle_t - \sigma^2)} \right) \right| \le \frac{\lambda^2}{2} \mathbf{E} \left| \langle X \rangle_t - \sigma^2 \right|$$

following from (7.7), it suffices to show that

$$\left| \mathbf{E} \left(e^{i\lambda X_t + \frac{\lambda^2}{2}(\langle X \rangle_t - \sigma^2)} - e^{-\frac{\lambda^2}{2}\sigma^2} \right) \right| \le c_1 |\lambda|^3 \mathbf{E} \sum_{0 < s \le t} |\Delta X_s|^3. \tag{7.9}$$

Denote

$$V_s = e^{i\lambda X_s}, \quad U_s = e^{\frac{\lambda^2}{2}\langle X \rangle_s}.$$

Then the left hand side of the inequality (7.9) has the form

$$\left| \mathbf{E} e^{-\frac{\lambda^2}{2}\sigma^2} (V_t U_t - 1) \right|,$$

besides by the assumption (7.7) we have

$$e^{-\frac{\lambda^2}{2}\sigma^2} U_s \le 1, \quad s \le t. \tag{7.10}$$

Let us calculate the product $V_t U_t$ by means of Ito's formula (Ch. 2, § 3). We have

$$V_t = 1 + i\lambda V_- \cdot X_t - \frac{\lambda^2}{2} V_- \circ \langle X^c\rangle_t + \sum_{0<s\le t} V_{s-}\left(e^{i\lambda\Delta X_s} - 1 - i\lambda\Delta X_s\right), \tag{7.11}$$

$$U_t = 1 + \frac{\lambda^2}{2} U_- \circ \langle X\rangle_t + \sum_{0<s\le t} U_{s-}\left(e^{\frac{\lambda^2}{2}\Delta\langle X\rangle_s} - 1 - \frac{\lambda^2}{2}\Delta\langle X\rangle_s\right). \tag{7.12}$$

Since $(U_s)_{s\ge 0}$ is a predictable process, then

$$V_t U_t = 1 + U \cdot V_t + V_- \circ U_t.$$

Therefore (7.11) and (7.12) give

$$\begin{aligned} V_t U_t = {} & 1 + i\lambda U V_- \cdot X_t - \frac{\lambda^2}{2} U V_- \circ \langle X^c\rangle_t \\ & + \sum_{0<s\le t} U_s V_{s-}\left(e^{i\lambda\Delta X_s} - 1 - i\lambda\Delta X_s\right) + \frac{\lambda^2}{2} U_- V_- \circ \langle X\rangle_t \\ & + \sum_{0<s\le t} U_{s-} V_{s-}\left(e^{\frac{\lambda^2}{2}\Delta\langle X\rangle_s} - 1 - \frac{\lambda^2}{2}\Delta\langle X\rangle_s\right). \end{aligned} \tag{7.13}$$

Observe that in virtue of Condition (7.7) and the inequality

$$| U_s V_s | \le U_s \le e^{\frac{\lambda^2}{2}\sigma^2}$$

(following from (7.7)) the equality

$$\mathbf{E}\, i\lambda U V_- \cdot X_t = 0 \tag{7.14}$$

holds. Besides

$$-\frac{\lambda^2}{2} U V_- \circ \langle X^c\rangle_t + \frac{\lambda^2}{2} U_- V_- \circ \langle X\rangle_t = \frac{\lambda^2}{2} U_- V_- \circ \langle X^d\rangle_t, \tag{7.15}$$

where $\langle X^c\rangle$ and $\langle X^d\rangle$ are the quadratic characterisrics of local square integrable martingales $(X_s^c, \mathcal{F}_s)$ and $(X_s^d, \mathcal{F}_s)$, involved in the decomposition $X = X^c + X^d$ (Theorem 1.7.2).

By (7.13)- (7.15) we get

$$\mathbf{E}\,(V_tU_t - 1) = \mathbf{E}\sum_{0<s\le t} U_sV_{s-}\,(e^{i\lambda\Delta X_s} - 1 - i\lambda\Delta X_s)$$

$$+\,\mathbf{E}\sum_{0<s\le t} U_{s-}V_{s-}\left(e^{\frac{\lambda^2}{2}\Delta\langle X\rangle_s} - 1 - \frac{\lambda^2}{2}\Delta\langle X\rangle_s\right) + \mathbf{E}\,\frac{\lambda^2}{2}U_-V_-\circ\langle X^d\rangle_t. \quad (7.16)$$

The process $\langle X^d\rangle$ presents the compensator of the process

$$\sum_s (\Delta X_s^d)^2 = \sum_s (\Delta X_s)^2$$

(Theorem 1.8.1). Therefore by Condition 1 to Theorem 1.6.3 and in view of the inequality (7.7) we get

$$\mathbf{E}\,\frac{\lambda^2}{2}U_-V_-\circ\langle X^d\rangle_t = \mathbf{E}\,\frac{\lambda^2}{2}\sum_{0<s\le t} U_{s-}V_{s-}\,(\Delta X_s)^2.$$

Consequently, the equality (7.16) can be rewritten in the following form

$$\mathbf{E}\,(V_tU_t - 1) = \mathbf{E}\sum_{0<s\le t} U_sV_{s-}\left(e^{i\lambda\Delta X_s} - 1 - i\lambda\Delta X_s + \frac{\lambda^2}{2}(\Delta X_s)^2\right)$$

$$-\,\mathbf{E}\,\frac{\lambda^2}{2}\sum_{0<s\le t}\Delta U_sV_{s-}\,(\Delta X_s)^2 + \mathbf{E}\sum_{0<s\le t} U_{s-}V_{s-}\left(e^{\frac{\lambda^2}{2}\Delta\langle X\rangle_s} - 1 - \frac{\lambda^2}{2}\Delta\,\langle X\rangle_s\right). \quad (7.17$$

Since

$$U_s = U_{s-}\,e^{\frac{\lambda^2}{2}\Delta\langle X\rangle_s}$$

then

$$\mathbf{E}\sum_{0<s\le t}\Delta U_sV_{s-}\,(\Delta X_s)^2 = \mathbf{E}\sum_{0<s\le t} U_{s-}V_{s-}\,(e^{\frac{\lambda^2}{2}\Delta\langle X\rangle_s} - 1)\,(\Delta X_s)^2$$

$$= \mathbf{E}U_-V_-\,(e^{\frac{\lambda^2}{2}\Delta\langle X\rangle} - 1)\circ\langle X^d\rangle_t = \mathbf{E}\sum_{0<s\le t} U_{s-}V_{s-}\,(e^{\frac{\lambda^2}{2}\Delta\langle X\rangle_s} - 1)\,\Delta\,\langle X\rangle_s.$$

Hence

$$\mathbf{E}\,(V_tU_t - 1) = \mathbf{E}\sum_{0<s\le t} U_sV_{s-}\left[e^{i\lambda\Delta X_s} - 1 - i\lambda\Delta X_s + \frac{\lambda^2}{2}(\Delta X_s)^2\right] \tag{7.18}$$

$$+\,\mathbf{E}\sum_{0<s\le t} U_{s-}V_{s-}\,[e^{\frac{\lambda^2}{2}\Delta\langle X\rangle_s} - 1 - \frac{\lambda^2}{2}\Delta\langle X\rangle_s - \frac{\lambda^2}{2}(e^{\frac{\lambda^2}{2}\Delta\langle X\rangle_s} - 1)\,\Delta\langle X\rangle_s].$$

Observe now that

$$U_{s-}\left[e^{\frac{\lambda^2}{2}\Delta\langle X\rangle_s} - 1 - \frac{\lambda^2}{2}\Delta\langle X\rangle_s - \frac{\lambda^2}{2}(e^{\frac{\lambda^2}{2}\Delta\langle X\rangle_s} - 1)\,\Delta\langle X\rangle_s\right]$$

$$= U_s\left[1 - e^{\frac{\lambda^2}{2}\Delta\langle X\rangle_s} - \frac{\lambda^2}{2}\Delta\langle X\rangle_s\right].$$

By this and (7.18) we get the following representation for $\mathbf{E}\,(V_tU_t - 1)$:

$$\mathbf{E}\,(V_tU_t - 1) = \mathbf{E}\sum_{0<s\le t} U_sV_{s-}\left[e^{i\lambda\Delta X_s} - 1 - i\lambda\Delta X_s + \frac{\lambda^2}{2}(\Delta X_s)^2\right]$$

$$+\,\mathbf{E}\sum_{0<s\le t} U_sV_{s-}\left[1 - e^{-\frac{\lambda^2}{2}\Delta\langle X\rangle_s} - \frac{\lambda^2}{2}\Delta\langle X\rangle_s\right], \tag{7.19}$$

in view of which the inequality

$$|\,\mathbf{E}e^{-\frac{\lambda^2}{2}\sigma^2}(V_tU_t - 1)\,| \le \mathbf{E}\sum_{0<s\le t}\left|e^{i\lambda\Delta X_s} - 1 - i\lambda\Delta X_s + \frac{\lambda^2}{2}(\Delta X_s)^2\right|$$

$$+\,\mathbf{E}\sum_{0<s\le t}\left(e^{-\frac{\lambda^2}{2}\Delta\langle X\rangle_s} - 1 + \frac{\lambda^2}{2}\Delta\langle X\rangle_s\right) \tag{7.20}$$

holds by (7.10).

Since

$$\left|e^{i\lambda\Delta X_s} - 1 - i\lambda\Delta X_s + \frac{\lambda^2}{2}(\Delta X_s)^2\right| \le \frac{|\lambda^3|}{6}\,|\Delta X_s|^3,$$

to complete the proof of the lemma it suffices to show only that

$$\mathbf{E}\sum_{0<s\le t}\left(e^{-\frac{\lambda^2}{2}\Delta\langle X\rangle_s}-1+\frac{\lambda^2}{2}\Delta\langle X\rangle_s\right)\le\frac{1}{2^{3/2}}|\lambda^3|\,\mathbf{E}\sum_{0<s\le t}|\Delta X_s|^3. \quad (7.21)$$

To prove this inequality, denote

$$g(x)=e^{-\frac{\lambda^2}{2}x}-1+\frac{\lambda^2}{2}x$$

for $x\ge 0$. Then the left-hand side of the inequality (7.21) may be rewritten as

$$\mathbf{E}\sum_{0<s\le t} g(\Delta\langle X\rangle_s).$$

Consider the increasing process

$$\sum_s g((\Delta X_s)^2).$$

Since $g(x)\le\lambda^2x^2$, then

$$\sum_{0<s\le t} g((\Delta X_s)^2)\le\lambda^2\sum_{0<s\le t}(\Delta X_s)^2.$$

Consequently

$$\mathbf{E}\sum_{0<s\le t} g((\Delta X_s)^2)\le\lambda^2\mathbf{E}\,[X,X]_t=\lambda^2\mathbf{E}\,\langle X\rangle_t\le\lambda^2\sigma^2.$$

Therefore by Theorem 1.6.3 and Corollary 2 to it ($a>0$)

$$\mathbf{E}\sum_{0<s\le t} g((\Delta X_s)^2)\,I(\Delta\langle X\rangle_s>a)=\mathbf{E}\sum_{0<s\le t}{}^{P}(g((\Delta X)^2)_s\,I(\Delta\langle X\rangle_s>a)$$

where ${}^{P}(g((\Delta X)^2))$ is the predictable projection of the process $g((\Delta X)^2)$.

The function $g(x)$ is concave and hence in view of Problem 1.3.9 we have

$${}^{P}(g((\Delta X)^2))\ge g({}^{P}((\Delta X)^2)),$$

while by Problem 1.8.12 we have

$${}^{P}((\Delta X)^2)=\Delta\langle X\rangle.$$

Consequently,

$$\mathbf{E}\sum_{0<s\le t} g(\Delta\langle X\rangle_s)=\lim_{a\to 0}\mathbf{E}\sum_{0<s\le t} g(\Delta\langle X\rangle_s)\,I(\Delta\langle X\rangle_s>a)$$

$$\le\lim_{a\to 0}\mathbf{E}\sum_{0<s\le t} g((\Delta X_s)^2)\,I(\Delta\langle X\rangle_s>a)$$

$$\le\mathbf{E}\sum_{0<s\le t} g((\Delta X_s)^2). \quad (7.22)$$

It is not hard to verify that the function g (x) possesses as $x \geq 0$ the following property

$$g(x) \leq \frac{|\lambda^3| x^{3/2}}{2^{3/2}}.$$

By taking into consideration this inequality and (7.22) we get

$$\mathbf{E} \sum_{0 < s \leq t} g(\Delta \langle X \rangle_s) \leq \frac{|\lambda^3|}{2^{3/2}} \mathbf{E} \sum_{0 < s \leq t} |\Delta X_s|^3.$$

The lemma is proved.

The assumption (7.7) introduced in Lemma 2 is restrictive. In this connection we establish the following result.

Lemma 3. *Let* $X = (X_s, \mathcal{F}_s)$ *be a locally square integrable martingale with* $X_0 = 0$ *and the quadratic characteristic* $\langle X \rangle$. *Then there can be found a square integrable martingale* $\tilde{X} = (\tilde{X}_t, \mathcal{F}_t)$ *such that*

$$\left| \mathbf{E} e^{i\lambda \tilde{X}_t} - e^{-\frac{\lambda^2}{2}\sigma^2} \right| \leq c_1 |\lambda^3| \mathbf{E} \sum_{0 < s \leq t} |\Delta X_s|^3 + \frac{\lambda^2}{2} \mathbf{E} |\langle X \rangle_t - \sigma^2|,$$

$$\lambda \in R, \tag{7.23}$$

and

$$\mathbf{E} (X_t - \tilde{X}_t)^2 \leq \mathbf{E} |\langle X \rangle_t - \sigma^2|. \tag{7.24}$$

Proof. Define a predictable Markov time $\tau = \inf (s \leq t: \langle X \rangle_s \geq \sigma^2)$ with $\tau = \infty$ if $\langle X \rangle_t < \sigma^2$ (Problem 1.3.11), and predictable processes $\alpha = (\alpha_s)_{s \geq 0}$ and $\beta = (\beta_s)_{s \geq 0}$ with

$$\alpha_s = I(\tau \geq s) \frac{\sigma^2 - \langle X \rangle_{s-}}{\Delta \langle X \rangle_s} \quad \left(\frac{0}{0} = 0 \right),$$

$$\beta_s = I(\tau > s) + \sqrt{\alpha_s}\, I(\tau = s).$$

Observe that $\beta^2 \circ \langle X \rangle \leq \sigma^2$, and consequently a locally square integrable martingale $\tilde{X} = (\tilde{X}_t, \mathcal{F}_t)$ is defined with $\tilde{X}_s = \beta \cdot X_s$ (see Ch. 2, § 2).

Let us show that

$$\langle \tilde{X} \rangle_t \leq \sigma^2.$$

By Theorem 2.2.2 and the remark to it we have

$$\langle \tilde{X} \rangle_t = \beta^2 \circ \langle X \rangle_t.$$

Consequently

$$\langle \tilde{X} \rangle_t = I(\tau > t) \langle X \rangle_t + I(\tau \le t) [\langle X \rangle_{\tau-} + \beta_\tau^2 \Delta \langle X \rangle_\tau] \le \sigma^2.$$

Therefore by Lemma 2 we have

$$\left| \mathbf{E} e^{i\lambda \tilde{X}_t} e^{-\frac{\lambda^2}{2}\sigma^2} \right| \le c_1 |\lambda^3| \mathbf{E} \sum_{0 < s \le t} |\Delta \tilde{X}_s|^3 + \frac{\lambda^2}{2} \mathbf{E} | \langle \tilde{X} \rangle_t - \sigma^2 |. \quad (7.25)$$

By Theorem 2.2.2 and the remark to it $\Delta \tilde{X} = \beta \Delta X$. Consequently on the set $\{\tau > t\}$ we have

$$\sum_{0 < s \le t} |\Delta \tilde{X}_s|^3 = \sum_{0 < s \le t} |\Delta X_s|^3,$$

and on the set $\{\tau \le t\}$

$$\sum_{0 < s \le t} |\Delta \tilde{X}_s|^3 = \sum_{0 < s < \tau} |\Delta X_s|^3 + \left(\sqrt{\frac{\sigma^2 - \langle X \rangle_{\tau-}}{\Delta \langle X \rangle_\tau}} \right)^3 |\Delta X_\tau|^3 I(\Delta \langle X \rangle_\tau > 0).$$

Observe now that on the set $\{\tau \le t\}$ we have

$$\frac{\sigma^2 - \langle X \rangle_{\tau-}}{\Delta \langle X \rangle_\tau} I(\Delta \langle X \rangle_\tau > 0) \le 1,$$

and hence on the set $\{\tau \le t\}$

$$\sum_{0 < s \le t} |\Delta \tilde{X}_s|^3 \le \sum_{0 < s \le \tau} |\Delta X_s|^3.$$

From this it follows that

$$\sum_{0 < s \le t} |\Delta \tilde{X}_s|^3 \le \sum_{0 < s \le t} |\Delta X_s|^3. \quad (7.26)$$

Next, observe that

$$| \langle \tilde{X} \rangle_t - \sigma^2 | = I(\tau > t) | \langle X \rangle_t - \sigma^2 | + I(\tau \le t) | \langle \tilde{X} \rangle_\tau - \sigma^2 |$$
$$= I(\tau > t) | \langle X \rangle_t - \sigma^2 |,$$

since on the set $\{\tau \le t\}$ we have

$$\langle \tilde{X} \rangle_\tau = \sigma^2.$$

This gives

$$| \langle \tilde{X} \rangle_t - \sigma^2 | \le | \langle X \rangle_t - \sigma^2 |. \quad (7.27)$$

The desired inequality (7.23) follows from (7.25) - (7.27).

To prove the inequality (7.24), we utilize the fact that $(X_s - \tilde{X}_s, \mathcal{F}_s)$ is a locally square integrable martingale with

$$X_s - \tilde{X}_s = (1-\beta)\cdot X_s,$$

and consequently

$$\langle X - \tilde{X}\rangle_t = (1-\beta)^2 \circ \langle X\rangle_t.$$

In view of Problem 1.8.6 we have

$$\mathbf{E}\,(X_t - \tilde{X}_t)^2 \le \mathbf{E}\,\langle X - \tilde{X}\rangle_t.$$

By the definition of β_s we have $0 \le \beta_s \le 1$. Therefore $(1-\beta_s)^2 \le 1-\beta_s^2$, and hence the inequality

$$\mathbf{E}\,(X_t - \tilde{X}_t)^2 \le \mathbf{E}\,(1-\beta^2)\circ \langle X\rangle_t \tag{7.28}$$

holds. On the set $\{\tau > t\}$ we have

$$(1-\beta^2)\circ\langle X\rangle_t = 0,$$

while on the set $\{\tau \le t\}$ we have

$$(1-\beta^2)\circ\langle X\rangle_t = \left(1 - \frac{\sigma^2 - \langle X\rangle_{\tau-}}{\Delta\langle X\rangle_\tau}\right)\Delta\langle X\rangle_\tau$$

$$= \langle X\rangle_\tau - \sigma^2 \le \langle X\rangle_t - \sigma^2.$$

Thus

$$(1-\beta^2)\circ\langle X\rangle_t \le |\,\langle X\rangle_t - \sigma^2\,|,$$

so that (7.24) follows from (7.28).

The lemma is proved.

Let $X = (X_s, \mathcal{F}_s)$ be a semimartingale with $X_0 = 0$ and with the decomposition

$$X_s = A_s + M_s, \tag{7.29}$$

where $A = (A_s)_{s\ge 0} \in \mathcal{V}$ and $M = (M_s, \mathcal{F}_s)$ is a locally square integrable martingale with the quadratic characteristic $\langle M\rangle$.

Let $\tilde{M} = (\tilde{M}_s, \mathcal{F}_s)$ be a locally square integrable martingale for which (see Lemma 3)

$$\left|\mathbf{E}e^{i\lambda\tilde{M}_t} - e^{-\frac{\lambda^2}{2}\sigma^2}\right| \le c_1|\lambda^3|\,\mathbf{E}\sum_{0<s\le t}|\Delta M_s|^3 + \frac{\lambda^2}{2}\mathbf{E}\,|\,\langle M\rangle_t - \sigma^2\,|,$$

$$\lambda \in R, \tag{7.30}$$

$$\mathbf{E}\,(M_t - \tilde{M}_t)^2 \le \mathbf{E}\,|\langle M\rangle_t - \sigma^2|. \tag{7.31}$$

Lemma 3 implies

Lemma 4. *Let* $X = (X_s, \mathcal{F}_s)$ *be a semimartingale with the decomposition* (7.29) *and* $\tilde{X} = (\tilde{X}_s, \mathcal{F}_s)$ *a semimartingale with the decomposition* $\tilde{X}_s = A_s + \tilde{M}_s$, *where* $\tilde{M} = (\tilde{M}_s, \mathcal{F}_s)$ *is a locally square integrable martingale with the properties* (7.30) *and* (7.31). *Then*

$$\left| \mathbf{E}e^{i\lambda \tilde{X}_t} - e^{-\frac{\lambda^2}{2}\sigma^2} \right|$$

$$\le |\lambda|\,\mathbf{E}\,|A_t| + c_1 |\lambda^3|\,\mathbf{E}\sum_{0<s\le t} |\Delta M_s|^3 + \frac{\lambda^2}{2}\mathbf{E}\,|\langle M\rangle_t - \sigma^2|, \quad \lambda \in R, \tag{7.32}$$

and

$$\mathbf{E}\,(X_t - \tilde{X}_t)^2 \le \mathbf{E}\,|\langle M\rangle_t - \sigma^2|. \tag{7.33}$$

The proof of the relation (7.32) follows from the inequalities

$$\left| \mathbf{E}e^{i\lambda \tilde{X}_t} - e^{-\frac{\lambda^2}{2}\sigma^2} \right| \le \mathbf{E}\,|e^{i\lambda A_t} - 1| + \left| \mathbf{E}e^{i\lambda \tilde{M}_t} - e^{-\frac{\lambda^2}{2}\sigma^2} \right|,$$

(7.30) and

$$|e^{i\lambda A_t} - 1| \le |\lambda|\,|A_t|.$$

The inequality (7.33) follows from (7.31) and the equality $X_t - \tilde{X}_t = M_t - \tilde{M}_t$.

4. Utilizing Lemmas 1 and 4 we will get an upper bound for the quantity

$$\sup_{x\in R} \left| P(X_t \le x) - \Phi\left(\frac{x}{\sigma}\right) \right|,$$

in case in which $X = (X_s, \mathcal{F}_s)$ is a semimartingale.

Theorem 1. *Let* $X = (X_s, \mathcal{F}_s)$ *be a semimartingale,* $X_0 = 0$, *with the decomposition* (7.29) *and* $\sigma > 0$. *Then*

$$\sup_{x\in R} \left| P(X_t \le x) - \Phi\left(\frac{x}{\sigma}\right) \right|$$

$$\le c_2 \left[(\mathbf{E}\,|A_t|)^{1/2} + (\mathbf{E}\,|\langle M\rangle_t - \sigma^2|)^{1/3} + \left(\mathbf{E}\sum_{0<s\le t} |\Delta M_s|^3 \right)^{1/4} \right]$$

with

$$c_2 = 2 + \left(\frac{2}{\pi} + \frac{24}{\sqrt{2\pi\sigma^2}}\right)\left(1 + \frac{1}{\sqrt{2\pi\sigma^2}}\right).$$

Proof. Let us first estimate the variable

$$\sup_{x \in R} \left| \mathsf{P}(\tilde{X}_t \le x) - \Phi\left(\frac{x}{\sigma}\right)\right|,$$

where $\tilde{X} = (\tilde{X}_s, \mathcal{F}_s)$ is a semimartingale defined in Lemma 4.

In virtue of Esseen's inequality (see (7.3)) we have to estimate the variable

$$\int_0^{z^{-1}} \frac{\left| \mathbf{E} e^{i\lambda \tilde{X}_t} - e^{-\frac{\lambda^2}{2}\sigma^2}\right|}{\lambda} d\lambda$$

and then to choose a parameter z appropriately.

By the estimate (7.32) for $\lambda > 0$ we have

$$\frac{\left| \mathbf{E} e^{i\lambda \tilde{X}_t} - e^{-\frac{\lambda^2}{2}\sigma^2}\right|}{\lambda}$$

$$\le \mathbf{E}\,|A_t| + c_1\lambda^2 \mathbf{E} \sum_{0<s\le t} |\Delta M_s|^3 + \frac{\lambda}{2}\mathbf{E}\,|\langle M\rangle_t - \sigma^2|. \tag{7.34}$$

Define

$$a = \mathbf{E}\,|A_t|, \quad b = \mathbf{E}\,|\langle M\rangle_t - \sigma^2|, \quad c = \mathbf{E}\sum_{0<s\le t} |\Delta M_s|^3 \tag{7.35}$$

and set

$$z = a^{1/2} + b^{1/3} + c^{1/4}. \tag{7.36}$$

Then by (7.34) we have

$$\frac{2}{\pi}\int_0^{z^{-1}} \frac{\left| \mathbf{E} e^{i\lambda \tilde{X}_t} - e^{-\frac{\lambda^2}{2}\sigma^2}\right|}{\lambda} d\lambda + \frac{24z}{\pi\sqrt{2\pi\sigma^2}}$$

$$\le \frac{2}{\pi}\int_0^{z^{-1}} \left(a + \frac{\lambda}{2}b + c_1 c\lambda^2\right) d\lambda + \frac{24z}{\pi\sqrt{2\pi\sigma^2}}$$

$$\frac{2}{\pi}\left[z^{-1}a + \frac{z^{-2}}{4}b + \frac{z^{-3}}{3}c_1 c\right] + \frac{24z}{\pi\sqrt{2\pi\sigma^2}}. \tag{7.37}$$

Since (see (7.36))

$$az^{-1} \le a^{1/2}, \quad bz^{-2} \le b^{1/3}, \quad cz^{-3} \le c^{1/4},$$

the right-hand side of the inequality (7.37) is bounded from above by the variable

$$L = \frac{2}{\pi}\left[a^{1/2} + \frac{1}{4}b^{1/3} + \frac{c_1}{3}c^{1/4}\right] + \frac{24\,(a^{1/2} + b^{1/3} + c^{1/4})}{\sqrt{2\pi\sigma^2}}$$

$$\le L_1\,(a^{1/2} + b^{1/3} + c^{1/4})$$

with

$$L_1 = \frac{24}{\sqrt{2\pi\sigma^2}} + \max\left(\frac{2}{\pi}, \frac{1}{2\pi}, \frac{2c_1}{3\pi}\right) = \frac{24}{\sqrt{2\pi\sigma^2}} + \frac{2}{\pi}.$$

Thus

$$\sup_{x \in R}\left|P(\tilde{X}_t \le x) - \Phi\left(\frac{x}{\sigma}\right)\right| \le \left(\frac{2}{\pi} + \frac{24}{\sqrt{2\pi\sigma^2}}\right)(a^{1/2} + b^{1/3} + c^{1/4}). \quad (7.38)$$

By Lemma 1

$$\sup_{x \in R}\left|P(X_t \le x) - \Phi\left(\frac{x}{\sigma}\right)\right|$$

$$\le \left(1 + \frac{1}{\sqrt{2\pi\sigma^2}}\right)\sup\left|P(\tilde{X}_t \le x) - \Phi\left(\frac{x}{\sigma}\right)\right| + 2\,(\mathbf{E}\,(X_t - \tilde{X}_t)^2)^{1/3}.$$

This, (7.39) and (7.38) give

$$\sup_{x \in R}\left|P(X_t \le x) - \Phi\left(\frac{x}{\sigma}\right)\right| \le c_2\,(a^{1/2} + b^{1/3} + c^{1/4}),$$

with

$$c_2 = 2 + \left(\frac{2}{\pi} + \frac{24}{\sqrt{2\pi\sigma^2}}\right)\left(1 + \frac{1}{\sqrt{2\pi\sigma^2}}\right)$$

and a, b and c defined by (7.35).

Theorem 2. *Let* $X = (X_s, \mathcal{F}_s)$ *be a semimartingale with* $X_0 = 0$ *and with the triplet* $T = (B, C, \nu)$.

Then

$$\sup_{x \in R} \left| P(X_t \le x) - \Phi\left(\frac{x}{\sigma}\right)\right|$$

$$\le c_3 [(\mathbf{E}(|B_t| + |x| I(|x| > 1) * \nu_t))^{1/2} + (\mathbf{E}|\langle M^1\rangle_t - \sigma^2|)^{1/3}$$
$$+ (\mathbf{E}|x|^3 I(|x| \le 1) * \nu_t)^{1/4}] \tag{7.39}$$

with

$$\langle M^1\rangle_t = C_t + x^2 I(|x| \le 1) * \nu_t - \sum_{0 < s \le t} \left(\int_{|x| \le 1} x\nu(\{s\}, dx) \right)^2. \tag{7.40}$$

Proof. The canonical representation for a semimartingale (see Ch. 4, § 1) implies that in the representation (7.29)

$$A_s = B_s + x I(|x| > 1) * \mu_s, \quad M_s = M_s^1,$$

where $\mu = \mu(dt, dx)$ is the jump measure of X,

$$M_s^1 = X_s^c + x I(|x| \le 1) * (\mu - \nu)_s \tag{7.41}$$

and the quadratic characteristic $\langle M^1\rangle$ is given by the formula (7.40).

Since

$$|A_t| \le |B| + |x| I(|x| > 1) * \mu_t$$

and

$$\mathbf{E}|x| I(|x| > 1) * \mu_t = \mathbf{E}|x| I(|x| > 1) * \nu_t$$

(see Ch. 3, § 2), then

$$\mathbf{E}|A_t| \le \mathbf{E}|B_t| + \mathbf{E}|x| I(|x| > 1) * \nu_t. \tag{7.42}$$

Let us show that

$$\mathbf{E} \sum_{0 < s \le t} |\Delta M_s^1|^3 \le 8\mathbf{E}|x|^3 I(|x| \le 1) * \nu_t. \tag{7.43}$$

From (7.41) it follows that

$$\Delta M_s^1 = \Delta X_s I(|\Delta X_s| \le 1) - \hat{x}_s$$

with

$$\hat{x}_s^1 = \int_{|x| \le 1} x\nu(\{s\}, dx).$$

Therefore by Hölder's inequality with $p = 3$ and $q = 3/2$, and by the fact that $\nu(\{s\} \times R_0) \le 1$ (Lemma 3.3.1) we get

$$|\Delta M_s^1|^3 \le 4\,(|\Delta X_s|^3\, I(|\Delta X_s| \le 1) + |\hat{x}_s^1|^3)$$

$$\le 4\,(|\Delta X_s|^3\, I(|\Delta X_s| \le 1) + \int_{|x| \le 1} |x|^3 \nu(\{s\}, dx)).$$

This gives

$$E \sum_{0 < s \le t} |\Delta M_s^1|^3 \le 4E\,(|x|^3\, I(|x| \le 1) * \mu_t + |x|^3\, I(|x| \le 1) * \nu_t),$$

and hence the estimate (7.43) holds.

The desired estimate (7.39) follows from the assertion of Theorem 1, and from (7.42) and (7.43).

The theorem is proved.

Problems

1. Prove the formulas (7.11) - (7.13).

2. Formulate the assertions of Theorems 1 and 2 for the case of "schemes of series" (see Subsection 5.13).

§ 8. A martingale method of proving the central limit theorem for strictly stationary sequences. Relation to mixing conditions

1. On a probability space $(\Omega, \mathcal{F}, P)$ let a strictly stationary sequence

$$Y = (Y_k)_{-\infty < k < \infty}$$

be given. Let $\mathcal{G}$ be a σ-algebra of invariant sets of the sequence Y and

$$\mathcal{F}_k = \sigma\{\dots, Y_{k-1}, Y_k\}.$$

Denote

$$X^n = \frac{1}{\sqrt{n}} \sum_{k=1}^{n} Y_k, \quad n \geq 1.$$

In this section we deal (by resorting to the methods of proving the central limit theorem exposed above) with conditions ensuring

$$X^n \xrightarrow{d} X \quad (\mathcal{G}\text{-stably})$$

where X is a random variable with the characteristic function

$$\phi(\lambda) = \mathbf{E} e^{-\frac{\lambda^2}{2}\sigma^2}, \quad \lambda \in R, \tag{8.1}$$

and $\sigma^2 = \sigma^2(\omega)$ is a certain $\mathcal{G}$-measurable random variable.

Denote

$$\gamma_k(p) = (\mathbf{E} \,|\, \mathbf{E}(Y_k \,|\, \mathcal{F}_0) \,|^{p/(p-1)})^{(p-1)/p}. \tag{8.2}$$

Theorem 1. *For a certain* $p \geq 2$ *let* $\mathbf{E}\,|\,Y_0\,|^p < \infty$ *and*

$$\sum_{k \geq 1} \gamma_k(p) < \infty. \tag{8.3}$$

Then

$$X^n \xrightarrow{d} X \quad (\mathcal{G}\text{-}stably),$$

where X *is a random variable with the characteristic function* (8.1) *and*

$$\sigma^2 = \mathbf{E}(Y_0^2 \,|\, \mathcal{G}) + 2 \sum_{k \geq 1} \mathbf{E}(Y_0 Y_k \,|\, \mathcal{G}). \tag{8.4}$$

2. The proof of this theorem is based on a number of auxiliary statements.

Lemma 1. *For a certain* $p \geq 2$ *let* $\mathbf{E}\,|\,Y_0\,|^p < \infty$. *Then*

$$\gamma_{k+1}(p) \leq \gamma_k(p), \quad k \geq 1.$$

Proof. Since Y is a strictly stationary sequence, the random variables $\mathbf{E}(Y_k \,|\, \mathcal{F}_0)$ and $\mathbf{E}(Y_0 \,|\, \mathcal{F}_{-k})$ coincide in distribution. Therefore

$$(\gamma_k(p))^{p/(p-1)} = \mathbf{E}\,|\,\mathbf{E}(Y_k \,|\, \mathcal{F}_0)\,|^{p/(p-1)} = \mathbf{E}\,|\,\mathbf{E}(Y_0 \,|\, \mathcal{F}_{-k})\,|^{p/(p-1)}.$$

By Jensen's inequality

$$\mathbf{E}\,(\,|\,\mathbf{E}\,(Y_0\,|\,\mathcal{F}_{-k})\,|^{p/(p-1)}\,|\,\mathcal{F}_{-(k+1)}) \ge |\,\mathbf{E}\,(Y_0\,|\,\mathcal{F}_{-(k+1)})\,|^{p/(p-1)}.$$

Consequently

$$(\gamma_k\,(p))^{p/(p-1)} \ge \mathbf{E}\,|\,\mathbf{E}\,(Y_0\,|\,\mathcal{F}_{-(k+1)})\,|^{p/(p-1)}$$

$$= \mathbf{E}\,|\,\mathbf{E}\,(Y_{k+1}\,|\,\mathcal{F}_0)\,|^{p/(p-1)} = (\gamma_{k+1}\,(p))^{p/(p-1)},$$

i.e. the desired inequality holds.

Lemma 2. *For a certain* $p \ge 2$ *let* $\mathbf{E}\,|\,Y_0\,|^2 < \infty$ *and*

$$\sum_{k \ge 1} \gamma_k\,(p) < \infty. \tag{8.5}$$

Then

1) $\mathbf{E}\,(Y_0\,|\,\mathcal{F}_{-\infty}) = 0$ (P-a.s.), *where* $\mathcal{F}_{-\infty} = \bigcap_{k > -\infty} \mathcal{F}_k$;

2) $\lim_k k\gamma_k\,(p) = 0$;

3) $\sum_{k \ge 1} \mathbf{E}\,|\,\mathbf{E}\,(Y_0 Y_k\,|\,\mathcal{G})\,| < \infty, \quad \sum_{k \ge 1} |\,\mathbf{E}\,(Y_0 Y_k)\,| < \infty$;

4) $\sum_{k \ge n} \mathbf{E}\,|\,\mathbf{E}\,(Y_k\,|\,\mathcal{F}_{n-1})\,| < \infty, \quad \sum_{k \ge n} \mathbf{E}\,|\,\mathbf{E}\,(Y_k\,|\,\mathcal{F}_n)\,| < \infty, \; n \ge 1.$

Proof. 1) By (8.5) we have $\lim_k \gamma_k\,(p) = 0$. Hence

$$\lim_k \mathbf{E}\,|\,\mathbf{E}\,(Y_k\,|\,\mathcal{F}_0)\,|^{p/(p-1)} = 0,$$

and since

$$\mathbf{E}\,|\,\mathbf{E}\,(Y_k\,|\,\mathcal{F}_0)\,|^{p/(p-1)} = \mathbf{E}\,|\,\mathbf{E}\,(Y_0\;\mathcal{F}_{-k})\,|^{p/(p-1)},$$

then

$$\mathbf{E}\,(Y_0\,|\,\mathcal{F}_{-k}) \xrightarrow{P} 0, \quad k \to \infty.$$

The desired relation follows from this by Levy's theorem [96], according to which

$$\mathbf{E}\,(Y_0\,|\,\mathcal{F}_{-k}) \to \mathbf{E}\,(Y_0\,|\,\mathcal{F}_{-\infty}) \quad \text{(P-a.s.)}.$$

2) Assume

$$\overline{\lim_k}\; k\gamma_k\,(p) = c > 0.$$

Then a number $N = N\,(c)$ and a sequence (i_k) with $i_k \ge N$ and $i_{k+1} \ge 2i_k$ can be found such that $i_k\gamma_{i_k}\,(p) \ge \frac{c}{2}$. By Lemma 1 and the fact that $\gamma_{i_{k+1}}\,(p) \ge c/\,(2i_{k+1})$, we get

$$\sum_{i\ge N}\gamma_i(p)\ge\sum_{\{k:\,i_k\ge N\}}\ \sum_{i=i_k+1}^{i_{k+1}}\gamma_i(p)\ge\sum_{\{k:\,i_k\ge N\}}\gamma_{i_{k+1}}(p)\,[i_{k+1}-i_k-1]$$

$$\ge\sum_{\{k:\,i_k\ge N\}}\frac{c}{2}\left[1-\frac{i_k}{i_{k+1}}-\frac{1}{i_{k+1}}\right].$$

By the inequality $i_{k+1}\ge 2i_k$ this gives

$$\sum_{i\ge N}\gamma_i(p)\ge\frac{c}{2}\sum_{\{k:\,i_k\ge N\}}\left(\frac{1}{2}-\frac{i}{i_{k+1}}\right)=\infty,$$

that contradicts to the assumption

$$\sum_{i\ge 1}\gamma_i(p)<\infty.$$

Hence, c = 0, i.e. the desired assertion holds.

3) It can be assumed without loosing generality that σ-algebras $\mathfrak{F}_k$, $-\infty<k<\infty$, are completed by sets in $\mathfrak{F}_\infty=\sigma\left(\bigcup_{k>-\infty}\mathfrak{F}_k\right)$ of zero probability. Then $\mathfrak{G}\subseteq\mathfrak{F}_k$, $-\infty<k<\infty$ ([260], Lemma 6.1, p. 225). Consequently

$$|\mathbf{E}(Y_0Y_k\mid\mathfrak{G})|=|\mathbf{E}(Y_0\mathbf{E}(Y_k\mid\mathfrak{F}_0)\mid\mathfrak{G})|.$$

Therefore by Hölder's inequality

$$|\mathbf{E}(Y_0Y_k\mid\mathfrak{G})|\le[\mathbf{E}(|Y_0|^p\mid\mathfrak{G})]^{1/p}\,[\mathbf{E}(|\mathbf{E}(Y_k\mid\mathfrak{F}_0)|^{p/(p-1)}\mid\mathfrak{G})]^{(p-1)/p}.$$

By this, applying once more Hölder's inequality, we get

$$\mathbf{E}\,|\mathbf{E}(Y_0Y_k\mid\mathfrak{G})|\le[\mathbf{E}\,|Y_0|^p]^{1/p}\,\gamma_k(p)$$

and

$$\sum_{k\ge 1}\mathbf{E}\,|\mathbf{E}(Y_0Y_k\mid\mathfrak{G})|\le[\mathbf{E}\,|Y_0|^p]^{1/p}\sum_{k\ge 1}\gamma_k(p)<\infty.$$

The second inequality is proved analogously, replacing the σ-algebra $\mathfrak{G}$ by a trivial σ-algebra (∅, Ω).

4) Utilize the fact that

$$\mathbf{E}\,|\mathbf{E}(Y_k\mid\mathfrak{F}_{n-1})|=\mathbf{E}\,|\mathbf{E}(Y_{k-n+1}\mid\mathfrak{F}_0)|.$$

Then by Hölder's inequality

$$\mathbf{E}\,|\mathbf{E}(Y_k\mid\mathfrak{F}_{n-1})|\le\gamma_{k-n+1}(p),$$

and hence

$$\sum_{k\ge n}\mathbf{E}\,|\mathbf{E}(Y_k\mid\mathfrak{F}_{n-1})|\le\sum_{k\ge n}\gamma_{k-n+1}(p)=\sum_{k\ge 1}\gamma_k(p)<\infty.$$

It is established analogously that

$$\sum_{k \geq n} \mathbf{E} \,|\, \mathbf{E}(Y_k \,|\, \mathfrak{F}_n) \,| \leq \sum_{k \geq 1} \gamma_k(p) + (\mathbf{E} \,|\, Y_0 \,|^{p/(p-1)})^{(p-1)/p}.$$

The lemma is proved.

For a certain $p \geq 2$ let $\mathbf{E} \,|\, Y_0 \,|^p < \infty$ and $\sum_{k \geq 1} \gamma_k(p) < \infty$. By Assertion 4) of Lemma 2 the random variable

$$\eta_1 = \sum_{k \geq 1} [\mathbf{E}(Y_k \,|\, \mathfrak{F}_0) - \mathbf{E}(Y_k \,|\, \mathfrak{F}_{-1})]$$

is defined.

Lemma 3. *If for a certain* $p \geq 2$ *we have* $\mathbf{E} \,|\, Y_0 \,|^p < \infty$ *and* $\sum_{k \geq 1} \gamma_k(p) < \infty$, *then*

$$\mathbf{E}\eta_1^2 = \mathbf{E}Y_0^2 + 2 \sum_{k \geq 1} \mathbf{E}Y_0 Y_k < \infty$$

and

$$\mathbf{E}(\eta_1^2 \,|\, \mathcal{G}) = \mathbf{E}(Y_0^2 \,|\, \mathcal{G}) + 2 \sum_{k \geq 1} \mathbf{E}(Y_0 Y_k \,|\, \mathcal{G}).$$

Proof. Let α be a $\mathcal{G}$-measurable random variable such that $\alpha = \alpha^2$, i.e. $\alpha = I_A(\omega)$ with $A \in \mathcal{G}$. We will show first that

$$\mathbf{E}\alpha\eta_1^2 = \mathbf{E}\alpha Y_0^2 + 2 \sum_{k \geq 1} \mathbf{E}\alpha Y_0 Y_k, \tag{8.6}$$

with

$$\sum_{k \geq 1} |\, \mathbf{E}(\alpha Y_0 Y_k) \,| < \infty$$

by Assertion 3) of Lemma 2.

Denote

$$\beta_i = \alpha [\mathbf{E}(Y_i \,|\, \mathfrak{F}_0) - \mathbf{E}(Y_i \,|\, \mathfrak{F}_{-1})], \quad b_{ij} = \mathbf{E}\beta_i\beta_j.$$

Under the condition $\sum_{i, j \geq 1} b_{ij} < \infty$ the series $\sum_{i \geq 1} \beta_i$ converge in mean-square and

$$\mathbf{E}\left(\sum_{i \geq 1} \beta_i\right)^2 = \sum_{i, j \geq 1} b_{ij},$$

i.e.

$$\mathbf{E}\alpha\eta_1^2 = \sum_{i, j \geq 1} b_{ij}. \tag{8.7}$$

It is not hard to verify that

$$b_{ij} = \mathbf{E}\,[\alpha Y_i \mathbf{E}\,(Y_j \mid \mathfrak{F}_0)] - \mathbf{E}\,[\alpha Y_i \mathbf{E}\,(Y_j \mid \mathfrak{F}_{-1})]. \tag{8.8}$$

Assuming that σ-algebras $\mathfrak{F}_{-k}$, $-\infty < k < \infty$ are completed by sets in $\mathfrak{F}_\infty$ of zero probability and taking into consideration that $\mathcal{G} \subseteq \mathfrak{F}_k$, $-\infty < k < \infty$, we get

$$\mathbf{E}\,[\alpha Y_i \mathbf{E}\,(Y_j \mid \mathfrak{F}_{-1})] = \mathbf{E}\,[\alpha Y_{i+1} \mathbf{E}\,(Y_{j+1} \mid \mathfrak{F}_0)].$$

Consequently

$$b_{ij} = \mathbf{E}\,[\alpha Y_i \mathbf{E}\,(Y_j \mid \mathfrak{F}_0)] - \mathbf{E}\,[\alpha Y_{i+1} \mathbf{E}\,(Y_{j+1} \mid \mathfrak{F}_0)].$$

Denote

$$\delta_j = \alpha \mathbf{E}\,(Y_j \mid \mathfrak{F}_0).$$

Then the equality (8.8) can be rewritten in the following manner:

$$b_{ij} = \mathbf{E}\delta_j\,(Y_i - Y_{i+1}) - \mathbf{E}Y_{i+1}\,(\delta_{j+1} - \delta_j).$$

This gives

$$\sum_{i,\,j=1}^{N} b_{ij} = r_1(N) + r_2(N)$$

with

$$r_1(N) = \sum_{j=1}^{N} \mathbf{E}\delta_j Y_1 + \sum_{i=1}^{N} \mathbf{E}Y_{i+1}\delta_1$$

and

$$r_2(N) = -\sum_{j=1}^{N} \mathbf{E}\delta_j Y_{N+1} - \sum_{i=1}^{N} \mathbf{E}Y_{i+1}\delta_{N+1}.$$

Since the variables δ_j are $\mathfrak{F}_0$-measurable, in view of Hölder's inequality we get

$$|\,\mathbf{E}\delta_j Y_{N+1}\,| = |\,\mathbf{E}\delta_j \mathbf{E}\,(Y_{N+1} \mid \mathfrak{F}_0)\,| = |\,\mathbf{E}\,(\alpha Y_j \mathbf{E}\,(Y_{N+1} \mid \mathfrak{F}_0))\,|$$
$$\le (\mathbf{E}\,|\,Y_j\,|^p)^{1/p}\,\gamma_{N+1}(p)$$

and analogously

$$|\,\mathbf{E}Y_{i+1}\delta_{N+1}\,| \le (\mathbf{E}\,|\,Y_{i+1}\,|^p)^{1/p}\,\gamma_{N+1}(p).$$

In view of these estimates we have the inequality

$$|\,r_2(N)\,| \le 2\,(\mathbf{E}\,|\,Y_0\,|^p)^{1/p}\,N\gamma_{N+1}(p).$$

Therefore by Assertion 2) of Lemma 2 we have

$$\lim_N r_2(N) = 0.$$

Let us show now that

$$\lim_N r_1(N) = \mathbf{E}\alpha Y_0^2 + 2\sum_{i \ge 1} \mathbf{E}\alpha Y_0 Y_i. \tag{8.9}$$

By the fact that

$$\mathbf{E}\delta_j Y_1 = \mathbf{E}\delta_{j-1} Y_0$$

and

$$\mathbf{E}Y_{i+1}\delta_1 = \mathbf{E}Y_i\delta_0$$

we have

$$r_1(N) = \sum_{j=1}^{N} \mathbf{E}\delta_{j-1}Y_0 + \sum_{i=1}^{N} \mathbf{E}Y_i\delta_0 = \sum_{i=0}^{N-1} \mathbf{E}\delta_i Y_0 + \sum_{i=1}^{N} \mathbf{E}Y_i\delta_0$$

$$= \sum_{i=0}^{N-1} \mathbf{E}\alpha Y_0 Y_i + \sum_{i=1}^{N-1} \mathbf{E}\alpha Y_0 Y_i + \mathbf{E}\alpha Y_N Y_0$$

$$= \mathbf{E}\alpha Y_0^2 + 2\sum_{i=1}^{N-1} \mathbf{E}\alpha Y_0 Y_i + \mathbf{E}\alpha Y_N Y_0.$$

By this and Assertion 3) of Lemma 2 we get the desired relation (8.9). Hence the equality (8.6) holds.

As $\alpha = 1$ by (8.6) we get the formula for $\mathbf{E}\eta_1^2$. The representation for $\mathbf{E}(\eta_1^2 \mid \mathcal{G})$ follows also from (8.6) since $\alpha = I_A(\omega)$, $A \in \mathcal{G}$ is arbitrary.

3. Proof of Theorem 1. By Assertion 4) of Lemma 2 random variables

$$v_n = \sum_{i\geq n} \mathbf{E}(Y_i \mid \mathcal{F}_{n-1}), \quad \eta_n = \sum_{i \geq n} [\mathbf{E}(Y_i \mid \mathcal{F}_n) - \mathbf{E}(Y_i \mid \mathcal{F}_{n-1})]$$

are defined. Besides

$$Y_n = \eta_n + v_n - v_{n+1}.$$

From this it follows, in particular, that

$$\frac{1}{\sqrt{n}}\sum_{i=1}^{n} Y_i = \frac{1}{\sqrt{n}}\sum_{i=1}^{n} \eta_i + \frac{v_1 - v_{n+1}}{\sqrt{n}}.$$

Since $(v_n)_{n\geq 1}$ is a strictly stationary sequence, for each $\varepsilon > 0$ we have

$$\mathrm{P}\left(\left|\frac{1}{\sqrt{n}}(v_1 - v_{n+1})\right| \geq \varepsilon\right) \leq 2\mathrm{P}\left(\frac{|v_1|}{\sqrt{n}} \geq \frac{\varepsilon}{2}\right) \to 0, \quad n \to \infty,$$

and hence, according to Problem 2.2 the convergence problem for the sequence of distributions of random variables

$$\frac{1}{\sqrt{n}}\sum_{i=1}^{n} Y_i, \quad n \geq 1,$$

is reduced to ensuring the weak convergence of the probability distributions of the sequence

$$\frac{1}{\sqrt{n}}\sum_{i=1}^{n} \eta_i, \quad n \geq 1.$$

Denote

$$X_t^n = \sum_{i=1}^{[n(t \wedge 1)]} \frac{\eta_i}{\sqrt{n}}, \qquad \mathcal{F}_t^n = \mathcal{F}_{[n(t \wedge 1)]}.$$

It is not hard to verify that the process $X^n = (X_t^n, \mathcal{F}_t^n)$ is a square integrable martingale (Problem 2). Besides by the strict stationarity of the sequence $(\eta_n)_{n \geq 1}$ and by Lemma 3

$$\mathbf{E}\langle X^n \rangle_1 = \frac{1}{n} \sum_{i=1}^{n} \mathbf{E}\eta_i^2 = \mathbf{E}\eta_1^2 = \mathbf{E}Y_0^2 + 2 \sum_{i \geq 1} \mathbf{E}Y_0 Y_i.$$

According to Theorem 5.4 (II)

$$X_1^n \xrightarrow{d} X \quad (\mathcal{G}\text{-stably}),$$

where X is a random variable with the characteristic function (8.1) and the quantity σ^2, defined by the formula (8.4), provided the following conditions are fullfilled:

(o) $\mathcal{G} \subseteq \bigcap_{n \geq 1} \mathcal{F}_0^n$,

(l_2) $\displaystyle\sum_{i=1}^{n} \mathbf{E}\left[\frac{\eta_i^2}{n} I\left(\frac{|\eta_i|}{\sqrt{n}} > \delta\right) \Big| \mathcal{F}_{i-1}^n\right] \xrightarrow{P} 0, \quad \delta > 0,$

(c_1) $\displaystyle\frac{1}{n} \sum_{i=1}^{n} \eta_i^2 \xrightarrow{P} \sigma^2.$

Condition (o) is fullfilled, since $\mathcal{F}_0^n = \mathcal{F}_0$ and since one may assume without loosing generality that σ-algebras $\mathcal{F}_{-k}$, $-\infty < k < \infty$, are completed by sets in $\mathcal{F}_\infty$ of zero probability, and hence $\mathcal{G} \subseteq \mathcal{F}_0$.

Condition (l_2) is fullfilled since $(\eta_n)_{n \geq 1}$ is a strictly stationary sequence and

$$\mathbf{E} \sum_{i=1}^{n} \mathbf{E}\left[\frac{\eta_i^2}{n} I\left(\frac{|\eta_i|}{\sqrt{n}} > \delta\right) \Big| \mathcal{F}_{i-1}\right] = \sum_{i=1}^{n} \mathbf{E}\frac{\eta_i^2}{n} I\left(\frac{|\eta_i|}{\sqrt{n}} > \delta\right)$$
$$= \mathbf{E}\eta_1^2 I(|\eta_1| > \delta\sqrt{n}) \to 0, \quad n \to \infty.$$

Condition (c_1) is fullfilled by the Birkhoff-Khintchine theorem, since (P-a.s.)

$$\lim_n \frac{1}{n} \sum_{i=1}^{n} \eta_i^2 = \mathbf{E}(\eta_1^2 \mid \mathcal{G}),$$

while by Lemma 3 we have

$$\mathbf{E}(\eta_1^2 \mid \mathcal{G}) = \sigma^2.$$

4. Let us relate the basic Condition (8.3) of Theorem 1 to various mixing conditions under which the central limit theorem for strictly stationary sequences $Y = (Y_k)_{-\infty < k < \infty}$ is usually proved. Here we assume $\mathbf{E}Y_0 = 0$.

Recall the definitions of the following three coefficients ϕ_k, α_k, and ρ_k.

1) The uniform strongly mixing coefficient

$$\phi_k = \sup | \mathsf{P}(B \mid A) - \mathsf{P}(B) |, \ k \geq 1, \quad \mathsf{P}(A) > 0,$$

where sup is taken over all sets $A \in \mathcal{F}_0$ and a set $B \in \mathcal{F}^k = \sigma\{Y_k, Y_{k+1}, \ldots\}$.

2) The strongly mixing coefficient

$$\alpha_k = \sup | \mathsf{P}(AB) - \mathsf{P}(A)\,\mathsf{P}(B) |, \ k \geq 1,$$

where sup is taken over all $A \in \mathcal{F}_0$ and $B \in \mathcal{F}^k$.

3) The maximal correlation coefficient

$$\rho_k = \sup \ \operatorname{cov}(\xi', \xi''), \ k \geq 1,$$

where sup is taken over all random variables ξ' and ξ'' with finite second moments such that ξ' is measurable relative to $\mathcal{F}_0$, and ξ'' relative to $\mathcal{F}^k$.

Lemma 4 [77, 205]. *The following inequalities hold:*

1) $\gamma_k(p) \leq 2\phi_k^{(p-1)/p} (\mathbf{E} \mid Y_0 \mid^p)^{1/p}, \ p \geq 2,$

2) $\gamma_k(p) \leq 2(2^{1/p} + 1)\, \alpha_k^{(p-2)/p} (\mathbf{E} \mid Y_0 \mid^p)^{1/p}, \ p > 2,$

3) $\gamma_k(2) \leq \rho_k^{1/2}.$

From these inequalities it follows immediately that under the assumption

$$\mathbf{E} \mid Y_0 \mid^p < \infty$$

any of the conditions

$$\sum_{k \geq 1} \phi_k^{(p-1)/p} < \infty, \ p \geq 2, \tag{8.10}$$

$$\sum_{k \geq 1} \alpha_k^{(p-2)/p} < \infty, \ p > 2, \tag{8.11}$$

guarantee the convergence of the series $\sum_{k \geq 1} \gamma_k(p)$ (see (8.3)). Besides

$$\sum_{k \geq 1} \rho_k^{1/2} < \infty \Rightarrow \sum_{k \geq 1} \gamma_k(2) < \infty. \tag{8.12}$$

Observe that if

$$\sum_{k \geq 1} \phi_k^{(p-1)/p} < \infty \ (p \geq 2)$$

and

$$\sum_{k\geq 1} \alpha_k^{(p-2)/p} < \infty \quad (p > 2),$$

then $\phi_k \to 0$ and $\alpha_k \to 0$ as $k \to \infty$. In these cases any set $A \in \mathcal{G}$ has probability 0 or 1 (see, for instance, [118]). Therefore Theorem 1 and Lemma 4 imply the following well known result (cf. [118]).

Theorem 2. *Let* $EY_0 = 0$ *and let any of the following pairs of conditions be fullfilled:*

$$\sum_{k\geq 1} \phi_k^{(p-1)/p} < \infty, \quad E\,|\,Y_0\,|^p < \infty, \; p \geq 2$$

or

$$\sum_{k>1} \alpha_k^{(p-2)/p} < \infty, \quad E\,|\,Y_0\,|^p < \infty, \; p > 2.$$

Then

$$\frac{1}{\sqrt{n}} \sum_{k=1}^{n} Y_k \xrightarrow{d} N(0, \sigma^2),$$

with

$$\sigma^2 = EY_0^2 + 2 \sum_{k\geq 1} EY_0Y_k.$$

Problems

1. Prove the inequalities 1) - 3) of Lemma 4.

2. Show that the process $X^n = (X_t^n, \mathcal{F}_t^n)$, introduced in the course of proving Theorem 1, is a square integrable martingale.

CHAPTER 6

THE SPACE D. RELATIVE COMPACTNESS OF PROBABILITY DISTRIBUTIONS OF SEMIMARTINGALES

§ 1. The space D. Skorohod's topology

1. In Chapter 5 the weak convergence has been studied for finite dimensional distributions of semimartingales only. On considering questions concerning the weak convergence of probability distributions of semimartingales, trajectories of which belong by definition to the space D, we will need certain known facts about the space D exposed here for convenience of the reader and for further references.

The space $D = D([0, \infty), R)$ is a space of functions $X = (X_t)_{t \ge 0}$, right continuous ($X_t = X_{t+}$) and having left-hand limits ($X_{t-} = \lim_{s \uparrow t} X_s$), with values in R ($X_t \in R$, $t \in R_+$). Sometimes we need to consider the space $D([0, \infty), S)$ too, where S is a complete separable metric space. For the sake of simplicity of the exposition, we consider only $D([0, \infty), R)$, all properties of which, modified in a natural way, extend to the space $D([0, \infty), S)$.

2. Recall how the metric is introduced in the space D.

First the space $D([0, 1]) = D([0, 1], R)$ is considered of right-continuous functions having left-hand limits. In this space the metric $d_0(.,.)$ is introduced in the following manner: for $X = (X_t)_{0 \le t \le 1}$ and $Y = (Y_t)_{0 \le t \le 1} \in D([0, 1])$

$$d_0(X, Y) = \inf_{\lambda \in \Lambda} \left\{ \sup_{0 \le t \le 1} | X_t - Y_{\lambda(t)} | + \sup_{0 \le s < t \le 1} \left| \log \frac{\lambda(t) - \lambda(s)}{t - s} \right| \right\} \tag{1.1}$$

with $\Lambda = \{\lambda = (\lambda(t))_{0 \le t \le 1} : \lambda$ is a strictly increasing continuous function, $\lambda(0) = 0$, $\lambda(1) = 1\}$.

Theorem 1 [14]. *The space* $D([0, 1])$ *is a complete separable metric space in the metric* d_0. *The metric* d_0 *is equivalent to Skorohod's metric*

$$d(X, Y) = \inf_{\lambda \in \Lambda} \{ \sup_{0 \le t \le 1} | X_t - Y_{\lambda(t)} | + \sup_{0 \le t \le 1} | \lambda(t) - t | \}. \tag{1.2}$$

Denote by $D^0([0, 1])$ a subset of $D([0, 1])$ with the following properties:

$$X \in D^0([0, 1]) \Rightarrow X_1 = X_{1-}.$$

The space $D^0([0, 1])$ is closed in Skorohod's topology, i.e. $D^0([0, 1])$ is a complete separable metric space in the metric d_0.

Analogously to $D^0([0, 1])$, let a subset D^0 of D exist with the property

$$X \in D^0 \Rightarrow \lim_{t \to \infty} X_t,$$

and let it be finite.

There exists a one to one correspondence between the sets $D^0([0, 1])$ and D^0 that is established by a change of time, determined by the function

$$\psi(t) = \begin{cases} -\log(1-t), & 0 \le t < 1, \\ \infty, & t = 1. \end{cases}$$

In D^0 define the metric ($X, Y \in D^0$)

$$\rho^0(X, Y) = d_0(X', Y') \tag{1.3}$$

with

$$X'_t = X_{\psi^{-1}(t)}$$

and

$$Y'_t = Y_{\psi^{-1}(t)}.$$

The space D^0 is, consequently, a complete separable metric space in the metric ρ^0.

Define now the function $g_k = g_k(t)$, $k \ge 1$ with

$$g_k(t) = \begin{cases} 1, & t \in [0, k], \\ k + 1 - t, & t \in (k, k+1], \\ 0, & t \in (k+1, \infty), \end{cases}$$

and associate with $X \in D$ the collection of functions

$$X^k = (X^k_t)_{t \ge 0}, \quad k \ge 1,$$

by setting

$$X^k_t = X_t g_k(t).$$

Obviously, the functions X^k, $k \ge 1$, belong to the space D^0. Hence one can associate with functions $X, Y \in D$ a sequence of numbers $\rho^0(X^k, Y^k)$, $k \ge 1$, which define in D the metric

$$\rho(X, Y) = \sum_{k=1}^{\infty} 2^{-k} \frac{\rho^0(X^k, Y^k)}{1 + \rho^0(X^k, Y^k)}, \tag{1.4}$$

introduced by Lindvall.

Theorem 2 [182]. *The space* D *is a complete separable metric space in the metric* ρ. *Besides the convergence in the metric* ρ ($\lim_n \rho(X^n, X) = 0$) *is equivalent* *to the fullfillment of the following condition: there exist functions* $\lambda_n \in \Lambda_\infty$, $n \ge 1$ (Λ_∞ *is a set of continuous strictly increasing functions* $\lambda_n = (\lambda_n, (t))_{t \ge 0}$ *with* $\lambda_n(0) = 0$) *such that for each* $L > 0$

$$\sup_{0 \le t \le L} |X_t - X^n_{\lambda_n(t)}| \to 0, \quad \sup_{0 \le t \le L} |\lambda_n(t) - t| \to 0, \ n \to \infty.$$

For a function $X \in D$ define

$$\Delta(X) = \{t > 0 : \Delta X_t = 0\} \cup \{0\}$$

and observe that $\Delta(X)$ is a dense subset in R_+, [14].

Theorem 2 has the following corollaries.

Corollary 1. *If* $\rho(X^n, X) \to 0$, *then* $X^n_t \to X_t$ *for each* $t \in \Delta(X)$.

Corollary 2. *If* $\rho(X^n, X) \to 0$, *then there exists a sequence* $(s_n)_{n \ge 1}$ *such that*

$$s_n \to t, \quad \Delta X^n_{s_n} \to \Delta X_t.$$

To prove these corollaries introduce the function μ_n, inverse to λ_n, where λ_n is the same function as in Theorem 2. Since $\lambda_n(t) \to t$, then $\mu_n(t) \to t$, and hence, as n is sufficiently large, the following inequalities hold:

$$\mu_n(t) \le t + 1, \quad \lambda_n(t+1) \le t + 2.$$

Corollary 2 follows from the inequalities (for sufficiently large n)

$$|X^n_t - X_t| \le |X^n_t - X_{\mu_n(t)}| + |X_{\mu_n(t)} - X_t|$$

$$\le \sup_{0 \le s \le t+1} |X^n_s - X_{\mu_n(s)}| + |X_{\mu_n(t)} - X_t|$$

$$\le \sup_{0 \le s \le \lambda_n(t+1)} |X^n_{\lambda_n(s)} - X_s| + |X_{\mu_n(t)} - X_t|,$$

from Theorem 2 and the fact that for $t \in \Delta(X)$ we have $X_{\mu_n(t)} \to X_t$.

The assertion of Corollary 2 holds with $s_n = \lambda_n(t)$ due to the estimate

$$|\Delta X^n_{\lambda_n(t)} - \Delta X_t| \le 2 \sup_{0 \le s \le t+1} |X^n_{\lambda_n(s)} - X_s|.$$

Observe an additional property of the metric ρ in the special case of nondecreasing functions in D, i.e. functions in V^+.

Theorem 3 [110]. *Let* X^n, $n \ge 1$, *and let* X *belong to* V^+. *The following properties are equivalent:*

1) $\rho(X^n, X) \to 0$,

projection of X in the space R^k.

From Corollary 1 to Theorem 2 it follows that as $t_j \in \Delta(X)$, $j = 1, \ldots, k$, the following implication holds:

$$\rho(X^n, X) \to 0 \Rightarrow \pi_{t_1, \ldots, t_k}(X^n) \to \pi_{t_1, \ldots, t_k}(X).$$

For $A \in R^k$ denote

$$\pi^{-1}_{t_1, \ldots, t_k}(A) = \{(X_{t_1}, \ldots, X_{t_k}) \in A\}$$

and

$$\mathcal{D}_\Delta = \{\pi^{-1}_{t_1, \ldots, t_k}(A): A \in B(R^k),\ t_j \in \Delta(X),\ j = 1, \ldots, k,\ k \geq 1\}$$

where $B(R^k)$ is the Borel σ-algebra in R^k.

Theorem 4 [14]. *Let* $\sigma(\mathcal{D}_\Delta)$ *be the smallest σ-algebra, containing* $\mathcal{D}_\Delta$, *and* $\mathcal{D}$ *the σ-algebra of Borel sets with the topology in the space* D *induced by the metric* ρ.

Then

$$\sigma(\mathcal{D}_\Delta) = \mathcal{D}.$$

4. For $X \in$ D denote

$$W_X[s, t) = \sup_{s \leq u,\, v < t} |X_u - X_v|,$$

$$W'_L(X, \sigma) = \inf_{t_j} \max_{0 < j \leq k} W_X[t_{j-1}, t_j),$$

where $(t_j)_{0 \leq j \leq k}$ are partition points of the interval

$$[0, L]: 0 = t_0 < t_1 < \ldots < t_k = L$$

with

$$t_j - t_{j-1} > \sigma,\ j = 1, \ldots, k - 1,$$

and

$$W''_L(X, \sigma) = \sup_{\substack{t_1, t_2 \in [0, L] \\ t_2 - t_1 \leq \sigma}} \sup_{t \in [t_1, t_2]} |X_t - X_{t_1}| \wedge |X_{t_2} - X_t|.$$

By utilizing Theorem 2, one can describe compact sets in D.

Theorem 5 [15, 182]. *Let* A *be a subset in* D. *Then* A *has a compact closure in the topology induced by the metric* ρ, *if and only if for each* L > 0

$$\sup_{X \in A} \sup_{0 \leq t \leq L} |X_t| < \infty$$

and if at least one of the following conditions is satisfied:

(1) $$\lim_{\sigma \to 0} \sup_{X \in A} W_L'(X, \sigma) = 0 \quad \forall L > 0,$$

or

(2) $$\begin{cases} \lim_{\sigma \to 0} \sup_{X \in A} W_L''(X, \sigma) = 0 & \forall L > 0, \\ \lim_{\sigma \to 0} \sup_{X \in A} W_X[0, \sigma) = 0. \end{cases}$$

5. Consider a measurable space $(D, \mathscr{D})$, where $\mathscr{D}$ is the σ-algebra of Borel sets in topology induced by the metric ρ. Let $(Q^n)_{n \geq 1}$ be a family of probability measures defined on a measurable space $(D, \mathscr{D})$.

A family of probability measures $(Q^n)_{n \geq 1}$ is called *relatively weakly compact* (or *relatively compact*), if any subsequence $(Q^{n'})_{n \geq 1}$ contains a weakly convergent subsequence $(Q^{n''})_{n \geq 1}$, i.e. a subsequence such that for any bounded function

$$f = f(X), \ X \in D,$$

taking values in R, and continuous in the topology induced by the metric ρ, we have

$$\lim_{n''} \int_D f(X)\, dQ^{n''} = \int_D f(X)\, dQ, \tag{1.5}$$

where Q is a certain probability measure on $(D, \mathscr{D})$.

Since a space D is a complete separable metric space in the metric ρ, the relative compactness of a family of probability measures $(Q^n)_{n \geq 1}$, according to Prohorov's theorem, is equivalent to the fact that this family is dense: for any $\varepsilon > 0$ one can choose a compact $K_\varepsilon \in \mathscr{D}$ depending on ε only, and such that

$$\sup_n Q^n(D \setminus K_\varepsilon) \leq \varepsilon. \tag{1.6}$$

We present now conditions for the density of a family of probability measures $(Q^n)_{n \geq 1}$, defined on a measurable space $(D, \mathscr{D})$, equipped by the metric ρ.

Theorem 6 [15, 182]. *A family of probability measures* $(Q^n)_{n \geq 1}$, *defined on a measuarble space* $(D, \mathscr{D})$, *is dense if and only if*

$$\lim_{a \to \infty} \overline{\lim_n}\, Q^n\big(\sup_{0 \leq t \leq L} |X_t| \geq a\big) = 0 \quad \forall L > 0,$$

and if at least one of the following conditions is fullfilled:

(1) $$\lim_{\sigma \to 0} \overline{\lim_n}\, Q^n(W_L'(X, \sigma) > \eta) = 0 \quad \forall \eta > 0, \ L > 0,$$

or

$$(2)\qquad \begin{cases} \lim\limits_{\sigma \to 0} \overline{\lim\limits_{n}}\, Q^n (W''_L (X, \sigma) > \eta) = 0 & \forall \eta > 0,\ L > 0, \\ \lim\limits_{\sigma \to 0} \overline{\lim\limits_{n}}\, Q^n (W_X [0, \sigma) > \eta) = 0 & \forall \eta > 0. \end{cases}$$

6. A sequence of probability measures $(Q^n)_{n \geq 1}$, defined on a measurable space $(D, \mathcal{D})$, equipped by the metric ρ, is called *weakly convergent* to a probability measure Q on $(D, \mathcal{D})$, if for each bounded and continuous in a metric ρ function f = f (X) taking values in R, we have

$$\lim_n \int_D f(X)\, dQ^n = \int_D f(X)\, dQ. \tag{1.7}$$

The weak convergence of a sequence $(Q^n)_{n \geq 1}$ to Q will be denoted by

$$Q^n \xrightarrow{w} Q.$$

If

$$Q^n \xrightarrow{w} Q,$$

then the relation (1.7) holds for a continuous in a metric ρ function f = f (X) taking values in R, under the condition of uniform integrability

$$\lim_N \sup_n \int_{\{|f(X)| > N\}} |f(X)|\, dQ^n = 0. \tag{1.8}$$

Theorem 7 [15]. *Let* f = f (X) *be a* $\mathcal{D}$*-measurable function taking values in* R, *satisfying the condition of uniform integrability* (1.8) *relative to a family of probability measures* $(Q^n)_{n \geq 1}$ *on* $(D, \mathcal{D})$, *and let*

$$A_f = \{X: \exists X^n,\ n \geq 1,\ \lim_n \rho(X^n, X) = 0,\ \overline{\lim_n}\, f(X^n) > \underline{\lim_n}\, f(X^n)\}.$$

If $Q^n \xrightarrow{w} Q$, *where* Q *is a probability measure on* $(D, \mathcal{D})$ *and* $Q(A_f) = 0$, *then*

$$\lim_n \int_D f(X)\, dQ^n = \int_D f(X)\, dQ.$$

7. Let $f = f(x_1, \ldots, x_k)$ be a bounded and continuous jointly in variables $(x_1, \ldots, x_k)$ function $(x_j \in R,\ j = 1, \ldots, k)$, taking values in R. If $(Q^n)_{n \geq 1}$ and Q are probability measures on $(D, \mathcal{D})$ such that

$$\lim_n \int_D f(X_{t_1}, \ldots, X_{t_k})\, dQ^n = \int_D f(X_{t_1}, \ldots, X_{t_k})\, dQ$$

for each function f with the indicated properties, then it is said that finite dimensional distributions (Q^n_S), $n \geq 1$, converge weakly to finite dimensional distributions Q_S, where $S = \{t_1, ..., t_k\}$. The weak convergence of finite dimensional distributions is denoted by

$$Q^n \xrightarrow{w_f(S)} Q.$$

As $S = R_+$ the notation

$$Q^n \xrightarrow{w_f} Q$$

will be used.

Denote

$$\Delta_Q = \{t > 0: Q(\Delta X_t = 0) = 1\} \cup \{0\}$$

and observe that Δ_Q is a dense subset in R_+.

Theorem 8 [15]. *Let* $(Q^n)_{n \geq 1}$ *and* Q *be probability measures on* $(D, \mathcal{D})$. *Then*

$$Q^n \xrightarrow{w} Q \Leftrightarrow \begin{cases} (Q^n)_{n \geq 1} \text{ is dense} \\ Q^n \xrightarrow{w_f(\Delta_Q)} Q. \end{cases}$$

Remark. If $\Delta_Q = R_+$, then the implication $(\Leftarrow)$ takes place with the convergence

$$\text{«} \xrightarrow{w_f(S)} \text{»}$$

instead of the convergence

$$\text{«} \xrightarrow{w_f(\Delta_Q)} \text{»}$$

where S is a dense set in R_+.

8. A remark concerning the metric ρ (see (1.4)) in case of the space $D([0, \infty), R^k)$, that is a space of functions $X_t = (X^1_t, ..., X^k_t)$ with $X^j = (X^j_t)_{t \geq 0} \in D$, $j = 1, ..., k$. In this case instead of the distance $d_0(.,.)$, defined by the formula (1.1), take the metric ($X_t = (X^1_t, ..., X^k_t)$, $Y_t = (Y^1_t, ..., Y^k_t)$)

$$d_0(X, Y) = \inf_{\substack{\lambda_j \in \Lambda \\ j = 1, ..., k}} \left\{ \sup_{0 \leq t \leq 1} \sum_{j=1}^{k} |X^j_t - Y^j_{\lambda_j(t)}| + \sup_{0 \leq s < t \leq 1} \sum_{j=1}^{k} \left| \log \frac{\lambda_j(t) - \lambda_j(s)}{t - s} \right| \right\} \tag{1.9}$$

Utilizing this metric $d_0(.,.)$ the corresponding metric $\rho(.,.)$ is defined in the same manner as in (1.4) in case $k = 1$.

Problems

1. Let $(Q^n)_{n \geq 1}$ and Q be probability measures on $(D, \mathscr{D})$ and $Q(C) = 1$. Show that

$$Q^n \xrightarrow{w} Q \Rightarrow Q^n \xrightarrow{w_f} Q.$$

2. Let $X^n = (X_t^n)_{t \geq 0}$, $Y^n = (Y_t^n)_{t \geq 0}$ and $Z^n = (Z_t^n)_{t \geq 0}$ be stochastic processes with trajectories in D, defined on a probability space $(\Omega^n, \mathscr{F}^n, P^n)$, $n \geq 1$, and Q^{X^n}, Q^{Y^n} and Q^{Z^n} the probability distributions of processes X^n, Y^n and Z^n, i.e. $Q^{X^n}(A) = P^n(X^n \in A)$, $A \in \mathscr{D}$ etc.

If

$$X^n = Y^n + Z^n, \; n \geq 1$$

and

$$\lim_n P^n(\sup_{t \leq L} |Z_t^n| \geq \varepsilon) = 0$$

for each $\varepsilon > 0$ and $L > 0$, then (Q is a probability measure on $(D, \mathscr{D})$)

$$Q^{Y^n} \xrightarrow{w} Q \Rightarrow \begin{cases} Q^{X^n} \xrightarrow{w} Q, \\ Q^{X^n} \xrightarrow{w_f(\Delta_Q)} Q. \end{cases}$$

3. Let X^n, $n \geq 1$, and X be stochastic processes defined on probability spaces $(\Omega^n, \mathscr{F}^n, P^n)$, $n \geq 1$, and $(\Omega, \mathscr{F}, P)$ respectively, with trajectories in D, and let Q^{X^n}, $n \geq 1$, and Q^X be probability distributions of these processes. Let $H = H(Y)$ be a continuous function in the metric ρ. Show that

$$Q^{X^n} \xrightarrow{w} Q^X \Rightarrow H(X^n) \xrightarrow{d} H(X).$$

4. Let α_n be a random variable, defined on a probability space $(\Omega^n, \mathscr{F}^n, P^n)$, $n \geq 1$, $a = \text{const}$. Show that

$$\alpha^n \xrightarrow{d} a \Leftrightarrow \lim_n P^n(|\alpha_n - a| \geq \varepsilon) = 0, \quad \varepsilon > 0.$$

5. Let $X \in C$, $X^n \in D$, $n \geq 1$, and $\lim_n \rho(X^n, X) = 0$. Show that for each $L > 0$

$$\lim_n \sup_{t \leq L} |X_t^n - X_t| = 0.$$

6. Let ρ be the metric defined by (1.4). For a fixed $Y \in D$ set $f(X) = \rho(X, Y)$, $X \in D$. Show that

$$|f(X') - f(X'')| \leq \rho(X', X'').$$

7. Let ρ be the metric defined by (1.4), and Γ a subset of D. Set $f(X) = \inf_{Y \in \Gamma} \rho(X, Y)$, $X \in D$. Show that

$$|f(X') - f(X'')| \leq \rho(X', X'').$$

§ 2. Continuous functions on $R_+ \times D$

1. On establishing the conditions for the relative compactness and weak convergence of measures corresponding to semimartingales we will need certain properties of functions $\phi = \phi(t, X)$, $t \in R_+$, $X \in D$, taking values in R, in particular the continuity property.

Definition 1. A function $\phi = \phi(t, X)$, $t \in R_+$, $X \in D$, taking values in R *is continuous at point* (t, X), if $\phi(t_n, X^n) \to \phi(t, X)$ for each sequence (t_n, X^n), $n \geq 1$, such that $t_n \to t$, $\rho(X^n, X) \to 0$, and *continuous on* (S, D), if it is continuous at each point (t, X), $X \in D$, $t \in S(X)$ ($S(X)$ is a nonempty set in R_+).

Example 1. The function

$$\phi(t, X) = X_t$$

is continuous on $(\Delta(X), D)$ with $\Delta(X) = \{t > 0: \Delta X_t = 0\} \cup \{0\}$.

Example 2. The function

$$\phi(t, X) = \sup_{s \leq t} |X_s|$$

is continuous on $(\Delta(X), D)$.

Example 3. The function

$$\phi(t, X) = \sup_{s \leq t} |\Delta X_s|$$

is continuous on $(\Delta(X), D)$.

Example 4. The function

$$\phi(t, X) = \sum_{0 < s \leq t} h(\Delta X_s) I(b < |\Delta X_s| < c)$$

with $b > 0$, $c \leq \infty$ and a continuous function $h = h(x)$, $x \in R$, is continuous on $(\Delta(X)$ D).

The proof of continuity of the above-mentioned functions is provided in the following manner. Let (t_n, X^n), $n \geq 1$, be a sequence with the properties $t_n \to t$ and $\rho(X^n, X) \to 0$. By Theorem 1.2 there exists a sequence $(\lambda_n)_{n \geq 1}$ of continuous strictly increasing functions with $\lambda_n(0) = 0$, $n \geq 1$, such that for each $L > 0$

$$\sup_{s \leq L} |X_s - X^n_{\lambda_n(s)}| \to 0, \quad \sup_{s \leq L} |\lambda_n(s) - s| \to 0.$$

Let μ_n be the function inverse to λ_n. Then

$$\sup_{s \leq L} |\mu_n(s) - s| \to 0$$

obviously.

Let us show that

$$\mu_n(t_n) \to t. \tag{2.1}$$

Since $t_n \to t$, one can assume $t_n \leq L$, $n \geq 1$, with a sufficiently large number L.

Then

$$|\mu_n(t_n) - t| \le |\mu_n(t_n) - t_n| + |t_n - t| \le \sup_{s \le L} |\mu_n(s) - s| + |t_n - t|,$$

and hence (2.1) holds.

Denote

$$\delta_n(L) = \sup_{s \le L} |X^n_{\lambda_n(s)} - X_s|. \tag{2.2}$$

In Example 1 we have

$$|\phi(t_n, X^n) - \phi(t, X)| = |X^n_{t_n} - X_t|,$$

and the following inequalities:

$$|X^n_{t_n} - X_t| \le |X^n_{\lambda_n(\mu_n(t_n))} - X_{\mu_n(t_n)}| + |X_{\mu_n(t_n)} - X_t|$$

$$\le \sup_{s \le L} |X^n_{\lambda_n(s)} - X_s| + |X_{\mu_n(t_n)} - X_t| = \delta_n(L) + |X_{\mu_n(t_n)} - X_t|,$$

take place, where L is chosen to be so large that $\mu_n(t_n) \le L$, $n \ge 1$ (this is allowed by (2.1) at least for sufficiently large n). But $\delta_n(L) \to 0$ for each $L > 0$, while $X_{\mu_n(t_n)} \to X_t$ by (2.1) for each $t \in \Delta(X)$, i.e.

$$|X^n_{t_n} - X_t| \to 0, \quad n \to \infty.$$

In Example 2

$$\phi(t_n, X^n) = \sup_{s \le t_n} |X^n_s| = \sup_{s \le t_n} |X^n_{\lambda_n(\mu_n(s))}|$$

$$\le \sup_{s \le \mu_n(t_n)} |X^n_{\lambda_n(s)} - X_s| + \sup_{s \le \mu_n(t_n)} |X_s|.$$

Choose L so large that $\mu_n(t_n) \le L$ (at least for large values of n). Then

$$\phi(t_n, X^n) \le \delta_n(L) + \phi(\mu_n(t_n), X). \tag{2.3}$$

For $t \in \Delta(X)$ we have $\phi(\mu_n(t_n), X) \to \phi(t, X)$ by (2.1). Therefore, from (2.3) it follows that

$$\overline{\lim_n}\ \phi(t_n, X^n) \le \phi(t, X). \tag{2.4}$$

On the other hand

$$\phi(\mu_n(t_n), X) = \sup_{s \le \mu_n(t_n)} |X_s| \le \sup_{s \le \mu_n(t_n)} |X_s - X^n_{\lambda_n(s)}| + \sup_{s \le \mu_n(t_n)} |X^n_{\lambda_n(s)}|$$

$$\le \delta_n(L) + \sup_{s \le t_n} |X^n_{\lambda_n(\mu_n(s))}| = \delta_n(L) + \phi(t_n, X^n)$$

where, as in the first case, L is sufficiently large. As $t \in \Delta(X)$ the last inequality implies

$$\phi(t, X) \le \underline{\lim}_n \phi(t_n, X^n). \tag{2.5}$$

The desired property

$$\lim_n \phi(t_n, X^n) = \phi(t, X)$$

follows from (2.4) and (2.5).

In Example 3

$$\phi(t_n, X^n) = \sup_{s \le t_n} |\Delta X_s^n| = \sup_{s \le t_n} |\Delta X^n_{\lambda_n(\mu_n(s))}|$$

$$\le \sup_{s \le t_n} |\Delta X^n_{\lambda_n(\mu_n(s))} - \Delta X_{\mu_n(s)}| + \sup_{s \le t_n} |\Delta X_{\mu_n(s)}|$$

$$\le 2\delta_n(L) + \sup_{s \le \mu_n(t_n)} |\Delta X_s| = 2\delta_n(L) + \phi(\mu_n(t_n), X)$$

with L chosen to be sufficiently large. Therefore, as $t \in \Delta(X)$, we have

$$\overline{\lim}_n \phi(t_n, X^n) \le \phi(t, X). \tag{2.6}$$

The inequality

$$\underline{\lim}_n \phi(t_n, X^n) \ge \phi(t, X) \tag{2.7}$$

follows from the following relations:

$$\phi(\mu_n(t_n), X) = \sup_{s \le \mu_n(t_n)} |\Delta X_s|$$

$$\le \sup_{s \le \mu_n(t_n)} |\Delta X_s - \Delta X^n_{\lambda_n(s)}| + \sup_{s \le t} |\Delta X^n_{\lambda_n(\mu_n(s))}|$$

$$\le 2\delta_n(L) + \sup_{s \le t_n} |\Delta X_s^n| = 2\delta_n(L) + \phi(t_n, X^n),$$

where L is a sufficiently large number.

Hence

$$\lim_n \phi(t_n, X^n) = \phi(t, X), \quad t \in \Delta(X).$$

In Example 4 we apply Corollary 2 to Theorem 1.2 according to which

$$\Delta X^n_{\lambda_n(s)} \to \Delta X_s \tag{2.8}$$

for each $s \in R_+$. In view of the indicated property it can be assumed that for each fixed $L > 0$ we have

$$I(b < |\Delta X_s| < c) = I(b < |\Delta X^n_{\lambda_n(s)}| < c)$$

as n is sufficiently large for each $s \in [0, L]$.

Therefore, as n is sufficiently large,

$$\phi(t_n, X^n) = \sum_{0 < s \le \mu_n(t_n)} h(\Delta X^n_{\lambda_n(s)})\, I(b < |\Delta X^n_{\lambda_n(s)}| < c)$$

$$= \sum_{0 < s \le \mu_n(t_n)} h(\Delta X^n_{\lambda_n(s)})\, I(b < |\Delta X_s| < c).$$

Consequently (for sufficiently large L),

$$\phi(t_n, X^n) - \phi(\mu_n(t_n), X)$$
$$\le \sum_{s \le \mu_n(t_n)} |h(\Delta X^n_{\lambda_n(s)}) - h(\Delta X_s)|\, I(b < |\Delta X_s| < c)$$
$$\le \sum_{s \le L} |h(\Delta X^n_{\lambda_n(s)}) - h(\Delta X_s)|\, I(b < |\Delta X_s| < c) \to 0, \quad n \to \infty$$

(see (2.8)).

Now, the desired relation

$$\lim_n \phi(t_n, X^n) = \phi(t, X), \quad t \in \Delta(X)$$

follows from this, (2.1) and the obvious equality

$$\lim_n \phi(\mu_n(t_n), X) = \phi(t, X), \quad t \in \Delta(X).$$

2. Let us establish a number of general properties of a function $\phi = \phi(t, X)$, $t \in R_+$, $X \in D$ taking values in R. Along with the notation $\phi = \phi(t, X)$ we will further on use the notation $\phi(X) = (\phi_t(X))_{t \ge 0}$, where for each (t, X) we have $\phi_t(X) = \phi(t, X)$.

Theorem 1. *Let the following conditions be fullfilled:*

(1) *for each* $t \in R_+$ *a function* $\phi(t, X)$ *is continuous in the metric* ρ;

(2) *for each* $X \in D$ *we have* $\phi(X) \in V$;

(3) *the function*

$$\left(\int_0^t (1 + h(\sup_{v \le s} |X_v|))\, dU_s - \mathrm{Var}(\phi(X))_t \right)_{t \ge 0} \in V^+$$

for each $X \in D$, *where* $h = h(x)$ *is a nonnegative monotone nondecreasing continuous function and* $U = (U_t)_{t \ge 0} \in V^+$.

Then $\phi(t, X)$ *is a continuous function of* $\Delta(U) \times D$ *with*

$$\Delta(U) = \{t > 0: \Delta U_t = 0\} \cup \{0\}.$$

Proof. Let (t_n, X^n), $n \ge 1$, be a sequence such that $t_n \to t$ and $\rho(X^n, X) \to 0$, n

$\to \infty$. By Condition (1) we have

$$\phi(t, X^n) \to \phi(t, X).$$

Therefore it suffices, due to the inequality

$$|\phi(t_n, X^n) - \phi(t, X)| \le |\phi(t_n, X^n) - \phi(t, X^n)| + |\phi(t, X^n) - \phi(t, X)|,$$

to verify that

$$|\phi(t_n, X^n) - \phi(t, X^n)| \to 0, \quad n \to \infty.$$

By Conditions (2) and (3) for a sufficiently large number $L \in \Delta(X)$ and $t \in \Delta(U)$ we have

$$|\phi(t_n, X^n) - \phi(t, X^n)| \le |\mathrm{Var}(\phi(X^n))_{t_n} - \mathrm{Var}(\phi(X^n))_t|$$

$$\le \int_{t_n \wedge t}^{t_n \vee t} (1 + h(\sup_{v \le s} |X_v^n|)\, dU_s$$

$$\le (1 + h(\sup_{s \le L} |X_s^n|))(U_{t_n \vee t} - U_{t_n \wedge t}) \to 0, \quad n \to \infty,$$

since as $t \in \Delta(U)$ we have $U_{t_n \vee t} - U_{t_n \wedge t} \to 0$, and in view of Problem 2 we have

$$\sup_{s \le L} |X_s^n| \to \sup_{s \le L} |X_s|.$$

Theorem 2. *Let* $\phi = \phi(t, X)$ *be a* $B(R_+) \otimes \mathcal{D}$*-measurable function such that*

(1) $|\phi(t, X)| \le c(1 + \sup_{s \le t} |X_s|^i)$, $i = 1, 2$;

(2) *for each* $t \in S$ ($S \in B(R_+)$) *a function* $\phi(t, X)$ *is continuous in the metric* ρ.

Let $u = (u_t)_{t \ge 0} \in V^+$ *and*

$$\int_{R_+ \setminus S} du_s = 0.$$

Then for each $t \in R_+$ *the function* $\psi(t, X)$ *with*

$$\psi(t, X) = \int_0^t \phi(s, X)\, du_s$$

is continuous in the metric ρ.

Proof. Let a sequence $(X^n)_{n \ge 1}$ be such that $\rho(X^n, X) \to 0$, $n \to \infty$. Then

$$\psi(t, X^n) = \int_0^t \phi(s, X^n)\, du_s = \int_{[0, t] \cap S} \phi(s, X^n)\, du_s. \tag{2.9}$$

By Condition (2) for each $s \in [0, t] \cap S$ we have $\phi(s, X^n) \to \phi(s, X)$. Therefore the desired assertion takes place, provided it is shown that the limit can be taken under the integral sign on the right-hand side of (2.9). By Condition (2) we have

$$|\phi(s, X^n)| \le c\,(1 + \sup_{s \le L} |X_s^n|^i)$$

with $L \ge t \ge s$.

Consequently, in view of Problem 2, for sufficiently large numbers n

$$\sup_{s \le t} |\phi(s, X^n)| \le 1 + c\,(1 + \sup_{s \le L} |X_s|^i), \quad i = 1, 2,$$

and the limit under the integral in

$$\int_{[0,\, t] \cap S} \phi(s, X^n)\, du_s$$

can be taken due to the Lebesgue dominated theorem.

Theorem 3. *Let* $\phi(X) = (\phi_t(X))_{t \ge 0} \in V^+$, $X \in D$ *and*

$$\tau_t(X) = \inf(s: \phi_s(X) + s \ge t).$$

If for every $t \in R_+$ *a function* $\phi_t(X)$ *is continuous in the metric* ρ, *then the following implication holds* ($t \in R_+$):

$$\rho(X^n, X) \to 0 \Rightarrow \sup_{s \le t} |\tau_s(X^n) - \tau_s(X)| \to 0.$$

Proof. First of all observe that for every $X \in D$ we have $(\tau_t(X))_{t \ge 0} \in V^+ \cap C$. Therefore it suffices, in view of Problem 5.3.2, to establish (for each $t \in R_+$)) the implication

$$\rho(X^n, X) \to 0 \Rightarrow \tau_t(X^n) \to \tau_t(X). \tag{2.10}$$

Observe that the definition of $\tau_t(X)$ implies the inequality $\tau_t(X) \le t$ for $X \in D$ and $t \in R_+$. This entails, in particular, that for each $X \in D$ and $t \in R_+$

$$\phi_{\tau_t(X)}(X) + \tau_t(X) \ge t, \quad \phi_{\tau_t(X)-}(X) + \tau_t(X) \le t. \tag{2.11}$$

Denote

$$c = \overline{\lim_n}\ \tau_t(X^n).$$

For each $\varepsilon > 0$ one can choose arbitrarily large numbers n such that

$$\tau_t(X^n) - \varepsilon < \overline{\lim_n}\ \tau_t(X^n) < \tau_t(X^n) + \varepsilon.$$

Consequently, by (2.11)

$$\phi_{c+\varepsilon}(X^n) + \tau_t(X^n) \ge t, \quad \phi_{(c-\varepsilon)\vee 0}(X^n) + \tau_t(X^n) \le t.$$

By taking the limit $\lim_{\varepsilon \to 0} \overline{\lim_n}$ in these relations we arrive at the inequalities

$$\phi_c(X) + c \geq t, \quad \phi_{c-}(X) + c \leq t. \tag{2.12}$$

Assume now $\tau_t(X) > c$. Then $\phi_c(X) \leq \phi_{\tau_t(X)-}(X)$ and by (2.11) and (2.12) we get

$$\phi_c(X) + c < \phi_{\tau_t(X)-}(X) + \tau_t(X) \leq t,$$

which contradicts the first inequality in (2.12), i.e. $\tau_t(X) \leq c$.

If $\tau_t(X) < c$, then by (2.11) and (2.12) we get

$$\phi_{c-}(X) + c > \phi_{\tau_t(X)}(X) + \tau_t(X) \geq t,$$

which contradicts the second inequality in (2.12), i.e. $\tau_t(X) = c$.

By an analogous construction for $c = \underline{\lim}_n \tau_t(X)$ we arrive at the equality $\tau_t(X) = c$.

3. Let Q be a probabiluty measure on $(D, \mathcal{D})$ and

$$\Delta_Q = \{t > 0: Q(\Delta X_t = 0) = 1\} \cup \{0\}.$$

Definition 2. A function $\phi(t, X)$, $t \in R_+$, $X \in D$ taking values in R and measurable relative to the σ-algebra $B(R_+) \otimes \mathcal{D}$, is called *continuous on* $S \times D$ *Q-a.s.* (S is a nonempty subset in R_+), if for each $t \in S$ and all sequences $(t_n)_{n \geq 1}$ and $(X^n(X))_{n \geq 1}$ where $X^n(X)_{n \geq 1}$ are $\mathcal{D}$-measurable functions taking values in D, such that $t^n \to t$ and $\rho(X^n(X), X) \to 0$, the following relation holds:

$$Q(X: \lim_n \phi(t_n, X^n(X)) = \phi(t, X)) = 1.$$

Theorem 4. *Let* a $B(R_+) \otimes \mathcal{D}$*-measurable function* $\phi = \phi(t, X)$ *be continuous on* $(\Delta(X), D)$, *and let a nonnegative* $\mathcal{D}$*-measurable function* $\tau(X)$ *be continuous in the metric* ρ, *such that*

$$Q(\tau(X) \in \Delta(X)) = 1.$$

Then the function $\phi(t \wedge \tau(X), X)$ *is continuous on* $\Delta_Q \times D$ *Q-a.s.*

Proof. Let $(t_n)_{n \geq 1}$ and $(X^n(X))_{n \geq 1}$ be sequences with the properties indicated in Definition 2. If t and $\tau(X)$ belong to $\Delta(X)$, then obviously $t \wedge \tau(X) \in \Delta(X)$. Next,

$$t_n \wedge \tau(X^n(X)) \to t \wedge \tau(X).$$

Therefore by the definition of continuity on $(\Delta(X), D)$ of functions $\phi(t, X)$ (see Definition 1)

$$\phi(t_n \wedge \tau(X^n(X)), X^n(X)) \to \phi(t \wedge \tau(X), X) \tag{2.13}$$

for $t \in \Delta(X)$, $\tau(X) \in \Delta(X)$, $X \in D$.

By the conditions of the theorem $\tau(X) \in \Delta(X)$ Q-a.s., while the definitions of sets $\Delta(X)$ and Δ_Q entail $t \in \Delta_Q \to Q(t \in \Delta(X)) = 1$.

Hence, as $t \in \Delta_Q$ the convergence in (2.13) takes place Q-a.s.

Theorem 5. *Let* $\phi = (\phi_t(X))_{t \geq 0} \in V^+ \cap C$, $X \in D$, *let* $\phi_t(X) = \phi(t, X)$ *be a*

$B(R_+) \otimes \mathcal{D}$*-measurable function,*

$$\tau_t(X) = \inf(s: \phi_s(X) + s \geq t)$$

and let $T_j(X)$, $j \geq 1$, *be a* $\mathcal{D}$*-measurable function such that* $Q(0 < T_j(X) < \infty) = 1$, $j \geq 1$ (Q *is a probability measure on* $(D, \mathcal{D})$).

Then there exists a nondecreasing sequence of numbers $(a_k)_{k \geq 1}$ *with* $a_1 \geq 0$, $\lim_k a_k = \infty$ *and*

$$Q(\tau_{a_k}(X) = T_j(X)) = 0, \ k \geq 1, \ j \geq 1. \tag{2.14}$$

Proof. By the assumption $\phi(X) \in V^+ \cap C$, $X \in D$, and the monotony of a function $\phi_t(X) + t$, $t \geq 0$, for each $X \in D$ the function $\tau_t(X)$, $t \geq 0$, itself is continuous and strictly monotone for each $X \in D$.

Define a $\mathcal{D}$-measurable function $\alpha_j(X)$, $j \geq 1$, by setting

$$\alpha_j(X) = \inf(t: \tau_t(X) = T_j(X)).$$

In accordance with the assumption $Q(0 < T_j(X) < \infty) = 1$ the analogous relation holds for $\alpha_j(X)$ too: $Q(0 < \alpha_j(X) < \infty) = 1$. Besides,

$$\{\tau_a(X) = T_j(X)\} = \{\alpha_j(X) = a\}$$

obviously, i.e.

$$Q(\tau_a(X) = T_j(X)) = Q(\alpha_j(X) = a), \ a \in R_+. \tag{2.15}$$

Denote by $F_{\alpha_j}(a)$ the distribution function for $\alpha_j(X)$:

$$F_{\alpha_j}(a) = Q(\alpha_j(X) \leq a).$$

Then $Q(\alpha_j(X) = a) = F_{\alpha_j}(a) - F_{\alpha_j}(a-)$. Since the set

$$A_j = \{a > 0: F_{\alpha_j}(a) - F_{\alpha_j}(a-) > 0\}$$

is at most countable, a sequence of numbers $(a_k)_{k \geq 1}$ with the indicated properties can be chosen within the set $R_+ \setminus \bigcup_{j \geq 1} A_j$; then (2.14) follows from (2.15).

Problems

1. Let $X \in D$ and

$$\phi(t, X) = \int_0^t X_{s-} ds.$$

Show that the function $\phi(t, X)$ is continuous on $R_+ \times D$.

2. Let $X \in D$ and

$$\phi(t, X) = \int_0^t X_{s-}\, du_s$$

with $u = (u_t)_{t \geq 0} \in V^+$ and

$$u_t^d = \sum_{0 < s \leq t} \Delta u_s > 0.$$

Show that at a fixed t the function $\phi(t, X)$ is, in general, not continuous in the metric ρ.

§ 3. Conditions on adapted processes sufficient for relative compactness of families of their distributions

1. Let a sequence of stochastic basises $(\Omega^n, \mathcal{F}^n, \mathbb{F}^n, P^n)$, $n \geq 1$, be given (see Ch. 1, § 1). For every $n \geq 1$ assume that a stochastic process $X^n = (X^n_t)_{t \geq 0}$, defined on a probability space $(\Omega^n, \mathcal{F}^n, P^n)$, is adapted to a family $\mathbb{F}^n$ with trajectories in D. We denote the probability distribution of a process X^n by Q^n. In other words, Q^n is a probability measure on the measurable space $(D, \mathcal{D})$ such that for each set $\Gamma \in \mathcal{D}$

$$Q^n(\Gamma) = P^n(X^n \in \Gamma).$$

In this section we establish sufficient conditions for relative compactness of families of measures (Q^n), $n \geq 1$, which are fit for treating the weak convergence of sequences of measures Q^n, $n \geq 1$, to a probability measure Q, in the practically most important case in which Q is associated with a left quasi-continuous process.

2. Theorem 1. *Let the following conditions be fullfilled:*

1) *for each* $L > 0$

$$\lim_{a \to \infty} \overline{\lim_n} P^n(\sup_{t \leq L} |X^n_t| \geq a) = 0;$$

2) *for each* $L > 0$ *and* $\eta > 0$

$$\lim_{\delta \to 0} \overline{\lim_n} \sup_{S \in T_L(\mathbb{F}^n)} P^n(\sup_{t \leq \delta} |X^n_{S+t} - X^n_S| \geq \eta) = 0,$$

where $T_L(\mathbb{F}^n)$ *is a family of stopping times taking values in* $[0, L]$.

Then the family of measures (Q^n), $n \geq 1$, *is relatively compact.*

Proof. In view of Theorem 1.6 it suffices to show that the implication

$$2) \Rightarrow \lim_{\sigma \to 0} \overline{\lim_n} P^n(W'_L(X^n, \sigma) > \eta) = 0 \tag{3.1}$$

holds.

To this end define the following Markov times:

$$S^n_0 = 0, \quad S^n_k = \inf(t > S^n_{k-1} : |X^n_t - X^n_{S_{k-1}}| \geq \eta), \ k \geq 1.$$

We will show that for given L, η, ε a positive number l can be chosen such that

$$\overline{\lim_n} P^n(S^n_l < L) \leq 2\varepsilon, \tag{3.2}$$

as well as a positive number σ such that

$$\underline{\lim_n} P^n(\{S^n_l \geq L\} \cap \bigcap_{k=1}^{l} \{(S^n_k - S^n_{k-1} \geq \sigma) \cup (S_{k-1} \geq L)\} \geq 1 - 3\varepsilon.$$

Observe first that the following sets coincide:

$$\{S^n_{k+1} \le L,\ S^n_{k+1} - S^n_k < \delta\}$$

$$= \{\sup_{t\le\delta} |X^n_{S^n_k + t} - X^n_{S^n_k}| \ge \eta,\ S^n_k \le L,\ S^n_{k+1} - S^n_k < \delta\}. \qquad (3.4)$$

By Condition 2) one can choose δ such that

$$\overline{\lim_n} \sup_{S \in T_L(\mathbb{F}^n)} P^n(\sup_{t\le\delta} |X^n_{S+t} - X^n_S| \ge \eta) \le \varepsilon.$$

Then by (3.4) for given δ we have

$$\overline{\lim_n} P^n(S^n_k \le L,\ S^n_{k+1} - S^n_k < \delta) \le \varepsilon \quad \forall k \ge 1. \qquad (3.5)$$

For a given δ we choose an integer l such that

$$l\delta > 2L. \qquad (3.6)$$

Next, we have ($\mathbf{E}^n$ is the mathematical expectation with respect to the measure P^n)

$$\sum_{k=1}^{l} \mathbf{E}^n (S^n_k - S^n_{k-1}) I(S^n_l < L) = \mathbf{E}^n \sum_{k=1}^{l} (S^n_k - S^n_{k-1}) I(S^n_l < L)$$

$$= \mathbf{E}^n S^n_l I(S^n_l < L) \le L P^n(S^n_l < L). \qquad (3.7)$$

On the other hand

$$\sum_{k=1}^{l} \mathbf{E}^n (S^n_k - S^n_{k-1}) I(S^n_l < L)$$

$$\ge \sum_{k=1}^{l} \mathbf{E}^n (S^n_k - S^n_{k-1}) I(S^n_l < L,\ S^n_k - S^n_{k-1} \ge \delta)$$

$$\ge \delta \sum_{k=1}^{l} P^n(S^n_l < L,\ S^n_k - S^n_{k-1} \ge \delta)$$

$$= \delta \sum_{k=1}^{l} [P^n(S^n_l < L) - P^n(S^n_l < L,\ S^n_k - S^n_{k-1} < \delta)]. \qquad (3.8)$$

By (3.8), (3.7) and (3.5) we get the inequality

$$L \overline{\lim_n} P^n(S^n_l < L) \ge \delta l \overline{\lim_n} P^n(S^n_l < L) - \delta l \varepsilon,$$

which along with (3.6) gives the desired inequality (3.2).

To prove (3.3) set ε'/l and, utilizing again the coincidence of the sets in (3.4), choose σ such that (c.f. (3.5))

$$\overline{\lim_n}\ P^n(S_k^n \le L,\ S_{k+1}^n - S_k^n < \sigma) \le \varepsilon' \quad \forall k \ge 1. \tag{3.9}$$

We have

$$P^n(\{S_1^n \ge L\} \cap \bigcap_{k=1} \{(S_k^n - S_{k-1}^n \ge \sigma) \cup (S_{k-1} \ge L)\}$$

$$= 1 - P^n(\{S_1^n < L\} \cup \bigcup_{k=1}^{l} \{S_k^n - S_{k-1}^n < \sigma,\ S_{k-1}^n < L\})$$

$$\ge 1 - P^n(S_1^n < L) - \sum_{k=1}^{l} P^n(S_k^n - S_{k-1}^n < \sigma,\ S_{k-1}^n < L). \tag{3.10}$$

In view of (3.2) and (3.9) this gives

$$\underline{\lim_n}\ P^n(\{S_1^n \ge L\} \cap \bigcap_{k=1}^{l} \{(S_k^n - S_{k-1}^n \ge \sigma) \cup (S_{k-1} \ge L)\})$$

$$\ge 1 - \overline{\lim_n}\ P^n(S_1^n < L) - \sum_{k=1}^{l} \overline{\lim_n}\ P^n(S_k^n - S_{k-1}^n < \sigma < L)$$

$$\ge 1 - 2\rho - l\varepsilon',$$

i. e. the inequality (3.3) holds, since $l\varepsilon' = \varepsilon$.

Let us establish now the first implication in (3.1).

To this end it suffices to show that for each $\zeta > 0$ one can choose σ such that

$$\overline{\lim_n}\ P^n(W'_L(X^n, \sigma) > \eta) \le \zeta. \tag{3.11}$$

Set $\zeta = 3\varepsilon$ and choose l and σ such that the inequality (3.3) holds. Let

$$\omega \in \{S_1^n \ge L\} \cap \bigcap_{k=1}^{l} \{S_k^n - S_{k-1}^n \ge \sigma\}.$$

For a given ω the points $(S_k^n)_{k \ge 0}$, belonging to $[0, L]$, partition the interval $0 = t_0 < t_1 < \dots < t_r \le L$ in such a way that $t_{p+1} - t_p \ge \sigma$, $0 \le p \le r-1$. Therefore

$$\sup_{t_p \le s < t < t_{p+1}} |X_t^n - X_s^n| \le \eta.$$

Consequently, for a given ω we have

$$W'_L(X^n, \sigma) \le \eta,$$

i.e.

$$\{S_1^n \ge L\} \cap \bigcap_{k=1}^{l} \{(S_k^n - S_{k-1}^n \ge \sigma) \cup (S_{k-1} \ge L)\} \subseteq \{W'_L(X^n, \sigma) \le \eta\}.$$

In view of (3.3) this gives

$$\overline{\lim_n}\ P^n(W_L'(X^n, \sigma) > \eta)$$

$$\le 1 - \underline{\lim_n}\ P^n(\{S_1^n \ge L\} \cap \bigcap_{k=1}^{l} \{(S_k^n - S_{k-1}^n \ge \sigma) \cup (S_{k-1} \ge L)\})$$

$$\le 3\varepsilon = \zeta,$$

i.e. (3.11) and (3.1) hold.

3. We give a sufficient condition for the relative compactness, useful in studying the weak convergence of a sequence of measures (Q^n), $n \ge 1$, to a probability measure Q, corresponding to a process which is not necessarily left quasi-continuous.

To this end we introduce the following

Definition. A sequence $T_i^n(u)$, $i \ge 0$, $u \in (0, \infty]$, $n \ge 1$ of $\mathbb{F}^n$-stopping times is called $\mathbb{F}^n$-*sieve*, if it satisfies the following conditions:

1) $T_0^n(u) = 0$, $T_i^n(\infty) = \infty$, $i \ge 1$,

2) $T_i^n(u) < \infty \Rightarrow T_i^n(u) < T_{i+1}^n(u)$,

3) $\lim_{i \to \infty} T_i^n(u) = \infty$,

4) $u' < u \Rightarrow \bigcup_{i \ge 1} [\![T_i^n(u)]\!] \subset \bigcup_{i \ge 1} [\![T_i^n(u')]\!]$.

We formulate the condition analogous to Condition 2) of Theorem 1.

Assume there exists a $\mathbb{F}^n$-sieve with the following properties:

2') for each $L > 0$ and $u \in (0, \infty]$

$$\lim_{\sigma \to 0} \overline{\lim_n}\ P^n(\bigcup_{i \ge 1} \{T_i^n(u) \le L,\ T_i^n(u) - T_{i-1}^n(u) < \sigma\}) = 0;$$

2'') there exists a function $q = q(L, \eta, \varepsilon)$ with the following properties: $0 < q(L, \eta, \varepsilon) \le \eta$, q is a nondecreasing function of ε, and for each $L > 0$, $\eta > 0$ and $\varepsilon > 0$ there exist $\delta > 0$ and an integer n_0 such that as $u \le q(L, \eta, \varepsilon)$

$$\sup_{n \ge n_0} \sup_{S \in T_L(\mathbb{F}^n)} P^n(\sup_{t \le \delta} |X_{S+t}^n(u) - X_S^n(u)| \ge \eta) \le \varepsilon,$$

with

$$X_t^n(u) = X_t^n - \sum_{i \ge 1} \Delta X_{T_i^n(u)}^n\, I_{[\![T_i^n(u), \infty[\![}(t);$$

2''') for each $L > 0$, $\eta > 0$, $u \in (0, \infty]$ and any sequence (S^n), $n \ge 1$ with $S^n \in T_L(\mathbb{F}^n)$

$$\lim_{\sigma \to 0} \overline{\lim_n} \, P^n \left(\bigcup_{i \geq 1} \{T_i^n(u) - \sigma < S^n < T_i^n(u) + \sigma, \ S^n \neq T_i^n(u), \right.$$

$$\left. |\Delta X^n_{S^n}| \geq \eta\} \right) = 0.$$

Theorem 1 is generalized as follows:

Theorem 2 [112]. *Let Condition* 1) *of Theorem 1 and Conditions* 2'), 2'') *and* 2''') *be fullfilled. Then the family of measures* (Q^n), $n \geq 1$, *is relatively compact.*

Remark. As $T_i^n(u) \equiv \infty$, $i \geq 1$, Conditions 2') and 2''') are automatically fullfilled, while Condition 2'') turns into Condition 2) of Theorem 1.

Problems

1. Let $X_t^n = \alpha^n t$, where α^n is a random vaiable on a probability space $(\Omega^n, \mathcal{F}^n, P^n)$, $n \geq 1$. Show that the family of distributions (Q^n), $n \geq 1$, of processes $X^n = (X_t^n)_{t \geq 0}$, $n \geq 1$, is relatively compact if

$$\lim_{a \to \infty} \overline{\lim_n} \, P^n (|\alpha^n| \geq a) = 0.$$

2. Let X^n be a $\mathbb{F}^n$-adapted process with trajectories in D, $T_0^n(u) = 0$ and $T_i^n(u) = \inf(t > T_{i-1}^n(u): |\Delta X_t^n| > u)$, $i \geq 1$. Show that $(T_i^n(u), i \geq 0, u \in (0, \infty]$ is a $\mathbb{F}^n$-sieve.

§ 4. Relative compactness of probability distributions of semimartingales

1. In this section all notations of § 3 are preserved, and probability measures Q^n, $n \geq 1$, will be here the probability distributions of semimartingales X^n, $n \geq 1$.

For each $n \geq 1$ let $T^n = (B^n, C^n, \nu^n)$ be the triplet of $\mathbb{F}^n$-predictable characteristics X^n.

The conditions for the weak convergence of finite dimensional distributions of semimartingales X^n, $n \geq 1$, have been formulated in terms determined by the triplets T^n, $n \geq 1$ (see Ch. 5). It is natural, therefore, to express in the same terms also conditions for relative compactness of distributions of X^n, $n \geq 1$, as this will allow us to formulate the conditions for the weak convergence of distributions of semimartingales (see Ch. 7 and 8 below) in terms corresponding to the "convergence" of the triplets.

2. The conditions for the relative compactness of distributions of semimartingales will be expressed in terms of their triplets in a form suitable for verifying Conditions 1) and 2) of Theorem 3.1.

The following version of the canonical representation for a semimartingale X^n with the triplet T^n of predictable characteristics (see Ch. 4, § 1, formula (1.9)) serves as a starting point in establishing conditions for the relative compactness of distributions of semimartingales:

$$X_t^n = X_0^n + B_t^{na} + X_t^{nc} + \int_0^t \int_{|x|>a} x \, d\mu^n + \int_0^t \int_{|x|\leq a} x \, d(\mu^n - \nu^n) \quad (4.1)$$

as $a \in (0, 1]$, with (Problem 4.1.5)

$$B_t^{na} = B_t^n - \int_0^t \int_{a<|x|\leq 1} x \, d\nu^n. \quad (4.2)$$

Denote

$$M_t^{n,a} = X_t^{nc} + \int_0^t \int_{|x|\leq a} x \, d(\mu^n - \nu^n). \quad (4.3)$$

Clearly,

$$M^{n,a} = (M_t^{n,a})_{t\geq 0} \in \mathcal{M}_{loc}^2 (\mathbb{F}^n, P^n)$$

and its quadratic characteristic (Theorem 3.5.1 and Lemma 3.5.1) is given by the formula:

$$\langle M^{n,a}\rangle_t = C^n_t + \int_0^t \int_{|x|\le a} x^2\, dv^n - \sum_{0<s\le t}\left(\int_{|x|\le a} x v^n(\{s\}, dx)\right)^2. \quad (4.4)$$

We give the four groups of assumptions used for formulating the basic result on the relative compactness of the family (Q^n), $n \ge 1$, of distributions of semimartingales X^n, $n \ge 1$, with the triplets T^n, $n \ge 1$.

Group I:

1) $\lim_{l\to\infty} \overline{\lim_n}\, P^n(|X^n_0| \ge l) = 0$;

2) for each $L > 0$ and $\varepsilon > 0$

$$\lim_{l\to\infty} \overline{\lim_n}\, P^n\left(\int_0^L \int_{|x|>l} dv^n \ge \varepsilon\right) = 0.$$

Group II:

1) There exists a stochastic process $\beta^{na} = (\beta^{na}_t)_{t\ge 0}$, $a \in (0, 1]$ with trajectories in V, $|\Delta\beta^{na}| \le 1$, $n \ge 1$, such that for each $L > 0$ and $\varepsilon > 0$

$$\lim_n P^n(\sup_{t\le L} |B^{na}_t - \beta^{na}_t| \ge \varepsilon) = 0;$$

2) there exists a $\mathbb{F}^n$-predictable process $\gamma^{na} = (\gamma^{na}_t)_{t\ge 0}$, $a \in (0, 1]$, with trajectories in V^+ such that for each $L > 0$ and $\varepsilon > 0$

$$\lim_{a\to 0} \overline{\lim_n}\, P^n(\sup_{t\le L} |\langle M^{n,a}\rangle_t - \gamma^{na}_t| \ge \varepsilon) = 0;$$

3) to every nonnegative continuous bounded function $g = g(x)$ with the property $g(x) = I(|x| > b)\, g(x)$ for a certain $b > 0$ there corresponds a $\mathbb{F}^n$-predictable process $\delta^{ng} = (\delta^{ng}_t)_{t\ge 0}$, $n \ge 1$, with trajectories in V^+ such that for each $L > 0$ and $\varepsilon > 0$

$$\lim_n P^n(\sup_{t\le L} |g * v^n_t - \delta^{ng}_t| \ge \varepsilon) = 0.$$

Group III:

There exists a stochastic process $G^n = (G^n_t)_{t\ge 0}$, $n \ge 1$, with trajectories in V^+, and a function

$$\overline{G}^n = (\overline{G}^n_t)_{t\ge 0} \in V^+, \ n \ge 1,$$

such that the processes

$$G_t^n + \int_0^t (1 + \sup_{u<s} |X_u^n|)\, d\overline{G}_s^n - \mathrm{Var}\,(\beta^{na})_t,$$

$$G_t^n + \int_0^t (1 + \sup_{u<s} |X_u^n|)\, d\overline{G}_s^n - \delta_t^{ng}$$

and

$$G_t^n + \int_0^t (1 + \sup_{u<s} (X_u^n)^2)\, d\overline{G}_s^n - \gamma_t^{na}$$

are increasing (i.e. belong to V^+).

Group IV:

1) Let $\overline{G}^n \equiv 0$ and let $\tilde{Q}^n$ be a probability measure on $(D, \mathcal{D})$ presenting the probability distribution of a process G^n. There exists a probability measure $\tilde{Q}$ on $(D, \mathcal{D})$ such that

$$\tilde{Q}\,(V^+ \cap C) = 1$$

and

$$\tilde{Q}^n \xrightarrow{w} \tilde{Q}.$$

2) Let G^n be a $\mathbb{F}^n$-predictable process such that $\lim_n P^n(G_t^n \geq \varepsilon) = 0$ for each $t \in R_+$ and $\varepsilon > 0$, and let a function $\overline{G} = (\overline{G}_t)_{t \geq 0} \in V^+ \cap C$ exist such that

$$\overline{G}_t^n \to \overline{G}_t$$

for every t in a certain set dense in R_+.

3) Let $\overline{G}^n \equiv 0$ and $(\Omega^n, \mathcal{F}^n, P^n) \equiv (\Omega, \mathcal{F}, P)$, while $\mathbb{F}^n$ may depend on n. There exists a stochastic process $G = (G_t)_{t \geq 0}$ defined on $(\Omega, \mathcal{F}, P)$, belonging to $V^+ \cap C$ and such that

$$G_t^n \xrightarrow{P} G_t$$

for every t in a certain set, dense in R_+.

Theorem 1. *Let the Groups* I, II *and* III *of conditions be fullfilled as well as any of Conditions* 1), 2) *and* 3) *of group* IV. *Then the family of distributions* (Q^n), $n \geq 1$, *of semimartingales* X^n, $n \geq 1$, *with the triplets* T^n, $n \geq 1$, *are relatively compact.*

Remark. If the conditions of the Groups I, II and III are fullfilled as well as any of

Conditions $IV_{1)}$ and $IV_{3)}$, then the assertion of the theorem is in force: the $\mathbb{F}^n$-predictability condition imposed on the processes γ^{na} and δ^{ng} in the assumptions $II_{2)}$ and $II_{3)}$ is superfluous.

3. The proof of Theorem 1 consists in checking Conditions 1) and 2) of Theorem 3.1.

We verify Condition 1) in several steps presented below as lemmas; the proof utilizes the decomposition

$$X^n = Y^n(a, b, c) + X^n(a, b, c), \ \ a < c, \tag{4.5}$$

where

$$Y_t^n(a, b, c) = X_0^n I(|X_0^n| > b) + (B_t^{na} - \beta_t^{na}) + h^c * \mu_t^n, \tag{4.6}$$

with

$$h^c(x) = xI(|x| > c) \tag{4.7}$$

and

$$X_t^n(a, b, c) = X_0^n I(|X_0^n| \le b) + \beta_t^{na} + h^{ac} * \mu_t^n + M_t^{n, a}, \tag{4.8}$$

with

$$h^{ac}(x) = xI(a < |x| \le c). \tag{4.9}$$

Lemma 1. *Let Conditions* I *and* $II_{1)}$ *be fullfilled. Then for each* $L > 0$ *and* $l > 0$, $a \in (0, 1]$

$$\lim_{b, c \to \infty} \overline{\lim_n} P^n(\sup_{t \le L} |Y_t^n(a, b, c)| \ge l) = 0.$$

Proof. From (4.6) it follows that

$$\sup_{t \le L} |Y_t^n(a, b, c)| \le |X_0^n| I(|X_0^n| > b) + \sup_{t \le L} |B_t^{na} - \beta_t^{na}| + |h^c| * \mu_L^n.$$

By Assumption $II_{1)}$ we have

$$\lim_n P^n\left(\sup_{t \le L} |B_t^{na} - \beta_t^{na}| \ge \frac{l}{3}\right) = 0, \ l > 0.$$

Next, in view of Assumption $I_{1)}$ we have

$$\overline{\lim_{b \to \infty}} \ \overline{\lim_n} P^n\left(|X_0^n| I(|X_0^n| > b) \ge \frac{l}{3}\right) \le \overline{\lim_{b \to \infty}} \ \overline{\lim_n} P^n(|X_0^n| > b) = 0.$$

Therefore it suffices to show that

$$\lim_{c \to \infty} \overline{\lim_n} P^n\left(|h^c| * \mu_L^n \ge \frac{l}{3}\right) = 0.$$

To this end note that as $c \ge 1$ we have

$$\left\{|h^c| * \mu_L^n \ge \frac{l}{3}\right\} \subseteq \{I(|x| > c) * \mu_L^n \ge 1\},$$

and hence it suffices that

$$\lim_{c \to \infty} \overline{\lim_n} P^n(I(|x| > c) * \mu_L^n \ge 1) = 0. \tag{4.10}$$

Since a process $I(|x| > c) * v^n$ presents the compensator of a process $I(|x| > c) * \mu^n$, by the Lenglart-Rebolledo inequality (Theorem 1.9.3) we have

$$P^n(I(|x| > c) * \mu_L^n \geq 1) \leq \varepsilon + P^n(I(|x| > c) * v_L^n \geq \varepsilon),$$

which leads to (4.10) as the limit $\lim_{\varepsilon \to 0} \overline{\lim}_{c \to \infty} \overline{\lim}_n$ is taken.

Lemma 2. *Let Conditions* $II_{2)}$, $II_{3)}$ *and* III *be fullfilled as well as any of Conditions* $IV_{2)}$ *and* $IV_{3)}$. *Then for each* $L > 0$

$$\lim_{a \to 0} \overline{\lim}_{b, c \to \infty} \overline{\lim}_{l \to \infty} \overline{\lim}_n P^n(\sup_{t \leq L} |X_t^n(a, b, c)| \geq l) = 0.$$

Proof. By (4.8) we get

$$\sup_{t \leq L} |X_t^n(a, b, c)| \leq b + \mathrm{Var}(\beta^{na})_L + |h^{ac}| * \mu_L + \sup_{t \leq L} |M_t^{n, a}|.$$

Denote by $\xi^n(a, b, c)$ any of the terms on the right-hand side of this inequality. Obviously the desired assertion takes place, provided

$$\lim_{a \to 0} \overline{\lim}_{b, c \to \infty} \overline{\lim}_{l \to \infty} \overline{\lim}_n P^n(\xi^n(a, b, c) \geq l) = 0. \tag{4.11}$$

Therefore the proof of the lemma is reduced to verifying (4.11).

If $\xi^n(a, b, c) = b$, then (4.11) is obviously satisfied.

Let $\xi^n(a, b, c) = \mathrm{Var}(\beta^{na})_L$. By the Assumptions $IV_{1)}$ and $IV_{3)}$ we have $\overline{G}^n \equiv 0$ and hence in view of Assumption III the inequality

$$\mathrm{Var}(\beta^{na})_L \leq G_L^n$$

holds. Consequently, (4.11) takes place, provided

$$\lim_{l \to \infty} \overline{\lim}_n P^n(G_L^n \geq l) = 0. \tag{4.12}$$

Let Assumption $IV_{1)}$ be satisfied. We utilize the fact that

$$P^n(G_L^n \geq l) = \tilde{Q}^n(X_L \geq l).$$

Since

$$\tilde{Q}^n \xrightarrow{w} \tilde{Q},$$

the family $(\tilde{Q}^n)$, $n \geq 1$, is relatively compact by Theorem 1.8, and hence by Theorem 1.6 we have

$$\lim_{l \to \infty} \overline{\lim}_n \tilde{Q}^n(X_L \geq l) = 0,$$

i.e. the desired relation (4.12) holds.

If Assumption $IV_{3)}$ is satisfied, then taking $L' > L$ with L' in $\{t > 0: P(\Delta G_t = 0) = 1\}$ if necessary, we set $L \in \{t > 0: P(\Delta G_t = 0) = 1\}$. Then the desired relation (4.12) follows from the following inequality ($P^n \equiv P$)

$$P(G_L^n \geq l) \leq P\left(G_L \geq \frac{l}{2}\right) + P\left(|G_L^n - G_L| \geq \frac{l}{2}\right)$$

as the limit $\lim_{l\to\infty} \overline{\lim_n}$ is taken.

Let

$$\xi^n(a, b, c) = |h^{ac}| * \mu_L^n.$$

Obviously one can choose a function $g = g(x)$, satisfying Assumption $II_{3)}$ and such that for fixed a and c one has $g \geq |h^{ac}|$. Then (4.11) takes place, provided

$$\lim_{l\to\infty} \overline{\lim_n} P^n(g * \mu_L^n \geq l) = 0. \tag{4.13}$$

To verify (4.13), observe that $g * \nu^n$ is the compensator of the process $g * \mu^n$ and hence by the Lenglart-Rebolledo inequality (Theorem 1.9.3)

$$P^n(g * \mu_L^n \geq l) \leq \frac{1}{\sqrt{l}} + P^n(g * \nu_L^n \geq \sqrt{l}).$$

Therefore (4.13) takes place, provided

$$\lim_{l\to\infty} \overline{\lim_n} P^n(g * \nu_L^n \geq l) = 0. \tag{4.14}$$

By Assumption $II_{3)}$ there exists a $\mathbb{F}^n$-predictable process $\delta^{ng} \in \mathcal{V}^+$, such that

$$\lim_n P^n(\sup_{t\leq L} |g * \nu_t^n - \delta_t^{ng}| \geq \varepsilon) = 0, \quad \varepsilon > 0.$$

Due to this fact it is not hard to deduce that (4.14) takes place, provided

$$\lim_{l\to\infty} \overline{\lim_n} P^n(\delta_L^{ng} \geq l) = 0.$$

But by the Assumptions III, $IV_{1)}$ and $IV_{3)}$ we have

$$\delta_L^{ng} \leq G_L^n.$$

Consequently, (4.14) holds since (4.12) is fullfilled.

Finally, let

$$\xi^n(a, b, c) = \sup_{t\leq L} |M_t^{n,a}|.$$

Since $M^{n,a}$ is independent of b and c, then (4.11) takes place, provided

$$\lim_{a\to 0} \lim_{l\to\infty} \overline{\lim_n} P^n(\sup_{t\leq L} |M_t^{n,a}| \geq l) = 0. \tag{4.15}$$

To verify (4.15) we utilize the fact that by the Lenglart-Rebolledo inequality (Theorem 1.9.3)

$$P^n(\sup_{t\leq L} |M_t^{n,a}| \geq l) \leq \frac{1}{l} + P^n(\langle M^{n,a}\rangle_L \geq l).$$

Therefore (4.15) is a consequence of the following relation:

$$\lim_{a \to 0} \overline{\lim_{l \to \infty}} \overline{\lim_{n}} P^n (\langle M^{n,a} \rangle_L \geq l) = 0. \tag{4.16}$$

By Assumption $II_{2)}$ there exists a $\mathbb{F}^n$-predictable process $\gamma^{na} \in \mathcal{V}^+$, such that

$$\lim_{a \to 0} \overline{\lim_{n}} P^n (\sup_{t \leq L} | \langle M^{n,a} \rangle_t - \gamma_t^{na} | \geq \varepsilon) = 0, \quad \varepsilon > 0.$$

Due to this fact, (4.16) takes place, provided

$$\lim_{a \to 0} \overline{\lim_{l \to \infty}} \overline{\lim_{n}} P^n (\gamma_L^{na} \geq l) = 0.$$

But by the assumptions III, $IV_{1)}$ and $IV_{3)}$ we have

$$\gamma_L^{na} \leq G_L^n$$

and hence (4.16) follows from (4.12).

The lemma is proved.

For fixed a, b, c and L define a Markov time

$$\tau_{abc}^n = \inf (t \leq L: \ | Y_t^n (a, b, c) | \ \geq 1). \tag{4.17}$$

Lemma 3. *Let the Assumptions* I, II, III *and* $IV_{2)}$ *be fullfilled. Then for each* L > 0

$$\lim_{a \to 0} \overline{\lim_{b, c \to \infty}} \overline{\lim_{l \to \infty}} P^n (\sup_{t \leq L \wedge \tau_{abc}^n} | X_t^n (a, b, c) | \geq l) = 0.$$

Proof. Let g = g (x) be a function satisfying Assumption $II_{3)}$ and such that

$$g \geq | h^{ac} | \vee (h^{ac})^2,$$

δ^{ng} the process corresponding to this function as indicated in Assumption $II_{3)}$, and g^{na} the process met in assumption $II_{2)}$.

We will show that for a fixed L > 0 and each n ≥ 1 there exists a $\mathbb{F}^n$-Markov time σ_{ac}^n such that

1) $$\sup_{t \leq L \wedge \sigma_{ac}^n} | g * \nu_t^n - \delta_t^{ng} | + \sup_{t \leq L \wedge \sigma_{ac}^n} (g * \nu_t^n - \delta_t^{ng})^2$$

$$+ \sup_{t \leq L \wedge \sigma_{ac}^n} | \langle M^{n,a} \rangle_t - \gamma_t^{na} | + G_{L \wedge \sigma_{ac}^n}^n \leq 1;$$

2) $$\lim_{a \to 0} \overline{\lim_{c \to \infty}} \overline{\lim_{n}} P^n (\sigma_{ac}^n \leq L) = 0.$$

To this end define the process $\phi^n (a, c) = (\phi_t^n (a, c))_{t \geq 0}$ with

$$\phi^n_t(a, c) = |\sup_{s \le t} g * v^n_s - \delta^{ng}_s| + \sup_{s \le t} (g * v^n_s - \delta^{ng}_s)^2$$

$$+ \sup_{s \le t} |\langle M^{n, a}\rangle_s - \gamma^{na}_s| + G^n_t,$$

and observe that by the assumptions $II_{2)}$, $II_{3)}$ and $IV_{2)}$ we have

$$\lim_{a \to 0} \overline{\lim_n} \, P^n(\phi^n_L(a, c) \ge \varepsilon) = 0, \quad \varepsilon > 0. \tag{4.18}$$

Define first a Markov time $\tilde{\sigma}^n_{ac} = \inf(t \le L: \phi^n_t(a, c) \ge 1)$ with $\inf(\varnothing) = \infty$. The process $\phi^n(a, c)$ is a $\mathbb{F}^n$-predictable process from $\mathcal{V}^+$. Therefore, by Problem 1.3.11 $\tilde{\sigma}^n_{ac}$ is a $\mathbb{F}^n$-predictable Markov time and consequently, by Theorem 1.3.4, one can choose a Markov time σ^n_{ac} such that

$$\sigma^n_{ac} < \tilde{\sigma}^n_{ac}, \quad P^n\left(\tilde{\sigma}^n_{ac} - \sigma^n_{ac} > \frac{1}{n}, \ \tilde{\sigma}^n_{ac} < \infty\right) \le \frac{1}{n}. \tag{4.19}$$

Obviously, this Markov time possesses the desired property 1), as well as property 2) since

$$\{\tilde{\sigma}^n_{ac} \le L\} = (\phi^n_L(a, c) \ge 1)$$

and

$$(\sigma^n_{ac} \le L\} \subseteq \left\{\sigma^n_{ac} \le L, \ \tilde{\sigma}^n_{ac} - \sigma^n_{ac} \le \frac{1}{n}\right\} \cup \left\{\tilde{\sigma}^n_{ac} - \sigma^n_{ac} > \frac{1}{n}, \ \tilde{\sigma}^n_{ac} < \infty\right\}$$

$$\subseteq \left\{\tilde{\sigma}^n_{ac} \le L + \frac{1}{n}\right\} \cup \left\{\tilde{\sigma}^n_{ac} - \sigma^n_{ac} > \frac{1}{n}, \ \tilde{\sigma}^n_{ac} < \infty\right\}$$

$$= \{\tilde{\sigma}^n_{ac} \le L\} \cup \left\{\tilde{\sigma}^n_{ac} - \sigma^n_{ac} > \frac{1}{n}, \ \tilde{\sigma}^n_{ac} < \infty\right\}$$

(observe that the definition of $\tilde{\sigma}^n_{ac}$ entails the alternative: $\tilde{\sigma}^n_{ac} \le L$, $\tilde{\sigma}^n_{ac} = \infty$), and hence

$$P^n(\sigma^n_{ac} \le L) \le P^n(\tilde{\sigma}^n_{ac} \le L) + \frac{1}{n} = P^n(\phi^n_L(a, c) \ge 1) + \frac{1}{n},$$

i.e. property 2) is a consequence of (4.18) and (4.19).

Denote

$$\delta^n_{abc} = \tau^n_{abc} \wedge \sigma^n_{ac}. \tag{4.20}$$

By property 2) of the Markov time σ^n_{ac} it suffices to prove the assertion of the lemma for Markov time δ^n_{abc} instead of τ^n_{abc}, i.e. it suffices to show that

$$\lim_{a\to 0}\ \overline{\lim_{b,c\to\infty}}\ \overline{\lim_{l\to\infty}}\ \overline{\lim_{n}}\ P^n\left(\sup_{t\le L\wedge\delta^n_{abc}} |X^n_t(a,b,c)|\ge l\right)=0. \tag{4.21}$$

To prove (4.21) we establish the inequality

$$\overline{\lim_{n}}\ E^n \sup_{t\le L\wedge\delta^n_{abc}} (X^n_t(a,b,c))^2 \le r(a,b,g), \tag{4.22}$$

with a constant r (a, b, g) depending on a, b and g.

Observe at once that (4.21) is a simple consequence of (4.22) since by Chebyshev's inequality and (4.22) we have

$$\overline{\lim_{n}}\ P^n\left(\sup_{t\le L\wedge\delta^n_{abc}} |X^n_t(a,b,c)|\ge l\right)\le \frac{r(a,b,g)}{l^2}\to 0,\quad l\to\infty.$$

Thus, to conclude the proof of the lemma it remains to establish (4.22).

Define Markov times

$$\delta^{nk}_{abc}=\inf(t\le L:\ |X^n_t(a,b,c)|\ge k)\wedge\delta^n_{abc},\quad k\ge 1, \tag{4.23}$$

and observe that

$$\lim_{k\to\infty}\delta^{nk}_{abc}=\delta^n_{abc}. \tag{4.24}$$

Denote

$$\xi^{nk}_t=\sup_{s\le t\wedge\delta^{nk}_{abc}}(X^n_s(a,b,c))^2 \tag{4.25}$$

and observe that since jumps of the process $X^n(a, b, c)$ are uniformly bounded due to (4.8), (4.3), (4.9) and assumption $II_{1)}$ which entail $|\Delta\beta^{na}|\le 1$, then $\xi^{nk}_t \le$ const.

Next, from (4.8) it follows as $t\le L\wedge\delta^{nk}_{abc}$ that

$$\xi^{nk}_t\le\left(b+\mathrm{Var}(\beta^{na})_t+|h^{ac}|*\mu_t+\sup_{s\le t}|M^{n,a}_s|\right)^2. \tag{4.26}$$

Define a process $m^n=(m^n_t)_{t\ge 0}$ with

$$m^n_t=|h^{ac}|*\mu^n_t-|h^{ac}|*\nu^n_t. \tag{4.27}$$

By Theorem 3.5.1 this process belongs to $\mathcal{M}^2_{loc}(\mathbb{F}^n)$. Its quadratic characteristic $\langle m^n\rangle$ possesses by Problem 3.5.8 the property

$$(h^{ac})^2*\nu^n-\langle m^n\rangle\in\mathcal{V}^+,$$

and hence in accordance with the definition of the function g = g (x) we have

$$g*\nu^n-\langle m^n\rangle\in\mathcal{V}^+. \tag{4.28}$$

Utilizing these facts as well as (4.27) and (4.26), for $t \le L \wedge \delta^{nk}_{abc}$ we get

$$\xi^{nk}_t \le 5\,(b^2 + \mathrm{Var}^2\,(\beta^{na})_t + (g * v^n_t)^2 + \sup_{s \le t}\,(m^n_s)^2 + \sup_{s \le t}\,(M^{n,\,a}_s)^2). \qquad (4.29)$$

In accordance with Problem 1.9.7 we have

$$\mathbf{E}^n \sup_{s \le t \wedge \delta^{nk}_{abc}} (m^n_s)^2 \le 4\mathbf{E}^n \langle m^n \rangle_{t \wedge \delta^{nk}_{abc}},$$

$$\mathbf{E}^n \sup_{s \le t \wedge \delta^{nk}_{abc}} (M^{n,\,a}_s)^2 \le 4\mathbf{E}^n \langle M^{n,\,a} \rangle_{t \wedge \delta^{nk}_{abc}}.$$

In view of these inequalities from (4.29) we deduce as $t \le L$ that

$$\mathbf{E}^n \xi^{nk}_t \le 5\,(b^2 + \mathbf{E}^n\,\mathrm{Var}^2\,(\beta^{na})_{t \wedge \delta^{nk}_{abc}} + \mathbf{E}^n\,(g * v^n_{t \wedge \delta^{nk}_{abc}})^2$$

$$+ 4\mathbf{E}^n \langle m^n \rangle_{t \wedge \delta^{nk}_{abc}} + 4\mathbf{E}^n \langle M^{n,\,a} \rangle_{t \wedge \delta^{nk}_{abc}}). \qquad (4.30)$$

We use now the property (4.28) of the quadratic characteristic $\langle m^n \rangle$. Then as $t \le L$

$$\langle m^n \rangle_{t \wedge \delta^{nk}_{abc}} \le g * v^n_{t \wedge \delta^{nk}_{abc}} \le \sup_{t \le L} |\, g * v^n_t - \delta^{ng}_t \,| + \delta^{ng}_{t \wedge \delta^{nk}_{abc}}. \qquad (4.31)$$

Obviously, the analogous estimates hold for

$$(g * v^n_{t \wedge \delta^{nk}_{abc}})^2 \text{ and } \langle M^{n,\,a} \rangle_{t \wedge \delta^{nk}_{abc}}$$

(as $t \le L$):

$$(g * v^n_{t \wedge \delta^{nk}_{abc}})^2 \le 2 \sup_{t \le L} (g * v^n_t - \delta^{ng}_t)^2 + 2\,(\delta^{ng}_{t \wedge \delta^{nk}_{abc}})^2, \qquad (4.32)$$

$$\langle M^{n,\,a} \rangle_{t \wedge \delta^{nk}_{abc}} \le \sup_{t \le L} |\langle M^{n,\,a} \rangle_t - \gamma^{na}_t\,| + \gamma^{na}_{t \wedge \delta^{nk}_{abc}}. \qquad (4.33)$$

By (4.30) - (4.33) we get, in view of property 1) of the Markov time σ^n_{ac} and the fact that $\delta^{nk}_{abc} \le \sigma^n_{ac}$, the following inequality for $\mathbf{E}^n\,\xi^{nk}_t$ as $t \le L$:

$$\mathbf{E}^n \xi^{nk}_t \le 20\,(b^2 + 1 + \mathbf{E}^n\,[\mathrm{Var}^2\,(\beta^{na})_{t \wedge \delta^{nk}_{abc}} + (\delta^{ng}_{t \wedge \delta^{nk}_{abc}})^2 + \gamma^{na}_{t \wedge \delta^{nk}_{abc}}]).$$

By this and the inequality $a \le 1 + a^2$, for $t \le L$ we get

$$\mathbf{E}^n\,\xi^{nk}_t \le 40\,(2 + b^2 + \mathbf{E}^n\,[\mathrm{Var}^2\,(\beta^{na})_{t \wedge \delta^{nk}_{abc}})\,(\delta^{ng}_{t \wedge \delta^{nk}_{abc}})^2 + \gamma^{na}_{t \wedge \delta^{nk}_{abc}}]). \qquad (4.34)$$

We utilize now the assumptions III and IV$_{2)}$, according to which as $t \le L \wedge \delta^{nk}_{abc}$ the Cauchy-Bunjakovski inequality gives

$$\mathrm{Var}^2(\beta^{na})_t \le \left(1 + \int_0^t (1 + \sup_{u<s} |X^u_n|)\, d\overline{G}^n_s\right)^2 \le 2 + 2\overline{G}^n_L \int_0^t \left(1 + \sup_{u<s} |X^n_u|\right)^2 d\overline{G}^n_s .$$

Next, by the decomposition (4.5) and the fact that $\delta^{nk}_{abc} \le \tau^n_{abc}$, we have (as $s \le L \wedge \delta^{nk}_{abc}$)

$$\sup_{u<s} (X^n_u)^2 \le 2 \sup_{u<s} (Y^n_u(a, b, c))^2 + 2 \sup_{u<s} (X^n_u(a, b, c))^2 \le 2\,(1 + \xi^{nk}_{s-}).$$

Consequently, as $t \le L$ we have

$$\mathrm{Var}^2(\beta^{na})_{t \wedge \delta^{nk}_{abc}} \le 2 + 12\,(\overline{G}^n_L)^2 + 8\overline{G}^n_L \int_0^t \xi^{nk}_{s-}\, d\overline{G}^n_s . \tag{4.35}$$

The same estimate holds for $(\delta^{ng}_{t \wedge \delta^{nk}_{abc}})^2$ too:

$$(\delta^{ng}_{t \wedge \delta^{nk}_{abc}})^2 \le 2 + 12\,(\overline{G}^n_L)^2 + 8\overline{G}^n_L \int_0^t \xi^{nk}_{s-}\, d\overline{G}^n_s . \tag{4.36}$$

In an analogous manner we establish that

$$\gamma^{na}_{t \wedge \delta^{nk}_{abc}} \le 1 + 3\overline{G}^n_L + \int_0^t \xi^{nk}_{s-}\, d\overline{G}^n_s . \tag{4.37}$$

Now, by the estimates (4.34) - (4.37) we get

$$E^n \xi^{nk}_t \le 40 \left[7 + b^2 + 24\,(\overline{G}^n_L)^2 + 3\overline{G}^n_L + (2 + 16\overline{G}^n_L) \int_0^t (E^n \xi^{nk}_{s-})\, d\overline{G}^n_s \right]. \tag{4.38}$$

Denote

$$R^{nk}_t = E^n \xi^{nk}_t, \quad r_1(x) = 40\,(7 + b^2 + 24x^2 + 3x), \quad r_2(x) = 2 + 16x.$$

Then the inequality (4.38) can be rewritten in the form

$$R^{nk}_t \le r_1(\overline{G}^n_L) + r_2(\overline{G}^n_L) \int_0^t R^{nk}_{s-}\, d\overline{G}^n_s, \quad t \le L,$$

where we made use of the fact that by the uniform boundedness of the variables ξ^{nk}_t, $t \ge 0$, we have $R^{nk}_{t-} = E^n \xi^{nk}_{t-}$. Therefore by Theorem 2.4.3 we estimate R^{nk}_L in the following manner:

$$R_L^{nk} \le r_1(\overline{G}_L^n) \exp [r_2(\overline{G}_L^n)\overline{G}_L^n].$$

Thus, in accordance with the definitions of R_L^{nk} and ξ_L^{nk} we have

$$E^n \sup_{t \le L \wedge \delta_{abc}^{nk}} (X_t^n(a, b, c))^2 \le r_1(\overline{G}_L^n) \exp [r_2(\overline{G}_L^n)\overline{G}_L^n]. \qquad (4.39)$$

The right-hand side of the inequality (4.39) is independent of k. Therefore, in view of the relation $\lim_{k \to \infty} \delta_{abc}^{nk} = \delta_{ab}^n$ and Lebesgue's theorem on the monotone limiting transition under the integral sign, taking the limit as $k \to \infty$ in (4.39) we arrive at the inequality

$$E^n \sup_{t \le L \wedge \delta_{abc}^{n}} (X_t^n(a, b, c))^2 \le r_1(\overline{G}_L^n) \exp [r_2(\overline{G}_L^n)\overline{G}_L^n]. \qquad (4.40)$$

By assumption $IV_{2)}$ and Problem 5.3.2

$$\overline{G}_L^n \to \overline{G}_L, \quad n \to \infty.$$

Consequently, taking the limit $\overline{\lim}_n$ in (4.40) we arrive at the desired inequality (4.22) with

$$r(a, b, g) = r_1(\overline{G}_L) \exp [r_2(\overline{G}_L)\overline{G}_L].$$

Lemma 4. *Let the assumptions of Groups* I, II *and* III *be fullfilled as well as any of Conditions* $IV_{1)}$, $IV_{2)}$ *and* $IV_{3)}$. *Then for each* $L > 0$

$$\lim_{l \to \infty} \overline{\lim_n} P^n(\sup_{t \le L} |X_t^n| \ge l) = 0.$$

Proof. By the decomposition (4.5)

$$\sup_{t \le L} |X_t^n| \le \sup_{t \le L} |Y_t^n(a, b, c)| + \sup_{t \le L} |X_t^n(a, b, c)|.$$

Consequently,

$$P^n(\sup_{t \le L} |X_t^n| \ge l + 1) \le P^n(\sup_{t \le L} |Y_t^n(a, b, c)| \ge 1)$$

$$+ P^n(\sup_{t \le L} |X_t^n(a, b, c)| \ge l). \qquad (4.41)$$

If any of the assumptions $IV_{1)}$ or $IV_{3)}$ is fullfilled, then by Lemma's 1 and 2 the desired assertion follows from (4.41) as the limit $\lim_{a \to 0} \overline{\lim}_{b, c \to \infty} \overline{\lim}_{l \to \infty} \overline{\lim}_n$ is taken.

If Condition $IV_{2)}$ is fullfilled, then instead of (4.41) we use the inequalities

$$P^n(\sup_{t\le L}|X_t^n|\ge l+1)$$

$$\le P^n(\sup_{t\le L}|Y_t^n(a,b,c)|\ge 1)+P^n(\sup_{t\le L\wedge\tau_{abc}^n}|X_t^n(a,b,c)|\ge 1)+P^n(\tau_{abc}^n<L),$$

$$P^n(\tau_{abc}^n<L)\le P^n(\sup_{t\le L}|Y_t^n(a,b,c)|\ge 1),$$

by which the desired assertion is established in virtue of Lemma's 1 and 3 as the limit

$\lim_{a\to 0}\ \overline{\lim}_{b,c\to\infty}\ \overline{\lim}_{l\to\infty}\ \overline{\lim}_{n}$ is taken.

4. Here we give some preliminary results needed for verifying Condition 2) of Theorem 3.1.

Let β^{na}, γ^{na} and δ^{ng} be the processes, met in assumption II. For $\eta>0$, $\zeta>0$ and $L>0$ denote

$$r_1^{na}=P^n(\sup_{t\le L+1}|B_t^{na}-\beta_t^{na}|\ge\eta/10),$$

$$r_2^{na}=P^n(\sup_{t\le L+1}|\langle M^{n,a}\rangle_t-\gamma_t^{na}|\ge\zeta/4),$$

$$r_3^{ng}=P^n(\sup_{t\le L+1}|g*v_t^n-\delta_t^{ng}|\ge\zeta/4),\tag{4.42}$$

$$r^{nag}=r_1^{na}+r_2^{na}+r_3^{ng};$$

$$p^{nl}=P^n(\sup_{t\le L+1}|X_t^n|\ge l),\tag{4.43}$$

$$q^{nc}=P^n(I(|x|>c)*v_{L+1}^n\ge\zeta);\tag{4.44}$$

$$\tilde G^n=G^n+\overline G^n.\tag{4.45}$$

Lemma 5. *Let* $T_L(\mathbb{F}^n)$ *be a set of stopping times taking values in* $[0,L]$. *If Condition* III *is fullfilled, then for* $\delta\in(0,1]$ *we have*

$$\sup_{S \in T_L(\mathbb{F}^n)} P^n \left(\sup_{t \le \delta} |X^n_{S+t} - X^n_S| \ge \eta \right)$$

$$\le r^{nag} + \zeta \left(1 + \frac{5}{\eta} + \frac{25}{\eta^2} \right) + 3p^{nl} + q^{nc}$$

$$+ 3 \sup_{S \in T_L(\mathbb{F}^n)} P^n \left(\tilde{G}^n_{S+\delta} - \tilde{G}^n_S \ge \frac{\eta \wedge \zeta}{5(2+1^2)} \right).$$

Proof. Utilizing the decomposition (4.5) and the representations (4.6) and (4.8) for $Y^n(a, b, c)$ and $X^n(a, b, c)$, we get for X^n the representation

$$X^n_t = X^n_0 + \beta^{na}_t + (B^{na}_t - \beta^{na}_t) + M^{n,a}_t + h^{ac} * \mu^n_t + h^c * \mu^n_t, \tag{4.46}$$

due to which for $S \in T_L(\mathbb{F}^n)$ and $\delta \in (0, 1]$ we have

$$\sup_{t \le \delta} |X^n_{S+t} - X^n_S| \le [\mathrm{Var}(\beta^{na})_{S+\delta} - \mathrm{Var}(\beta^{na})_S]$$

$$+ 2 \sup_{t \le L+1} |B^{na}_t - \beta^{na}_t| + \sup_{t \le \delta} |M^{n,a}_{S+t} - M^{n,a}_S|$$

$$+ [\,|h^{ac}| * \mu^n_{S+\delta} - |h^{ac}| * \mu^n_S] + 2\,|h^c| * \mu^n_{L+1} = \sum_{j=1}^{5} I^n_j.$$

By this inequality we get

$$\sup_{S \in T_L(\mathbb{F}^n)} P^n \left(\sup_{t \le \delta} |X^n_{S+t} - X^n_S| \ge \eta \right) \le \sum_{j=1}^{5} \sup_{S \in T_L(\mathbb{F}^n)} P^n (I^n_j \ge \eta/5). \tag{4.47}$$

We will estimate now each term on the right-hand side of (4.47).

By definition

$$I^n_1 = \mathrm{Var}(\beta^{na})_{S+\delta} - \mathrm{Var}(\beta^{na})_S.$$

Therefore by assumption III

$$I^n_1 \le G^n_{S+\delta} - G^n_S + \int_S^{S+\delta} (1 + \sup_{u<s} |X^n_u|)\, d\overline{G}^n_s$$

and consequently on the set

$$\{ \sup_{t \le L+1} |X^n_t| < 1 \}$$

we have

$$I^n_1 \le (1+1)\,[\tilde{G}^n_{S+\delta} - \tilde{G}^n_S] \le (2+1^2)\,[\tilde{G}^n_{S+\delta} - \tilde{G}^n_S],$$

i.e. (see (4.43))

$$\sup_{S \in T_L(\mathbb{F}^n)} P^n(I_1^n \geq \eta/5) \leq p^{nl} + \sup_{S \in T_L(\mathbb{F}^n)} P^n\left(\tilde{G}^n_{S+\delta} - \tilde{G}^n_S \geq \frac{\eta}{5(2+l^2)}\right). \tag{4.48}$$

Since

$$I_2^n = 2 \sup_{t \leq L+1} |B_t^{na} - \beta_t^{na}|,$$

in view of the definitions (4.42) we have

$$\sup_{S \in T_L(\mathbb{F}^n)} P^n(I_2^n \geq \eta/5) \leq r_1^{na}. \tag{4.49}$$

To estimate the third term on the right-hand side of the inequality (4.47), observe that a process $\hat{M}^{na} = (\hat{M}_t^{na})_{t \geq 0}$ with

$$\hat{M}_t^{na} = M^{n,a}_{S+t} - M^{n,a}_S, \quad S \in T_L(\mathbb{F}^n) \tag{4.50}$$

is a locally square integrable martingale relative to the family $\mathbb{F}^{n,S} = (\mathcal{F}_t^{n,S})_{t \geq 0}$ with $\mathcal{F}_t^{n,S} = \mathcal{F}^n_{S+t}$ (Problem 1.8.13) and has the quadratic characteristic

$$\langle \hat{M}^{na} \rangle_t = \langle M^{n,a} \rangle_{S+t} - \langle M^{n,a} \rangle_S.$$

In view of (4.50) we have

$$I_3^n = \sup_{t \leq \delta} |\hat{M}_t^{na}|.$$

By Problem 1.8.6 we have

$$E^n (\hat{M}_\tau^{na})^2 \leq E^n \langle \hat{M}^{na} \rangle_\tau$$

for each stopping time τ relative to $\mathbb{F}^{n,S}$. Therefore by the Lenglart-Rebolledo inequality

$$P^n(I_3^n \geq \eta/5) \leq \frac{25\zeta}{\eta^2} + P^n(\langle M^{n,a} \rangle_{S+\delta} - \langle M^{n,a} \rangle_S \geq \zeta).$$

We use now the fact that for $\delta \in (0, 1]$

$$\langle M^{n,a} \rangle_{S+\delta} - \langle M^{n,a} \rangle_S \leq 2 \sup_{t \leq L+1} |\langle M^{n,a} \rangle_t - \gamma_t^{na}| + |\gamma^{na}_{S+\delta} - \gamma_S^{na}|,$$

as well as the fact that by assumption III on the set $\{\sup_{t \leq L+1} |X_t^n| < 1\}$

$$\gamma^{na}_{S+\delta} - \gamma^n_S \leq (1 + l^2)(\tilde{G}^n_{S+\delta} - \tilde{G}^n_S).$$

This implies

$$\sup_{S \in T_L(\mathbb{F}^n)} P^n(I_3^n \geq \eta/5)$$
$$\leq \frac{25\zeta}{\eta^2} + r_2^{na} + p^{nl} + \sup_{S \in T_L(\mathbb{F}^n)} P^n\left(\tilde{G}^n_{S+\delta} - \tilde{G}^n_S \geq \frac{\zeta}{2(2+l^2)}\right). \quad (4.51)$$

On estimating the fourth term in the inequality (4.47) we use the fact that a process $A^n = (A_t^n)_{t \geq 0}$ with

$$A_t^n = |h^{ac}| * \mu^n_{S+t} - |h^{ac}| * \mu^n_S$$

belongs to $\mathcal{A}^+_{loc}(\mathbb{F}^{n,S})$ and has the compensator

$$\tilde{A}_t^n = |h^{ac}| * \nu^n_{S+t} - |h^{ac}| * \nu^n_S$$

with respect to the family $\mathbb{F}^{n,S}$ (Problem 1.6.10). Since $I_4^n = A_\delta^n$, by the Lenglart-Rebolledo inequality (Theorem 1.9.3)

$$P^n(I_4^n \geq \eta/5) \leq \frac{5\zeta}{\eta} + P^n(\tilde{A}_\delta^n \geq \zeta).$$

Let $g = g(x)$ be a continuous nonnegative bounded function equal to zero in a certain neighbourhood of the origin and such that $g \geq |h^{ac}|$. Then for $\delta \in (0, 1]$

$$\tilde{A}_\delta^n \leq g * \nu^n_{S+\delta} - g * \nu^n_S \leq 2 \sup_{t \leq L+1} |g * \nu_t^n - \delta^{ng}| + \delta^{ng}_{S+\delta} - \delta^{ng}_S.$$

Consequently, by assumptions III on the set $\{\sup_{t \leq L+1} |X_t^n| < 1\}$ we have

$$\tilde{A}_\delta^n \leq 2 \sup_{t \leq L+1} |g * \nu_t^n - \delta_t^{ng}| + (2 + l^2)(\tilde{G}^n_{S+\delta} - \tilde{G}^n_S),$$

which gives

$$\sup_{S \in T_L(\mathbb{F}^n)} P^n(I_4^n \geq \eta/5) \leq \frac{5\zeta}{\eta} + r_3^{ng} + p^{nl}$$
$$+ \sup_{S \in T_L(\mathbb{F}^n)} P^n\left(\tilde{G}^n_{S+\delta} - \tilde{G}^n_S \geq \frac{\zeta}{2(2+l^2)}\right). \quad (4.52)$$

The last term in the inequality (4.47) is estimated as follows: firstly

$$\{|h^c| * \mu^n_{L+1} \geq \eta/10\} \subseteq \{I(|x| > c) * \mu^n_{L+1} \geq 1\},$$

secondly by the Lenglart-Rebolledo inequality (Theorem 1.9.3)

$$P^n(I(|x| > c) * \mu^n_{L+1} \geq 1) \leq \zeta + P^n(I(|x| > c) * \nu^n_{L+1} \geq \zeta).$$

Consequently,

$$\sup_{S \in T_L(\mathbb{F}^n)} P^n (I_5^n \geq \eta/5) = P^n (|h^c| * \mu_{L+1}^n \geq \eta/10) \leq \zeta + q^{nc}. \qquad (4.53)$$

The proof of the desired assertion follows now from (4.47) and the estimates (4.48) - (4.53).

5. Proof of Theorem 1. Condition 1) of Theorem 3.1 is verified in Lemma 4. Utilizing Lemma 5, we will show that Condition 2) of Theorem 3.1 is fullfilled as well.

First assume

$$\lim_{\delta \to 0} \overline{\lim_n} \sup_{S \in T_L(\mathbb{F}^n)} P^n \left(\tilde{G}_{S+\delta}^n - \tilde{G}_S^n \geq \frac{\eta \wedge \zeta}{5(2+l^2)} \right) = 0 \qquad (4.54)$$

for each $\eta > 0$, $\zeta > 0$, $l > 0$, $L > 0$, $a \in (0, 1]$ and g. Let us show that in this case Condition 2) of Theorem 3.1 is fullfilled. In fact

$$\mathrm{II}_{1)} \Rightarrow \lim_n r_1^{na} = 0, \quad \mathrm{II}_{2)} \Rightarrow \lim_{a \to 0} \overline{\lim_n} r_2^{na} = 0, \quad \mathrm{II}_{3)} \Rightarrow \lim_n r_3^{ng} = 0,$$

and hence

$$\mathrm{II} \Rightarrow \lim_{a \to 0} \overline{\lim_n} r^{nag} = 0$$

(see (4.42)). Besides,

$$\mathrm{I}_{2)} \Rightarrow \lim_{c \to \infty} \overline{\lim_n} q^{nc} = 0,$$

while by Lemma 4

$$\lim_{l \to \infty} \overline{\lim_n} p^{nl} = 0,$$

i.e. by taking the limit $\lim_{\zeta \to 0} \overline{\lim}_{a \to 0} \overline{\lim}_{c \to \infty} \overline{\lim}_{l \to \infty} \overline{\lim}_{\delta \to 0} \overline{\lim}_n$ in the equality established in Lemma 5, we arrive at the desired Condition 2) of Theorem 3.1:

$$\lim_{\delta \to 0} \overline{\lim_n} \sup_{S \in T_L(\mathbb{F}^n)} P^n (\sup_{t \leq \delta} |X_{S+t}^n - X_S^n| \geq \eta) = 0.$$

Thus, it remains for accomplishing the proof to verify Condition (4.54).

Aiming at this, we set

$$\varepsilon = \frac{\eta \wedge \zeta}{5(2+l^2)}$$

and observe that

$$\sup_{S \in T_L(\mathbb{F}^n)} P^n (\tilde{G}_{S+\delta}^n - \tilde{G}_S^n \geq \varepsilon) \leq P^n (\sup_{t \leq L} |\tilde{G}_{t+\delta}^n - \tilde{G}_t^n| \geq \varepsilon),$$

i.e. it suffices to show that

$$\lim_{\delta \to 0} \overline{\lim_{n}} P^n \left(\sup_{t \le L} | \tilde{G}^n_{t+\delta} - \tilde{G}^n_t | \ge \varepsilon \right) = 0. \tag{4.55}$$

The relation (4.55) will be verified separately under each of the assumptions $IV_{1)}$, $IV_{2)}$ and $IV_{3)}$.

If assumption $IV_{1)}$ is fullfilled, then $\tilde{G}^n = G^n$ and

$$P^n \left(\sup_{t \le 1} | G^n_{t+\delta} - G^n_t | \ge \varepsilon \right) = \tilde{Q}^n \left(\sup_{t \le L} [X_{t+\delta} - X_t] \ge \varepsilon \right),$$

where

$$X \in V^+ \quad \tilde{Q}^n\text{-a.s.}$$

For each $X \in V^+$ denote

$$H_\delta(X) = \sup_{t \le L} [X_{t+\delta} - X_t], \quad \delta \in (0, 1].$$

Then

$$\tilde{Q}^n \left(\sup_{t \le L} [X_{t+\delta} - X_t] \ge \varepsilon \right) = \int_{V^+} I(H_\delta(X) \ge \varepsilon)\, d\tilde{Q}^n \le \int_{V^+} 1 \wedge \frac{H_\delta(X)}{\varepsilon} d\tilde{Q}^n.$$

The function

$$1 \wedge \frac{H_\delta(X)}{\varepsilon}$$

is bounded and continuous in $V^+ \cap C$, and possesses obviously the following property:

$$\lim_{\delta \to 0} \left(1 \wedge \frac{H_\delta(X)}{\varepsilon} \right) = 0, \qquad X \in V^+ \cap C.$$

Consequently, in view of Theorem 1.7 we get

$$\overline{\lim_{\delta \to 0}} \; \overline{\lim_{n}} P^n \left(\sup_{t \le L} (\tilde{G}^n_t - \tilde{G}^n_t) \ge \varepsilon \right)$$

$$\le \overline{\lim_{\delta \to 0}} \; \overline{\lim_{n}} \int_{V^+} 1 \wedge \frac{H_\delta(X)}{\varepsilon} d\tilde{Q}^n$$

$$= \overline{\lim_{\delta \to 0}} \int_{V^+ \cap C} 1 \wedge \frac{H_\delta(X)}{\varepsilon} d\tilde{Q}$$

$$= \int_{V^+ \cap C} \lim_{\delta \to 0} \left(1 \wedge \frac{H_\delta(X)}{\varepsilon} \right) d\tilde{Q} = 0,$$

i.e. the desired relation (4.55) holds.

If assumption $IV_{2)}$ is fullfilled, then

$$P^n(\sup_{t\le L}[\tilde{G}^n_{t+\delta}-\tilde{G}^n_t]\ge\varepsilon)\le I\left(\sup_{t\le L}[\overline{G}^n_{t+\delta}-\overline{G}^n_t]\ge\frac{\varepsilon}{2}\right)+P^n\left(G^n_{L+\delta}\ge\frac{\varepsilon}{2}\right).$$

But

$$\lim_n P^n(G^n_{L+\delta}\ge\varepsilon/2)=0,$$

and as $\delta\in(0,1]$

$$\sup_{t\le L}[\overline{G}^n_{t+\delta}-\overline{G}^n_t]\le 2\sup_{t\le L+1}|\overline{G}^n_t-\overline{G}_t|+\sup_{t\le L}[\overline{G}_{t+\delta}-\overline{G}_t].$$

Consequently, the relation (4.55) holds by the continuity of the function $\overline{G}$ and Problem 5.3.2, in view of which

$$\sup_{t\le L'}|\overline{G}^n_t-G_t|\to 0,\quad n\to\infty,$$

for each $L'>0$.

If assumption $IV_{3)}$ is fullfilled, then $P^n\equiv P$ and $\tilde{G}^n=G^n$. As $\delta\in(0,1]$ we have

$$P(\sup_{t\le L}[G^n_{t+\delta}-G^n_t]\ge\varepsilon)$$

$$\le P\left(2\sup_{t\le L+1}|G^n_t-G_t|\ge\frac{\varepsilon}{2}\right)+P\left(\sup_{t\le L}[G_{t+\delta}-G_t]\ge\frac{\varepsilon}{2}\right).$$

Consequently, (4.55) holds by Problem 5.3.2 in view of which

$$\sup_{t\le L'}|G^n_t-G_t|\xrightarrow{P}0,\quad L'>0,$$

and also by the continuity of trajectories of the process G.

The remark to Theorem 1 is in force, since the $\mathbb{F}^n$-predictability of the processes γ^{na} and δ^{ng} is used only in the course of proving Lemma 3, in connection with assumption $IV_{2)}$.

6. Let $X^n=(X^n_t,\mathcal{F}^n_t)$ be semimartingales with the triplets $T^n=(B^n,C^n,\nu^n)$, $n\ge 1$. For each $n\ge 1$ let $\hat{\tau}_n=(\hat{\tau}_n(t))_{t\ge 0}\in V^+$, and for each $t\in R_+$ let $\hat{\tau}_n(t)\in \mathbf{T}(\mathbb{F}^n)$, i.e. $\hat{\tau}_n$ is a random change of time. We consider a semimartingale

$$\hat{X}^n=(\hat{X}^n_t,\hat{\mathcal{F}}^n_t)$$

with

$$\hat{X}^n_t=X^n_{\hat{\tau}_n(t)},\qquad \hat{\mathcal{F}}^n_t=\mathcal{F}^n_{\hat{\tau}_n(t)}$$

(see Ch. 4, § 7), and we formulate the theorem analogous to Theorem 1, concerning the family of distributions $(\hat{Q}^n)$ of semimartingales $\hat{X}^n$, $n\ge 1$. Aiming at this we formulate conditions analogous to the groups of Conditions I - IV.

Group $\hat{\mathrm{I}}$:

1) $\lim\limits_{l \to \infty} \overline{\lim\limits_{n}} \, \mathsf{P}^n (|X_0^n| \geq l) = 0,$

2) for each $L > 0$ and $\varepsilon > 0$

$$\lim_{l \to \infty} \overline{\lim_{n}} \, \mathsf{P}^n \left(\int_0^{\hat{\tau}_n(L)} \int_{|x| > l} dv^n \geq \varepsilon \right) = 0.$$

Group $\hat{\mathrm{II}}$:

1) There exists a stochastic process $\beta^{na} = (\beta_t^{na})_{t \geq 0}$ with $|\Delta\beta_{na}| \leq 1$, $a \in (0, 1]$, $n \geq 1$ and trajectories in V, such that for each $L > 0$ and $\varepsilon > 0$

$$\lim_{n} \mathsf{P}^n \left(\sup_{t \leq \hat{\tau}_n(L)} |B_t^{na} - \beta_t^{na}| \geq \varepsilon \right) = 0;$$

2) there exists a stochastic process $\gamma^{na} = (\gamma_t^{na})_{t \geq 0}$, $n \geq 1$, $a \in (0, 1]$ with trajectories in V^+ such that for each $L > 0$ and $\varepsilon > 0$

$$\lim_{a \to 0} \overline{\lim_{n}} \, \mathsf{P}^n \left(\sup_{t \leq \hat{\tau}_n(L)} |\langle M^{n,a} \rangle_t - \gamma_t^{na}| \geq \varepsilon \right) = 0;$$

3) for any nonnegative continuous bounded function $g = g(x)$ such that

$$g(x) = I(|x| > b)\, g(x)$$

for a certain $b > 0$, there exists a stochastic process $\delta^{ng} = (\delta_t^{ng})_{t \geq 0}$, $n \geq 1$, in V^+ such that for each $L > 0$ and $\varepsilon > 0$

$$\lim_{n} \mathsf{P}^n \left(\sup_{t \leq \hat{\tau}_n(L)} |g * v_t^n - \delta_t^{ng}| \geq \varepsilon \right) = 0.$$

Group $\hat{\mathrm{III}}$:

There exists a stochastic process $G^n = (G_t^n)_{t \geq 0}$ with trajectories in V^+, $n \geq 1$, such that

$$G^n - \mathrm{Var}\,(\beta^{na}) - \delta^{ng} - \gamma^{na} \in V^+, \quad n \geq 1.$$

Group $\hat{\mathrm{IV}}$:

1) Let $\tilde{Q}^n$ be a probability measure on $(D, \mathcal{D})$, presenting the distribution of a

process $\hat{G}^n = (\hat{G}^n_t)_{t \geq 0}$ with $\hat{G}^n_t = G^n_{\hat{\tau}^n(t)}$. There exists a probability measure $\tilde{Q}$ on (D, $\mathcal{D}$) such that

$$\tilde{Q}(V^+ \cap C) = 1$$

and

$$\tilde{Q}^n \xrightarrow{w} \tilde{Q}.$$

2) Let $(\Omega^n, \mathcal{F}^n, P^n) \equiv (\Omega, \mathcal{F}, P)$, and allow $\mathbb{F}^n$ to depend on n. There exists a process $G \in V^+ \cap C$, defined on $(\Omega, \mathcal{F}, P)$ such that

$$G^n_{\hat{\tau}_n(t)} \xrightarrow{P} G_t$$

for every t from a set dense in R_+.

Theorem 2. *Let the assumptions of Groups* $\hat{\mathrm{I}}$, $\hat{\mathrm{II}}$ *and* $\hat{\mathrm{III}}$ *be fullfilled, as well as any of Conditions* $\hat{\mathrm{IV}}_{1)}$ *and* $\hat{\mathrm{IV}}_{2)}$*. Then the family of distributions* $(\hat{Q}^n)$, $n \geq 1$, *is relatively compact.*

We do not provide the proof of this theorem as it is analogous to the proof of Theorem 1.

7. We give yet another result concerning the relative compactness of distributions of semimartingales X^n, $n \geq 1$, which is expressed in terms of the triplets $T^n = (B^n, C^n, \nu^n)$, $n \geq 1$, and proved by applying Theorem 3.2. Aiming at this we formulate the following conditions.

Group III':

There exists a $\mathbb{F}^n$-predictable process $G^n = (G^n_t)_{t \geq 0}$, $n \geq 1$, with trajectories in V^+, such that a process

$$G^n - \mathrm{Var}(B^n) - C^n - (x^2 \wedge 1) * \nu^n$$

is increasing (i.e. belongs to V_+).

Group IV':

1) Let Q^{G^n} be a probability measure on (D, $\mathcal{D}$), presenting the distribution of the process G^n. There exists a probability measure Q^G on (D, $\mathcal{D}$), concentrated on one of the functions in the class V^+ (i.e. Q^G is a Dirac measure) and

$$Q^{G^n} \xrightarrow{w} Q^G.$$

2) Let $(\Omega^n, \mathcal{F}^n, P^n) \equiv (\Omega, \mathcal{F}, P)$, and allow $\mathbb{F}^n$ to depend on n. There exists a process $G = (G_t)_{t \geq 0}$ on $(\Omega, \mathcal{F}, P)$ with trajectories in V^+, which is $\mathbb{F}$-predictable ($\mathbb{F} = \bigcap_{n \geq 1} \mathbb{F}^n$) and such that

$$G_t^n \xrightarrow{P} G_t, \quad \sum_{0<s\le t} (\Delta G_s^n)^2 \xrightarrow{P} \sum_{0<\sigma'' \tau} (\Delta G_s)^2$$

for each $t \in \{t > 0: \mathsf{P}(\Delta G_t = 0) = 1\}$.

Theorem 3 [112]. *Let the assumptions* I *and* III' *be fullfilled as well as any of the assumptions* $IV'_{1)}$ *and* $IV'_{2)}$. *Then the family of distributions* (Q^n), $n \ge 1$, *of semimartingales* X^n, $n \ge 1$, *with triplets* T^n, $n \ge 1$, *is relatively compact.*

8. Let us formulate the assumptions analogous to the assumptions of Groups I - IV and respectively the theorem analogous to Theorem 1, in case in which a probability measure Q^n presents the probability distribution of a vector-valued semimartingale $X^n = (X^{n1}, ..., X^{nk})$, $k > 1$, with the triplet $T^n = (B^n, C^n, \nu^n)$ of predictable characteristics (see Ch. 4, Subsection 1.3), defined on a probability measure $(\Omega^n, \mathcal{F}^n, \mathsf{P}^n)$ with a distinguished family $\mathbb{F}^n$ satisfying Conditions (a) and (b) (Ch. 1, Subsection 1).

The norm $|x|$ of a vector $x = (x^1, ..., x^k)$ is considered to be the Euclidean norm

$$|x| = \left(\sum_{j=1}^{k} (x^j)^2\right)^{1/2}.$$

Set $(a \in (0, 1])$

$$B_t^{na} = B_t^n - \int_0^t \int_{a<|x|\le 1} x\, d\nu^n, \quad M_t^{n,a} = X_t^{nc} + \int_0^t \int_{|x|\le a} x\, d(\mu^n - \nu^n),$$

$$\langle M^{n,a}\rangle_t$$

$$= C_t^n + \int_0^t \int_{|x|\le a} xx^*\, d\nu^n - \sum_{0<s\le t} \left(\int_{|x|\le a} x\nu^n(\{s\}, dx)\right)\left(\int_{|x|\le a} x\nu^n(\{s\}, dx)\right)^*,$$

where all vectors are considered as column vectors and * is the transposition sign.

The following conditions are analogous to Conditions I - IV.

Group I^k:

1) $\lim_{l\to\infty} \overline{\lim_n}\, \mathsf{P}^n(|X_0^n| \ge l) = 0$;

2) for each $L > 0$ and $\varepsilon > 0$

$$\lim_{l \to \infty} \overline{\lim_{n}} \, P^n\left(\int_0^L \int_{|x|>l} dv^n \geq \varepsilon \right) == 0.$$

Group II^k:

1) There exists a vector-valued stochastic process

$$\beta^{na} = (\beta_t^{na})_{t \geq 0}, \quad \beta_t^{na} = (\beta_t^{na,\,1}, \ldots, \beta_t^{na,\,k}), \quad a \in (0, 1],$$

each component of which belongs to V and

$$|\Delta\beta^{na,\,j}| \leq 1, \; j = 1, \ldots, k, \; n \geq 1,$$

such that for each $L > 0$ and $\varepsilon > 0$

$$\lim_{n} P^n(\sup_{t \leq L} |B_t^{na} - \beta_t^{na}| \geq \varepsilon) = 0;$$

2) there exists a $\mathbb{F}^n$-predictable vector-valued process

$$\gamma^{na} = (\gamma_t^{na})_{t \geq 0}, \quad \gamma_t^{na} = (\gamma_t^{na,\,1}, \ldots, \gamma_t^{na,\,k}), \quad a \in (0, 1], \; n \geq 1,$$

each component of which belongs to V^+, such that for each $L > 0$ and $\varepsilon > 0$

$$\lim_{a \to 0} \overline{\lim_{n}} \, P^n\left(\sum_{j=1}^{k} \sup_{t \leq L} |\langle M^{n,\,a}\rangle_t^{jj} - \gamma_t^{na,\,j}| \geq \varepsilon \right) = 0,$$

where $\langle M^{n,\,a}\rangle^{jj}$, $j = 1, \ldots, k$, are the diagonal elements of the matrix $\langle M^{n,\,a}\rangle$;

3) for any bounded nonnegative function $g = g(x)$ of a vector-valued argument $x = (x^1, \ldots, x^k)$, jointly continuous in variables and such that

$$g(x) = I(|x| > b)\, g(x)$$

for a certain $b > 0$, there exists a $\mathbb{F}^n$-predictable process

$$\delta^{ng} = (\delta_t^{ng})_{t \geq 0}, \; n \geq 1$$

in V^+ such that for each $L > 0$ and $\varepsilon > 0$

$$\lim_{n} P^n(\sup_{t \leq L} |g * v_t^n - \delta_t^{ng}| \geq \varepsilon) = 0.$$

Group III^k:

There exists a stochastic process $G^n = (G_t^n)_{t \geq 0}$ with trajectories in V^+ and a function

$$\overline{G}^n = (\overline{G}^n)_{t \geq 0} \in V^+$$

such that processes

$$G_t^n + \int_0^t (1 + \sup_{u<s} |X_u^n|)\, d\bar{G}_s^n - \sum_{j=1}^k \mathrm{Var}\,(\beta^{na,j})_t,$$

$$G_t^n + \int_0^t (1 + \sup_{u<s} |X_u^n|)\, d\bar{G}_s^n - \delta_t^{ng},$$

$$G_t^n + \int_0^t (1 + \sup_{u<s} |X_u^n|^2)\, d\bar{G}_s^n - \sum_{j=1}^k \gamma_t^{na,j}$$

are increasing (i.e. belong to V^+).

Group $IV^k = IV$:

The following theorem is analogous to Theorem 1.

Theorem 4. *Let the assumptions of Groups* I^k, II^k *and* III^k *be fullfilled, as well as any of Conditions* $IV^k_{1)}$, $IV^k_{2)}$ *and* $IV^k_{3)}$. *Then the family of distributions* (Q^n), $n \geq 1$, *of vector-valued semimartingales* X^n, $n \geq 1$, *with the triplets* T^n, $n \geq 1$, *is relatively compact.*

Remark. If the assumptions of Groups I^k, II^k and III^k are fullfilled as well as any of Conditions $IV^k_{1)}$ and $IV^k_{3)}$, then the assertion of the theorem is still in force: the $\mathbb{F}^n$-predictability condition imposed on the processes γ^{na} and δ^{ng} in the assumptions $II^k_{2)}$ and $II^k_{3)}$, is superfluous.

Problems

1. Let $X_t^n = \alpha^n A_t^n$, where α^n is a random variable on a probability space $(\Omega^n, \mathcal{F}^n, P^n)$, $n \geq 1$, and let $(A_t^n)_{t \geq 0}$ be a function in V^+. Show that the family of distributions (Q^n), $n \geq 1$, of processes X^n, $n \geq 1$, is relatively compact if $\sup_n |\alpha^n| \leq c$ and if there exists a function $A = (A_t)_{t \geq 0}$ in V^+ such that

$$A_t^n \to A_t$$

and

$$\sum_{0<s\leq t} (\Delta A_s)^2$$

for each $t \in \{t > 0: \Delta A_t = 0\}$.

2. Let α^n and A^n be such as in Problem 1, while a function A belongs to $V^+ \cap C$.

Show that the family of distributions (Q^n), $n \geq 1$, of processes X^n, $n \geq 1$, with $X_t^n = a^n A_t^n$ are relatively compact, if

$$\lim_{c \to \infty} \overline{\lim_n} \, P^n(|\alpha^n| \geq c) = 0$$

and

$$A_t^n \to A_t$$

for each t in a subset dense in R_+.

3. Prove Theorem 2.
4. Prove Theorem 4.
5. Formulate and prove a vector-valued version of Theorem 2.
6. Formulate a vector-valued version of Theorem 3.

§ 5. Conditions necessary for the weak convergence of probability distributions of semimartingales

1. Let $(\Omega^n, \mathcal{F}^n, \mathbb{F}^n, P^n)$ be a sequence of stochastic basises. For every $n \geq 1$ let X^n be a semimartingale defined on a probability space $(\Omega^n, \mathcal{F}^n, P^n)$, adapted to the family $\mathbb{F}^n$, with the triplets $T^n = (B^n, C^n, \nu^n)$ of predictable characteristics and with a probability distribution Q^n.

Suppose there exists a probability measure Q on a measurable space $(D, \mathcal{D})$ such that

$$Q^n \xrightarrow{w} Q. \tag{5.1}$$

We establish certain consequences of the fact of the convergence (5.1).

By Theorem 1.8 we have

$$X_0^n \xrightarrow{d} X_0,$$

i.e. the sequence of random variables X_0^n, $n \geq 1$, converges weakly in distribution (to a random variable X_0).

A more interesting consequence of the convergence (5.1) is formulated in the following theorem.

Theorem 1. *Let the convergence* (5.1) *take place. In order that*

$$Q(C) = 1,$$

it is necessary and sufficient that for each $L > 0$, $l > 0$ *and* $\varepsilon > 0$

$$\lim_n P^n\left(\int_0^L \int_{|x|>l} d\nu^n \geq \varepsilon\right) = 0. \tag{5.2}$$

Proof. Denote

$$\phi_L(X) = \sup_{t \leq L} |\Delta X_t|, \quad X \in D.$$

If $Q(C) = 1$, then $Q(\phi_L(X) = 0) = 1$ for each $L > 0$, and hence

$$\int_D e^{-\phi_L(X)} dQ = 1.$$

The set of discontinuities of the function $\phi_L(X)$ has zero Q-measure (see § 2, Example 3). Therefore, by Theorem 1.7

$$\lim_n \int_D e^{-\phi_L(X)} dQ^n = \int_D e^{-\phi_L(X)} dQ = 1.$$

Due to Chebyshev's inequality, this gives

$$P^n(\phi_L(X^n) \geq \varepsilon) = P^n(1 - e^{-\phi_L(X^n)} \geq 1 - e^{-\varepsilon})$$

$$\leq \frac{1}{1 - e^{-\varepsilon}} \int_{\Omega^n} (1 - e^{-\phi_L(X^n)})\, dP^n$$

$$= \frac{1}{1 - e^{-\varepsilon}} \int_D (1 - e^{-\phi_L(X)})\, dQ^n \to 0, \quad n \to \infty,$$

i.e.

$$\lim_n P^n(\sup_{t \leq L} |\Delta X_t^n| \geq \varepsilon) = 0, \tag{5.3}$$

which is equivalent to (5.2) by Lemma 5.5.1 and the remark to it.

If Condition (5.2) is fullfilled, then by Lemma 5.5.1 and the remark to it the relation (5.3) holds. Let L belong to a set $\Delta_Q = \{t > 0:\ Q(\Delta X_t) = 0) = 1\} \cup \{0\}$, dense in R_+. In accordance with Problem 3 in § 2, for such L the function $\phi_L(X)$ has a set of discontinuity points of zero Q-measure. Therefore, taking into consideration Theorem 1.7, we get

$$\lim_n \int_D e^{-\phi_L(X)}\, dQ^n = \int_D e^{-\phi_L(X)}\, dQ, \quad L \in \Delta_Q. \tag{5.4}$$

On the other hand we have the estimate

$$\int_D e^{-\phi_L(X)}\, dQ^n \geq e^{-\varepsilon} + \int_{\{\phi_L(X) \geq \varepsilon\}} e^{-\phi_L(X)}\, dQ^n,$$

where by (5.3)

$$\int_{\{\phi_L(X) \geq \varepsilon\}} e^{-\phi_L(X)}\, dQ^n \leq Q^n(\phi_L(X) \geq \varepsilon)$$

$$= P^n(\sup_{t \leq L} |\Delta X_t^n| \geq \varepsilon) \to 0, \quad n \to \infty.$$

As $L \in \Delta_Q$, this and (5.4) give

$$\int_D e^{-\phi_L(X)}\, dQ \geq e^{-\varepsilon}$$

for each $\varepsilon > 0$. This means that

$$\int_D e^{-\phi_L(X)}\, dQ \geq 1,$$

and since the converse inequality

$$\int_D e^{-\phi_L(X)}\, dQ \le 1$$

is obvious, then

$$\int_D e^{-\phi_L(X)}\, dQ = 1, \quad L \in \Delta_Q. \tag{5.5}$$

By taking the limit in (5.5) as $L_k \uparrow \infty$, $L_k \in \Delta_Q$, $k \ge 1$, we arrive at the inequality

$$\int_D \exp\left(-\sup_{t>0} |\Delta X_t|\right) dQ = 1.$$

Consequently, Q (C) = 1.

2. Let the convergence (5.1) take place with Q (C) = 1. On a measurable space (D, $\mathcal{D}^Q$) (where a σ-algebra $\mathcal{D}^Q$ is a completion of $\mathcal{D}$ relative to the measure Q) define a family of σ-algebras $\mathbb{D}^Q = (\mathcal{D}^Q_t)_{t \ge 0}$ with

$$\mathcal{D}^Q_t = \bigcap_{u>t} \sigma\{X_s,\ s \le u\} \vee \mathfrak{N},$$

where $\mathfrak{N}$ is a system of sets in $\mathcal{D}^Q$ of zero Q-measure.

Let there exist a $\mathbb{D}^Q$-predictable process $B(X) = (B_t(X))$ with trajectories in $V \cap C$ such that a process $M(X) = (M_t(X))_{t \ge 0}$ with

$$M_t(X) = X_t - X_0 - B_t(X) \tag{5.6}$$

presents a local martingale relative to ($\mathbb{D}^Q$, Q), the quadratic characteristic of which we denote by $\langle M(X)\rangle = (\langle M(X)\rangle_t)_{t \ge 0}$. In this case it is said that a measure Q with Q (C) = 1 solves *the martingale problem of a diffusion type* with the parameters (B (X), ⟨M (X)⟩).

Let $M^{n,a}$ be a local square integrable martingale defined by the formula (4.3), i.e.

$$M^{n,a}_t = X^{nc}_t + \int_0^t \int_{|x| \le a} x\, d(\mu^n - \nu^n), \quad a \in (0, 1].$$

As a consequence of the convergence (5.1) ($Q^n \xrightarrow{w} Q$), we establish relations between ⟨M (X)⟩, $[M^{n,a}, M^{n,a}]$ and $\langle M^{n,a}\rangle$.

Theorem 2. *Let the convergence* (5.1) *take place with* Q (C) = 1 *and let the measure* Q *solve the martingale problem of a diffusion type with the parameters* (B (X) *and* ⟨M (X)⟩).

Let the following conditions be fullfilled:

(α) *for every* t *in a set dense in* R_+ *functions* $B_t(X)$, $\langle M(X)\rangle_t$ *are continuous in the metric* ρ (§ 1) *at each point* $X \in \mathbb{C}$;

(sup B)
$$\lim_n \mathsf{P}^n(\sup_{t \le L} |B_t^n - B_t(X^n)| \ge \varepsilon) = 0$$

for each $L > 0$ *and* $\varepsilon > 0$;

(K)
$$\lim_{l \to \infty} \overline{\lim_n} \mathsf{P}^n(\mathrm{Var}(B(X^n))_L \ge l) = 0,$$

$$\lim_{\sigma \to 0} \overline{\lim_n} \mathsf{P}^n(\sup_{t \le L} [\mathrm{Var}(B(X^n))_{t+\sigma} - \mathrm{Var}(B(X^n))_t] \ge \varepsilon) = 0,$$

$$\lim_{\sigma \to 0} \overline{\lim_n} \mathsf{P}^n(\sup_{t \le L} [\langle M(X^n)\rangle_{t+\sigma} - \langle M(X^n)\rangle_t] \ge \varepsilon) = 0$$

for each $L > 0$ *and* $\varepsilon > 0$.

Then for each $a \in (0, 1]$, $L > 0$ *and* $\varepsilon > 0$

$$\lim_n \mathsf{P}^n(\sup_{t \le L} |[M^{n,a}, M^{n,a}]_t - \langle M(X^n)\rangle_t| \ge \varepsilon) = 0,$$

$$\lim_n \mathsf{P}^n(\sup_{t \le L} |\langle M^{n,a}\rangle_t - \langle M(X^n)\rangle_t| \ge \varepsilon) = 0.$$

The proof of Theorem 2 is based on a number of auxiliary results formulated below as lemmas.

For the sake of brevity the relation

$$\lim_n \mathsf{P}^n(|\alpha^n| \ge \varepsilon) = 0, \quad \varepsilon > 0,$$

is denoted by

$$\alpha^n \xrightarrow{\mathsf{P}^n} 0.$$

3. Lemma 1. *Let Condition* (sup B) *of Theorem* 2 *be fullfilled and let the convergence* (5.1) *take place with* $\mathsf{Q}(\mathsf{C}) = 1$.

Then for each $a \in (0, 1]$ *and* $L > 0$

$$\sup_{t \le L} |M_t^{n,a} - M_t(X^n)| \xrightarrow{\mathsf{P}^n} 0.$$

Proof. Use the fact that

$$M_t^{n,a} = X_t^n - X_0^n - B_t^n + \int_0^t \int_{a < |x| \le 1} x\, d\nu^n - \int_0^t \int_{|x| > a} d\, x\mu^n$$

(see (4.1) - (4.3)). This and (5.6) give

$$\sup_{t \le L} | M_t^{n,a} - M_t(X^n) |$$

$$\le \sup_{t \le L} | B_t^n - B_t(X^n) | + \int_0^L \int_{|x|>a} d\nu^n + \int_0^L \int_{|x|>a} |x| \, d\mu^n. \qquad (5.7)$$

By Condition (sup B) we have

$$\sup_{t \le L} | B_t^n - B_t(X^n) | \overset{P^n}{\to} 0,$$

and by Theorem 1 we have

$$\int_0^L \int_{|x|>a} d\nu^n \overset{P^n}{\to} 0.$$

Therefore by the inequality (5.7) the desired assertion holds, provided

$$\int_0^L \int_{|x|>a} |x| \, d\mu^n \overset{P^n}{\to} 0. \qquad (5.8)$$

Since

$$\left\{ \int_0^L \int_{|x|>a} |x| \, d\mu^n \ge \varepsilon \right\} \subseteq \left(\int_0^L \int_{|x|>a} d\mu^n \ge 1 \right),$$

then (5.8) is satisfied, provided

$$\lim_n P^n \left(\int_0^L \int_{|x|>a} d\mu^n \ge 1 \right) = 0. \qquad (5.9)$$

To establish (5.9), we use the Lenglart-Rebolledo inequality (Theorem 1.9.3) in view of which

$$P^n \left(\int_0^L \int_{|x|>a} d\mu^n \ge 1 \right) \ge b + P^n \left(\int_0^L \int_{|x|>a} d\nu^n \ge b \right). \qquad (5.10)$$

By Theorem 1 we have

$$\int_0^L \int_{|x|>a} d\nu^n \overset{P^n}{\to} 0.$$

Consequently, we arrive at (5.9) as in (5.10) the limit $\lim_{b \to 0} \overline{\lim_n}$ is taken.

4. Denote by $Q^{M(X^n)}$ and $Q^{M^{n,a}}$ the probability distributions of processes $M(X^n)$ and $M^{n,a}$, with (cf. (5.6))

$$M_t(X^n) = X_t^n - X_0^n - B_t(X^n). \tag{5.11}$$

Lemma 2. *Let the conditions of Theorem 2 be fullfilled. Then*

$$Q^{M(X^n)} \xrightarrow{w} Q', \quad Q^{M^{n,a}} \xrightarrow{w} Q' \quad (a \in (0, 1],$$

where Q' *is a probability measure on* $(D, \mathcal{D})$ *with* Q' (C) = 1, *presenting the distribution of* M (X) *(see (5.6)), defined on* $(D, \mathcal{D}, Q)$.

Proof. By Condition (a) for each t in a set, dense in R_+, the measure Q of a set of discontinuity points of the function $M_t(X)$ is equal to zero. Therefore, in view of Theorem 1.7 for each $\lambda_j \in R$ and t_j in a set, dense in R_+, $j = 1, \ldots, k$, $k \geq 1$ ($i = \sqrt{-1}$) we get

$$\lim_n \int_D \exp\left(i \sum_{j=1}^{k} \lambda_j M_{t_j}(X) \right) dQ^{M(X^n)}$$

$$= \lim_n \int_D \exp\left(i \sum_{j=1}^{k} \lambda_j M_{t_j}(X) \right) dQ^n = \int_C \exp\left(i \sum_{j=1}^{k} \lambda_j M_{t_j}(X) \right) dQ,$$

i.e. finite dimensional distributions of processes $M(X^n)$, $n \geq 1$, weakly converge on a dense set to finite dimensional distributions of a process M (X), defined on a probability space $(D, \mathcal{D}, Q)$.

Since $\Delta_{Q'} = R_+$, by Theorem 1.8 it suffices for the weak convergence

$$Q^{M(X^n)} \xrightarrow{w} Q'$$

to show that the family of measures $(Q^{M(X^n)})$, $n \geq 1$, is relatively compact. To this end it suffices by Theorem 1.6 to verify the conditions

$$\lim_{b \to \infty} \overline{\lim_n}\, Q^{M(X^n)}\left(\sup_{t \leq L} |X_t| \geq b \right) = 0, \tag{5.12}$$

$$\lim_{\sigma \to 0} \overline{\lim_n}\, Q^{M(X^n)}\left(W_L'(X, \sigma) \geq \eta \right) = 0 \tag{5.13}$$

for each $L > 0$ and $\eta > 0$.

By (5.11)

$$\sup_{t \leq L} |M(X^n)| \leq 2 \sup_{t \leq L} |X_t^n| + \mathrm{Var}\,(B(X^n))_L.$$

Therefore

$$Q^{M(X^n)}(\sup_{t \le L} |X_t| \ge b) = P^n(\sup_{t \le L} |M_t(X^n)| \ge b)$$

$$\le P^n\left(2 \sup_{t \le L} |X^n_t| \ge \frac{b}{2}\right) + P^n\left(\text{Var}(B(X^n))_L \ge \frac{b}{2}\right), \tag{5.14}$$

and hence (5.12) is a consequence of Condition (K), (5.1) and Theorem 1.6 in view of which

$$\lim_{b \to \infty} \overline{\lim_n} P^n(\sup_{t \le L} |X^n_t| \ge b) = 0.$$

To establish (5.13), observe that by (5.11) the estimate

$$W'_L(M(X^n), \sigma)$$
$$\le W'_L(X^n, \sigma) + \sup_{t \le L}[\text{Var}(B(X^n))_{t+\sigma} - \text{Var}(B(X^n))_t] \tag{5.15}$$

takes place which gives

$$Q^{M(X^n)}(W'_L(X, \sigma) > \eta) = P^n(W'_L(M(X^n), \sigma) \ge \eta)$$
$$\le P^n(W'_L(X^n, \sigma) \ge \eta/2) + P^n(\sup_{t \le L}[\text{Var}(B(X^n))_{t+\sigma} - \text{Var}(B(X^n))_t] \ge \eta/2), \tag{5.16}$$

i.e. (5.13) is a consequence of Condition (K), (5.1) and Theorem 1.6 in view of which

$$\lim_{\sigma \to 0} \overline{\lim_n} P^n(W'_L(X^n, \sigma) \ge \eta) = 0, \quad \eta > 0.$$

Thus

$$Q^{M(X^n)} \xrightarrow{w} Q'.$$

The convergence

$$Q^{M^{n,a}} \xrightarrow{w} Q'$$

follows from this by Lemma 1 and Problem 1.2. Next, Q' (C) = 1, since Q' is the distribution of a continuous process M (X).

5. The processes $M^{n,a}$, $n \ge 1$, are locally square integrable martingales, relative to $(\mathbb{F}^n, P^n)$, $n \ge 1$, and as $a \in (0, 1]$ they possess the property $|\Delta M^{n,a}| \le 2$.

We will need below the following properties (for each L > 0) of the processes $M^{n,a}$, $n \ge 1$:

$$\lim_{l \to \infty} \overline{\lim_n} P^n([M^{n,a}, M^{n,a}]_L \ge l) = 0, \tag{5.17}$$

$$\lim_{l \to \infty} \overline{\lim_n} P^n(\langle M^{n,a} \rangle_L \ge l) = 0. \tag{5.18}$$

These properties follow from the general fact of independent interest, formulated here as the following lemma.

Lemma 3. Let $(\Omega^n, \mathcal{F}^n, \mathbb{F}^n, P^n)$, $n \geq 1$, *be a sequence of stochastic basises and for every* $n \geq 1$ *let* $M^n \in \mathcal{M}^2_{loc}(\mathbb{F}^n, P^n)$ with

$$\sup_n \mathbf{E}^n \sup_{t \leq L} (\Delta M^n_t)^2 < \infty, \quad L > 0$$

($\mathbf{E}^n$ *is the mathematical expectation with respect to a measure* P^n).

The following conditions are equivalent:

1) $\lim_{l \to \infty} \overline{\lim_n} P^n(\sup_{t \leq L} |M^n_t| \geq l) = 0,$

2) $\lim_{l \to \infty} \overline{\lim_n} P^n([M^n, M^n]_L \geq l) = 0,$

3) $\lim_{l \to \infty} \overline{\lim_n} P^n(\langle M^n \rangle_L \geq l) = 0.$

Proof. Recall the following relations (τ is a $\mathbb{F}^n$-stopping time)

$$\mathbf{E}^n (M^n_\tau)^2 \leq \mathbf{E}^n [M^n, M^n]_\tau = \mathbf{E}^n \langle M^n \rangle_\tau \leq \mathbf{E}^n \sup_{t \leq \tau} |M^n_t|^2$$

(see Problems 1.8.6 and 1.8.7). Therefore by the relations

$$\Delta [M^n, M^n] = (\Delta M^n)^2, \quad \Delta \sup_{s \leq t} (M^n_s)^2 \leq \sup_{s \leq t} (\Delta M^n_s)^2, \quad t \in R_+$$

and the Lenglart-Rebolledo inequality (Theorem1.9.3) we get

$$P^n(\sup_{t \leq L} |M^n_t| \geq l)$$

$$\leq \frac{b + \sup_n \mathbf{E}^n \sup_{t \leq L} (\Delta M^n_t)^2}{l^2} + P^n([M^n, M^n]_L \geq b), \tag{5.19}$$

$$P^n([M^n, M^n]_L \geq l) \leq \frac{b}{l} + P^n(\langle M^n \rangle_L \geq b), \tag{5.20}$$

$$P^n(\langle M^n \rangle_L \geq l)$$

$$\leq \frac{b + \sup_n \mathbf{E}^n \sup_{t \leq L} (\Delta M^n_t)^2}{l} + P^n([M^n, M^n]_L \geq b), \tag{5.21}$$

$$P^n([M^n, M^n]_L \geq l)$$

$$\leq \frac{b + \sup_n E^n \sup_{t \leq L} (\Delta M_t^n)^2}{l} + P^n(\sup_{t \leq L} (M_t^n)^2 \geq b). \tag{5.22}$$

By taking the limit $\lim_{b \to \infty} \overline{\lim}_{l \to \infty} \overline{\lim}_n$, we arrive at (5.19) ⇒ «2) ⇒ 1)», (5.20) ⇒ «3) ⇒ 2)», (5.21) ⇒ «2) ⇒ 3)», (5.22) ⇒ «1) ⇒ 2)».

The lemma is proved.

Thus the properties (5.17) and (5.18) of the processes $M^{n, a}$, $n \geq 1$, follow from Lemma 3, since by Lemma 2 we have

$$Q^{M^{n, a}} \overset{w}{\to} Q',$$

while due to Theorem 1.6 in this case

$$\lim_{l \to \infty} \overline{\lim}_n P^n(\sup_{t \leq L} |M_t^{n, a}| \geq l) = 0, \quad L > 0.$$

6. Let $L > 0$ and (t_j^n), $j = 0, 1, \ldots, k$, $k \geq 1$, be a sequence of condensing partitions of an interval $[0, L]$ with

$$0 \equiv t_0^k < t_1^k < \ldots < t_k^k \equiv L, \quad \max_{1 \leq j \leq k} (t_j^k - t_{j-1}^k) \to 0, \ k \to \infty.$$

We establish yet another property of the processes $M^{n, a}$, $n \geq 1$, namely:

$$\lim_{k \to \infty} \overline{\lim}_n P^n(\sup_{t \leq L} |[M^{n, a}, M^{n, a}]_t - \sum_{j=0}^{k-1} (M^{n, a}_{t^k_{j+1} \wedge t} - M^{n, a}_{t^k_j \wedge t})^2| \geq \varepsilon) = 0$$

for each $\varepsilon > 0$ and $L > 0$. This property follows from the following assertion of an independent interest.

Lemma 4. *Let* $(\Omega^n, \mathcal{F}^n, \mathbb{F}^n, P^n)$, $n \geq 1$, *be a sequence of stochastic basises and for every* $n \geq 1$ *let* $M^n \in \mathcal{M}^2_{loc}(\mathbb{F}^n, P^n)$ *with*

$$\sup_n E^n \sup_{t \leq L} (\Delta M_t^n)^2 < \infty, \quad L > 0.$$

If a sequence of distributions Q^{M^n}, $n \geq 1$, *of processes* M^n, $n \geq 1$, *converges weakly to a certain probability measure* Q' *on* $(D, \mathcal{D})$ *with* $Q'(C) = 1$, *then*

$$\lim_{k \to \infty} \overline{\lim}_n P^n(\sup_{t \leq L} |[M^n, M^n]_t - \sum_{j=0}^{k-1} (M^n_{t^k_{j+1} \wedge t} - M^n_{t^k_j \wedge t})^2| \geq \varepsilon) = 0, \ \varepsilon > 0.$$

Corollary. *Let* $(\Omega^n, \mathcal{F}^n, \mathbb{F}^n, P^n) \equiv (\Omega, \mathcal{F}, \mathbb{F}, P)$ *and* $M \in \mathcal{M}^c_{loc}(\mathbb{F}, P)$. *Then for each* $L > 0$

$$\sup_{t \le L} \left| \langle M \rangle_t - \sum_{j=0}^{k-1} (M_{t^k_{j+1} \wedge t} - M_{t^k_j \wedge t})^2 \right| \xrightarrow{P} 0, \quad k \to \infty.$$

Proof of Lemma 4. By Ito's formula (Ch. 2, § 3) we see as $t > s$ that

$$(M^n_t - M^n_s)^2 = 2 \int_s^t (M^n_{u-} - M^n_s)\, dM^n_u + [M^n, M^n]_t - [M^n, M^n]_s. \tag{5.24}$$

Denote

$$f^k_u(M^n) = 2 \sum_{j=0}^{k-1} (M^n_{u-} - M^n_{t^k_j})\, I_{]\!]t^k_j, t^k_{j+1}]\!]}(u). \tag{5.25}$$

By using the representation (5.24) and the notation (5.25), as $t \in [0, L]$ we get

$$\sum_{j=0}^{k-1} (M^n_{t^k_{j+1} \wedge t} - M^n_{t^k_j \wedge t})^2 = f^k(M^n) \cdot M^n_t + [M^n, M^n]_t.$$

Thus the assertion of the lemma takes place, provided

$$\lim_{k \to \infty} \overline{\lim_n}\, P^n\left(\sup_{t \le L} | f^k(M^n) \cdot M_t | \ge \varepsilon \right) = 0, \quad \varepsilon > 0. \tag{5.26}$$

Observe that trajectories of the process $(f^k_u(M^n))_{0 \le u \le t}$ are left-continuous having right-hand limits; consequently they are locally bounded (for instance for a localizing sequence $\tau_m = \inf(u \le L: |f^k_u(M^n)| \ge m)$, $m \ge 1$, the inequalities

$$|f^k_{u \wedge \tau_m}(M^n)| \le m, \quad m \ge 1,$$

hold). Therefore

$$(f^k(M^n))^2 \circ \langle M^n \rangle \in \mathcal{A}^+_{loc}(P^n),$$

and hence

$$f^k(M^n) \cdot M^n \in \mathcal{M}^2_{loc}(\mathbb{F}^n, P^n)$$

(see Theorem 2.2.2) with the quadratic characteristics

$$\langle f^k(M^n) \cdot M^n \rangle = (f^k(M^n))^2 \circ \langle M^n \rangle.$$

By taking into consideration Problem 1.8.6 and by the Lenglart-Rebolledo inequality (Theorem 1.9.3) this gives

$$P^n\left(\sup_{t \le L} | f^k(M^n) \cdot M^n_t | \ge \varepsilon \right) \le \frac{b}{\varepsilon^2} + P^n\left((f^k(M^n))^2 \circ \langle M^n \rangle_L \ge b \right).$$

By taking the limit $\lim_{b\to 0}\ \overline{\lim_{k\to\infty}}\ \overline{\lim_{n}}$ it is not hard to deduce from this that (5.26) takes place, provided

$$\lim_{k\to\infty}\ \overline{\lim_{n}}\ P^n\left((f^k(M^n))^2 \circ \langle M^n\rangle_L \ge b\right) = 0 \tag{5.27}$$

for each $b > 0$.

For $X \in D$ denote

$$H^k(X) = \max_{0\le j\le k-1}\ \sup_{t_j^k < s \le t_{j+1}^k} (X_{s-} - X_{t_j^k})^2$$

and observe that $H^k(X)$ is a continuous function in the metric ρ at the point $X \in C$ and that $(f^k(M^n))^2 \circ \langle M^n\rangle_L \le \langle M^n\rangle_L H^k(M^n)$. Consequently, (5.27) takes place, provided

$$\lim_{k\to\infty}\ \overline{\lim_{n}}\ P^n\left(\langle M^n\rangle_L H^k(M^n) \ge b\right) = 0 \tag{5.28}$$

for each $b > 0$.

To prove (5.28), we use the estimate

$$P^n\left(\langle M^n\rangle_L H^k(M^n) \ge b\right) \le P^n\left(H^k(M^n) \ge b/l\right) + P^n\left(\langle M^n\rangle_L \ge l\right). \tag{5.29}$$

Since by assumption $Q^{M^n} \xrightarrow{w} Q'$, where Q' is a probability measure on $(D, \mathcal{D})$, by Theorem 1.6 we have

$$\lim_{l\to\infty}\ \overline{\lim_{n}}\ P^n\left(\sup_{t\le L} |M_t^n| \ge l\right) = 0.$$

By Lemma 3 this gives

$$\lim_{l\to\infty}\ \overline{\lim_{n}}\ P^n\left(\langle M^n\rangle_L \ge l\right) = 0.$$

Hence, we arrive at the desired relation (5.28) as in (5.29) the limit $\lim_{l\to\infty}\ \overline{\lim_{k\to\infty}}\ \overline{\lim_{n}}$ is taken, provided

$$\lim_{k\to\infty}\ \overline{\lim_{n}}\ P^n\left(H^k(M^n) \ge \varepsilon\right) = 0 \tag{5.30}$$

for each $\varepsilon > 0$.

Let $g = g(x)$ be a nonnegative continuous bounded function with $g(0) = 0$ and $g(x) \ge I(|x| \ge \varepsilon)$. Then

$$P^n\left(H^k(M^n) \ge \varepsilon\right) \le \mathbf{E}^n g\left(H^k(M^n)\right) = \int_D g\left(H^k(X)\right) dQ^{M^n}. \tag{5.31}$$

Since

$$Q^{M^n} \xrightarrow{w} Q'$$

the set of discontinuity points of function $g(H^k(X))$ has zero Q'-measure. Therefore by Theorem 1.7

$$\lim_n \int_D g(H^k(X))\, dQ^{M^n} = \int_C g(H^k(X))\, dQ' \to 0, \quad k \to \infty.$$

The desired relation follows from this and (5.31) in an obvious manner.

The lemma is proved.

The property (5.23) of the processes $M^{n,a}$, $n \geq 1$, follows from Lemmas 2 and 4.

7. The following result, generalizing Problem 5.3.2, is also used in the course of proving Theorem 2.

Lemma 5. *Let* $(\Omega^n, \mathcal{F}^n, \mathbb{F}^n, P^n)$, $n \geq 1$, *be a sequence of probability spaces. For every* $n \geq 1$ *let* A^n *and* C^n *be stochastic processes, defined on* $(\Omega^n, \mathcal{F}^n, P^n)$ *with trajectories in* V^+.

If for every t in a set Δ, *dense in* R_+,

$$A_t^n - C_t^n \xrightarrow{P^n} 0 \tag{5.32}$$

and for each $L > 0$ *and* $\varepsilon > 0$

$$\lim_{\sigma \to 0} \overline{\lim_n}\, P^n(\sup_{t \leq L} [C_{t+\sigma}^n - C_t^n] \geq \varepsilon) = 0, \tag{5.33}$$

then for each $L > 0$ *and* $\varepsilon > 0$

$$\lim_n P^n(\sup_{t \leq L} | A_t^n - C_t^n | \geq \varepsilon) = 0.$$

Proof. One may assume $L \in \Delta$ without loosing generality. Let (t_j^k), $j = 0, 1, \ldots, k$, $k \geq 1$, be a sequence of partitions of the interval $[0, L]$, $0 \equiv t_0^k < t_1^k < \ldots < t_k^k \equiv L$, such that

$$t_j^k \in \Delta, \ j = 1, 2, \ldots, k, \quad \max_{1 \leq j \leq k} (t_j^k - t_{j-1}^k) \to 0, \quad k \to \infty.$$

Then

$$\sup_{t \leq L} | A_t^n - C_t^n | = \max_{0 \leq j \leq k-1} \sup_{t_j^k \leq t < t_{j+1}^k} | A_t^n - C_t^n |.$$

Since A^n and C^n are increasing processes

$$\sup_{t_j^k \leq t < t_{j+1}^k} | A_t^n - C_t^n | \leq | A_{t_{j+1}^k}^n - C_{t_j^k}^n | + | A_{t_j^k}^n - C_{t_{j+1}^k}^n |$$

$$\leq 2 | C_{t_{j+1}^k}^n - C_{t_j^k}^n | + | A_{t_{j+1}^k}^n - C_{t_{j+1}^k}^n | + | A_{t_j^k}^n - C_{t_j^k}^n |.$$

Consequently,

$$\sup_{t \le L} | A^n_t - C^n_t | \le 2 \sum_{j=1}^{k} | A^n_{t^k_j} - C^n_{t^k_j} | + 2 \max_{0 \le j \le k-1} [C^n_{t^k_{j+1}} - C^n_{t^k_j}].$$

By Condition (5.32) we have

$$\sum_{j=1}^{k} | A^n_{t^k_j} - C^n_{t^k_j} | \xrightarrow{P^n} 0.$$

Therefore it suffices to show for each $\varepsilon > 0$ that

$$\lim_{k \to \infty} \overline{\lim_{n}} \, P^n (\max_{0 \le j \le k-1} [C^n_{t^k_j} - C^n_{t^k_j}] \ge \varepsilon) = 0.$$

For each $k \ge 1$ denote

$$\delta_k = \max_{0 \le j \le k-1} (t^k_{j+1} - t^k_j).$$

Then

$$\{ \max_{0 \le j \le k-1} [C^n_{t^k_{j+1}} - C^n_{t^k_j}] \ge \varepsilon\} \subseteq \{ \sup_{t \le L} | C^n_{t+\delta_k} - C^n_t] \ge \varepsilon\},$$

and hence by Condition (5.33)

$$P^n (\max_{0 \le j \le k-1} [C^n_{t^k_{j+1}} - C^n_{t^k_j}] \ge \varepsilon) \le P^n (\sup_{t \le L} [C^n_{t+\delta_k} - C^n_t] \ge \varepsilon) \to 0$$

as the limit $\lim_{k \to \infty} \overline{\lim_{n}}$ is taken, since $\delta_k \to 0$ as $k \to \infty$.

8. Proof of Theorem 2. Fix $L > 0$ and introduce a partition sequence

$$\{t^k_j\}, \quad j = 0, 1, \ldots, k, \quad k \ge 1,$$

$$t^k_j \in S, \; j = 0, 1, \ldots, k, \quad \max_{1 \le j \le k} (t^k_j - t^k_{j-1}) \to 0, \; k \to \infty,$$

of the interval $[0, L]$ such as in Lemma 4. Then by Corollary to Lemma 4

$$\sup_{t \le L} | \langle M(X) \rangle_t - \sum_{j=0}^{k-1} (M_{t^k_{j+1} \wedge t}(X) - M_{t^k_j \wedge t}(X))^2 \xrightarrow{Q} 0. \tag{5.34}$$

Since $Q(C) = 1$, then for every t in a subset, dense in R_+, a function

$$h^k_t(X) = | \langle M(X) \rangle_t - \sum_{j=0}^{k-1} (M_{t^k_{j+1} \wedge t}(X) - M_{t^k_j \wedge t}(X))^2 |$$

has the set of discontinuity points of zero Q-measure by assumption (α) of Theorem 5.2 and by (5.6). Therefore the convergence $Q^n \xrightarrow{w} Q$ and Theorem 1.7 entail

$$\lim_n \int_D e^{-h_t^k(X)} dQ^n = \int_C e^{-h_t^k(X)} dQ.$$

Consequently in view of (5.34)

$$\lim_{k \to \infty} \lim_n E^n e^{-h_t^k(X^n)} = 1, \quad t \le L.$$

It is not hard to deduce from this that for each $\varepsilon > 0$ and $t \le L$

$$\lim_{k \to \infty} \overline{\lim_n} P^n\left(|\langle M(X^n)\rangle_t - \sum_{j=0}^{k-1} (M_{t_{j+1}^k \wedge t}(X^n) - M_{t_j^k \wedge t}(X^n))^2 \ge \varepsilon\right) = 0,$$

and hence in accordance with Lemma 1 as $t \le L$ and $\varepsilon > 0$

$$\lim_{k \to \infty} \overline{\lim_n} P^n\left(|\langle M(X^n)\rangle_t - \sum_{j=0}^{k-1} (M^{n,a}_{t_{j+1}^k \wedge t} - M^{n,a}_{t_j^k \wedge t})^2| \ge \varepsilon\right) = 0. \tag{5.35}$$

From (5.35) and (5.23) it follows that as $t \le L$ and $\varepsilon > 0$

$$\lim_n P^n(|\langle M(X^n)\rangle_t - [M^{n,a}, M^{n,a}]_t| \ge \varepsilon) = 0. \tag{5.36}$$

Therefore the first assertion of the theorem is a consequence of Lemma 5 with $A^n = [M^{n,a}, M^{n,a}]$ and $C^n = \langle M(X^n)\rangle$, the conditions of which are satisfied by (5.36) and the third relation in Condition (K) of Theorem 2.

To establish the second assertion of Theorem 2, we use the inequality (as $\varepsilon \in (0, a)$)

$$|\langle M^{n,a}\rangle_t - \langle M(X^n)\rangle_t| \le |\langle M^{n,a}\rangle_t - \langle M^{n,\varepsilon}\rangle_t$$

$$+ |\langle M^{n,\varepsilon}\rangle_t - [M^{n,\varepsilon}, M^{n,\varepsilon}]_t| + |[M^{n,\varepsilon}, M^{n,\varepsilon}]_t - \langle M(X^n)\rangle_t|. \tag{5.37}$$

Since $Q^n \xrightarrow{w} Q$ with $Q(C) = 1$, by Theorem 1 the relation (5.2) holds. Hence in view of Remark to Lemma 5.5.1

$$\sup_{t \le L} |\Delta X_t^n| \xrightarrow{P^n} 0. \tag{5.38}$$

Then by an obvious generalization of Lemma 5.5.2 for fixed a and ε we get

$$\sup_{t \le L} |\langle M^{n,a}\rangle_t - \langle M^{n,\varepsilon}\rangle_t| \xrightarrow{P^n} 0. \tag{5.39}$$

Besides, under the assumption (5.38) the obvious generalization of (5.23) (Ch. 5, § 5) holds by Problem 6:

$$\lim_{\varepsilon \to 0} \overline{\lim_n} P^n(\sup_{t \le L} |\langle M^{n,\varepsilon}\rangle_t - [M^{n,\varepsilon}, M^{n,\varepsilon}]_t| \ge b) = 0, \quad b > 0. \tag{5.40}$$

Consequently, the second assertion of the theorem is deduced from the first one by the inequality (5.37) and the relations (5.39) and (5.40).

Problems

1. Let $(\Omega^n, \mathcal{F}^n, \mathbb{F}^n, P^n)$, $n \geq 1$, be a sequence of stochastic basises and for every $n \geq 1$ let

$$M^n \in \mathcal{M}_{loc}(\mathbb{F}^n, P^n)$$

and

$$\sup_n E^n \sup_{t \leq L} |\Delta M_t^n| < \infty, \quad L > 0,$$

(E^n is the mathematical expectation with respect to the measure P^n).

Show that the following conditions are equivalent:

1) $\lim\limits_{l \to \infty} \overline{\lim\limits_n} P^n (\sup\limits_{t \leq L} |M_t^n| \geq l) = 0,$

2) $\lim\limits_{l \to \infty} \overline{\lim\limits_n} P^n ([M^n, M^n]_L \geq l) = 0.$

2. Let processes M^n, $n \geq 1$, be such as in Problem 1, and

$$\lim_n E^n \sup_{t \leq L} |\Delta M_t^n| = 0.$$

Show that the following conditions are equivalent:

1) $\sup\limits_{t \leq L} |M_t^n| \xrightarrow{P^n} 0,$

2) $[M^n, M^n]_L \xrightarrow{P^n} 0.$

3. Let $(\Omega^n, \mathcal{F}^n, \mathbb{F}^n, P^n)$, $n \geq 1$, be a sequence of stochastic basises, and for every $n > 1$ let $M^n \in \mathcal{M}^2_{loc}(\mathbb{F}^n, P^n)$ and $\langle M \rangle_L \xrightarrow{P^n} 0$. Show that

$$\sup_{t \leq L} |M_t^n| \xrightarrow{P^n} 0.$$

4. Let the conditions of Theorem 2 be fullfilled except the third relation in Condition (K). Show that in this case for each $t > 0$ in a set, dense in R_+, and $\varepsilon > 0$

$$\lim_n P^n (|[M^{n,a}, M^{n,a}]_t - \langle M(X^n) \rangle_t| \geq \varepsilon) = 0,$$

$$\lim_n P^n (|\langle M^{n,a} \rangle_t - \langle M(X^n) \rangle_t| \geq \varepsilon) = 0.$$

5. Let the stochastic processes A^{nk} and $C^n \in V^+$, $n \geq 1$, and for each $\varepsilon > 0$ and t in a set Δ, dense in R_+,

$$\lim_{k \to \infty} \overline{\lim_{n}} \; P^n (|A_t^{nk} - C_t^n| \geq \varepsilon) = 0.$$

Show that if

$$\lim_{\sigma \to 0} \overline{\lim_{n}} \; P^n (\sup_{t \leq L} [C_{t+\sigma}^n - C_t^n] \geq \varepsilon) = 0, \quad L > 0, \quad \varepsilon > 0,$$

then for each $L > 0$ and $\varepsilon > 0$ we have

$$\lim_{k \to \infty} \overline{\lim_{n}} \; P^n (\sup_{t \leq L} |A_t^{nk} - C_t^n| \geq \varepsilon) = 0.$$

6. Let the conditions of Theorem 2 be fullfilled. Show that for each $L > 0$ and $0 < \varepsilon \leq 1$

$$\lim_{l \to \infty} \overline{\lim_{n}} \; P^n (\sup_{t \leq L} |M_t^{n, \varepsilon}| \geq l) = 0,$$

$$\lim_{l \to \infty} \overline{\lim_{n}} \; P^n (\langle M^{n, \varepsilon} \rangle_L \geq l) = 0,$$

$$\lim_{l \to \infty} \overline{\lim_{n}} \; P^n ([M^{n, \varepsilon}, M^{n, \varepsilon}]_L \geq l) = 0.$$

CHAPTER 7

WEAK CONVERGENCE OF DISTRIBUTIONS OF SEMIMARTINGALES TO DISTRIBUTIONS OF PROCESSES WITH CONDITIONALLY INDEPENDENT INCREMENTS

§ 1. The functional central limit theorem (invariance principle)

1. Using the method of stochastic exponentials in Chapter 5 we have studied conditions for the weak convergence of finite dimensional distributions of semimartingales to distributions of processes with conditionally independent increments. Sufficient conditions of such convergence were established for limiting processes presenting left quasi-continuous semimartingales, point processes and Gaussian semimartingales. Utilizing these results and results on relative compactness of probability distribution families of semimartingales (Ch. 6), we study in the present chapter conditions for weak convergence of distributions of semimartingales to distributions of processes (of indicated type) with conditionally independent increments.

Throughout this chapter we assume to be given a stochastic basis $(\Omega, \mathcal{F}, \mathbb{F}, P)$, and a sequence $\mathbb{F}^n$, $n \geq 1$, which satisfy Conditions (a) and (b) (Ch. 1, Subsection 1.1) where $\mathbb{F}^n = (\mathcal{F}^n_t)_{t \geq 0}$ is a family of σ-algebras $\mathcal{F}^n_t \subseteq \mathcal{F}$, $t \geq 0$, for every $n \geq 1$. Assume to be given also a σ-albegra

$$\mathcal{G} \subseteq \mathcal{F}_0.$$

Let $X^n = (X^n_t, \mathcal{F}^n_t)$ be a semimartingale, defined on a probability space $(\Omega, \mathcal{F}, P)$ and adapted to a family $\mathbb{F}^n$. The probability distribution of the process X^n is denoted by Q^n, and its triplet by $T^n = (B^n, C^n, \nu^n)$. A "limiting process" $X = (X_t, \mathcal{F}_t)$ is assumed to be a semimartingale with $\mathcal{G}$-conditionally independent increments and with the ($\mathcal{G}$-measurable) triplet $T = (B, C, \nu)$ (see Ch. 4, § 4).

Everywhere below the weak convergence $Q^n \xrightarrow{w} Q$, where Q is the probability distribution of the "limiting" process $X = (X_t, \mathcal{F}_t)$, is denoted for clarity by $X^n \xrightarrow{d} X$ and $X^n \xrightarrow{d} X$ ($\mathcal{G}$-stably), in case the convergence $X^n \xrightarrow{d} X$ takes place along with the convergence

$$X^n \xrightarrow{d_f(S)} X \ (\mathcal{G}\text{-stably}),$$

where S is a set dense in R_+.

Observe that the definition of $X^n \xrightarrow{d} X$ ($\mathcal{G}$-stably) is equivalent to to the following definition:

$$\lim_n \mathbf{E}\, \xi f(X^n) = \mathbf{E}\, \xi f(X)$$

for each bounded and $\mathcal{G}$-measurable random variable ξ and each continuous in Skorohod's topology (Ch. 6, § 1) and bounded function

$$f = f(X), \ X \in D.$$

Obviously, one needs to check only the implication

$$X^n \xrightarrow{d} X \ (\mathcal{G}\text{-stably}) \Rightarrow \lim_n \mathbf{E}\, \xi f(X^n) = \mathbf{E}\, \xi f(X)$$

for a nonnegative random variable ξ. Without loosing generality one may assume $\xi \le a$ with $a > 1$ and $\mathbf{E}\, \xi = 1$.

Define the probability measure $\tilde{P}$ by setting $d\tilde{P} = \xi dP$, and denote by $\tilde{\mathbf{E}}$ the mathematical expectation relative to the measure $\tilde{P}$. Then it suffices to prove that

$$X^n \xrightarrow{d} X \ (\mathcal{G}\text{-stably}) \Rightarrow \lim_n \tilde{\mathbf{E}}\, f(X^n) = \tilde{\mathbf{E}}\, f(X),$$

i.e. distributions of X^n, $n \ge 1$, relative to $\tilde{P}$ converge weakly to distributions X, relative to $\tilde{P}$. The corresponding convergence of finite dimensional distributions is guaranteed by the convergence

$$X^n \xrightarrow{d_f(S)} X \ (\mathcal{G}\text{-stably}).$$

Therefore it suffices to show that the family of distributions of X^n, $n \ge 1$, relative to $\tilde{P}$ is tight. Since $X^n \xrightarrow{d} X$, the family of distributions of X^n, $n \ge 1$, relative to P is tight, i.e. for each $\varepsilon > 0$ one can choose a compact K_ε such that

$$\sup_n P(X^n \in D \setminus K_\varepsilon) \le \frac{\varepsilon}{a}$$

(see Ch. 6, formula (1.6)).

Then by the absolute continuity $\tilde{P} << P$ and the condition $\xi \le a$ we have

$$\sup_n \tilde{P}(X^n \in D \setminus K_\varepsilon) \le a \sup_n P(X^n \in D \setminus K_\varepsilon) \le \varepsilon,$$

i.e. the family of distributions X^n, $n \ge 1$, relative to $\tilde{P}$, is tight.

On investigating conditions for the convergences $X^n \xrightarrow{d} X$ and $X^n \xrightarrow{d} X$ ($\mathcal{G}$-stably) we consider first the case, important from the point of view of applications, when $X = (X_t, \mathcal{F}_t)$ is a Gaussian martingale or a martingale with $\mathcal{G}$-conditionally independent and $\mathcal{G}$-conditionally Gaussian increments.

2. We begin with the consideration of the functional analogies of Theorems 5.5.1 - 5.5.3.

As in Ch. 5, § 5, we assume that $X = (X_t, \mathcal{F}_t)$ is a locally square integrable martingale with $\mathcal{G}$-conditionally Gaussian and $\mathcal{G}$-conditionally independent increments and $X_0 = 0$, i.e. X is a semimartingale with the triplet $T = (B, C, \nu)$ which is $\mathcal{G}$-measurable and possesses the following properties:

$$B = 0, \quad \nu^c (R_+ \times R_0) = 0,$$

$$\int_{R_0} e^{i\lambda x} \nu (\{t\}, dx) = I(a_t > 0) \exp\left(-\frac{\lambda^2}{2} \int_{R_0} x^2 \nu (\{t\}, dx) \right).$$

The quadratic characteristic of the process X is given here by the formula (see Theorem 4.9.6)

$$\langle X \rangle_t = C_t + \sum_{0 < s \le t} \int_{R_0} x^2 \nu (\{s\}, dx), \tag{1.1}$$

and the stochastic exponential related to the process X by the formula

$$\mathcal{E}_t (G(\lambda)) = \exp\left(-\frac{\lambda^2}{2} \langle X \rangle_t \right). \tag{1.2}$$

For each $n \ge 1$ let a process $X^n = (X^n_t, \mathcal{F}^n_t)$ be a semimartingale with the triplet $T^n = (B^n, C^n, \nu^n)$ and for $a \in (0, 1]$ let

$$\langle M^{n,a} \rangle_t = C^n_t + x^2 I(|x| \le a) * \nu^n_t - \sum_{0 < s \le t} \left(\int_{|x| \le a} x \nu^n(\{s\}, dx) \right)^2 \tag{1.3}$$

be the quadratic characteristic of a locally square integrable martingale $M^{n,a} = (M^{n,a}_t, \mathcal{F}^n_t)$ with

$$M^{n,a}_t = X^{nc}_t + xI(|x| \le a) * (\mu^n - \nu^n)_t. \tag{1.4}$$

Theorem 1. *Let*

$$\sum_{s > 0} \nu (\{s\} \times R_0) = 0,$$

i.e. $\langle X \rangle = C$, *and let* S *be a set dense in* R_+.

1) *If the following conditions are fullfilled:*

(o) $\mathcal{G} \subseteq \bigcap_{n \geq 1} \mathcal{F}_0^n$,

(a) $\nu^n((0, t] \times \{ |x| > a\}) \xrightarrow{P} 0, \ a \in (0, 1], \ t \in S$,

(sup b) $\sup_{s \leq t} |B_s^n| \xrightarrow{P} 0, \ t \in S$,

(c) $\langle M^{n, a} \rangle_t \xrightarrow{P} C_t, \ a \in (0, 1], \ t \in S$,

(e) $X_0^n \xrightarrow{P} 0$,

then

$$X^n \xrightarrow{d} X \ (\mathcal{G}\text{-}stably).$$

2) *If* $X^n \xrightarrow{d} X$ *and Condition* (sup b) *holds, then Conditions* (a) *and* (e) *are fullfilled. If, in addition,* C *is a deterministic function, then Condition* (c) *is fullfilled as well as the condition*

$$[M^{n, a}, M^{n, a}]_t \xrightarrow{P} C_t, \ t \in S.$$

Consider also the case of semimartingales $\hat{X}^n = (\hat{X}_t^n, \hat{\mathcal{F}}_t^n)$ with $\hat{X}_t^n = X^n_{\hat{\tau}_n(t)}$ and $\hat{\mathcal{F}}_t^n = \mathcal{F}^n_{\hat{\tau}_n(t)}$, where $\hat{\tau}_n = (\hat{\tau}_n(t))_{t \geq 0} \in V^+$, $n \geq 1$, and $\hat{\tau}_n(t) \in \mathbf{T}(\mathbb{F}^n)$ for fixed $t \in R_+$, i.e. $\hat{\tau}_n$ is a random change of time.

Theorem 2. *Let*

$$\sum_{s > 0} \nu(\{s\} \times R_0) = 0,$$

i.e. $\langle X \rangle = C$, *and let* S *be a set dense in* R_+.

If the following conditions are fullfilled:

(o) $\mathcal{G} \subseteq \bigcap_{n \geq 1} \mathcal{F}_0^n$,

(a) $\nu^n((0, \hat{\tau}_n(t)] \times \{ |x| > a\}) \xrightarrow{P} 0, \ a \in (0, 1], \ t \in S$,

(sup b) $\sup_{s \leq t} |B^n_{\hat{\tau}_n(s)}| \xrightarrow{P} 0, \ t \in S$,

(c) $\langle M^{n, a} \rangle_{\hat{\tau}_n(t)} \xrightarrow{P} C_t, \ a \in (0, 1], \ t \in S$,

(e) $X_0^n \xrightarrow{P} 0$,

then

$$\hat{X}^n \xrightarrow{d} X \quad (\mathcal{G}\text{-}stably)$$

The case of nested families is considered in the following

Theorem 3. *Let*

$$\sum_{s>0} \nu(\{s\} \times R_0) = 0,$$

i.e. $\langle X \rangle = C$, *and let* S *be a set dense in* R_+.

1) *If the following conditions are fullfilled:*

(o) *there exists a sequence of numbers* $(s_n)_{n\ge 1}$, $s_n \downarrow 0$, $n \to \infty$, *such that*

$$\mathcal{F}^n_{s_n} \subseteq \mathcal{F}^{n+1}_{s_{n+1}}, \quad \sigma\left(\bigcup_{n\ge 1} \mathcal{F}^n_{s_n}\right) = \sigma\left(\bigcup_{n\ge 1} \mathcal{F}^n_\infty\right), \quad \mathcal{G} \subseteq \sigma\left(\bigcup_{n\ge 1} \mathcal{F}^n_\infty\right);$$

(a) $\nu^n((0, t] \times \{|x| > a\}) \xrightarrow{P} 0, \ a \in (0, 1], \ t \in S;$

(sup b) $\sup_{s\le t} |B^n_s| \xrightarrow{P} 0, \ t \in S;$

(c) $\langle M^{n,a} \rangle_t \xrightarrow{P} C_t, \ t \in S;$

(e) $X^n_0 \xrightarrow{P} 0,$

then

$$X^n \xrightarrow{d} X \quad (\mathcal{G}\text{-}stably).$$

2) *If* $X^n \xrightarrow{d} X$ *and Condition* (sup b) *holds, then Conditions* (a) *and* (e) *are fullfilled. Under the assumption that* C *is a deterministic function, Condition* (c) *is fullfilled too as well as the condition*

$$[M^{n,a}, M^{n,a}]_t \xrightarrow{P} C_t, \ t \in S.$$

Proof of Theorems 1 - 3. Conditions (o) - (e) of Theorems 1 - 3 ensure validity of the conditions of Theorems 5.5.1 - 5.5.3, according to which the weak convergence of finite dimensional distributions

$$X^n \xrightarrow{d_f(S)} X \quad (\mathcal{G}\text{-stably})$$

holds for a set S dense in R_+. Therefore, to prove the convergence

$$X^n \xrightarrow{d} X \quad (\mathcal{G}\text{-stably})$$

it suffices, in virtue of $\langle X \rangle = C$ and $\nu = 0$, to establish the relative compactness of the family of distributions (Q^n), $n \ge 1$, of semimartingales X^n, $n \ge 1$. To prove the relative compactness of this family we apply Theorems 6.4.1 and 6.4.2. Aiming at this, we will show that the assumptions stipulated in these theorems are fullfilled in virtue of Conditions (o) - (e) of Theorems 1 - 3.

We consider first the conditions of Theorems 1 and 3, and we verify Conditions I, II, III and $IV_{3)}$ of Theorem 6.4.1.

Assumption $I_{1)}$ is fullfilled by Condition (e), and assumption $I_{2)}$ by Condition (a). Set

$$\beta_t^{na} = B_t^{na} - B_t^{n}$$

and observe that assumption $II_{1)}$ is fullfilled by Condition (sup b). The assumptions $II_{2)}$ and $II_{3)}$ are obviously fullfilled as $\gamma^{na} = \langle M^{n,a}\rangle$ and $\delta^{ng} = g * \nu^n$. Assumption III is fullfilled as $\overline{G}^n \equiv 0$ and $G^n = \mathrm{Var}\,(\beta^{na}) + \gamma^{na} + \delta^{ng}$.

Let us show now that assumption $IV_{3)}$ is fullfilled as $G = C$.

Since $\gamma^{na} = \langle M^{n,a}\rangle$, by Condition (c) we have

$$\gamma_t^{na} \xrightarrow{P} C_t, \quad a \in (0, 1], \quad t \in S.$$

Therefore it suffices to show that

$$\mathrm{Var}\,(\beta^{na})_t \xrightarrow{P} 0, \qquad \delta_t^{ng} \xrightarrow{P} 0, \quad t \in S.$$

This relation holds by Condition (a) and the estimates (see Ch. 6, § 4, formula (4.2))

$$\mathrm{Var}\,(\beta^{na})_t \le \int_0^t \int_{a<|x|\le 1} |x|\, d\nu^n \le \nu^n\,((0, t] \times \{\,|x| > a\})$$

and

$$\delta_t^{ng} = g * \nu_t^n \le \max_x g\,(x)\, \nu^n\,((0, t] \times \{\,|x| > a\})$$

for a certain $a > 0$ such that $I\,(\,|x| > a)\, g\,(x) = g\,(x)$.

Thus in case of Theorems 1 and 3 the relative compactness of the family (Q^n), $n \ge 1$, is established.

In case of Theorem 2 the relative compactness of the distributions of the family of semimartingales $\hat{X}^n$, $n \ge 1$, is established analogously by utilizing Theorem 6.4.2.

Assertion 2) of Theorems 1 and 3 is established in the following way. Since under the assumption

$$\sum_{s>0} \nu\,(\{s\} \times R_0) = 0$$

the process X has trajectories in C, i.e. Q (C) = 1, then by Theorem 6.1.8 we have

$$X_0^n \xrightarrow{d} 0$$

and by Problem 6.1.4 Condition (e) is fullfilled. Condition (a) is fullfilled in view of Theorem 6.5.1. Finally, Condition (c) is fullfilled by Theorem 6.5.2, since $B\,(X) = 0$, $\langle M\,(X)\rangle = C$ and consequently Conditions (α) and (K) of Theorem 6.5.2 are obviously fullfilled. Condition (sup B) of Theorem 6.5.2 is fullfilled by Condition (sup b) of Theorems 1 and 3.

Theorems 1 - 3 are proved.

3. Consider now the case in which $X^n = (X^n_t, \mathcal{F}^n_t)$ is a locally square integrable martingale for every $n \geq 1$ with the quadratic variation $[X^n, X^n]$, the quadratic characteristic $\langle X^n \rangle$ and $X^n_0 = 0$.

Theorem 4. *Let*

$$\sum_{s>0} \nu(\{s\} \times R_0) = 0,$$

i.e. $\langle X \rangle = C$, *let* S *be a set dense in* R_+ *and* $X^n \in \mathcal{M}^2_{loc,\,0}(\mathbb{F}^n)$.

1) *if Condition* (o) *of Theorems* 1 *and* 3 *is fullfilled, as well as any of the conditions*

$$(c_1) \qquad [X^n, X^n]_t \xrightarrow{P} C_t, \quad t \in S,$$

or

$$(c_2) \qquad \langle X^n \rangle_t \xrightarrow{P} C_t, \quad t \in S,$$

and also the condition

$$(L_2) \qquad x^2 I(|x| > a) * \nu^n_t \xrightarrow{P} 0, \quad a \in (0, 1], \quad t \in S,$$

then

$$X^n \xrightarrow{d} X \quad (\mathcal{G}\text{-}stably).$$

2) *If* $X^n \xrightarrow{d} X$, *Condition* (L_2) *is fullfilled and the fuction* C *is deterministic, then Conditions* (c_1) *and* (c_2) *hold.*

If $X^n = (X^n, \mathcal{F}^n)$ is a local square integrable martingale and

$$\hat{\tau}_n = (\hat{\tau}_n(t))_{t \geq 0} \in V^+, \quad \hat{\tau}_n(t) \in \mathbf{T}(\mathbb{F}^n), \quad t \in R_+,$$

then the process $\hat{X}^n = (\hat{X}^n_t, \hat{\mathcal{F}}^n_t)$ with

$$\hat{X}^n_t = X^n_{\hat{\tau}_n(t)}, \quad \hat{\mathcal{F}}^n_t = \mathcal{F}^n_{\hat{\tau}_n(t)}$$

may fail to be even a local martingale (see Ch. 4, § 7). However, the following theorem for $\hat{X}_n$, $n \geq 1$, holds, analogously to Theorem 4.

Theorem 5. *Let*

$$\sum_{s>0} \nu(\{s\} \times R_0) = 0,$$

i.e. $\langle X\rangle = C$, *let* S *be a set dense in* R_+ *and* $X^n \in \mathcal{M}^2_{loc}(\mathbb{F}^n)$, $n \ge 1$.

If Condition (o) *of Theorem* 2 *is fullfilled as well as any of the conditions*

$(\hat{c}_1)$ $$[X^n, X^n]_{\hat{\tau}_n(t)} \xrightarrow{P} C_t, \ t \in S,$$

or

$(\hat{c}_2)$ $$\langle X^n\rangle_{\hat{\tau}_n(t)} \xrightarrow{P} C_t, \ t \in S,$$

and also the condition

$(\hat{L}_2)$ $$x^2 I(|x| > a) * \nu^n_{\hat{\tau}_n(t)} \xrightarrow{P} 0, \ a \in (0, 1], \ t \in S,$$

then

$$\hat{X}^n \xrightarrow{d} X \ (\mathcal{G}\text{-}stably).$$

The proof of Theorems 4 and 5 is based on the fact that the conditions of these theorems ensure validity of the conditions of Theorems 1 - 3 (for more details see the proofs of Theorems 5.5.4 and 5.5.6).

4. In the statement of Theorems 4 and 5 (as well as Theorems 1 - 3) the condition

$$\sum_{s > 0} \nu(\{s\}, \times R_0) = 0$$

is stipulated, according to which the limiting process has continuous trajectories and $\langle X\rangle = C$. This condition is dropped below. Recall that

$$\langle X\rangle_t = C_t + \sum_{0 < s \le t} \int_{R_0} x^2 \nu(\{s\}, dx).$$

For each $n \ge 1$, let $X^n = (X^n_t, \mathcal{F}^n_t)$ be a local square integrable martingale with $X^n_0 = 0$, the triplet of $\mathbb{F}^n$-predictable characteristics $T^n(B^n, C^n, \nu^n)$ and the quadratic characteristic $\langle X^n\rangle$. Denote

$$F^n_t(x) = I(a^n_t > 0)\,[\nu^n(\{t\} \times (-\infty, x] \cap R_0) + I(x \ge 0)(1 - a^n_t)] \qquad (1.5)$$

and

$$\Phi(x) = \frac{1}{\sqrt{2\pi}} \int_{-\infty}^{x} e^{-\frac{y^2}{2}}\, dy. \qquad (1.6)$$

Theorem 6. *Let*

$$S = \{t > 0:\ P(\Delta \langle X\rangle_t = 0) = 1\}$$

and

$$X^n \in \mathcal{M}^2_{loc,0}(\mathbb{F}^n),\ n \geq 1.$$

If the following conditions are fullfilled:

(o) $\mathcal{G} \subseteq \bigcap_{n \geq 1} \mathcal{F}^n_0$,

(L^c_2) $x^2 I(|x| > a) * \nu^{nc}_t \xrightarrow{P} 0$, $a \in (0, 1]$, $t \in S$,

$$(\Lambda_2) \quad \sum_{s \in \{s \leq t:\ a^n_s > 0\}} \int_{|x| > a} |x| \left| F^n_s(x) - \Phi\left(\frac{x}{\sqrt{\Delta \langle X^n\rangle_s}}\right)\right| dx \xrightarrow{P} 0,$$

$$a \in (0, 1],\ t \in S,$$

(c_2) $\langle X^n\rangle_t \xrightarrow{P} \langle X\rangle_t$, $t \in S$,

(k_2) $\sum_{0 < s \leq t} (\Delta \langle X^n\rangle_s)^2 \xrightarrow{P} \sum_{0 < s \leq t} (\Delta \langle X\rangle_s)^2$, $t \in S$,

then

$$X^n \xrightarrow{d} X \ (\mathcal{G}\textit{-stably}).$$

Proof. The weak convergence of finite dimensional distributions $X^n \xrightarrow{d_f(S)} X$ ($\mathcal{G}$-stably) is established in Theorem 5.6.1. Therefore by Theorem 6.1.8 it suffices to verify that the family (Q^n), $n \geq 1$, of distributions of the processes X^n, $n \geq 1$, is relatively compact. We verify this by applying Theorem 6.4.3.

Assumption $I_{1)}$ of this theorem is fullfilled, since $X^n_0 = 0$, $n \geq 1$. Assumption $I_{2)}$ is fullfilled too, since (see the definition of $F^n_t(x)$ in (1.5))

$$\int_0^L \int_{|x| > 1} d\nu^n = \int_0^L \int_{|x| > 1} d\nu^{nc} + \sum_{0 < s \leq L} \int_{|x| > 1} dF^n_s(x),$$

and by Chebyshev's inequality (see also Ch. 5, § 6)

$$\int_{|x| > 1} dF^n_s(x) \leq \frac{1}{1^2} \int_R x^2 dF^n_s(x) = \frac{\Delta \langle X^n\rangle_s}{1^2},$$

i.e. as $l \geq 1$

$$\int_0^L \int_{|x|>l} d\nu^n \leq \int_0^L \int_{|x|>l} x^2 d\nu^{nc} + \frac{\langle X^n \rangle_L}{l^2} \tag{1.7}$$

and, consequently,

$$\lim_{l \to \infty} \overline{\lim_n} \, P\left(\int_0^L \int_{|x|>l} d\nu^n \geq \varepsilon \right) = 0, \ \varepsilon > 0$$

by the estimate (1.7) and Conditions (L_2^c) and (c_2) of the theorem.

Let us verify now assumption III' of Theorem 6.4.3. To this end observe that by Problem 4.1.10 we have

$$B_t^n = -\int_0^t \int_{|x|>1} x d\nu^n.$$

Consequently, the following processes are also increasing:

$$|x| I(|x| > 1) * \nu^n - \operatorname{Var}(B^n),$$

$$x^2 I(|x| > 1) * \nu^n - |x| I(|x| > 1) * \nu^n,$$

$$x^2 * \nu^n - x^2 I(|x| > 1) * \nu^n, \ \langle X^n \rangle - x^2 * \nu^n.$$

Hence the process $\langle X^n \rangle - \operatorname{Var}(B^n)$ is increasing. It is not hard to deduce from this that assumption III' is fullfilled, with the process $G^n = 3 \langle X^n \rangle$.

Assumption $IV'_{2)}$ of Theorem 6.4.3 is fullfilled with the process $G = 3 \langle X \rangle$ by Conditions (c_2) and (k_2) of Theorem 6.1.3.

5. Let now $X^n = (X_t^n, \mathcal{F}_t^n)$, $n \geq 1$, be a local martingale. In this case the quadratic characteristic $\langle X^n \rangle$ is undefined and the following theorems, analogous to Theorems 4 and 5, are formulated in terms of the quadratic variation $[X^n, X^n]$ only.

Theorem 7. *Let*

$$\sum_{s>0} \nu(\{s\} \times R_0) = 0,$$

i.e. $\langle X \rangle = C$, *let* S *be a set dense in* R_+ *and* $X^n \in \mathcal{M}_{loc,0}(\mathbb{F}^n)$, $n \geq 1$.

1) *If Condition* (o) *of Theorems* 1 *and* 3, *and the condition*

(c_1) $$[X^n, X^n]_t \xrightarrow{P} C_t, \ t \in S,$$

are fullfilled, as well as any of the conditions

(L_1) $$|x| I(|x| > a) * \nu_t^n \xrightarrow{P} 0, \ a \in (0, 1], \ t \in S,$$

or

(ρ) *the family of random variables*

$$(\sup_{s\le t} |\Delta X^n_s|),\ n\ge 1,$$

is uniformly integrable for every $t \in R_+$,
then

$$X^n \xrightarrow{d} X \ (\mathcal{G}\text{-}stably).$$

2) *If* $X^n \xrightarrow{d} X$, *the function* C *is deterministic and any of Conditions* (L_1) *or* (ρ) *is fullfilled, then Condition* (c_1) *holds.*

Let $\hat{X}^n_t = X^n_{\hat{\tau}_n(t)}$ and $\hat{\mathcal{F}}^n_t = \mathcal{F}^n_{\hat{\tau}_n(t)}$, with $\hat{\tau}_n = (\hat{\tau}_n(t))_{t\ge 0} \in V^+$, $\hat{\tau}_n(t) \in T(\mathbb{F}^n)$, $t \in R$, i.e. $\hat{\tau}_n$ is a random change of time. In case in which $X^n = (X^n_t, \mathcal{F}^n_t)$ is a local martingale, the following theorem is valid.

Theorem 8. *Let*

$$\sum_{s>0} \nu(\{s\}, \times R_0) = 0,$$

i.e. $\langle X\rangle = C$, *let* S *be a set dense in* R_+ *and* $X^n \in \mathcal{M}_{loc,0}(\mathbb{F}^n)$, $n \ge 1$.

If Condition (o) *of Theorem* 2 *is fullfilled, as well as the condition*

($\hat{c}_1$) $$[X^n, X^n]_{\hat{\tau}_n(t)} \xrightarrow{P} C_t,\ t\in S,$$

and any of the following conditions

($\hat{L}_1$) $$|x| I(|x|>a) * \nu^n_{\hat{t}_n(t)} \xrightarrow{P} 0,\ a\in(0,1],\ t\in S,$$

or

($\hat{\rho}$) *the family of random variables*

$$(\sup_{s\le \hat{\tau}_n(t)} |\Delta X^n_s|),\ n\ge 1,$$

is uniformly integrable for every $t \in R_+$,
then

$$X^n \xrightarrow{d} X \ (\mathcal{G}\text{-}stably).$$

The proof of Theorems 7 and 8 (just as the proof of Theorems 4 and 5) is based on the fact that their conditions satisfy the conditions of Theorems 1 - 3 (for more details see the proof of Theorems 5.5.4 and 5.5.6).

We present now the theorem analogous to Theorem 6, concerning the case in which processes $X^n = (X_t^n, \mathcal{F}_t^n)$, $n \geq 1$, are local martingales.

Let functions $F_t^n(x)$ and $\Phi(x)$ be defined by the formulas (1.5) and (1.6), and a process $\Delta^{n1} = (\Delta_t^{n1})_{t \geq 0}$ by the relation (6.10) (Ch. 5, § 6).

Theorem 9. *Let*

$$S = \{t > 0\colon \mathsf{P}(\Delta \langle X \rangle_t = 0) = 1\}$$

and

$$X^n \in \mathcal{M}_{\mathrm{loc},0}(\mathbb{F}^n),\ n \geq 1.$$

If the following conditions are fullfilled

(o) $\mathcal{G} \subseteq \bigcap_{n \geq 1} \mathcal{F}_0^n$,

(L_1^c) $|x| I(|x| > a) * \nu_t^{nc} \xrightarrow{\mathsf{P}} 0,\ a \in (0, 1],\ t \in S,$

$$(L_1)\quad \sum_{s \in \{s \leq t:\, a_s^n > 0\}} \int_{|x| > a} \left| F_s^n(x) - \Phi\left(\frac{x}{\sqrt{\Delta_s^{n1}}}\right) \right| dx \xrightarrow{\mathsf{P}} 0,\ a \in (0, 1],$$

$t \in S,$

$$(c_1^\Delta)\quad C_t^n + x^2 I(|x|, a) * \nu_t^{nc} + \sum_{0 < s \leq t} \Delta_s^{n1} \xrightarrow{\mathsf{P}} C_t,\ a \in (0, 1],\ t \in S,$$

$$(k_2^\Delta)\quad \sum_{0 < s \leq t} (\Delta_s^{n1})^2 \xrightarrow{\mathsf{P}} \sum_{0 < s \leq t} (\Delta \langle X \rangle_s)^2,\ t \in S,$$

then

$$X^n \xrightarrow{d} X\ (\mathcal{G}\text{-}\textit{stably}).$$

Proof. The weak convergence of finite dimensional distributions

$$X^n \xrightarrow{d_f(S)} X\ (\mathcal{G}\text{-stably})$$

is established in Theorem 5.6.2. Therefore, as in the course of proving Theorem 6, we will verify the relative compactness of distributions of local martingales X^n, $n \geq 1$, by applying Theorem 6.4.3, i.e. verifying the validity of the assumptions I, III' and IV' of this theorem.

Assumption $I_{1)}$ of this theorem is fullfilled, since $X_0^n = 0$, $n \geq 1$.

To verify the rest of the assumptions of Theorem 6.4.3, we will need also the following estimates:

$$\int_{|x|>l} dF_s^n(x) \le \frac{\Delta_s^{n1}}{2l-1}, \quad l \ge 1, \tag{1.8}$$

$$\int_{|x|>1} |x|\, dF_s^n(x) \le \Delta_s^{n1} \tag{1.9}$$

and

$$\int_R (x^2 \wedge 1)\, dF_s^n(x) \le \Delta_s^{n1}. \tag{1.10}$$

To prove (1.8) we use Chebyshev's inequality with the function

$$h(x) = \frac{1}{2} x^2 I(|x| \le 1) + \left(|x| - \frac{1}{2}\right) I(|x| > 1),$$

according to which

$$\int_{|x|>l} dF_s^n(x) \le \frac{1}{h(l)} \int_R h(x)\, dF_s^n(x). \tag{1.11}$$

Since

$$\frac{dh(x)}{dx} = xI(|x| \le 1) + I(|x| > 1) \operatorname{sign} x = g_1(x)$$

(see Ch. 5, (6.7)), in view of the definition of the process Δ^{n1} (see Ch. 5, (6.10)) we get, integrating by parts, that

$$\int_R h(x)\, dF_s^n(x) = \int_0^\infty g_1(x)\, [(1 - F_s^n(x)) + F_s^n(-x)]\, dx$$

$$= \int_0^\infty g_1(x) \left[\left(1 - \Phi\left(\frac{x}{\sqrt{\Delta_s^{n1}}} \right) \right) + \Phi\left(-\frac{x}{\sqrt{\Delta_s^{n1}}} \right) \right] dx$$

$$= \int_R h(x)\, d\Phi\left(\frac{x}{\sqrt{\Delta_s^{n1}}} \right) \le \frac{1}{2} \int_R x^2 d\Phi\left(\frac{x}{\sqrt{\Delta_s^{n1}}} \right) = \frac{1}{2} \Delta_s^{n1}. \tag{1.12}$$

The estimate (1.8) follows obviously from this and (1.11). The estimate (1.9) follows from (1.12), since

$$\int_{|x|>1} |x|\, dF_s^n(x) \le 2 \int_{|x|>1} h(x)\, dF_s^n(x) \le 2 \int_R h(x)\, dF_s^n(x),$$

while the estimate (1.10) follows from (1.12) and the following inequalities

$$\int_R (x^2 \wedge 1)\, dF_s^n(x) \le \int_R (x^2 \wedge |x|)\, dF_s^n(x) \le \int_R 2h(x)\, dF_s^n(x).$$

Let us verify now assumptions $I_{2)}$, III' and $IV'_{2)}$ of Theorem 6.4.3. Denote

$$V_t^{n1} = C_t^n + x^2 I(|x| \le 1) * v_t^{nc} + \sum_{0 < s \le t} \Delta_s^{n1}. \tag{1.13}$$

In view of the inequality (1.8), for $l \ge 1$ we get

$$\int_0^L \int_{|x|>1} dv^n = \int_0^L \int_{|x|>1} dv^{nc} + \sum_{0<s\le t} \int_{|x|>1} dF_s^n(x)$$

$$\le \int_0^L \int_{|x|>1} |x|\, dv^{nc} + \frac{V_L^{n1}}{2l-1}.$$

Consequently, assumption $I_{2)}$ is fullfilled by Conditions (L_1^c) and (c_1^Δ) of the theorem.

In accordance with Problem 4.1.10 we have

$$B_t^n = -\int_0^t \int_{|x|>1} x dv^n.$$

Therefore a process

$$|x| I(|x| > 1) * v^n - \mathrm{Var}(B^n)$$

is increasing and hence by the inequalities (1.9) and (1.13) such is a process

$$|x| I(|x| > 1) * v^{nc} + V^{n1} - \mathrm{Var}(B^n).$$

Consequently, assumption III' holds with a process

$$G^n = 2|x| I(|x| > 1) * n^{nc} + 2V^{n1},$$

and assumption $IV'_{2)}$ with a process $G = 2\langle X\rangle$ by Conditions (L_1^c), (c_1^Δ) and (k_2^Δ) of Theorem 6.1.3.

6. We apply Theorem 4 to a number of simple examples.

Example 1. Let a Poisson process $\pi = (\pi_t)_{t\ge 0}$ with $\mathbf{E}\pi_t \equiv t$ be given on a certain probability space. We set

$$X_t^n = \frac{1}{\sqrt{n}}(\pi_{nt} - nt), \quad n \ge 1,$$

and we show that

$$X^n \xrightarrow{d} W,$$

where W is a Wiener process.

Since we are interested in a convergence in distribution, we may assume that all considerations concern one and the same stochastic basis $(\Omega, \mathcal{F}, \mathbb{F}, P)$. $W = (W_t, \mathcal{F}_t)$ and $\pi = (\pi_t, \mathcal{F}_t)$ are Wiener and Poisson processes, besides the Poisson process has the compensator $A_t \equiv t$. For each $n \geq 1$ define a family $\mathbb{F}^n = (\mathcal{F}^n_t)_{t \geq 0}$ with $\mathcal{F}^n_t = \mathcal{F}_{nt}$. Then the process $X^n = (X^n_t, \mathcal{F}^n_t)$ is a square integrable martingale with the quadratic characteristic $\langle X^n \rangle \equiv t$ (see Problem 3.4.5).

Apply now Theorem 4. For a Wiener process we have $C_t \equiv t$. Assume $\mathcal{G} = \{\varnothing, \Omega\}$. Then Condition (o) of Theorems 1 and 3 is fullfilled automatically. Condition (c_2) is fullfilled, since $\langle X^n \rangle_t \equiv t$. Condition (L_2) is fullfilled, since for each $t \in R_+$

$$\mathbf{E}x^2 I(|x| > a) * \nu^n_t = \mathbf{E}x^2 I(|x| > a) * \mu^n_t = \mathbf{E} \sum_{0 < s \leq t} \frac{(\Delta\pi_{ns})^2}{n} I\left(\frac{\Delta\pi_{ns}}{\sqrt{n}} > a\right)$$
$$= \frac{1}{n}\mathbf{E}I\left(\frac{1}{\sqrt{n}} > a\right) \circ \pi_{nt} = tI\left(\frac{1}{\sqrt{n}} > a\right) \to 0, \; n \to \infty.$$

Example 2. Let a stochastic basis $(\Omega, \mathcal{F}, \mathbb{F}, P)$ be given and $(\Theta_t, \mathcal{F}_t)$ is a Markov process with two states (0, 1), right-continuous trajectories and the matrix of transition intensities

$$\begin{pmatrix} -\lambda_1 & \lambda_1 \\ \lambda_2 & -\lambda_2 \end{pmatrix}$$

with $\lambda_1 > 0$ and $\lambda_2 > 0$. Define the stochastic process

$$M_t = \Theta_t - \Theta_0 - \int_0^t (\lambda_1 - (\lambda_1 + \lambda_2)\,\Theta_s)\, ds. \tag{1.14}$$

We set

$$X^n_t = \frac{M_{nt}}{\sqrt{n}}, \quad n \geq 1,$$

and we will show that

$$X^n \xrightarrow{d} aW, \quad a^2 = \frac{2\lambda_1\lambda_2}{\lambda_1 + \lambda_2}$$

where W is a Wiener process.

The process $M = (M_t, \mathcal{F}_t)$ is a square integrable martingale (Lemma 9.2 in [188]). Consequently, $X^n = (X^n_t, \mathcal{F}^n_t)$ is a square integrable martingale too, where $\mathcal{F}^n_t = \mathcal{F}_{nt}$.

To verify the condition of Theorem 4, it is necessary to determine the quadratic characteristic $\langle M \rangle$. To this end observe that by (1.14) $(\Theta_t, \mathcal{F}_t)$ is a special semimartingale with the decomposition:

$$\Theta_t = \Theta_0 + A_t + M_t,$$

where M is a square integrable martingale and

$$A_t = \int_0^t (\lambda_1 - (\lambda_1 + \lambda_2)\,\Theta_s)\,ds.$$

Since $\Theta_t^2 = \Theta_t$, then $(\Theta_t^2, \mathcal{F}_t)$ is a special semimartingale with just the same decomposition.

Therefore by Problem 4.1.8

$$\langle M \rangle_t = \int_0^t [\lambda_1\,(1 - \Theta_s) + \lambda_2 \Theta_s]\,ds. \tag{1.15}$$

Let us verify now the conditions of Theorem 4. As

$$a^2 = \frac{2\lambda_1\lambda_2}{\lambda_1 + \lambda_2}$$

we have

$$C_t = \langle aW \rangle_t = \frac{2\lambda_1\lambda_2}{\lambda_1 + \lambda_2}\,t.$$

Set

$$\mathcal{G} = \{\varnothing, \Omega).$$

Then it suffices to verify Conditions (c_2) and (L_2), i.e. to show that as $t \in R_+$

$$\langle X^n \rangle_t \xrightarrow{P} \frac{2\lambda_1\lambda_2}{\lambda_1 + \lambda_2}\,t \tag{1.16}$$

and

$$x^2 I(|x| > a) * \nu_t^n \xrightarrow{P} 0, \quad a \in (0, 1]. \tag{1.17}$$

From (1.15) and the definition of X_t^n it follows that

$$\langle X^n \rangle_t = \frac{1}{n} \int_0^{nt} [\lambda_1\,(1 - \Theta_s) + \lambda_2 \Theta_s]\,ds.$$

Therefore (1.16) is fullfilled, provided

$$\frac{1}{n} \int_0^{nt} \Theta_s\,ds \xrightarrow{P} \frac{\lambda_1 t}{\lambda_1 + \lambda_2}$$

or equivalently

$$\frac{1}{n}\int_0^{nt}(\lambda_1-(\lambda_1+\lambda_2)\Theta_s)\,ds\xrightarrow{P}0.$$

The last equation is equivalent by (1.16) to the relation

$$\frac{M_{nt}}{n}\xrightarrow{P}0, \tag{1.18}$$

which holds since (see (1.15))

$$\mathbf{E}\left(\frac{M_{nt}}{n}\right)^2=\mathbf{E}\frac{\langle M\rangle_{nt}}{n^2}\le\frac{1}{n^2}\int_0^{nt}(\lambda_1+\lambda_2)\,ds\to 0,\ \ n\to\infty.$$

Thus the relation (1.16) is fullfilled.

Next, for each $t\in R_+$ and $a\in(0,1]$

$$\mathbf{E}x^2 I(|x|>a)*\nu_t^n=\mathbf{E}x^2 I(|x|>a)*\mu_t^n$$

$$=\mathbf{E}\sum_{0<s\le t}(\Delta X_s^n)^2 I(|\Delta X_s^n|>a)$$

$$=\mathbf{E}\sum_{0<s\le t}\frac{(\Delta M_{ns})^2}{n}I\left(\frac{|\Delta M_{ns}|}{\sqrt{n}}>a\right).$$

Since $|\Delta M|=|\Delta\Theta|\le 1$, this gives

$$\mathbf{E}x^2 I(|x|>a)*\nu_t^n\le I\left(\frac{1}{\sqrt{n}}>a\right)\frac{\mathbf{E}\langle M\rangle_{nt}}{n}$$

$$\le I\left(\frac{1}{\sqrt{n}}>a\right)(\lambda_1+\lambda_2)\,t\to 0,\ \ n\to\infty,$$

and consequently the relation (1.17) is fullfilled.

Example 3. Let $(\Omega, \mathcal{F}, \mathbb{F}, P)$ be a stochastic basis, $\tilde{W}=(\tilde{W}_t)_{t\ge 0}$ a Wiener process and $\xi=(\xi_t,\mathcal{F}_t)$ a stochastic process, defined by Ito's stochastic equation

$$\xi_t=\int_0^t(1-a\xi_s)\,ds+\int_0^t\xi_s\,d\tilde{W}_s. \tag{1.19}$$

We set

$$X_t^n = \frac{1}{\sqrt{n}} \int_0^{nt} \xi_s d\tilde{W}_s, \quad n \geq 1,$$

and we will show that (W is a Wiener process) as $a > \frac{3}{2}$

$$X^n \xrightarrow{d} \sqrt{\frac{2}{a(2a-1)}}\, W.$$

Since $X_n = (X_t^n, \mathcal{F}_t^n)$ with $\mathcal{F}_t^n = \mathcal{F}_{nt}$ is a square integrable martingale with continuous trajectories and the quadratic characteristic

$$\langle X^n \rangle_t = \frac{1}{n} \int_0^{nt} \xi_s^2 \, ds,$$

it suffices to show that

$$\frac{1}{n} \int_0^{nt} \xi_s^2 \, ds \xrightarrow{P} \frac{2t}{a(2a-1)}. \tag{1.20}$$

To establish (1.20), we need two auxiliary facts:

$$\sup_{t \geq 0} E\xi_t^2 \leq \frac{2}{a(2a-1)}, \quad \sup_{t \geq 0} E\xi_t^4 \leq \frac{2}{a(2a-1)(a-1)\left(a-\frac{3}{2}\right)} \tag{1.21}$$

and

$$\frac{\xi_{nt}}{\sqrt{n}} \xrightarrow{P} 0, \quad t \in R_+. \tag{1.22}$$

To establish (1.21) we utilize the following representation for ξ_t^m, $m = 2, 3, 4,$ which are easily obtained by using Ito's formula (Ch. 2, § 3):

$$\begin{aligned}
\xi_t^2 &= \int_0^t [2\xi_s(1 - a\xi_s) + \xi_s^2] \, ds + 2\int_0^t \xi_s^2 \, d\tilde{W}_s, \\
\xi_t^3 &= \int_0^t [3\xi_s^2(1 - a\xi_s) + 3\xi_s^3] \, ds + 3\int_0^t \xi_s^3 \, d\tilde{W}_s, \\
\xi_t^4 &= \int_0^t [4\xi_s^3(1 - a\xi_s) + 6\xi_s^4] \, ds + 4\int_0^t \xi_s^4 \, d\tilde{W}_s.
\end{aligned} \tag{1.23}$$

By Theorem 4.6 in [188] we have

$$\mathbf{E}\,|\,\xi_t\,|^m < \infty,\ \ m \geq 1,\ \ t \in R_+.$$

Therefore, by (1.23) we get

$$\mathbf{E}\xi_t^2 = \int_0^t [2\mathbf{E}\xi_s - (2a - 1)\,\mathbf{E}\xi_s^2]\,ds,$$

$$\mathbf{E}\xi_t^3 = \int_0^t [3\mathbf{E}\xi_s^2 - 3\,(a - 1)\,\mathbf{E}\xi_s^3]\,ds,$$

and

$$\mathbf{E}\xi_t^4 = \int_0^t \left[\,4\mathbf{E}\xi_s^3 - 4\left(a - \frac{3}{2}\right)\mathbf{E}\xi_s^4\,\right] ds.$$

These equations for $\mathbf{E}\xi_t^m$, $m = 2,3,4$, together with the equation

$$\mathbf{E}\xi_t = \int_0^t (1 - a\mathbf{E}\xi_s)\,ds$$

deduced from (1.19), admit the following estimates (as $a > 3/2$ and $t \in R_+$):

$$0 \leq \mathbf{E}\xi_t \leq 1/a, \quad \mathbf{E}\xi_t^2 \leq \frac{2}{a\,(2a - 1)}, \quad 0 \leq \mathbf{E}\xi_t^3 \leq \frac{2}{a\,(2a - 1)\,(a - 1)},$$

$$\mathbf{E}\xi_t^4 \leq \frac{2}{a\,(2a - 1)\,(a - 1)\,(a - 3/2)},$$

which prove the inequality (1.21).

Next, by (1.19)

$$\frac{\xi_{nt}}{\sqrt{n}} = \frac{e^{-nat}}{\sqrt{n}}\left(\int_0^{nt} e^{as}\,ds + \int_0^{nt} e^{as}\xi_s\,d\tilde{W}_s\right).$$

Consequently, (1.22) holds, provided

$$\frac{e^{-nat}}{\sqrt{n}}\int_0^{nt} e^{as}\xi_s\,d\tilde{W}_s \overset{P}{\to} 0.$$

But

$$\mathbf{E}\left(\frac{e^{-nat}}{\sqrt{n}}\int_0^{nt} e^{as}\xi_s d\tilde{W}_s\right)^2 = \frac{e^{-2nat}}{n}\int_0^{nt} e^{2as}\mathbf{E}\xi_s^2 ds$$

$$\le \frac{1}{a^2(2a-1)n}(1-e^{-2nat}) \to 0,\ n\to\infty,$$

i.e. (1.22) holds.

We establish now the desired relation (1.20). From (1.19) it follows that

$$\frac{1}{n}\int_0^{nt}(1-a\xi_s)\,ds = \frac{\xi_{nt}}{n} - \frac{1}{n}\int_0^{nt}\xi_s d\tilde{W}_s. \tag{1.24}$$

By (1.22) we have

$$\frac{\xi_{nt}}{n}\xrightarrow{P} 0.$$

Besides, in view of (1.21)

$$\mathbf{E}\left(\frac{1}{n}\int_0^{nt}\xi_s d\tilde{W}_s\right)^2 = \frac{1}{n^2}\int_0^{nt}\mathbf{E}\xi_s^2 ds \le \frac{2t}{a(2a-1)n} \to 0,\ n\to 0.$$

Thus from (1.24) we deduce

$$\frac{1}{n}\int_0^{nt}\xi_s ds \xrightarrow{P} \frac{t}{a}. \tag{1.25}$$

Analogously, from the first relation in (1.23) we obtain

$$\frac{1}{n}\int_0^{nt}[2\xi_s-(2a-1)\,\xi_s^2]\,ds = \frac{\xi_{nt}^2}{n} - \frac{2}{n}\int_0^{nt}\xi_s^2 d\tilde{W}_s. \tag{1.26}$$

By (1.22) we have

$$\frac{\xi_{nt}^2}{n}\xrightarrow{P} 0.$$

Besides,

$$\mathbf{E}\left(\frac{2}{n}\int_0^{nt}\xi_s^2 d\tilde{W}_s\right)^2 = \frac{4}{n^2}\int_0^{nt}\mathbf{E}\xi_s^4 ds \le \frac{8t}{a(2a-1)(a-1)\left(a-\frac{3}{2}\right)} \to 0,$$

as $n\to\infty$. Therefore (1.20) follows obviously from (1.25) and (1.26).

6. Restricting our attention to the important special case of semimartingales $X^n = (X^n_t, \mathcal{F}^n_t)$, $n \ge 1$, given in the following form:

$$X^n_t = \sum_{k=1}^{[nt]} \xi_{nk}, \quad \mathcal{F}^n_t = \mathcal{H}^n_{[nt]}$$

(see Ch. 5, Subsection 1.6), we will formulate the conditions under which the functional central limit theorem holds. The triplet of a semimartingale X^n is given by the formula (1.25) (Ch. 5, § 1).

Theorem 1 implies directly

Theorem 10. *Let*

$$\sum_{s>0} \nu(\{s\} \times R_0) = 0,$$

i.e. $\langle X \rangle = C$, *and let* S *be a subset, dense in* R_+.

If the following conditions are fullfilled:

(o) $\mathcal{G} = \{\varnothing, \Omega\}$,

(a) $\displaystyle\sum_{k=1}^{[nt]} \mathsf{P}(|\xi_{nk}| > a \mid \mathcal{H}^n_{k-1}) \xrightarrow{\mathsf{P}} 0, \; a \in (0, 1], \; t \in S,$

(sup b) $\displaystyle\sup_{s \le t} \left| \sum_{k=1}^{[ns]} \mathbf{E}(\xi_{nk} I(|\xi_{nk}| \le 1) \mid \mathcal{H}^n_{k-1} \right| \xrightarrow{\mathsf{P}} 0, \; t \in S,$

(c) $\displaystyle\sum_{k=1}^{[nt]} [\mathbf{E}(\xi^2_{nk} I(|\xi_{nk}| \le a) \mid \mathcal{H}^n_{k-1}) - (\mathbf{E}(\xi_{nk} I(|\xi_{nk}| \le a) \mid \mathcal{H}^n_{k-1}))^2] \xrightarrow{\mathsf{P}} C_t,$

$a \in (0, 1], \; t \in S,$

then

$$X^n \xrightarrow{d} X.$$

Observe that Condition (o) is a consequence of the assumption

$$\mathcal{F}^n_0 = (\varnothing, \Omega), \; n \ge 1,$$

and Condition (e) is dropped since $X^n_0 = 0, \; n \ge 1$.

Consider now the case in which $\xi^n = (\xi_{nk}, \mathcal{H}^n_k)$ is a martingale difference, $n \ge 1$. From Theorems 4 and 7 we deduce the following result.

Theorem 11. *Let*

$$\sum_{s>0} \nu(\{s\} \times R_0) = 0,$$

i.e. $\langle X \rangle = C$, *let* S *be a set dense in* R_+ *and* $\mathcal{G} = \{\emptyset, \Omega\}$.

1) *If* $\xi^n = (\xi_{nk}, \mathcal{H}^n_k)$, $n \geq 1$, *are square integrable martingale differences and if any of the following conditions is fullfilled:*

$$(c_1) \quad \sum_{k=1}^{[nt]} \xi^2_{nk} \xrightarrow{P} C_t, \; t \in S,$$

or

$$(c_2) \quad \sum_{k=1}^{[nt]} E\,(\xi^2_{nk} \mid \mathcal{H}^n_{k-1}) \xrightarrow{P} C_t, \; t \in S,$$

as well as the condition

$$(L_2) \quad \sum_{k=1}^{[nt]} E\,(\xi^2_{nk} I\,(\,|\,\xi_{nk}\,| > a) \mid \mathcal{H}^n_{k-1}) \xrightarrow{P} 0, \;\; a \in (0, 1], \;\; t \in S,$$

then

$$X^n \xrightarrow{d} X.$$

2) *If* $\xi^n = (\xi_{nk}, \mathcal{H}^n_k)$, $n \geq 1$, *are martingale differences, and Condition* (c_1) *is fullfilled as well as any of the following conditions:*

$$(L_1) \quad \sum_{k=1}^{[nt]} E\,(\,|\,\xi_{nk}\,| I\,(\,|\,\xi_{nk}\,| > a) \mid \mathcal{H}^n_{k-1}) \xrightarrow{P} 0, \;\; a \in (0, 1], \;\; t \in S,$$

or

(ρ) *the family of random variables*

$$\Big(\max_{1 \leq k \leq [nt]} |\,\xi_{nk}\,| \Big), \quad n \geq 1,$$

is uniformly integrable for each $t \in R_+$,
then

$$X^n \xrightarrow{d} X.$$

Theorems 3, 4 and 7 give the following result.

Theorem 12. *Suppose that the family* $(\mathcal{H}^n_k)_{k>0}$, $n \geq 1$, *is nested:*

$$\mathcal{H}^n_k \subseteq \mathcal{H}^{n+1}_k, \quad k \leq n, \; n \geq 1,$$

and the conditions of Theorems 10 *or* 11 *(Conditions* 1) *or* 2)) *are fullfilled except the condition* $\mathcal{G} = (\emptyset, \Omega)$ *which is replaced by the condition*

$$\mathcal{G} \subseteq \sigma\Big(\bigcup_{n \geq 1} \mathcal{H}^n_\infty \Big).$$

Then

$$X^n \xrightarrow{d} X \ (\mathcal{G}\text{-}stably\ ).$$

Suppose now

$$\hat{X}^n_t = \sum_{k=1}^{[n\hat{\tau}_n(t)]} \xi_{nk}, \quad \hat{\mathcal{F}}^n_t = \mathcal{H}^n_{[n\hat{\tau}_n(t)]},$$

with

$$\hat{\tau}_n = (\hat{\tau}_n(t))_{t \geq 0} \in V^+$$

where $\hat{\tau}_n(t)$ is a Markov time for every $t \in R_+$, relatively to the family

$$\mathbb{F}^n = (\mathcal{F}^n_t)_{t \geq 0} \text{ with } \mathcal{F}^n_t = \mathcal{H}^n_{[nt]},$$

i.e. $\hat{\tau}_n$ is a random change of time.

Then Theorems 10 and 11 are reformulated as the following assertion, the validity of which is guaranteed by Theorems 2, 5 and 8.

Theorem 13. *If the conditions of Theorems* 10 *or* 11 *(Conditions* 1) *or* 2)) *are fullfilled as* [nt] *is replaced by* $[n\hat{\tau}_n(t)]$, *then*

$$\hat{X}_n \xrightarrow{d} X.$$

We give now the assertion analogous to Theorems 6 and 9. Denote by $F^n_k(x)$ the conditional distribution function of a random variable ξ_{nk} under the condition $\mathcal{H}^n_{k-1}$.

Theorem 14. *Let*

$$S = \{t > 0: P(\Delta \langle X\rangle_t = 0) = 1\}$$

and $\mathcal{G} = \{\varnothing, \Omega\}$.

1) *If* $\xi^n = (\xi_{nk}, \mathcal{H}^n_k)$, $n \geq 1$, *are square integrable martingale differences with*

$$\Delta^n_k = \mathbf{E}(\xi^2_{nk} \mid \mathcal{H}^n_{k-1})$$

and the following conditions are fullfilled:

$$(L_2) \quad \sum_{k=1}^{[nt]} \int_{|x|>a} |x| \left| F^n_k(x) - \Phi\left(\frac{x}{\sqrt{\Delta^n_k}}\right)\right| dx \xrightarrow{P} 0, \ a \in (0, 1], \ t \in S,$$

$$(c_2) \quad \sum_{k=1}^{[nt]} \Delta^n_k \xrightarrow{P} \langle X\rangle_t, \ t \in S,$$

$$(\mathrm{k}_2)\ \sum_{k=1}^{[nt]} (\Delta_k^n)^2 \xrightarrow{P} \sum_{0<s\le t} (\Delta \langle X\rangle_s)^2, \quad t \in S,$$

then

$$X^n \xrightarrow{d} X.$$

2) *If* $\xi^n = (\xi_{nk}, \mathcal{H}_k^n)$, $n \ge 1$, *are martingale differences*, Δ_k^{n1} *is defined by the relation*

$$\int_R h(x)\, dF_k^n(x) = \int_R h(x)\, d\Phi\left(\frac{x}{\sqrt{\Delta_k^{n1}}}\right)$$

with

$$h(x) = \frac{1}{2} x^2 I(|x| \le 1) + \left(|x| - \frac{1}{2}\right) I(|x| > 1),$$

and the following conditions are fullfilled:

$$(\Lambda_1)\ \sum_{k=1}^{[nt]} \int_{|x|>a} \left| F_k^n(x) - \Phi\left(\frac{x}{\sqrt{\Delta_k^{n1}}}\right)\right| dx \xrightarrow{P} 0, \quad a \in (0, 1],\ t \in S,$$

$$(\overset{\Delta}{c_1})\ \sum_{k=1}^{[nt]} \Delta_k^{n1} \xrightarrow{P} \langle X\rangle_t, \quad t \in S,$$

$$(\overset{\Delta}{k_2})\ \sum_{k=1}^{[nt]} (\Delta_k^{n1})^2 \xrightarrow{P} \sum_{0<s\le t} (\Delta \langle X\rangle_s)^2, \quad t \in S,$$

then

$$X^n \xrightarrow{d} X.$$

Example 4 (Donsker's theorem, [16]). Let $(\xi_k)_{k\ge 1}$ be a sequence of independent and identically distributed random variables with $\mathbf{E}\xi_1 = 0$ and $\mathbf{E}\xi_1^2 = \sigma^2$. Denote

$$X_t^n = \frac{1}{\sqrt{n}} \sum_{k=1}^{[nt]} \xi_k.$$

Let $W = (W_t)_{t\ge 0}$ be a Wiener process.

Donsker's theorem tells us that

$$X^n \xrightarrow{d} \sigma W, \quad n \to \infty.$$

To prove Donsker's theorem apply Theorem 11. Denote

$$\mathcal{H}_k = \sigma\{\xi_1, \dots, \xi_k\}.$$

The sequence $(\frac{1}{\sqrt{n}}\xi_k, \mathcal{H}_k)$ is a sequence of square integrable martingale differences.

Besides, Conditions (c_1) and (L_2) of Theorem 11 are satisfied. Indeed,

$$(c_1)\ \frac{1}{n}\sum_{k=1}^{[nt]} \xi_k^2 \to \sigma^2 t \ \text{(P-a.s.)},\ n \to \infty,\ \forall\, t > 0$$

by the Birkhof-Khintchine theorem ([289, [332]), and

$$(L_2)\ \frac{1}{n}\sum_{k=1}^{[nt]} E\,(\xi_k^2 I\,(\frac{1}{\sqrt{n}}\,|\,\xi_k\,| > a)\,|\,\mathcal{H}_{k-1}) = \frac{[nt]}{n} E\xi_1^2 I\,(\,|\,\xi_1\,| > a\sqrt{n}\,) \to 0,$$

$$n \to \infty,\ \forall\, t > 0.$$

7. We will formulate the vector-valued version of the assertions of Theorems 1 and 3.

Let $X_t = (X_t^1, \dots, X_t^k)$ and $X = (X_t, \mathcal{F}_t)$ be a locally square integrable vector-valued martingale with $\mathcal{G}$-conditionally Gaussian increments and the quadratic characteristic

$$\langle X\rangle_t = C_t,\ t \geq 0.$$

Let $X^n = (X_t^n, \mathcal{F}_t^n)$, $n \geq 1$, be vector-valued semimartingales $(X_t^n = (X_t^{n1}, \dots, X_t^{nk}))$ with the triplets of predictable characteristics $T^n = (B^n, C^n, \nu^n)$, $n \geq 1$, (see Ch. 4, Subsection 1.3) and $M^{n,a} = (M_t^{n,a}, \mathcal{F}_t^n)$, $n \geq 1$, local square integrable martingales with

$$M_t^{n,a} = X_t^{nc} + \int_0^t \int_{|x| \leq a} x\, d\,(\mu^n - \nu^n)$$

$(\,|\,x\,| = (\sum_{j=1}^k (x^j)^2)^{1/2}$ and the quadratic characteristic

$$\langle M^{n,a}\rangle_t = C_t^n + \int_0^t \int_{|x| \leq a} xx^*\, d\nu^n - \sum_{0 < s \leq t} \hat{x}_s^{na} (\hat{x}_s^{na})^*,$$

where

$$\hat{x}_s^{na} = \int_{|x| \le a} x\nu^n(\{s\}, dx)$$

(all vectors here are column-vectors, and * is the transposition sign).

Theorem 15. *Let* S *be a set dense in* R_+. *If one of the following conditions is fullfilled:*

(o') $\mathcal{G} \subseteq \bigcap_{n \ge 1} \mathcal{F}_0^n$,

(o'') *there exists a sequence of numbers* $(s_n)_{n \ge 1}$, $s_n \downarrow 0$, $n \to \infty$, *such that*

$$\mathcal{F}_{s_n}^n \subseteq \mathcal{F}_{s_{n+1}}^{n+1}, \ n \ge 1, \quad \sigma(\bigcup_{n \ge 1} \mathcal{F}_{s_n}^n) = \sigma(\bigcup_{n \ge 1} \mathcal{F}_\infty^n), \quad \mathcal{G} \subseteq \sigma(\bigcup_{n \ge 1} \mathcal{F}_\infty^n)$$

as well as the conditions

(a) $\nu^n((0, t] \times \{|x| > a\}) \xrightarrow{P} 0$, $a \in (0, 1]$, $t \in S$,

(sup b) $\sup_{s \le t} |B_s^n| \xrightarrow{P} 0$, $t \in S$,

(c) $\langle M^{n, a} \rangle_t \xrightarrow{P} C_t$, $a \in (0, 1]$, $t \in S$,

(e) $X_0^n \xrightarrow{P} 0$,

then

$$X^n \xrightarrow{d} X \ (\mathcal{G}\text{-}stably).$$

Problems

1. Let $(\theta_t, \mathcal{F}_t)$ be the Markov process introduced in Example 2 and

$$X_t^n = \frac{1}{\sqrt{n}} \int_0^{nt} (\Theta_s - E\Theta_s)\, ds, \ n \ge 1.$$

Show that then

$$X^n \xrightarrow{d} aW,$$

where $a^2 = 2\lambda_1\lambda_2/(\lambda_1 + \lambda_2)^3$ and W is a Wiener process.

2. Let $\xi = (\xi_t, \mathcal{F}_t)$ be the stochastic process defined by Ito's stochastic equation (1.19) (see Example 3) and

$$X_t^n = \frac{1}{\sqrt{n}} \int_0^{nt} (\xi_s - E\xi_s)\, ds, \ n \ge 1.$$

Show that as $a > \frac{3}{2}$

$$X^n \xrightarrow{d} bW, \quad b^2 = \frac{2}{(2a - 1)\, a^3},$$

where W is a Wiener process.

3. Prove Theorem 15.

4. Formulate and prove vector-valued versions of Theorems 2 and 4 - 9.

§ 2. Weak convergence of distributions of semimartingales to distributions of point processes

1. In this section we consider functional versions of Theorems 5.3.1 and 5.3.2. Here the process $X = (X_t, \mathcal{F}_t)$ is assumed to be a semimartingale with $\mathcal{G}$-conditionally independent increments that is a point (counting) process with the compensator $A = (A_t)_{t \geq 0}$, besides $X_t < \infty$ (P-a.s.), $t \in R_+$.

Theorem 1. *Let* $X^n = (X^n_t, \mathcal{F}^n_t)$, $n \geq 1$, *be semimartingales that are counting processes with the compensators* $A^n = (A^n_t)_{t \geq 0}$, $n \geq 1$, *and let*

$$S = \{t > 0 \colon P(\Delta A_t = 0) = 1\}.$$

If the following conditions are fullfilled:

(o) $\mathcal{G} \subseteq \bigcap_{n \geq 1} \mathcal{F}^n_0$,

(a) $A^n_t \xrightarrow{P} A_t$, $t \in S$,

(a$_2$) $\sum_{0 < s \leq t} (\Delta A^n_s)^2 \xrightarrow{P} \sum_{0 < s \leq t} (\Delta A_s)^2$, $t \in S$,

then

$$X^n \xrightarrow{d} X \quad (\mathcal{G}\text{-}stably).$$

Remark. If $A \in \mathcal{V}^+ \cap C$, i.e. $\{t > 0 \colon P(\Delta A_t = 0) = 1\} \cup \{0\} = R_+$, then the assertion of Theorem 1 holds true under Conditions (o) and (a), as for S one may take any set dense in R_+.

Proof of Theorem 1. The convergence

$$X^n \xrightarrow{d_f(S)} X \quad (\mathcal{G}\text{-stably})$$

is established in Theorem 5.3.1 (see also Corollary 1 to this theorem). Therefore, by Theorem 6.1.8 it suffices to show that the family of distributions of counting processes X^n, $n \geq 1$, is relatively compact. To this end we verify the assumptions of Theorem 6.4.3.

Observe first that the representation for the semimartingale X^n has the following form:

$$X^n_t = B^n_t + \int_0^t \int_{\{1\}} x \, d(\mu^n - \nu^n),$$

with $B^n = A^n$ and $I(x = 1) * \nu^n = A^n$, besides $\nu^n(R_+ \times \{x \neq 1\}) = 0$ (see Ch. 3, § 4).

Assumption I is obviously fullfilled, assumption III' is fullfilled with $G^n = 2A^n$, since

$$\mathrm{Var}\,(B^n) + C^n + (x^2 \wedge 1) * \nu^n = 2A^n,$$

and assumption $IV'_{2)}$ is fullfilled with $G = 2A$ by Conditions (a) and (a_2) of the theorem.

2. Consider now the case in which $X^n = (X^n_t, \mathcal{F}^n_t)$, $n \geq 1$, are semimartingales (not necessarily presenting counting processes) with the triplets $T^n = (B^n, C^n, \nu^n)$, $n \geq 1$.

Theorem 2. *Let* $X^n = (X^n_t, \mathcal{F}^n_t)$ *be a semimartingale with the triplet* $T^n = (B^n, C^n, \nu^n)$ *of predictable characteristics,* $n \geq 1$, *and let* $0 < \delta < 1/2$,

$$S = \{t > 0 \colon \mathsf{P}\,(\Delta A_t = 0) = 1\}$$

and

$$\Delta_\delta = \{(|x| > \delta) \cap (|x - 1| > \delta)\}.$$

If the following conditions are fullfilled:

(o) $\mathcal{G} \subseteq \bigcap_{n \geq 1} \mathcal{F}^n_0$,

(a) $I(|x - 1| \leq \delta) * \nu^n_t \xrightarrow{\mathsf{P}} A_t$, $t \in S$,

(a_2) $\sum_{0 < s \leq t} (\nu^n(\{s\} \times \{|x - 1| \leq \delta\})^2 \xrightarrow{\mathsf{P}} \sum_{0 < s \leq t} (\Delta A_s)^2$, $t \in S$,

(sup b) $\sup_{s \leq t} |B^n_s - xI(\delta < |x| \leq 1) * \nu^n_s| \xrightarrow{\mathsf{P}} 0$, $t \in S$,

(c) $C^n_t + x^2 I(|x| \leq \delta) * \nu^n_t - \sum_{0 < s \leq t} \left(\int_{|x| \leq \delta} x\nu^n(\{s\}, dx) \right)^2 \xrightarrow{\mathsf{P}} 0$, $t \in S$,

(d) $\nu^n((0, t] \times \Delta_\delta) \xrightarrow{\mathsf{P}} 0$, $t \in S$,

(f) $X^n_0 \xrightarrow{\mathsf{P}} 0$,

then

$$X^n \xrightarrow{d} X \ (\mathcal{G}\text{-}stably).$$

Remark. If $A \in \mathcal{V}^+ \cap C$, then the assertion of the theorem remains true even if (a_2) is not assumed, and as for S one can take any set dense in R_+.

Proof of Theorem 2. Denote

$$Y_t^{n\delta} = I(|x - 1| \le \delta) * \mu_t^n.$$

Obviously, $Y^{nd} = (Y_t^{n\delta}, \mathcal{F}_t^n)$ is a counting process with the compensator

$$A_t^n = I(|x - 1| \le \delta) * \nu_t^n.$$

By Theorem 1 and Conditions (o), (a) and (a_2)

$$Y^{n\delta} \xrightarrow{d} X \ (\mathcal{G}\text{-stably}).$$

Therefore, in view of Problem 6.1.2, it suffices for the convergence $X^n \xrightarrow{d} X$ to show that for each $L > 0$

$$\sup_{t \le L} |X_t^n - Y_t^{n\delta}| \xrightarrow{P} 0. \tag{2.1}$$

In the course of proving Theorem 5.3.2 the inequality (see (3.25) in Ch. 5, § 3)

$$|X_t^n - Y_t^{n\delta}| \le |X_0^n| + |B_t^n - xI(\delta < |x| \le 1) * \nu_t^n| + |M_t^{n,\delta}| + I(x \in \Delta_\delta)|x| * \mu_t^n$$

has been established with

$$M_t^{n,\delta} = X_t^{nc} + \int_0^t \int_{|x| \le \delta} x\, d(\mu^n - \nu^n).$$

This gives

$$\sup_{t \le L} |X_t^n - Y_t^{n\delta}| \le |X_0^n| + \sup_{t \le L} |B_t^n - xI(\delta < |x| \le 1) * \nu_t^n|$$

$$+ \sup_{t \le L} |M_t^{n,\delta}| + I(x \in \Delta_\delta)|x| * \mu_L^n. \tag{2.2}$$

By utilizing this inequality we will show that the desired relation (2.1) holds.

By Condition (f) we have

$$|X_0^n| \xrightarrow{P} 0,$$

by Condition (sup b) the second term on the right-hand side of the inequality (2.2) tends to zero in probability as $n \to \infty$, and by Condition (d) the last term on the right-hand side of the inequality (2.2) tends to zero in probability as $n \to \infty$ (this fact has been established in the course of proving Theorem 5.3.2; see the implication (3.27) in Ch. 5, § 3). Therefore it suffices to show that

$$\sup_{t \le L} |M_t^{n,\delta}| \xrightarrow{P} 0, \quad n \to \infty.$$

The process $M^{n,\delta} = (M_t^{n,\delta}, \mathcal{F}_t^n)$ is a locally square integrable martingale with the quadratic characteristic

$$\langle M^{n,\delta}\rangle_t = C_t^n + x^2 I(|x| \le \delta) * \nu_t^n - \sum_{0 < s \le t} \left(\int_{|x| \le \delta} x\nu^n(\{s\}, dx) \right)^2$$

(see Theorem 3.5.1 and Lemma 3.5.1).

By Condition (c) we have

$$\langle M^{n,\delta}\rangle_L \xrightarrow{P} 0.$$

Therefore, in view of Problem 1.9.2, we have

$$\sup_{t \le L} |M_t^{n,\delta}| \xrightarrow{P} 0, \quad n \to \infty.$$

Thus the relation (2.1) is established and hence

$$X^n \xrightarrow{d} X.$$

Finally, the convergence $X^n \xrightarrow{d} X$ ($\mathcal{G}$-stably) follows from the convergence $X^n \xrightarrow{d} X$ and the convergence $X^n \xrightarrow{d_f(S)} X$ ($\mathcal{G}$-stably), proved already in Theorem 5.3.2.

Problem

1. Let $X^n = (X_t^n, \mathcal{F}_t^n)$, $n \ge 1$, and $X = (X_t, \mathcal{F}_t)$ be point (counting) processes ($X_t^n < \infty$, $X_t < \infty$ (P-a.s.), $t \in R_+$, $n \ge 1$) with the deterministic compensators A^n, $n \ge 1$ and A respectively.

Show that the implication

$$\rho(A^n, A) \to 0 \Rightarrow X^n \xrightarrow{d} X$$

holds, with the metric ρ introduced in Ch. 6, § 1.

§ 3. Weak convergence of distributions of semimartingales to the distribution of a left quasi-continuous semimartingale, with conditionally independent increments

1. As in Ch. 5, § 4, suppose that $X = (X_t, \mathcal{F}_t)$ is a semimartingale with $\mathcal{G}$-conditionally independent increments ($\mathcal{G} \subseteq \mathcal{F}_0$) and the triplet $T = (B, C, \nu)$ such that

$$\sum_{t > 0} \nu(\{t\}, \times R_0) = 0$$

(hence $\Delta B = 0$; see Corollary to Theorem 4.1.1).

Let $X^n = (X_t^n, \mathcal{F}_t^n)$, $n \geq 1$, in turn be semimartingales with the triplets T^n (B^n, C^n, ν^n), $n \geq 1$. To each semimartingale X^n relate a locally integrable martingale $M^{n,a} = (M_t^{n,a}, \mathcal{F}_t^n)$ with

$$M_t^{n,a} = X_t^{nc} + \int_0^t \int_{|x| \leq a} x\, d(\mu^n - \nu^n), \quad a \in (0, 1] \tag{3.1}$$

and the quadratic characteristic

$$\langle M^{n,a} \rangle_t = C_t^n + x^2 I(|x| \leq a) * \nu_t^n - \sum_{0 < s \leq t} \left(\int_{|x| \leq a} x \nu^n(\{s\}, dx) \right)^2. \tag{3.2}$$

Theorem 1. *Let* S *be a set dense in* R_+, *and let any of the following conditions be fullfilled:*

(o') $\mathcal{G} \subseteq \bigcap_{n \geq 1} \mathcal{F}_0^n$,

(o'') *the family of* σ*-algebras* $\mathbb{F}^n$, $n \geq 1$, *is nested, i.e. there exists a sequence of numbers* $(s_n)_{n \geq 1}$, $s_n \downarrow 0$, $n \to \infty$, *such that*

$$\mathcal{F}_{s_n}^n \subseteq \mathcal{F}_{s_{n+1}}^{n+1}, \quad n \geq 1, \qquad \sigma\left(\bigcup_{n \geq 1} \mathcal{F}_{s_n}^n \right) = \sigma\left(\bigcup_{n \geq 1} \mathcal{F}_\infty^n \right)$$

and

$$\mathcal{G} \subseteq \left(\bigcup_{n \geq 1} \mathcal{F}_\infty^n \right).$$

Let the following conditions be fullfilled:

(a) *for any continuous on* $R_0 \setminus (\{-1\} \cup \{1\})$ *and bounded function* $g = g(x)$ *with the property* $g(x) = I(|x| > \varepsilon)\, g(x)$ *for a certain* $\varepsilon > 0$ *the relation*

$$g * \nu_t^n \xrightarrow{P} g * \nu_t, \quad t \in S$$

holds;

(sup b) $\sup_{s \le t} |B^n_s - B_s| \xrightarrow{P} 0, \ t \in S;$

(c) $\lim_{a \to 0} \overline{\lim_n} \, P(|\langle M^{n,a}\rangle_t - C_t| \ge \varepsilon) = 0, \ \varepsilon > 0, \ t \in S;$

(e) $X^n_0 \xrightarrow{d} X_0$ ($\mathcal{G}$*-stably*).

Then

$$X^n \xrightarrow{d} X \ (\mathcal{G}\text{-}stably).$$

Proof. The convergence

$$X^n \xrightarrow{d_f(S)} X \ (\mathcal{G}\text{-stably})$$

is proved in Theorems 5.4.1 and 5.4.3 and Remarks to them. Therefore, it is necessary in view of Theorem 6.1.8 to establish the relative compactness of the family of semimartingales X^n, $n \ge 1$. To this end we will verify the assumptions of Theorem 6.4.1.

Assumption $I_{1)}$ is satisfied obviously by Condition (e).

Assumption $I_{2)}$ is valid by Condition (a). Indeed, if

$$g_l = g_l(x) = \left[0 \vee \left(\frac{2}{l} |x| - 1\right)\right] \wedge 1,$$

then

$$I(|x| > l) * \nu^n_L \le g_l * \nu^n_L,$$

and as $L \in S$ we get

$$P(I(|x| > l) * \nu^n_L \ge \varepsilon) \le P(g_l * \nu^n_L \ge \varepsilon)$$

$$\le P\left(g_l * \nu_L \ge \frac{\varepsilon}{2}\right) + P\left(|g_l * \nu_L - g_L * \nu^n_L| \ge \frac{\varepsilon}{2}\right).$$

Therefore by Condition (a) we have

$$\overline{\lim_n} \, P(|x| > l) * \nu^n_L \ge \varepsilon) \le P\left(g_l * \nu_L \ge \frac{\varepsilon}{2}\right),$$

and consequently assumption $I_{2)}$ is satisfied, since

$$g_l * \nu_L \xrightarrow{P} 0, \ l \to \infty.$$

Assumption $II_{1)}$ is satisfied with

$$\beta^{na} = B - xI(a < |x| \le 1) * \nu^n$$

by Condition (sup b), since

$$B^{na} = B^n - xI(a < |x| \le 1) * \nu^n$$

(see (4.2) in Ch. 6, § 4) and consequently

$$\sup_{s \le t} | B_s^{na} - \beta_s^{na} | = \sup_{s \le t} | B_s^n - B_s |.$$

Assumption $II_{2)}$ is satisfied with $\gamma^{na} = C$ by Condition (c) and Problem 1; assumption $II_{3)}$ is satisfied with $\delta^{ng} = g * \nu^n$. Obviously, assumption III is satisfied with $\overline{G}^n \equiv 0$ and

$$G^n = \mathrm{Var}\,(B) + C + g * \nu^n + g^a * \nu^n,$$

where $g^a = g^a(x)$ is the function involved in Condition (a), with the property

$$g^a(x) \ge | x | I(a < | x | \le 1).$$

Defining G^n in this way we get by Condition (a) that assumption $IV_{3)}$ is satisfied with

$$G = \mathrm{Var}\,(B) + C + g * \nu + g^a * \nu.$$

2. Consider now the case of semimartingales $\hat{X}^n = (\hat{X}_t^n, \hat{\mathcal{F}}_t^n)$, $n \ge 1$, with $\hat{X}_t^n = X^n_{\hat{\tau}_n(t)}$ and $\hat{\mathcal{F}}_t^n = \mathcal{F}^n_{\hat{\tau}_n(t)}$, where $\hat{\tau}_n = (\hat{\tau}_n(t))_{t \ge 0} \in V^+$ with $\hat{\tau}_n(t) \in (\mathbb{F}^n)$, $n \ge 1$, for every $t \in R_+$, while $X^n = (X_t^n, \mathcal{F}_t^n)$ is a semimartingale with the triplet $T^n = (B^n, C^n, \nu^n)$, $n \ge 1$. Applying Theorems 5.4.2 and 6.4.2 it is not hard to establish the following result.

Theorem 2. *Let S be a subset dense in* R_+.

If the following conditions are fullfilled:

(o) $\mathcal{G} \subseteq \bigcap_{n \ge 1} \mathcal{F}_0^n$,

(a) *for each continuous on* $R_0 \setminus (\{-1\} \cup \{1\})$ *and bounded function* $g = g(x)$ *with the property* $g(x) = I(|x| > \varepsilon)\, g(x)$ *for a certain* $\varepsilon > 0$, *the relation*

$$g * \nu^n_{\hat{\tau}_n(t)} \xrightarrow{P} g * \nu_t, \quad t \in S$$

holds,

(sup b) $\sup_{s \le t} | B^n_{\hat{\tau}_n(s)} - B_s | \xrightarrow{P} 0, \ t \in S,$

(c) $\lim_{a \to \infty} \overline{\lim_n}\, P(| \langle M^{n,a} \rangle_{\hat{\tau}_n(t)} - C_t | \ge \varepsilon) = 0, \ \varepsilon > 0, \ t \in S,$

(e) $X_0^n \xrightarrow{d} X_0$ ($\mathcal{G}$-*stably*),

then

$$\hat{X}^n \xrightarrow{d} X \;((\mathcal{G}\text{-}stably)).$$

Problems

1. Let $(a_k)_{k \geq 1}$ be a sequence of numbers and for every $k \geq 1$ let $A^n(a_k) = (A^n_t(a_k))_{t \geq 0}$ be a stochastic process with trajectories in V^+. Show that under the condition

$$\lim_{k \to \infty} \overline{\lim_{n}} \, P(|A^n_t(a_k) - A_t| \geq \varepsilon) = 0$$

for each $\varepsilon > 0$ and t in a set, dense in R_+, where $A = (A_t)_{t \geq 0}$ is a stochastic process with trajectories in $V^+ \cap C$, the relation (for each $\varepsilon > 0$)

$$\lim_{k \to \infty} \overline{\lim_{n}} \, P(\sup_{s \leq t} |A^n_s(a_k) - A_s| \geq \varepsilon) = 0, \quad t \in R_+$$

holds.

2. Let $X^n = (X^n_t, \mathcal{F}^n_t)$, $n \geq 1$, be semimartingales with the triplets $T^n = (B^n, C^n, \nu^n)$, $n \geq 1$, and the following conditions are fullfilled:

(a) for each continuous on $R_0 \setminus (\{-1\} \cup \{1\})$ and bounded function $g = g(x)$ with the property $g(x) = I(|x| > \varepsilon)\, g(x)$ for a certain $\varepsilon > 0$, the relation

$$g * \nu^n_t \xrightarrow{P} 0, \quad t \in R_+$$

holds,

(sup b) $\sup_{s \leq t} |B^n_s| \xrightarrow{P} 0, \quad t \in R_+,$

(c) $\lim_{a \to \infty} \overline{\lim_{n}} \, (\langle M^{n,a} \rangle_t \geq \varepsilon) = 0, \quad \varepsilon > 0, \quad t \in R_+,$

(e) $X^n_0 \xrightarrow{P} 0.$

Show that then

$$\sup_{s \leq t} |X^n_s| \xrightarrow{P} 0$$

for each $t \in R_+$.

3. Prove Theorem 2.

CHAPTER 8

WEAK CONVERGENCE OF DISTRIBUTIONS OF SEMIMARTINGALES TO THE DISTRIBUTION OF A SEMIMARTINGALE

§ 1. Convergence of stochastic exponentials and weak convergence of distributions of semimartingales

1. In the present chapter we study the question concerning the weak convergence

$$Q^n \xrightarrow{w} Q$$

of distributions Q^n of semimartingales X^n to the distribution Q of a semimartingale X as $n \to \infty$ in a sufficiently wide setting of the problem. Observe that since we avoid here special assumptions concerning the 'limiting" process X, such as, say, that it is a process with conditionally independent increments, the procedure leading in Ch. 7 to the weak convergence:

$$\left\{\begin{array}{l} Q^n \xrightarrow{w_f(S)} Q, \ \text{S is dense in } R_+ \\ \text{and a family } (Q^n)_{n\geq 1} \text{ is relatively compact} \end{array}\right\} \Rightarrow Q^n \xrightarrow{w} Q,$$

becomes, generally speaking, inapplicable, because in the general case the method of stochastic exponentials does not allow us to verify the weak convergence

$$Q^n \xrightarrow{w_f(S)} Q.$$

The method used below for establishing the weak convergence

$$Q^n \xrightarrow{w} Q$$

avoids the question on the weak convergence of finite dimensional distributions — it departs from the initial assumption that the family $(Q^n)_{n\geq 1}$ is relatively compact.

Let then $(Q^{n'})_{n'\geq 1}$ be an arbitrary weakly convergent subsequence of the sequence $(Q^n)_{n\geq 1}$, and Q' its weak limit. The conditions imposed on the triplets T^n and T of processes X^n and X are such that they allow us to show (Theorem 1) the coincidence of the triplet of the process X relative to the measure Q' with the triplet T of this process relative to the measure Q. This stage of the proof is accomplished by utilizing properties of the stochastic exponentials and characterizing Theorem 4.3.2. Next, under the assumption that the triplet defines uniquely the probability distribution of a semimartingale X, we establish the desired fact of the convergence

$$Q^n \xrightarrow{w} Q$$

(Theorem 2).

The question whether the family of measures $(Q^n)_{n \geq 1}$, corresponding to semimartingales is relatively compact, has been studied in Ch. 6, § 4 (see also §§ 2 - 4 below).

2. Let us turn to the precise description of the objects considered.

Suppose that the "limiting" semimartingale $X = (X_t, \mathcal{D}_t^Q)$ is given on a stochastic basis $(D, \mathcal{D}^Q, \mathbb{D}^Q = (\mathcal{D}_t^Q)_{t \geq 0}, Q)$ (see Ch. 6, Subsection 5.2).

Also, let

$$(\Omega^n, \mathcal{F}^n, \mathbb{F}^n = (\mathcal{F}_t^n)_{t \geq 0}, P^n)_{n \geq 1}$$

be a sequence of stochastic basises, $X^n = (X_t^n, \mathcal{F}_t^n)$ semimartingales with the triplets $T^n = (B^n, C^n, \nu^n)$ and Q^n, $n \geq 1$, their probability distributions in the space $(\mathbb{D}, \mathcal{D})$.

Denote by $T = (B(X), C(X), \nu(X))$ the triplet of a semimartingale X, satisfying throughout this chapter the following assumptions:

$$B(X) \in \mathcal{V} \cap \mathcal{P}(\mathbb{D}^Q), \;\; C(X) \in \mathcal{V}^+ \cap C$$

and $\nu(X) = \nu(X; dt, dx)$ is a σ-finite measure on $(R_+ \times R_0, B(R_+) \otimes B(R_0))$ with the properties:

$$\nu(X; \{0\} \times R_0) = 0,$$

$$\int_0^t \int_{R_0} (1 \wedge x^2)\, \nu(X; ds, dx) < \infty, \;\; t > 0,$$

$$\nu(X; \{t\} \times R_0) = \Delta B_t(X)$$

and for each $B(R_0)$-measurable function $f = f(x)$ with

$$|f(x)| \leq c(1 \wedge x^2)$$

the function

$$\int_0^t \int_{R_0} f(x)\, \nu(X; ds, dx)$$

is $\mathcal{P}(\mathbb{D}^Q)$-measurable.

Let $\mathcal{E}(G(X; \lambda))$ and $\mathcal{E}(G^n(\lambda))$ be the stochastic exponentials (see Ch. 4, § 2), associated with the cumulants

$$G_t(X;\lambda) = i\lambda B_t(X) - \frac{\lambda^2}{2} C_t(X) + \int_0^t \int_{R_0} (e^{i\lambda x} - 1 - i\lambda x I(|x| \le 1))\, \nu(X; ds, dx),$$
$$G_t^n(\lambda) = i\lambda B_t^n - \frac{\lambda^2}{2} C_t^n + \int_0^t \int_{R_0} (e^{i\lambda x} - 1 - i\lambda x I(|x| \le 1))\, \nu^n(ds, dx) \qquad (1.1)$$

of processes X and X^n respectively, $n \ge 1$.

Theorem 1. *Let the following conditions be fullfilled:*

(a) *the family* $(Q^n)_{n \ge 1}$ *is relatively compact;*

(b) *there exists a sequence* $(\tau_k(X))_{k \ge 1}$ *of* $\mathbb{D}^Q$*-Markov times and a sequence of constants* $(c_k(\lambda))_{k \ge 1}$*, depending on* $\lambda \in R$*, such that*

1) $\tau_k(X) \uparrow \infty$, $k \to \infty$, $X \in D$;

2) $|\mathcal{E}_{\tau_k(X)}(G(X;\lambda))| \ge c_k(\lambda) > 0$, $k \ge 1$, $\lambda \in R$, $X \in D$;

3) $\tau_k(X^n)$ *is a* $\mathbb{F}^n$*-Markov time,* $k \ge 1$;

4) *if* Q' *is a weak limit of a converging subsequence* $(Q^{n'})_{n' \ge 1}$ *of the sequence* $(Q^n)_{n \ge 1}$, *then for every* $k \ge 1$ *and any* $\lambda \in R$ *and* t *in a subset* $\Delta_{Q'}$*, dense in* R_+*, the functions*

$$X_{t \wedge \tau_k(X)}$$

and

$$\mathcal{E}_{t \wedge \tau_k(X)}(G(X;\lambda))$$

are continuous (Q'*-a.s.*) *in the metric* ρ *(see Ch. 6) and for each* $\varepsilon > 0$ *and* $\lambda \in R$

$$\overline{\lim_n}\, P^n(|\mathcal{E}_{t \wedge \tau_k(X^n)}(G^n(\lambda)) - \mathcal{E}_{t \wedge \tau_k(X^n)}(G(X^n;\lambda))| \ge \varepsilon) = 0. \quad (1.2)$$

Then the process $X = (X_t, \mathcal{D}_t^{Q'})$ *is a semimartingale relative to the measure* Q' *and its triplet* T' *coincides with the triplet* $T = (B(X), C(X), \nu(X))$.

Proof. Define the process

$$Z(X) = (Z_t(X))_{t \ge 0}$$

with

$$Z_t(X) = \exp(i\lambda(X_t - X_0))\, \mathcal{E}_t^{-1}(G(X;\lambda)). \qquad (1.3)$$

In view of Theorem 4.3.2 the desired assertion takes place, provided it is shown that $(Z_t(X), \mathcal{D}_t^{Q'})$ is a local martingale relative to the measure Q'.

As for a localizing sequence for $(Z_t(X), \mathcal{D}_t^{Q'})$ we can take the sequence $(\tau_k(X))_{k\geq 1}$, since by assumption (b_2) we have

$$|Z_{t\wedge\tau_k(X)}(X)| \leq c_k^{-1}(\lambda) < \infty.$$

Hence for any s and t with s < t and any $\mathcal{D}_s^{Q'}$-measurable random variable $\phi(X)$ with $|\phi(X)| \leq 1$ we need the equality

$$\int_D Z_{t\wedge\tau_k(X)}(X)\,\phi(X)\,dQ' = \int_D Z_{s\wedge\tau_k(X)}(X)\,\phi(X)\,dQ' \tag{1.4}$$

for each $k \geq 1$.

In fact it suffices to verify this equality only for s and t in a set, dense in R_+. Indeed, if it holds for such s and t, then it holds for all s and t in R_+ as well, since by the right-continuity in t of the function $Z_{t\wedge\tau_k(X)}(X)$ (for every $k \geq 1$ and $X \in D$)) and its uniform boundedness we have

$$\int_D X_{t_j\wedge\tau_k(X)}(X)\,\phi(X)\,dQ' \to \int_D Z_{t\wedge\tau_k(X)}(X)\,\phi(X)\,dQ',\ j\to\infty,$$

where $(t_j)_{j\geq 1}$ is a sequence in a set

$$\Delta_{Q'} = \{t > 0: Q'(\Delta X_t = 0) = 1\} \cup \{0\}, \tag{1.5}$$

dense in R_+, such that $t_j \downarrow t,\ j \to 0$.

Next, it suffices to verify the equality (1.14) for functions $\phi(X)$ of type

$$\phi(X) = \psi(X_{s_1}, \ldots, X_{s_l}) \tag{1.6}$$

only, where $l \geq 1,\ 0 \leq s_1 < s_2 < \ldots < s_l \leq s$ with $s_j \in \Delta_{Q'}$ and $\psi = \psi(x_1, \ldots, x_l)$ is a function continuous in all variables jointly.

Observe also that by Example 1 in Ch. 6, § 2, a function $\psi(x_1, \ldots, x_l)$, continuous jointly in all variables, possesses the following property: $\psi(X_{s_1}, \ldots, X_{s_l})$ is a continuous function in the metric ρ (Q'-a.s.), if $s_j \in \Delta_{Q'}\ j = 1$.

Thus it is assumed that $\psi(x_1, \ldots, x_l)$ is a continuous function in all variables, $s_j \in \Delta_{Q'},\ j = 1, \ldots, l,\ s, t \in \Delta_{Q'}$.

Then, in view of Conditions (b_2) and (b_4) and the definition of Z (X) the function

$$Z_{t\wedge\tau_k(X)}(X)\,\phi(X)$$

with $\phi(X) = \psi(X_{s_1}, \ldots, X_{s_l})$ is bounded and continuous in the metric ρ (Q'-a.s.). Consequently,

$$\int_D Z_{t\wedge\tau_k(X)}(X)\,\phi(X)\,dQ' = \lim_{n'} \int_D Z_{t\wedge\tau_k(X)}(X)\,\phi(X)\,dQ^{n'}$$

$$= \lim_{n'} \mathbf{E}^{n'} Z_{t\wedge\tau_k(X^{n'})}(X^{n'})\,\phi(X^{n'}). \qquad (1.7)$$

The analogous relation takes place with s instead of t.

Suppose for time being that there exists for every n and k a uniformly integrable (relative to the measure P^n) martingale $(Z_t^{n,k}, \mathfrak{F}_t^n)$ such that for each $t \in \Delta_{Q'}$

$$\lim_n \mathbf{E}^n \,|\, Z_t^{n,k} - Z_{t\wedge\tau_k(X^n)}(X^n) \,| = 0. \qquad (1.8)$$

We will show that under this assumption the desired equality (1.4) takes place.

Indeed, in view of $|\phi| \le 1$, by (1.7) and (1.8) we get

$$\left| \int_D Z_{t\wedge\tau_k(X)}(X)\,\phi(X)\,dQ' - \int_D Z_{s\wedge\tau_k(X)}(X)\,\phi(X)\,dQ' \right|$$

$$= \left| \lim_{n'} \mathbf{E}^{n'} [(Z_{t\wedge\tau_k(X^{n'})}(X^{n'}) - Z_{s\wedge\tau_k(X^{n'})}(X^{n'}))\,\phi(X^{n'})] \right|$$

$$\le \left| \lim_{n'} \mathbf{E}^{n'} [(Z_t^{n',k} - Z_s^{n',k})\,\phi(X^{n'})] \right| + \lim_{n'} \mathbf{E}^{n'} \left| Z_t^{n',k} - Z_{t\wedge\tau_k(X^{n'})}(X^{n'}) \right|$$

$$+ \lim_{n'} \mathbf{E}^{n'} \left| Z_s^{n',k} - Z_{s\wedge\tau_k(X^{n'})}(X^{n'}) \right| = 0.$$

Hence the proof of the theorem is completed by constructing uniformly integrable martingales $(Z_t^{n,k}, \mathfrak{F}_t^n)$ and by verifying the relation (1.8).

Aiming at this, we introduce for every $k \ge 1$ a stopping time

$$\tilde{T}^{n,k} = \inf\left(t: \left| \mathfrak{C}_{t\wedge\tau_k(X^n)}(G^n(\lambda)) \right| \le \frac{1}{2} c_k(\lambda)) \right) \wedge n$$

(relative to $\mathbb{F}^n$) with constants $c_k(\lambda)$ involved in Condition (b_2). A stopping time $\tilde{T}^{n,k}$ is $\mathbb{F}^n$-predictable, since $(\,|\,\mathfrak{C}_t(G^n(\lambda))\,|\,)_{t\ge 0}$ is a $\mathbb{F}^n$-predictable process with trajectories in D (Problem 1.3.11). Then by Theorem 1.3.4 there can be found a stopping time $T^{n,k}$ with the properties $T^{n,k} < \tilde{T}^{n,k}$ and

$$P^n\left(\tilde{T}^{n,k} - T^{n,k} > \frac{1}{n}\right) \leq \frac{1}{n}. \tag{1.9}$$

Define a stochastic process $Z^{n,k} = (Z_t^{n,k})_{t \geq 0}$, by setting

$$Z_t^{n,k} = \exp(i\lambda(X^n_{t\wedge\tau_k(X^n)\wedge T^{n,k}} - X_0^n))\,\mathcal{E}^{-1}_{t\wedge\tau_k(X^n)\wedge T^{n,k}}(G^n(\lambda)).$$

The definition of the stopping time $T^{n,k}$ entails

$$|Z^{n,k}| \leq 2/c_k(\lambda).$$

On the other hand by Condition (b_3) of Theorem 4.3.1 $(Z_t^{n,k}, \mathcal{F}_t^n)$ is a local martingale. Hence, $(Z_t^{n,k}, \mathcal{F}_t^n)$ is a uniformly integrable martingale.

We verify now the relation (1.8).

By Condition (b_2) and the estimate $|Z_t^{n,k}| \leq 2/c_k(\lambda)$ we have

$$|Z_t^{n,k} - Z_{t\wedge\tau_k(X^n)}(X^n)| \leq 3/c_k(\lambda).$$

Therefore for each $\varepsilon \in (0, 3/c_k(\lambda))$ we have

$$E^n|Z_t^{n,k} - Z_{t\wedge\tau_k(X^n)}(X^n)| \leq \varepsilon + \frac{3}{c_k(\lambda)} P^n(|Z_t^{n,k} - Z_{t\wedge\tau_k(X^n)}(X^n)| \geq \varepsilon).$$

Consequently, taking in this inequality the limit $\lim_{\varepsilon\to 0}\overline{\lim_n}$ we get the desired relation (1.8), provided

$$\lim_n P^n(|Z_t^{n,k} - Z_{t\wedge\tau_k(X^n)}(X^n)| \geq \varepsilon) = 0 \tag{1.10}$$

for each $\varepsilon > 0$

To establish (1.10), observe that

$$I(t \leq T^{n,k})\,|Z_t^{n,k} - Z_{t\wedge\tau_k(X^n)}(X^n)|$$

$$\leq I(t \leq T^{n,k})\,|\mathcal{E}^{-1}_{t\wedge\tau_k(X^n)}(G^n(\lambda)) - \mathcal{E}^{-1}_{t\wedge\tau_k(X^n)}(G(X^n;\lambda))|$$

$$\leq \frac{2}{c_k^2(\lambda)}|\mathcal{E}_{t\wedge\tau_k(X^n)}(G^n(\lambda)) - \mathcal{E}_{t\wedge\tau_k(X^n)}(G^n(X^n;\lambda))|.$$

Therefore, in view of Condition (b_4), the desired relation (1.10) takes place, provided

$$\lim_n P^n(T^{n,k} < t) = 0. \tag{1.11}$$

To prove (1.11) we apply the obvious relation:

$$P^n(T^{n,k} < t) \le P^n\left(T^{n,k} < t, \tilde{T}^{n,k} - T^{n,k} \le \frac{1}{n}\right) + P^n\left(\tilde{T}^{n,k} - T^{n,k} > \frac{1}{n}\right)$$

$$\le P^n\left(\tilde{T}^{n,k} < t + \frac{1}{n}\right) + \frac{1}{n}$$

$$\le P^n(\tilde{T}^{n,k} < t + c) + \frac{1}{n},$$

with a constant $c > 1$ chosen in such a way that $t + c \in \Delta_{Q'}$.

Thus instead of (1.11) it suffices to establish

$$\lim_n P^n(\tilde{T}^{n,k} < t) = 0 \tag{1.12}$$

for each $t \in \Delta_{Q'}$.

By the definition of the stopping time $\tilde{T}^{n,k}$ and by the fact that

$$|\mathcal{E}_{t \wedge \tau_k(X^n)}(G^n(\lambda))|$$

is a nonincreasing function of t (Lemma 4.2.1) we get

$$\{\tilde{T}^{n,k} < t\} \subseteq \left\{ |\mathcal{E}_{t \wedge \tau_k(X^n)}(G^n(\lambda))| \le \frac{1}{2} c_k(\lambda) \right\}.$$

On the other hand, by Condition (b_2) we have

$$P^n(|\mathcal{E}_{t \wedge \tau_k(X^n)}(G(X^n, \lambda))| \ge c_k(\lambda)) = 1.$$

Consequently, due to the inequality

$$||a| - |b|| \le |a - b|$$

we have

$$P^n(\tilde{T}^{n,k} < t)$$

$$\le P^n\left(|\mathcal{E}_{t \wedge \tau_k(X^n)}(G^n(\lambda))| \le \frac{1}{2} c_k(\lambda), |\mathcal{E}_{t \wedge \tau_k(X^n)}(G(X^n; \lambda))| \ge c_k(\lambda) \right)$$

$$\le P^n\left(|\mathcal{E}_{t \wedge \tau_k(X^n)}(G^n(\lambda)) - \mathcal{E}_{t \wedge \tau_k(X^n)}(G(X^n; \lambda))| \ge \frac{1}{2} c_k(\lambda) \right),$$

i.e. (1.12) is a consequence of Condition (b_4).

The theorem is proved.

3. Let a process $X = (X_t)_{t \ge 0}$ belong to $S(\mathbb{D}^Q, Q)$ and have the triplet of predictable characteristics $T = (B(X), C(X), \nu(X))$.

We will say that the initial distribution Q_0 of a random variable X_0 and the triplet T *determine the measure* Q *uniquely* in the sense that if $\tilde{Q}$ is a probability measure on $(D, \mathcal{D}^{\tilde{Q}})$ with $\tilde{Q}_0 = Q_0$ and $X = (X_t)_{t \geq 0}$ is a semimartingale relative to $(\mathbb{D}^{\tilde{Q}}, \tilde{Q})$ with just the same triplet T, then $\tilde{Q} = Q$.

Theorem 2. *Let the conditions of Theorem* 1 *be fullfilled and*

$$Q_0^n \overset{w}{\to} Q_0$$

(Q_0^n *and* Q_0 *are the restrictions of the measures* Q^n *and* Q *to* $\mathcal{D}_0$). *Also, let the measure* Q_0 *and the triplet* T = (B (X), C (X), ν (X)) *determine uniquely the distribution of a semimartingale* $(X_t, \mathcal{D}_t^Q)$.

Then

$$Q^n \overset{w}{\to} Q.$$

Proof. Since the conditions of Theorem 1 are fullfilled, there exists a weakly convergent subsequence $(Q^{n'})_{n' \geq 1}$ of the sequence $(Q^n)_{n \geq 1}$ with a weak limit Q', such that in virtue of Theorem 1 a process $(X_t, \mathcal{D}_t^{Q'})$ is a semimartingale relative to a measure Q' with the triplet T = (B (X), C (X), ν (X)).

We have required that

$$Q_0^{n'} \overset{w}{\to} Q_0$$

and that the triplet T, together with Q_0 determine uniquely the distribution of the semimartingale $(X_t, \mathcal{D}_t^Q)$, i.e. Q' = Q. In other words, for each bounded function f = f (X), continuous in the metric ρ (see Ch. 6)

$$\lim_{n'} \int_D f(X)\, dQ^{n'} = \int_D f(X)\, dQ.$$

We will show that in fact

$$\lim_{n} \int_D f(X)\, dQ^{n} = \int_D f(X)\, dQ. \tag{1.13}$$

Let

$$c = \overline{\lim_{n}} \int_D f(X)\, dQ^n$$

and let (n'') be a subsequence of the sequence (n) such that

$$\lim_{n''} \int_D f(X)\, dQ^{n''} = c.$$

By the assumed relative compactness of the family $(Q^n)_{n \geq 1}$ one may choose a weakly convergent subsequence $(Q^{\tilde{n}})$ of a subsequence $(Q^{n''})_{n'' \geq 1}$ with a weak limit $\tilde{Q}$. In view of Theorem 1 the process $(X_t, \mathcal{D}_t^{\tilde{Q}})$ is a semimartingale relative to the measure $\tilde{Q}$ with the triplet $T = (B(X),\ C(X),\ \nu(X))$ and by the condition of Theorem 2 we have

$$\tilde{Q} = Q.$$

Hence,

$$c = \lim_{n''} \int_D f(X)\, dQ^{n''} = \lim_{\tilde{n}} \int_D f(X)\, dQ^{\tilde{n}} = \int_D f(X)\, dQ,$$

i.e.

$$\overline{\lim_{n}} \int_D f(X)\, dQ^n = \int_D f(X)\, dQ.$$

It is established analogously that

$$\underline{\lim_{n}} \int_D f(X)\, dQ^n = \int_D f(X)\, dQ,$$

and hence the desired relation (1.13) holds.

Theorem 2 is proved.

Problem

1. Show that Condition (1.2) follows from the following condition (for each t in a set $\Delta_{Q'}$, dense in R_+):

$$\lim_n P^n\left(\sup_{s \leq t} |\mathcal{E}_s(G^n(\lambda)) - \mathcal{E}_s(G(X^n, \lambda))| \geq \varepsilon \right) = 0, \ \varepsilon > 0.$$

§ 2. Weak convergence to the distribution of a left quasi-continuous semimartingale

1. Suppose that the triplet $T = (B(X),\ C(X),\ \nu(X))$ of a Q-semimartingale

$$X = (X_t, \mathcal{D}_t^Q)$$

is such that

$$B_t(X) = \int_0^t b(s, X)\, du_s,\quad t \in R_+,\quad X \in D, \tag{2.1}$$

and

$$C_t(X) = \int_0^t c(s, X)\, du_s,\quad t \in R_+,\quad X \in D, \tag{2.2}$$

where $u = (u_t)_{t \geq 0}$ is a function in $V^+ \cap C$, and $b(t, X)$ and $c(t, X)$ are $\mathcal{P}(\mathbb{D}^Q)$-measurable functions, $c(t, X) \geq 0$, $t \in R_+$, $X \in D$, besides $|b| \circ u$, and $c \circ u \in V^+$.

We assume

$$\nu(X; dt, dx) = K(X, t; dx)\, du_t, \tag{2.3}$$

where for each $X \in D$ and $t \in R_+$ the function $K(X, t; dx)$ is a finite measure on $(R_0, B(R_0))$ and $K(X, t; A)$ is a nonnegative $\mathcal{P}(\mathbb{D}^Q)$-measurable function for each $A \in B(R_0)$, besides

$$\int_0^t \int_{R_0} (1 \wedge x^2)\, K(X, s; dx)\, du_s < \infty,\quad X \in D,\quad t \in R_+.$$

A Q-semimartingale $X = (X_t, \mathcal{D}_t^O)$ with the triplet $T = (B(X),\ C(X),\ \nu(X))$ is a left quasi-continuous process (Theorem 4.1.1). In accordance with the definition of the left quasi-continuity (see Definition 4 in Ch. 1, § 3) we have here that

$$Q\Big(X: \sup_{\tau \in T_p(\mathbb{D}^Q)} |\Delta X_\tau| = 0\Big) = 1.$$

The stochastic exponential $\mathcal{E}(G(X, \lambda))$, related to the triplet

$$T = (B(X),\ C(X),\ \nu(X)),$$

is given by the formula (see Ch. 4, § 2)

$$\mathcal{E}_t (G (X; \lambda)) = \exp (G_t (X; \lambda)),$$

$$G_t (X; \lambda) = i\lambda \int_0^t b (s, X)\, du_s - \frac{\lambda^2}{2} \int_0^t c (s, X)\, du_s$$

$$+ \int_0^t \int_{R_0} (e^{i\lambda x} - 1 - i\lambda x I (|x| \le 1)) K (X, s; dx)\, du_s. \qquad (2.4)$$

Along with (2.1) - (2.4), we assume the following properties of $b (t, X)$, $c (t, X)$ and $K (X, t; dx)$:

(T_1) for every t in a set S, dense in R_+, such that

$$\int_{R_+ \setminus S} du_s = 0,$$

the functions

$$b (t, X), \ c (t, X) \text{ and } a^g (t, X) = \int_{R_0} g (x) K (X, t; dx)$$

($g = g (x)$ is a continuous function on $R_0 \setminus (\{-1\} \cup \{1\})$ such that $0 \le g (x) \le c (1 \wedge x^2)$ with a constant c) are continuous in the metric ρ (see Ch. 6).

(T_2)
$$| b (t, X) | \le L (t) (1 + \sup_{s < t} | X_s |),$$

$$c (t, X) \le L (t) (1 + \sup_{s < t} (X_s)^2),$$

$$a^g (t, X) \le L^g (t) (1 + \sup_{s < t} | X_s |),$$

where $g = g (x)$ is a function involved in Condition (T_1) and

$$\int_0^t [L (s) + L^g (s)]\, ds < \infty, \ t > 0.$$

2. Let $X^n = (X_t^n, \mathcal{F}_t^n)$ be a P^n-semimartingale with the triplet $T^n = (B^n, C^n, \nu^n)$ of predictable characteristics, and $M^{n, a} = (M_t^{n, a}, \mathcal{F}_t^n)$ a locally integrable martingale with

$$M_t^{n, a} = X_t^{nc} + \int_0^t \int_{|x| \le a} x\, d (\mu^n - \nu^n), \ a \in (0, 1]$$

and the quadratic characteristic

$$\langle M^{n,a}\rangle_t = C_t^n + \int_0^t \int_{|x|\le a} x^2 dv^n - \sum_{0<s\le t}\left(\int_{|x|\le a} xv^n(\{s\}, dx)\right)^2 .$$

We assume the following conditions concerning the triplets T^n, $n \ge 1$ (for each $L > 0$ and $\varepsilon > 0$):

(U_1)
$$\lim_{l\to\infty} \overline{\lim_n}\, P^n\left(\int_0^L \int_{|x|>l} dv^n \ge \varepsilon\right) = 0,$$

$$\lim_{\delta\to 0} \overline{\lim_n}\, P^n\left(\int_0^L \int_{|x|\le\delta} x^2 K(X^n, s; dx)\, du_s \ge \varepsilon\right) = 0;$$

(U_2) if $g = g(x)$ is a nonnegative, bounded and continuous function on

$$R_0 \setminus (\{-1\} \cup \{1\})$$

with the property

$$g(x) = I(|x| > \delta)\, g(x)$$

for a certain $\delta > 0$ and if

$$a^g(t, X) = \int_{R_0} g(x) K(X, t; dx),$$

then

$$\lim_n P^n\left(\sup_{t\le L} \left| g * v_t^n - \int_0^t a^g(s, X^n)\, du_s\right| \ge \varepsilon\right) = 0;$$

(U_3)
$$\lim_n P^n\left(\sup_{t\le L} \left| B_t^n - \int_0^t b(s, X^n)\, du_s\right| \ge \varepsilon\right) = 0;$$

(U_4)
$$\lim_{a\to 0} \overline{\lim_n}\, P^n\left(\sup_{t\le L} \left| \langle M^{n,a}\rangle_t - \int_0^t c(s, X^n)\, du_s\right| \ge \varepsilon\right) = 0.$$

Remark. 1) The implication

$$\lim_{l\to\infty} \overline{\lim_n}\, P^n\left(\int_0^L \int_{|x|>l} dv^n \ge \varepsilon\right) = 0 \Leftrightarrow \lim_{l\to\infty} \overline{\lim_n}\, P^n\left(\sup_{t\le L} |\Delta X_t^n| \ge l\right) = 0, \tag{2.5}$$

takes place, i.e. the first Condition (U_1) is equivalent to the condition

$$\lim_{l \to \infty} \overline{\lim_{n}} \; P^n \left(\sup_{t \le L} |\Delta X_t^n| \ge l \right) = 0 \tag{2.6}$$

(Problem 1).

2) If for each $X \in D$ and for the function g involved in Condition (U_2) we have

$$a^g (t, X) \le L^g (t)$$

and

$$\int_0^t L^g (s) \, du_s < \infty, \quad t > 0,$$

then Condition (U_2) is equivalent to the condition (for each $L > 0$ and $\varepsilon > 0$)

$$\lim_{n} P^n \left(\left| g * \nu_L^n - \int_0^L a^g (s, X^n) \, du_s, \right| \ge \varepsilon \right) = 0 \tag{2.7}$$

(Problem 2).

3) If for each $X \in D$ we have

$$c (t, X) \le L (t)$$

and

$$\int_0^t L (s) \, du_s < \infty, \quad t > 0,$$

then Condition (U_4) is equivalent to the condition (for each $L > 0$, $\varepsilon > 0$)

$$\lim_{a \to 0} \overline{\lim_{n}} \; P^n \left(\left| \langle M^{n, a} \rangle_L - \int_0^L c (s, X^n) \, du_s \right| \ge \varepsilon \right) = 0 \tag{2.8}$$

(Problem 2).

3. Theorem 1. *Let* Q^n *be the probability distribution of a* P^n*-semimartingale* $X^n = (X_t^n, \mathcal{F}_t^n)$ *with the triplet of predictable characteristics* $T^n = (B^n, C^n, \nu^n)$, $n \ge 1$, *and* Q *the probability distribution of a semimartingale* $X = (X_t, \mathcal{D}_t^Q)$ *with the triplet* $T = (B (X), \; C (X), \; \nu (X))$ *possessing the properties* (2.1) - (2.3).

(a) *If Conditions* (T_2) (*on the triplet* $T = (B (X), \; C (X), \; \nu (X))$ *and* (U_1) - (U_4) (*on the triplets* $T^n = (B^n, C^n, \nu^n)$, $n \ge 1$) *are fullfilled as well as the condition*

$$\lim_{l \to \infty} \overline{\lim_{n}} \; P^n (|X_0^n| \ge l) = 0, \tag{2.9}$$

then the family $(Q^n)_{n \ge 1}$ *is relatively compact.*

(b) *If the assumptions of assertion* (a) *are fullfilled, as well as Condition* (T_1) *with* $S = \Delta_{Q'}$, *where* Q' *is the limit of a certain weak convergent subsequence* $(Q^{n'})$ *of the sequence* $(Q^n)_{n \geq 1}$, *then the process* $X = (X_t, \overset{Q'}{\mathcal{D}_t})$ *is a* Q'*-semimartingale with the triplet* T = (B (X), C (X), ν (X)) *of predictable characteristics.*

(c) *If the assumptions of assertion* (b) *are fullfilled, with* (2.9) *replaced by*

$$X_0^n \xrightarrow{d} X_0, \tag{2.10}$$

and Q_0 (*the restriction of the measure* Q *on the σ-algebra* $\overset{Q}{\mathcal{D}_0}$) *and the triplet* T = (B (X), C (X), ν (X)) *determine uniquely the probability distribution* Q *of the semimartingale* $X = (X_t, \overset{Q}{\mathcal{D}_t})$, *then*

$$Q^n \xrightarrow{w} Q.$$

Remark. The assertion of the theorem remains valid if the measure Q' involved in Condition (b) of the theorem possesses the property

$$Q'(\tilde{D}) = 1, \quad \tilde{D} \subset D,$$

while the functions involved in Condition (T_1) are continuous in the metric ρ for each $X \in \tilde{D}$.

4. **Proof** of assertion (a) in Theorem 1.

Let us verify the assumptions of Theorem 6.4.1. Assumption $I_{1)}$ of Theorem 6.4.1 is fullfilled by (2.9) and assumption $I_{2)}$ of the same theorem by Condition (U_1) of the present theorem. Assumption $II_{1)}$ of Theorem 6.4.1 is satisfied with

$$\beta_t^{na} = \int_0^t b(s, X^n)\, du_s - xI(a < |x| \leq 1) * \nu_t^n,$$

since in this case

$$B_t^{na} - \beta_t^{na} = B_t^n - \int_0^t b(s, X^n)\, du_s$$

and by Condition (U_3)

$$\lim_n P^n\left(\sup_{t \leq L} \left| B_t^n - \int_0^t b(s, X^n)\, du_s \right| \geq \varepsilon \right) = 0$$

for each L > 0 and ε > 0.

Assumptions $II_{2)}$ and $II_{3)}$ of Theorem 6.4.1 are satisfied with

$$\gamma_t^{na} = \int_0^t c(s, X^n)\, du_s, \qquad \delta_t^{ng} = \int_0^t a^g(s, X^n)\, du_s$$

by Conditions (U_4) and (U_2) respectively.

Let us verify now assumption III of Theorem 6.4.1.

Let $\bar{g} = \bar{g}(x)$ be a function involved in Condition (U_2) such that

$$\bar{g}(x) \geq |x|\, I(a < |x| \leq 1).$$

Obviously, the process

$$\int_0^t |b(s, X^n)|\, du_s + \bar{g} * v_t^n - \mathrm{Var}(\beta^{na})_t, \quad t \geq 0,$$

is increasing, as well as the process

$$\int_0^t |b(s, X^n)|\, du_s + \int_0^t a^{\bar{g}}(s, X^n)\, du_s$$

$$+ \sup_{s \leq t} \left| \bar{g} * v_s^n - \int_0^s a^{\bar{g}}(r, X^n)\, du_r \right| - \mathrm{Var}(\beta^{na})_t, \quad t \geq 0.$$

Finally, by Condition (T_2) the process

$$\int_0^t (1 + \sup_{r < s} |X_r^n|)(L(s) + L^{\bar{g}}(s))\, du_s$$

$$+ \sup_{s \leq t} \left| \bar{g} * v_s^n - \int_0^s a^{\bar{g}}(r, X^n)\, du_r \right| - \mathrm{Var}(\beta^{na})_t, \quad t \geq 0,$$

is increasing. Analogously, using Condition (T_2) one may verify that the processes

$$\int_0^t (1 + \sup_{r < s} |X_r^n|)\, L^g(s)\, du_s - \delta_t^{ng}, \quad t \geq 0,$$

and

$$\int_0^t (1 + \sup_{r < s} (X_r^n)^2)\, L(s)\, du_s - \gamma_t^{na}, \quad t \geq 0,$$

are increasing.

Therefore Condition III of Theorem 6.4.1 is satisfied, with

$$G_t^n = \sup_{s \le t} \left| \bar{g} * \nu_s^n - \int_0^s \bar{a}^g (r, X^n)\, du_r \right|, \quad \overline{G}_t^n = (L^g + \bar{L}^g + L) \circ u_t.$$

Consequently, assumption $IV_{2)}$ of Theorem 6.4.1 is satisfied by Condition (U_2) with

$$\overline{G}_t = (L^g + \bar{L}^g + L) \circ u_t.$$

Thus, the assumptions of Theorem 6.4.1 are fullfilled and consequently the family $(Q^n)_{n \ge 1}$ is relatively compact.

5. Proof of assertion (b) in Theorem 1. It suffices to verify the assumptions of Theorem 1.1. Assumption (a) of this theorem on the relative compactness of the family $(Q^n)_{n \ge 1}$ is satisfied in virtue of assertion (a) of the present theorem proved already.

Let us verify assertion (b) of Theorem 1.1.

For $X \in D$, $\lambda \in R$ and $t \in R_+$ denote

$$\phi_t (X, \lambda) = \frac{\lambda^2}{2} c (t, X) + \int_{R_0} (1 - \cos \lambda x) K (X, t; dx) \tag{2.11}$$

and

$$\psi_t (X, \lambda) = \int_0^t \phi_s (X, \lambda)\, du_s. \tag{2.12}$$

By Conditions (T_1) and (T_2), and Theorem 6.2.2 the function $\psi_t (X, \lambda)$ is continuous for each $t \in R_+$ and $\lambda \in R$ in the metric ρ.

Define Markov times

$$\sigma_a (X) = \inf (t\colon \psi_t (X, \lambda) + t \ge a), \quad a \ge 0$$

By Theorem 6.2.3 the function $\sigma_a (X)$ is continuous in the metric ρ uniformly in a which belongs to an arbitrary finite interval (for each fixed $\lambda \in R$). Obviously, for each $\lambda \in R$ and $X \in D$

$$\lim_{a \to \infty} \sigma_a (X) = \infty. \tag{2.13}$$

Let $T_j (X)$, $j \ge 1$, be jump times of a function $X \in D$. By Theorem 6.2.5 one may choose a nonnegative sequence of numbers a_k, $k \ge 1$, such that (for every fixed $\lambda \in R$)

$$Q' (\sigma_{a_k} (X) = T_j (X) < \infty) = 0, \; k \ge 1, \; j \ge 1, \; a_k \uparrow \infty, \; k \to \infty, \tag{2.14}$$

where Q' is the weak limit of a weakly convergent subsequence $(Q^{n'})$ of the sequence $(Q^n)_{n \ge 1}$.

Set

$$\tau_k (X) = \sigma_{a_k} (X). \tag{2.15}$$

Assumption (b_1) of Theorem 1.1 is fullfilled by (2.13), and assumption (b_2) by (2.2) - (2.4), (2.11) and (2.12), since

$$| \mathcal{E}_{\tau_k (X)} (G (X; \lambda)) | = \exp (- \psi_{\tau_k (X)} (X, \lambda)) \geq \exp (- a_k), \; k \geq 1, \; \lambda \in R, \; X \in D.$$

Assumption (b_3) is also fullfilled, since

$$\{\tau_k (X^n) \leq t\} = \{\sigma_{a_k} (X^n) \leq t\} = \{\psi_t (X^n, \lambda) + t \geq a_k\} \in \mathcal{F}^n_t, \; t \in R_+.$$

We turn now to verifying assumption (b_4). We will show that for each $t \in \Delta_{Q'}$ the functions $X_{t \wedge \tau_k (X)}$ and $\mathcal{E}_{t \wedge \tau_k} (X) (G (X; \lambda))$ are continuous, Q'-a.s., $k \geq 1$, in the metric ρ. As $t \in \Delta_{Q'}$ by (2.14) we have

$$Q' (t \wedge \tau_k (X) \in \Delta_{Q'}) = 1.$$

Therefore, by Example 1 (Ch. 6, § 2), $X_{t \wedge \tau_k (X)}$ possesses the desired property.

Next, by the representation (2.4) for $\mathcal{E}$ (G (X; λ)) it suffices to show that $G_{t \wedge \tau_k (X)} (X; \lambda)$ is a continuous function in the metric ρ. By (2.1) - (2.3), Theorem 6.2.2, Condition (T_2) and Condition (T_1), which holds for the chosen set S, we have that $G_t (X; \lambda)$ is a function, continuous in the metric ρ for every $t \in R_+$ (λ is fixed).

Denote

$$g (x) = | e^{i\lambda x} - 1 - i\lambda x I (| x | \leq 1) |.$$

Let L (t) and L^g (t) be the functions involved in Condition (T_2), where L^g (t) is associated with this function g = g(x). Set

$$U_t = \int_0^t \left[\left(\lambda + \frac{\lambda^2}{2} \right) L (s) + L^g (s) \right] du_s. \tag{2.16}$$

Then by Condition (T_2) it is not hard to deduce from the representation (2.4) for G_t (X; λ) that the process

$$\int_0^t (1 + \sup_{r < s} | X_r | + \sup_{r < s} (X_r)^2) \, dU_s - \mathrm{Var} \, (G (X; \lambda))_t, \; t \geq 0,$$

is increasing. Thus, for G_t (X; λ) the conditions of Theorem 6.2.1 are satisfied with

$$h (x) = x \vee 0 + (x \vee 0)^2,$$

and consequently G_t (X, λ) is (for fixed λ) a continuous function on $R_+ \times D$ in all variables. Hence $G_{t \wedge \tau_k (X)} (X; \lambda)$ is a continuous function in the metric ρ, for each $t \in R$.

Thus, to verify assumption (b_4) it remains to establish the relation (1.2) (see § 1) by utilizing the problem in § 1, i.e. by verifying the following relations ($t \in S$):

$$\lim_n P^n\big(\sup_{s \le t} |\mathcal{E}_s(G^n(\lambda)) - \mathcal{E}_s(G(X^n;\lambda))| \ge \varepsilon\big) = 0, \ \varepsilon > 0. \tag{2.17}$$

First of all we note that in view of Lemma 5.4.1

$$\mathcal{E}_t(G^n(\lambda)) = \mathcal{E}_t(G^{na}(\lambda))\, e^{i\lambda\Delta_t^{na}} \tag{2.18}$$

with

$$\Delta_t^{na} = \sum_{0<s\le t} \hat{x}_s^{na}, \quad \hat{x}_s^{na} = \int_{|x|\le a} x\nu^n(\{s\}, dx), \tag{2.19}$$

$$G_t^{na}(\lambda) = i\lambda(B_t^n - \Delta_t^{na}) - \frac{\lambda^2}{2} C_t^n$$

$$+ \int_0^t \int_{R_0} [e^{i\lambda(x - \hat{x}_s^{na})} - 1 - i\lambda(xI(|x|\le 1) - \hat{x}_s^{na})]\, d\nu^n$$

$$+ \sum_{0<s\le t} (e^{-i\lambda\hat{x}_s^{na}} - 1 + i\lambda\hat{x}_s^{na})(1 - a_s^n) \tag{2.20}$$

and

$$a_s^n = \nu^n(\{s\} \times R_0). \tag{2.21}$$

According to the definition of the stochastic exponential (Ch. 2, § 3)

$$\mathcal{E}_t(G^{na}(\lambda)) = \exp(G_t^{na}(\lambda)) \prod_{0<s\le t} (1 + \Delta G_s^{na}(\lambda))\, e^{-\Delta G_s^{na}(\lambda)}.$$

Therefore (2.17) takes place, provided

$$\lim_{a\to 0} \overline{\lim_n} P^n\left(\sup_{s\le t}\left|\prod_{0<u\le s}(1 + \Delta G_u^{na}(\lambda))\, e^{-\Delta G_u^{na}(\lambda)} - 1\right| \ge \varepsilon\right) = 0 \tag{2.22}$$

and

$$\lim_{a\to 0} \overline{\lim_n} P^n\left(\sup_{s\le t} |G_s^{na}(\lambda) + i\lambda\Delta_s^{na} - G_s(X^n;\lambda)| \ge \varepsilon\right) = 0 \tag{2.23}$$

for each $\lambda \in R$, $t \in R_+$ and $\varepsilon > 0$. Observe that

$$G_t^{na}(\lambda) + i\lambda\Delta_t^{na} = i\lambda B_t^n + R_t^{na}(\lambda) + Q_t^{na}(\lambda) \tag{2.24}$$

with

$$R_t^{na}(\lambda) = -\frac{\lambda^2}{2} C_t^n + \int_0^t \int_{|x| \le a} [e^{i\lambda (x - \hat{x}_s^{na})} - 1 - i\lambda (x - \hat{x}_s^{na})] \, d\nu^n$$

$$+ \sum_{0 < s \le t} (e^{-i\lambda \hat{x}_s^{na}} - 1 + i\lambda \hat{x}_s^{na}) (1 - a_s^{na}), \tag{2.25}$$

$$a_s^{na} = \nu^n (\{s\} \times \{ |x| \le a\}), \tag{2.26}$$

$$Q_t^{na}(\lambda) = \int_0^t \int_{|x| > a} [e^{i\lambda (x - \hat{x}_s^{na})} - 1 - i\lambda (xI(|x| \le 1) - \hat{x}_s^{na})] \, d\nu^n$$

$$+ \sum_{0 < s \le t} (e^{-i\lambda \hat{x}_s^{na}} - 1 + i\lambda \hat{x}_s^{na}) (a_s^{na} - a_s^n) \tag{2.27}$$

as $a \in (0, 1]$, and that

$$G_t(X; \lambda) = i\lambda \int_0^t b(s, X) \, du_s + R_t(X; \lambda) + Q_t^a(X; \lambda), \tag{2.28}$$

with

$$R_t^a(X; \lambda) = -\frac{\lambda^2}{2} \int_0^t c(s, X) \, du_s$$

$$+ \int_0^t \int_{|x| \le a} (e^{i\lambda x} - 1 - i\lambda x) K(X, s; dx) \, du_s, \tag{2.29}$$

and

$$Q_t^a(X; \lambda) = \int_0^t \int_{|x| > a} (e^{i\lambda x} - 1 - i\lambda x I(|x| \le 1)) K(X, s; dx) \, du_s \tag{2.30}$$

as $a \in (0, 1]$.

Denote

$$\tilde{Q}_t^{na}(\lambda) = \int_0^t \int_{|x| > a} (e^{i\lambda x} - 1 \, , \, i\lambda x I(|x| \le 1)) \, d\nu^n. \tag{2.31}$$

Since by Condition (U_3) for each $t \in R_+$ and $\varepsilon > 0$

$$\lim_n P^n\left(\sup_{s\le t}\left|B_s^n - \int_0^s b(r, X^n)\,du_r\right| \ge \varepsilon\right) = 0,$$

then (2.23) is a consequence of the following relations (for each $\lambda \in R$, $t \in R_+$ and $\varepsilon > 0$):

$$\lim_{a\to 0}\ \overline{\lim_n}\ P^n(\sup_{s\le t} |R_s^{na}(\lambda) - R_s^a(X^n;\lambda)| \ge \varepsilon) = 0, \tag{2.32}$$

$$\lim_{a\to 0}\ \overline{\lim_n}\ P^n(\sup_{s\le t} |Q_t^{na}(\lambda) - \tilde{Q}_s^{na}(\lambda)| \ge \varepsilon) = 0, \tag{2.33}$$

and

$$\lim_{a\to 0}\ \overline{\lim_n}\ P^n(\sup_{s\le t} |\tilde{Q}_s^{na}(\lambda) - Q_s^a(X^n;\lambda)| \ge \varepsilon) = 0. \tag{2.34}$$

Consequently the proof of this part of the theorem is completed by verifying the validity of the relations (2.22) and (2.32) - (2.34).

In the course of verifying these relations we will need a number of general facts which are formulated below as lemmas.

Lemma 1. *Let Condition* (U_2) *be fullfilled. Then for each* $t \in R_+$, $a \in (0, 1]$ *and* $\varepsilon > 0$

$$\lim_n P^n(\sup_{s\le t} \nu^n(\{s\}\times\{|x| > a\}) \ge \varepsilon) = 0.$$

Proof. Let $g = g(x)$ be a function involved in Condition (U_2) such that

$$g(x) \ge I(|x| > a).$$

Since for each $X \in D$ we have

$$\left(\int_0^t a^g(s, X)\,du_s\right)_{t\ge 0} \in C,$$

then

$$\begin{aligned}\sup_{s\le t} \nu^n(\{s\}\times\{|x|>a\}) &\le \sup_{s\le t} \Delta(g * \nu_s^n)\\ &\le 2\sup_{s\le t}\left|g*\nu_s^n - \int_0^s a^g(r, X^n)\,du_r\right|,\end{aligned}$$

and hence the desired assertion is a simple consequence of Condition (U_2).

Lemma 2. *Let Condition* (U_2) *be fullfilled. Then for each* $t \in R_+$, $a \in (0, 1]$ *and* $\varepsilon > 0$

$$\lim_n P^n(\sup_{s\le t} |\hat{x}_s^{na}| \ge \varepsilon) = 0.$$

Proof. In accordance with the definition of $\hat{x}^{na}_s$ (see (2.19)), as $0 < \delta < a$ we have

$$\sup_{s \le t} |\hat{x}^{na}_s| \le \delta + \sup_{s \le t} \nu^n(\{s\} \times \{|x| > \delta\}).$$

Therefore as $\delta < \varepsilon$

$$P^n(\sup_{s \le t} |\hat{x}^{na}_s| \ge \varepsilon) \le P^n\left(\sup_{s \le t} \nu^n(\{s\} \times \{|x| > \delta\}) \ge \frac{\varepsilon - \delta}{a}\right), \quad (2.35)$$

and consequently the desired assertion takes place by Lemma 1.

Lemma 3. *Let the conditions, stipulated in assertion* (a) *of Theorem* 1 *be fullfilled. Then for each* $t \in R_+$ *and* $a \in (0, 1]$

$$\lim_{l \to \infty} \overline{\lim_{n}} P^n(\nu^n((0, t] \times \{|x| > a\}) \ge l) = 0.$$

Proof. Let $g = g(x)$ be a function involved in Condition (U_2) such that

$$g(x) \ge I(|x| > a).$$

Then

$$\begin{aligned}\nu^n((0, t] \times \{|x| > a\}) &\le g * \nu^n_t \\ &\le \int_0^t a^g(s, X^n)\, du_s + \sup_{s \le t}\left| g * \nu^n_s - \int_0^s a^g(r, X^n)\, du_r \right| \\ &\le \int_0^t L^g(s)\, du_s\,(1 + \sup_{s \le t} |X^n_s|) + \sup_{s \le t}\left| g * \nu^n_s - \int_0^s a^g(r, X^n)\, du_r \right|. \quad (2.36)\end{aligned}$$

Since under the stipulated assumptions the family $(Q^n)_{n \ge 1}$ is relatively compact, the relation

$$\lim_{l \to \infty} \overline{\lim_{n}} P^n(\sup_{s \le t} |X^n_s| \ge l) = 0 \quad (2.37)$$

takes place. Consequently, by the estimate (2.36) the desired relation is a consequence of (2.37) and Condition (U_2).

Lemma 4. *Let the conditions stipulated in assertion* (b) *of Theorem* 1 *be fullfilled and let* $g = g(x)$ *be a function involved in Condition* (T_1). *Then for each* $t \in R_+$ *and* $\varepsilon > 0$

$$\lim_{\delta \to 0} \overline{\lim_{n}} P^n\left(\sup_{s \le t}\left| \int_0^s \int_{|x| > \delta} g(x)\, d\nu^n - \int_0^s \int_{|x| > \delta} g(x) K(X^n, r; dx)\, du_r \right| \ge \varepsilon\right) = 0.$$

Proof. As $\delta \in (0, 1]$ set

$$h(x) = \begin{cases} 0, & |x| \le \delta/2 \\ 2g(\delta)\,|x|/\delta - g(\delta), & \delta/2 < |x| \le \delta, \\ g(x) & |x| > \delta, \end{cases} \tag{2.38}$$

$$f(x) = \begin{cases} h(x), & |x| \le \delta, \\ 2g(\delta) - g(\delta)\,|x|/\delta, & \delta < |x| \le 2\delta, \\ 0, & |x| > 2\delta \end{cases} \tag{2.39}$$

and observe that

$$|g(x) I(|x| > \delta) - h(x)| \le h(x) I(|x| \le \delta), \tag{2.40}$$

$$f(x) \le cx^2 I(|x| \le 2\delta), \tag{2.41}$$

with a certain constant c.

In view of (2.40) and (2.41) we have

$$\sup_{s \le t} \left| \int_0^s \int_{|x|>\delta} g(x)\, d\nu^n - \int_0^s \int_{|x|>\delta} g(x) K(X^n, r; dx)\, du_r \right|$$

$$\le \sup_{s \le t} \left| h * \nu_s^n - \int_0^s a^h(r, X^n)\, du_r \right|$$

$$+ hI(|x| \le \delta) * \nu_t^n + \int_0^t \int_{|x| \le \delta} h(x) K(X^n, r; dx)\, du_r$$

$$\le \sup_{s \le t} \left| h * \nu_s^n - \int_0^s a^h(r, X^n)\, du_r \right| + f * \nu_t^n + \int_0^t \int_{R_0} f(x) K(X^n, r; dx)\, du_r$$

$$\le \sup_{s \le t} \left| h * \nu_s^n - \int_0^s a^h(r, X^n)\, du_r \right| + \sup_{s \le t} \left| f * \nu_s^n - \int_0^s a^f(r, X^n)\, du_r \right|$$

$$+ 2\int_0^t a^f(r, X^n)\, du_r \le \sup_{s \le t} \left| h * \nu_s^n - \int_0^s a^h(r, X^n)\, du_r \right|$$

$$+ \sup_{s \le t} \left| f * \nu_s^n - \int_0^s a^f(r, X^n)\, du_r \right| + 2c \int_0^t \int_{|x| \le 2\delta} x^2 K(X^n, r; dx)\, du_r. \tag{2.42}$$

The functions h and f satisfy Condition (U_2). Therefore the desired assertion is a consequence of the inequality (2.42) and Conditions (U_1) and (U_2).

The lemma is proved.

We turn now to verifying the relations (2.22), (2.32) - (2.34).

Proof (2.22). For simplicity of the exposition we suppress the symbol λ. Denote

$$\Gamma_t^{na} = \left(\sup_{s \le t} |\Delta G_s^{na}| \ge 1/2 \right), \quad J_s^{na} = \left| \prod_{0 < u \le s} (1 + \Delta G_u^{na}) e^{-\Delta G_u^{na}} - 1 \right|.$$

Then

$$P^n\left(\sup_{s \le t} J_s^{na} \ge \varepsilon \right) \le P^n\left(\sup_{s \le t} J_s^{na} \ge \varepsilon, \ \Omega^n \setminus \Gamma_t^{na} \right) + P^n(\Gamma_t^{na})$$

$$\le P^n\left(\sup_{s \le t} J_s^{na} \ge \varepsilon, \ \Omega^n \setminus \Gamma_t^{na}, \ \sup_{s \le t} |X_s^n| < l \right)$$

$$+ P^n\left(\sup_{s \le t} |X_s^n| \ge l \right) + P^n(\Gamma_t^{na}). \tag{2.43}$$

Analogously to the proof of Lemma 5.4.4 it is established, in view of Lemma 1, that for each $t \in R_+$ and $\varepsilon > 0$

$$\lim_n P^n\left(\sup_{s \le t} |\Delta G_s^{na}| \ge \varepsilon \right) = 0, \tag{2.44}$$

$$\lim_n P^n(\Gamma_t^{na}) = 0.$$

Consequently, by (2.37) and (2.43) it suffices to show that for each $l > 0$, $t \in R_+$ and $\varepsilon > 0$

$$\lim_{a \to 0} \overline{\lim_n} \, P^n\left(\sup_{s \le t} J_s^{na} \ge \varepsilon, \ \Omega^n \setminus \Gamma_t^{na}, \ \sup_{s \le t} |X_s^n| < l \right) = 0. \tag{2.45}$$

We have the following estimate for $\sup_{s \le t} J_t^{na}$ on the set $\Omega^n \setminus \Gamma_t^{na}$, which has been established in the course of proving Lemma 5.4.5, with a instead of δ and Ω^n instead of Ω (see Ch. 5, § 4, (4.21)):

$$\sup_{s \le t} J_s^{na} \le \exp\left(\sup_{s \le t} |\Delta G_s^{na}| \left[\frac{\lambda^2}{2} \langle M^{n,a} \rangle_t + 4I(|x| > a) * \nu_t^n \right] \right) - 1.$$

From this estimate and (2.44) it follows that the relation (2.45) takes place provided for each $t \in R_+$ and $l > 0$

$$\lim_{b \to \infty} \overline{\lim_{a \to 0}} \, \overline{\lim_n} \, P^n\left(\langle M^{n,a} \rangle_t \ge b, \ \sup_{s \le t} |X_s^n| < l \right) = 0, \tag{2.46}$$

$$\lim_{b\to\infty}\ \overline{\lim_{n}}\ P^n\left(g * \nu^n_t \geq b,\quad \sup_{s\leq t} |X^n_s| < 1\right) = 0, \tag{2.47}$$

where $g = g(x)$ is a function involved in Condition (U_2) such that $I(|x| > a) \leq g(x)$.

The desired relations (2.46) and (2.47) hold by the estimates

$$\langle M^{n,a}\rangle_t \leq \sup_{s\leq t} |\langle M^{n,a}\rangle_s - \int_0^s c(r, X^n)\, du_r| + \int_0^t c(r, X^n)\, du_r,$$

$$g * \nu^n_t \leq \sup_{s\leq t} \left| g * \nu^n_s - \int_0^s a^g(r, X^n)\, du_r \right| + \int_0^t a^g(r, X^n)\, du_r,$$

since by Conditions (U_2) and (U_4) for each $t \in R_+$ and $\varepsilon > 0$

$$\lim_{a\to 0}\ \overline{\lim_{n}} \left(\langle M^{n,a}\rangle_s - \int_0^s c(r, X^n)\, du_r \,|\geq \varepsilon \right) = 0,$$

while by Condition (T_2)

$$\int_0^t c(s, X^n)\, du_s \leq (1 + l^2) \int_0^t L(s)\, du_s$$

and

$$\int_0^t a^g(s, X^n)\, du_s \leq (1 + l) \int_0^t L^g(s)\, du_s$$

on the set $\{\sup_{s\leq t} |X^n_s| < 1\}$.

Proof of (2.32). In view of (4.33) (Ch. 5, § 4) with a instead of δ we have

$$\left| R^{na}_s(\lambda) + \frac{\lambda^2}{2} \langle M^{n,a}\rangle_s \right| \leq \frac{|\lambda^3|\, a}{3} \langle M^{n,a}\rangle_s,$$

and by (2.29) we have

$$\left| R^a_s(\lambda) + \frac{\lambda^2}{2} \int_0^s c(r, X^n)\, du_r \right| \leq \frac{\lambda^2}{2} \int_0^s \int_{|x|\leq a} x^2 K(X^n, r; dx)\, du_r.$$

Consequently,

$$\sup_{s\le t} | R_s^{na}(\lambda) - R_s^a(X^n;\lambda) | \le \frac{|\lambda^3| a}{3} \langle M^{n,a}\rangle_t + \sup_{s\le t} \frac{\lambda^2}{2} \left| \langle M^{n,a}\rangle_s - \int_0^s c(r, X^n)\, du_r \right| + \frac{\lambda^2}{2} \int_0^t \int_{|x|\le a} x^2 K(X^n, r; dx)\, du_r.$$

By Condition (U_1)

$$\lim_{a\to 0} \overline{\lim_n} P^n \left(\int_0^t \int_{|x|\le a} x^2 K(X^n, r; dx)\, du_r \ge \varepsilon \right) = 0, \ \varepsilon > 0,$$

and by Condition (U_4)

$$\lim_{a\to 0} \overline{\lim_n} P^n \left(\sup_{s\le t} \left| \langle M^{n,a}\rangle_s - \int_0^s c(r, X^n)\, du_r \right| \ge \varepsilon \right) = 0, \ \varepsilon > 0.$$

Therefore it suffices to show that

$$\lim_{a\to 0} \overline{\lim_n} P^n (a \langle M^{n,a}\rangle_t \ge \varepsilon) = 0, \ \varepsilon > 0. \tag{2.48}$$

We have

$$P^n (a \langle M^{n,a}\rangle_t \ge \varepsilon) \le P^n (a \langle M^{n,a}\rangle_t \ge \varepsilon, \ \langle M^{n,a}\rangle_t < b) + P^n (\langle M^{n,a}\rangle_t \ge b)$$

$$\le I(ab \ge \varepsilon) + P^n (\langle M^{n,a}\rangle_t \ge b, \ \sup_{s\le t} |X_s^n| < l) + P^n (\sup_{s\le t} |X_s^n| \ge l). \tag{2.49}$$

Consequently, we arrive at the desired relation (2.48) by taking the limit

$$\lim_{l\to\infty} \overline{\lim_{b\to\infty}} \ \overline{\lim_{a\to 0}} \ \overline{\lim_n}$$

in the inequality (2.49), by (2.46) and (2.37).

Proof of (2.33). In view of (4.35) (Ch. 5, § 4) with a instead of δ we have

$$\sup_{s\le t} | Q_s^{na}(\lambda) - \tilde{Q}_s^{na}(\lambda) | \le 2|\lambda| \sup_{s\le t} |\hat{x}_s^{na}| \, \nu^n((0, t] \times \{ |x| > a\}).$$

The desired assertion follows from this inequality by Lemmas 2 and 3.

Proof of (2.34). This relation follows from Lemma 4 applied to the real and imaginary parts of the function

$$g(x) = e^{i\lambda x} - 1 - i\lambda x I(|x| \le 1).$$

6. Proof of assertion (c). This assertion follows from Theorem 1.2.

Problems

1. Prove the validity of the implication (2.5). (Hint: use the method of proving Lemma 5.5.1.)

2. Show the equivalence of Conditions (U_2) and (2.7), as well as (U_4) and (2.8). (Hint: see Lemma 6.5.5 and Problem 6.5.5.)

§ 3. Diffusion approximation

1. Let the triplet $T = (B(X),\ C(X),\ \nu(X))$ of a Q-semimartingale $X = (X_t, \mathcal{D}_t^Q)$ posses the following properties ($t \in R_+$ and $X \in D$):

$$B_t(X) = \int_0^t b(s, X)\, ds, \tag{3.1}$$

$$C_t(X) = \int_0^t c(s, X)\, ds, \tag{3.2}$$

and

$$\nu(X; R_+ \times R_0) = 0, \tag{3.3}$$

where $b(t, X)$ and $c(t, X)$ are $\mathcal{P}(\mathcal{D}^Q)$-measurable functions and $c(t, X) \geq 0$.

If $\mu(X) = \mu(X; dt, dx)$ is the jump measure of X, then in accordance with the definition of the compensator (Ch. 3, § 2)

$$\mathbf{E}_Q \mu(X; P_+ \times R_0) = \mathbf{E}_Q \nu(X; R_+ \times R_0)$$

($\mathbf{E}_Q$ is the mathematical expectation with respect to the measure Q). Consequently, the Q-semimartingale $X = (X_t, \mathcal{D}_t^Q)$ has continuous trajectories, i.e.

$$Q(C) = 1. \tag{3.4}$$

The canonical representation for such a martingale is of the type

$$X_t = X_0 + \int_0^t b(s, X)\, ds + X_t^c, \tag{3.5}$$

where $(X_t^c, \mathcal{D}_t^Q)$ is a local Q-martingale with the quadratic characteristic

$$\left(\int_0^t c(s, X)\, ds\right)_{t \geq 0}.$$

If $c(t, X) > 0$ for every $t \in R_+$ and $X \in D$, then the stochastic integral

$$W_t = \int_0^t c^{-1/2}(s, X)\, dX_s^c, \quad t \geq 0,$$

is defined. Besides, the continuous local Q-martingale $(W_t, \mathcal{D}_t^Q)$ has the quadratic characteristic $\langle W \rangle_t \equiv t$, i.e. $(W_t, \mathcal{D}_t^Q)$ is a Wiener process ([162, 188]). Hence the Q-semimartingale $X = (X_t, \mathcal{D}_t^Q)$ admits the representation in the form

$$X_t = X_0 + \int_0^t b\,(s, X)\,ds + \int_0^t c^{1/2}\,(s, X)\,dW_s \tag{3.6}$$

as well.

In the general case $(c\,(t, X) \geq 0)$ consider a stochastic basis

$$(D \times \tilde{D},\ \mathcal{D}^Q \otimes \tilde{\mathcal{D}}^{Q^W},\ \mathbb{D}^Q \otimes \tilde{\mathbb{D}}^{Q^W},\ Q \times Q^W),$$

where Q^W is a Wiener measure.

A local $Q \times Q^W$-martingale $(W'_t,\ \mathcal{D}_t^Q \times \tilde{\mathcal{D}}_t^{Q^W})$ with

$$W'_t = \int_0^t I\,(c\,(s, X) > 0)\,c^{-1/2}\,(s, X)\,dX_s^c + \int_0^t I\,(c\,(s, X) = 0)\,d\tilde{X}_s$$

is a Wiener process, and a Q-semimartingale $X = (X_t, \mathcal{D}_t^Q)$, considered as a stochastic process on a probability space

$$(D \times \tilde{D},\ \mathcal{D}^Q \otimes \tilde{\mathcal{D}}^{Q^W},\ Q \times Q^W),$$

can be represented in the following form (Problem 1):

$$X_t = X_0 + \int_0^t b\,(s, X)\,ds + \int_0^t c^{1/2}\,(s, X)\,dW'_s. \tag{3.7}$$

In view of the representations (3.6) and (3.7) it seems natural that under the assumptions (3.1) - (3.3) the semimartingale $X = (X_t, \mathcal{D}_t^Q)$ is called a process of *diffusion type*.

Along with (3.1) and (3.2) we assume that the conditions of "linear growth" of the functions $b\,(t, X)$ and $c\,(t, X)$:

$$|\,b\,(t, X)\,| \leq L\,(t)\,(1 + \sup_{s<t} |\,X_s\,|), \tag{3.8}$$

$$c\,(t, X) \leq L\,(t)\,(1 + \sup_{s<t} (X_s)^2) \tag{3.9}$$

are fullfilled with

$$\int_0^t L(s)\,ds < \infty, \quad t \in R_+. \tag{3.10}$$

2. For every $n \geq 1$ let $X^n = (X_t^n, \mathcal{F}_t^n)$ be a P^n-semimartingale with the triplet of predictable characteristics $T^n = (B^n, C^n, \nu^n)$, and let $M^{n,a} = (M_t^{n,a}, \mathcal{F}_t^n)$ be a locally square integrable P^n-martingale with

$$M_t^{n,a} = X_t^{nc} + \int_0^t \int_{|x| \leq a} x\, d(\mu^n - \nu^n), \quad a \in (0, 1]$$

and the quadratic characteristic

$$\langle M^{n,a} \rangle_t = C_t^n + \int_0^t \int_{|x| \leq a} x^2 d\nu^n - \sum_{0 < s \leq t} \left(\int_{|x| \leq a} x\nu^n(\{s\}, dx) \right)^2.$$

The probability distribution of a semimartingale $X^n = (X_t^n, \mathcal{F}_t^n)$ is denoted by Q^n.

Let us formulate the conditions imposed on the triplets T^n, $n \geq 1$.

For each $L > 0$, $\varepsilon > 0$ and $a \in (0, 1]$

(A) $$\lim_n P^n\left(\int_0^L \int_{|x| > a} d\nu^n \geq \varepsilon \right) = 0,$$

(sup B) $$\lim_n P^n\left(\sup_{t \leq L} \left| B_t^n - \int_0^t b(s, X^n)\, ds \right| \geq \varepsilon \right) = 0,$$

(sup C) $$\lim_n P^n\left(\sup_{t \leq L} \left| \langle M^{n,a} \rangle_t - \int_0^t c(s, X^n)\, ds \right| \geq \varepsilon \right) = 0.$$

Observe that Condition (A) is equivalent to the condition (see Remark to Lemma 5.5.1)

(A*) $$\lim_n P^n(\sup_{t \leq L} |\Delta X_t^n| \geq \varepsilon) = 0,$$

while Condition (sup C) is equivalent in case

$$c(t, X) \le L(t), \quad \int_0^t L(s)\, ds < \infty, \quad t \in R_+$$

to the condition

$$\lim_n P^n\left(\left| \langle M^{n,a}\rangle_t - \int_0^t c(s, X^n)\, ds \right| \ge \varepsilon\right) = 0$$

for each $t > 0$ (see § 2, Remark 3) to Condition (U_4).

3. Theorem 1. *Let the conditions of "linear growth"* (3.8) - (3.10) *be fullfilled as well as Conditions* (A), (sup B), (sup C) *and*

$$\lim_{l \to \infty} \overline{\lim_n} P^n(|X_0^n| \ge l) = 0. \tag{3.11}$$

Then the family of measures $(Q^n)_{n \ge 1}$ *is relatively compact and the limit* Q' *of any weak convergent subsequence* $(Q^{n'})$ *of the sequence* $(Q^n)_{n \ge 1}$ *possesses the following property:*

$$Q'(C) = 1. \tag{3.12}$$

(b) *Let the conditions stipulated in assertion* (a) *be fullfilled. For each* $t \in S$ (S *is a subset dense in* R_+,

$$\int_{R_+ \setminus S} dt = 0)$$

let the functions $b(t, X)$ *and* $c(t, X)$ *be continuous in the metric* ρ *at each point* $X \in C$.

Then the process $X = (X_t, \mathcal{D}_t^{Q'})$ *is a* Q'*-semimartingale with the triplet* $T = (B(X), C(X), 0)$, *where* $B(X)$ *and* $C(X)$ *are defined by the formulas* (3.1) *and* (3.2), *while* Q' *is the limit of any weak convergent subsequence* $(Q^{n'})$ *of the sequence* $(Q^n)_{n \ge 1}$.

(c) *If the conditions stipulated in assertion* (b) *are fullfilled with*

$$X_0^n \xrightarrow{d} X_0,$$

instead of (3.11), *and the distribution of* X_0 *and the functions* $b(t, X)$ *and* $c(t, X)$ *uniquely determine the measure* Q, *then*

$$Q^n \xrightarrow{w} Q.$$

(d) *If*

$$Q^n \xrightarrow{w} Q$$

and the conditions of "linear growth" (3.8) - (3.10) *are fullfilled as well as the continuity conditions on* $b(t, X)$ *and* $c(t, X)$ *stipulated in assertions* (b) *and* (sup B), *then Conditions* (A) *and* (sup C) *hold, and also*

$$\lim_n P^n\left(\sup_{t\le L}\left|[M^{n,a},M^{n,a}]_t-\int_0^t c(s,X^n)\,ds\right|\ge\varepsilon\right)=0 \tag{3.13}$$

for each L > 0 *and* ε > 0.

Proof. (a) We will show that the conditions stipulated in assertion (a) of Theorem 2.1 are fullfilled.

In fact, (3.8) - (3.10) guarantee the validity of Condition (T_2) of Theorem 2.1, Condition (A) entails Conditions (U_1) and (U_2) of Theorem 2.1, Conditions (sup B) and (sup C) coincide with Conditions (U_3) and (U_4) as $u_t \equiv t$. Consequently, the desired assertion concerning the relative compactness of the family $(Q^n)_{n\ge1}$ follows by Theorem 2.1.

Let Q' be the limit of a certain weak convergent subsequence $(Q^{n'})$ of the sequence $(Q^n)_{n\ge1}$. To establish (3.12) it suffices to show that

$$\int_D \exp\left(-\sup_{t>0}|\Delta X_t|\right)dQ'=1. \tag{3.14}$$

Since the set

$$\{t>0\colon Q'(\Delta X_t=0)=1\}$$

is dense in R_+, there exists in this set an increasing sequence L_k, $k\ge1$, such that

$$\lim_k L_k=\infty.$$

Therefore to prove (3.14) it suffices to show that

$$\int_D \exp\left(-\sup_{t\le L_k}|\Delta X_t|\right)dQ'=1,\ \ k\ge1. \tag{3.15}$$

In view of Example 3 (Ch. 6, § 2), the function

$$H(X)=\sup_{t\le L_k}|\Delta X_t|$$

is continuous in the metric ρ Q'-a.s. Hence

$$\int_D \exp\left(-\sup_{t\le L_k}|\Delta X_t|\right)dQ'=\lim_{n'}\int_D \exp\left(-\sup_{t\le L_k}|\Delta X_t|\right)dQ^{n'}$$

$$=\lim_{n'}E^{n'}\exp\left(-\sup_{t\le L_k}|\Delta X_t^{n'}|\right),\ \ k\ge1,$$

where $E^{n'}$ is the mathematical expectation with respect to the measure $P^{n'}$.

Since Condition (A) is equivalent to Condition (A*), then

$$\overline{\lim_{n}}\ \mathbf{E}^{n'} I\,(\sup_{t \le L_k} |\Delta X_t^{n'}| \ge \varepsilon) \exp(-\sup_{t \le L_k} |\Delta X_t^{n'}|)$$

$$\le \overline{\lim_{n}}\ \mathsf{P}^{n}\,(\sup_{t \le L_k} |\Delta X_t^{n}| \ge \varepsilon) = 0.$$

Consequently

$$\overline{\lim_{n'}}\ \mathbf{E}^{n'} \exp(-\sup_{t \le L_k} |\Delta X_t^{n'}|)$$

$$\ge \overline{\lim_{n'}}\ \mathbf{E}^{n'} I\,(\sup_{t \le L_k} |\Delta X_t^{n'}| < \varepsilon) \exp(-\sup_{t \le L_k} |\Delta X_t^{n'}|)$$

$$\ge e^{-\varepsilon}\,(1 - \overline{\lim_{n'}}\ \mathsf{P}^{n'}\,(\sup_{t \le L_k} |\Delta X_t^{n'}| \ge \varepsilon)) = e^{-\varepsilon} \to 1, \ \varepsilon \to 0,$$

i.e. the desired relation (3.15) holds, since the reverse inequality

$$\lim_{n'} \mathbf{E}^{n'} \exp(-\sup_{t \le L_k} |\Delta X_t^{n'}|) \le 1$$

is obvious.

(b), (c) These assertions take place by the corresponding assertions of Theorem 2.1 and the fact that for every $t \in S$ the functions b (t, X) and c (t, X) are continuous in the metric ρ Q'-a.s. (see 3.12)).

(d) Condition (A) is satisfied by Theorem 6.5.1. To establish the validity of Conditions (sup C) and (3.13), we will apply Theorem 6.5.2. We will show that the conditions of this theorem are fullfilled.

Condition (α) of this theorem is satisfied by the assumption concerning the continuity of the functions b (t, X) and c (t, X) for every t in a set S, dense in R_+,

$$\int_{R_+ \setminus S} dt = 0,$$

in the metric ρ, at each point $X \in C$ (i.e. Q-a.s. continuity in the metric ρ of the functions b (t, X) and c (t, X) for every t in S), and by Theorem 6.2.2, since for B_t (X) and $\langle M(X)\rangle_t \equiv C_t(X)$ the representations (3.1) and (3.2) take place and the conditions of "linear growth" (3.8) - (3.10) are satisfied.

Conditions (sup B) of Theorem 6.5.2 and the present theorem coincide.

To verify Condition (K) of Theorem 6.5.2 we remark first that since by (3.1), (3.2) (3.8) and (3.9) the following estimates take place:

$$\mathrm{Var}\,(B\,(X^n))_{L'} \le (1 + \sup_{t \le L'} |X_t^n|) \int_0^{L'} L\,(s)\,ds,$$

$$\sup_{t \le L'} [\mathrm{Var}\,(B\,(X^n))_{t+\sigma} - \mathrm{Var}\,(B\,(X^n))_t] \le (1 + \sup_{t \le L'} |X_t^n|) \sup_{t \le L'} \int_t^{t+\sigma} L\,(s)\,ds,$$

$$\sup_{t \le L'} [C_{t+\sigma}\,(X^n) - C_t\,(X^n)] \le (1 + \sup_{t \le L'} (X_t^n)^2) \sup_{t \le L'} \int_t^{t+\sigma} L\,(s)\,ds$$

and by (3.10) (for any $L' > 0$)

$$\lim_{\sigma \to 0} \sup_{t \le L'} \int_t^{t+\sigma} L\,(s)\,ds = 0, \tag{3.16}$$

then in virtue of the weak convergence

$$Q^n \xrightarrow{w} Q$$

we have (for any $L' > 0$)

$$\lim_{l \to \infty} \overline{\lim_n}\, P^n\,(\sup_{t \le L'} |X_t^n| \ge l) = 0. \tag{3.17}$$

Hence the first condition in (K) is a simple consequence of (3.10) and (3.17), and the second and third of (3.16) and (3.17).

4. We present an example of applying Theorem 1 on a diffusion approximation to a problem of the queuing theory.

Let there be one finite source of orders of size N (N is an integer) and one servant. If the source of orders contains N_1 orders ($N_1 \le N$), then it generates an order with the intensity λN_1, where λ is a positive constant. An order being served returns to the source. The servant follows the rule: "the first order is served first", and if there is a queue of length $N - N_1$, then it serves with the intensity

$$Nf\left(\frac{N - N_1}{N}\right),$$

where $f = f\,(x)$ is a certain nonnegative function. Besides, the duration of generating orders and of service are assumed to be independent, and the probability of generating or serving two or more orders within a time interval $[t,\ t + \delta t]$ is of order $o\,(\delta t)$.

We will denote by Q_t the length of the queue at time t, and we will give the description of the queue, useful for applying Theorem 1.

It is assumed that a stochastic basis $(\Omega, \mathcal{F}, \mathbb{F}, P)$ is fixed, and the queue $(Q_t)_{t \ge 0}$ is a $\mathbb{F}$-adapted process with trajectories in D and with

$$Q_t = Q_0 + A_t - D_t, \tag{3.18}$$

where $A = (A_t)_{t \ge 0}$ and $D = (D_t)_{t \ge 0}$ are counting processes with the compensators $\tilde{A} = (\tilde{A}_t)_{t \ge 0}$ and $\tilde{D} = (\tilde{D}_t)_{t \ge 0}$. In accordance with the description given above

$$\tilde{A}_t = \int_0^t \lambda (N - Q_s)\, ds, \quad \tilde{D}_t = \int_0^t I(Q_s > 0)\, Nf\left(\frac{Q_s}{N}\right) ds, \tag{3.19}$$

$$\sum_{t>0} \Delta A_t \Delta D_t = 0 \quad \text{(P-a.s.)}.$$

If N is a large number, then the calculation of one or another probability characteristic of a queue is a considerably hard problem. In view of this it is natural to utilize the asymptotic description (as $N \to \infty$) of a queue, which, as we will see, is much more useful for the study of its characteristcs. To this end we introduce a stochastic process $q^N = (q_t^N)_{t \geq 0}$ with

$$q_t^N = \frac{Q_t}{N}, \tag{3.20}$$

which is naturally called a normalized queue, and we consider the differential equation

$$\dot{q}_t = \lambda (1 - q_t) - f(q_t) \tag{3.21}$$

with the initial condition $q_0 \in (0, 1]$.

The first step towards the asymptotic description of a queue $(Q_t)_{t \geq 0}$ as $N \to \infty$ we present in the following lemma.

Lemma 1. *Let a nonnegative function* f (x) *satisfy Lipschitz' condition, and let the differential equation* (3.21) *have a solution with the property*

$$\inf_{s \leq t} q_s > 0, \quad t \in R_+.$$

Then under the condition $q_0^N \to q_0$, $N \to \infty$, *the following convergence holds (for each* $t \in R_+$*):*

$$\sup_{s \leq t} |q_s^N - q_s| \xrightarrow{P} 0, \quad N \to \infty.$$

Proof. Denote

$$M^A = A - \tilde{A}, \quad M^D = D - \tilde{D}$$

and

$$M^N = \frac{1}{N}(M^A - M^D).$$

In view of Problem 3.4.5 M^A and M^D are locally square integrable martingales with the quadratic characteristics $\langle M^A \rangle = \tilde{A}$ and $\langle M^D \rangle = \tilde{D}$. Besides, by the assumption

$$\sum_{t>0} \Delta A_t \Delta D_t = 0 \quad \text{(P-a.s.)}$$

M^A and M^D are such that the mutual quadratic characteristic $\langle M^A, M^D \rangle = 0$. Consequently, M^N is a locally square integrable martingale with the quadratic characteristic

$$\langle M^N \rangle = \frac{\tilde{A} + \tilde{D}}{N^2},$$

i.e.

$$\langle M^N \rangle_t = \frac{1}{N} \int_0^t [\lambda (1 - q_s^N) + I (q_s^N > 0) f (q_s^N)] \, ds. \tag{3.22}$$

The definition of q_t^N gives

$$q_t^N = q_0^N + \int_0^t \lambda (1 - q_s^N) \, ds - \int_0^t I (q_s^N > 0) f (q_s^N) \, ds + M_t^N. \tag{3.23}$$

Introduce the Markov time $\tau_N = \inf (t: q_t^N \le 2/N)$. Since

$$| \Delta q_t^N | \le \frac{1}{N},$$

then

$$q_{t \wedge \tau_N}^N > 0, \;\; t \in R_+,$$

and hence

$$q_{t \wedge \tau_N}^N = q_0^N + \int_0^{t \wedge \tau_N} [\lambda (1 - q_s^N) - f (q_s^N)] \, ds + M_{t \wedge \tau_N}^N. \tag{3.24}$$

Denote

$$U_t^N = | q_t^N - q_t | .$$

By (3.24) and (3.21) it is verified, in view of Lipschitz' condition

$$| f (x) - f (y) | \le c \, | x - y |$$

that as $t \le L$

$$U_{t \wedge \tau_N}^N \le U_0^N + \int_0^t (\lambda + c) \, U_{s \wedge \tau_N}^N \, ds + \sup_{s \le L} | M_s^N |.$$

By Theorem 2.4.3 this gives

$$U_{t \wedge \tau_N}^N \le (U_0^N + \sup_{s \le L} | M_s^N |) \, e^{(\lambda + c) t}$$

and hence as $t \le L$

$$\sup_{s \le t \wedge \tau_N} U_s^N \le (U_0^N + \sup_{s \le L} | M_s^N |) \, e^{(\lambda + c) t}. \tag{3.25}$$

By condition $U_0^N \xrightarrow{P} 0$ as $N \to \infty$. Next, by Problem 1.9.2 we have

$$\sup_{s \le L} |M_s^N| \xrightarrow{P} 0$$

as $N \to \infty$, since, according to (3.22),

$$\langle M^N \rangle_L \le \frac{L}{N} (\lambda + \max_{0 \le x \le 1} f(x)) \to 0, \ N \to \infty.$$

Thus by the estimate (3.25) we get

$$\lim_N P\left(\sup_{s \le t \wedge \tau_N} U_s^N \ge \varepsilon \right) = 0 \tag{3.26}$$

for each $\varepsilon > 0$.

To prove the desired assertion, we note that for each $\varepsilon > 0$

$$P\left(\sup_{s \le t} U_s^N \ge \varepsilon \right) \le P(\tau_N < t) + P\left(\sup_{s \le t \wedge \tau_N} U_s^N \ge \varepsilon \right),$$

and consequently it suffices by (3.26) to show that

$$\lim_N P(\tau_N < t) = 0. \tag{3.27}$$

We have

$$\{\tau_N < t\} \subseteq \{\tau_N \le t\} \subseteq \left\{ \inf_{s \le t \wedge \tau_N} q_s^N \le 2/N \right\}.$$

On the other hand on the set $\{s \le t \wedge \tau_N\}$ we have

$$q_s^N \ge q_s - |q_s - q_s^N| \ge q_s - \sup_{s \le t \wedge \tau_N} U_s^N,$$

and hence

$$\inf_{s \le t \wedge \tau_N} q_s^N \ge \inf_{s \ge t} q_s - \sup_{s \le t \wedge \tau_N} U_s^N.$$

Thus we get

$$P(\tau_N < t) \le P\left(\sup_{s \le t \wedge \tau_N} U_s^N \ge \inf_{s \le t} q_s - 2/N \right),$$

and consequently (3.27) follows from (3.26).

The lemma is proved.

We consider a stochastic process $X^N = (X_t^N)_{t \ge 0}$ with

$$X_t^N = \sqrt{N}\,(q_t^N - q_t) \tag{3.28}$$

and we will show that a sequence of processes X^N, $N \ge 1$, admits the diffusion approximation.

Theorem 2. *Let* $f = f(x)$ *be a nonnegative continuously differentiable function* *the derivative of which* f' = f' (x) *satisfies Lipschitz' condition, and the differential equation* (3.21) *has a solution with the property*

$$\inf_{s \le t} q_s > 0, \ t \in R_+.$$

If the following conditions are fullfilled

(a) $q_0^N \xrightarrow{P} q_0, \ N \to \infty$,

(b) *the sequence of distributions or random variables* X_0^N, $N \ge 1$, *converges weakly to a distribution* F,

then the sequence of probability distributions of processes X^N, $N \ge 1$, *converges weakly to the distribution of the diffusion process* $X = (X_t)_{t \ge 0}$, *defined by the stochastic equation*

$$dX_t = b(t) X_t dt + c^{1/2}(t) dW_t \tag{3.29}$$

with a Wiener process $W = (W_t)_{t \ge 0}$, *with the initial condition* X_0 *having the distribution* F *and with*

$$b(t) = -(\lambda + f'(q_t)), \quad c(t) = \lambda(1 - q_t) + f(q_t). \tag{3.30}$$

Proof. It suffices to verify the condition of statement (c) of Theorem 1.

The semimartingale X has the triplet T = (B (X), C (X), 0) with

$$B_t(X) = \int_0^t b(s) X_s ds, \quad C_t(X) = \int_0^t c(s) ds.$$

Consequently, Conditions (3.1), (3.2), and (3.8) - (3.10) are fullfilled. The function c (t, X) is independent of X, and the function b (t, X) = b (t) X_t satisfies the continuity condition in the metric ρ at each point X ∈ C, due to Problem 6.1.5.

The distribution of X_0 together with B (X), C (X), determine uniquely the probability distribution of the process $X = (X_t)_{t \ge 0}$. The convergence

$$X_0^n \xrightarrow{d} X_0$$

is ensured by the conditions of the theorem.

Therefore it remains to verify Conditions (A), (sup B) and (sup C) only.

Observe that for each $t \in R_+$

$$|\Delta X_t^N| = \sqrt{N}\, |\Delta q_t^N| = \frac{1}{\sqrt{N}} |\Delta Q_t| \le \frac{1}{\sqrt{N}}.$$

This implies Condition (A*), and hence Condition (A) too.

To verify Conditions (sup B) and (sup C) the triplet $T^N = (B^N, C^N, \nu^N)$ has to be calculated.

Aiming at this we note first that

$$\Delta X_t^N = \frac{\Delta Q_t}{\sqrt{N}} = \frac{1}{\sqrt{N}}(\Delta A_t - \Delta D_t),$$

and hence

$$\nu^N\left(R_+ \times R_0 \setminus \left(\left\{-\frac{1}{\sqrt{N}}\right\} \cup \left\{\frac{1}{\sqrt{N}}\right\}\right)\right) = 0,$$

$$\nu^N\left((0, t] \times \left\{\frac{1}{\sqrt{N}}\right\}\right) = \tilde{A}_t, \quad \nu^N\left((0, t] \times \left\{-\frac{1}{\sqrt{N}}\right\}\right) = \tilde{D}_t. \tag{3.31}$$

Next, by the representation (3.23) for q_t^N and the equation (3.21) for q_t we obtain

$$X_t^N = X_0^N + \int_0^t [-\lambda X_s^N - \sqrt{N}\ (I(q_s^N > 0) f(q_s^N) - f(q_s))]\, ds + \sqrt{N}\ M_t^N. \tag{3.32}$$

This gives

$$B_t^N = \int_0^t [-\lambda X_s^N - \sqrt{N}\ (I(q_s^N > 0) f(q_s^N) - f(q_s))]\, ds. \tag{3.33}$$

The local martingale M^N has trajectories in $\mathcal{A}_{loc}$. Therefore by Theorem 1.7.3 and Corollary to it M^N does not possess the continuous martingale component. Consequently

$$C^N = 0 \tag{3.34}$$

and for $a \in (0, 1]$

$$\langle M^{N, a}\rangle_t = \int_0^t \int_{|x| \le a} x^2 d\nu^N = I\left(\frac{1}{\sqrt{N}} \le a\right)\frac{\tilde{A}_t + \tilde{D}_t}{N}$$

$$= I\left(\frac{1}{\sqrt{N}} \le a\right)\int_0^t [\lambda(1 - q_s^N) + I(q_s^N > 0) f(q_s^N)[\, ds. \tag{3.35}$$

We will show that for each $t \in R_+$

$$\lim_N \mathsf{P}\left(\int_0^t I(q_s^N = 0)\, ds > 0\right) = 0. \tag{3.36}$$

We apply here the inclusion

$$\left\{ \int_0^t I\,(q_s^N = 0) > 0 \right\} \subseteq \left\{ \inf_{s \le t} q_s^N = 0 \right\}$$

and the inequality

$$\inf_{s \le t} q_s^N \ge \inf_{s \le t} q_s - \sup_{s \le t} | q_s^N - q_s | .$$

Therefore

$$P\left(\int_0^t (q_s^N = 0)\, ds > 0 \right) \le P\left(\sup_{s \le t} | q_s^N - q_s | \ge \inf_{s \le t} q_s \right),$$

and the desired relation holds by Lemma 1.

On verifying Conditions (sup B) and (sup C) it suffices, by (3.36), to replace B^N by $\overline{B}^N$ with

$$\overline{B}_t^N = \int_0^t [-\lambda X_s^N - \sqrt{N}\ (f\,(q_s^N) - f\,(q_s))]\, ds,$$

and $\langle M^{N,a}\rangle$ by $\overline{\langle M^{N,a}\rangle}$ with

$$\overline{\langle M^{N,a}\rangle} = I\left(\frac{1}{\sqrt{N}} \le a \right) \int_0^t [\lambda\,(1 - q_s^N) + f\,(q_s^N)]\, ds.$$

We will verify first Condition (sup C). We have (c is a constant involved in Lipschitz' condition) as $N > 1/a^2$

$$\sup_{s \le t} \left| \overline{\langle M^{N,a}\rangle}_s - \int_0^s c\,(u)\, du \right| \le \int_0^t [\lambda\, | q_s^N - q_s | + | f\,(q_s^N) - f\,(q_s) |]\, ds$$

$$\le t\,(\lambda + c) \sup_{s \le t} | q_s^N - q_s | .$$

Hence Condition (sup C) is a consequence of Lemma 1.

Condition (sup B) is verified analogously (c is a constant involved in Lipschitz' condition, $\theta_s \in [0, 1]$):

$$\sup_{s \le t} \left| \overline{B}_s^N - \int_0^s b(u) X_u^N \, du \right| \le \int_0^t | \sqrt{N} \, (f(q_s^N) - f(q_s)) - f'(q_s) X_s^N | \, ds$$

$$= \int_0^t | f'(q_s + \theta_s (q_s^N - q_s)) \sqrt{N} \, (q_s^N - q_s) - f'(q_s) X_s^N | \, ds$$

$$= \int_0^t | X_s^N | \, | f'(q_s + \theta_s (q_s^N - q_s)) - f'(q_s) | \, ds$$

$$\le \sup_{s \le t} | X_s^N | \sup_{s \le t} | q_s^N - q_s | \, ct.$$

This inequality gives ($t > 0$)

$$P\left(\sup_{s \le t} \left| \overline{B}_s^N - \int_0^s b(u) X_u^N \, du \right| \ge \varepsilon \right)$$

$$\le P\left(\sup_{s \le t} | X_s^N | \ge l \right) + P\left(\sup_{s \ge t} | q_s^N - q_s | \ge \frac{\varepsilon}{lct} \right).$$

By this and Lemma 1 Condition (sup B) takes place, provided

$$\lim_{l \to \infty} \overline{\lim_{N}} \, P\left(\sup_{s \le L} | X_s^N | \ge l \right) = 0 \tag{3.37}$$

for each $L \in R_+$.

To verify (3.37) we apply the representation (3.32) for X_t^N, by which as $t \le L$

$$| X_t^N | \le | X_0^N | + \int_0^t (\lambda + \sup_{0 \le x \le 1} | f'(x) |) | X_s^N | \, ds$$

$$+ \sqrt{N} \sup_{0 \le x \le 1} f(x) \int_0^L I(q_s^N = 0) \, ds + \sqrt{N} \sup_{t \le L} | M_t^N | .$$

By Theorem 2.4.3 this gives

$$| X_t^N | \le \left[| X_0^N | + \sqrt{N} \sup_{0 \le x \le 1} f(x) \int_0^L I(q_s^N = 0) \, ds + \sqrt{N} \sup_{t \le L} | M_t^N | \right]$$

$$\exp [(\lambda + \sup_{0 \le x \le 1} | f'(x) |) L]. \tag{3.38}$$

Consequently, $\sup_{t \le L} | X_t^N |$ does not exceed the right hand side of the inequality (3.38). Therefore (3.37) takes place, provided

$$\lim_{l \to \infty} \overline{\lim_N} \, P (| X_0^N | \ge l) = 0, \tag{3.39}$$

$$\lim_{l \to \infty} \overline{\lim_N} \left(\sqrt{N} \int_0^L I (q_s^N = 0) \, ds \ge l \right) = 0 \tag{3.40}$$

and

$$\lim_{l \to \infty} \overline{\lim_N} \, P \left(\sqrt{N} \sup_{t \le L} | M_t^N | \ge l \right) = 0. \tag{3.41}$$

The relation (3.39) holds since a sequence of random variables X_0^N, $N \ge 1$, converges in distribution to a certain random variable, and (3.40) holds by (3.36). By Doob's inequality (Theorem 1.9.1) and by (3.22)

$$P (\sqrt{N} \sup_{t \le L} | M_t^N | \ge l) \le \frac{N}{l^2} \mathbf{E} \langle M^N \rangle_L$$

$$= \frac{1}{l^2} \mathbf{E} \int_0^L [\lambda - q_s^N) + I (q_s^N > 0) f (q_s^N)] \, ds$$

$$\le l^{-2} (\lambda + \sup_{0 \le x \le 1} f (x)) L,$$

and hence (3.41) holds too.

Theorem 2 is proved.

5. As the diffusion approximation is so important for applications, we present a vector-valued version of the assertion of Theorem 1.

Suppose that a vector-valued semimartingale $(X_t, \mathcal{D}_t^Q)$, $(X_t = (X_t^1, \ldots, X_t^k))$ has the triplet $T = (B(X), C(X), 0)$, with

$$B_t (X) = \int_0^t b (s, X) \, ds \tag{3.42}$$

and

$$C_t (X) = \int_0^t c (s, X) \, ds, \tag{3.43}$$

where a vector $b (t, X)$ and a matrix $c (t, X)$ possess elements $b_j (t, X)$ and $c_{ij} (t, X)$, i,

$j = 1, \ldots, k$, which are predictable functions at each $t \in R_+$ and $X \in D(R^k)$, and a matrix $c(t, X)$ is symmetric and nonnegative definite. As in the scalar case, $(X_t)_{t \geq 0}$ admits the representation (possibly on an enlarged probability space)

$$X_t = X_0 + \int_0^t b(s, X)\,ds + \int_0^t c^{1/2}(s, X)\,dW_s$$

with a vector-valued Wiener process $(W_t)_{t \geq 0}$ $(W_t = (W_t^1, \ldots, W_t^k))$ consisting of independent components.

Let us formulate the conditions imposed on elements of $b(t, X)$ and $c(t, X)$:

$$\sum_{j=1}^{k} |b_j(t, X)| \leq L(t)(1 + \sup_{s \leq t} |X_s|), \tag{3.44}$$

$$\sum_{j=1}^{k} c_{jj}(t, X) \leq L(t)(1 + \sup_{s \leq t} |X_s|^2), \tag{3.45}$$

$$\int_0^t L(s)\,ds < \infty, \quad t \in R_+; \tag{3.46}$$

all vectors here and below are considered as column-vectors, and

$$|x| = \left(\sum_{j=1}^{k} (x^j)^2 \right)^{1/2}.$$

Let $X^n = (X_t^n, \mathcal{F}_t^n)$ be a P^n-semimartingale $(X_t^n = (X_t^{n1}, \ldots, X_t^{nk}))$ with the triplet of predictable characteristics $T^n = (B^n, C^n, \nu^n)$, and $M^{n,a} = (M_t^{n,a}, \mathcal{F}_t^n)$ a local square integrable martingale with

$$M_t^{n,a} = X_t^{nc} + \int_0^t \int_{|x| \leq a} x\,d(\mu^n - \nu^n), \quad a \in (0, 1],$$

and with the quadratic characteristic

$$\langle M^{n,a} \rangle_t = C_t^n + \int_0^t \int_{|x| \leq a} xx^*\,d\nu^n - \sum_{0 < s \leq t} \hat{x}_s^{na} (\hat{x}_s^{na})^*,$$

where

$$\hat{x}_t^{na} = \int_{|x| \le a} x\nu^n(\{t\}, dx),$$

and * is the transposition sign.

For each $L > 0$, $\varepsilon > 0$ and $a \in (0, 1]$ let

$$\text{(A)} \quad \lim_n P^n\left(\int_0^L \int_{|x|>a} d\nu^n \ge \varepsilon\right) = 0,$$

$$\text{(sup B)} \quad \lim_n P^n\left(\sup_{t \le L}\left|B_t^n - \int_0^t b(s, X^n)\,ds\right| \ge \varepsilon\right) = 0,$$

$$\text{(sup C)} \quad \lim_n P^n\left(\sup_{t \le L}\left|\langle M^{n,a}\rangle_t(ij) - \int_0^t c_{ij}(s, X^n)\,ds\right| \ge \varepsilon\right) = 0, \quad i, j = 1, \ldots, k.$$

Theorem 3. (a) *Let the conditions of "linear growth"* (3.44) *and* (3.45) *be fullfilled, as well as Conditions* (A), (sup B), (sup C) *and*

$$\lim_{l \to \infty} \overline{\lim_n}\, P^n(|X_0^n| \ge l) = 0. \tag{3.47}$$

Then the family of probability distributions $(Q^n)_{n \ge 1}$ *of semimartingales* $X^n = (X_t^n, \mathcal{F}_t^n)$, $n \ge 1$, *is relatively compact and the limit* Q' *of any weak convergent subsequence* $(Q^{n'})$ *of the sequence* $(Q^n)_{n \ge 1}$ *possesses the following property:*

$$Q'(C(R_k)) = 1.$$

(b) *Let the conditions stipulated in assertion* (a) *be fullfilled, and for each* $t \in S$ (S *is a subset dense in* R_+,

$$\int_{R_+ \setminus S} dt = 0)$$

let elements of the vector $b(t, X)$ *and the matrix* $c(t, X)$ *be continuous in Skorohod's metric at each point* $X \in C(R^k)$.

Then the process $(X_t, \mathcal{D}_t^{Q'})$ *is a* Q'*-semimartingale with the triplet* $T = (B(X), c(X), 0)$, *where* $B_t(X)$ *and* $C_t(X)$ *are determined by the formulas* (3.42) *and* (3.43), *while* Q' *is the limit of any weak convergent subsequence* $(Q^{n'})$ *of the sequence* $(Q^n)_{n \ge 1}$.

(c) *If the conditions stipulated in assertion* (b) *are fullfilled with*

$$X_0^n \xrightarrow{d} X_0$$

instead of (3.47) *and the distribution of* X_0, *the vector* b (t, X) *and the matrix* c (t, X) *uniquely determine the probability distribution* Q *of a semimartingale* $(X_t, \mathcal{D}_t^Q)$, *then*

$$Q^n \xrightarrow{w} Q$$

where Q^n *is the probability distribution of a semimartingale* X^n, $n \geq 1$.

Remark. The functions b (s, X) and c (s, X) involved in the definition of B_t (X) and C_t (X) (see (3.42) and (3.43)), may eventually be nonpredictable. It suffices to require that they are $B(R_+) \otimes \mathcal{D}_\infty$-measurable and $\mathcal{D}_s$-measurable for each $s \geq 0$.

Problems

1. Prove the representation (3.7).
2. Prove Theorem 3.
3. Let X^n be a semimartingale with the decomposition

$$X_t^n = X_0^n + A_t^n + M_t^n$$

where $A^n \in \mathcal{V} \cap C$ and $M^n \in \mathcal{M}^2_{loc}$, and let

$$L^{2,\varepsilon}(M^n) = \left(L_t^{2,\varepsilon}(M^n)\right)_{t \geq 0}$$

be the Lindeberg process:

$$L_t^{2,\varepsilon}(M^n) = \sum_{s \leq t} (\Delta M_s^n)^2 I(|\Delta M_s^n| > \varepsilon)$$

with the compensator $\tilde{L}^{2,\varepsilon}(M^n)$.

Show that Conditions (A), (sup B) and sup (C) in Theorem 1 are satisfied if for each $T > 0$

1) $\tilde{L}_T^{2,\varepsilon}(M^n) \xrightarrow{P} 0$,

2) $\sup_{t \leq T} \left| A_t^n - \int_0^t b(s, X_s^n)\, ds \right| \xrightarrow{P} 0$,

3) $\sup_{s \leq t} \left| \langle M^n \rangle_t - \int_0^t c(s, X_s^n)\, ds \right| \xrightarrow{P} 0$.

§ 4. Weak convergence to a distribution of a point process with a continuous compensator

1. Let $D\{1\}$ be a set of piecewise constant functions $X = (X_t)_{t \geq 0}$ in D with $X_0 = 0$ and with unit jumps. Assume a probability measure Q on $(D, \mathcal{D})$ is such that

$$Q(D\{1\}) = 1. \tag{4.1}$$

Obviously, then a stochastic process $(X_t, \mathcal{D}_t)$ presents (up to sets of zero Q-measure) a point (counting) process.

In this section we will treat the question concerning the weak convergence

$$Q^n \xrightarrow{w} Q, \tag{4.2}$$

where a measure Q possesses the property (4.1), while Q^n presents the probability distribution of a P^n-semimartingale $X^n = (X^n_t, \mathcal{F}^n_t)$ with the triplet of predictable characteristics $T^n = (B^n, C^n, \nu^n)$, $n \geq 1$.

On proving characteristic statements concerning the convergence (4.2) we will need the following general fact.

Lemma 1. 1) *The space* $D\{1\}$ *is closed in the metric* ρ *(see Ch.* 6).

2) *If* Q^n, $n \geq 1$, *are probability measures on* $(D, \mathcal{D})$ *such that*

$$Q^n(D\{1\}) = 1, \ n \geq 1$$

and

$$Q^n \xrightarrow{w} Q',$$

then

$$Q'(D\{1\}) = 1.$$

Proof. 1) Let $Y^n \in D\{1\}$, $n \geq 1$ and $Y \in D$. We will show that

$$\rho(Y^n, Y) \to 0 \Rightarrow Y \in D\{1\}.$$

In view of Corollary 1 to Theorem 6.1.2 we have $Y^n_0 \to Y_0$. Hence $Y_0 = 0$. By Corollary 2 to the same theorem one can choose a subsequence $(s_n)_{n \geq 1}$ such that for each $t \in R_+$ we have $s_n \to t$ and $\Delta Y^n_s \to \Delta Y_t$. Hence a function Y has jumps of unite size only. To each function Y we associate a function $Z(Y) = (Z_t(Y))_{t \geq 0}$ with

$$Z_t(Y) = Y_t - \sum_{0 < s \leq t} \Delta Y_s.$$

Obviously $Z(Y)$ admits the following representation:

$$Z_t(Y) = Y_t - \sum_{0 < s \leq t} I\left(|\Delta Y_s| > \frac{1}{2} \right).$$

In view of Example 4 (Ch. 6, § 2) a function $Z_t(Y)$ is continuous for each $t \in \Delta(Y) = \{t > 0: \Delta Y_t = 0\} \cup \{0\}$ in the metric ρ. Hence for each $t \in \Delta(Y)$

$$Z_t(Y) = \lim_n Z_t(Y^n) = \lim_n \left[Y_t^n - \sum_{0 < s \le t} I\left(|\Delta Y_s^n| > \frac{1}{2} \right) \right] = 0.$$

Hence Y_t admits for each $t \in \Delta(Y)$ the representation

$$Y_t = \sum_{0 < s \le t} \Delta Y_s. \tag{4.3}$$

The representation (4.3) for Y remains true for every $t \in R_+$, since $Y \in D$ and $\Delta(Y)$ is a set dense in R_+.

2) Define a function $f = f(X)$, $X \in D$ by setting

$$f(X) = \inf_{Y \in D\{1\}} \rho(X, Y).$$

In view of Problem 6.1.7 the function $f = f(X)$ is continuous in the metric ρ. Therefore

$$\lim_n \int_D (1 - e^{-f(X)})\, dQ^n = \int_D (1 - e^{-f(X)})\, dQ'.$$

Since $f(X) = 0$, as $X \in D\{1\}$, then for every $n \ge 1$

$$\int_D (1 - e^{-f(X)})\, dQ^n = \int_{D\{1\}} (1 - e^{-f(X)})\, dQ^n = 0,$$

i.e.

$$\int_D (1 - e^{-f(X)})\, dQ' = 0,$$

and hence

$$\int_{D \setminus D\{1\}} (1 - e^{-f(X)})\, dQ' = 0.$$

The last equality takes place only if

$$Q'(D \setminus D\{1\} = 0,$$

since the space $D\{1\}$ is close in the metric ρ, and thus for each $X \in D \setminus D\{1\}$ the inequality $f(X) > 0$ holds.

The lemma is proved.

2. Let the measure Q possess the property (4.1), i.e. $(X_t, \mathcal{D}_t^Q)$ is a counting process with the compensator $B(X) = (B_t(X))_{t \ge 0}$ such that

$$B_t(X) = \int_0^t b(s, X)\, du_s, \tag{4.4}$$

where $b(t, X)$ is a $\mathcal{P}(\mathbb{D})$-measurable nonnegative function with the following property: for each $t \in R_+$ and $X \in D$

$$b(t, X) \le L(t)(1 + \sup_{s \le t} |X_s|); \quad \int_0^t L(s)\, du_s < \infty, \; t \in R_+, \qquad (4.5)$$

while a function

$$u = (u_t)_{t \ge 0} \in V^+ \cap C.$$

Theorem 1. *Let Conditions* (4.4) *and* (4.5) *be fullfilled, and for every* t *in a set* S, *dense in* R_+ *let* b (t, X) *be a function, continuous in the metric* ρ *at each point* $X \in D\{1\}$.

For each $n \ge 1$ *let a* P^n*-semimartingale* $X^n = (X_t^n, \mathcal{F}_t^n)$ *present a point (counting) process with the probability distribution* Q^n *and the compensator* $B^n = (B_t^n)_{t \ge 0}$ *such that for each* $L > 0$ *and* $\varepsilon > 0$

$$\lim_n P^n\left(\sup_{t \le L} \left| B_t^n - \int_0^t b(s, X^n)\, du_s \right| \ge \varepsilon \right) = 0. \qquad (4.6)$$

Then

$$Q^n \xrightarrow{w} Q.$$

Remark. If $b(t, X) \le L(t)$ and the second inequality in (4.5) is fullfilled, then Condition (4.6) is equivalent to the condition (for each $L > 0$ and $\varepsilon > 0$)

$$\lim_n P^n\left(\left| B_L^n - \int_0^L b(s, X^n)\, du_s \right| \ge \varepsilon \right) = 0. \qquad (4.7)$$

Proof of Theorem 1. Observe that the triplets $T = (B(X), C(X), \nu(X))$ and $T^n = (B^n, C^n, \nu^n)$, $n \ge 1$, of semimartingales $(X_t, \mathcal{D}_t^Q)$ and $(X_t^n, \mathcal{F}_t^n)$ possess the following properties:

$$C(X) = 0, \quad C^n = 0, \; n \ge 1,$$
$$\nu(X; (0, t] \times \{1\}) = B_t(X), \quad \nu(X; R_+ \times R_0 \setminus \{1\}) = 0,$$
$$\nu^n((0, t] \times \{1\}) = B_t^n, \quad \nu^n(R_+ \times R_0 \setminus \{1\}) = 0, \; n \ge 1.$$

It is not hard to verify that all conditions of assertion (a) of Theorem 2.1 are fullfilled. Therefore the family $(Q^n)_{n \ge 1}$ is relatively compact.

If Q' is the limit of any weak convergent subsequence $(Q^{n'})$ of the sequence $(Q)_{n \ge 1}$, then by Lemma 1 we have

$$Q'(D\{1\}) = 1.$$

Consequently, as $t \in S$ the function $b(t, X)$ is continuous in the metric ρ Q'-a.s., i.e. by assertion (b) of Theorem 2.1 the process $(X_t, \mathcal{D}_t^{Q'})$ presents a point process relative to the measure Q' with the compensator $B(X)$. Since $X_0^n = 0$, $n \geq 1$, and $X_0 = 0$, and the compensator $B(X)$ of a point process defines uniquely the probability distribution (Proposition 3.4.2), then in accordance with assertion (c) of Theorem 2.1 the convergence

$$Q^n \xrightarrow{w} Q$$

takes place.

3. Theorem 2. *Let Conditions* (4.4) *and* (4.5) *be satisfied and for each* t *from a set* S *dense in* R_+ *let* $b(t, X)$ *be a continuous function in the metric* ρ *at each point* $X \in D\{1\}$.

For every $n \geq 1$ *let a* P^n*-semimartingale* $X^n = (X_t^n, \mathcal{F}_t^n)$ *with the probability distribution* Q^n *and with the triplet of predictable characteristics* $T^n = (B^n, C^n, \nu^n)$ *be such that for any* $L > 0$, $\varepsilon > 0$ *and* $\delta \in (0, 1/2)$ *the following conditions are fullfilled:*

$$\text{(sup a)} \quad \lim_n P^n\left(\sup_{t \leq L}\left| I(|x-1| \leq \delta) * \nu_t^n - \int_0^t b(s, X^n)\, du_s \right| \geq \varepsilon\right) = 0,$$

$$\text{(sup b)} \quad \lim_n P^n\left(\sup_{t \leq L} | B_t^n - xI(\delta < |x| \leq 1) * \nu_t^n | \geq \varepsilon\right) = 0,$$

$$\text{(c)} \quad \lim_n P^n\left(C_L^n + x^2 I(|x| \leq \delta) * \nu_L^n - \sum_{0 < s \leq L}\left(\int_{|x| \leq \delta} x\nu^n(\{s\}, dx)\right)^2 \geq \varepsilon\right) = 0,$$

$$\text{(d)} \quad \lim_n P^n(\nu^n((0, L] \times \Delta_\delta) \geq \varepsilon) = 0, \quad \Delta_\delta = \{x : |x| > \delta,\ |x-1| > \delta\},$$

$$\text{(f)} \quad X_0^n \xrightarrow{d} 0.$$

Then

$$Q^n \xrightarrow{w} Q.$$

Remark. If $b(t, X) \leq L(t)$ and the second inequality (4.5) is fullfilled, then Condition (sup a) is equivalent to the condition (for any $L > 0$ and $\varepsilon > 0$)

$$\lim_n \mathsf{P}^n\left(\left| I(|x-1|\le\delta) * \nu_L^n - \int_0^L b(s, X^n)\, du_s \right| \ge \varepsilon\right) = 0.$$

Proof of Theorem 2. Let $\mu^n = \mu^n(dt, dx)$ be the jump measure of X^n. Denote

$$Y_t^{n\delta} = I(|x-1|\le\delta) * \mu_t^n.$$

Obviously, $(Y_t^{n\delta}, \mathcal{F}_t^n)$ is a P^n-semimartingale presenting a point process with the compensator

$$A_t^n = I(|x-1|\le\delta) * \nu_t^n.$$

Let $\overline{\mathsf{Q}}^n$ be the probability distribution $Y^{n\delta} = (Y_t^{n\delta}, \mathcal{F}_t^{n\delta})$. By Condition (sup a) and Theorem 1

$$\overline{\mathsf{Q}}^n \xrightarrow{w} \mathsf{Q}.$$

Conditions (sup b), (c), (d) and (f) of the theorem guarantee the following "approximation" of the processes X^n and $Y^{n\delta}$:

$$\lim_n \mathsf{P}^n\left(\sup_{t\le L} |X_t^n - Y_t^{n\delta}| \ge \varepsilon\right) = 0$$

for each $L > 0$ and $\varepsilon > 0$. This statement is proved so as the relation (2.1) (Ch. 7, § 2), replacing P by P^n.

Therefore by Problem 6.2.1

$$\mathsf{Q}^n \xrightarrow{w} \mathsf{Q}.$$

4. Example. Let a P^n-semimartingale $X^n = (X_t^n, \mathcal{F}_t^n)$ present a point process with the compensator

$$B_t^n = \int_0^1 (1 + X_s^n \Theta_s^n)\, ds,$$

where $(\Theta_t^n, \mathcal{F}_t^n)$ is a P^n-Markov process with two states (0, 1) and the transition intensity matrix

$$n\begin{pmatrix} -\lambda_1 & \lambda_1 \\ \lambda_2 & -\lambda_2 \end{pmatrix}.$$

Let us show that the sequence of probability distributions Q^n, $n \ge 1$, of the point processes X^n, $n \ge 1$, converges weakly to the probability measure Q which presents the probability distribution of a point process with the compensator

$$B_t(X) = \int_0^t \left(1 + X_{s-} \frac{\lambda_1}{\lambda_1 + \lambda_2}\right) ds.$$

Since

$$\int_0^\infty |\Delta X_s| \, ds = 0$$

for each $X \in D$, it suffices to verify by Theorem 1 that for each $L > 0$ and $\varepsilon > 0$

$$\lim_n P^n \left(\sup_{t \le L} \left| \int_0^t X^n_{s-} \left(\Theta^n_s - \frac{\lambda_1}{\lambda_1 + \lambda_2} \right) ds \right| \ge \varepsilon \right) = 0. \tag{4.8}$$

On considering Example in Ch. 5, § 3 it has been indicated that (see (3.30))

$$\Theta^n_t = \Theta^n_0 + n \int_0^t [\lambda_1 - (\lambda_1 + \lambda_2)\, \Theta^n_s] \, ds + M^n_t, \tag{4.9}$$

where $(M^n_t, \mathcal{F}^n_t)$ is a locally square integrable martingale with the quadratic characteristics

$$\langle M^n \rangle_t = n \int_0^t [\lambda_1 (1 - \Theta^n_s) + \lambda_2 \Theta^n_s] \, ds. \tag{4.10}$$

By Ito's formula (Ch. 2, § 3) we get, taking into consideration (4.9), that

$$X^n_t \Theta^n_t = n \int_0^t X^n_{s-} [\lambda_1 - (\lambda_1 + \lambda_2)\, \Theta^n_s] \, ds + \int_0^t X^n_{s-} \, dM^n_s + \int_D^t \Theta^n_s \, dX^n_s.$$

This gives

$$\sup_{t \le L} \left| \int_0^t X^n_{s-} \left(\Theta^n_s - \frac{\lambda_1}{\lambda_1 + \lambda_2} \right) ds \right| \le \frac{2X^n_L}{n(\lambda_1 + \lambda_2)} + \frac{\sup_{t \le L} \left| \int_0^t X^n_{s-} \, dM^n_s \right|}{n(\lambda_1 + \lambda_2)}.$$

Set

$$\tau_{nl} = \inf(t: X^n_t \ge l), \quad l \ge 1.$$

Then

$$\sup_{t \le L \wedge \tau_{nl}} \left| \int_0^t X^n_{s-} \left(\Theta^n_s - \frac{\lambda_1}{\lambda_1 + \lambda_2} \right) ds \right| \le \frac{2l}{n(\lambda_1 + \lambda_2)} + \frac{\sup_{t \le L \wedge \tau_{nl}} \left| \int_0^t X^n_{s-}\, dM^n_s \right|}{n(\lambda_1 + \lambda_2)} .$$

By Doob's inequality (Theorem 1.9.1) and (4.10)

$$P^n \left(\frac{1}{n} \sup_{t \le L \wedge \tau_{nl}} \left| \int_0^t X^n_{s-}\, dM^n_s \right| \ge \varepsilon \right) \le \frac{1}{(n\varepsilon)^2} E^n \int_0^{L \wedge \tau_{nl}} (X^n_{s-})^2\, d\langle M^n \rangle_s$$

$$\le \frac{l^2 (\lambda_1 + \lambda_2) L}{n\varepsilon^2} \to 0, \; n \to \infty.$$

Therefore

$$\lim_n P^n \left(\sup_{t \le L \wedge \tau_{nl}} \left| \int_0^t X^n_{s-} \left(Q^n_s - \frac{\lambda_1}{\lambda_1 + \lambda_2} \right) ds \right| \ge \varepsilon \right) = 0,$$

and hence the desired relation (4.8) takes place, provided

$$\lim_{l \to \infty} \overline{\lim_n}\, P^n (\tau_{nl} < L) = 0 \tag{4.11}$$

for each $L > 0$.

But

$$\{\tau_{nl} < L\} \subseteq \{X^n_L \ge l\}.$$

Consequently, it suffices to show that for each $L > 0$

$$\lim_{l \to \infty} \overline{\lim_n}\, P^n (X^n_L \ge l) = 0. \tag{4.12}$$

By the definition of the compensator B^n we have

$$E^n X^n_{t \wedge \tau_{nl}} = E^n \int_0^{t \wedge \tau_{nl}} (1 + X^n_s \Theta^n_s)\, ds \le \int_0^t (1 + E^n X^n_{s \wedge \tau_{nl}})\, ds.$$

By Theorem 2.4.3 this gives

$$E^n X^n_{L \wedge \tau_{nl}} \le L e^L.$$

In view of $t_{nl} \uparrow \infty$ as $l \to \infty$, by the monotone limiting transition under the Lebesgue integral sign we get

$$E^n X^n_L = \lim_{l \to \infty} E^n X^n_{L \wedge \tau_{nl}} \le L e^L.$$

Hence by Chebyshev's inequality

$$\overline{\lim_n}\ \mathsf{P}^n (X^n_L \geq l) \leq \frac{1}{l}\ \overline{\lim_n}\ \mathbf{E}^n X^n_L \leq \frac{Le^L}{l} \to 0,\ l \to \infty,$$

i.e. (4.12) holds.

Problem

1. Let $X^n = (X^n_t, \mathcal{F}^n_t)$, $n \geq 1$, be a semimartingale, presenting a point process with the compensator

$$B^n_t = \int_0^t (1 + X^n_s f(\Theta^n_s))\, ds,$$

where $f = f(x)$ is a nonnegative bounded and continuous function, while Θ^n_t is the solution of Ito's stochastic equation

$$d\Theta^n_t = -\ ,\ (\Theta^n_t - a_t)\, dt + dW^n_t$$

with the initial condition Θ^n_0 and a Wiener process $(W^n_t, \mathcal{F}^n_t)$, where a_t, $t \geq 0$ is a continuous function.

Show that the sequence of probability distributions Q^n of point processes X^n, $n \geq 1$, converges weakly to the probability distribution Q of a point process with the compensator

$$B_t(X) = \int_0^t (1 + X_{s-} f(a_s))\, ds.$$

§ 5. Weak convergence of invariant measures

1. For every $x \in R$ let

$$D_x = \{X \in D: X_0 = x\}, \tag{5.1}$$

and let Q_x be a probability measure on $(D, \mathcal{D})$ such that

$$Q_x(D_x) = 1, \quad x \in R, \tag{5.2}$$

and for each $\Gamma \in \mathcal{D}$ a function $Q_x(\Gamma)$ is $B(R)$-measurable.

To a probability measure $r = r(dx)$ on $(R, B(R))$ we relate a probability measure Q^r on $(D, \mathcal{D})$ by setting

$$Q^r(\Gamma) = \int_R Q_x(\Gamma)\, r(dx), \quad \Gamma \in \mathcal{D}. \tag{5.3}$$

Definition. A probability measure $r = r(dx)$ is called *invariant* for a family Q_x, $x \in R$, if for each nonnegative continuous bounded function $f = f(x)$ the function

$$\int_D f(X_t)\, dQ^r, \quad t \geq 0,$$

is a constant (independent of t).

2. Let $(Q_x^n)_{n \geq 1}$ be a family of probability measures on $(D, \mathcal{D})$, where a measure Q_x^n for every $n \geq 1$ possesses the properties indicated above.

Let $r = r(dx)$ and $r^n = r^n(dx)$, $n \geq 1$, be measures invariant for families Q_x, $x \in R$, and Q_x^n, $x \in R$, $n \geq 1$. There arises the natural question which conditions ensure

$$r^n \xrightarrow{w} r.$$

The answer to this question is presented in

Theorem 1. *Let the following conditions be fullfilled:*

1) *for each* $x \in R$ *and each sequence* x^n, $n \geq 1$, *such that*

$$\lim_n x^n = x,$$

the weak convergence

$$Q^n_{x^n} \xrightarrow{w} Q_x$$

takes place;

2) *the family of invariant measures* $(r^n)_{n \geq 1}$ *is relatively compact;*

3) *for each nonnegative continuous and bounded function* $f = f(x)$ *and each* $t \in R_+$ *the function*

$$g_t(x) = \int_D f(X_t)\, dQ_x$$

is continuous in x;

4) *the measure* $r = r(dx)$ *invariant for the family* Q_x, $x \in R$, *is unique.*

Then

$$r^n \xrightarrow{w} r.$$

Proof. Since $(r^n)_{n \geq 1}$ is the relatively compact family of probability measures, there exists a weak convergent subsequence $(r^{n'})$ of the sequence $(r^n)_{n \geq 1}$ with the weak limit r'.

We will show that r' is a measure invariant for the family Q_x, $x \in R$. By the definition of the measure $Q^{r'}$ it follows that for each nonnegative bounded B (R)-measurable function $f = f(x)$ and each $t \in R_+$

$$\int_D f(X_t)\, dQ^{r'} = \int_R \int_D f(X_t)\, dQ_x r'(dx) = \int_R g_t(x)\, r'(dx), \tag{5.4}$$

where

$$g_t(x) = \int_D f(X_t)\, dQ_x. \tag{5.5}$$

Let $f = f(x)$ be a continuous nonnegative and bounded function. In accordance with the definition of an invariant measure, it suffices to show that for each function $f = f(x)$ with the indicated properties we have

$$\int_R g_t(x)\, r'(dx) = \int_R g_s(x)\, r'(dx) \tag{5.6}$$

for each $s, t \in R$.

We utilize now assumption 3), in view of which $g_t(x)$ is a continuous function in x for every $t \in R_+$. Then by the weak convergence

$$r^{n'} \xrightarrow{w} r'$$

for each $t \in R_+$ the equality

$$\lim_{n'} \int_R g_t(x)\, r^{n'}(dx) = \int_R g_t(x)\, r'(dx) \tag{5.7}$$

takes place, and consequently

$$\int_R [g_t(x) - g_s(x)]\, r'(dx) = \lim_{n'} \int_R [g_t(x) - g_s(x)]\, r^{n'}(dx). \tag{5.8}$$

For a given function $f = f(x)$ denote

$$g_t^n(x) = \int_D f(X_t)\, dQ_x^n. \tag{5.9}$$

Since the measure r^n is invariant for the family Q_x^n, $x \in R$, then

$$\int_R g_t^n(x)\, r^n(dx) = \int_R g_s^n(x)\, r^n(dx), \quad t,\ s \in R,\ n \geq 1. \tag{5.10}$$

In view of (5.8) and (5.10) we get

$$\left| \int_R [g_t(x) - g_s(x)]\, r'(dx) \right|$$

$$\leq \overline{\lim_{n'}} \int_R |g_t(x) - g_t^{n'}(x)|\, r^{n'}(dx) + \overline{\lim_{n'}} \int_R |g_s(x) - g_s^{n'}(x)|\, r^{n'}(dx).$$

By this estimate it follows that (5.6) takes place provided

$$\lim_n \int_R |g_t(x) - g_t^n(x)|\, r^n(dx) = 0 \tag{5.11}$$

for ech $t \in R_+$.

To establish (5.11), we note that for each $l > 0$ the inequality

$$\int_R |g_t(x) - g_t^n(x)|\, r^n(dx) \leq 2 \sup_{x \in R} f(x) \int_{|x| \geq l} r^n(dx) + \sup_{|x| \leq l} |g_t(x) - g_t^n(x)|$$

takes place. Since the family of measures $(r^n)_{n \geq 1}$ is relatively compact, in virtue of Prohorov's theorem [248] the family $(r^n)_{n \geq 1}$ is dense, i.e.

$$\lim_{l \to \infty} \overline{\lim_n} \int_{|x| \geq l} r^n(dx) = 0,$$

and consequently it suffices to show that for each $l > 0$

$$\lim_n \sup_{|x| \leq l} |g_t(x) - g_t^n(x)|. \tag{5.12}$$

Fix $l > 0$ and $\varepsilon > 0$. By the definition of sup one can choose for each $n \geq 1$ a number x^n with the properties: $|x^n| \leq l$,

$$\sup_{|x| \leq l} |g_t(x) - g_t^n(x)| \leq \varepsilon + |g_t(x^n) - g_t^n(x^n)|. \tag{5.13}$$

Choose a subsequence $(\tilde{n})$ of the sequence (n) such that

$$\overline{\lim_n} \sup_{|x| \leq l} |g_t(x) - g_t^n(x)| = \lim_{\tilde{n}} \sup_{|x| \leq l} |g_t(x) - g_t^{\tilde{n}}(x)|$$

and

$$\lim_{\tilde{n}} x^{\tilde{n}} = \tilde{x},$$

where $\tilde{x}$ is a certain point of the interval $[-1, 1]$. Then by (5.13) we have

$$\overline{\lim_{n}} \sup_{|x| \le 1} |g_t(x) - g_t^{n}(x)| \le \varepsilon + \overline{\lim_{\tilde{n}}} |g_t(x^{\tilde{n}}) - g_t^{\tilde{n}}(x^{\tilde{n}})|$$

$$\le \varepsilon + \overline{\lim_{\tilde{n}}} |g_t(x^{\tilde{n}}) - g_t(\tilde{x})| + \overline{\lim_{\tilde{n}}} |g_t^{\tilde{n}}(x^{\tilde{n}}) - g_t(\tilde{x})| . \tag{5.14}$$

By Condition 3) we have

$$g_t(x^{\tilde{n}}) \to g_t(\tilde{x}),$$

while by Condition 1) we have

$$g_t^{\tilde{n}}(x^{\tilde{n}}) \to g_t(\tilde{x})$$

for each

$$t \in \{t > 0: Q_{\tilde{x}}(\Delta X_t = 0) = 1\} \cup \{0\} = \Delta_{Q_{\tilde{x}}}$$

(Theorem 6.1.8), since

$$g_t^{\tilde{n}}(x^{\tilde{n}}) = \int_D f(X_t)\, dQ^{\tilde{n}}_{x^{\tilde{n}}}, \quad g_t(\tilde{x}) = \int_D f(X_t)\, dQ_{\tilde{x}}.$$

Thus it is deduced from (5.14) that as $t \in \Delta_{Q_{\tilde{x}}}$

$$\overline{\lim_{n}} \sup_{|x| \le 1} |g_t(x) - g_t^{n}(x)| \le \varepsilon \to 0, \ \varepsilon \to 0,$$

and hence the equality (5.6) takes place for each s, $t \in \Delta_{Q_{\tilde{x}}}$, i.e. the function

$$\int_D f(X_t)\, dQ^{r'} = \int_R g_t(x)\, r'(dx), \ t \ge 0,$$

is a constant on $\Delta_{Q_{\tilde{x}}}$.

On the other hand the function

$$\int_D f(X_t)\, dQ^{r'}, \ t \ge 0,$$

belongs to D, while $\Delta_{Q_{\tilde{x}}}$ is a set, dense in R_+. Consequently,

$$\int_D f(X_t)\, dQ^{r'} = \text{const}, \ t \in R_+,$$

i.e. r' is a measure invariant for the family Q_x, $x \in R$.

In view of the uniqueness of measures invariant for the family Q_x, $x \in R$, we have

$$r' = r.$$

We will establish now the weak convergence

$$r^n \xrightarrow{w} r.$$

Let $g = g(x)$ be a bounded continuous function, (n') a subsequence of the sequence (n) such that

$$\overline{\lim_n} \int_R g(x)\, r^n(dx) = \lim_{n'} \int_R g(x)\, r^{n'}(dx)$$

and $(r^{n''})$ a weak convergent subsequence of the sequence $(r^{n'})_{n' \ge 1}$ with the weak limit r''. As it has been proved

$$r'' = r.$$

Then

$$\overline{\lim_n} \int_R g(x)\, r^n(dx) = \lim_{n''} \int_R g(x)\, r^{n''}(dx) = \int_R g(x)\, r(dx).$$

It is proved analogously that

$$\underline{\lim_n} \int_R g(x)\, r^n(dx) = \int_R g(x)\, r(dx).$$

Hence the desired assertion

$$r^n \xrightarrow{w} r$$

takes place.

3. Remark. It is not simple, in general, to verify the conditions of Theorem 1. However, if Q_x and $Q^n_{x^n}$, $n \ge 1$, are the probability distributions of semimartingales with the initial values x and x^n, $n \ge 1$, then Condition 1) of Theorem 1 can be established under the same assumptions as in Theorems 2.1, 3.1, 4.1 and 4.2. Besides, the compactness of the family $(r^n)_{n \ge 1}$ takes place, provided

$$\lim_{l \to \infty} \overline{\lim_n} \overline{\lim_{t \to \infty}} Q^n_x(|X_t| \ge l) = 0 \tag{5.15}$$

for each $x \in R$.

Conditions 3) and 4) are often easily verified, for instance, in case in which Q_x is the distribution of a Markov (for example, diffusion) process.

4. We consider the same example from the queueing theory as in Subsection 3.4 in order to show how Theorem 1 is utilized for determining conditions for the weak convergence of invariant measures.

Let $(Q_t)_{t \ge 0}$ be a queue defined by (3.18). By the representation of the compensators $\tilde{A}$ and $\tilde{D}$ (see (3.19)) a Markov process $(Q_t)_{t \ge 0}$, possessing the intensity matrix $||\Lambda_{ij}||$ of transition probabilities with the elements

$$\Lambda_{i,\,i+1} = \lambda\,(N - i),\ i \le N - 1, \quad \Lambda_{i,\,i-1} = Nf\left(\frac{i}{N}\right),\ i \ge 1, \tag{5.16}$$
$$\Lambda_{ij} = 0,\ |i - j| \ge 2, \quad \Lambda_{ii} = -\sum_{j \ne i} \Lambda_{ij},\ i, j = 0, 1, \ldots, N,$$

takes values in the set $\{0, 1, \ldots, N\}$. The function $f = f(x)$ is involved also in the expression on the right-hand side of the differential equation

$$q_t = \lambda\,(1 - q_t) - f\,(q_t) \tag{5.17}$$

(see (3.21)).

Let us formulate the conditions imposed on the function $f = f(x)$ which we will need below:

(α) $f = f(x)$ is a nonnegative differential function the derivate of which $f' = f'(x)$ satisfies Lipschitz' condition and the condition

$$\inf_{0 \le x \le 1} f'(x) > -\lambda; \tag{5.18}$$

(β) the equation

$$\lambda\,(1 - x) = f\,(x) \tag{5.19}$$

has the unique positive solution q ($q > 0$, $\lambda\,(1 - q) = f(q)$);

(γ) for every $N \ge 1$ a Markov process $(Q_t)_{t \ge 0}$ has the invariant (stationary) distribution.

Let

$$q_t^N = \frac{Q_t}{N}$$

(see (3.20)) and

$$X_t^N = \sqrt{N}\,(q_t^N - q), \tag{5.20}$$

where q is the solution of the equation (5.19). Let Q_x^N be the probability distribution of the process $(X_t^N)_{t \ge 0}$ with $X_0^N = x$, where x takes values in the set

$$J^N = \{i/\sqrt{N} - q\sqrt{N},\ i = 0, 1, \ldots, N\}.$$

In view of Condition (γ) for the family of measures Q_x^N, $x \in J^n$, there exists the invariant measure $r^N = r^N(dx)$ with the support J^N, $N \ge 1$.

Let Q_x be the probability distribution of a Gaussian stochastic process $(X_t)_{t \ge 0}$ with $X_0 = x$ that is defined by Ito's stochastic differential equation

$$dX_t = -(\lambda + f'(q))\,X_t\,dt + \sqrt{\lambda\,(1 - q) + f(q)}\ dW_t \tag{5.21}$$

with a Wiener process $(W_t)_{t \ge 0}$. Observe that by (β) we have

$$\lambda\,(1 - q) + f(q) = 2\lambda\,(1 - q),$$

while by (5.18) the unique measure $r = r(dx)$ invariant for the family Q_x, $x \in R$, is presented by

$$r(dx) = \frac{1}{\sqrt{2\pi\sigma^2}} \exp\left(-\frac{x^2}{2\sigma^2} \right) dx \tag{5.22}$$

with

$$\sigma^2 = \frac{f(q)}{\lambda + f'(q)} . \tag{5.23}$$

Theorem 2. *Let Conditions* (α) - (γ) *be fullfilled. Then*

$$r^N \xrightarrow{w} r, \ N \to \infty.$$

Proof. It suffices to verify Conditions 1) - 3) of Theorem 1.

If $x^N \to x$ as $N \to \infty$, then by the condition $X_0^N = x^N$ and (5.20) we get

$$q_0^N - q = \frac{x^N}{\sqrt{N}} \to 0,$$

as $N \to \infty$. This, (α) and (β) entail the conditions of Theorem 3.2, according to which

$$Q^N_{x^N} \xrightarrow{w} Q_x .$$

Thus Condition 1) of Theorem 1 is fullfilled.

Let us verify now the relative compactness of the family $(r^N)_{N \geq 1}$. Aiming at this we will verify the property (5.15), i.e. we will show that

$$\lim_{l \to \infty} \overline{\lim_{N}} \ \overline{\lim_{t \to \infty}} \ P(|X_t^N| \geq l) = 0 \tag{5.24}$$

as $X_0^N = x, \ x \in J^N$.

The desired relation obviously takes place, provided

$$\overline{\lim_{t \to \infty}} \ \mathbf{E}(X_t^N)^2 \leq c \tag{5.25}$$

with a constant c independent of N.

Let us calculate $\mathbf{E}(X_t^N)^2$. We need for this the representation for $(X_t^N)^2$. By (3.32) X_t^N admits the representation

$$X_t^N = x + \int_0^t [-\lambda X_s^N - \sqrt{N}\,(I(q_s^N > 0)\, f(q_s^N) - f(q))]\, ds + \sqrt{N}\, M_t^N ,$$

where $x \in J^N$ and q is the solution of the equation (5.19). By Ito's formula (Ch. 2, § 3) we get

$$(X_t^N)^2 = x^2 + 2\int_0^t X_s^N [-\lambda X_s^N - \sqrt{N}\ (I(q_s^N > 0) f(q_s^N) - f(q))]\, ds$$
$$+ 2\sqrt{N}\int_0^t X_{s-}^N\, dM_s^N + N[M^N, M^N]_t.$$

Consequently

$$\mathbf{E}(X_t^N)^2 = x^2 + 2\int_0^t \mathbf{E}[-\lambda (X_s^N)^2 - \sqrt{N}\ X_s^N (I(q_s^N > 0) f(q_s^N) - f(q))]\, ds$$
$$+ N\mathbf{E}\langle M^N\rangle_t. \tag{5.26}$$

Observe now that (assume $\frac{0}{0} = 0$)

$$\sqrt{N}\ X_s^N (I(q_s^N > 0) f(q_s^N) - f(q))$$
$$= \sqrt{N}\ X_s^N (f(q_s^N) - f(q)) - \sqrt{N}\ X_s^N I(q_s^N = 0) f(q_s^N)$$
$$= (X_s^N)^2 \frac{f(q_s^N) - f(q)}{q_s^N - q} + Nqf(0)\, I(q_s^N = 0). \tag{5.27}$$

From (5.26), (5.27) and the representation for $\langle M^N\rangle_t$ (see (3.22)) it follows that $V_t^N = \mathbf{E}(X_t^N)^2$ satisfies the relation

$$V_t^N = x^2 - 2\int_0^t (\lambda + \inf_{0 \le x \le 1} f'(x))\, V_s^N\, ds + \int_0^t (a_s^N + b_s^N)\, ds \tag{5.28}$$

with

$$a_s^N = 2\mathbf{E}\left[\left(\inf_{0 \le x \le 1} f'(x) - \frac{f(q_s^N) - f(q)}{q_s^N - q}\right)(X_s^N)^2\right] - 2Nqf(0)\, \mathsf{P}(q_s^N = 0)$$

and

$$b_s^N = \mathbf{E}[\lambda(1 - q_s^N) + I(q_s^N > 0) f(q_s^N)].$$

Denote

$$p = 2\lambda + \inf_{0 \le x \le 1} f'(x)$$

and observe the following properties of p, a_s^N and b_s^N: $p > 0$ by (5.18); $a_s^N < 0$ by construction; $b_s^N \le \lambda + \sup_{0 \le x \le 1} f(x) = p'$. In view of these properties we deduce from (5.28) that

$$V_t^N = x^2 e^{-pt} + \int_0^t e^{-p(t-s)} (a_s^N + b_s^N)\, ds \le x^2 e^{-pt} + p' \int_0^t e^{-p(t-s)}\, ds$$

$$= x^2 e^{-pt} + p'(1 - e^{-pt}).$$

Consequently

$$\overline{\lim_{t \to \infty}}\, V_t^N \le p' < \infty,$$

i.e. (5.25) takes place and hence Condition 2) of Theorem 1 is fullfilled.

Condition 3) of the theorem is fullfilled too, since in our case

$$g_t(x) = \int_R f(y) \frac{1}{\sqrt{2\pi\gamma_t}} \exp\left(-\frac{(y - m_t(x))^2}{2\gamma_t} \right) dy,$$

with

$$m_t(x) = x \exp(-(\lambda + f'(q))\, t)$$

and

$$\gamma_t = \frac{f(q)}{\lambda + f'(q)} (1 - \exp(-2(\lambda + f'(q))\, t).$$

Thus the conditions of Theorem 1 are verified and hence Theorem 2 is proved.

Problem

1. Let X^x be an element of the set D_x (see (5.1)), and let a measure Q^r be defined by (5.3). Show that if for every $x \in R$ the process $(X_t^x, \mathcal{D}_t^{Q_x})$ is a Q_x-semimartingale with the triplet of predictable characteristics $T^x = (B(X^x), C(X^x), \nu(X^x))$, then $(X_t, \mathcal{D}_t^{Q^r})$ is a Q^r-semimartingale with the triplet of predictable characteristics $T = (B(X), C(X), \nu(X))$.

PART III

CHAPTER 9

INVARIANCE PRINCIPLE AND DIFFUSION APPROXIMATION FOR MODELS GENERATED BY STATIONARY PROCESSES

§ 1. Generalization of Donsker's invariance principle

1. Let $(\xi_k)_{k \geq 1}$ be a sequence of independent and identically distributed random variables with $\mathbf{E}\xi_1 = 0$ and $\mathbf{E}\xi_1^2 = \sigma^2$.

Denote

$$X^n = (X_t^n)_{0 \leq t \leq 1}, \quad X_t^n = \frac{1}{\sqrt{n}} \sum_{k=1}^{[nt]} \xi_k. \tag{1.1}$$

Donsker's theorem (see Ch. 7, Subsection 1.6, Example 4) tells us that

$$X^n \xrightarrow{d} \sigma W, \tag{1.2}$$

where $W = (W_t)_{0 \leq t \leq 1}$ is a Wiener process.

We present first a simple generalization of this result.

Let $(\xi_k)_{k \geq 1}$ be a sequence of random variables and $(\mathcal{F}_k)_{k \geq 1}$ a family of σ-algebras $\mathcal{F}_k = \sigma\{\xi_1, \ldots, \xi_k\}$.

Theorem 1. *Let* $(\xi_k)_{k \geq 1}$ *be a strictly stationary ergodic sequence with* $\mathbf{E}\xi_1 = 0$ *and* $\mathbf{E}\xi_1^2 = \sigma^2$, *that is a sequence of martingale differences relative to* $(\mathcal{F}_k)_{k \geq 1}$:

$$\mathbf{E}(\xi_k \mid \mathcal{F}_{k-1}) = 0 \quad (\text{P-a.s.}).$$

For processes X^n, $n \geq 1$, *defined by the formula* (1.1), *the weak convergence* (1.2) *holds.*

Proof. Set

$$\mathcal{F}_t^n = \mathcal{F}_{[nt]}.$$

Then $(X_t^n, \mathcal{F}_t^n)$ is a square integrable martingale for each $n \geq 1$. Therefore, we can apply Theorem 7.1.11, according to which it suffices to verify the following conditions: for each $t \in [0, 1]$

$$\frac{1}{n}\sum_{k=1}^{[nt]} \xi_k^2 \xrightarrow{P} \sigma^2 t,$$

$$\frac{1}{n}\sum_{k=1}^{[nt]} \mathbf{E}\left(\xi_k^2 I\left(\frac{|\xi_k|}{\sqrt{n}} \geq \varepsilon\right) | \mathcal{F}_{k-1}\right) \xrightarrow{P} 0.$$

The first condition holds in virtue of the Birkhof-Khintchine theorem (see [289], [332]), since P-a.s.

$$\frac{1}{[nt]}\sum_{k=1}^{[nt]} \xi_k^2 \to \mathbf{E}\xi_1^2 = \sigma^2, \ n \to \infty,$$

and, besides,

$$\lim_n \ [nt] / n = t.$$

The second condition is satisfied in virtue of the following relations:

$$\mathbf{E}\frac{1}{n}\sum_{k=1}^{[nt]} \mathbf{E}\left(\xi_k^2 I\left(\frac{|\xi_k|}{\sqrt{n}} \geq \varepsilon\right) | \mathcal{F}_{k-1}\right) = \frac{[nt]}{n}\mathbf{E}\xi_1^2 I\left(|\xi_1| \geq \sqrt{n}\ \varepsilon\right)$$

and

$$\lim_n \ \mathbf{E}\xi_1^2 I\left(|\xi_1| \geq \sqrt{n}\ \varepsilon\right) = 0.$$

2. The processes X^n, $n \geq 1$, involved in Donsker's theorem and Theorem 1, are square integrable martingales and processes with strictly stationary increments.

In view of these properties of the processes X^n, $n \geq 1$, we will consider the more sophisticated generalization of Donsker's theorem.

Let $M = (M_t)_{t \geq 0}$ be a square integrable martingale that is a process with strictly stationary increments, and

$$X_t^n = \frac{1}{\sqrt{n}} M_{nt}. \tag{1.3}$$

It will be established below that in this case the weak convergence of type (1.2) takes place for the processes $X^n = (X_t^n)_{t \geq 0}$, $n \geq 1$.

To formulate this result we assume a stochastic basis $(\Omega, \mathcal{F}, \mathbb{F}^P = (\mathcal{F}_t^P)_{t \in R}, P)$ is given, which has been defined in Ch. 4, § 11 by means of a group $\theta = (\theta_t)_{t \in R}$ of measure preserving transformations related to a σ-algebra of invariant sets J.

We assume also that on a given stochastic basis a square integrable martingale M is defined that is a helix (Ch. 4, § 11).

Along with a process X^n, given by (1.3), we define also a family $\mathbb{F}^n = (\mathcal{F}^n_t)_{t \geq 0}$ of σ-algebras $\mathcal{F}^n_t = \mathcal{F}^P_{[nt]}$. Obviously, this definition of X^n_t and $\mathcal{F}^n_t$ leads to a process $(X^n_t, \mathcal{F}^n_t)$ that is a square integrable martingale helix.

Theorem 2. *Let* $M = (M_t)_{t \geq 0}$, *be a square integrable martingale defined above that is a helix*

$$X^n_t = \frac{1}{\sqrt{n}} M_{nt}.$$

Then the weak convergence

$$X^n \xrightarrow{d} \sqrt{\eta^2}\ W \ (J\text{-}stably),$$

takes place where $W = (W_t)_{t \geq 0}$ *is a Wiener process independent of the random variable* $\eta^2 = E\,(M_1^2 \mid J)$.

Proof. In view of Problem 4.9.15 it suffices to show that

$$X^n \xrightarrow{d} Z \ (J\text{-stably}),$$

where $Z = (Z_t)_{t \geq 0}$ is a martingale with J-conditionally Gaussian and J-conditionally independent increments, and with the quadratic characteristic $\langle Z \rangle_t = \eta^2 t$. To this end it suffices, according to Theorem 7.1.4, to verify the conditions

$$J \subseteq \bigcap_{n \geq 0} \mathcal{F}^n_0,$$

$$[X^n, X^n]_t \xrightarrow{P} \eta^2 t, \ t > 0, \tag{1.4}$$

$$x^2 I(|x| > a) * \nu^n_t \xrightarrow{P} 0, \ t > 0, \ a \in (0, 1],$$

where $\nu^n = \nu^n(dt, dx)$ is the compensator of the jump measure of the process X^n.

The first condition in (1.4) is satisfied, since $\mathcal{F}^n_0 \equiv \mathcal{F}^P_0$ and $J \subseteq \mathcal{F}^P_0$ (Problem 4.11.1).

To verify the second condition in (1.4), observe that

$$[X^n, X^n]_t = \frac{1}{n}[M, M]_{nt} = \frac{1}{n}\sum_{k=1}^{n}([M, M]_{kt} - [M, M]_{(k-1)t}).$$

Since M is a helix, [M, M] is a helix too by Theorem 4.11.2, and hence

$$\lim_n \frac{1}{n}\sum_{k=1}^{n}([M, M]_{kt} - [M, M]_{(k-1)t}) = E\,([M, M]_t \mid J) \ \ (P\text{-a.s.})$$

by the Birkhof-Khintchine theorem (see [289, 332]).

In view of Problem 1 and 4.11.3 the following equality holds (P-a.s.):

$$\mathbf{E}([M, M]_t \mid J) = \mathbf{E}(M_t^2 \mid J) = \eta^2 t.$$

Consequently the second condition in (1.4) holds as well.

The last condition in (1.4) is verified in the following manner. Denote

$$Y_t^n = \frac{1}{n} \sum_{0 < s \le nt} (\Delta M_s)^2 I(|\Delta M_s| > \sqrt{n}\, a)$$

and observe that

$$\mathbf{E} x^2 I(|x| > a) * \nu_t^n = \mathbf{E} \sum_{0 < s \le t} (\Delta X_s^n)^2 I(|\Delta X_s^n| > a) = \mathbf{E} Y_t^n.$$

Due to Theorem 4.11.5 the process $Y^n = (Y_t^n)_{t \ge 0}$ is a nondecreasing helix. Therefore

$$\mathbf{E} Y_t^n = nt\, \mathbf{E} Y_{1/n}^n, \quad n \ge 1$$

in view of Problem 4.11.2, and hence

$$\mathbf{E} Y_t^n = t\, \mathbf{E} \sum_{0 < s \le 1} (\Delta M_s)^2 I(|\Delta M_s| > \sqrt{n}\, a)$$

$$\le t\, \mathbf{E}\, [M, M]_1\, I([M, M]_1 > na^2) \to 0, \quad n \to \infty.$$

3. We will formulate the invariance principle concerning a square integrable martingale with strictly stationary increments, in the following special situation.

Let $\xi = (\xi_t)_{t \in R}$ be a strictly stationary process, given on a complete probability space $(\Omega, \mathcal{F}, \mathsf{P})$ with trajectories in $(\mathbf{X}, \mathfrak{X})$ ($\mathfrak{X}$ is a σ-algebra of cylindric sets). To a stationary process ξ we relate a σ-algebra of invariant sets J^ξ, i.e. the collection of sets A in $\mathcal{F}$, which possesses the following property: there exists $B \in \mathfrak{X}$ such that for each $t \in R$

$$A = \{\omega: (\xi_s(\omega))_{s \ge t} \in B\},$$

and it is completed by sets in $\mathcal{F}$ of P-measure zero. Denote also

$$G_t^\xi = \sigma\{\xi_s, -\infty < s \le t\}, \quad \mathcal{F}_t^\xi = \bigcap_{\varepsilon > 0} G_{t+\varepsilon}^\xi \vee \mathcal{N},$$

where $\mathcal{N}$ is a system of sets in $\mathcal{F}$ of P-measure zero.

The following result is implied by Theorem 2.

Theorem 3. *Let* $M = (M_t)_{t \ge 0}$ *be a square integrable martingale that is a process with strictly stationary increments, given on a stochastic basis*

$$(\Omega, \mathcal{F}, \mathbf{F}^\xi = (\mathcal{F}_t^\xi)_{t \ge 0}, \mathsf{P})$$

and

$$X^n = (X_t^n)_{t > 0}$$

with

$$X_t^n = \frac{1}{\sqrt{n}} M_{nt}.$$

Then

$$X^n \xrightarrow{d} \sqrt{\eta^2}\, W \quad (J^{\xi}\text{-}stably),$$

where $W = (W_t)_{t \geq 0}$ *is a Wiener process independent of* $\eta^2 = \mathbf{E}\,(M_1^2 \mid J^{\xi})$.

Problems

1. Let M be a square integrable martingale helix, [M, M] its quadratic variation and J the σ-algebra of invariant sets. Show that

$$\mathbf{E}\,([M, M]_t \mid J) = \mathbf{E}\,(M_t^2 \mid J) \quad (\text{P-a.s.}).$$

2. Let M be a square integrable martingale that is a process with strictly stationary increments and $M_0 = 0$. Show that

$$\mathbf{E}M_1^2 = \lim_{t \to \infty} t\,\mathbf{E}M_{1/t}^2.$$

3. Prove Theorem 3 (Hint: apply the result in Ch. 4, Subsection 11.8).

§ 2. Invariance principle for strictly stationary processes

1. Let $\xi = (\xi_t)_{t \in R}$ be a measurable with respect to (t, ω), strictly stationary process, given on a complete probability space $(\Omega, \mathcal{F}, P)$, and J^ξ a σ-algebra of invariant sets of a process ξ (see the definition of J^ξ in Subsection 1.3).

Denote

$$G_t^\xi = \sigma\{\xi_s : -\infty < s \le t\}$$

and

$$\mathcal{F}_t^\xi = \bigcap_{\varepsilon > 0} G_{t+\varepsilon}^\xi \vee \mathcal{N},$$

where $\mathcal{N}$ is the family of sets in $\mathcal{F}$ of P-measure zero.

In this section we assume $\mathbf{E}\xi_0^2 < \infty$ and $\mathbf{E}\xi_0 = 0$. Consider a stochastic process

$$Y_t = \int_0^t \xi_s ds$$

defined in view of $\mathbf{E}\,|\,\xi_s\,| = \mathbf{E}\,|\,\xi_0\,| < \infty$.

Applying the results of Ch. 5 and 7 we establish below the central limit theorem (functional as well) for the sequence of distributions of the processes

$$Y^n = (Y_t^n)_{t \ge 0}, \quad n \ge 1, \quad Y_t^n = \frac{1}{\sqrt{n}} Y_{nt}. \tag{2.1}$$

2. First we dwell on certain assumptions concerning processes ξ.

For a random variable ξ we denote

$$\|\xi\|_q = \begin{cases} (\mathbf{E}\,|\,\xi\,|^q)^{1/q}, & q \ge 1 \\ \text{ess sup}\,|\,\xi\,|, & q = \infty. \end{cases}$$

The central rôle below is played by the following conditions: for certain $q \ge 2$

$$\|\xi_0\|_q < \infty, \tag{2.2}$$

$$\int_0^\infty \|\mathbf{E}(\xi_t \,|\, G_0^\xi)\|_{q/(q-1)}\, dt < \infty; \tag{2.3}$$

as $q = \infty$ we assume[1] $q/(q-1) = 1$.

---

[1]Observe that in the standard notations "(1/p) + (1/q) = 1" the expression q/q − 1 involved in (2.3) is exactly p.

In view of (2.2) we have

$$||\xi_0||_2 = (\mathbf{E}\xi_0^2)^{1/2} < \infty.$$

Therefore, the process ξ possesses the correlation function

$$R(t) = \mathbf{E}\xi_t\xi_0.$$

In virtue of the representation

$$R(t) = \mathbf{E}\xi_0\mathbf{E}(\xi_t \mid G_0^\xi)$$

and Hölder's inequality the estimate

$$|R(t)| \le ||\xi_0||_q \, ||\mathbf{E}(\xi_t \mid G_0^\xi)||_{q/(q-1)}$$

takes place, and hence by Conditions (2.2) and (2.3) we have

$$\int_0^\infty |R(t)|\, dt < \infty. \tag{2.4}$$

It is known (see [118]), that

$$\mathbf{E}\left(\int_0^T \xi_t dt\right)^2 = \int_{-T}^{T} (T - |t|) R(t)\, dt.$$

Besides, it is easily deduced from (2.4) (for instance, by Theorem 2.6.9) that

$$\lim_{T\to\infty} \frac{1}{T} t R(t)\, dt = 0.$$

Consequently

$$\int_0^\infty R(t)\, dt = \lim_{T\to\infty} \frac{1}{2}\mathbf{E}\left(\frac{1}{\sqrt{T}}\int_0^T \xi_t dt\right)^2.$$

It is established analogously that the random variable

$$\eta = 2\int_0^\infty \mathbf{E}(\xi_t\xi_0 \mid J^\xi)\, dt \tag{2.5}$$

is nonnegative (P-a.s.) (Problem 1).

3. Theorem 1. *Let processes* Y^n, $n \ge 1$, *be defined by* (2.1) *and for certain* $q \in [2, \infty]$ *let Conditions* (2.2) *and* (2.3) *be satisfied.*

Then

$$Y^n \xrightarrow{d_f} \sqrt{\eta}\, W \quad (J^\xi\text{-}stably),$$

where $W = (W_t)_{t\ge 0}$ *is a Wiener process independent of the nonnegative random variable* η *defined by* (2.5).

If $q = 2$, then the stronger result holds

$$Y^n \xrightarrow{d} \sqrt{\eta}\, W \quad (J^{\xi}\text{-stably}).$$

4. In the course of proving Theorem 1 and all auxiliary facts below one may assume that the underlying process ξ is given on a coordinate stochastic basis

$$(\mathbf{X}, \mathfrak{X}^Q, \mathbb{F}^Q = (\mathcal{F}^Q_t)_{t \in R}, Q)$$

(see Ch.4., Subsection 11.8) with the group $\theta = (\theta_t)_{t \in R}$ of measure preserving transformations that is related to the σ-algebra of invariant sets $J = J^{\xi}$.

The following result is of independent interest.

Lemma 1. *Let for certain* $q \in [2, \infty]$ *Conditions* (2.2) *and* (2.3) *be satisfied. Then the process* $Y = (Y_t)_{t \geq 0}$ *with*

$$Y_t = \int_0^t \xi_s ds \tag{2.6}$$

admits the representation of the form

$$Y_t = X_t - X_0 + M_t, \tag{2.7}$$

where

1) $X = (X_t)_{t \geq 0}$ *is a* $(\mathbb{F}^Q, Q)$*-semimartingale that is a modification of the strictly stationary process*

$$\left(-\int_t^{\infty} \pi_t(\xi_s)\, ds \right)_{t \geq 0},$$

where $\pi(\xi_s) = (\pi_t(\xi_s))_{t \geq 0}$ *is the* $\mathbb{F}^Q$*-optional projection of the random variable* ξ_s (*see Theorem* 1.3.13 *and Ch.* 4, § 11);

2) $(X_t, \xi_t)_{t \geq 0}$ *is a strictly stationary process*;

3) $X_0 = -\int_0^{\infty} E(\xi_s \mid \mathcal{F}^Q_0)\, ds, \quad EX_0^2 < \infty,$ *as* $q = 2$;

4) $M = (M_t)_{t \geq 0}$ *is a* $(\mathbb{F}^Q, Q)$*-square integrable martingale that is a helix such that*

$$E(M_1^2 \mid J) = 2 \int_0^{\infty} E(\xi_s \xi_0 \mid J)\, ds.$$

Proof. Due to the corollary to Theorem 4.11.1 there exists a modification of the process $\pi(\xi_s)$ such that $\pi_t(\xi_s)(\omega)$, as a function of (ω, s, t), is $\mathfrak{X}^Q \otimes \mathcal{B}_R \otimes \mathcal{B}_{R_+}$-measurable. This modification will be considered further on.

We will show first that

$$\int_0^\infty \mathbf{E} \mid \pi_t (\xi_s) \mid ds < \infty$$

for each $t \geq 0$. This will guarantee, in particular, the existence of the process

$$\left(\int_t^\infty \pi_t (\xi_s) \, ds \right)_{t \geq 0} .$$

To this end, observe that in virtue of Lemma 4.11.1 and Hölder's inequality

$$\mathbf{E} \mid \pi_t (\xi_s (\omega) \mid = \mathbf{E} \mid \mathbf{E} (\xi_s \mid \overset{Q}{\mathfrak{F}}_t) (\omega) \mid = \mathbf{E} \mid \mathbf{E} (\xi_s \mid \overset{Q}{\mathfrak{F}}_t) (\theta_{-t} \omega) \mid$$

$$= \mathbf{E} \mid \mathbf{E} (\xi_s (\theta_{-t} \omega) \mid \overset{Q}{\mathfrak{F}}_0) (\omega) \mid$$

$$= \mathbf{E} \mid \mathbf{E} (\xi_{s-t} \mid \overset{Q}{\mathfrak{F}}_0) (\omega) \mid$$

$$= I (s \leq t) \, \mathbf{E} \mid \xi_{s-t} \mid + I (s > t) \, \mathbf{E} \mid \mathbf{E} (\xi_{s-t} \mid \overset{Q}{\mathfrak{F}}_0) \mid$$

$$\leq I (s \leq t) \, \mathbf{E} \mid \xi_0 \mid + I (s > t) \mid\mid \mathbf{E} (\xi_{s-t} \mid \overset{Q}{\mathfrak{F}}_0) \mid\mid_{q/(q-1)} .$$

Therefore, assuming Condition (2.3) with a given $q \geq 2$, we get

$$\int_0^\infty \mathbf{E} \mid \pi_t (\xi_s) \mid ds \leq t \, \mathbf{E} \mid \xi_0 \mid + \int_0^\infty \mid\mid \mathbf{E} (\xi_s \mid \overset{Q}{\mathfrak{F}}_0) \mid\mid_{q/(q-1)} ds < \infty.$$

Define the process

$$M' = (M_t')_{t \in R}$$

with

$$M_t' = \int_0^\infty [\pi_t (\xi_u) - \pi_0 (\xi_u)] \, du.$$

The process M' possesses the following properties:

$$\mathbf{E} \mid M_t' \mid < \infty,$$

$$\mathbf{E} (M_t' \mid \overset{Q}{\mathfrak{F}}_s) = M_s' \quad \text{Q-a.s.,} \quad t > s,$$

$$M_{t+s}' (\omega) - M_s' (\omega) = M_t' (\theta_s \omega) \quad \text{Q-a.s.,} \quad t, s \in R,$$

where the last condition is established by means of Lemma 4.11.1. Therefore, in virtue of the remark at the end of § 4 in Ch. 1 and Lemma 4.11.3, there exists a modification $M = (M_t)_{t \geq 0}$ of the process $M' = (M_t')_{t \geq 0}$ such that $M \in \overline{\mathfrak{M}}$ is a helix.

Observe now that for each $t \geq 0$

$$\int_0^t \xi_s \, ds - M_t' - \int_0^\infty \pi_0(\xi_s)\, ds = -\int_t^\infty \pi_t(\xi_s)\, ds.$$

Obviously this implies that the process X, defined by the decomposition (2.7) with the given martingale M, is a modification of the process

$$X' = \left(\int_t^\infty \pi_t(\xi_s)\, ds \right)_{t \geq 0}.$$

Let us show now that X is a strictly stationary process. It suffices to show that X' is such a process. By Lemma 4.11.1 we have

$$-X_t'(\theta_h\omega) = \int_t^\infty \pi_t(\xi_u)(\theta_h\omega)\, du = \int_t^\infty \pi_{t+h}(\xi_u(\theta_h\omega))\, du$$

$$= \int_t^\infty \pi_{t+h}(\xi_{u+h})\, du = \int_{t+h}^\infty \pi_{t+h}(\xi_u)\, du = -X_{t+h}'(\omega),$$

i.e. X' is a strictly stationary process.

Let us show now that actually $M \in \overline{\mathfrak{M}}^2$. Apply here Theorem 4.12.2, which gives

$$\mathbf{E}M_t^2 = -2\mathbf{E}\int_0^t X_s\xi_s\, ds,$$

provided

$$\int_0^t \mathbf{E}\,|X_s\xi_s|\, ds < \infty.$$

As the bivariate process (X, ξ) is stationary, we have

$$\mathbf{E}\,|X_s\xi_s| = \mathbf{E}\,|X_0\xi_0| = \mathbf{E}\,|X_0'\xi_0|.$$

Consequently, by this equality and in view of Hölder's inequality

$$\mathbf{E}\,|X_s\xi_s| = \mathbf{E}\left|\int_0^\infty \pi_0(\xi_s)\, ds\, \xi_0\right| \leq \int_0^\infty \mathbf{E}\,|\pi_0(\xi_s)\,\xi_0|\, ds$$

$$\leq ||\xi_0||_q \int_0^\infty ||\mathbf{E}(\xi_s \mid \mathcal{F}_0^Q)||_{q/(q-1)}\, ds < \infty,$$

i.e.

$$\int_0^t \mathbf{E}\,|X_s\xi_s|\,ds \le t\,||\xi_0||_q \int_0^\infty ||\mathbf{E}\,(\xi_s\,|\,\mathcal{F}^Q_0)||_{q/(q-1)}\,ds < \infty$$

due to Condition (2.3). Hence

$$\mathbf{E}M_t^2 = -2\int_0^t \mathbf{E}\,(X_s\xi_s)\,ds = -2t\,\mathbf{E}\,(X_0\xi_0)$$

$$= 2t\,\mathbf{E}\left(\int_0^\infty \pi_0\,(\xi_s)\,ds \cdot \xi_0\right) = 2t\int_0^\infty \mathbf{E}\,(\xi_s\xi_0)\,ds.$$

According to the remark to Theorem 4.12.2

$$\mathbf{E}\,(M_1^2\,|\,J) = 2\int_0^\infty \mathbf{E}\,(\xi_s\xi_0\,|\,J)\,ds \quad (Q\text{-a.s.}).$$

Let $q = 2$. Then in virtue of the estimate

$$\mathbf{E}X_0^2 \le \left(\int_0^\infty ||\mathbf{E}\,(\xi_s\,|\,\mathcal{F}^Q_0)||_2\,ds\right)^2$$

and Condition (2.3) which holds as $q = 2$, the desired assertion $\mathbf{E}X_0^2 < \infty$ holds.

5. Proof of Theorem 1. Let $X = (X_t)_{t\ge 0}$ and $M = (M_t)_{t\ge 0}$ be the processes defined in Lemma 1. Set

$$X_t^n = \frac{1}{\sqrt{n}} X_{nt}, \quad M_t^n = \frac{1}{\sqrt{n}} M_{nt}, \quad \mathcal{F}^n_t = \mathcal{F}^Q_{nt}.$$

Then in virtue of the representation (2.7) for X_t^n and the representation (2.1) for Y_t^n we get

$$Y_t^n = X_t^n - \frac{1}{\sqrt{n}} X_0 + M_t^n.$$

By Lemma 1 the process $M^n = (M_t^n)_{t\ge 0}$ is a square integrable martingale helix (relative to $\mathbb{F}^n = (\mathcal{F}^n_t)_{t\ge 0}$). Therefore by Theorem 1.1

$$M^n \xrightarrow{d} \sqrt{\eta}\,W \quad (J\text{-stably}),$$

where $W = (W_t)_{t\ge 0}$ is a Wiener process independent of

$$\eta = \mathbf{E}\,(M_1^2\,|\,J) = 2\int_0^\infty \mathbf{E}\,(\xi_s\xi_0\,|\,J)\,ds.$$

The stochastic process $X^n = (X^n_t)_{t\geq 0}$ is strictly stationary. Therefore

$$Q(|X^n_t| > \varepsilon) = Q\left(\frac{|X_{nt}|}{\sqrt{n}} > \varepsilon\right) = Q\left(\frac{|X_0|}{\sqrt{n}} > \varepsilon\right) \to 0, \ n \to \infty.$$

Since

$$\frac{|X_0|}{\sqrt{n}} \to 0, \ n \to \infty,$$

for each $t \geq 0$ we have

$$|Y^n_t - M^n_t| \xrightarrow{P} 0, \ n \to \infty.$$

Hence the weak convergence

$$Y^n \xrightarrow{d_f} \sqrt{\eta}\, W \quad \text{(J-stably)}$$

takes place, due to Problem 5.2.2.

Consider now the case in which $q = 2$. To establish the weak convergence

$$Y^n \xrightarrow{d} \sqrt{\eta}\, W \quad \text{(J-stably)}$$

it suffices to show that

$$\sup_{s\leq t} |Y^n_s - M^n_s| \xrightarrow{P} 0, \ \ n \to \infty, \ \ t > 0.$$

Obviously, it suffices that

$$\sup_{s\leq t} |X^n_s| = \sup_{s\leq t} \left|\frac{X_{ns}}{\sqrt{n}}\right| \xrightarrow{P} 0, \ \ n \to \infty. \tag{2.8}$$

The process X is a semimartingale and a strictly stationary process admitting the decomposition

$$X_t = X_0 + A_t + M_t$$

with

$$A_t = \int_0^t \xi_s \, ds$$

due to (2.7).

By Lemma 1 we have $\mathbf{E}X_0^2 < \infty$ as $q = 2$. Besides, $\mathbf{E}\,\mathrm{Var}^2\,(A)_t < \infty$ and $\mathbf{E}M_t^2 < \infty$. Therefore, the relation (2.8) holds in virtue of Theorem 4.12.1.

Thus, Theorem 1 is proved in case in which the process ξ is given on a coordinate basis with $J = J^{\xi}$. The originally formulated assertion of the theorem is deduced here in a standard manner.

6. Example 1. Let $\xi = (\xi_t)_{t \in R}$ be a stationary ergodic process and $f = f(x)$ a measurable function such that $\mathbf{E}f^2(\xi_0) < \infty$ and $\mathbf{E}f(\xi_0) = 0$.

Set $Y^n = (Y_t^n)_{t \geq 0}$ with

$$Y_t^n = \frac{1}{\sqrt{n}} \int_0^{nt} f(\xi_s)\, ds.$$

If the condition

$$\int_0^{\infty} \| \mathbf{E}(f(\xi_t) \mid G_0^{\xi}) \|_2\, dt < \infty$$

is satisfied, then

$$Y^n \xrightarrow{d} \left(2 \int_0^{\infty} \mathbf{E}(f(\xi_t) f(\xi_0))\, dt \right)^{1/2} W,$$

where $W = (W_t)_{t \geq 0}$ is a Wiener process.

To prove this result it suffices to consider the strictly stationary process $\zeta = (\zeta_t)_{t \in R}$ with $\zeta_t = f(\xi_t)$ and to observe that in virtue of the inclusion of σ-algebras $G_t^{\zeta} \subseteq G_t^{\xi}$, $t \in R$, we have the inequality

$$\| \mathbf{E}(\zeta_t \mid G_0^{\zeta}) \|_2 \leq \| \mathbf{E}(f(\xi_t) \mid G_0^{\xi}) \|_2.$$

Example 2. Let $Z = (Z_t)_{t \in R}$ be a process with independent increments and a helix with trajectories, having left-hand limits. Besides, let

$$\mathbf{E}(Z_t - Z_s) = 0, \quad \mathbf{E}(Z_t - Z_s)^2 = | t - s |.$$

Define a strictly stationary process $\xi = (\xi_t)_{t \in R}$ with

$$\xi_t = \int_{-\infty}^{t} b(t - s)\, dZ_s, \tag{2.9}$$

where $b = b(t)$ is a measurable function $b(t) = b(-t)$ with

$$\int_0^{\infty} b^2(t)\, dt < \infty$$

and the integral in (2.9) is the stochastic integral with respect to a square integrable martingale (for more details see Ch. 4, Subsection 11. 6). The processes

$$Y^n = (Y_t^n)_{t \geq 0}$$

are defined as in (2.1). Then under the condition

$$\int_0^{\infty}\left(\int_t^{\infty} b^2(s)\,ds\right)^{1/2} dt < \infty \tag{2.10}$$

the weak convergence

$$Y^n \xrightarrow{d} \left(2\int_0^{\infty} b(u) \int_u^{\infty} b(s)\,ds\,du\right)^{1/2} W$$

takes place, where W is a Wiener process.

To prove this convergence, it suffices to verify Conditions (2.2) and (2.3) as $q = 2$.

In view of Problem 4.11.4

$$\mathbf{E}\xi_0^2 = \int_{-\infty}^{0} b^2(-s)\,ds = \int_0^{\infty} b^2(s)\,ds < \infty.$$

Next, as $s > 0$

$$\mathbf{E}(\xi_t \mid G_0^{\xi}) = \int_{-\infty}^{0} b(t-s)\,dZ_s$$

and consequently

$$\| \mathbf{E}(\xi_t \mid G_0^{\xi}) \|_2 = \left(\int_{-\infty}^{0} b^2(t-s)\,ds\right)^{1/2} = \left(\int_t^{\infty} b^2(s)\,ds\right)^{1/2},$$

i.e. (2.10) implies (2.3). Therefore it remains to show that

$$2\int_0^{\infty} \mathbf{E}(\xi_s\xi_0 \mid J^{\xi})\,ds = \left(\int_0^{\infty} b(s)\,ds\right)^2. \tag{2.11}$$

Denote

$$G_{-\infty}^{\xi} = \bigcap_{t \le 0} G_t^{\xi}.$$

In view of Problem 4.11.1

$$J^{\xi} \subseteq G_{-\infty}^{\xi}.$$

But

$$\mathbf{E}(\xi_s\xi_0 \mid G_{-\infty}^{\xi}) = \int_{-\infty}^{0} b(s-u)\,b(-u)\,du \quad \text{(P-a.s.)}.$$

Consequently,

$$2\int_0^\infty \mathbf{E}\,(\xi_s\xi_0 \mid J^\xi)\,ds = 2\int_0^\infty\int_0^\infty b\,(s+u)\,b\,(u)\,du\,ds = 2\int_0^\infty b\,(u)\int_u^\infty b\,(s)\,ds\,du,$$

i.e. the representation (2.11) holds.

Example 3 (cf. [352]). Let $\xi = (\xi_t)_{t\in R}$ be a strictly stationary ergodic process with $\mathbf{E}\xi_0^2 < \infty$ and $\mathbf{E}\xi_0 = 0$, and let Condition (2.3) be satisfied as $q = 2$, i.e.

$$\int_0^\infty \| \mathbf{E}\,(\xi_t \mid G_0^\xi) \|_2\,dt < \infty.$$

Then by Theorem 1 the sequence of processes

$$Y^n = (Y_t^n)_{t\geq 0},\quad n \geq 1,$$

with

$$Y_t^n = \sqrt{n}\int_0^t \xi_{ns}\,ds$$

converges weakly to $\sqrt{\eta}\ W$, where $W = (W_t)_{t\geq 0}$ is a Wiener process and

$$\eta = 2\int_0^\infty \mathbf{E}\xi_t\xi_0\,dt.$$

Consider now the situation in which[2]

$$Y_t^n = \sqrt{n}\int_0^t \xi_{Y_s^n + ns}\,ds,\quad |\xi_t| \leq C,\ t\in R.$$

In this case the weak convergence

$$Y^n \xrightarrow{d} Y$$

takes place with

$$Y_t = \sqrt{\eta}\ W_t - \mathbf{E}\xi_0^2\,t.$$

In fact, as ξ_t is bounded for $n \geq 4C^2$, the inequality

$$\frac{1}{\sqrt{n}}\xi_t + 1 \geq 1/2$$

holds. Therefore for such values of n the change of variables $u = Y_s^n + ns$ leads to the following representation

---

[2]The equation with respect to Y^n has a solution, provided there exists a bounded derivative $\dot{\xi}_t$.

$$Y_t^n = \frac{1}{\sqrt{n}} \int_0^{Y_t^n + nt} \xi_u \Big/ \left(\frac{1}{\sqrt{n}} \xi_u + 1 \right) du.$$

Denote

$$Z_t^n = \frac{1}{\sqrt{n}} \int_0^{nt} \xi_u \Big/ \left(\frac{1}{\sqrt{n}} \xi_u + 1 \right) du.$$

Then

$$Y_t^n = Z^n_{Y_t^n / n + t}.$$

The definition of Y_t^n implies

$$\sup_{t \le T} \frac{1}{n} | Y_t^n | \le \frac{CT}{\sqrt{n}}.$$

Consequently, the desired weak convergence takes place, provided

$$Z^n \xrightarrow{d} Y.$$

Observe now that

$$Z_t^n = \frac{1}{\sqrt{n}} \int_0^{nt} \xi_u \, du - \frac{1}{n} \int_0^{nt} \xi_u^2 \, du + \frac{1}{n^{3/2}} \int_0^{nt} \frac{\xi_u^3}{\frac{1}{\sqrt{n}} \xi_u + 1} du$$

$$= Z_t^{n1} - Z_t^{n2} + Z_t^{n3}.$$

Besides, by Theorem 1 we have

$$Z^{n1} \xrightarrow{d} \sqrt{\eta}\, W$$

and by the Birkhof-Khintchine theorem (Corollary to Theorem 4.11.9)

$$Z_t^{n2} \to \mathbf{E}\xi_0^2 t \quad (\text{P-a.s.}),$$

and hence by Problem 5.3.2

$$\sup_{t \le T} | Z_t^{n2} - \mathbf{E}\xi_0^2 t | \to 0 \quad (\text{P-a.s.})$$

for each $T > 0$; finally, for $n \ge 4C^2$

$$\sup_{t \le T} | Z_t^{n3} | \le \frac{C^3 T}{\sqrt{n} \left(1 - \frac{C}{\sqrt{n}} \right)} \to 0, \ n \to \infty.$$

This and Problem 6.1.2 give the desired weak convergence.

7. The invariance principle for stationary processes traditionally is proved under conditions expressed in terms of strong and uniformly strong mixing conditions (see, for instance, [118]). In this regard it is natural to give conditions which guarantee (2.3), expressed in terms of the behaviour of corresponding mixing coefficients. Aiming at this we will estimate the quantity

$$\| \mathbf{E}(\xi_s \mid G_0) \|_{q/q-1}$$

by means of mixing coefficients.

Introduce σ-algebras

$$\mathcal{G}_t^{\xi} = \sigma\{\xi_s,\ s \geq t\},\ t \in R.$$

Definition 1. The function $\alpha = \alpha(t)$, $t \geq 0$, with

$$\alpha(t) = \sup_{A \in G_0^{\xi},\ B \in \mathcal{G}_t^{\xi}} | P(AB) - P(A)P(B) |$$

is called a *strong mixing coefficient.*

Definition 2. The function $\phi = \phi(t)$, $t \geq 0$ with

$$\phi(t) = \sup_{A \in G_0^{\xi},\ B \in \mathcal{G}_t^{\xi}} | P(B \mid A) - P(B) |$$

is called a *uniformly strong mixing coefficient.*

Obviously, α and φ are nonincreasing functions and

$$\alpha(t) \leq \phi(t).$$

Let us estimate

$$\| \mathbf{E}(\xi_s \mid G_0^{\xi}) \|_{q/(q-1)}$$

by means of α (s) and φ (s).

Lemma 2. 1) *If* $|\xi_s| \leq C$ *and* $q \in (0, \infty]$, *then*

$$\| \mathbf{E}(\xi_s \mid G_0^{\xi}) \|_{q/(q-1)} \leq 4C(\alpha(s))^{1-1/q}.$$

2) *If* $q > 2$ *and* $\| \xi_0 \|_q < \infty$, *then*

$$\| \mathbf{E}(\xi_s \mid G_0^{\xi}) \|_{q/(q-1)} \leq 5 \| \xi_0 \|_q (\alpha(s))^{1-2/q}.$$

Proof. 1) First we show that the estimate

$$\mathbf{E} | \mathbf{E}(\xi_s \mid G_0^{\xi}) | \leq 4C\alpha(s) \tag{2.12}$$

holds. To this end, denote

$$\beta_1 = \text{sign}\, \mathbf{E}(\xi_s \mid G_0^{\xi}),$$

$$\beta_2 = \text{sign}\,[\mathbf{E}(\beta_1 \mid \mathcal{G}_s^{\xi}) - \mathbf{E}\beta_1]$$

and observe (taking into consideration the equality $\mathbf{E}\xi_s = 0$) that

$$\mathbf{E}\,|\,\mathbf{E}\,(\xi_s \,|\, G_0^{\xi})\,| = \mathbf{E}\beta_1\mathbf{E}\,(\xi_s \,|\, G_0^{\xi}) = \mathbf{E}\beta_1\xi_s = \mathbf{E}\xi_s\mathbf{E}\,(\beta_1 \,|\, \mathcal{G}_s^{\xi})$$

$$= \mathbf{E}\xi_s\,[\mathbf{E}\,(\beta_1 \,|\, \mathcal{G}_s^{\xi}) - \mathbf{E}\beta_1] \le \mathbf{E}\,|\,\xi_s\,|\,|\,\mathbf{E}\,(\beta_1 \,|\, \mathcal{G}_s^{\xi}) - \mathbf{E}\beta_1\,|$$

$$\le C\,\mathbf{E}\beta_2\,[\mathbf{E}\,(\beta_1 \,|\, \mathcal{G}_s^{\xi}) - \mathbf{E}\beta_1] = C\,[\mathbf{E}\beta_1\beta_2 - \mathbf{E}\beta_1\mathbf{E}\beta_2].$$

Therefore it suffices to show that

$$\mathbf{E}\beta_1\beta_2 - \mathbf{E}\beta_1\mathbf{E}\beta_2 \le 4\alpha\,(s).$$

This inequality holds, since for $A = \{\beta_1 = 1\}$ and $B = \{\beta_2 = 1\}$ we have

$$\mathbf{E}\beta_1\beta_2 - \mathbf{E}\beta_1\mathbf{E}\beta_2 = P\,(AB) - P\,(\bar{A}B) - P\,(A\bar{B}) + P\,(\bar{A}\,\bar{B})$$

$$- P\,(A)\,P\,(B) + P\,(\bar{A})\,P\,(B) - P\,(A)\,P\,(\bar{B}) - P\,(\bar{A})\,P\,(\bar{B}),$$

besides the set

$$A \in G_0^{\xi}$$

and the set

$$B \in \mathcal{G}_s^{\xi}.$$

Hence the inequality (2.12) is established. Applying it we will deduce the first assertion of the lemma. We have

$$\|\,\mathbf{E}\,(\xi_s \,|\, G_0^{\xi})\,\|_{q/(q-1)} = C\,\|\,\mathbf{E}\left(\frac{\xi_s}{C} \,\Big|\, G_0^{\xi}\right)\|_{q/(q-1)}$$

$$\le C\,\mathbf{E}\,|\,\mathbf{E}\,(\frac{\xi_s}{C} \,|\, G_0^{\xi})\,|^{(q-1)/q} = C^{1/q}\,\mathbf{E}\,|\,\mathbf{E}\,(\xi_s \,|\, G_0^{\xi})\,|^{(q-1)/q}.$$

This and the inequality (2.12) imply

$$\|\,\mathbf{E}\,(\xi_s \,|\, G_0^{\xi})\,\|_{q/(q-1)} \le C\,(4\alpha\,(s))^{1-1/q},$$

i.e. the desired assertion holds.

2) Set

$$C = (\alpha\,(s))^{-1/q}\,\|\,\xi_0\,\|_q$$

and define the random variables

$$\xi_s^{'} = \xi_s I\,(\,|\,\xi_s\,| \le C)$$

and

$$\xi_s^{''} = \xi_s I\,(\,|\,\xi_s\,| > C).$$

Since

$$\xi_s = \xi_s' + \xi_s'',$$

then

$$||\mathbf{E}(\xi_s \mid G_0^\xi)||_{q/(q-1)} \le ||\mathbf{E}(\xi_s' \mid G_0^\xi)||_{q/(q-1)} + ||\mathbf{E}(\xi_s'' \mid G_0^\xi)||_{q/(q-1)}.$$

In view of assertion 1) of the lemma and the inequality

$$|\xi_s'| \le C$$

the estimate

$$||\mathbf{E}(\xi_s' \mid G_0^\xi)||_{q/(q-1)} \le 4C\,(\alpha(s))^{1-1/q}$$

holds. Due to the definition of C, this gives

$$||\mathbf{E}(\xi_s' \mid G_0^\xi)||_{q/(q-1)} \le 4\,||\xi_0||_q\,(\alpha(s))^{1-2/q}.$$

Consequently, it remains to show that

$$||\mathbf{E}(\xi_s'' \mid G_0^\xi)||_{q/(q-1)} \le ||\xi_0||_q\,(\alpha(s))^{1-2/q}. \tag{2.13}$$

The last inequality may be established in the following manner. In virtue of Jensen's inequality for conditional mathematical expectations

$$||\mathbf{E}(\xi_s'' \mid G_0^\xi)||_{q/(q-1)} \le ||\xi_s''||_{q/(q-1)} = ||\xi_0''||_{q/(q-1)}.$$

Next, we have

$$||\xi_0''||_{q/(q-1)} = \left(\mathbf{E}\,|\xi_0|^{q/(q-1)} I(|\xi_0| > C)\right)^{(q-1)/q}$$

$$\le \left(\frac{\mathbf{E}\,|\xi_0|^q}{C^{q(q-2)/(q-1)}}\right)^{(q-1)/q} = \frac{\left(\mathbf{E}\,|\xi_0|^q\right)^{(q-1)/q}}{C^{q-2}}$$

$$= \frac{\left(\mathbf{E}\,|\xi_0|^q\right)^{(q-1)/q}}{||\xi_0||_q^{q-2}}\,(\alpha(s))^{1-q/2},$$

where the right-hand side of the last equality coincides with that of the inequality (2.13).

Lemma 3. *Let* $q \in (1, \infty)$ *and let the condition*

$$||\xi_0||_{q/(q-1)} < \infty$$

be satisfied. Then

$$||\mathbf{E}(\xi_s \mid G_0^\xi)||_{q/(q-1)} \le 2\,||\xi_0||_{q/(q-1)}\,(\phi(s))^{1/q}.$$

Proof. Denote by

$$F = F(x)$$

and

$$F_{G_0^\xi} = F_{G_0^\xi}(x)$$

the distribution and conditional distribution (under the condition G_0^ξ) functions of the random variables ξ_s.

Also, denote

$$\phi_1(t) = \operatorname{ess\,sup}_{\omega} \sup_{B \in \mathfrak{S}_t^\xi} | P(B \mid G_0^\xi) - P(B) |, \quad t \geq 0$$

and observe that

$$\operatorname{Var}(F_{G_0^\xi} - F)(\infty) \leq 2\,\phi_1(s). \tag{2.14}$$

Set $p = q/(q-1)$ and utilize the fact that $\mathbf{E}\xi_s = 0$.
Then

$$\| \mathbf{E}(\xi_s \mid G_0^\xi) \|_p = \left(\mathbf{E} \left| \int_R x \, d[(F_{G_0^\xi} - F)(x)] \right|^p \right)^{1/p}$$

$$\leq \left(\mathbf{E} \left[\int_R |x| \, d\operatorname{Var}(F_{G_0^\xi} - F)(x) \right]^p \right)^{1/p}.$$

Due to Hölder's inequality and (2.14) we get

$$\left[\int_R |x| \, d\operatorname{Var}(F_{G_0^\xi} - F)(x) \right]^p \leq \int_R |x|^p \, d\operatorname{Var}(F_{G_0^\xi} - F)\,[2\phi_1(s)]^{p-1}$$

$$\leq [2\phi_1(s)]^{p-1} \int_R |x|^p \, d(F_{G_0^\xi} + F)(x)$$

$$= [2\phi_1(s)]^{p-1} \left(\mathbf{E}(|\xi_s|^p \mid G_0^\xi) + \mathbf{E}|\xi_s|^p \right).$$

This implies the inequality

$$\| \mathbf{E}(\xi_s \mid G_0^\xi) \|_p \leq 2 \| \xi_0 \|_p (\phi_1(s))^{1-1/p}.$$

To establish the required in Lemma 3 inequality, it suffuces to show that $\phi_1(s) = \phi(s)$.

For $B \in \mathfrak{S}_s^\xi$ and $A \in G_0^\xi$ we have the following relations:

$$| P(B \mid A) - P(B) | = \frac{1}{P(A)} \left| \int_A (P(B \mid G_0^\xi) - P(B)) \, dP \right| \leq \phi_1(s),$$

i.e. $\phi(s) \leq \phi_1(s)$.

The definition of ϕ_1 (s) implies also that for a fixed $\varepsilon > 0$ one can choose sets $B_\varepsilon \in \mathcal{G}^\xi_s$ and $A_\varepsilon \in G^\xi_0$ such that

$$P(B_\varepsilon \mid G^\xi_0) - P(B_\varepsilon) > \phi_1(s) - \varepsilon$$

for $\omega \in A_\varepsilon$ with $P(A_\varepsilon) > 0$. Integrating this inequality over the set A_ε, we get

$$P(A_\varepsilon B_\varepsilon) - P(A_\varepsilon) P(B_\varepsilon) > (\phi_1(s) - \varepsilon) P(A_\varepsilon)$$

and hence

$$\phi(s) > \phi_1(s) - \varepsilon.$$

As ε is arbitrary, we get

$$\phi(s) \geq \phi_1(s),$$

i.e.

$$\phi(s) = \phi_1(s).$$

8. The estimates for

$$\| E(\xi_s \mid G^\xi_0) \|_{q/(q-1)},$$

presented in Lemmas 2 and 3, allow us to formulate sufficient conditions under which the functional central limit theorem holds for strictly stationary processes, in terms of the behaviour of the mixing coefficients α and ϕ.

Indeed, Theorem 1 and Lemmas 2 and 3 imply the following result.

Theorem 2. *Let at least one of the following conditions be satisfied:*

1) $|\xi_0| \leq C, \quad \int_0^\infty \alpha^{1/2}(s)\, ds < \infty,$

2) $\|\xi_0\|_2 < \infty, \quad \int_0^\infty \phi^{1/2}(s)\, ds < \infty.$

Then

$$Y^n \xrightarrow{d} \left(2 \int_0^\infty E(\xi_s \xi_0)\, ds \right)^{1/2} W,$$

where W *is a Wiener process.*

Remark. Under Conditions 1) and 2) of Theorem 2

$$\lim_{s \to \infty} \alpha(s) = 0.$$

This means that each set from the σ-algebra of invariant sets J^ξ has measure 0 or 1 (Problem 2) and consequently the condition J^ξ, involved in (2.5) may be omitted.

Problems

1. Show that under Condition (2.3) the random variable

$$\eta = 2\int_0^\infty \mathbf{E}\,(\xi_t \xi_0 \mid J^\xi)\,dt$$

is defined and $\eta \geq 0$ (P-a.s.).

2. Let

$$\lim_{t\to\infty} \alpha(t) = 0,$$

where $\alpha = \alpha(t)$ is the strong mixing coefficient of the process ξ. Show that here each set from the σ-algebra of invariant sets J^ξ has measure 0 or 1.

3. The function $\rho = \rho(t)$

$$\rho(t) = \sup \frac{|\mathrm{Cov}(\eta, \zeta)|}{\sqrt{D\eta\, D\zeta}},$$

where sup is taken over all $\eta \in G_0^\xi$ and $\zeta \in \mathcal{G}_t^\xi$ with $\mathbf{E}\eta^2 < \infty$ and $\mathbf{E}\zeta^2 < \infty$, is called the maximal correlation coefficient. Show that

$$\|\mathbf{E}\,(\xi_s \mid G_0^\xi)\|_2 \leq \|\xi_0\|_2\, \rho(s).$$

4. Let

$$\xi = (\xi_t)_{t\in R} = (\xi_1(t), \ldots, \xi_k(t))_{t\in R}$$

be a strictly stationary ergodic process such that for $i = 1, \ldots, k$

$$\mathbf{E}\xi_i^2(0) < \infty, \qquad \mathbf{E}\xi_i(0) = 0,$$

$$\int_0^\infty \|\mathbf{E}\,(\xi_i(t) \mid G_0^\xi)\|_2\, dt < \infty,$$

with

$$G_0^\xi = \sigma\{\xi_s, -\infty < s \leq 0\}.$$

Set

$$Y^n = (Y_t^n)_{t\geq 0}, \quad Y_t^n = \frac{1}{\sqrt{n}}\int_0^{nt} \xi_s\, ds.$$

Show that

$$Y^n \xrightarrow{d} \Lambda^{1/2} W,$$

where $W = (W_t)_{t\geq 0} = (W_1(t), \ldots, W_k(t))_{t\geq 0}$ is a Wiener process with independent components and Λ is a nonnegative matrix, defined by the formula

$$\Lambda = \int_0^\infty \mathbf{E}\,(\xi_t \xi_0^* + \xi_0 \xi_t^*)\, dt$$

(* is the transposition sign, and all vectors are vector-columns).

§ 3. Invariance principle for a Markov process

1. Let $(\Omega, \mathcal{F}, \mathbb{F} = (\mathcal{F}_t)_{t \geq 0}, \mathsf{P})$ be a stochastic basis on which a progressively measurable homogeneous Markov process $\xi = (\xi_t)_{t \geq 0}$ is defined with a phase space $(E, \mathcal{E})$.

Let $f = f(x)$ be a ξ-measurable function with values in R such that

$$\int_0^t |f(\xi_s)|\, ds < \infty \quad (\mathsf{P}\text{-a.s.}).$$

Define the process $Y = (Y_t^0)_{t \geq 0}$ with

$$Y_t = \int_0^t f(\xi_s)\, ds \tag{3.1}$$

and the processes $Y^n = (Y_t^n)_{t \geq 0}$, $n \geq 1$, with

$$Y_t^n = \frac{1}{\sqrt{n}} Y_{nt}. \tag{3.2}$$

In this section we will establish conditions under which the functional central limit theorem holds for the sequence Y^n, $n \geq 1$.

2. Here some facts are presented from the theory of Markov processes which we will need in the sequel. Here and elsewhere below all necessary notions and definitions are reproduced from [98].

Denote the transition function of a Markov process by $p(t, x, dy)$, and the distribution of the random variable ξ_0 by $q = q(dx)$.

To the transition function $p(t, x, dy)$ we relate the operator T_t which operates according to the formula

$$T_t h(x) = \int_E h(y)\, p(t, x, dy)$$

for each $\mathcal{E}$-measurable function $h = h(y)$ such that

$$\int_{E \times E} |h(y)|\, p(t, x, dy)\, q(dx) < \infty.$$

The family T_t, $t > 0$, presents a semigroup with the operation

$$T_t T_s = T_{t+s}.$$

Let $h = h(y)$ be a function with the property indicated above. The Markov characterization of a process ξ, defined on a stochastic basis $(\Omega, \mathcal{F}, \mathbb{F} = (\mathcal{F}_t)_{t \geq 0}, \mathsf{P})$ is formulated in terms of the semigroup T_t, $t > 0$, in the following manner:

$$\mathbf{E}(h(\xi_t) \mid \mathcal{F}_s) = T_{t-s} h(\xi_s) \ (\mathsf{P}\text{-a.s.}), \ t > s.$$

A probability measure $r = r(dx)$ on $(E, \mathcal{E})$ is called invariant if for each set $\Gamma \in \mathcal{E}$

$$\int_E p(t, x, \Gamma)\, r(dx) = r(\Gamma), \ t > 0,$$

if for each $\mathcal{E}$-measurable function $h = h(x)$ such that

$$\int_E |h(x)|\, r(dx) < \infty$$

the following equality holds:

$$\int_E T_t h(x)\, r(dx) = \int_E h(x)\, r(dx), \ t > 0.$$

If $q = r$, then a Markov process ξ is strictly stationary.

Let $L_2(E, \mathcal{E}, r)$ be a Hilbert space of functions $h = h(y)$ with the norm

$$\|h\|_2 = \left(\int_E h^2(y)\, r(dy) \right)^{1/2}.$$

Denote by B_0 the centre of the semigroup T_t, $t > 0$:

$$B_0 = \{h \in L_2(E, \mathcal{E}, r): \lim_{t \to 0} \|T_t h - h\|_2 = 0\}.$$

An operator A, defined on the region

$$D_A = \{h \in B_0: \exists\, g \in B_0, \lim_{t \to 0} \left\| \frac{T_t h - h}{t} - g \right\|_2 = 0\}$$

by the relation

$$\lim_{t \to 0} \left\| AR - \frac{T_t h - h}{t} \right\|_2 = 0,$$

is called an *infinitesimal operator* of a semigroup T_t, $t > 0$, relative to $L_2(E, \mathcal{E}, r)$.

Denote by R_A the set of values of an operator A:

$$R_A = \{Ah: h \in D_A\}.$$

For a function $h \in D_A$ and for each $t > 0$ the following equality holds:

$$T_t h(x) = h(x) + \int_0^t T_u Ah(x)\, du \ \ (r\text{-a.s.}). \tag{3.3}$$

Lemma 1. *Let* $h = h(x) \in D_A$ *and* $q = r$. *Then there exists a square integrable martingale* $M = (M_t)_{t \geq 0} \in \overline{\mathcal{M}}^2(\mathbb{F}, \mathsf{P})$, *which is a process with strictly stationary increments and with*

$$\mathbf{E}M_t^2 = -2t \int_E h(x)\, Ah(x)\, r(dx), \tag{3.4}$$

such that the semimartingale $X = (X_t)_{t \ge 0} \in S(\mathbb{F}, P)$, *defined by the formula*

$$X_t = h(\xi_0) + \int_0^t Ah(\xi_u)\,du + M_t, \tag{3.5}$$

presents a modification of the process $h(x) = (h(x_t))_{t \ge 0}$.

Proof. Define the process $M' = (M'_t)_{t \ge 0}$ by setting

$$M'_t = h(\xi_t) - h(\xi_0) - \int_0^t Ah(\xi_u)\,du. \tag{3.6}$$

Since the Markov process ξ is stationary as $q = r$ and $h \in D_A$, the functions h and Ah belong to $L_2(E, \mathcal{E}, r)$. It is not hard to deduce from this that

$$\mathbf{E}(M'_t)^2 < \infty, \quad t > 0.$$

Let us show that M' satisfies the martingale equality:

$$\mathbf{E}(M'_t \mid \mathcal{F}_s) = M'_s \quad (P\text{-a.s.}), \quad t > s.$$

In fact by (3.3), (3.6) and the martingale characterization of the Markov process ξ we get

$$\mathbf{E}(M'_t \mid \mathcal{F}_s) - M'_s = \mathbf{E}(h(\xi_t) \mid \mathcal{F}_s) - h(\xi_s) - \int_s^t \mathbf{E}(Ah(\xi_u) \mid \mathcal{F}_s)\,du$$

$$= T_{t-s}h(\xi_s) - h(\xi_s) - \int_s^t T_{u-s}Ah(\xi_s)\,du$$

$$= T_{t-s}h(\xi_s) - h(\xi_s) - \int_0^{t-s} T_u Ah(\xi_s)\,ds = 0 \quad (P\text{-a.s.}).$$

In view of the remark at the end of § 4 in Ch. 1, the process M' possesses a modification

$$M = (M_t)_{t \ge 0} \in \overline{\mathcal{M}}^2(\mathbb{F}, P).$$

Inserting this modification in (3.5), we see that the process X is a modification of $h(\xi)$.

Also, by (3.6) it follows that M' is a process with strictly stationary increments. Therefore such is the process M.

Let us establish now the representation (3.4). In virtue of Theorem 4.12.2

$$\mathbf{E}M_t^2 = -2\int_0^t X_u\,Ah(\xi_u)\,du,$$

provided

$$\int_0^t \mathbf{E}\,|\,X_u\,Ah(\xi_u)\,|\,du < \infty, \quad t > 0.$$

As X is a modification of h (ξ) and ξ is a stationary process, we have

$$\mathbf{E}\,|\,X_u\,Ah(\xi_u)\,| = \mathbf{E}\,|\,h(\xi_u)\,Ah(\xi_u)\,| \le \|h\|_2\,\|Ah\|_2 < \infty.$$

Hence

$$\mathbf{E}M_t^2 = -2\int_0^t \mathbf{E}\,X_u\,Ah(\xi_u)\,du = -2\int_0^t \mathbf{E}\,h(\xi_u)\,Ah(\xi_u)\,du$$

$$= -2t\int_E h(x)\,Ah(x)\,r(dx).$$

3. Let $f \in R_A$. Then under this assumption it follows, as is not hard to see, that

1) $$\int_E f^2(x)\,r(dx) < \infty, \qquad \int_E f(x)\,r(dx) = 0,$$

2) the equation Ag = f with a function g = g (x) such that

$$\int_E g^2(x)\,r(dx) < \infty,$$

has a solution.

3) under the condition

$$\int_0^\infty \|T_t f\|_2\,dt < \infty$$

there exists a solution of the equation Ag = f and it has the form:

$$g(x) = -\int_0^\infty T_t f(x)\,dt.$$

Lemma 2. *Let* $f \in R_A$ *and let* 0 *be the simple eigen-value of the operator* A. *Then under the condition*

$$\int_E g(x)\,r(dx) = 0$$

the equation

$$Ag = f \tag{3.7}$$

has a unique solution with the minimal norm $\|g\|_2$.

Proof. If h = const, then $T_t\,h = h$ and the definition of A implies A h = 0. In other words, h = const is the solution of the equation Ah = 0. If g is the solution of the equation (3.7) and

$$\bar{g} = \int_E g(x)\, r(dx),$$

then obviously $g - \bar{g}$ is the solution of the same equation.

If g_i, $i = 1, 2$ are the solutions of the equation (3.7) and $h = g_1 - g_2$, then $h =$ const. Besides, if g_i, $i = 1, 2$ are the solutions of the equation (3.7) such that

$$\int_E g_i(x)\, r(dx) = 0, \quad i = 1, 2,$$

then $h = 0$.

Let $g^0 = g^0(x)$ be the solution of the equation (3.7) and

$$\int_E g^0(x)\, r(dx) = 0,$$

and let $g = g(x)$ be an arbitrary solution of the equation (3.7). Then

$$g = g^0 + h$$

with $h =$ const. Besides,

$$\|g\|_2^2 = \|g^0\|_2^2 + \|h\|_2^2$$

since

$$\int_E g^0(x)\, h\, r(dx) = h \int_E g^0(x)\, r(dx) = 0,$$

i.e.

$$\|g^0\|_2 \le \|g\|_2 .$$

Lemma 3. *For each* $x \in E$ *let*

$$\lim_{t \to \infty} \mathrm{Var}\,(p(t, x, \cdot) - r(\cdot)) = 0.$$

Then 0 *is the simple eigen-value of the operator* A.

Proof. Let $h \in D_A$ and $A h = 0$. Then by (3.3) we have $T_t h = h$ (r-a.s.), $t > 0$. We will show that

$$\lim_{t \to \infty} T_t h = \int_E h(y)\, r(dy).$$

We have

$$|T_t h - \int_E h(y)\, r(dy)| \le \int_E |h(y)|\, |p(t, x, dy) - r(dy)|$$

$$\le \{\mathrm{Var}\,(p(t, x \cdot) - r(\cdot)) \int_E h^2(y)\, [p(t, x, dy) + r(dy)]\}^{1/2} \to 0, \quad t \to \infty.$$

This implies h = const, i.e. the desired assertion holds.

Lemma 4. *Let* q = r *and let* 0 *be the simple eigen-value of the operator* A.

Then the Markov process ξ *is ergodic, i.e. the* σ*-algebra* $\mathfrak{J}^{\xi}$ *of invariant sets includes sets of probability* 0 *or* 1.

Proof. Let $B \in \mathfrak{J}^{\xi}$. Define a $\mathcal{E}$-measurable function h = h (x) such that

$$h(\xi_0) = \mathbf{E}(I_B \mid \xi_0).$$

As ξ possesses the Markov property relative to $\mathbb{F}$, we get (P-a.s.)

$$h(\xi_0) = \mathbf{E}[\mathbf{E}(I_B \mid \mathcal{F}_t) \mid \xi_0] = \mathbf{E}[\mathbf{E}(I_B \mid \xi_t) \mid \xi_0] = \mathbf{E}[h(\xi_t) \mid \xi_0].$$

Hence $T_t h(x) = h(x)$ (r-a.s.), t > 0. Consequently Ah = 0 and by Lemma 2 h = const. By this

$$\mathbf{E}(I_B \mid \mathcal{F}_t) = h(\xi_t) = \text{const} \quad (\text{P-a.s.}).$$

Since

$$B \in \sigma\Big(\bigcup_{t \geq 0} \mathcal{F}_t\Big),$$

and

$$\mathbf{E}(I_B \mid \mathcal{F}_t) \to \mathbf{E}\Big(I_B \mid \sigma\Big(\bigcup_{t \geq 0} \mathcal{F}_t\Big)\Big) = I_B \quad (\text{P-a.s.}),$$

by Theorem 1.4.1, then I_B = const P-a.s. Hence P (B) = 0 or P (B) = 1.

4. We now formulate the invariance principle in the stationary, as well as in the nonstationary case.

Theorem 1. *Let* q = r *and let the function* f, *involved in the definition of* Y^n, n ≥ 1, (*see* (3.1) *and* (3.2)), *belong to* R_A, *that is the set of values of the infinitesimal operator* A *relative to* $L_2(E, \mathcal{E}, r)$.

If 0 *is the simple eigen-value of the operator* A, *then*

$$Y^n \xrightarrow{d} aW,$$

where $W = (W_t)_{t \geq 0}$ *is a Wiener process,*

$$a = \Big(-2 \int_E g(x)\, f(x)\, r(dx)\Big)^{1/2}$$

and g = g (x) *is the solution of the equation* Ag = f.

Theorem 2. *Let* q ≠ r, $f \in R_A$ *and*

$$\int_0^t |f(\xi_s)|\, ds < \infty \quad (\text{P-a.s.}).$$

The assertion of Theorem 1 *remains valid if for each* x ∈ E

$$\lim_{t \to \infty} \operatorname{Var}(p(t, x\, \cdot\,) - r(\,\cdot\,)) = 0. \tag{3.8}$$

Proof of Theorem 1. Since $f \in R_A$, the equation $Ag = f$ has a solution. Next, by Lemma 1 the process $g(\xi)$ is a modification of the semimartingale X with the decomposition:

$$X_t = g(\xi_0) + \int_0^t f(\xi_u)\,du + M_t, \tag{3.9}$$

where

$$M = (M_t)_{t \geq 0} \in \overline{\mathcal{M}}^2(\mathbb{F}, P)$$

is a process with strictly stationary increments. Besides, in view of the equality $Ag = f$ we have

$$EM_t^2 = -2 \int_E g(x)\, f(x)\, r(dx).$$

By Lemma 4 ξ is an ergodic process. Hence

$$E(M_1^2 \mid J^{\xi}) = EM_1^2 = -2 \int_E g(x)\, f(x)\, r(dx) \quad (P\text{-a.s.}).$$

Denote $M^n = (M_t^n)_{t \geq 0}$ with

$$M_t^n = \frac{1}{\sqrt{n}} M_{nt}.$$

From (3.1), (3.2) and (3.9) it follows that

$$Y_t^n = \frac{X_{nt}}{\sqrt{n}} - \frac{g(\xi_0)}{\sqrt{n}} - M_t^n.$$

By Theorem 1.3

$$M^n \xrightarrow{d} aW,$$

where W is a Wiener process and

$$a = \left(-2 \int_E g(x)\, f(x)\, r(dx) \right)^{1/2} = (EM_1^2)^{1/2}.$$

By Theorem 4.12.1

$$\sup_{s \leq t} \frac{|X_{ns}|}{\sqrt{n}} \xrightarrow{P} 0, \quad n \to \infty, \quad t > 0,$$

since X is a strictly stationary process,

$$Eg^2(\xi_0) < \infty, \quad \int_0^t E f^2(\xi_s)\, ds < \infty \text{ and } EM_t^2 < \infty, \quad t > 0.$$

Consequently

$$\sup_{s \le t} n^{-1/2} | X_{ns} - g(\xi_0) | \xrightarrow{P} 0, \; n \to \infty, \; t > 0,$$

and hence in view of Problem 6.2.2 we have

$$Y^n \xrightarrow{d} aW.$$

Remark. To prove Theorem 1 one may apply the results of § 2 (see Subsection 6), since $f(\xi)$ is a strictly stationary process. In particular, by Example 1 in § 2

$$Y^n \xrightarrow{d} aW,$$

provided ξ is an ergodic process and

$$\int_0^\infty || T_t f ||_2 \, dt < \infty,$$

with

$$a = \left(2 \int_0^\infty \mathbf{E} f(\xi_t) f(\xi_0) \, dt \right)^{1/2}.$$

Observe that under the condition

$$\int_0^\infty || T_t f ||_2 \, dt < \infty$$

the equation $Ag = f$ is solved by the function

$$g(x) = - \int_0^\infty T_t f(x) \, dt,$$

and besides

$$\int_0^\infty \mathbf{E} f(\xi_t) f(\xi_0) \, dt = \int_0^\infty \mathbf{E} T_t f(\xi_0) f(\xi_0) \, dt = \mathbf{E} f(\xi_0) \int_0^\infty T_t f(\xi_0) \, dt$$

$$= - \mathbf{E} f(\xi_0) g(\xi_0) = - \int_E f(x) g(x) r(dx).$$

Proof of Theorem 2. Along with the processes Y^n, $n \ge 1$, we consider the processes

$$\tilde{Y}^n = (\tilde{Y}^n_t)_{t \ge 0}, \; n \ge 1$$

with

$$\tilde{Y}_t^n = \frac{1}{\sqrt{n}} \int_{n^{1/4}}^{n^{1/4}+nt} f(\xi_s)\, ds.$$

If q = r, then ξ is a strictly stationary process and obviously

$$Y^n \overset{d}{=} \tilde{Y}^n, \quad n \geq 1.$$

In view of Condition (3.8) and Lemmas 3 and 4 the process ξ is ergodic. Therefore,

$$\tilde{Y}^n \overset{d}{\to} aW$$

by Theorem 1.

Assume now $q \neq r$. We will prove that

$$\tilde{Y}^n \overset{d}{\to} aW$$

in this case too. Let $\psi = \psi(Y)$, where $Y \in D$ is a bounded function continuous in the Skorohod topology (see Ch. 6, § 1) and $|\psi| \leq k$. Denote by $\mathbf{E}_q$ the mathematical expectation with respect to q, that is the distribution of the random variable ξ_0. For each n define a $\mathcal{E}$-measurable function $\alpha_n = \alpha_n(x)$ such that

$$\alpha_n(\xi_s) = \mathbf{E}\,[\psi(\tilde{Y}^n) \mid \xi_s], \quad \text{P-a.s.}, \quad s \leq n^{1/4}.$$

Clearly, α_n may be chosen such that $|\alpha_n| \leq k$.

Then

$$\mathbf{E}_q \psi(\tilde{Y}^n) = \int_E \int_E \alpha_n(y)\, P(n^{1/4}, x, dy)\, q(dx).$$

On the other hand

$$\mathbf{E}_r \psi(\tilde{Y}^n) = \int_E \alpha_n(y)\, r(dy) = \int_E \int_E \alpha_n(y)\, r(dy)\, q(dx).$$

Consequently

$$|\mathbf{E}_q \psi(\tilde{Y}^n) - \mathbf{E}_r \psi(\tilde{Y}^n)| \leq k \operatorname{Var}(P(n^{1/4}, x, \cdot\,) - r(\,\cdot\,)) \to 0, \quad n \to \infty,$$

and hence

$$\lim_n \mathbf{E}_q \psi(\tilde{Y}^n) = \lim_n \mathbf{E}_r \psi(\tilde{Y}^n) = \mathbf{E}\,\psi(aW),$$

i.e.

$$\tilde{Y}^n \overset{d}{\to} aW$$

as $q \neq r$.

Denote

$$V_t^n = (t - n^{-3/4}) \vee 0$$

and observe that

$$\sup_{s \le t} | V^n_s - s | \to 0, \ n \to \infty.$$

Set

$$\hat{Y}^n_t = (\hat{Y}^n_t)_{t \ge 0}, \quad \hat{Y}^n_t = \tilde{Y}^n_{V^n_t}.$$

This implies

$$\hat{Y}^n \xrightarrow{d} aW$$

(Problem 3). Therefore, to accomplish the proof it remains to show that

$$\sup_{s \le t} | Y^n_s - \hat{Y}^n_s | \xrightarrow{P} 0, \ n \to 0, \ t > 0$$

(Problem 6.2.2). Observe that

$$\sup_{t > 0} | Y^n_t - \hat{Y}^n_t | \le \frac{1}{\sqrt{n}} \int_0^{n^{1/4}} | f(\xi_s) | \, ds.$$

If q = r, then

$$\mathbf{E} \frac{1}{\sqrt{n}} \int_0^{n^{1/4}} | f(\xi_s) | \, ds = \frac{1}{n^{1/4}} \int_E | f(x) | \, r(dx) \to 0, \ n \to \infty.$$

If q ≠ r, then

$$\frac{1}{n^{1/2}} \int_0^T | f(\xi_s) | \, ds \to 0 \ \text{(P-a.s.)}, \ n \to \infty.$$

Therefore, as q ≠ r, it suffices to show that

$$\lim_{T \to \infty} \overline{\lim_{n}} \, \mathsf{P} \left(\frac{1}{\sqrt{n}} \int_T^{n^{1/4}} | f(\xi_s) | \, ds \ge \varepsilon \right) = 0, \ \varepsilon > 0. \tag{3.10}$$

To this end denote

$$\beta_{n, T} = I \left(\frac{1}{\sqrt{n}} \int_T^{n^{1/4}} | f(\xi_s) | \, ds \ge \varepsilon \right).$$

Then, as $n^{1/4} > T$,

$$|E_q\beta_{n,T} - E_r\beta_{n,T}| \le \mathrm{Var}(P(T, x, \cdot) - r(\cdot)) \to 0, \ T \to \infty,$$

and

$$E_r\beta_{n,T} \le \frac{1}{\varepsilon}\frac{n^{1/4} - T}{n^{1/2}} \to 0, \ n \to \infty,$$

as $n^{1/4} > T$, i.e. the desired relation holds.

Problems

1. Let the σ-algebra J^{ξ} of invariant sets of a strictly stationary Markov process contain sets of probability 0 or 1. Show that 0 is a simple eigen-value of the operator A.

2. Let $f \in R_A$ and let g_i, $i = 1, 2$ be the solutions of the equation $Ag = f$. If 0 is a simple own value of the operator A, then

$$\int_E f(x)\, g_1(x)\, r(dx) = \int_E f(x)\, g_2(x)\, r(dx).$$

3. Let X^n, $n \ge 1$, and X be stochastic processes with trajectories in D and C respectively,

$$V^n = (V_t^n)_{t \ge 0} \in V$$

and

$$\hat{X}_t^n = X^n_{V_t^n}.$$

Show that under the conditions

$$\sup_{s \le t} |V_s^n - s| \to 0, \ n \to \infty, \ t > 0$$

and

$$X^n \xrightarrow{d} X$$

the following convergence takes place:

$$\hat{X}^n \xrightarrow{d} X.$$

§ 4. Diffusion approximation for systems with a "broad bandwidth noise" (scalar case)

1. The treatment of certain physical systems often leads to the study of properties of the stochastic process defined by the ordinary differential equation

$$\dot{x}_t = a(t, x_t) + b(t, x_t)\,\eta_t,$$

where (η_t) is a stationary process with a constant spectral density over a wide frequency range (a physical white noise).

In the present section the problem is considered of approximating the process (x_t) by a certain diffusion process defined by Ito's stochastic equation. The treatment of this problem is justified by the fact that, as a rule, the properties of a diffusion process are easier to study then the properties of the process (x_t).

It seems natural to pose the following problem.

Let $X^n = (X^n_t)_{t \geq 0}$, $n \geq 1$, be a sequence of continuous stochastic processes such that X^n_t solves the differential equation

$$\dot{X}^n_t = a(t, X^n_t) + b(t, X^n_t)\,\eta^n_t \tag{4.1}$$

with an initial condition X^n_0 for each n. Here X^n_0 and $(\eta^n_t)_{t \geq 0}$ are independent random quantities and

$$\int_0^t |\eta^n_s|\, ds < \infty \quad (\text{P-a.s.}), \quad t > 0.$$

Define the stochastic process $Y^n = (Y^n_t)_{t \geq 0}$ with

$$Y^n_t = \int_0^t \eta^n_s\, ds \tag{4.2}$$

and suppose

$$(X^n_0, Y^n) \xrightarrow{d} (X_0, \sigma W), \tag{4.3}$$

where σ is a constant and $W = (W_t)_{t \geq 0}$ a Wiener process.

The problem consists in imposing conditions on $a = a(t, x)$ and $b = b(t, x)$, which ensure the weak convergence

$$X^n \xrightarrow{d} X, \tag{4.4}$$

where $X = (X_t)_{t \geq 0}$ is a certain diffusion process. It is necessary also to specify the differential equation for X relative to X_0 and W.

Observe that $\sqrt{n}\ \xi_{nt}$ may serve as an example of η_t^n where $(\xi_t)_{t \in R}$ is a strictly stationary ergodic process with $E\xi_0^2 < \infty$ and $E\xi_0 = 0$, which satisfies the condition of the weak dependence

$$\int_0^\infty (E\,[E\,(\xi_t \mid G_0^\xi)]^2)^{1/2}\, dt < \infty$$

(see Theorem 2.1).

2. One of the solutions of the problem formulated above is given in the following theorem.

Theorem 1. *Let* 1) $a = a\,(t, x)$ *be a continuous function satisfying Lipschitz' condition and the condition of linear growth:*

$$|\,a\,(t, x') - a\,(t, x'')\,| \le K\,|\,x' - x''\,|,$$

$$|\,a\,(t, x)\,| \le K\,(\,1 + |\,x\,|\,);$$

2) $b = b\,(t, x)$ *is a continuous function having continuous derivatives* $b_t = b_t\,(t, x)$ *and* $b_x = b_x\,(t, x)$;

3) $b_t\,(t, x)$, $b_x\,(t, x)$ *and* $b\,(t, 0)$ *are uniformly bounded functions* (*by a constant* K);

4) b_x *satisfies Lipschitz' condition*

$$|\,b_x\,(t, x') - b_x\,(t, x'')\,| \le K\,|\,x' - x''\,|.$$

Under Condition (4.3), *the weak convergence* (4.4) *holds with the limiting process defined by Ito's equation*

$$X_t = X_0 + \int_0^t [a\,(s, X_s) + \frac{\sigma^2}{2}\, b_x\,(s, X_s)\, b\,(s, X_s)]\, ds + $$
$$+ \int_0^t \sigma\, b\,(s, X_s)\, dW_s. \tag{4.5}$$

3. The proof of the theorem will be based on the change of variables used by Krasnosel'skii and Pokrovskii [353] in the theory of vibrocorrect differential equations.

Following [353], we will show that there exists a function $Q = Q\,(s, s_0, y_0, t)$ continuous in all variables, such that X^n may be represented in the form

$$X_t^n = Q\,(Y_t^n, 0, Z_t^n, t), \tag{4.6}$$

where the process (Z_t^n) is such that

$$\dot{Z}_t^n = a\,(t, Q\,(Y_t^n, 0, Z_t^n, t)), \quad Z_0^n = X_0^n. \tag{4.7}$$

The required function Q is determined in the following manner:

$$Q(s, s_0, y_0, t) = y_s,$$

where y_s is the solution of the differential equation

$$\frac{dy_s}{ds} = b(t, y_s), \quad y_{s_0} = y_0$$

(the solution of this equation exists and it is unique due to the assumption on the function b; t plays the rôle of a parameter).

The function Q defined in this manner satisfies the following identities:

$$Q(s_0, s_0, y_0, t) = y_0, \tag{4.8}$$

$$Q(s_2, s_1, Q(s_1, s_0, y_0, t), t) = Q(s_2, s_0, y_0, t) \tag{4.9}$$

and, as a consequence of (4.8) and (4.9) the identity

$$Q(s', s'', Q(s'', s', z, t), t) = z. \tag{4.10}$$

Due to the assumption on the function b and to the general results concerning the differentiability of the solutions of differential equations relative to the initial quantities and parameters, the function Q is continuously differentiable in all variables.

Denote by Q_i the partial derivative of the function Q with respect to the i-th argument, $i = 1, ..., 4$. We present the facts concerning the functions Q_i, $i = 1, ..., 4$, which we will need in the sequel.

Obviously,

$$Q_1(s', s'', z, t) = b(t, Q(s', s'', z, t)) \tag{4.11}$$

and hence by (4.10)

$$Q_1(s', s'', Q(s'', s', z, t), t) = b(t, z). \tag{4.12}$$

Differentiating both sides of the identity (4.10) with respect to s'', in view of (4.11) we get

$$Q_2(s', s'', Q(s'', s', z, t), t)$$
$$+ Q_3(s', s'', Q(s'', s', z, t), t)\, b(t, Q(s'', s', z, t)) = 0. \tag{4.13}$$

Next, differentiating both sides of the identity (4.10) with respect to t and z, we get

$$Q_4(s', s'', Q(s'', s', z, t), t)$$
$$+ Q_3(s', s'', Q(s'', s', z, t), t)\, Q_4(s'', s', z, t) = 0 \tag{4.14}$$

and

$$Q_3(s', s'', Q(s'', s', z, t), t)\, Q_3(s'', s', z, t) = 1 \tag{4.15}$$

respectively. Observe also that

$$Q_3(0, s, Q(s, 0, z, t), t) = 1 \tag{4.16}$$

and

$$Q_4(0, s, Q(s, 0, z, t), t) = 0. \tag{4.17}$$

Let X_t^n be the solution of the differential equation (4.1).

Set

$$Z_t^n = Q(0, Y_t^n, X_t^n, t). \tag{4.18}$$

Then by (4.10)

$$X_t^n = Q(Y_t^n, 0, Z_t^n, t). \tag{4.19}$$

By (4.18), (4.1) and (4.2) we get

$$\dot{Z}_t^n = Q_2(0, Y_t^n, X_t^n, t)\,\eta_t^n + Q_3(0, Y_t^n, X_t^n, t)\,[a(t, X_t^n) + b(t, X_t^n)\,\eta_t^n]$$

$$+ Q_4(0, Y_t^n, X_t^n, t). \tag{4.20}$$

By (4.13) and (4.19)

$$Q_2(0, Y_t^n, X_t^n, t) + Q_3(0, Y_t^n, X_t^n, t)\, b(t, X_t^n) = 0. \tag{4.21}$$

Besides, by (4.19) and (4.16)

$$Q_3(0, Y_t^n, X_t^n, t) = Q_3(0, Y_t^n, Q(Y_t^n, 0, Z_t^n, t), t) = 1. \tag{4.22}$$

Consequently, taking into consideration (4.21) and (4.22), by (4.20) we get

$$\dot{Z}_t^n = a(t, X_t^n)$$

or, taking into consideration (4.19), we see that Z_t^n is the solution of the equation (4.7), where by (4.18) and (4.10)

$$Z_0^n = Q(0, 0, X_0^n, 0) = X_0^n.$$

Thus, to establish the representations (4.6) and (4.7), it suffices to show that the equation (4.7) has a unique solution. Aiming at this, we note that by the definition of the function Q

$$\frac{\partial Q_3(s, 0, y_0, t)}{\partial s} = b_x(t, Q_3(s, 0, y_0, t)),$$

$$Q_3(0, 0, y_0, t) = 1.$$

Therefore, in accordance with the condition on the function b_x, the following estimate holds:

$$|Q_3(s, 0, y_0, t)| \le 1 + K|s|.$$

By this the function

$$\tilde{a}_n(t, z) = a(t, Q(Y_t^n, 0, z, t))$$

satisfies the condition of linear growth:

$$|\tilde{a}_n(t, z)| = |a(t, Q(Y_t^n, 0, z, t))| \le K(1 + |Q(Y_t^n, 0, z, t)|$$

$$\le K(2 + |Q(Y_t^n, 0, 0, t)| + K|Y_t^n||z|) \tag{4.23}$$

and Lipschitz' condition

$$|\tilde{a}_n(t, z') - \tilde{a}_n(t, z'')| = |a(t, Q(Y^n_t, 0, z', t)) - a(t, Q(Y^n_t, 0, z'', t))|$$

$$\le (1 + K|Y^n_t|)|z' - z''|, \tag{4.24}$$

and hence the equation (4.7) has a unique solution.

Let us show now that the following weak convergence takes place:

$$(X^n, Z^n) \xrightarrow{d} (X, Z), \tag{4.25}$$

where the random variables X and Z are determined in the following manner:

$$X_t = Q(\sigma W_t, 0, Z_t, t), \tag{4.26}$$

$$\dot{Z}_t = a(t, Q(\sigma W_t, 0, Z_t, t)),$$

$$Z_0 = X_0. \tag{4.27}$$

Indeed, for each t X^n_t and Z^n_t are continuous functions of X^n_0 and Y^n (the continuity in Y^n is understood in Skorohod's metric; Ch. 6, § 1). Therefore, (4.3) implies the convergence

$$(X^n, Z^n) \xrightarrow{d_f} (X, Z)$$

(see Theorem 6.1.7). Consequently, by Theorem 6.1.8 the weak convergence holds, provided the family of distributions of X^n, Z^n, $n \ge 1$, is relatively compact. By the representation (4.6) for X^n and the properties of the function Q it suffices to establish here that the family of distributions of Z^n, $n \ge 1$, is relatively compact. For this, in turn, it suffices to verify the following conditions (Theorem 6.3.1):

$$\lim_{a \to \infty} \overline{\lim_n} \, P\left(\sup_{s \le L} |Z^n_t| > a\right) = 0, \ \forall L > 0,$$

$$\lim_{\delta \to 0} \overline{\lim_n} \sup_{T^n \in \Gamma_L} P\left(\sup_{t \le \delta} |Z^n_{T^n + t} - Z^n_{T^n}| > \eta\right) = 0, \ \forall L \ge 0, \ \eta \ge 0. \tag{4.28}$$

Both of these conditions are satisfied in virtue of the fact that (Z^n_t) presents the solution of the differential equation (4.7), the right-hand side of which satisfies the condition of linear growth (4.23) and Lipschitz' condition (4.24), and also in virtue of the weak convergence of the sequence of processes Y^n, $n \ge 1$, and the continuity of the function Q (Problem 2).

Let us establish now the equation (4.5) for the limiting process X.

By (4.26), (4.27) and Ito's formula we get

$$dX_t = Q_1(\sigma W_t, 0, Z_t, t)\, \sigma W_t + \frac{1}{2}\frac{\partial}{\partial s} Q_1(s, 0, Z_t, t)\,|_{s=\sigma W_t}\, \sigma^2 dt$$
$$+ Q_3(\sigma W_t, 0, Z_t, t)\, a(t, Q(\sigma W_t, 0, Z_t, t))\, dt. \tag{4.29}$$

By (4.10) and (4.26)

$$Z_t = Q(0, \sigma W_t, X_t, t). \tag{4.30}$$

Therefore in view of (4.11)

$$Q_1(\sigma W_t, 0, Z_t, t) = Q_1(\sigma W_t, 0, Q(0, \sigma W_t, X_t, t), t)$$
$$= b(t, Q(\sigma W_t, 0, Z_t, t)) = b(t, X_t). \tag{4.31}$$

Hence

$$\frac{\partial}{\partial s} Q_1(s, 0, Z_t, t)\,|_{s=\sigma W_t} = \frac{\partial}{\partial s} b(t, Q(s, 0, Z_t, t))\,|_{s=\sigma W_t}$$
$$= b_x(t, X_t)\, Q_1(\sigma W_t, 0, Z_t, t)$$
$$= b_x(t, X_t)\, b(t, X_t). \tag{4.32}$$

Finally, in view of (4.29) and (4.16)

$$Q_3(\sigma W_t, 0, Z_t, t) = Q_3(\sigma W_t, 0, Q(0, \sigma W_t, X_t, t), t) = 1. \tag{4.33}$$

The equation (4.5) is obtained by (4.29), (4.31) - (4.33) in an obvious manner.

Problems

1. Let the conditions of Theorem 1 be satisfied with

$$X_0^n, Y^n \overset{d}{\to} X_0, M$$

instead of (4.3), where M is a continuous Gaussian martingale with $\langle M\rangle_t = \mathbf{E}M_t^2$. Show that the assertion of the theorem remains valid with the limiting process

$$X_t = X_0 + \int_0^t a(s, X_s)\, ds + \frac{1}{2}\int_0^t b_x(s, X_s)\, b(s, X_s)\, d\langle M\rangle_s + \int_0^t b(s, X_s)\, dM_s.$$

2. Verify the relations (4.28).

§ 5. Diffusion approximation with a "broad bandwidth noise" (vector case)

1. In this section the diffusion approximation is studied for sequences of vector-valued processes $X^n = (X_t^n)_{t \geq 0} = (X_1^n(t), \ldots, X_k^n(t))$, $n \geq 1$, determined for each n by the differential equation

$$\dot{X}_t^n = a(t, X_t^n) + b(t, X_t^n)\,\eta_t^n, \tag{5.1}$$

where $a = a(t, x) = (a_1(t, x), \ldots, a_k(t, x))$ is a vector-valued function, $b = b(t, x)$ is a $(k \times k)$-matrix with elements $b_{ij}(t, x)$, $x = (x_1, \ldots, x_k)$, and

$$(\eta_t^n)_{t \in R} = (\eta_1^n(t), \ldots, \eta_k^n(t))_{t \in R}$$

is a vector-valued stochastic process with

$$\int_0^t \sum_{i=1}^k |\eta_i^n(s)|\, ds < \infty \quad (\text{P-a.s.}), \quad t > 0,$$

independent of the initial vector X_0^n.

Define the vector-valued stochastic process $Y^n = (Y_t^n)_{t \geq 0}$ with

$$Y_t^n = \int_0^t \eta_s^n\, ds. \tag{5.2}$$

(Here and elsewhere below the vectors are column-vectors).

The method of proving Theorem 4.1 which concerns the diffusion approximation (in the scalar case, i.e. in case $k = 1$) of the sequence X^n, $n \geq 1$, as applied to the case $k > 1$, requires an additional, special condition (the so-called Frobenius condition; see [353]). Therefore, in this section another proof of the diffusion approximation is presented, which is adapted to the case $k \geq 1$. However, as $k = 1$ this proof leads to to superfluous (as compared with the conditions of Theorem 4.1) restrictions on the process $\eta^n = (\eta_t^n)_{t \in R}$.

2. First of all, let us dwell on the description of properties of the process η^n. Assume

$$\eta_t^n = \sqrt{n}\,\xi_{nt}, \tag{5.3}$$

where $\xi = (\xi_t)_{t \in R} = (\xi_1(t), \ldots, \xi_k(t))_{t \in R}$ is a strictly stationary ergodic stochastic process such that for $i = 1, \ldots, k$

$$\mathbf{E}\xi_i^2(0) < \infty, \quad \mathbf{E}\xi_i(0) = 0$$

and

$$\int_0^\infty (\mathbf{E}\,[\mathbf{E}\,(\xi_i\,(t)\mid G_0^\xi)]^2)^{1/2}\,dt < \infty \tag{5.4}$$

with

$$G_t^\xi = \sigma\,\{\xi_s,\ -\infty < s \le t\}.$$

In accordance with Problem 2.4

$$Y^n \xrightarrow{d} \Lambda^{1/2}\,W, \tag{5.5}$$

where $W = (W_t)_{t \geq 0} = (W_1\ (t), \ldots, W_k\ (t))$ is a vector-valued Wiener process with independent components, and Λ a nonnegative definite symmetric matrix, given by the formula

$$\Lambda = \int_0^\infty \mathbf{E}\,(\xi_t\xi_0^* + \xi_0\xi_t^*)\,dt \tag{5.6}$$

($*$ is the transposition sign).

Denote

$$b_{ij}^{(p)}\,(t, x) = \frac{\partial}{\partial x_p}\,b_{ij}\,(t, x) \tag{5.7}$$

(assuming that all derivatives exist).

Define the vector-valued function $D\,(t, x) = (D_1\,(t, x), \ldots, D_k\,(t, x))$ by setting

$$D_i\,(t, x) = \sum_{jpq} b_{ij}^{(p)}\,(t, x)\,b_{pq}\,(t, x)\,\lambda_{qj}, \tag{5.8}$$

where λ_{qj} are the elements of the matrix Λ.

3. **Theorem 1.** *Denoting by* g (t, x) *and* h (t, x) *any of the elements of the vector-valued function* a (t, x) *and matrix* b (t, x) *respectively, suppose that the following conditions are satisfied:*

1) g (t, x) *is a continuous function,*

$$|\,g\,(t, x)\,| \le L\,(1 + \sum_{i=1}^k |\,x_i\,|),$$

$$|\,g\,(t, x') - g\,(t, x'')\,| \le L \sum_{i=1}^k |\,x_i' - x_i''\,|,$$

2) h (t, x) *is a continuous function, it has continuous derivatives* h_t, h_{x_i}, $h_{x_i t}$, $h_{x_i x_j}$, *and it is bounded by a constant* L, *together with these derivatives.*

Let the conditions (5.4) *and*

$$X_0^n \xrightarrow{d} X_0$$

also be satisfied.

Then

$$X^n \xrightarrow{d} X,$$

where $X = (X_t)_{t \geq 0}$ *is the diffusion process defined by Ito's stochastic equation*

$$X_t = X_0 + \int_0^t [a(s, X_s) + \frac{1}{2} D(s, X_s)]\, ds + \int_0^t b(s, X_s)\, \Lambda^{1/2}\, dW_s. \quad (5.9)$$

4. To avoid tedious manipulations, we will illustrate the idea of proving this theorem in case $k = 1$. Observe that Λ is a constant as $k = 1$, and

$$D(t, x) = \Lambda\, b_x(t, x)\, b(t, x).$$

The proof of the theorem is preceded by a number of auxiliary constructions.

Define a family of σ-algebras $\mathbb{F} = (\mathcal{F}_t)_{t \geq 0}$ satisfying the "usual" conditions (see Ch. 1, § 1, Definition 1), by departing from the σ-algebras $G_t^\xi = \sigma(\xi_s, s \leq t)$.

Denote

$$Y_t = \int_0^t \xi_s\, ds. \quad (5.10)$$

Under Conditions (5.4), there exists by Lemma 2.1 a strictly stationary process $Z = (Z_t)_{t \geq 0}$, that is a semimartingale (relative to $\mathbb{F}$) with the decomposition

$$Z_t = Z_0 + Y_t + M_t, \quad (5.11)$$

where $M = (M_t)_{t \geq 0}$ is a square integrable martingale which is a process with strictly stationary increments. Here

$$Z_0 = \int_0^\infty \mathbf{E}(\xi_t \mid G_0^\xi)\, dt$$

and hence $\mathbf{E}Z_0^2 < \infty$ by (5.4). Besides, $(Z_t, \xi_t)_{t \geq 0}$ is a strictly stationary ergodic process with $\mathbf{E}\,|Z_0\xi_0| < \infty$ and

$$\mathbf{E}Z_0\xi_0 = \int_0^\infty \mathbf{E}\,\xi_t\xi_0\, dt = \frac{\Lambda}{2} \quad (5.12)$$

(see (5.6) with $k = 1$).

Obviously

$$Y_t^n = \int_0^t \sqrt{n}\ \xi_{ns}\, ds = \frac{1}{\sqrt{n}} Y_{nt}.$$

Therefore $Z_t^n = \frac{1}{\sqrt{n}} Z_{nt}$ may be decomposed as follows:

$$Z_t^n = Z_0^n + Y_t^n + M_t^n \tag{5.13}$$

with

$$M_t^n = \frac{1}{\sqrt{n}} M_{nt}.$$

Denote

$$\mathcal{F}_t^n = \mathcal{F}_{nt} \vee \sigma(X_0^n). \tag{5.14}$$

As X_0^n and ξ are independent, the processes $Z^n = (Z_t^n)_{t \ge 0}$ and $M^n = (M_t^n)_{t \ge 0}$ are a martingale and, respectively, a square integrable martingale relative to $\mathbb{F}^n = (\mathcal{F}_t^n)_{t \ge 0}$.

By Theorem 4.12.1

$$\sup_{t \le T} |Z_t^n| \overset{P}{\to} 0, \ n \to \infty, \ \forall\, T > 0. \tag{5.15}$$

By Theorem 1.2 and 1.3

$$\lim_n \mathbf{E} \sum_{t \le T} (\Delta M_t^n)^2 I(|\Delta M_t^n| > \varepsilon) = 0, \ \forall\, T > 0, \ \varepsilon > 0,$$

$$\lim_n [M^n, M^n]_t = \Lambda t \ \ (\text{P-a.s.}), \ \forall\, t > 0, \tag{5.16}$$

where a nonnegative constant Λ is determined by the formula (5.6) with $k = 1$. Next, by the corollary to Lemma 5.5.5

$$\langle M^n \rangle_t \overset{P}{\to} \Lambda t, \ \forall\, t > 0. \tag{5.17}$$

Finally, in view of Problem 5.3.2

$$\sup_{t \le T} |[M^n, M^n]_t - \Lambda t| \overset{P}{\to} 0,$$

$$\sup_{t \le T} |\langle M^n \rangle_t - \Lambda t| \overset{P}{\to} 0, \ \forall\, T > 0. \tag{5.18}$$

Consider now the process $(Z_t^n \eta_t^n)_{t \ge 0}$. Since

$$Z_t^n \eta_t^n = \frac{1}{\sqrt{n}} Z_{nt} \sqrt{n}\ \xi_{nt} = Z_{nt} \xi_{nt},$$

we have

$$\mathbf{E}\,|\,Z_t^n\eta_t^n\,| = \mathbf{E}\,|\,Z_0\xi_0\,|$$

as the process $(Z_t\xi_t)_{t\geq 0}$ is stationary, and

$$\lim_{n\to\infty}\ \sup_{t\leq T}\left|\int_0^T Z_s^n\eta_s^n\,ds - \frac{\Lambda t}{2}\right| = 0\ \ (\text{P-a.s.})$$

and

$$\lim_{n}\ \sup_{t\leq T}\left|\int_0^t |\,Z_s^n\eta_s^n\,|\,ds - t\,\mathbf{E}\,|\,Z_0\xi_0\,|\right| = 0\ \ (\text{P-a.s.}),\ \ \forall\, T>0, \tag{5.19}$$

as it is ergodic, due to the formula (5.13) and Theorem 6.1. Utilizing the decomposition (5.13) for Z_t^n and (5.3), the equation (5.1) may be rewritten in the following form:

$$X_t^n = X_0^n + \int_0^t a\,(s, X_s^n)\,ds + \int_0^t b\,(s, X_s^n)\,dM_s^n - \int_0^t b\,(s, X_s^n)\,dZ_s^n, \tag{5.20}$$

where the integrals M^n and Z^n are stochastic integrals with respect to a square integrable martingale and a semimartingale.

Along with (5.20), we will need yet another representation for X_t^n.

By Ito's formula (Ch. 2, § 3)

$$b\,(t, X_t^n)\,Z_t^n = b\,(0, X_0^n)\,Z_0^n + \int_0^t b\,(s, X_s^n)\,dZ_s^n$$

$$+ \int_0^t Z_s^n\,[b_t\,(s, X_s^n) + b_x\,(s, X_s^n)\,a\,(s, X_s^n)]\,ds$$

$$+ \int_0^t b_x\,(s, X_s^n)\,b\,(s, X_s^n)\,Z_s^n\eta_s^n\,ds.$$

This and (5.20) give the following representation for X^n:

$$X_t^n = X_0^n + \int_0^t a\,(s, X_s^n)\,ds + \int_0^t b_x\,(s, X_s^n)\,b\,(s, X_s^n)\,Z_s^n\eta_s^n\,ds$$

$$+ \int_0^t b\,(s, X_s^n)\,dM_s^n + \alpha_t^n \tag{5.21}$$

with

$$\alpha_t^n = b(0, X_0^n)Z_0^n - b(t, X_t^n)Z_t^n$$

$$+ \int_0^t Z_s^n [b_t(s, X_s^n) + b_x(s, X_s^n)\, a(s, X_s^n)]\, ds. \qquad (5.22)$$

5. We will formulate a number of the auxiliary statements.

Lemma 1. *Let the conditions of Theorem* 1 *be satisfied. Then the family of distributions of* X^n, $n \geq 1$, *is relatively compact.*

Corollay. *For each* $t > 0$ *and* $\varepsilon > 0$

$$\lim_N \overline{\lim_n} \, P\left(\sup_{s \leq t} \left| X_s^n - X_{\frac{[Ns]}{N}}^n \right| \geq \varepsilon \right) = 0,$$

where [a] *is the integer part of* a.

Proof. In view of Theorem 6.3.1 it suffices for each $l > 0$ and $\rho > 0$ that

$$\lim_{c \to \infty} \overline{\lim_n} \, P(\sup_{t \leq l} |X_t^n| \geq c) = 0, \qquad (5.23)$$

$$\lim_{\delta \to 0} \overline{\lim_n} \sup_{S \in T_l(\mathbb{F}^n)} P(\sup_{t \leq \delta} |X_{S+t}^n - X_S^n| \geq \rho) = 0. \qquad (5.24)$$

To establish (5.23) denote

$$\beta_t^n = \sup_{s \leq t} |X_s^n|, \quad \gamma_t^n = \sup_{s \leq t} |Z_s^n|, \quad \delta_t^n = \sup_{s \leq t} \left| \int_0^s b(u, X_u^n)\, dM_u^n \right| .$$

By (5.21), (5.22) and the conditions of Theorem 1 the following inequality holds (for each $t \leq l$):

$$\beta_t^n \leq L_1^n + L_2^n \int_0^t \beta_s^n \, ds$$

with

$$L_1^n = \beta_0^n + 2L\gamma_l^n + Ll(1 + \gamma_l^n(1 + L)) + \delta_l^n + L^2 \int_0^l |Z_s^n \eta_s^n| \, ds$$

and

$$L_2^n = L(1 + \gamma_l^n L),$$

which by the Gronwall-Bellman inequality gives

$$\beta_l^n \leq L_1^n \exp(L_2^n l).$$

Since

$$L_2^n \xrightarrow{P} L_2$$

(see (5.15)), (5.23) is valid, provided

$$\lim_{c\to\infty} \overline{\lim_n}\, P(L_1^n \ge c) = 0. \tag{5.25}$$

Note that the relation (5.25) holds, provided it holds for each term involved in L_1^n. The relation of type (5.25) is satisfied by β_0^n and $2L\gamma_1^n + Ll(1 + \gamma_1^n(1 + l))$ since

$$\beta_0^n = |X_0^n|, \quad X_0^n \xrightarrow{d} X_0 \text{ and } \gamma_1^n \xrightarrow{P} 0$$

(see (5.15)). The relation (5.25) is satisfied by δ_1^n in virtue of the Lenglart-Rebolledo inequality (Theorem 1.9.3), which gives

$$P(\delta_1^n \ge c) \le \frac{b}{c^2} + P\left(\int_0^1 b^2(u, X_u^n)\, d\langle M^n\rangle_u \ge b\right)$$

and hence

$$P(\delta_1^n \ge c) \le \frac{b}{c^2} + P(L^2 \langle M^n\rangle_1 \ge b) \le \frac{b}{c^2} + \frac{L^2}{b}\mathbf{E}\langle M^n\rangle_1$$

$$= \frac{b}{c^2} + \frac{L^2}{b}\mathbf{E}(M_1^n)^2 = \frac{b}{c^2} + \frac{L^2}{b}\mathbf{E}M_1^2 \to 0$$

as the limit $\lim_{b\to\infty} \overline{\lim_{c\to\infty}}\, \overline{\lim_n}$ is taken. The last term of L_1^n satisfies the relation (5.25) in virtue of the following estimate

$$P\left(L^2 \int_0^1 |Z_s^n \eta_s^n|\, ds \ge c\right) \le \frac{L^2 1}{c}\mathbf{E}|Z_0\xi_0|.$$

We will establish now the validity of (5.24). We have (see (5.21))

$$X_{S+t}^n - X_S^n = \int_S^{S+t} a(u, X_u^n)\, du + \int_S^{S+t} b_x(u, X_u^n)\, b(u, X_u^n)\, Z_u^n \eta_u^n\, du$$

$$+ \int_S^{S+t} b(u, X_u^n)\, dM_u^n + (\alpha_{S+t}^n - \alpha_S^n). \tag{5.26}$$

Obviously (5.24) is valid if each term on the right-hand side of (5.26) satisfies (5.24).

For the first term the following estimate holds:

$$\sup_{t\le\delta}\left|\int_S^{S+t} a(u, X_u^n)\,du\right| \le \int_S^{S+\delta} L\,(1+\sup_{v\le u}|X_v^n|)\,du$$

$$\le L\delta\left(1+\sup_{u\le 1+\delta}|X_u^n|\right).$$

By this estimate and (5.23) we verify the validity of (5.24) for the first term. The second term is estimated as follows:

$$\sup_{t\le\delta}\left|\int_S^{S+t} b_x(u, X_u^n)\, b(u, X_u^n)\, Z_u^n\eta_u^n\,du\right| \le L^2\int_S^{S+t}|Z_s^n\eta_s^n|\,ds$$

$$\le L^2\delta\,\mathbf{E}\,|Z_0\xi_0| + L^2\sup_{t\le 1+\delta}\left|\int_0^t|Z_s^n\eta_s^n|\,ds - t\,\mathbf{E}\,|Z_0\eta_0|\right|,$$

and by this and (5.19) it satisfies (5.24). To estimate the third term we use the Lenglart-Rebolledo inequality (Theorem 1.9.3):

$$\mathsf{P}\left(\sup_{t\le\delta}\int_S^{S+t} b(u, X_u^n)\,dM_u^n \ge \rho\right) \le \frac{q}{\rho^2} + \mathsf{P}\left(\int_S^{S+t} b^2(u, X_u^n)\,d\langle M^n\rangle_u \ge q\right).$$

Consequently,

$$\mathsf{P}\left(\sup_{t\le\delta}\left|\int_S^{S+t} b(u, X_u^n)\,dM_u^n\right| \ge \rho\right) \le \frac{q}{\rho^2} + \mathsf{P}\left(\langle M^n\rangle_{S+\delta} - \langle M^n\rangle_S \ge q/3L^2\right)$$

$$\le \frac{q}{\rho^2} + 2\mathsf{P}\left(\sup_{t\le 1+\delta}|\langle M^n\rangle_t - \Lambda t| \ge q/3L^2\right) + I\,(\Lambda\delta > 3/3L^2.$$

We obtain the desired relation (5.24) by passing here to the limit $\lim_{q\to 0}\ \overline{\lim}_{\delta\to 0}\ \overline{\lim}_{n}$.

To verify the validity of (5.24) for the last term it suffices to show that

$$\sup_{t\le T}|\alpha_t^n| \xrightarrow{\mathsf{P}} 0,\ n\to\infty,\ \forall\, T>0. \tag{5.27}$$

The definition of α_t^n (see (5.22)) gives

$$\sup_{t\le T}|\alpha_t^n| \le 2L\sup_{t\le T}|Z_t^n| + T\{\sup_{t\le T}|Z_t^n|\,[L\,(2+\sup_{t\le T}|X_t^n|)]\,\}.$$

Therefore (5.27) takes place in virtue of (5.15) and (5.23).

Lemma 2. *Let*

$$A^n = (A^n_t)_{t\geq 0} \in \mathcal{V}(\mathbb{F}^n),\ n \geq 1,$$

and let the following conditions be satisfied:

$$\sup_{t \leq T} |A^n_t| \xrightarrow{P} 0,\ n \to \infty,\ \forall\, T > 0$$

and

$$\lim_{c \to \infty} \overline{\lim_{n}}\ P(\mathrm{Var}\,(A^n)_T \geq c) = 0,\ \forall\, T > 0.$$

If the conditions of Theorem 1 *are satisfied and the function* h = h (t, x) *satisfies Lipschitz' condition*

$$|h(t', x') - h(t'', x'')| \leq L(|t' - t''| + |x' - x''|),$$

then

$$\sup_{t \leq T} \left| \int_0^t h(s, X^n_s)\, dA^n_s \right| \xrightarrow{P} 0,\ n \to \infty,\ \forall\, T > 0.$$

Proof. The following estimate takes place:

$$\sup_{t \leq T} \left| \int_0^t h(s, X^n_s)\, dA^n_s \right| \leq \sup_{t \leq T} \left| \int_0^t h\left(\frac{[Ns]}{N}, X^n_{\frac{[Ns]}{N}}\right) dA^n_s \right|$$
$$+ L\left(\frac{1}{N} + \sup_{t \leq T} \left| X^n_t - X^n_{\frac{[Nt]}{N}} \right| \right) \mathrm{Var}\,(A^n)_T.$$

Therefore the desired assertion is obtained in an obvious manner due to the corollary to Lemma 1.

6. Proof of Theorem 1. Define the process $\overline{X}^n = (\overline{X}^n_t)_{t > 0}$ by

$$\overline{X}^n_t = X^n_t - \alpha^n_t.$$

By (5.27)

$$\sup_{t \leq T} |X^n_t - \overline{X}^n_t| \xrightarrow{P} 0,\ n \to \infty,\ \forall\, T > 0. \tag{5.28}$$

Therefore, in view Problem 6.1.2 it suffices to show that

$$\overline{X}^n \xrightarrow{d} X.$$

From (5.21) it follows that X^n_t admits the representation:

$$\overline{X}^n_t = X^n_0 + \int_0^t a(s, X^n_s)\, ds + \int_0^t b_x(s, X^n_s)\, b(s, X^n_s)\, Z^n_s \eta^n_s\, ds + \int_0^t b(s, X^n_s)\, dM^n_s.$$

Utilize now assertion c) of Theorem 8.3.1 and Problem 8.3.3. According to these assertions it suffices to verify the following conditions (for each $T > 0$):

1) $\lim_n \mathrm{E} \sum_{t \le T} b^2(t, X_t^n)(\Delta M_t^n)^2 I(|b(t, X_t^n)\Delta M_t^n| \ge \varepsilon) = 0, \ \varepsilon > 0,$

2) $\sup_{t \le T} \left| \int_0^t a(s, X_s^n)\, ds - \int_0^t a(s, \overline{X}_s^n)\, ds \right| \xrightarrow{P} 0,$

3) $\sup_{t \le T} \left| \int_0^t b_x(s, X_s^n)\, b(s, X_s^n) Z_s^n \eta_s^n\, ds - \frac{1}{2}\int_0^t b_x(s, X_s^n)\, b(s, X_s^n)\, \Lambda\, ds \right| \xrightarrow{P} 0,$

4) $\sup_{t \le T} \left| \int_0^t b^2(s, X_s^n)\, d\langle M^n\rangle_s - \int_0^t b^2(s, X_s^n)\, \Lambda\, ds \right| \xrightarrow{P} 0.$

Condition 1) follows from (5.16) and the uniform boundedness of the function $b(t, x)$; Condition 2) follows from (5.28) and Lipschitz' condition (relative to x) on the function $a(t, x)$; Condition 3) is verified by the estimate

$$\sup_{t \le T} \left| \int_0^t b_x(s, X_s^n)\, b(s, X_s^n) Z_s^n \eta_s^n\, ds - \frac{1}{2}\int_0^t b_x(s, \overline{X}_s^n)\, b(s, \overline{X}_s^n)\, \Lambda\, ds \right|$$

$$\le \frac{\Lambda}{2} \int_0^T | b_x(s, X_s^n)\, b(s, X_s^n) - b_x(s, \overline{X}_s^n)\, b(s, \overline{X}_s^n) |\, ds$$

$$+ \sup_{t \le T} \left| \int_0^t b_x(s, X_s^n)\, b(s, X_s^n) \left(Z_s^n \eta_s^n - \frac{\Lambda}{2} \right) ds \right|,$$

by Lipschitz' condition relative to t and x on the function $b_x(t, x)\, b(t, x)$, by (5.28), (5.19) and Lemma 2; Condition 4) is verified by the estimate

$$\sup_{t \le T} \left| \int_0^t b^2(s, X_s^n)\, d\langle M^n\rangle_s - \int_0^t b^2(s, \overline{X}_s^n)\, \Lambda\, ds \right|$$

$$\le \int_0^T | b^2(s, X_s^n) - b^2(s, \overline{X}_s^n) |\, \Lambda\, ds + \sup_{t \le T} \left| \int_0^t b^2(s, X_s^n)\, d(\langle M^n\rangle_s - \Lambda s) \right|,$$

by Lipschitz' condition relative to t and x on the function $b^2(t, x)$, by (5.28), (5.18) and Lemma 2.

Theorem 1 is proved.

Problem

1. Prove Theorem 1 in the multivariate case.

§ 6. Ergodic theorem and invariant principle in case of nonhomogeneous time avaraging

1. Let x_t, $t \geq 0$, be a continuous function $((x_t)_{t\geq 0} \in C)$ and $\xi = (\xi_t)_{t\in R}$ a strictly stationary process. Denote

$$G_t^{\xi} = \sigma\{\xi_s, -\infty < s \leq t\}, \ \ t \in R.$$

Also, denote by J^{ξ} the σ-algebra of invariant sets of ξ.

Let a $B_R \otimes B_R$-measurable function $b = b(x, y)$ be given, such that

$$\mathbf{E}\,|\,b(x, \xi_0)\,| < \infty, \ \ x \in R. \tag{6.1}$$

In this section the limiting (as $n \to \infty$) properties of the processes

$$\int_0^t b(x_s, \xi_{ns})\,ds$$

and

$$Y_t^n = \sqrt{n}\int_0^t [b(x_s, \xi_{ns}) - \mathbf{E}(b(x_s, \xi_0)\,|\,J^{\xi})]\,ds$$

are studied, which then are applied (in §7) to establish a stochastic version of Bogoljubov's averaging principle.

Observe that as $x_t \equiv c$ the equality

$$\int_0^t b(c, \xi_{ns})\,ds = \frac{1}{n}\int_0^{nt} b(c, \xi_s)\,ds$$

holds, and by the Birkhof-Khintchine theorem (see Corollary to Theorem 4.11.9)

$$\int_0^t b(c, \xi_{ns})\,ds \to t\,\mathbf{E}(b(c, \xi_0)\,|\,J^{\xi}) \ \ (\text{P-a.s.}) \tag{6.2}$$

as $n \to \infty$.

Next, under the conditions

$$||\,b(c, \xi_0)\,||_2 < \infty$$

and

$$\int_0^{\infty} ||\,\mathbf{E}(b(c, \xi_s)\,|\,G_0^{\xi}) - \mathbf{E}(b(c, \xi_0)\,|\,J^{\xi})\,||_2\,ds < \infty$$

with

$$||\,\alpha\,||^p = (\mathbf{E}\,|\,\alpha\,|^p)^{1/p}, \ \ p \geq 1,$$

the sequence of the processes $Y^n = (Y^n_t)_{t \ge 0}$, $n \ge 1$ (as $x_t \equiv c$) converges weakly to $\sqrt{\eta}\, W$ as $n \to \infty$, where $W = (W_t)_{t \ge 0}$ is a Wiener process independent of

$$\eta = 2 \int_0^\infty \{\mathbf{E}\,[b\,(c, \xi_s)\, b\,(c, \xi_0) \mid J^\xi] - [\mathbf{E}\,(b\,(c, \xi_0) \mid J^\xi]^2\}\, ds$$

(see Ch. 9, § 2, Example 1).

In the present section we extend these results on the case $x_t \equiv$ const.

2. Theorem 1. *Let Condition* (6.1) *be satisfied, and let the function* $b = b\,(x, y)$ *satisfy Lipschitz' condition with respect to* x:

$$| b\,(x', y) - b\,(x'', y) | \le L\, | x' - x'' |$$

with a constant L, *independent of* y. *Then* (P-a.s.)

$$\lim_n \sup_{0 \le t \le T} \left| \int_0^t b\,(x_s, \xi_{ns})\, ds - \int_0^t \mathbf{E}\,(b\,(x_s, \xi_0) \mid J^\xi)\, ds \right| = 0$$

for each $T > 0$.

Proof. We will show first that for each $t > 0$ (P-a.s.)

$$\lim_n \int_0^t b\,(x_s, \xi_{ns})\, ds = \int_0^t \mathbf{E}\,(b\,(x_s, \xi_0) \mid J^\xi)\, ds. \tag{6.3}$$

To this end denote

$$x^N_t = x_{[Nt]/N}, \quad N = 1, 2, \ldots \tag{6.4}$$

By (6.2) for each $N \ge 1$ we have P-a.s.

$$\lim_n \int_0^t b\,(x^N_s, \xi_{ns})\, ds = \int_0^t \mathbf{E}\,(b\,(x^N_s, \xi_0) \mid J^\xi)\, ds.$$

Therefore

$$\overline{\lim_n} \left| \int_0^t b\,(x_s, \xi_{ns})\, ds - \int_0^t \mathbf{E}\,(b\,(x_s, \xi_0) \mid J^\xi)\, ds \right|$$

$$\le \overline{\lim_n} \int_0^t | b\,(x_s, \xi_{ns}) - b\,(x^N_s, \xi_{ns}) |\, ds + \overline{\lim_n} \int_0^t | \mathbf{E}\,(b\,(x_s, \xi_0) - b\,(x^N_s, \xi_0) \mid J^\xi) |\, ds$$

$$\le 2L \int_0^t | x_s - x^N_s |\, ds \le 2Lt \sup_{0 \le s \le t} | x_s - x^N_s | \to 0, \ N \to \infty.$$

To establish the relation asserted in the theorem, denote by g (x, y) any of the functions $b^+(x, y) = b(x, y) \vee 0$ or $b^-(x, y) = -(b(x, y) \wedge 0)$. Evidently, the function $g(x, y)$ satisfies Lipschitz' condition relative to x with the same constant L. Therefore, for each $t > 0$ (P-a.s.)

$$\lim_n \int_0^t g(x_s, \xi_{ns})\, ds = \int_0^t \mathbf{E}(g(x_s, \xi_0) \mid J^\xi)\, ds$$

by (6.3). Since $g(x, y)$ is a nonnegative function, in view of Problem 5.3.2 P-a.s. we have

$$\lim_n \sup_{0 \le t \le T} \left| \int_0^t g(x_s, \xi_{ns})\, ds - \int_0^t \mathbf{E}(g(x_s, \xi_0) \mid J^\xi)\, ds \right| = 0, \ \forall T > 0.$$

This gives the desired assertion in an obvious manner.

Theorem 1 is proved.

3. Denote

$$a(\omega, x, y) = b(x, y) - \mathbf{E}(b(x, \xi_0) \mid J^\xi)$$

(for brevity we will supress the symbol ω) and

$$Y_t^n = \sqrt{n} \int_0^t a(x_s, \xi_{ns})\, ds. \tag{6.5}$$

Introduce the following conditions: as $p > 2$

$$\sup_x \| b(x, \xi_0) \|_p < \infty \tag{6.6}$$

and

$$\int_0^\infty \operatorname{ess\,sup}_x \| \mathbf{E}(a(x, \xi_s) \mid G_0^\xi) \|_p\, ds < \infty, \tag{6.7}$$

where $\sup_x$ and $\operatorname{ess\,sup}_x$ is taken over all x's such that

$$|x| \le \sup_{t \ge 0} |x_t|.$$

If Conditions (6.6) and (6.7) are satisfied as $p > 2$, then the function

$$\eta(x) = \eta(\omega, x) = 2\int_0^\infty \mathbf{E}(a(x, \xi_s)\, a(x, \xi_0) \mid J^\xi)\, ds \tag{6.8}$$

may be defined.

Theorem 2. *Let Condition* (6.1) *be satisfied and let the function* $b = b(x, y)$ *satisfy Lipschitz' condition relative to* x:

$$|b(x', y) - b(x'', y)| \le L\, |x' - x''|$$

with a constant L, *independent of* y. *If, in addition, Conditions* (6.6) *and* (6.7) *are satisfied for a certain* p > 2, *then*

$$Y^n \xrightarrow{d} Y \quad (\mathcal{J}^{\xi}\textit{-stably}),$$

where $Y = (Y_t)_{t\geq 0}$ *is the stochastic integral*:

$$Y_t = \int_0^t \eta^{1/2}(x_s)\, dW_s$$

with respect to a Wiener process $W = (W_t)_{t\geq 0}$, *independent of events of the* σ-*algebra* $\mathcal{J}^{\xi}$.

4. To prove this theorem, we will need a number of auxiliary facts.

Without loosing generality we may assume that the process ξ is defined on a coordinate stochastic basis (see Ch. 4, Subsection 11.8), denoted by

$$(\Omega, \mathcal{F}, (\mathcal{F}_t)_{t\in R}, P).$$

For each n define the stochastic process $\overline{Y}^n = (\overline{Y}^n_t)_{t\geq 0}$ with

$$\overline{Y}^n_t = \int_0^t a\,(x_{s/n}, \xi_s)\, ds. \tag{6.9}$$

Obviously

$$Y^n_t = \frac{1}{\sqrt{n}}\,\overline{Y}^n_{nt}. \tag{6.10}$$

Assume that Conditions (6.6) and (6.7) are satisfied as p = 2. Then, analogously to Lemma 2.1, one can verify that there exists a square integrable martingale

$$\overline{M}^n = (\overline{M}^n_t, \mathcal{F}_t)$$

that is a modification of the stochastic process

$$\int_0^\infty [\pi_t(a\,(x_{s/n}, \xi_s)) - \pi_0(a\,(x_{s/n}, \xi_s))]\, ds,$$

where $\pi\,(a\,(x_{s/n}, \xi_s)) = (\pi_t(a\,(x_{s/n}, \xi_s)))_{t\geq 0}$ is the optional projection of the random variable $a\,(x_{s/n}, \xi_s)$ with respect to the family $\mathbb{F} = (\mathcal{F}_t)_{t\geq 0}$. Besides, the stochastic process $\overline{X}^n = (\overline{X}^n_t)_{t\geq 0}$ with

$$\overline{X}_0 = -\int_0^\infty \pi_0(a\,(x_{s/n}, \xi_s))\, ds$$

and

$$\overline{X}^n_t = \overline{X}^n_0 + \overline{Y}^n_t - \overline{M}^n_t \tag{6.11}$$

is a modification of the process

$$\left(-\int_t^{\infty} \pi_t(a\,(x_{s/n}, \xi_s)\,ds \right)_{t \geq 0}$$

Along with the processes Y^n, $n \geq 1$, consider the processes $Y^{n,N} = (Y_t^{n,N})_{t \geq 0}$ with

$$Y_t^{n,N} = \sqrt{n} \int_0^t a\,(x_s^N, \xi_{ns})\,ds, \tag{6.12}$$

where the function x_s^N is defined by (6.4).

The process $Y_t^{n,N}$ may be represented as

$$Y_t^{n,N} = \frac{1}{\sqrt{n}} \overline{Y}_{nt}^{n,N}, \tag{6.13}$$

where $\overline{Y}_t^{n,N}$ is defined by the formula (6.9) with $x_{s/n}^N$ instead of $x_{s/n}$. Besides the representation of type (6.11) holds:

$$\overline{X}_t^{n,N} = \overline{X}_0^{n,N} + \overline{Y}_t^{n,N} - \overline{M}_t^{n,N}, \tag{6.14}$$

where $\overline{X}_t^{n,N}$ and $\overline{M}_t^{n,N}$ are defined analogously to $\overline{X}_t^n$ and $\overline{M}_t^n$ with $x_{s/n}^N$ instead of $x_{s/n}$.

Lemma 1. *Let Conditions* (6.6) *and* (6.7) *be satisfied as* $p = 2$. *Then for each* $t > 0$

$$\lim_N \sup_n \mathbf{E}\,(Y_t^n - Y_t^{n,N})^2 = 0.$$

Besides, if (6.6) *and* (6.7) *are satisfied for a certain* $p > 2$, *then for each* $T > 0$

$$\lim_N \overline{\lim_n} \mathbf{E} \sup_{0 \leq t \leq T} (Y_t^n - Y_t^{n,N})^2 = 0.$$

Proof. Denote

$$r^{n,N}(s, \xi_s) = a\,(x_{s/n}, \xi_s) - a\,(x_{s/n}^N, \xi_s).$$

Then by (6.10) and (6.13) the equality

$$(Y_t^n - Y_t^{n,N})^2 = \frac{2}{n} \int_0^{nt} \int_0^{ns} r^{n,N}(s, \xi_s)\, r^{n,N}(u, \xi_u)\,du\,ds$$

holds and, consequently,

$$\mathbf{E}\,(Y_t^n - Y_t^{n,N})^2 = \frac{2}{n}\int_0^{nt}\int_0^{ns} \mathbf{E}\left[r^{n,N}(u,\xi_u)\,\mathbf{E}\,(r^{n,N}(s,\xi_s)\mid \mathcal{F}_u)\right] du\, ds$$

$$\le \frac{4L}{n}\int_0^{nt}\int_0^{ns} |\, x_{u/n} - x_{u/n}^N \,|\, \sup_x \|\,\mathbf{E}\,(a\,(x,\xi_s)\mid \mathcal{F}_u)\,\|_2\, du\, ds.$$

Obviously,

$$\|\,\mathbf{E}\,(a\,(x,\xi_s)\mid \mathcal{F}_u)\,\|_2 = \|\,\mathbf{E}\,(a\,(x,\xi_{s-u})\mid G_0^{\xi})\,\|_2$$

and hence

$$\mathbf{E}\,(Y_t^n - Y_t^{n,N})^2 \le \frac{4L}{n}\int_0^{nt}\int_{nu}^{nt} \sup_x \|\,\mathbf{E}\,(a\,(x,\xi_{s-u})\mid G_0^{\xi})\,\|_2\; |\, x_{u/n} - x_{u/n}^N \,|\, ds\, du$$

$$= \frac{4L}{n}\int_0^{nt}\int_0^{n(t-u)} \sup_x \|\,\mathbf{E}\,(a\,(x,\xi_s)\mid G_0^{\xi})\,\|_2\; |\, x_{u/n} - x_{u/n}^N \,|\, ds\, du$$

$$\le 4L\int_0^{\infty} \sup_x \|\,\mathbf{E}\,(a\,(x,\xi_s)\mid G_0^{\xi})\,\|_2\, ds \int_0^t |\, x_u - x_u^N \,|\, du.$$

This implies the first assertion of the lemma in an obvious manner, since

$$\int_0^t |\, x_u - x_u^N \,|\, du \le t \sup_{u\le t} |\, x_u - x_u^N \,| \to 0, \quad N\to\infty.$$

To prove the second assertion of the lemma, utilize the decompositions (6.11) and (6.14) which give

$$\sup_{t\le T}\,(Y_t^n - Y_t^{n,N})^2 = \frac{1}{\sqrt{n}} \sup_{t\le nT}\,(\overline{Y}_t^n - \overline{Y}_t^{n,N})^2$$

$$\le \frac{3}{n}\left[2\sup_{t\le nT}\,(\overline{X}_t^n)^2 + 2\sup_{t\le nT}\,(\overline{X}_t^{n,N})^2 + \sup_{t\le nT}\,(\overline{M}_t^n - \overline{M}_t^{n,N})^2\right].$$

By Doob's inequality (Theorem 1.9.2)

$$\mathbf{E}\sup_{t\le nT}\,(\overline{M}_t^n - \overline{M}_t^{n,N})^2 \le 4\mathbf{E}\,(\overline{M}_{nT}^n - \overline{M}_{nT}^{n,N})^2.$$

In view of (6.11) and (6.14) this gives the following inequality:

$$\mathbf{E}\sup_{t\le T}(Y_t^n - Y_t^{n,N})^2 \le \frac{c}{n}\left[\mathbf{E}\sup_{t\le nT}(\overline{X}_t^n)^2 + \mathbf{E}\sup_{t\le nT}(\overline{X}_t^{n,N})^2\right] + c\mathbf{E}(Y_T^n - Y_T^{n,N})^2.$$

Therefore, in view of the first assertion of the lemma, proved already, it suffices to show that

$$\lim_n \frac{1}{n}\mathbf{E}\sup_{t\le nT}(\overline{X}_t^n)^2 = 0, \quad \lim_n \frac{1}{n}\mathbf{E}\sup_{t\le nT}(\overline{X}_t^{n,N})^2 = 0. \tag{6.15}$$

To establish the first relation in (6.15), denote

$$\zeta_k^n = \sup_{(k-1)T < t \le kT} |\overline{X}_t^n|.$$

Then as $p > 2$

$$\mathbf{E}\frac{1}{n}\sup_{t\le nT}(\overline{X}_t^n)^2 = \mathbf{E}\frac{1}{n}\max_{k\le n}(\zeta_k^n)^2 \le \left\{\mathbf{E}\frac{1}{n^{p/2}}\max_{k\le n}|\zeta_k^n|^p\right\}^{2/p}$$

$$\le \left\{\frac{1}{n^{p/2}}\sum_{k=1}^n \mathbf{E}|\zeta_k^n|^p\right\}^{2/p},$$

and the desired assertion takes place, provided

$$\mathbf{E}|\zeta_k^n|^p \le c,$$

with a constant c independent of k and n.

The definition of the process $\overline{X}_t^n$ implies that

$$\zeta_k^n = \sup_{(k-1)T < t \le kT}\left|\int_t^\infty \pi_t(a(x_{s/n}, \xi_s))\,ds\right|$$

$$= \sup_{(k-1)T < t \le kT}\left|\mathbf{E}\left(\int_{kT}^\infty \pi_{kT}(a(x_{s/n}, \xi_s))\,ds + \int_t^{kT} a(x_{s/n}, \xi_s)\,ds \mid \mathcal{F}_t\right)\right|$$

$$\le \sup_{(k-1)T < t \le kT}\mathbf{E}\left\{\int_{kT}^\infty |\pi_{kT}(a(x_{s/n}, \xi_s))|\,ds + \int_{(k-1)T}^{kT} |a(x_{s/n}, \xi_s)|\,ds \mid \mathcal{F}_t\right\}$$

$$\le \sup_{t\le kT}\mathbf{E}\left\{\int_{kT}^\infty |\pi_{kT}(a(x_{s/n}, \xi_s))|\,ds + \int_{(k-1)T}^{kT} |a(x_{s/n}, \xi_s)|\,ds \mid \mathcal{F}_t\right\}.$$

Consequently, by Doob's inequality (Theorem 1.9.2) as $p > 2$

$$\mathbf{E}\,|\zeta_k^n|^p \le \left(\frac{p}{p-1}\right)^p \mathbf{E}\left(\int_{kT}^{\infty} |\pi_{kT}(a(x_{s/n}, \xi_s))|\,ds + \int_{(k-1)T}^{kT} |a(x_{s/n}, \xi_s)|\,ds\right)^p$$

and hence

$$\mathbf{E}\,|\zeta_k^n|^p \le \left(\frac{p}{p-1}\right)^p \left(\int_{kT}^{\infty} ||\pi_{kT}(a(x_{s/n}, \xi_s))||_p\,ds + \int_{(k-1)T}^{kT} ||a(x_{s/n}, \xi_s)||_p\,ds\right)^p.$$

This gives the desired estimate (6.16) in an obvious manner, since

$$\int_{kT}^{\infty} ||\pi_{kT}(a(x_{s/n}, \xi_s))||_p\,ds \le \int_{kT}^{\infty} \sup_x ||\pi_{kT}(a(x, \xi_s))||_p\,ds$$

$$= \int_0^{\infty} \sup_x ||\mathbf{E}(a(x, \xi_s) \mid G_0^{\xi})||_p\,ds$$

and

$$\int_{(k-1)T}^{kT} ||a(x_{s/n}, \xi_s)||_p\,ds \le \int_{(k-1)T}^{kT} \sup_x ||a(x, \xi_s)||_p\,ds$$

$$= \int_{(k-1)T}^{kT} \sup_x ||a(x, \xi_0)||_p\,ds$$

$$= T \sup_x ||a(x, \xi_0)||_p.$$

Thus (6.16) takes place, and hence the first relation in (6.15) is valid. The second relation in (6.15) is established analogously.

The following lemma allows us to establish the structure of the limiting process involved in the assertion of Theorem 2.

Lemma 2. *Let Condition* (6.1) *be satisfied, as well as Conditions* (6.6) *and* (6.7) *as* $p = 2$. *Then for each* $N \ge 1$

$$Y^{n,N} \xrightarrow{d} Z^n \quad (J^{\xi}\text{-stably}),$$

where $Z^n = (Z_t^n)_{t \ge 0}$ *is the stochastic integral:*

$$Z_t^N = \int_0^t \eta^{1/2}(x_s^N)\, dW_s$$

with respect to a Wiener process $W = (W_t)_{t \geq 0}$, independent of events of the σ-algebra J^{ξ}.

Proof. By (6.13) and (6.9) (with x_s^N instead of x_s) the process $Y^{n,N}$ admits the representation

$$Y_t^{n,N} = \frac{1}{\sqrt{n}} \int_0^{nt} a\,(x_{\frac{[Ns/n]}{N}}, \xi_s)\, ds$$

$$= \frac{1}{\sqrt{n}} \int_0^{nt} \sum_{j \geq 0} a\,(x_{\frac{jn}{N}}, \xi_s)\, I_{]\frac{jn}{N}, \frac{(j+1)n}{N}]}(s)\, ds. \qquad (6.17)$$

Define the stochastic process $\overline{Y}^{n,N,j} = (\overline{Y}_t^{n,N,j})_{t \geq 0}$ with

$$\overline{Y}_t^{n,N,j} = \int_0^t a\,(x_{\frac{jn}{N}}, \xi_s)\, ds. \qquad (6.18)$$

Since Conditions (6.6) and (6.7) are satisfied as $p = 2$, there exists, in view of Lemma 2.1, a square integrable martingale $\overline{M}^{n,N,j} = (\overline{M}_t^{n,N,j})_{t \geq 0}$, that is a helix such that the semimartingale $\overline{X}^{n,N,j} = (\overline{X}_t^{n,N,j})_{t \geq 0}$ with

$$\overline{X}_0^{n,N,j} = \int_0^{\infty} \mathbf{E}\,(a\,(x_{\frac{jn}{N}}, \xi_s) \mid G_0^{\xi})\, ds$$

and with the decomposition

$$\overline{X}_t^{n,N,j} = \overline{X}_0^{n,N,j} + \overline{Y}_t^{n,N,j} - \overline{M}_t^{n,N,j}, \qquad (6.19)$$

presents a strictly stationary process. It has been established in the course of proving Theorem 2.1 that as $n \to \infty$

$$\sup_{t \leq nT} \frac{1}{\sqrt{n}} | \overline{X}_t^{n,N,j} | \xrightarrow{P} 0, \ T > 0, \qquad (6.20)$$

$$\frac{1}{n} [\overline{M}^{n,N,j}, \overline{M}^{n,N,j}]_t \to t\eta\,(x_{\frac{jn}{N}}) \ \text{(P-a.s.)} \qquad (6.21)$$

and

$$\frac{1}{n} E \sum_{s \le nt} (\Delta \overline{M}_s^{n, N, j})^2 I(|\Delta \overline{M}_s^{n, N, j}| > \sqrt{n}\, a) \to 0, \ t > 0. \tag{6.22}$$

The following representation for $Y_t^{n, N}$ is easily deduced from (6.17) and (6.18):

$$Y_t^{n, N} = M_t^{n, N} + X_t^{n, N} \tag{6.23}$$

with

$$M_t^{n, N} = \frac{1}{\sqrt{n}} \int_0^{nt} \sum_{j \ge 0} I_{]\frac{jn}{N}, \frac{(j+1)n}{N}]}(s)\, d\overline{M}_s^{n, N, j} \tag{6.24}$$

and

$$X_t^{n, N} = \frac{1}{\sqrt{n}} \int_0^{nt} \sum_{j \ge 0} I_{]\frac{jn}{N}, \frac{(j+1)n}{N}]}(s)\, d\overline{X}_s^{n, N, j}. \tag{6.25}$$

Utilizing Problem 6.1.2 it can be shown that the desired assertion takes place, provided

$$\sup_{t \le T} |X_t^{n, N}| \xrightarrow{P} 0, \ T > 0 \tag{6.26}$$

and

$$M^{n, N} \xrightarrow{d} Z^n \ (J^{\xi}\text{-stably}) \tag{6.27}$$

as $n \to \infty$.

To verify (6.26) observe that

$$X_t^{n, N} = \frac{1}{\sqrt{n}} \sum_{j \ge 0} \left[\overline{X}^{n, N, j}_{(nt) \wedge \left(\frac{(j+1)n}{N}\right)} - \overline{X}^{n, N, j}_{(nt) \wedge \left(\frac{jn}{N}\right)} \right]$$

and, consequently,

$$\frac{1}{\sqrt{n}} \sup_{t \le T} |X_t^{n, N}| \le \frac{2}{\sqrt{n}} \sum_{j=0}^{NT-1} \sup_{t \le nT} |\overline{X}_t^{n, N, j}|.$$

This gives the desired relation (6.26) in virtue of (6.20).

To prove (6.27), observe first that $(M_t^{n, N}, \mathcal{F}_{nt})$ is a square integrable martingale, that is, one can utilize Theorem 7.1.4. According to Theorem 7.1.4 it suffices to verify the conditions:

1) $J^{\xi} \subseteq \mathcal{F}_0$

and as $n \to \infty$

$$2)\ [M^{n,N}, M^{n,N}]_t \xrightarrow{P} \int_0^t \eta(x_s^N)\, ds,\ t > 0$$

and

$$3)\ \mathbf{E} \sum_{s \le t} (\Delta M_s^{n,N})^2 I(|\Delta M_s^{n,N}| > a) \to 0,\ t > 0,\ a > 0.$$

The first condition is satisfied due to Problem 4.11.1. Let us verify the second condition. In virtue of the representation (6.24) and the properties of the stochastic integrals with respect to martingales (Ch. 2, § 2)

$$[M^{n,N}, M^{n,N}]_t = \frac{1}{n} \int_0^{nt} \sum_{j \ge 0} I_{]\frac{jn}{N}, \frac{(j+1)n}{N}]}(s)\, d[\overline{M}^{n,N,j}, \overline{M}^{n,N,j}]_s$$

$$= \frac{1}{n} \sum_{j \ge 0} \left([\overline{M}^{n,N,j}, \overline{M}^{n,N,j}]_{(nt) \wedge \left(\frac{(j+1)n}{N}\right)} - [\overline{M}^{n,N,j}, \overline{M}^{n,N,j}]_{(nt) \wedge \left(\frac{jn}{N}\right)} \right).$$

Therefore Condition 2) follows from (6.21).

Condition 3) follows from (6.22), since

$$\mathbf{E} \sum_{s \le t} (\Delta M_s^{n,N})^2 I(|\Delta M_s^{n,N}| > a)$$

$$= \frac{1}{n} \mathbf{E} \sum_{s \le t} \sum_{j \ge 0} I_{]\frac{jn}{N}, \frac{(j+1)n}{N}]}(s)\, (\Delta \overline{M}_s^{n,N,j})^2 I(|\Delta \overline{M}_s^{n,N,j}| > \sqrt{n}\ a).$$

Finally, the assertion of Lemma 2 is established in virtue of Problem 4.9.15.

5. Proof of Theorem 2. Let Y and Z^n be the stochastic integrals involved in the assertions of Theorem 2 and Lemma 2. Without loosing generality we may assume that the one and the same Wiener process W defines Y and Z^n, $N \ge 1$.

Under this assumption

$$\lim_N \mathbf{E} \sup_{t \le T} (Y_t - Z_t^N)^2 = 0. \tag{6.28}$$

Indeed, by Doob's inequality (Theorem 1.9.2)

$$\mathbf{E} \sup_{t \le T} (Y_t - Z_t^N)^2 \le 4\mathbf{E} \int_0^T (\eta^{1/2}(x_s) - \eta^{1/2}(x_s^N))^2 \, ds$$

$$\le 4 \int_0^T \mathbf{E} \, | \, \eta (x_s) - \eta (x_s^N) \, | \, ds.$$

It is easily verified (Problem 1) that the function $\eta(x) = \eta(\omega, x)$, defined by the formula (6.8), is continuous in x for each ω. Besides,

$$\mathbf{E} \, | \, \eta (x_s) - \eta (x_s^N) \, |^2 \le 2\mathbf{E} \, \eta^2(x_s) + 2\mathbf{E} \, \eta^2(x_s^N)$$

$$\le 32 \sup_x \| \, a (x, \xi_0) \, \|_2^2 \left(\int_0^\infty \sup_x \| \, \mathbf{E} \, (a (x, \xi_s) \, | \, G_0^\xi) \, \|_2 \, ds \right)^2 < \infty.$$

Consequently,

$$\lim_N \int_0^T \mathbf{E} \, | \, \eta (x_s) - \eta (x_s^N) \, | \, ds = 0,$$

i.e. the desired relation (6.28) holds.

Now, let α be a J^ξ-measurable and bounded random variable and $f = f(y_1, ..., y_k)$ a continuous function. Then

$$\mathbf{E}\alpha f (Y_{t_1}^n, ..., Y_{t_k}^n) = \mathbf{E}\alpha f (Y_{t_1}^{n, N}, ..., Y_{t_k}^{n, N})$$

$$= \mathbf{E}\alpha \, [f (Y_{t_1}^n, ..., Y_{t_k}^n) - f (Y_{t_1}^{n, N}, ..., Y_{t_k}^{n, N})].$$

From (6.28) it follows that

$$\lim_N \mathbf{E}\alpha f (Z_{t_1}^N, ..., Z_{t_k}^N) = \mathbf{E}\alpha f (Y_{t_1}, ..., Y_{t_k}).$$

Therefore by Lemmas 1 and 2

$$\lim_n \mathbf{E}\alpha f (Y_{t_1}^n, ..., Y_{t_k}^n) = \mathbf{E}\alpha f (Y_{t_1}, ..., Y_{t_k}).$$

This implies

$$Y^n \xrightarrow{d_f} Y \quad (J^\xi\text{-stably}).$$

Therefore to conclude the proof it suffices to show that the family of distributions of Y^n, $n \ge 1$, is relatively compact. According to Theorem 6.1.6 it suffices to establish

$$\lim_{a \to \infty} \overline{\lim_n} \, \mathsf{P} \left(\sup_{t \le L} | \, Y_t^n \, | \ge a \right) = 0, \; L > 0 \tag{6.29}$$

and

$$\lim_{\sigma \to 0} \overline{\lim_{n}} P(W'_L(Y^n, \sigma) \geq \varepsilon) = 0, \ L > 0, \ \varepsilon > 0 \tag{6.30}$$

with the function W'_L defined in Ch. 6, Subsection 4.1.

Since

$$Y^{n, N} \xrightarrow{d} Z^N$$

as $n \to \infty$ for each $N \geq 1$ (Lemma 2), by the same Theorem 6.1.6

$$\lim_{a \to \infty} \overline{\lim_{n}} P\left(\sup_{t \leq L} |Y_t^{n, N}| \geq a\right) = 0, \ L > 0, \ N \geq 1 \tag{6.31}$$

and

$$\lim_{\sigma \to 0} \overline{\lim_{n}} P(W'_L(Y^{n, N}, \sigma) \geq \varepsilon) = 0, \ L > 0, \ \varepsilon > 0, \ N \geq 1. \tag{6.32}$$

Thus the desired relations (6.29) and (6.30) follow from (6.31), (6.32) and Lemma 1. Indeed,

$$\sup_{t \leq L} |Y_t^n| \leq \sup_{t \leq L} |Y_t^{n, N}| + \sup_{t \leq L} |Y_t^n - Y_t^{n, N}|,$$

and hence for $a > 1$

$$P\left(\sup_{t \leq L} |Y_t^n| \geq a\right) \leq P\left(\sup_{t \leq L} |Y_t^{n, N}| \geq a - 1\right) + P\left(\sup_{t \leq L} |Y_t^n - Y_t^{n, N}| \geq 1\right) \to 0$$

as the limit $\lim_{N} \overline{\lim_{a \to \infty}} \overline{\lim_{n}}$ is taken.

Analogously, in view of the inequality

$$W'_L(Y^n, \sigma) \leq W'_L(Y^{n, N}, \sigma) + \sup_{t \leq L} |Y_t^n - Y_t^{n, N}|$$

the following estimate holds:

$$P(W'_L(Y^n, \sigma) \geq \varepsilon) \leq P\left(W'_L(Y^{n, N}, \sigma) \geq \frac{\varepsilon}{2}\right) + P\left(\sup_{t \leq L} |Y_t^n - Y_t^{n, N}| \geq \frac{\varepsilon}{2}\right),$$

the right-hand side of which tends to zero as the limit $\lim_{N} \overline{\lim_{\sigma \to 0}} \overline{\lim_{n}}$ is taken. This proves (6.30).

Hence the relative compactness of Y^n, $n \geq 1$, is established.

Theorem 2 is proved.

Problems

1. Show that under Conditions (6.6) and (6.7) as $p > 2$ the function $\eta(x) = \eta(\omega, x)$, defined by the formula (6.8), is a continuous function of x for each fixed ω.

2. Let $b = b(x, y)$ be a uniformly bounded function and let the strong mixing coefficient $\alpha = \alpha(t)$ of a strictly stationary process ξ possess the following property:

$$\int_0^\infty \alpha^{1/p}(t)\, dt < \infty$$

for a certain $p > 2$. Show that the assertion of Theorem 2 holds, the σ-algebra of invariant sets of J^ξ consists of sets of probability 0 or 1, and

$$\eta(x) = 2\int_0^\infty \mathbf{E}\,(a(x, \xi_s)\, a(x, \xi_0))\, ds$$

(cf. (6.8)).

3. Prove Theorem 2 for

$$Y_t^n = \sqrt{n}\int_0^t [b(x, \xi_{ns}) - \mathbf{E}\, b(x_s, \xi_{ns})]\, ds$$

under Conditions (6.1), (6.6), (6.7) and Lipschitz' condition:

$$|\, b(x', y) - b(x'', y)\,| \le L\,|\, x' - x''\,| .$$

§ 7. Stochastic version of Bogoljubov's averaging principle

1. Let $\xi = (\xi_t)_{t \in R}$ be a strictly stationary ergodic process and $b = b(x, y)$ a measurable function such that for each $n \geq 1$ the differential equation

$$\dot{X}_t^n = b(X_t^n, \xi_{nt}) \tag{7.1}$$

with the initial condition $X_0^n = x_0$ has a unique solution on each finite time interval.

According to Bogoljubov's averaging principle (see, e.g. [361]), the function $X^n = (X_t^n)_{t \geq 0}$, that is the solution of the equation (7.1), is approximated for a large value of n by the function $X = (X_t)_{t \geq 0}$, defined by the differential equation

$$\dot{X}_t = B(X_t) \tag{7.2}$$

with $X_0 = x_0$, where $B(x)$ is the "averaged right-hand side" of the equation (7.1), i.e.

$$B(x) = \mathbf{E} b(x, \xi_0) = \mathbf{E} b(x, \xi_{nt}). \tag{7.3}$$

In this section we present the conditions under which X^n is approximated by the function X for suffieciently large values of n, and we establish the diffusion approximation for the sequence of processes $Y^n = (Y_t^n)_{t \geq 0}$ with

$$Y_t^n = \sqrt{n}\,(X_t^n - X_t). \tag{7.4}$$

2. **Theorem 1.** *Let* $\xi = (\xi_t)_{t \in R}$ *be a strictly stationary ergodic process, and let the function* $b = b(x, y)$ *satisfy the following conditions:*

1) $b(x, y)$ *is measurable in both of the variables;*

2) $b(x, y)$ *satisfies Lipschitz' condition in* x:

$$| b(x', y) - b(x'', y) | \leq L | x' - x'' |$$

with a constant L, *independent of* y;

3) $\mathbf{E} | b(x, \xi_0) | < \infty, \; x \in R.$

Then

$$\lim_n \sup_{t \leq T} | X_t^n - X_t | = 0 \quad (\text{P-a.s.}), \quad T > 0.$$

Proof. Observe that Conditions 1) - 3) imposed on the function $b = b(x, y)$ guarantee the existence and uniqueness of solutions of the equations (7.1) and (7.2) on each finite time interval, with the following properties:

$$\sup_{0 \leq s \leq t} | X_s | < \infty \text{ and } \sup_{0 \leq s \leq t} | X_s^n | < \infty \quad (\text{P-a.s.}), \quad t > 0, \; n \geq 1.$$

By (7.1) and (7.2) we have

$$X_t^n - X_t = \int_0^t [b(X_s^n, \xi_{ns}) - B(X_s)]\,ds$$

$$= \int_0^t [b(X_s, \xi_{ns}) - B(X_s)]\,ds + \int_0^t [b(X_s^n, \xi_{ns}) - b(X_s, \xi_{ns})]\,ds. \quad (7.5)$$

Denote

$$\alpha^n = \sup_{t \le T} \left| \int_0^t [b(X_s, \xi_{ns}) - B(X_s)]\,ds \right|, \quad \Delta_t^n = |X_t^n - X_t|. \quad (7.6)$$

Then (7.5) and (7.6) entail the inequality

$$\Delta_t^n \le \alpha^n + L \int_0^t \Delta_s^n\,ds, \quad t \le T,$$

which gives

$$\Delta_t^n \le \alpha^n \exp(LT), \quad t \le T.$$

Hence

$$\sup_{t \le T} |X_t^n - X_t| \le \alpha^n \exp(LT).$$

This implies the desired assertion, since

$$\lim_n \alpha^n = 0 \quad (P\text{-a.s.})$$

by Theorem 6.1.

3. We will study here the problem on the diffusion approximation of the sequences of processes Y^n, $n \ge 1$, defined by the relation (7.4).

Denote

$$a(x, y) = b(x, y) - B(x).$$

For a certain $p > 2$ let the following conditions be satisfied:

$$\sup_x \| b(x, \xi_0) \|_p < \infty \quad (7.7)$$

and

$$\int_0^\infty \operatorname{ess\,sup}_x \| \mathbf{E}(a(x, \xi_s) \mid G_0^\xi) \|_p\,ds < \infty, \quad (7.8)$$

where

$$\| \alpha \|_p = (\mathbf{E} |\alpha|^p)^{1/p}, \quad G_0^\xi = \sigma\{\xi_s, -\infty < s \le 0\}$$

and $\sup_x$ and $\operatorname{ess\,sup}_x$ is taken over the set

$$\{x : |x| \leq \sup_{t \geq 0} |X_t|\}.$$

Under Conditions (7.7) and (7.8) the function

$$\sigma(x) = 2 \int_0^{\infty} \mathbf{E}(a(x, \xi_s)\, a(x, \xi_0))\, ds \tag{7.9}$$

is defined.

If the function $b = b(x, y)$ is differentiable with respect to x and

$$\mathbf{E}\, |b'_x(x, \xi_0)| < \infty, \quad x \in \mathbf{R},$$

then set

$$A(x) = \mathbf{E} b'_x(x, \xi_0). \tag{7.10}$$

Theorem 2. *Let* $\xi = (\xi_t)_{t \in \mathbf{R}}$ *be a strictly stationary ergodic process, and let the function* $b = b(x, y)$ *satisfy the following conditions:*

1) $b(x, y)$ *is measurable in both of the variables;*

2) *for each* y *the function* $b(x, y)$ *is differrentiable with respect to* x *and*

$$|b'_x(x, y)| \leq L;$$

3) *the function* $b'_x = b'_x(x, y)$ *satisfies Lipschitz' condition in* x:

$$|b'_x(x^1, y) - b'_x(x^2, y)| \leq L\, |x^1 - x^2|$$

with a constant L, *independent of* y;

4) *for a certain* $p > 2$ *Conditions* (7.7) *and* (7.8) *are satisfied.*

Then

$$Y^n \xrightarrow{d} Y,$$

where $Y = (Y_t)_{t \geq 0}$ *is a diffusion process defined by Ito's differential equation*

$$Y_t = \int_0^t A(X_s)\, Y_s\, ds + \int_0^t \sigma^{1/2}(X_s)\, dW_s \tag{7.11}$$

with respect to a Wiener process $W = (W_t)_{t \geq 0}$, *where the functions* $A(x)$ *and* $\sigma(x)$ *are given by the formulas* (7.9) *and* (7.10), *and* X_s *is the solution of the equation* (7.2)

Proof. From (7.1), (7.2) and (7.4) it follows that

$$Y^n_t = \sqrt{n} \int_0^t [b(X^n_s, \xi_{ns}) - B(X_s)]\, ds$$

$$= \sqrt{n} \int_0^t [b(X^n_s, \xi_{ns}) - b(X_s, \xi_{ns})]\, ds \tag{7.12}$$

$$+ \sqrt{n} \int_0^t [b(X_s, \xi_{ns}) - B(X_s)]\, ds.$$

Denote $Z^n = (Z_t^n)_{t \geq 0}$ with

$$Z_t^n = \sqrt{n} \int_0^t [b(X_s, \xi_{ns}) - B(X_s)]\, ds.$$

Since ξ is a strictly stationary ergodic process and the conditions of Theorem 6.2 are satisfied (see (6.7) and (6.8)), then

$$Z^n \xrightarrow{d} Z$$

where $Z = (Z_t)_{t \geq 0}$ is the stochastic integral:

$$Z_t = \int_0^t \sigma^{1/2}(X_s)\, dW_s$$

with respect to a Wiener process $W = (W_t)_{t \geq 0}$, where the function $\sigma = \sigma(x)$ is defined by the formula (7.9) and X_s is the solution of the equation (7.2).

From (7.12) and Condition 2) of the theorem it follows that

$$|Y_t^n| \leq L \int_0^t |Y_s^n|\, ds + \sup_{t \leq T} |Z_t^n|$$

as $t \leq T$. This gives

$$|Y_t^n| \leq \sup_{t \leq T} |Z_t^n| \exp(LT),$$

i.e.

$$\sup_{t \leq T} |Y_t^n| \leq \sup_{t \leq T} |Z_t^n| \exp(LT).$$

In virtue of the weak convergence

$$Z^n \xrightarrow{d} Z$$

the family of distributions of Z^n, $n \geq 1$, is relatively compact, and consequently

$$\lim_{a \to \infty} \overline{\lim_{n}}\, \mathsf{P}\left(\sup_{t \leq T} |Z_t^n| \geq a \right) = 0$$

(Theorem 6.1.6). Therefore

$$\lim_{a \to \infty} \overline{\lim_{n}}\, \mathsf{P}\left(\sup_{t \leq T} |Y_t^n| \geq a \right) = 0. \tag{7.13}$$

Consider the stochastic process $\hat{Y}^n = (\hat{Y}_t^n)_{t \geq 0}$, defined by the linear stochastic equation

$$\hat{Y}^n_t = \int_0^t A(X_s)\,\hat{Y}^n_s\,ds + Z^n_t. \tag{7.14}$$

By Theorem 4.1 we have

$$\hat{Y}^n \xrightarrow{d} Y.$$

Therefore, to prove the statement

$$Y^n \xrightarrow{d} Y,$$

it suffices, in view of Problem 6.1.2, to show

$$\sup_{t \le T} |Y^n_t - \hat{Y}^n_t| \xrightarrow{P} 0,\ n \to \infty,\ T > 0. \tag{7.15}$$

By (7.12) and (7.14) we get

$$Y^n_t - \hat{Y}^n_t = \int_0^t [\sqrt{n}\,(b(X^n_s, \xi_{ns}) - b(X_s, \xi_{ns})) - A(X_s)\,\hat{Y}^n_s]\,ds$$

$$= \int_0^t [\sqrt{n}\,(b(X^n_s, \xi_{ns}) - b(X_s, \xi_{ns})) - b'_x(X_s, \xi_{ns})\,(X^n_s - X_s)]\,ds$$

$$+ \int_0^t b'_x(X_s, \xi_{ns})\,(Y^n_s - \hat{Y}^n_s)\,ds + \int_0^t \hat{Y}^n_s[b'_x(X_s, \xi_{ns}) - A(X_s)]\,ds. \tag{7.16}$$

Denote

$$\Delta^n_t = |Y^n_t - \hat{Y}^n_t|.$$

Taking into account Conditions 2) and 3) of the theorem, by (7.16) we obtain the following inequality for Δ^n_t, $t \le T$:

$$\Delta^n_t \le L\left(\sup_{t \le T}\frac{|Y^n_t|}{\sqrt{n}} + 1\right)\int_0^t \Delta^n_s\,ds + \sup_{t \le T}\left|\int_0^t \hat{Y}^n_s\,[b'_x(X_s, \xi_{ns}) - A(X_s)]\,ds\right|$$

and hence

$$\sup_{t \le T}|Y^n_t - \hat{Y}^n_t|$$

$$\le \sup_{t \le T}\left|\int_0^t \hat{Y}^n_s\,[b'_x(X_s, \xi_{ns}) - A(X_s)]\,ds\right| \exp\left(T\left(1 + \sup_{t \le T}\frac{|Y^n_t|}{\sqrt{n}}\right)\right).$$

This and (7.13) entail the desired relation (7.15), provided the following convergence takes place:

$$\sup_{t \le T} \left| \int_0^t \hat{Y}_s^n \,[b_x'(X_s, \xi_{ns}) - A(X_s)]\, ds \right| \xrightarrow{P} 0, \quad n \to \infty. \tag{7.17}$$

To simplify the notations, we set

$$g(x, y) = b_x'(x, y) - A(x).$$

Then

$$g(X_s, \xi_{ns}) = b_x'(X_s, \xi_{ns}) - A(X_s). \tag{7.18}$$

Under the assumptions of the theorem we have

$$|g(x, y)| \le 2L.$$

Besides, by Theorem 6.1,

$$\sup_{t \le T} \left| \int_0^t g(X_s, \xi_{ns})\, ds \right| \to 0 \quad (\text{P-a.s.}). \tag{7.19}$$

To prove (7.17), we will show first that for each $N \ge 1$

$$\sup_{t \le T} \left| \int_0^t \hat{Y}^n_{\frac{[sN]}{N}}\, g(X_s, \xi_{ns})\, ds \right| \xrightarrow{P} 0, \quad n \to \infty. \tag{7.20}$$

To prove (7.20) it suffices, in view of the relation

$$\sup_{t \le T} \left| \int_0^t \hat{Y}^n_{\frac{[sN]}{N}}\, g(X_s, \xi_{ns})\, ds \right| = \sup_{t < T} \left| \sum_{j=0}^{N[t]} \int_{\frac{j}{N}}^{\frac{j+1}{N}} \hat{Y}^n_{\frac{j}{N}}\, g(X_s, \xi_{ns})\, ds \right|$$

$$\le 2N(T+1) \sup_{t \le T} |\hat{Y}_t^n| \sup_{t \le T} \left| \int_0^t g(X_s, \xi_{ns})\, ds \right|,$$

to show that

$$J_1^n = \sup_{t \le T} |\hat{Y}_t^n| \sup_{t \le T} \left| \int_0^t g(X_s, \xi_{ns})\, ds \right| \xrightarrow{P} 0, \quad n \to \infty. \tag{7.21}$$

But

$$J_1^n\, I\left(\sup_{t \le T} |\hat{Y}_t^n| < c \right) \to 0 \quad (\text{P-a.s.}), \quad n \to \infty$$

in virtue of (7.19). Hence to prove (7.21) one needs to show that

$$\lim_{c \to \infty} \overline{\lim_{n}}\, P\left(\sup_{t \le T} |\hat{Y}_t^n| \ge c \right) = 0.$$

The last relation holds as the family of distributions of $\hat{Y}^n$, $n \geq 1$, is relatively compact (see Theorem 6.6.1), by the weak convergence

$$\hat{Y}^n \xrightarrow{d} Y.$$

In accordance with the relation (7.20), to prove (7.15) it remains to show that

$$J_2^{n,N} = \int_0^T \left| \hat{Y}_s^n - \hat{Y}_{\frac{[sN]}{N}}^n \right| \, |g(X_s, \xi_{ns})| \, ds \xrightarrow{P} 0$$

as the limit $\lim\limits_N \overline{\lim\limits_n}$ is taken. Obviously,

$$J_2^{n,N} \leq 2LT \sup_{t \leq T} \left| \hat{Y}_t^n - \hat{Y}_{\frac{[tN]}{N}}^n \right| .$$

Therefore one needs to establish that

$$\sup_{t \leq T} \left| \hat{Y}_t^n - \hat{Y}_{\frac{[tN]}{N}}^n \right| \xrightarrow{P} 0 \tag{7.22}$$

as the limit $\lim\limits_N \overline{\lim\limits_n}$ is taken.

Let a be a positive number that is a continuity point for the distribution function of the random variable

$$\sup_{t \leq T} \left| Y_t - Y_{\frac{[tN]}{N}} \right|,$$

where $Y = (Y_t)_{t \geq 0}$ is the limiting stochastic process involved in the statement of Theorem 2. Then, as trajectories of Y are continuous, we have

$$\lim_N \overline{\lim_n} \, P\left(\sup_{t \leq T} \left| \hat{Y}_t^n - \hat{Y}_{\frac{[tN]}{N}}^n \right| \geq a \right) = \lim_N P\left(\sup_{t \leq T} \left| Y_t - Y_{\frac{[tN]}{N}} \right| \geq a \right) = 0,$$

and this proves (7.22).

Theorem 2 is proved.

Problem

1. Formulate and prove the vector version of Theorems 1 and 2.

CHAPTER 10

DIFFUSION APPROXIMATION FOR SEMIMARTINGALES WITH A NORMAL REFLECTION IN A CONVEX REGION

§ 1. Skorohod's problem on normal reflection

1. Let $D = D_{[0,\infty)}(R^d)$ ($C = C_{[0,\infty)}(R^d)$) be Skorohod's space of right-continuous (continuous) vector-valued functions[1] with left-hand limits

$$X = (X_t)_{t \geq 0}, \quad X_t = (X_1(t), \ldots, X_d(t)).$$

Denote

$$D(O) = \{X \in D: X_t \in \bar{O},\ t \geq 0\}$$

where $\bar{O}$ is the closure of a region $O \in R^d$ (∂O is the boundary of a region O). Obviously,

$$C(O) = C \cap D(O).$$

A function $Y = (Y_t)_{t \geq 0} \in D(O)$ is called the solution of Skorohod's problem on normal reflection within the region O for $X \in D$ with $X_0 \in \bar{O}$ if the function $\phi = Y - X$ possesses the following properties:

1) $\phi = (\phi_t)_{t \geq 0} \in D$, $\phi_0 = 0$,

2) $\mathrm{Var}(\phi)_t < \infty$, $t > 0$,

3) $\displaystyle\int_0^\infty I_O(Y_s)\, d\,\mathrm{Var}(\phi)_s = 0$,

4) $d\phi_t / d\,\mathrm{Var}(\phi)_t \in \mathfrak{N}_{Y_t}(O)$, $d\,\mathrm{Var}(\phi)$-a.e., where $\mathfrak{N}_y(O)$ is the set of all inner vectors at point $y \in \partial O$,

5) the function ϕ is adapted to the family $\mathbb{D} = (\mathcal{D}_t)_{t \geq 0}$ of σ-algebras

$$\mathcal{D}_t = \bigcap_{\varepsilon > 0} \sigma\{X_s, s \leq t + \varepsilon\},$$

and it is $B(R_+) \otimes \mathcal{D}_\infty$-measurable.

Observe that Condition 3) is equivalent to condition

3') $\displaystyle\int_0^t (f(Y_s), d\phi_s) = 0$, $t > 0$

[1]Here and elsewhere below all vectors are column vectors.

for each continuous and bounded function

$$f: \bar{O} \to R^d$$

with f (y) = 0, $y \in \partial O$, where (. , .) is the scalar product, while Condition 4) is equivalent to condition

4') the function

$$U_t = \int_0^t (\tilde{Y}_s - Y_s, d\phi_s), \ t \geq 0$$

is nondecreasing for each function

$$\tilde{Y} = (\tilde{Y}_t)_{t \geq 0} \in D(O).$$

It is said that the function ϕ with properties 1) - 5) is associated with the function Y $\in D(O)$.

2. To solve the problem on normal reflection we will need certain properties of the vector

$$[x]_\partial = \arg \min_{y \in \bar{O}} |x - y|,$$

where $|\cdot|$ is the Euclidean norm (for vectors and matrices).

Firstly, observe that the vector $[x]_\partial$ is uniquely determined and secondly, that $[x]_\partial$ is a continuous function of x in the metric $|\cdot|$. Indeed, let $y_i \in \bar{O}$, $i = 1, 2$, $y_1 \neq y_2$ and $|x - y_1| = |x - y_2| = \min_{y \in \bar{O}} |x - y|$. Define the vector

$$y_3 = \frac{1}{2}(y_1 + y_2).$$

By construction

$$(x - y_3, y_1 - y_2) = 0.$$

Hence

$$|x - y_1| = |x - y_3 + y_3 - y_1| = |x - y_3 + \frac{1}{2}(y_2 - y_1)|$$

$$= \left(|x - y_3|^2 + \frac{1}{4}|y_2 - y_1|^2 \right)^{1/2} > |x - y_3|.$$

On the other hand we have

$$y_3 \in \bar{O}$$

as $\bar{O}$ is convex and, consequently,

$$|x - y_3| \geq |x - y_1|.$$

The obtained contradiction shows that the assumption $y_1 \neq y_2$ is invalid.

Next, by the triangle inequality

$$|x - y| \leq |x - z| + |z - y|.$$

Therefore for each $y \in \bar{O}$ the inequality

$$|x - [x]_\partial| \le |x - y| \le |x - z| + |z - y|$$

holds and hence

$$|x - [x]_\partial| \le |x - z| + |z - [z]_\partial| .$$

Exchanging x and z we get

$$|z - [z]_\partial| \le |x - z| + |x - [x]_\partial| .$$

Therefore

$$||x - [x]_\partial| - |z - [z]_\partial|| \le |x - z| .$$

It is easily deduced from this inequality, as the vector $[x]_\partial$ is unique, that

$$|[x]_\partial - [z]_\partial| \to 0$$

when $|x - z| \to 0$ (Problem 1).

Let us dwell now on a simple case when the problem on normal reflection has a solution.

Lemma 1. *For a stepwise function* $X \in D$ *with* $X_0 \in \bar{O}$ *there exists a solution of the problem on normal reflection.*

Proof. Set $T_0 = 0$, $\phi_0 = 0$ and

$$T_n = \inf (t \ge T_{n-1}: X_t + \phi_{T_{n-1}} \notin \bar{O}), \ n \ge 1.$$

Then

$$Y_t = \begin{cases} X_t + \phi_{T_{n-1}}, & T_{n-1} \le t < T_n, \\ [X_{T_n} + \phi_{T_{n-1}}]_\partial, & t = T_n. \end{cases}$$

Here

$$\phi_t = \begin{cases} \phi_{T_{n-1}}, & T_{n-1} \le t < T_n, \\ [X_{T_n} + \phi_{T_{n-1}}]_\partial - X_{T_n}, & t = T_n. \end{cases}$$

On proving the existence and uniqueness of a solution of the problem on normal reflection, we will need a number of properties of the functions Y and ϕ, which is presented in the following lemmas.

Lemma 2. *Let* $X, \tilde{X} \in D$ *with* $X_0, \tilde{X}_0 \in \bar{O}$ *and let* Y *and* $\tilde{Y}$ *be solutions of the problems on normal reflection with the associated functions* ϕ *and* $\tilde{\phi}$ *respectively. Then*

$$1) \ |Y_t - \tilde{Y}_t|^2 = |X_t - \tilde{X}_t|^2 + 2\int_0^t (X_t - \tilde{X}_t - X_s + \tilde{X}_s, d(\phi_s - \tilde{\phi}_s))$$

and

$$2)\ |Y_t - Y_s|^2 \le |X_t - X_s|^2 + 2\int_s^t (X_t - X_u, d\phi_u),\ t > s.$$

Corollary 1. *A solution of the problem on normal reflection is unique.*

Corollary 2. *Let* $X \in C$, $X_0 \in \overline{O}$ *and let* $Y = X + \phi$ *be a solution of the problem on normal reflection. Then*

$$Y \in C\,(O).$$

Proof. We have

$$|Y_t - \tilde{Y}_t|^2 = |X_t - \tilde{X}_t|^2 + 2\,(X_t - \tilde{X}_t,\ \phi_t - \tilde{\phi}_t) + |\phi_t - \tilde{\phi}_t|^2. \tag{1.1}$$

By Ito's formula (Ch. 2, § 3)

$$|\phi_t - \tilde{\phi}_t|^2 = 2\int_0^t (\phi_{s-} - \tilde{\phi}_{s-},\ d\,(\phi_s - \tilde{\phi}_s)) + \sum_{s \le t} |\Delta\phi_s - \Delta\tilde{\phi}_s|^2$$

$$\le 2\int_0^t (\phi_s - \tilde{\phi}_s,\ d\,(\phi_s - \tilde{\phi}_s)).$$

Utilize now the fact that

$$\phi_s - \tilde{\phi}_s = (Y_s - \tilde{Y}_s) - (X_s - \tilde{X}_s),$$

and also property 4') of an associated function which gives

$$\int_0^t (Y_s - \tilde{Y}_s,\ d\,(\phi_s - \tilde{\phi}_s)) = \int_0^t (Y_s - \tilde{Y}_s,\ d\,\phi_s) + \int_0^t (\tilde{Y}_s - Y_s,\ d\,\tilde{\phi}_s) \le 0.$$

These and (1.1) imply the desired inequality.

2) By the obvious equality

$$|Y_t - Y_s|^2 = |X_t - X_s|^2 + 2\,(X_t - X_s, \phi_t - \phi_s) + |\phi_t - \phi_s|^2,$$

Ito's formula

$$|\phi_t - \phi_s|^2 = 2\int_s^t (\phi_{u-} - \phi_s,\ d\,\phi_u) + \sum_{s < u \le t} |\Delta\phi_u|^2 \le 2\int_s^t (\phi_u - \phi_s,\ d\,\phi_u)$$

and the inequality

$$\int_s^t (Y_u - Y_s,\ d\,\phi_u) \le 0,$$

which is deduced from property 4') of an associated function, we have

$$|Y_t - Y_s|^2 \le |X_t - X_s|^2 + 2(X_t - X_s, \phi_t - \phi_s) - 2\int_s^t (X_u - X_s, d\phi_u).$$

This gives the desired inequality in an obvious manner.

Lemma 3. *Let* $X^n \in D$ *with* $X_0^n \in \overline{O}$, *and let* $Y^n = X^n + \phi^n$ *be a solution of the problem on normal reflection with an associated function* ϕ^n, $n \ge 1$. *Let the following conditions be satisfied:*

1) *there exists a function* $X \in D$ *with* $X_0 \in \overline{O}$ *such that for each* $T > 0$

$$\lim_n \sup_{t \le T} |X_t^n - X_t| = 0,$$

2) $\sup_n \operatorname{Var}(\phi^n)_T \le K_T$

with a constant K_T *depending, possibly, on* T *and* X.

Then there exists a solution of the problem on normal reflection for $X : Y = X + \phi$ *such that for each* $T > 0$

$$\lim_n \sup_{t \le T} |\phi_t^n - \phi_t| = 0.$$

Proof. By assumption 1) we have

$$\lim_{n, m} \sup_{t \le T} |X_t^n - X_t^m| = 0$$

for each $T > 0$, and by assumption 2) and assertion 1) of Lemma 2

$$\lim_{n, m} \sup_{t \le T} |Y_t^n - Y_t^m| = 0.$$

Hence

$$\lim_{n, m} \sup_{t \le T} |\phi_t^n - \phi_t^m| = 0,$$

i.e. there exists $Y \in D(O)$, $\phi \in D$ with $\phi_0 = 0$ such that

$$\lim_n \left(\sup_{t \le T} |Y_t - Y_t^n| + \sup_{t \le T} |\phi_t - \phi_t^n| \right) = 0, \quad T > 0.$$

We will show now that

$$\operatorname{Var}(\phi)_t < \infty$$

for each $t > 0$. When the interval $[0, t]$ is partitioned: $0 = t_0 < t_1 < \dots < t_k = t$, we have

$$\sum_{j=1}^{k} |\phi_{t_j} - \phi_{t_{j-1}}| \le \sum_{j=1}^{k} |\phi_{t_j}^n - \phi_{t_{j-1}}^n| + 2\sum_{j=1}^{k} |\phi_{t_j} - \phi_{t_j}^n|$$

$$\le \sup_n \operatorname{Var}(\phi^n)_t + 2\sum_{j=1}^{k} |\phi_{t_j} - \phi_{t_j}^n| \to \sup_n \operatorname{Var}(\phi^n)_t, \quad n \to \infty.$$

This gives

$$\mathrm{Var}\,(\phi)_t \le K_t.$$

To complete the proof, it remains to show that ϕ satisfies Conditions 3') and 4') (see properties of an associated function).

Condition 3') is verified in the following manner. Let $f = f(y)$ be the function, involved in Condition 3'). Then

$$\int_0^t (f(Y_s^n), d\phi_s^n) = 0, \quad t > 0, \quad n \ge 1.$$

By Lemma 1 ([14], Ch. 3, § 14) one may choose the sequence of stepwise functions Z_s^N, $s \le t$, $N \ge 1$, such that

$$\sup_{s \le t} | Y_s - Z_s^N | \to 0, \quad N \to \infty.$$

Then

$$\left| \int_0^t (f(Y_s), d\phi_s) \right| = \left| \int_0^t (f(Y_s), d\phi_s) - \int_0^t (f(Y_s^n), d\phi_s^n) \right|$$

$$\le \left| \int_0^t (f(Y_s) - f(Y_s^n), d\phi_s^n) \right| + \left| \int_0^t (f(Z_s^N), d(\phi_s - \phi_s^n)) \right|$$

$$+ \left| \int_0^t (f(Y_s) - f(Z_s^N), d\phi_s) \right| + \left| \int_0^t (f(Y_s) - f(Z_s^N), d\phi_s^n) \right| .$$

Obviously, the right-hand side of this inequality tends to zero as the limit $\lim_N \overline{\lim_n}$ is taken, i.e. 3') is satisfied.

Condition 4') is verified analogously. It suffices to show that for each $s < t$ and each $\tilde{Y} \in D(O)$ the inequality

$$\int_s^t (\tilde{Y}_u - Y_u, d\phi_u) \ge 0$$

holds. Since

$$\int_s^t (\tilde{Y}_u - Y_u^n, d\phi_u^n) \ge 0, \quad n \ge 1,$$

it suffices that

$$\lim_n \left| \int_s^t (\tilde{Y}_u - Y_u, d\phi_u) - \int_s^t (\tilde{Y}_u - Y_u^n, d\phi_u^n) \right| = 0. \tag{1.2}$$

By Lemma 1 ([14], Ch. 3, § 14) there exists a sequence Z_s^N, $s \le t$, $N \ge 1$, of stepwise functions, approximating

$$\tilde{Y}_u - Y_u,\ u \le t: \sup_{u \le t} |\tilde{Y}_u - Y_u - Z_u^N| \to 0,\ N \to \infty.$$

Then the relation (1.2) holds, provided the inequality

$$\left| \int_s^t (\tilde{Y}_u - Y_u, d\phi_u) - \int_s^t (\tilde{Y}_u - Y_u^n, d\phi_u^n) \right|$$

$$\le \left| \int_s^t (\tilde{Y}_u - Y_u^n, d\phi_u^n) \right| + \left| \int_s^t (Z_u^N, d(\phi_u - \phi_u^n)) \right| \tag{1.3}$$

$$+ \left| \int_s^t (\tilde{Y}_u - Y_u - Z_u^N, d\phi_u) \right| + \left| \int_s^t (\tilde{Y}_u - Y_u - Z_u^N, d\phi_u^n) \right|$$

holds and the right-hand side of (1.3) tends to zero as the limit $\lim_N \overline{\lim_n}$ is taken.

Further on we will need the following additional condition on a region O.

Condition (α). There exists a unitary vector e and a constant $c > 0$ such that $(e, n) \ge c$ for each vector

$$n \in \bigcup_{y \in \partial O} \mathfrak{N}_y(O).$$

Lemma 4. *Let* $Y = X + \phi$ *be a solution of the problem on normal reflection for* $X \in D$ *with* $X_0 \in \overline{O}$. *If a region* O *satisfies Condition* (α), *then as* $s < t$

$$|Y_t - Y_s| + \mathrm{Var}(\phi)_t - \mathrm{Var}(\phi)_s \le K \sup_{s \le t_1 < t_2 \le t} |X_{t_2} - X_{t_1}|$$

with K *depending on the constant* c *involved in Condition* (α). *In particular*

$$\mathrm{Var}(\phi)_t \le 2K \sup_{s \le t} |X_s|.$$

Proof. For $s < t$ denote

$$K_{s,t} = \mathrm{Var}(\phi)_t - \mathrm{Var}(\phi)_s, \quad \Delta_{s,t} = \sup_{s \le t_1 < t_2 \le t} |X_{t_2} - X_{t_1}|.$$

By Condition (α)

$$(e, \phi_t - \phi_s) \geq c\, K_{s,t}.$$

Hence

$$(e, Y_t - Y_s) = (e, X_t - X_s) + (e, \phi_t - \phi_s) \geq (e, X_t - X_s) + c\, K_{s,t}.$$

This gives, by the Cauchy-Bunjakovski inequality, the following estimate for $K_{s,t}$:

$$K_{s,t} \leq c^{-1}(\,|\,Y_t - Y_s\,| + \Delta_{s,t}). \tag{1.4}$$

On the other hand, using assertion 2) of Lemma 1, we have

$$|\,Y_t - Y_s\,|^2 \leq \Delta_{s,t}^2 + 2K_{s,t}\Delta_{s,t}. \tag{1.5}$$

Denote $a = |\,Y_t - Y_s\,|$. By (1.4) and (1.5) we get the following inequality for a:

$$a^2 \leq 2c^{-1}\Delta_{s,t}\, a + (1 + 2c^{-1})\, \Delta_{s,t}^2,$$

which gives

$$a \leq (c^{-1} + \sqrt{c^{-2} + 4\,(1 + 2c^{-1})}\,)\, \Delta_{s,t}.$$

This and (1.4) entail the desired inequality with the constant

$$K = 1 + (1 + c^{-1})\,(c^{-1} + \sqrt{c^{-2} + 4\,(1 + 2c^{-1})}\,).$$

The second assertion of the lemma follows from the first one in an obvious manner.

3. Let a $B(R_+) \otimes \mathcal{D}_\infty$-measurable function $\Phi_t(X)$ taking values in R^d possess the following properties:

1) $\Phi(X) = (\Phi_t(X))_{t \geq 0} \in D$, for each $X \in D$,

2) $\Phi_t(X)$ is $\mathcal{D}_t$-measurable for each $t \in R_+$.

It is said that $\Phi(X)$ is continuous in Skorohod's topology at point $X \in D$, if for each sequence X^k, $k \geq 1$, with $X^k \in D$ such that

$$\lim_k \rho\,(X^k, X) = 0$$

we have

$$\lim_k \rho\,(\Phi(X^k), \Phi(X)) = 0,$$

where ρ is Skorohod's metric (see Ch. 6, § 1).

Theorem 1. *Let a convex region* O ($O \subseteq R^d$) *satisfy Condition* (α). *Then for each function* $X \in D$ *with* $X_0 \in \overline{O}$ *there exists the unique solution of the problem on normal reflection:*

$$Y = X + \phi,$$

where the function $\phi = \phi(X)$ *associated with* $Y \in D(O)$ *is continuous in Skorohod's topology at each point* $X \in D$, *and*

$$\mathrm{Var}\,(\phi\,(X))_t \le L \sup_{s \le t} |X_s|,$$

with L *depending only on the constant* c *involved in* (α).

Proof. For the given function $X \in D$ with $X_0 \in \overline{O}$ one can choose the sequence of stepwise functions $(X^n_t(X))_{t \ge 0}$, $n \ge 1$, such that $X^n_t(X)$ is a $B(R_+) \otimes \mathcal{D}_\infty$-measuarble function and $\mathcal{D}_t$-measurable for each $t \ge 0$ and $n \ge 1$, and

$$\sup_{t \le T} |X^n_t(X) - X_t| \to 0, \ n \to \infty,$$

for each T > 0 (Lemma 1, [14], Ch. 3, § 14). Besides, there exists by Lemma 1 the solution of the problem on normal reflection:

$$Y^n = X^n(X) + \phi^n,$$

and by Lemma 4

$$\mathrm{Var}\,(\phi^n)_T \le 2K \sup_{t \le T} |X^n_t(X)|.$$

Hence

$$\sup_n \mathrm{Var}\,(\phi^n)_T \le 2K \left(\sup_{t \le T} |X_t| + \sup_n \sup_{t \le T} |X_t - X^n_t(X)| \right),$$

i.e. Condition 2) of Lemma 3 is satisfied. Therefore, there exists by Lemma 3 the solution of the problem on normal reflection:

$$Y = X + \phi.$$

This solution is unique by Corollary 1 to Lemma 2. The function $\phi_t(X)$ is $B(R_+) \otimes \mathcal{D}_\infty$-measurable and $\mathcal{D}_t$-measurable for each $t \ge 0$, since such are the approximating functions ϕ^n_t (see the definition of ϕ^n in Lemma 1).

We will show now that $\phi(X) = (\phi_t(X))_{t \ge 0}$ is continuous in Skorohod's topology. Let $X^k \in D$, $X^k_0 \in \overline{O}$ and let $Y^k = X^k + \phi(X^k)$ be the solution of the problem on normal reflection for X^k. Denote by $X^k_j(t)$, j = 1, ..., d the elements of the vector X^k_t (respectively, $X_j(t)$, j = 1, ..., d are elements of the vector X_t). The convergence of a sequence X^k, $k \ge 1$, to X in Skorohod's topology $\rho(X^k, X) \to 0$, $k \to \infty$, is equivalent to the following type of convergence. There exists a sequence $\lambda^k = (\lambda^k_t)_{t \ge 0}$ of vector-valued functions with elements $\lambda^k_j(t)$, j = 1, ..., d, where $\lambda^k_j(t)$ is a continuous nondecreasing function $\lambda^k_j(0) = 0$ such that for each T > 0

$$\lim_k \sum_{j=1}^d \sup_{t \le T} |\lambda^k_j(t) - t| = 0,$$

$$\lim_k \sum_{j=1}^{d} \sup_{t \le T} | X_j^k(\lambda_j^k(t)) - X_j(t) | = 0$$

(see Ch. 6, § 1).

Denote by $\overline{X}_t^k$ the vector with elements

$$\overline{X}_j^k(t) = X_j^k(\lambda_j^k(t)),$$

by $\overline{Y}_t^k$ the vector with elements

$$\overline{Y}_j^k(t) = Y_j^k(\lambda_j^k(t)),$$

and by $\overline{\phi}_t^k$ the vector with elements

$$\overline{\phi}_j^k(t) = \phi_j^k(\lambda_j^k(t)).$$

Obviously,

$$\overline{Y}^k = \overline{X}^k + \overline{\phi}^k$$

is the solution of the problem on normal reflection for $\overline{X}^k$ with the associated function $\overline{\phi}^k$. Since Condition (α) is satisfied, we have

$$\mathrm{Var}\,(\overline{\phi}^k)_t \le 2K \sup_{s \le t} | \overline{X}_s^k |$$

by Lemma 4. Therefore, by the convergence

$$\sup_{t \le T} | \overline{X}_t^k - X_t | \to 0, \ k \to \infty,$$

for each $T > 0$ one may choose a constant K_T, depending on T and X, such that

$$\sup_k \mathrm{Var}\,(\overline{\phi}^k)_T \le K_T.$$

Hence,

$$\sup_{t \le T} | \overline{\phi}_t^k - \phi_t | \to 0, \ k \to \infty$$

by Lemma 3, i.e. for each $T > 0$

$$\lim_k \sum_{j=1}^{d} \sup_{t \le T} | \phi_j^k(\lambda_j^k(t)) - \phi_j(t) | = 0,$$

which is equivalent to

$$\lim_k \rho\,(\phi\,(X^k), \phi\,(X)) = 0.$$

4. The following condition on a region O is weaker then Condition (α).

Condition (β). There exist constants $\varepsilon > 0$ and $\delta > 0$, which define at each point x

$\in \partial O$ an open ball $B_\varepsilon(x_0)$ with the centre at x_0 as $|x - x_0| \le \delta$:

$$B_\varepsilon(x_0) = \{y \in R^d : |y - x_0| < \varepsilon\}$$

such that

$$B_\varepsilon(x_0) \subset O.$$

Remark. Each bounded region O satisfies Condition (β).

Theorem 2. *Let a convex region* O ($O \subseteq R^d$) *satisfy Condition* (β). *Then for each function* $X \in D$ *with* $X_0 \in \overline{O}$ *there exists the unique solution of the problem on normal reflection:*

$$Y = X + \phi,$$

where the function $\phi = \phi\,(X)$, *associated with* $Y \in D\,(O)$, *is continuous in Skorohod's topology at each point* $X \in D$.

Proof. Let ε and δ be the constants involved in Condition (β). Define at point $x \in \partial O$ the ball

$$B\,(x) = \{y \in R^d : |y - x| < \frac{\varepsilon}{2}\}$$

($\overline{B}\,(x)$ is the closure of $B\,(x)$). Denote by $\mathcal{H}_y(O)$ the set of all hyperplanes of support of O at point $y \in \partial O$, and by H (O) the closed half-space with the basic hyperplane H as a boundary, containing the region O. Define the region O_x as the interior of the set

$$\bigcap_{y \in \partial O \cap \overline{B}\,(x)} \; \bigcap_{H \in \mathcal{H}_y\,(O)} H\,(O).$$

The region O_x is convex and satisfies Condition (α) with the vector

$$e = (x_0 - x) / |x_0 - x|$$

and the constant $c = \varepsilon/2\delta$.

Define two sequences of numbers $(T_m)_{m \ge 0}$ and $(t_m)_{m \ge 1}$ such that $t_m \le T_m$, $m \ge 1$, and as $t_{m'} = \infty$ (for a certain m') the subsequent t_{m+1}, T_m are undefined when $m \ge m'$.

Set

$$T_0 = \inf\,(t \ge 0 : X_t \notin \overline{O}).$$

As $0 < T_0 < \infty$, define

$$Y_t^0 = \begin{Bmatrix} X_t,\ 0 \le t < T_0, \\ [X_{T_0}]_\partial,\ t + T_0. \end{Bmatrix}$$

For $m \ge 1$ set

$$\hat{X}_t^m = Y_{T_{m-1}}^{m-1} + X_{t+T_{m-1}} - X_{T_{m-1}}$$

with

$$Y^{m-1}_{T_{m-1}} \in \partial O.$$

By Theorem 1 there exists the solution of the problem on normal reflection for

$$\hat{X}^m = (\hat{X}^m_t)_{t \geq 1}$$

in the region

$$O_{Y^{m-1}_{T_{m-1}}} : \hat{Y}^m = \hat{X}^m + \hat{\phi}^m.$$

Set

$$t_m = \inf\left(t \geq T_{m-1} : | \hat{Y}^m_{t-T_{m-1}} - \hat{Y}^m_0 | \geq \frac{\varepsilon}{2} \right).$$

As $T_{m-1} < t_m < \infty$ define on $[T_{m-1}, t_m]$ the function

$$Y^m_t = \begin{cases} \hat{Y}^m_{t-T_{m-1}}, & T_{m-1} \leq t < t_m, \\ [\hat{Y}^m_{t_m - T_{m-1}}]_\partial, & t = t_m. \end{cases}$$

Next, let

$$T_m = \inf\,(t \geq t_m : Y^m_{t_m} + X_t - X_{t_m} \notin \bar{O})$$

and for $t_m < T_m < \infty$ define on $[t_m, T_m]$ the function

$$Y^m_t = \begin{cases} Y^m_{t_m} + X_t - X_{t_m}, & t_m \leq t < T_m, \\ [Y^m_{t_m} + X_{T_m} - X_{t_m}]_\partial, & t = T_m. \end{cases}$$

Set

$$T_\infty = \lim_m T_m.$$

Obviously, if $T_\infty < \infty$, then there exists the solution of the problem on normal reflection given by

$$Y_t = \sum_{m \geq 1} I\,(T_{m-1} \leq t < T_m)\, Y^m_t.$$

We will show now that $T_\infty = \infty$. By the definition of t_m we have

$$|\hat{Y}^m_{t_m - T_{m-1}} - \hat{Y}^m_0| \geq \frac{\varepsilon}{2}.$$

By Lemma 4

$$|\hat{Y}^m_{t_m - T_{m-1}} - \hat{Y}^m_0| \leq K \sup_{0 \leq s_1 < s_2 < t_m - T_{m-1}} |\hat{X}^m_{s_2} - \hat{X}^m_{s_1}|$$

$$= K\Delta_{T_{m-1}, t_m} \leq K\Delta_{T_{m-1}, T_m}$$

with

$$\Delta_{s,t} = \sup_{s \leq s_1 < s_2 \leq t} |X_{s_2} - X_{s_1}|.$$

This entails

$$\Delta_{T_{m-1}, T_m} \geq \frac{\varepsilon}{2}.$$

Let T > 0. By Lemma 1 ([14], Ch. 3, § 14) there exists a partition of the interval [0, T] at points $\{t_i\}$ with

$$\inf_i (t_i - t_{i-1}) > 0$$

and

$$\inf_{\{t_i\}} \max_i \Delta_{t_{i-1}, t_i} \leq \varepsilon/4K.$$

This means that each interval $[t_{i-1}, t_i]$ contains at most one element of the sequence T_m, $m \geq 1$, i.e. [0, T] contains at most $(1 + 4\varepsilon^{-1}KT)$ elements of the sequence T_m, $m \geq 1$, and hence, as T is arbitrary, $T_\infty = \infty$ holds.

Thus, under Condition (β) there exists the solution of the problem on normal reflection in the region O. This solution is unique by Corollary 1 to Lemma 2.

As in Theorem 1, it will be proved that ϕ (X) is continuous in Skorohod's topology. To this end, it suffices that

$$\mathrm{Var}\,(\phi\,(X))_T \leq K_T(X)$$

for each T> 0, where $K_T(X)$ is a continuous function of X in the uniform convergence topology:

$$\lim_k \sup_{t \leq T} |X^k_t - X_t| = 0 \Rightarrow \lim_k K_T(X^k) = K_T(X).$$

By construction

$$\mathrm{Var}\,(\phi\,(X))_T = \sum_{m \geq 1} I\,(T_{m-1} \leq T < t_m)\, \mathrm{Var}\,(\hat{\phi}^m)_{T - T_{m-1}}$$

$$+ \sum_{m \geq 1} I\,(t_m \leq T < T_m)\, |Y^m_{t_m} - X_{t_m}|.$$

By Theorem 1

$$\mathrm{Var}\,(\hat{\phi}^m)_{t-T_{m-1}} \le L \sup_{t \le T - T_{m-1}} |\hat{X}^m_t|$$

$$\le L\left(\left| Y^{m-1}_{T_{m-1}} - X_{T_{m-1}} \right| + \sup_{t \le T} |X_t| \right).$$

Hence

$$\mathrm{Var}\,(\phi\,(X))_T$$

$$\le L\left(\sum_{m \ge 1} \left[I\,(T_{m-1} \le T < t_m)\,(\,| Y^{m-1}_{T_{m-1}} - X_{T_{m-1}} | + \sup_{t \le t_m} |X_t| \right] \right)$$

$$+ \sum_{m \ge 1} I\,(t_m \le T < T_{m-1})\,| Y^m_{t_m} - X_{t_m} | = K_T\,(X).$$

The function $K_T\,(X)$ just defined is continuous in the uniform convergence topology, provided such are the functions

$$Y^{m-1}_{T_{m-1}} = Y^{m-1}_{T_{m-1}}\,(X) \text{ and } Y^m_{t_m} = Y^m_{t_m}\,(X).$$

Observe that

$$Y^0_{T_0} = [X_{T_0}]_\partial$$

possesses the desired property as $[x]_\partial$ is a continuous function of

$$\hat{Y}^1_{t_1 - T_0} = \hat{X}^1_{t_1 - T_0} + \hat{\phi}^1_{t_1 - T_0}.$$

The continuity of $\hat{Y}^1_{t_1 - T_0}$ follows from the continuity of $\hat{\phi}^1$, established in Theorem 1. As $m \ge 2$ the desired property is established by induction.

Problems

1. Let $x^n \in R^d$, $n \ge 1$, $x \in R^d$, $[x]_\partial = \arg\min_{y \in \overline{O}} |x - y|$, where $|\cdot|$ is the Eucledian norm, O a convex region in R^d, and $\overline{O}$ its closure. Show that

$$\lim_n |x^n - x| = 0$$

implies

$$\lim_n |[x^n]_\partial - [x]_\partial| = 0.$$

2. Let $d = 1$ and $O = \{x > 0\}$. Show that for $X \in D$ with $X_0 \ge 0$ the associated

function is presented by

$$\phi_t(X) = -\inf_{s \leq t} (X_s \wedge 0).$$

§ 2. Semimartingale with normal reflection

1. Let a convex region $O \subseteq R^d$ satisfy one of Conditions (α) or (β). Then for each $X \in D$ with $X_0 \in \overline{O}$ there exists the solution of the problem on normal reflection (Theorems 1.1 and 1.2):

$$Y_t = X_t + \phi_t(X). \tag{2.1}$$

Let $X = (X_t)_{t \geq 0}$ be a semimartingale, defined on a stochastic basis $(\Omega, \mathcal{F}, \mathbb{F}, P)$. Denote by $\mathbb{F}^X_+ = (\mathcal{F}^X_{t+})_{t \geq 0}$ the family of σ-algebras

$$\mathcal{F}^X_{t+} = \bigcap_{\varepsilon > 0} \sigma\{X_s, s \leq t + \varepsilon\} \vee \mathcal{N},$$

where $\mathcal{N}$ is the collection of sets in $\mathcal{F}$ of P-measure zero. By properties of the function $\phi(X) = (\phi_t(X))_{t \geq 0}$ it follows that $\phi(X)$ is a $\mathbb{F}^X_+$-adapted process, and hence $\phi(X)$ and $Y = (Y_t)_{t \geq 0}$ are semimartingales relative to $\mathbb{F}$ and $\mathbb{F}^X_+$.

Define by $\mathbb{F}^Y_+ = (\mathcal{F}^Y_{t+})_{t \geq 0}$ the family of σ-algebras

$$\mathcal{F}^Y_{t+} = \bigcap_{\varepsilon > 0} \sigma\{Y_s, s \leq t + \varepsilon\} \vee \mathcal{N}$$

($\mathcal{N}$ is the collection of sets in $\mathcal{F}$ of P-measure zero).

Assume X is a continuous semimartingale. Then it is decomposed as

$$X_t = X_0 + A_t + M_t \tag{2.2}$$

with the continuous processes $A = (A_t)_{t \geq 0}$ of local bounded variation and $M = (M_t)_{t \geq 0}$ a local martingale (Theorem 2.1.2). Besides, $\phi_t(X)$ is a continuous function of t, by Corollary 2 to Lemma 1.2. From (2.1) and (2.2) it follows that

$$Y_t = X_0 + A_t + M_t + \phi_t(X). \tag{2.3}$$

Since the process Y is a $\mathbb{F}^Y_+$-semimartingale and $\mathbb{F}^Y_+ \subset \mathbb{F}$, the process Y admits, by the vector version of Theorem 4.6.1, the representation

$$Y_t = X_0 + \overline{A}_t + \overline{M}_t, \tag{2.4}$$

where $\overline{A} = (\overline{A}_t)_{t \geq 0}$ is a $\mathbb{F}^Y_+$-adapted continuous process of locally bounded variation, and $\overline{M} = (\overline{M}_t)_{t \geq 0}$ is a $\mathbb{F}^Y_+$-continuous local martingale with

$$\langle \overline{M} \rangle = \langle M \rangle. \tag{2.5}$$

Suppose now that A is a $\mathbb{F}^Y_+$-adapted process, and define

$$\overline{\phi} = A - \overline{A}. \tag{2.6}$$

In this case the process Y admits the decomposition

$$Y_t = X_0 + A_t + \overline{M}_t + \overline{\phi}_t. \tag{2.7}$$

Theorem 1. *The stochastic process* $\overline{\phi} = (\overline{\phi}_t)_{t \geq 0}$, *defined by* (2.6), *is the process associated with* Y *in the problem on normal reflection for the process* $\overline{X} = (\overline{X}_t)_{t \geq 0}$ *with*

$$\overline{X}_t = X_0 + A_t + \overline{M}_t.$$

Proof. It suffices to verify Conditions 3') and 4') concerning an associated process.

Let $f = f(y)$ be the function involved in Condition 3'). We will show that

$$\int_0^t (f(Y_s), d\overline{\phi}_s) = 0 \quad \text{P-a.s.}, \quad t > 0. \tag{2.8}$$

Define the process $Z = (Z_t)_{t \geq 0}$ with

$$Z_t = \int_0^t (f(Y_s), d\phi_s) + \int_0^t (f(Y_s), dM_s)$$

and observe that by the equality

$$\phi_t + M_t = \overline{\phi}_t + \overline{M}_t, \quad t > 0 \tag{2.9}$$

which follows from (2.3) and (2.7), the process Z admits also the representation

$$Z_t = \int_0^t (f(Y_s), d\overline{\phi}_s) + \int_0^t (f(Y_s), d\overline{M}_s).$$

The processes

$$\left(\int_0^t (f(Y_s), dM_s) \right)_{t \geq 0}$$

and

$$\left(\int_0^t (f(Y_s), d\overline{M}_s) \right)_{t \geq 0}$$

are continuous local martingales. Besides,

$$\int_0^t (f(Y_s), d\phi_s) \equiv 0,$$

since ϕ is the function associated with Y. Consequently, (2.8) holds, in view of Problem 4.6.3.

We will verify Condition 4'). Let

$$\tilde{Y} = (\tilde{Y}_t)_{t \ge 0} \in D(O).$$

Denote

$$U_t = \int_0^t (\tilde{Y}_s - Y_s, d\phi_s), \quad \overline{U}_t = \int_0^t (\tilde{Y}_s - Y_s, d\overline{\phi}_s),$$

$$L_t = \int_0^t (\tilde{Y}_s - Y_s, dM_s), \quad \overline{L}_t = \int_0^t (\tilde{Y}_s - Y_s, d\overline{M}_s).$$

Obviously $L = (L_t)_{t \ge 0}$ and $\overline{L} = (\overline{L}_t)_{t \ge 0}$ are continuous local martingales and $U = (U_t)_{t \ge 0}$ an increasing process in the sense of the definition in Ch. 1, § 8.

By the equality

$$U_t + L_t = \overline{U}_t + \overline{L}_t,$$

which is a consequence of (2.9), and by Problem 4.6.3 it follows that $\overline{U} = (\overline{U}_t)_{t \ge 0}$ is an increasing process (see Ch. 1, § 6). Hence, Condition 4') is satisfied.

2. We will turn now to the description of the martingale problem of diffusion type with normal reflection in a region O.

Consider a vector b (t, X) and a symmetric nonnegative definite matrix c (t, X), $t \in R_+$, $X \in D$ with $B(R_+) \otimes \mathcal{D}_\infty$-measurable elements $b_i(t, X)$, $c_{ij}(t, X)$, $i, j = 1, \ldots, d$, which are $\mathcal{D}_t$-measurable for each $t \in R_+$. Let Q be a probability measiure on $(D, \mathcal{D}_\infty)$, $\mathcal{D}^Q_\infty$ the completion of $\mathcal{D}_\infty$ relative to the measure Q, $\mathcal{D}^Q_t = \mathcal{D}_t \vee \mathcal{N}$ where $\mathcal{N}$ is the collection of sets in $\mathcal{D}^Q_\infty$ of Q-measure zero, and $\mathbb{D}^Q = (\mathcal{D}^Q_t)_{t \ge 0}$. On a stochastic basis $(D, \mathcal{D}^Q_\infty, \mathbb{D}^Q, Q)$ consider a stochastic process $Y = (Y_t)_{t \ge 0}$ with trajectories in D (O), and define the probability measure Q^Y (distribution of Y):

$$Q^Y(\Gamma) = Q(Y \in \Gamma), \quad \Gamma \in \mathcal{D}^Q_\infty.$$

Obviously

$$Q^Y(D(O)) = 1.$$

Define the family $\mathbb{F}^Y_+ = (\mathcal{F}^Y_t)_{t \ge 0}$ of σ-algebras

$$\mathcal{F}^Y_{t+} = \bigcap_{\varepsilon > 0} \sigma\{Y_s, s \le t + \varepsilon\} \vee \mathcal{N},$$

where $\mathcal{N}$ is the collection of sets in $\mathcal{D}^Q_\infty$ of Q-measure zero.

Definition. A probability measure Q solves the martingale problem of diffusion type with the coefficient of drift b (t, Y), the diffusion coefficient c (t, Y), and the normal reflection in a region O, if

$$\int_0^t \sum_{i=1}^d (|b_i(s, Y)| + c_{ii}(s, Y))\, ds < \infty, \quad \text{Q-a.s.}, \ t > 0$$

and the stochastic process $M = (M_t)_{t \geq 0}$ with

$$M_t = Y_t - Y_0 - \int_0^t b(s, Y)\, ds - \phi_t, \tag{2.10}$$

where $\phi = (\phi_t)_{t \geq 0}$ is the function associated with $Y \in D(O)$, is a continuous local martingale relative to $(\mathbb{F}^Y_+, Q)$, with the quadratic characteristic

$$\langle M \rangle_t = \int_0^t c(s, Y)\, ds. \tag{2.11}$$

By the definition of M it follows that ϕ is a $\mathbb{F}^Y_+$-adapted process. By Doob's theorem (see [96]) and by (2.10) and (2.11) the stochastic process $Y = (Y_t)_{t \geq 0}$ defined on a basis, larger then $(D, \mathcal{D}^Q_\infty, \mathbb{D}^Q, Q)$ if necessary, presents a weak solution of the stochastic differential equation

$$Y_t = Y_0 + \int_0^t b(s, Y)\, ds + \int_0^t c^{1/2}(s, Y)\, dW_s + \phi_t \tag{2.12}$$

with respect to a Wiener process $W = (W_t)_{t \geq 0}$ independent of Y_0, with $W_t \in R^d$ having independent components.

We will say that the present martingale problem has a unique solution if the set of probability measures, solving this problem with the parameters b (t, Y) and c (t, Y) and with one and the same restrictions on the σ-algebra $\mathcal{D}_0$, consists of one point. The uniqueness of the solution of the martingale problem is equivalent to the weak uniqueness of the solution of the equation (2.12).

Let the equation (2.12) have a weak solution. Define the stochastic process $Z = (Z_t)_{t \geq 0}$ with

$$Z_t = Y_0 + \int_0^t b(s, Y)\, ds + \int_0^t c^{1/2}(s, Y)\, dW_s.$$

Then

$$Y_t = Z_t + \Phi_t(Z) \tag{2.13}$$

where $\Phi(Z) = (\Phi_t(Z))_{t \geq 0}$ is the function associated with Y in the problem on normal reflection for the function Z in a region O, i.e.

$$\Phi(Z) = \phi.$$

Therefore Z is the weak solution of the stochastic equation

$$Z_t = Y_0 + \int_0^t b(s, Z + \Phi(Z))\, ds + \int_0^t c^{1/2}(s, Z + \Phi(Z))\, dW_s. \tag{2.14}$$

The converse statement is also true: if $Z = (Z_t)_{t \geq 0}$ is the weak solution of the stochastic equation (2.14), then the process Y, defined by (2.13), is the weak solution of the stochastic equation (2.12). Besides, both equations (2.12) and (2.14) are simultaneously weak.

3. Let $g(t, Y)$ denote any of the elements of $b(t, Y)$ and $c(t, Y)$, $t \geq 0$, $Y \in D(O)$. We say that the coefficients of drift $b(t, Y)$ and diffusion $c(t, Y)$ satisfy the condition of linear growth and Lipschitz' condition, if

$$g^2(t, Y) \leq \int_0^t (1 + |Y_s|^2)\, dK(s) + k(1 + |Y_t|^2),$$

$$|g(t, Y') - g(t, Y'')|^2 \leq \int_0^t |Y'_s - Y''_s|\, dK(s) + k\,|Y'_t - Y''_t| \tag{2.15}$$

with $k \geq 0$ and $K(t) \in V^+ \cap C$.

Theorem 2 (Problem 1). *If the condition of linear growth and Lipschitz' condition* (2.15) *are satisfied, then the equation* (2.12) *has the unique strong solution.*

Remark. If the first of Conditions (2.15) is satisfied, then

$$g^2(t, Y) \leq L(t)\,(1 + \sup_{s \leq t} |Y_s|^2), \qquad \int_0^t L(s)\, ds < \infty. \tag{2.16}$$

Problems

1. Prove Theorem 2.

2. Let $O \subset R$, $O = \{x: x > 0\}$, let $X = (X_t)_{t \geq 0}$ be a Wiener process and $Y = (Y_t)_{t \geq 0}$ the solution of the problem on normal reflection for X. Show that

$$Y \overset{d}{=} |X|.$$

§ 3. Diffusion approximation with normal reflection

1. Let $X^n = (X_t^n)_{t \geq 0}$ be a semimartingale, defined on a stochastic basis $(\Omega^n, \mathcal{F}^n, \mathbb{F}^n, P^n)$, with $X_t^n \in R^d$ and the triplet of predictable characteristics $T^n = (B^n, C^n, \nu^n)$, related to the canonical representation (Ch. 4, § 1):

$$X_t^n = X_0^n + B_t^n + X_t^{nc} + \int_0^t \int_{|x|>1} x \, d\mu^n + \int_0^t \int_{|x| \leq 1} x \, d(\mu^n - \nu^n) \tag{3.1}$$

and

$$X_0^n \in \bar{O}, \tag{3.2}$$

$n \geq 1$. For each $n \geq 1$ and $a \in (0, 1]$ we define a locally square integrable martingale $M^{na} = (M_t^{na})_{t \geq 0}$ with

$$M_t^{na} = X_t^{nc} + \int_0^t \int_{|x| \leq a} x \, d(\mu^n - \nu^n) \tag{3.3}$$

and the quadratic characteristic

$$\langle M^{na} \rangle_t = C_t^n + \int_0^t \int_{|x| \leq a} xx^* d\nu^n - \sum_{s \leq t} \hat{x}_s^{na} (\hat{x}_s^{na})^*, \tag{3.4}$$

with

$$\hat{x}_s^{na} = \int_{|x| \leq a} x\nu^n(ds, dx).$$

Suppose that a convex region O satisfies one of Conditions (α) or (β).

Then for X^n there exists the solution of the problem on normal reflection (Theorems 1.1. and 1.2):

$$Y_t^n = X_t^n + \phi_t(X^n).$$

Consider the problem concerning the weak convergence of the sequence $Y^n = (Y_t^n)_{t \geq 0}$, $n \geq 1$, to a process of diffusion type with normal reflection $Y = (Y_t)_{t \geq 0}$ in the region O, defined by the stochastic differential equation (2.12) (§ 2). We will need the following conditions on elements of b (t, Y) and c (t, Y):

$$\sum_{i=1}^d |b_i(t, Y)| \leq L(t) (1 + \sup_{s \leq t} |Y_s|),$$

$$\sum_{i=1}^{d} c_{ii}(t, Y) \le L(t)\,(1 + \sup_{s \le t} |Y_s|^2), \tag{3.5}$$

$$\int_0^t L(s)\,ds < \infty, \quad t > 0, \quad Y \in D(O).$$

Theorem 1. *Let a region* O *satisfy Condition* (α) . *Let the coefficient of drift* b (t, Y) *and the diffusion coefficient* c (t, Y) *satisfy Conditions* (3.5), *for each* $t \in S$ (S *is a dense set in* R_+, $\int_{R_+ \setminus S} dt = 0$) *let elements of* b (t, Y) *and* c (t, Y) *be continuous in Skorohod's metric at each point* $Y \in C(O)$, *and let the equation* (2.12) (§ 2) *have the unique weak solution.*

If

(0) $\quad Y_0^n \xrightarrow{d} Y_0$,

and for each $t > 0$, $\varepsilon > 0$ *and* $0 < a \le 1$

(A) $$\lim_n P^n\left(\int_0^t \int_{|x| > a} dv^n \ge \varepsilon\right) = 0,$$

(sup B) $$\lim_n P^n\left(\sup_{s \le t}\left|B_s^n - \int_0^s b(u, Y^n)\,du\right| \ge \varepsilon\right) = 0,$$

(sup C) $$\lim_n P^n\left(\sup_{s \le t}\left|\langle M^{na}\rangle_s - \int_0^s c(u, Y^n)\,du\right| \ge \varepsilon\right) = 0,$$

then

$$Y^n \xrightarrow{d} Y.$$

Theorem 2. *Let the conditions of Theorem* 1 *be satisfied, however with Condition* (α) *replaced by Condition* (β) *and* (3.5) *by the condition that elements of* b (t, Y) *and* c (t, Y) *are bounded. Then the assertion of Theorem* 1 *is still in force.*

Corollary. *If* O *is a bounded region and the conditions of Theorem* 1 *are satisfied with Condition* (β) *instead of* (α), *then the assertion Theorem* 1 *is still in force.*

2. Proof of Theorem 1. For $X \in D$ and $X_0 \in \overline{O}$ let $Y = X + \phi(X)$ be a solution of the problem on normal reflection. Let Q be a probability measure on $(D, \mathcal{D}_\infty)$, which solves the martingale problem (2.10) and (2.11) (§ 2) with normal reflection in a region O. Then $X = Y - \phi(X)$ admits the representation (cf. (2.14) in § 2)

$$X_t = X_0 + \int_0^t b(s, X + \phi(X))\, ds + M_t, \tag{3.6}$$

where M is a continuous local martingale relative to $(\mathbb{D}^Q, Q)$ with the quadratic characteristic

$$\langle M \rangle_t = \int_0^t c(s, X + \phi(X))\, ds. \tag{3.7}$$

Let Q^{X^n} be the distribution of a semimartingale X^n:

$$Q^{X^n}(\Gamma) = P^n(X \in \Gamma).$$

According to Theorem 1.2

$$Y^n = X^n + \phi(X^n)$$

is a solution of the problem on normal reflection in a region O. Then the distribution of Y^n is given in the following manner:

$$Q^{Y^n}(\Gamma) = Q^{X^n}(Y^n \in \Gamma).$$

As $\phi(X)$ is continuous in Skorohod's topology, as well as $Y = Y(X) = X + \phi(X)$, the weak convergence

$$Q^{X^n} \xrightarrow{w} Q$$

implies the weak convergence

$$Q^{Y^n} \xrightarrow{w} Q^Y \quad (Y^n \xrightarrow{d} Y).$$

For the weak convergence

$$Q^{X^n} \xrightarrow{w} Q$$

it suffices to verify the following conditions (Theorem 8.3.3 and Remark to it):

1) $\displaystyle\sum_{i=1}^d |b_i(t, X + \phi(X))| \le L(t)\,(1 + \sup_{s \le t} |X_s|)$,

$$\sum_{i=1}^d c_{ii}(t, X + \phi(X)) \le L(t)\,(1 + \sup_{s \le t} |X_s|^2),$$

$$\int_0^t L(s)\,ds < \infty, \quad t > 0;$$

2) for each $t \in S$ (S is a set dense in R_+, $\int_{R_+\setminus S} dt = 0$), elements of $b(t, X + \phi(X))$ and $c(t, X + \phi(X))$ are continuous in Skorohod's metric at each point $x \in C$;

3) the equation (3.6) possesses a unique weak solution;

for each $t > 0$, $\varepsilon > 0$ and $0 < a \le 1$

$$\text{(a)} \qquad \lim_n P^n\left(\int_0^t \int_{|x|>a} dv^n \ge \varepsilon\right) = 0,$$

$$\text{(sup b)} \quad \lim_n P^n\left(\sup_{s\le t}\left| B^n_s - \int_0^s b(u, X^n + \phi(X^n))\,du\right| \ge \varepsilon\right) = 0,$$

$$\text{(sup c)} \quad \lim_n P^n\left(\sup_{s\le t}\left| \langle M^{na}\rangle_s - \int_0^s c(u, X^n + \phi(X^n))\,du\right| \ge \varepsilon\right) = 0.$$

These conditions are satisfied in view of the assumptions, stipulated in Theorem 1 and Theorem 1.1, the representation (3.8) for Y^n, and the one to one correspondence between the solutions of the equations (2.12) (§ 2) and (3.6).

The proof of Theorem 2 is carried out analogously; it is based on Theorem 1.2.

Problem

1. Let $X^n = (X^n_t)_{t\ge 0}$ be a locally square integrable martingale, $n \ge 1$, and

$$\text{1)} \quad E^n \sum_{s\le t}^{n} (\Delta X^n_s)^2 I(|\Delta X^n_s| > \varepsilon) \to 0, \quad n \to \infty, \quad t > 0, \quad \varepsilon > 0,$$

$$\text{2)} \quad \langle X^n\rangle_t \xrightarrow{P} t, \quad t > 0.$$

If Y^n is a solution of the problem on normal reflection in the region $O = \{x\colon x > 0\}$, then

$$Y^n \xrightarrow{d} |W|$$

where W is a Wiener process. (Hint: utilize Problem 2.3.)

§ 4. Diffusion approximation with reflection for queueing models with autonomious service

1. A queueing model is characterized by a queue Q_t at time $t \geq 0$, defined by the relation:

$$Q_t = A_t - \int_0^t I(Q_{s-} > 0)\, dD_s, \tag{4.1}$$

where $A = (A_t)_{t \geq 0}$ is a process of demands and $D = (D_t)_{t \geq 0}$ a process of service. It is assumed here that A and D are point processes on a certain stochastic basis $(\Omega, \mathfrak{F}, \mathbb{F}, P)$ with the $\mathbb{F}$-compensators $\tilde{A} = (\tilde{A}_t)_{t \geq 0}$ and $\tilde{D} = (\tilde{D}_t)_{t \geq 0}$:

$$\tilde{A}_t = \int_0^t \lambda(\varepsilon Q_s)\, ds, \quad \tilde{D}_t = \int_0^t \mu(\varepsilon Q_s)\, ds, \tag{4.2}$$

where ε is a small parameter and $\lambda(x)$ and $\mu(x)$ are nonnegative functions, besides, jumps of the processes A and D do not coincide:

$$\sum_{t>0} \Delta A_t \Delta D_t = 0. \tag{4.3}$$

If λ and μ are constant intensities, i.e. A and D are Poisson processes, and

$$\lambda = \mu, \tag{4.4}$$

then the stationary distribution of the process $Q = (Q_t)_{t \geq 0}$ does not exist. Moreover, as $t \to \infty$, a queue Q_t can be unboundedly increasding. Apparently, an analogous situation may take place in case of general intensities $\lambda(\varepsilon Q_t)$ and $\mu(\varepsilon Q_t)$, if

$$\lambda(0) = \mu(0).$$

Taking this into account, we assume $\varepsilon = 1/n$, $n \geq 1$, and we consider the stochastic process $Y^n = (Y_t^n)_{t \geq 0}$ with

$$Y_t^n = \frac{1}{\sqrt{n}} Q_{nt}. \tag{4.5}$$

We will show that under certain conditions the sequence of processes Y^n, $n \geq 1$, converges weakly to a nonnegative diffusion process $Y = (Y_t)_{t \geq 0}$ with normal reflection at zero.

2. Theorem 1. *If the functions* $\lambda(x)$ *and* $\mu(x)$ *are differentiable with the uniformly bounded and uniformly continuous derivatives* $\lambda'(x)$ *and* $\mu'(x)$, *and* $\lambda(0) = \mu(0)$, *then*

$$Y^n \xrightarrow{d} Y,$$

where $Y = (Y_t)_{t \geq 0}$ *is a nonnegative diffusion process with normal reflection at zero, determined by the stochastic equation*

$$Y_t = \int_0^t [\lambda'(0) - \mu'(0)]\, Y_s ds + \int_0^t \sqrt{2\lambda(0)}\, dW_t + \phi_t, \tag{4.6}$$

where $W = (W_t)_{t \geq 0}$ *is a Wiener process and* $\phi = (\phi_t)_{t \geq 0}$ *a function associated with* Y.

Remark. We have the equality

$$Y \overset{d}{=} |X|, \tag{4.7}$$

where $X = (X_t)_{t \geq 0}$ is a diffusion process determined by the stochastic equation (cf. (4.6))

$$X_t = \int_0^t [\lambda'(0) - \mu'(0)]\, X_s ds + \int_0^t \sqrt{2\lambda(0)}\, d\tilde{W}_t$$

relative to a Wiener process $\tilde{W} = (\tilde{W}_t)_{t \geq 0}$ (Problem 1).

3. In the course of proving this theorem we will need certain properties of the process $q^n = (q_t^n)_{t \geq 0}$ with

$$q_t^n = \frac{1}{n} Q_{nt}. \tag{4.8}$$

Lemma 1. *Let the conditions of Therorem* 1 *be satisfied. Then for each* $t > 0$

$$\lim_n \mathbf{E} \sup_{s \leq t} q_s^n = 0.$$

Proof. We will show that

$$\sup_n \mathbf{E} \sup_{s \leq t} q_s^n \leq K,$$

where a constant K can depend on t only.

By (4.1) and (4.8) we have

$$\sup_{s \leq t} q_s^n \leq \frac{1}{n} A_{nt}.$$

First, assume $EA_t < \infty$ for each $t > 0$. Then

$$\mathbf{E}\left(\sup_{s \leq t} q_s^n\right) \leq \frac{1}{n} \mathbf{E}A_{nt} = \frac{1}{n} \mathbf{E}\tilde{A}_{nt} = \frac{1}{n} \int_0^{nt} \mathbf{E}\lambda\left(\frac{1}{n} Q_s\right) ds$$

$$\leq t\,\lambda(0) + \sup_{x \geq 0} |\lambda'(x)| \int_0^t \mathbf{E}\left(\sup_{u \leq s} q_u^n\right) ds.$$

This gives the desired inequality by the Gronwall-Bellman inequality with

$$K = t\lambda(0)\exp(\sup_{x \geq 0} |\lambda'(x)|\, t).$$

In the general case when $\mathbf{E}A_t \leq \infty$, $t > 0$, we get the desired inequality with the same constant K by utilizing a localizing sequence for A and Fatou's lemma.

Observe next that the process Q, defined by the relation (4.1), is a solution of the problem on normal reflection for the process $X = (X_t)_{t \geq 0}$ with

$$X_t = A_t - D_t \tag{4.9}$$

in the region $O = \{x: x > 0\}$. Besides, by Problem 1.2

$$Q_t = X_t - \inf_{s \leq t} X_s. \tag{4.10}$$

By (4.10) and (4.8) it follows that

$$q_t^n = x_t^n - \inf_{s \leq t} x_s^n \tag{4.11}$$

with

$$x_t^n = \frac{1}{n} X_{nt}.$$

Therefore the desired assertion holds, provided

$$\lim_n \mathbf{E} \sup_{s \leq t} |x_s^n| = 0. \tag{4.12}$$

By (4.9) and (4.2) for $\varepsilon = 1/n$ we get

$$\begin{aligned} x_t^n &= \frac{1}{n}\int_0^{nt}\left[\lambda\left(\frac{1}{n}Q_s\right) - \mu\left(\frac{1}{n}Q_s\right)\right] ds + \frac{1}{n}[(A_{nt} - \tilde{A}_{nt}) - (D_{nt} - \tilde{D}_{nt})] \\ &= \int_0^t [\lambda(q_s^n) - \mu(q_s^n)]\, ds + M_t^n, \end{aligned} \tag{4.13}$$

where $M^n = (M_t^n)_{t \geq 0}$ with

$$M_t^n = \frac{1}{n}[(A_{nt} - \tilde{A}_{nt}) - (D_{nt} - \tilde{D}_{nt})]$$

is a locally square integrable martingale relative to $\mathbb{F}^n = (\mathcal{F}_{nt})_{t \geq 0}$ with the quadratic characteristic (Problem 3.4.5)

$$\langle M^n \rangle_t = \frac{1}{n^2}(\tilde{A}_{nt} + \tilde{D}_{nt}) = \frac{1}{n}\int_0^t [\lambda(q_s^n) + \mu(q_s^n)]\, ds.$$

Since

$$\sup_n \mathbf{E} \sup_{s \leq t} q_s^n \leq K,$$

this gives

$$\lim_n \mathbf{E} \langle M^n \rangle_t = 0$$

for each $t > 0$, and hence by Doob's inequality (Theorem 1.9.1)

$$\mathbf{E} \sup_{s \le t} (M_s^n)^2 \le 4\mathbf{E} \langle M^n \rangle_t$$

the following relation holds:

$$\lim_n \mathbf{E} \sup_{s \le t} |M_s^n| = 0, \quad t > 0. \tag{4.14}$$

By (4.11) we have the inequality

$$\sup_{s \le t} q_s^n \le 2 \sup_{s \le t} |x_s^n|,$$

in view of which (4.13) entails

$$\sup_{s \le t} |x_s^n| \le 4c \int_0^t \sup_{u \le s} |x_u^n| \, du + \sup_{s \le t} |M_s^n|$$

with

$$c = \sup_{x \ge 0} (|\lambda'(x)| \vee |\mu'(x)|).$$

Therefore, by the Gronwall-Bellman inequality

$$\sup_{s \le t} |x_s^n| \le \sup_{s \le t} |M_s^n| e^{4ct}, \quad t > 0.$$

Hence, (4.12) is a consequence of (4.14).

4. Proof of Theorem 1. From (4.10) it follows that

$$Y_t^n = \frac{1}{\sqrt{n}} (X_{nt} - \inf_{s \le t} X_{ns}),$$

i.e. in view of Problem 1.2 the process Y^n is a solution of the problem on normal reflection for the process $X^n = (X_t^n)_{t \ge 0}$ with

$$X_t^n = \frac{1}{\sqrt{n}} X_{nt}.$$

The process X^n is a semimartingale relative to $\mathbb{F}^n = (\mathcal{F}_{nt})_{t \ge 0}$ with the decomposition

$$X_t^n = \int_0^t \sqrt{n} \, [\lambda(q_s^n) - \mu(q_s^n)] \, ds + L_t^n, \tag{4.15}$$

where $L^n = (L_t^n)_{t \ge 0}$ is a locally square integrable martingale relative to $\mathbb{F}^n$ with

$$L_t^n = \frac{1}{\sqrt{n}} [(A_{nt} - \tilde{A}_{nt}) - (D_{nt} - \tilde{D}_{nt})]$$

and the quadratic characteristic

$$\langle L^n \rangle_t = \frac{1}{n} (\tilde{A}_{nt} + \tilde{D}_{nt}) = \int_0^t [\lambda(q_s^n) + \mu(q_s^n)] \, ds. \tag{4.16}$$

Let us verify now the conditions of Theorem 1.3. Since O = {x: x > 0}, the region O satisfies Condition (α). The coefficients of drift and diffusion in the equation (4.6) satisfy Conditions (3.5) (§ 2), and and they are continuous in Skorohod's metric for each $t \in R_+$ at each point $Y \in C$ (O). The equation (4.6) possesses a unique strong solution (Theorem 2.2).

Condition (0) is satisfied in view of $Y_0^n \equiv 0$ and $Y_0 = 0$, and Condition (A) in view of $|\Delta X_t^n| \le \frac{1}{\sqrt{n}}$, $t > 0$. Condition (sup B) is satisfied since

$$\sup_{s \le t} \left| \int_0^s \sqrt{n}\, [\lambda (q_u^n) - \mu (q_u^n)]\, du - \int_0^s [\lambda'(0) - \mu'(0)]\, Y^n du \right|$$

$$\le \int_0^t \left| \sqrt{n}\, [\lambda (q_u^n) - \mu (q_u^n)] - [\lambda'(0) - \mu'(0)]\, Y_u^n \right| du \xrightarrow{P} 0, \quad n \to \infty$$

by Lemma 1.

Condition (sup C) is verified in the following manner. From (4.16) and Lemma 1 it follows that

$$\langle L^n \rangle_t \xrightarrow{P} 2\lambda (0)\, t, \quad t > 0.$$

Therefore, in view of Problem 5.3.2, we have

$$\sup_{s \le t} \left| \langle L^n \rangle_s - 2\lambda (0)\, s \right| \xrightarrow{P} 0, \quad n \to \infty$$

for each t > 0. Next, by taking into consideration that

$$|\Delta L_t^n| = |\Delta X_t^n| \le \frac{1}{\sqrt{n}}$$

for a sufficiently large value of n, depending on a, we get the equality

$$\langle M^{na} \rangle = \langle L^n \rangle.$$

Hence, Condition (sup C) is satisfied.

Problem

1. Verify the equality (4.7).

HISTORIC-BIBLIOGRAPHICAL NOTES[1]

Chapter 1

In gambling, a system of a game which consists in doubling the stakes in the case of losing and in stopping in the case of winning, is called the martingale. Just from that stems the mathematical notion of "martingale", first introduced, apparently by Ville [36]. (In the French dictionary Claude et Paul Auge "Nouveau Petit Larousse", Librairie Larousse Paris-VI, 1956 the following interpretation of the word Martingale is given: 1) Courroie qui empêche le cheval de donner de la tête; 2) Languette de buffle ou d'étoffe; 3) Cordage servant de sousbarbe aux bouts-dehors du foc; 4) Système de jeu, qui prétend assurer un bénéfice par une augmentation progressive de la mise. In the English dictionary "Standard College Dictionary", Funk and Wagmalls Company, New York, 1957 the following interpretation of the term Martingale is given: 1) A forked strap that prevents a horse from rearing its head, connecting the head gear with the bellyband; 2) A dolphin striker; 3) A system of gambling in which one doubles the stakes to recover previous losses).

The random series, called martingales nowadays, have appeared in the works of Bernstein [12] and Levy [177]. The first systematic exposition of the ("classical") martingale theory has been given in Doob's book [96].

The case of martingales with discrete time has been discussed in many monographs and textbooks: Loéve [203], Breiman [28], Meyer [217], Neveu [231], Liptser and Shiryaev [188]-[190], Shiryaev [332], Gihman and Skorohod [47]. In the present monograph we intend to expose the modern martingale theory in case of *continuous* time.

§§ **1-3.** The material exposed here comprises the background of the "general theory of stochastic processes" and is based mainly on the results of the Strassburg group of probabilists (P.-A. Meyer, C. Dellacherie, C. Doléans-Dade, ...). Much of the related material is contained in the books by Meyer [217], Dellacherie and Meyer [83]-[85], Dellacherie [81], Jacod [103], Métivier [219], Elliot [335], Gihman and Skorohod [50]. In the proceedings of Séminaire de Probabilités, published by Springer-Verlag in the series *Lecture Notes in Mathematics* one can find original as well as survey papers concerning the general theory of stochastic processes and martingales.

§ **4.** The notion of a local martingale has been introduced in the paper by Ito and

[1] As the results of that or another paper have already entered the books, we have preferred to refer to the last ones.

Watanabe [124].

§ **5.** In the work by Kunita and Watanabe [162] the rôle played in the martingale theory by square integrable martingales has in many respects been clarified.

§ **6.** In the monographs by Dellacherie [81] and Meyer [217] increasing processes have been treated systematically as an important separate class. Concerning the proof of the Doob-Meyer decomposition, see Meyer [217], Dellacherie and Meyer [83]. In the original proof of the Doob-Meyer decomposition for submartingales of class $(\mathcal{D})$, given by Meyer, the notion of "natural" processes has been used instead of predictability. Doléans-Dade [92] has established that the notions of predictable increasing and natural increasing processes coincide.

In the present monograph we prefer to use (in accordance with [129]) the term "compensator", instead of "dual predictable projection"

§ **7.** The first of the decompositions of a local martingale in case of discrete time is called the "Gandy decomposition". Meyer [213] established the second decomposition and studied its properties. Concerning the structure of local martingales, see also [103].

§ **8.** The quadratic characteristic $\langle M \rangle$ for square integrable martingales has been introduced by Meyer [210]. The mutual quadratic characteristic or the quadratic covariance $\langle M, N \rangle$ has been introduced by Kunita and Watanabe [162]. The quadratic variation [M, M] for a locale martingale M has been introduced by Meyer [213]. Theorem 3 belongs to Chou [327] and Lepingle [179].

§ **9.** In this section the proofs are given for the basic inequalities concerning local martingales. The inequalities (9.1), (9.2), (9.9) belong to Doob [96]. The notion of "domination" has been introduced by Lenglart [178]. The inequalities (9.13) and (9.14) are presented in the works [178], [253], [255].

Concerning the Davis inequality ($p = 1$) and the Burkholder-Gandy inequality ($p > 1$), see the paper by Burkholder, Davis and Gandy [29], and also [213]. The Burkholder-Davis-Gandy inequalities present generalizations of the famous Khintchine and Marcinkiewicz-Zygmund inequalities (for details see [323], Ch. VII).

Chapter 2

§ **1.** Semimartingales have been singled out as a separate class in Doléans-Dade and Meyer's work [97]. Special semimartingales have been introduced by Yoeurp [126] and Meyer [213]. Among the books discussing properties of martingales are Dellacherie and Meyer [83]-[85], Jacod [103], Métivier [219], Elliot [335], Gihman and Skorohod [50]. The notion of a quasimartingale has been introduced by Fisk [303] and Orey [235], see also Métivier [219], Jacod [103], Rao [250].

§ **2.** The first construction of the stochastic integral with respect to a Wiener process and with a deterministic integrand has been given in 1923 by N. Wiener. The general definition of the stochastic integral with respect to a Wiener process belongs to Ito [121]. The integration with respect to square integrable martingales has been treated by Doob [96], Meyer [213], Courrége [163], Kunita and Watanabe [162].

The construction of a stochastic integral with respect to a local martingale based on the second decomposition

$$M = M^c + M^d, \quad M^c \in \mathcal{M}^c_{loc}, \quad M^d \in \mathcal{M}^d_{loc},$$

is given by Jacod [103]. The exposed construction basically follows the papers by Doléans-Dade and Meyer [94], Meyer [213]. See also the books by Métivier [219], Jacod [103], Dellacherie and Meyer [84], Elliot [335]. Lemma 1 is proved in the paper by Kabanov, Liptser and Shiryaev [132].

An interesting characterization of semimartingales by properties of the stochastic integral is given by Dellacherie (see [103], Ch.9 and [84]).

§ **3.** A rule for calculating the differential df (t, X_t) in case of semimartingales of type

$$X_t = \int_0^t b(s, \omega)\, ds + \int_0^t a(s, \omega)\, dW_s,$$

where $(W_s)_{s \geq 0}$ is a Wiener process, has been established in Ito's work [122].

The formula (3.1), called the Ito formula, is a generalization of the result from this work by Ito to the case of arbitrary semimartingales. Concerning its proof, see e.g., the books by Dellacherie and Meyer [84], Gihman and Skorohod [50], Elliot [335].

§ **4.** Theorem 1 belongs to Doléans-Dade [93].

§ **5.** The works by Ito and Watanabe [124], Meyer [210], [211], Jacod [103] (Ch.VI) are devoted to the multiplicative decomposition of nonnegative special semimartingales.

§ **6.** The exposition in the sections 1-6 basically follows the work [132] by Kabanov, Liptser and Shiryaev. Liptser's work [187] is devoted to the strong law of large numbers for local martingales.

Chapter 3

§§ **1-4.** The exposition of the theory of random measures and their compensators basically follows the scheme proposed by Jacod [100]. In Notes to Chapter III of the book [103], there are references on works which have stimulated the introduction of the notion of the dual predictable projection or compensator of a random measure. The formula (4.7) has been obtained in Jacod's work [100] in case of a multiplicative point process, and in the work [129] by Kabanov, Liptser and Shiryaev in case of counting processes. See also Ch.18, 19 in the English version of Liptser and Shiryaev's monograph [190].

§ 5. On constructing the stochastic integral with respect to an integer-valued martingale measure, we follow the works by Kabanov, Liptser and Shiryaev [132] and Jacod [103].

§ 6. See Ch.III of the monograph [103] by Jacod for more general versions of the formula (6.3), in case in which special semimartingales are considered instead of a local martingale M.

Chapter 4

§ 1. The triplets of predictable characteristics for locally infinitely divisible processes have been introduced by Grigelionis [65], [66]. The triplets were systematically treated by Jacod and Mémin [108] and, in connection with questions concerning the absolutely continuous change of measures, by Kabanov, Liptser and Shiryaev [132]. Concerning the canonical representation, see the monographs by Dellacherie and Meyer [84], Jacod [103] and the paper [132].

§ 2. The stochastic exponentials, constructed by the triplets of semimartingales, and their properties have been treated in the works by Grigelionis [67], Jacod and Mémin [108], Jacod [100], Liptser and Shiryaev [191].

§§ 3-4. A martingale characterization of stochastic processes with independent and conditionally independent increments has been studied by Grigelionis [67], Jacod and Mémin [109], Jacod [105].

§ 5. The notion of the local absolute continuity $\tilde{P} \overset{loc}{\ll} P$ has been introduced in the work [132] by Kabanov, Liptser and Shiryaev. The assertion of Theorem 2 in case of $M \equiv W$ - a Wiener process - presents nothing else then the famous Girsanov theorem [45] (for details, see [188], Ch.6). The formulation of Theorem 2 given here belongs to Van Schuppen and Wong [32]. The change of a probabilistic measure and the transformation of the triplets of semimartingales under the locally absolutely continuous change of a measure, has been studied to a various extent in the works by Grigelionis [58], [59], [63], Kabanov, Liptser and Shiryaev [132], Jacod and Mémin [108], and in the monographs by Liptser and Shiryaev [188], Jacod [103], Dellacherie and Meyer [84]. Our exposition is based on the results of the works mentioned above.

§ 6. The basic Theorem 1 belongs to Stricker [294]. Chapter IX of the monograph [103] by Jacod and Chapter VII of the monograph [84] by Dellacherie and Meyer contain extended material devoted to the questions on preserving the semimartingale property under the restriction as well as under the extension of a basic flow of σ-algebras.

§ 7. The exposition is based on the results given in the monograph [83] by Dellacherie and Meyer. For additional details concerning a random change of time, see Chapter X of the monograph [103] by Jacod.

§ 8. The exposition is based on the works by Jacod [102] and Liptser [184]

§ 9. Lemmas 1 and 2 are taken over from Fernique's paper [302]. The Theorems 1-3 belong to Stricker [295], [296]. Theorem 4 has been communicated to the authors by Butov.

§ 10. The theory of the optimal nonlinear filtration for the case in which the observed process X is presented by a process of diffusion type or by a point process, is exposed in the author's monograph [188], [190]. The given Theorem 1 presents a modification of the results by Khadzhiev [307] and Vetrov [35].

§ 11. Properties of helix semimartingales are treated by Protter [355]. Theorem 1 is proved by De Sam Lazaro and Meyer [358]. Theorems 8 and 9 belong to the authors.

§ 12. Theorem 1 is proved by Chikin [363], and Theorem 2 belongs to the authors.

§ 13. In this section the results known for sums of independent random variables, or for certain martingales, for instance, square integrable martingales, are extended on the case of semimartingales. In this connection see [366], Vol. I, pp. 34, 102-103 and [367], Vol. II, pp. 74-75, 169-170, 247, 251-252. The monographs [344], [364] and [365] are devoted to large deviations for random (basically) Markov processes.

Chapter 5

§§ 1-2. The method for proving the weak convergence of finite dimensional distributions (of point processes) which in the present book is called *the method of stochastic exponentials*, has been used first by Kabanov, Liptser and Shiryaev [134]. This method has been applied and further developed in the works by Liptser and Shiryaev [191], Jacod, Klopotowski and Mémin [107]. In the present sections the exposition is based on the works just mentioned. Concerning the earlier works on martingale limit theorems, we refer the reader to the book [321] by Hall and Heyde.

§ 3. Theorem 1 is contained in the work [134] by Kabanov, Liptser and Shiryaev. See its special cases in Brown [22] - [24]. Theorem 2 is proved in the works by Liptser and Shiryaev [193] and by Jacod and Mémin [109].

§ 4. The exposition is based on the work [193] by Liptser and Shiryaev and on the work [107] by Jacod, Klopotowski and Mémin.

§ 5. The exposition follows the works [191], [192], [195] by Liptser and Shiryaev. Concerning the theorems for nested σ-fields see also Hall and Heyde's book [321].

§ **6.** A "nonclassical" version of the central limit theorem for semimartingales has been considered by Liptser and Shiryaev [197], who were inspired by the works by Zolotarev [114] - [117] and Rotar' [265] - [270] (for the scheme of series of independent random variables).

§ **7.** The exposition reproduces the work [194] by Liptser and Shiryaev. (On writing this work the authors were influenced by Môri's work [225]).

§ **8.** The basic method by S. N. Bernstein [12] for proving the central limit theorem for the sum of dependent random variables, consists in approximating the sums of independent random variables, and then using one or another type of mixing. The summarizing exposition is given by Ibragimov and Linnik [118]. Gordin [53] applied the method of approximating the sum of dependent random variables by martingales. Theorem 1 is proved by Gordin [53] in case of a trivial σ-algebra of invariant sets. The presented formulation has been communicated to the authors by Chikin.

Chapter 6

§§ **1-2.** The basic information concerning the space D, Skorohod's topology and continuous functions on $R_+ \times D$, is contained in Billingsley's book [14]. See also the papers by Lindvall [182] and Stone [291].

§ **3.** Here sufficient conditions are given for the relative compactness of the family of semimartingale distributions, useful for studying the weak convergence to the distribution of left quasicontinuous semimartingales. The important Theorem 1 belongs to Aldous [3], [4]. Theorem 2 is proved in the work [112] by Jacod, Mémin and Métivier. The relative compactness of families of probability measures on D has been considered in Grigelionis' work [62].

§ **4.** The "triplet" conditions for the relative compactness of probability distributions of semimartingales have been considered by Liptser and Shiryaev [191], [193], [196], [197], and Jacod, Klopotowski and Mémin [107].

§ **5.** The necessary conditions for the weak convergence of semimartingale distributions have been considered in the works by Liptser and Shiryaev [192], [195], Rebolledo [255], Jacod [104], [105]. The necessary conditions for the so-called extended weak convergence have been studied in the works [160], [161] by Kubilius and Mikulevicius.

Chapter 7

§ **1.** The Theorems 1, 4 and 6, 7, 9 are essentially reproduced from Liptser and

Shiryaev's works [191], [192], [195]. The validity of the functional central limit theorem for semimartingales has been studied in the works by Jacod, Klopotowski and Mémin [107], Grigelionis and Mikulevicius [71], [72].

§§ 2-3. The exposition in these sections is essentially based on the works by Kabanov, Liptser and Shiryaev [191], Jacod, Klopotowski and Mémin [107], Grigelionis and Mikulevicius [71], [72], Liptser and Shiryaev [193].

Chapter 8

§ 1. The method of proving the weak convergence by avoiding questions on the weak convergence of finite dimensional distributions, has been developed in the works by Stroock and Varadan [297], Grigelionis and Mikulevicius [71], [72].

§§ 2-3. The exposition presents a modification of the works by Liptser and Shiryaev [195] and Butov [30]. Concerning different conditions for the weak convergence, see the works by Grigelionis and Mikulevicius [71], [72], Borovkov [20], Gihman and Skorohod [50]. The example from the queueing theory is borrowed from the work [144] by Kogan, Liptser and Smorodinski.

§ 4. See the works by Grigelionis and Mikulevicius [71], [72], Kabanov, Liptser and Shiryaev [134].

§ 5. The weak convergence of invariant measures has been studied in the works by Kurtz [164], Costantini, Gerardi and Nappo [152] concerning Markov processes. The formulation of the corresponding results is given in Theorem 1, without the Markov assumptions on the prelimiting and limiting processes. The example from the queueing theory is borrowed from Smorodinski's work [286].

Chapter 9

§ 1. Theorem 1 is proved by a number of authors: Billingsley [13], Ibragimov [350], Rosén [356]. Theorem 2 belongs to Chikin [363].

§ 2. The result formulated in Theorem 1 concerning an ergodic stationary process, is classical. Its various versions can be found in the works by Rosenblatt [262], Volkonskii and Rosanov [39], Statulyavichus [287], Serfling [359], Gordin [53], Mc Leish [205]. Additional bibliographical information is presented in the books by Ibragimov and Linnik [118], and Hall and Heyde [321].

The proof of Theorem 1 presented here belongs to Chikin [363] (see also Dürr and Goldstein [349]). The proofs of Lemmas 2 and 3 are borrowed from [118], [77], [205] and [359].

§ 3. On exposing the results of this section, we basically follow Bhattacharya

[343] (see also Touat [299]). The analogous result concerning the discrete time case is contained in Gordin and Lifsic [54].

§ **4.** The proof of Theorem 1 is based on the Krasnosel'skii and Pokrovskii transformation, utilized in investigating vibrocorrect differential equations [353]. See also Doss [347].

§ **5.** In this section we formulate a special case of Theorem 1 in § 4. The proof is based on Lemma 1 in § 1, proved by Chikin [363]; it is extended to the vector case. As for replacing the weak convergence by the uniform convergence in probability, the analogous results can be found in Ikeda and Watanabe [351], Mackevicius[354] and Gyöngy [348].

§ **6.** Theorem 1 belongs to the authors. Theorem 2, which is analogous to Theorem 1 in § 1, has been communicated to the authors by Chikin.

§ **7.** The model to which Bogoljubov's averaging principle is applicable, is borrowed from Wentzell and Freidlin [344]. The proof of Theorem 1 belongs to the authors, the proof of Theorem 2 is communicated by Chikin.

Chapter 10

§ **1.** Skorohod's problem has been formulated originally for the univariate case, in connection with stochastic differential equations on the halfline $[0, \infty)$ [360]. For multivariate convex regions it has been solved by Tanaka [361]. Theorem 2 generalizes Theorem 2.1 in [361]; it has been obtained in collaboration with Anulova [342].

§ **2.** The definition of the solution of the martingale problem with reflection and the equivalence to weak solutions belongs to Stroock and Varadan [297]. Theorem 2 is proved by Tanaka [361] for an arbitrary convex region with the linear growth of coefficients.

§ **3.** Limit theorems for semimartingales in regions have been studied by Grigelionis and Mikulevicius under boundary conditions of a general type [345]. Theorems 1 and 2, extending the results of Grigelionis and Mikulevicius to the case of normal reflection and fairly arbitrary convex regions, have been obtained in collaboration with Anulov [342].

§ **4.** The diffusion approximation with reflection, as applied to problems of the queueing theory, has been studied by Borovkov [20], De Zelicourt [113], Grigelionis and Mikulevicius [346].

BIBLIOGRAPHY

1. Aalen O., *Weak convergence of stochastic integrals related to counting processes*. Z. Wahrsch. verw. Geb., 1977, Bd. 38, pp. 261-277.
2. Aleskevicius G.J., *On a central limit problem for sums of random variables defined on a Markov chain*. Liet. Matem. Rink., 1966, Vol. 6, N. 1, pp. 15-21 (Russian).
3. Aldous D.J., *Stopping time and tightness*. Ann. Probab., 1978, Vol. 6, N. 2, pp. 335-340.
4. Aldous D.J., *Weak convergence and the general theory of processes: Incomplete draft of monograph*. Berkeley: Department of Statistics, University of California, Berkeley, 1981.
5. Aldous D.J., Eagleson G.K., *On mixing and stability of limit theorems*. Ann. Probab., 1978, Vol. 6, pp. 325-331.
6. Al-Hussaini A.N., Elliott R.J., *Semimartingales and empirical distribution*. Preprint, 1983.
7. Anisimov V.V., *Limit Theorems for Stochastic Processes and their Applications to Discrete Summation Schemes*. Visca Skola, Kiev, 1976 (Russian).
8. Anisimov V.V., Vojna A.A., *Limit theorems for summation schemes on random processes with an arbitrary state space*.Theor. Probab. and Math. Statist., 1980, N. 19, pp. 7-18.
9. Baxter J.R., Chacon R.V., *Compactness of stopping time*. Z. Wahrsch. verw. Geb., 1977, Bd. 40, pp. 169-182.
10. Béska M., Klopotowski A., Slominski L., *Limit theorem for random sums of dependent d-dimensional vectors*. Z. Wahrsch. verw. Geb., 1982, Bd. 61, pp. 43-57.
11. Bernard A., Maissonneuve B., *Decomposition atomique des martingales de la classe* $\mathcal{H}^1$. Séminaire de Probabilités XI, Berlin, Springer, 1977, pp. 303-323. (Lect. Notes Math., Vol. 581).
12. Bernstein S.N., *Sur l'extension du théorème limite du calcul des probabilités aux sommes de quantités dependantes*. Math. Ann., 1926, Vol. 97, pp. 1-59
13. Billingsley P., *The Lindeberg-Levy theorem for martingales*. Proc. Amer. Math. Soc., 1961, Vol. 12, pp. 788-792.
14. Billingsley P., *Convergence of Probability Measures*. New York, Wiley, 1968.
15. Borovkov A.A., *Convergence of weakly dependent processes to the Wiener process*. Theory Probab. Appl., 1967, Vol. XII, N. 2, pp. 159-186.
16. Borovkov A.A., *Theorems on the convergence to Markov diffusion process*. Z.

Wahrsch. verw. Geb., 1970, Bd. 18, pp. 47-76.
17. Borovkov A.A., *The convergence of distributions of functionals on stochastic processes*. Russian Math. Surveys, 1972, Vol. 27, N. 1, pp. 1-42.
18. Borovkov A.A., *Convergence of measures and random processes*. Russian Math. Surveys, 1976, Vol. 31, N. 2, pp. 1-69.
19. Borovkov A.A., *Rate of convergence and large deviations in invariance principle*. Proc. of Int. Congress Math. Helsinki, Helsinki, 1978, pp. 725-731.
20. Borovkov A.A., *Asymptotic Methods in Queueing Theory*. New York, Wiley, 1984.
21. Brown B.M., *Martingale central limit theorems*. Ann. Math. Statist., 1971, Vol. 42, pp. 59-66.
22. Brown T.C., *A martingale approach to the Poisson convergence of simple point processes*. Ann. Probab., 1978, Vol. 6, pp. 615-628.
23. Brown T.C., *Some Poisson approximation using compensators*. Ann. Probab., 1983, Vol. 11, pp. 726-744.
24. Brown T.C., *Poisson approximation and the definition of the Poisson process*. Amer. Math. Monthly, 1984, Vol. 91, N. 2, pp. 116-123.
25. Brown B.M., Eagleson G.K., *Martingale convergence to infinitely devisible laws with finite variances*. Trans. Amer. Math. Soc., 1971, Vol. 162, pp. 449-453.
26. Brown T.C., Pollett P.K., *Some distribution approximation for Markov queueing networks*. Adv. Appl. Probab., 1982, Vol. 14, pp. 654-671.
27. Brémaud P., *Point Processes and Queues. Martingale Dynamics*. New York, Springer, 1981.
28. Breiman L., *Probability*. Reading (MA), Addison-Wesley, 1968.
29. Burkholder D., Davis B., Gundy R., *Integral inequalities for convex functions of operators on martingales*. Proc. Sixth Berkeley Symp. Math. Statist. Probab., Berkeley, Univ. of California Press, 1972, Vol. 2, pp. 223-240
30. Butov A.A., *On the problem of weak convergence of a sequence of semimartingales to a process of diffusion type*. Uspekhi Math. Nauk, 1983, Vol. 38, N. 5, pp. 135-136.
31. Valkeila E., *Studies in Distributional Properties of Counting Processes*. Preprint, Helsinki, Department of Mathematics, University of Helsinki, 1984.
32. Van Schuppen J.H., Wong E., *Transformation of local martingales under a change of law*. Ann. Probab., 1974, Vol. 2, pp. 879-888.
33. Weber N.C., *Rates of convergence for backward martingale arrays*. Liet. Matem. Rink., 1982, Vol. 22, N. 2, pp. 20-27
34. Ventcel A.D., *Additive functionals of a multidimentional Wiener process*. Soviet

Math. Dokl., 1961, Vol. 2, N. 4, pp. 848-851.

35. Vetrov L.G., *On filtering interpolation and extrapolation of semimartingales*. Theory Probab. Appl., 1982, Vol. 27, N. 1, pp. 24-36.

36. Ville J., *Etude critique de la notion de collectif*. Paris, Gauthier-Villars, 1939.

37. Vladimirova A.I., *Limit theorems for a solution of recurrent equations with random parameters*. Vychislitel'naya i Prikladnaya Matematika, 1982, Vol. 47, pp. 126-130 (Russian)

38. Volkonskii V.A., *A multidimentional limit theorem for homogeneous Markov chains with a countable number of states*. Theory Probab. Appl., 1957, Vol. II, N. 2, pp. 221-244.

39. Volkonskii V.A., Rozanov Yu.A., *Some limit theorems for random functions. I; II*. Theory Probab. Appl., 1959, Vol. 4, N. 2, pp. 178-197; 1961, Vol. 6, N. 2, pp. 186-198.

40. Vostrikova L.Ju., *Functional limit theorems for the disorder problem*. Stochastics, 1983, Vol. 9, N. 1-2, pp. 103-104.

41. Wolfson D.P., *Limit theorems for sums of a sequence of random variables defined on a Markov chain*. J. Appl. Probab., 1977, Vol. 14, N. 3, pp. 614-620.

42. Galtchouk L.I., *Représentation des martingales engendrées par un processus `a accroisements indépendants (cas de martingales de carré intégrable)*. Ann. Inst. H. Poincaré, 1976, Vol. 12, N. 3, pp. 199-211.

43. Gal'chuk L.I., *Gaussian semimartingales*, in: "Statistics and Control of Stochastic processes". Eds. N.V. Krylov, R.Sh. Liptser, A.A. Novikov, Berlin, Springer, 1985, pp. 102-121.

44. Girko V.L., *Limit Theorems for Functions of Random Variables*. Kiev, Visca Skola, 1983, 207 p. (Russian)

45. Girsanov I.V., *On transforming a certain class of stochastic processes by absolutely continuous substitution of measures*. Theory Probab. Appl., 1960, Vol. V, N. 3, pp. 285-301.

46. Gikhman I.I., Skorokhod A.V., *Introduction to the Theory of Random Processes*. Scripta Technica, Philadelphia, Saunders, 1969.

47. Gihman I.I., Skorohod A.V., *The Theory of Stochastic Processes, I*. Berlin, Springer, 1974.

48. Gihman I.I., Skorohod A.V., *The Theory of Stochastic Processes, II*. Berlin, Springer, 1975.

49. Gihman I.I., Skorohod A.V., *The Theory of Stochastic Processes, III*. Berlin, Springer, 1979.

50. Gihman I.I., Skorohod A.V., *Stochastic Differential Equations*. Berlin, Springer, 1972.

51. Gine E., Marcus M.B., *The central limit theorem for stochastic integrals with respect to Levy processes*. Ann. Probab., 1983, Vol. 11, N. 1, pp. 58-77.
52. Gnedenko B.V., Kolmogorov A.N., *Limit Distributions for Sums of Independent Random Variables*. Cambridge (MA), Addison-Wesley, 1954.
53. Gordin M.I., *The central limit theorem for stationary processes*. Soviet Math. Dokl., 1969, Vol. 10, N. 5, pp. 1174-1176.
54. Gordin M.I., Lifsic B.A., *The central limit theorem for stationary Markov processes*. Soviet Math. Dokl., 1978, Vol. 19, N. 2, pp. 392-394.
55. Grinblat L.S., *A limit theorem for measurable random processes and its applications*. Proc. Amer. Math. Soc., 1976, Vol. 61, pp. 371-376.
56. Greenwood P., Shiryayev A., *Contiguity and Statistical Invariance Principle*. New York, Gordon and Breach, 1958, (Statistics Monographs, Vol. 1).
57. Grigelionis B., *On a Markov property of Markov processes*. Liet. Matem. Rink., 1968, Vol. 8, N. 3, pp. 489-502 (Russian)
58. Grigelionis B., *On absolutely continous change of measures and Markov property of stochastic processes*. Liet. Matem. Rink., 1969, Vol. 9, N. 1, pp. 57-71 (Russian).
59. Grigelionis B., *On the absolute continuity of measures corresponding to stochastic processes*. Liet. Matem. Rink., 1971, Vol. 11, N. 3, pp. 783-794 (Russian).
60. Grigelionis B., *On representation of integer-valued random measures by means of stochastic integrals with respect to the Poisson measure*. Liet. Matem. Rink., 1971, Vol. 11, N. 3, pp. 93-108 (Russian)
61. Grigelionis B., *On stochastic equations for nonlinear filtering problem of stochastic processes*. Liet. Matem. Rink., 1972, Vol. 12, N. 4, pp. 37-51 (Russian).
62. Grigelionis B., *On the relative compactness of sets of probability measures in* $D[0, \infty)$ (X). Lithuanian Math. Journal, 1973, Vol. 13, N. 4, pp. 576-586.
63. Grigelionis B., *Structure of densities of measures corresponding to stochastic processes*. Lithuanian Math. Journal, 1973, Vol. 13, N. 1, pp. 48-52.
64. Grigelionis B., *Representation by stochastic integrals of square-integrable martingales*. Lithuanian Math. Journal, 1974, Vol. 14, N. 4, pp. 573-584.
65. Grigelionis B., *Random point processes and martingales*. Lithuanian Math. Journal, 1975, Vol. 15, N. 3, pp. 444-453.
66. Grigelionis B., *Characterization of stochastic processes with conditionally independent increments*. Lithuanian Math. Journal, 1975, Vol. 15, N. 4, pp. 562-567.
67. Grigelionis B., *Martingale characterization of random processes by independent*

increments. Lithuanian Math. Journal, 1977, Vol. 17, N. 1, pp. 52-60.

68. Grigelionis B., *Additive Markov processes*. Lithuanian Math. Journal, 1978, Vol. 18, N. 3, pp. 340-342.
69. Grigelionis B., *Theory of nonlinear estimation and semimartingales*. Izv. Academy Uzbek SSR, 1981, Vol. 3, pp. 17-22 (Russian)
70. Grigelionis B., Mikulevicius R., *Weak convergence of semimartingales*. Lithuanian Math. Journal, 1981, Vol. 21, N. 3, pp. 213-224.
71. Grigelionis B., Mikulevicius R., *Weak convergence of stochastic point processes*. Lithuanian Math. Journal, 1981, Vol. 21, N. 4, pp. 297-301.
72. Grigelionis B., Mikulevicius R., *On stably weak convergence of semimartingales and of point processes*. Theory Probab. Appl., 1983, Vol. 28, N. 2, pp. 337-350.
73. Grigoresku S., Oprisan G., *Limit theorems for J - X processes with a general state space*. Z. Wahrsch. verw. Geb., 1976, Bd. 35, N. 1, pp. 65-73.
74. Guess H.A., Gilespie J.H., *Diffusion approximation to linear difference equations with stationary coefficients*. J. Appl. Probab., 1977, Vol. 14, pp. 58-74.
75. Gänssler P., Strobel J., Stute W., *On central limit theorems for martingale triangular arrays*. Acta Math. Acad. Sci. Hungar., 1978, Vol. 31, pp. 205-216.
76. Gänssler P., Häusler E., *Remarks on the functional central limit theorem for martingales*. Z. Wahrsch. verw. Geb., 1980, Bd. 50, pp. 237-243.
77. Davydov Yu.A., *Convergence of distributions generated by stationary stochastic processes*. Theory Probab. Appl., 1968, Vol. 23, N. 1, pp. 691-696.
78. Davydov Yu.A., *The invariance principle for stationary processes*. Theory Probab. Appl., 1970, Vol. 25, N. 3, pp. 487-498.
79. Dvoretsky A., *Asymptotic normality for sums of dependent random variables*. Proc. Sixth Berkeley Symp. Math. Statist., Probability, Berkeley, Univ. of California Press, 1972, pp. 513-535.
80. Dvoretsky A., *Asymptotic normality of sums of dependent random vectors*. Multivariate analysis IV, Ed. P.R. Krishnaiah, Amsterdam, North Holland, 1977, pp. 23-24.
81. Dellacherie C., *Capacités et processus stochastiques*. Berlin, Springer, 1972.
82. Dellacherie C., *Intégrales stochastiques par rapport aux processus de Wiener ou de Poisson*. Séminaire de Probabilités VIII, Berlin, Springer, 1974, pp. 25-26 (Lect. Notes Math., Vol. 381).
83. Dellacherie C., Meyer P., *Probabilités et potentiel, I*. Paris, Hermann, 1975.
84. Dellacherie C., Meyer P., *Probabilités et potentiel, II*. Paris, Hermann, 1980.
85. Dellacherie C., Meyer P., *Probabilités et potentiel, III*. Paris, Hermann, 1983.

86. Jeganathan P., *A solution of the martingale central limit problem, Part I.* Sankhyâ, Ser. A., 1982, Vol. 44, pp. 299-318.

87. Jeganathan P., *A solution of the martingale central limit problem, Part II.* Sankhyâ, Ser. A., 1982, Vol. 44, pp. 319-340.

88. Jeganathan P., *A solution of the martingale central limit problem, Part III.* Invariance principles with mixtures of Levy process limits, Sankhyâ, Ser. A., 1983, Vol. 45, pp. 125-140.

89. Jain N.G., Monrad D., *Gaussian quasimartingales.* Z. Wahrsch. verw. Geb., 1982, Bd. 59, pp. 139-159.

90. Dobrushin R.L., *Limit laws for Markov chains.* Izv. Academy SSSR, Ser. Math., 1953, Vol. 17, pp. 291-330 (Russian)

91. Dobrushin R.L., *Central limit theorem for nonstationary Markov chains. I; II.* Theory Probab. Appl., 1956, Vol. 1, N. 1, pp. 65-80; N. 4, pp. 329-383.

92. Doleans-Dade C., *Processus croissants naturels et processus croissants très bien measurables.* C. R. Acad. Sci. Paris, 1967, Vol. 264, pp. 874-876.

93. Doleans-Dade C., *Quelques applications de la formule de changement de variable pour les semimartingales.* Z. Wahrsch. verw. Geb., 1970, Bd. 16, pp. 181-194.

94. Doleans-Dade C., Meyer P.-A., *Integrales stochastiques par rapport aux martingales locales.* Séminaire de Probabilités IV, Berlin, Springer, 1970, pp. 77-107 (Lect. Notes Math. Vol., 124).

95. Drogin R., *An invariance principle for martingales.* Ann. Math. Statistics, 1972, Vol. 43, pp. 602-620.

96. Doob J.L., *Stochastic Processes.* New York, Chapman & Hall, 1953.

97. Durrett R., Resnick S.I., *Functional limit theorems for dependent variables.* Ann. Probab., 1978, Vol 6, pp. 829-846.

98. Dynkin E.B., *Markov Processes.* Berlin, Springer, 1965.

99. Davis M.H.A., *The representation of martingales of jump processes.* SIAM J. Control and Optimization, 1976, Vol. 14, pp. 623-638.

100. Jacod J., *Multivariate point processes; Predictable projection, Radon - Nikodym derivatives, representation of martingales.* Z. Wahrsch. verw. Geb., 1975, Bd. 31, pp. 235-253.

101. Jacod J., *Un théorème de répresentation pour les martingales discontinues.* Z. Wahrsch. verw. Geb., 1976, Bd. 34, pp. 225-244.

102. Jacod J., *A general theorem of representation for martingales.* Proc. of Symp. in Pure Math., 1977, Vol. 31, pp. 101-104.

103 Jacod J., *Calcul stochastique et problèmes de martingales.* Berlin, Springer, 1979 (Lect. Notes Math., Vol. 714).

104. Jacod J., *Convergence en loi de semimartingales et variation quadratique.* Séminaire de Probabilités XV, Berlin, Springer, 1981, pp. 547-560 (Lect. Notes

Math., Vol. 850).

105. Jacod J., *Processus à accroissements indépendents: Une condition nécessary et suffisante de convergence en loi.* Z. Wahrsch. verw. Geb., 1983, Bd. 63, pp. 109-136.

106. Jacod J., Yor M., *Etudes des solutions extrémales et reprèsentation integrable des solutions pour certain problèmes de martingales.* Z. Wahrsch. verw. Geb., 1977, Bd. 38, pp. 83-125.

107. Jacod J., Klopotowski A., Mémin J., *Théorème de la limite centrale et convergence fonctionelle vers un processus à accroissements indépendants: la méthode des martingales.* Ann. Inst. H. Poincaré, 1982, Vol. 28, N. 1, pp. 1-45.

108. Jacod J., Mémin J., *Caractéristique locales et conditions de continuité absolute pour les semimartingales.* Z. Wahrsch. verw. Geb., 1976, Bd. 36, pp. 1-37.

109. Jacod J., Mémin J., *Sur la convergence des semimartingales vers un processus à accroissements independants.* Séminaire de Probabilitees XIII, Berlin, Springer, 1979, pp. 227-248 (Lect. Notes Math., Vol. 784).

110. Jacod J., Mémin J., *Un nouveau critère de compacité relative pour une suite de processus.* Preprint, 1980.

111. Jacod J., Mémin J., *Sur un type de convergence intermédiaire entre la convergence en loi la convergence en probabilité.* Séminaire de Probabilités XV, Berlin, Springer, 1981, pp. 529-546 (Lect. Notes Math., Vol. 850).

112. Jacod J., Mémin J., Métivier M., *On tightness and stopping times.* Stoch. Proc. and Appl., 1983, Vol. 14, pp. 109-146.

113. De Zelicourt C., *Une méthode de martingales la convergence d'une suite de processus de sauts markoviens vers une diffusion associée à une condition frontière. Application aux systèmes de files.* Ann. Inst. H. Poincaré, 1981, Bd. 17, N. 4, pp. 351-375.

114. Zolotarev V.M., *On the closeness of distributions of two sums of independent random variables.* Theory Probab. Appl., 1965, Vol. X, N. 3, pp. 472-479.

115. Zolotarev V.M., *Théorèmes limites pour les sommes de variable aléatoires independants qui ne sont pas infinitesimales.* C. R. Acad. Sci. Paris, 1967, Vol. 264, pp. 799-800.

116. Zolotarev V.M., *A generalization of the Lindberg-Feller theorem.* Theory Probab. Appl., 1967, Vol. XII, N. 4, pp. 6608-618.

117. Zolotarev V.M., *Théorèmes limites generaux pour les sommes de variable aléatories independants.* C. R. Acad. Sci. Paris, 1970, Vol. 270, pp. 889-902.

118. Ibragimov I.A., Linnik Yu. V., *Independent and Stationary Sequences of Random Variables.* Groningen, Wolters - Noordhoff, 1971.

119. Ibragimov I.A., *A note on the central limit theorem for dependent random variables*. Theory Probab. Appl., 1975, Vol. XX, N. 1, pp. 135-141.
120. Ibragimov I.A., Has'minskii, *Statistical Estimation: Asymptotic Theory*. New York, Springer, 1981.
121. Ito K., *Stochastic integral*. Proc. Imp. Acad. Tokyo, 1944, Vol. 20, pp. 519-524.
122. Ito K., *On a formula concerning stochastic differentials*. Nagoya Math. J., 1951, Vol. 3, pp. 55-65.
123. Ito K., *Multiple Wiener integral*. J. Math. Soc. Japan, 1951, Vol. 3, pp. 157-169.
124. Ito K., Watanabe S., *Transformation of Markov processes by multiplicative functionals*. Ann. Inst. Fourier, 1965, Vol. 15, pp. 15-30.
125. Yor M., *Reprèsentation intégrable des martingales de carré integrable*. C. R. Acad. Sci. Paris, 1976, Ser. A - B, Vol. 282, pp. 889-901.
126. Yoeurp Ch., *Decompositions des martingales locales et formules exponentielles*. Séminaire de Probabilités X, Berlin, Springer, 1976, pp. 432-480 (Lect. Notes Math., Vol. 511).
127. Kabanov Yu.M., *Representation of functionals of Wiener and Poisson processes in the form of stochastic integrals*. Theory Probab. Appl., 1973, Vol. 28, N. 2, pp. 362-365.
128. Kabanov Yu.M., *Integral representation of functionals of processes with independent increments*. Theory Probab. Appl., 1974, Vol. 29, N. 4, pp. 853-857.
129. Kabanov Ju.M., Liptser P.Sh., Shiryayev A.N., *Martingale methods in the theory of point processes*. Proc. School and Sem. Theory of Stochastic Processes, Druskininkai, 1974. Part II, Vilnius, 1975, pp. 269-354 (Russian).
130. Kabanov Ju.M., Lipcer R.S., Sirjaev A.N., *"Predictable" criteria for absolute continuity and singularity of probability measures (the continuous time case)*. Soviet Math. Dokl., 1977, Vol. 18, N. 6, pp. 1515-1518.
131. Kabanov Ju.M., Lipcer R.S., Sirjaev A.N., *On the question of absolute continuity and singularity of probability measures*. Math. USSR Sbornik, 1977, Vol. 33, N. 2, pp. 203-221.
132. Kabanov Ju.M., Lipcer R.S., Sirjaev A.N., *Absolute continuity and singularity of locally absolutely continuous probability distributions. I; II*. Math. USSR Sbornik, 1979, Vol. 35, N. 5, pp. 631-680; 1980, Vol. 36, N. 1, pp. 31-58.
133. Kabanov Ju.M., Lipcer R.S., Sirjaev A.N., *On the representation of integral-valued random measures and local martingales by means of random measures with deterministic compensators*. Math. USSR Sbornik, 1981, Vol. 39, N. 2, pp. 267-280.

134. Kabanov Yu.M., Liptser R.Sh., Shiryaev A.N., *Some limit theorems for simple point processes (a martingale approach)*. Stochastics, 1980, Vol. 3, pp. 203-216.

135. Kallenberg O., *Random Measures*. Berlin, Akademic Verlag, 1975.

136. Kallianpur G., Wolpert F., *Weak convergence of solution of stochastic differential equations with applications to neuronal models*. Depart. of Statist. Univer. of North Carolina Technical Report, 1984, Vol. 60, 23 p.

137. Kaplan E.I., *Limit theorems for exit times of random sequences with mixing*. 1980, N. 21, pp. 59-65.

138. Kaplan E.I., Sil'vestrov D.C., *Theorems of invariance principle type for semi-Markov processes with an arbitrary set of states*. Analytical Methods in Probability Theory, Kiev, Naukova Dumka, 1979, pp. 121-125 (Russian).

139. Kato Y., *Rates of convergence in central limit theorem for martingale difference*. Bull. Math. Statist., 1979, Vol. 18, pp. 1-8.

140. Keilson J., Wishart D., *A central limit theorem for processes defined on a finite Markov chain*. Proc. Cambr. Phil. Soc., 1964, Vol. 60, pp. 547-567.

141. Kembleton S.R., *A stable limit theorems for Markov chain*. Ann. Math. Statist., 1969, Vol. 40, pp. 1467-1473.

142. Clark J.M.C., *The representation of functionals of Brownian motion by stochastic integrals*. Ann. Math. Statist., 1970, Vol. 41, pp. 1282-1295.

143. Klopotowski A., *Mixtures of infinitely divisible distributions as limit laws for sums of dependent random variables*. Z. Wahrsch. verw. Geb., 1980, Bd. 51, pp. 101-115.

144. Kogan Ya.A., Liptser R.Sh., Smorodinskii A.V., *Gaussian diffusion approximation of closed Markov models of computer networks*. Problems Inf. Transmission, 1986, Vol. 22, N. 1, pp. 38-51.

145. Cogburn R., *The central limit theorem for Markov processes*. Proc. Sixth Berkeley Symp. Math. Stat. Probab., Berkeley, Univ. of California Press, 1972, Vol. 2, pp. 485-512.

146. Kolmogorov, A.N., *Foundations of the Theory of Probability*. New York, Chelsea, 1956.

147. Kolmogorov A.N., Rozanov Yu.A., *On strong mixing conditions for stationary Gaussian processes*. Theory Probab. Appl., 1960, Vol. V, N. 2, pp. 204-208.

148. Kolomiets E.I., *Relations between triplets of local characteristics of semimartingales*. Russian Math. Surveys, 1984, Vol. 39, N. 4, pp. 123-124.

149. Komlos J., Major P., Tusnady G., *Weak convergence and embedding*. Limit theorems in probability theory, Ed. P. Revesz, Amsterdam, North-Holland, 1975, pp. 117-135.

150. Korolyuk V.S., Lebedintseva E.P., *Limit theorem for sojourn time of a semi-Markov process in a subset of states*. Ukrainian Math. Journal, 1978, Vol. 30, N. 5, pp. 513-516.
151. Korolyuk V.S., Polishchuk L.I., Tomusyak A.A., *A limit theorem for semi-Markov processes*. Cybernetics, 1969, Vol. 5, N. 4, pp. 524-526.
152. Costantini C., Gerardi A., Nappo G., *On the convergence of sequences of stationary jump Markov processes*. Stat. and Probab. Letters, 1983, N. 1, pp. 155-160.
153. Cremers H., Kadelka D., *On weak convergence of stochastic processes with Lusin path space*. Manuscripta Math., 1984, Vol. 45, pp. 115-125.
154. Kruglov C. M., *Limit theorems for sums of independent random variables with values in Hilbert space*. Theory Probab. Appl., 1972, Vol. 17, N. 2, pp. 199-217.
155. Kruglov V.M., *Central limit theorem and law of large numbers in the mean*. Theory Probab. Appl., 1981, Vol. 26, N. 4, pp. 813-815.
156. Kubilius K., *Asymptotics of distributions of martingales with a continuous parameter*. Lithuanian Math. Journal, 1979, Vol. 19, N. 4, pp. 524-533.
157. Kubilius K., *Asymptotics of distributions of martingales*. Lithuanian Math. Journal, 1981, Vol. 21, N. 3, pp. 227-240.
158. Kubilius K., *Necessary and sufficient conditions for the convergence of semimartingales to processes with conditionally independent increments*. Lithuanian Math. Journal, 1984, Vol. 24, pp. 130-142.
159. Kubilius K., *Rate of convergence in the functional central limit theorem for semimartingales*. Lithuanian Math. Journal, 1985, Vol. 25, N. 1, pp. 44-52.
160. Kubilius K., Mikulevicius R., *On necessary and sufficient conditions for the convergence of semimartingales*. Séminaire de Probabilités XVII, Berlin, Springer, 1983, pp. 338-351 (Lect. Notes Math., Vol. 1021).
161. Kubilius K., Mikulevicius R., *Necessary and sufficient conditions for convergence of semimartingales and point processes. I; II*. Lithuanian Math. Journal, 1984, Vol. 24, N. 3, pp. 270-276; N. 4, pp. 346-357.
162. Kunita H., Watanabe S., *On square integrable martingales*. Nagoya Math. J., 1967, Vol. 30, pp. 209-245.
163. Courrège Ph., *Intégrales stochastiques associées à une martingale de carré intégrable*. CRAS, Paris, 1963, Vol. 256, pp. 867-870.
164. Kurtz T.G., *Limit theorems for sequences of jump Markov processes approximating ordinary differential processes*. J. Appl. Probab., 1971, Vol. 8, pp. 344-356.
165. Xu C., *Remarks on the martingale invariance principles*, Fudan J. Nat. Sci., 1982, Vol. 21, N. 2, pp. 192-205.

166. Kubacki K.S., Szynal D., *Weak convergence of martingales with random indices to infinitely divisible laws.* Acta Math. Hung., 1983, Vol. 42, N. 1-2, pp. 143-151.

167. Kurtz T.G., *Semigroups of conditioned shifts and approximation of Markov processes,* Ann. Probab., 1975, Vol. 3, pp. 618-642.

168. Kutoyants Yu.A., *Locally asymptotic normality for processes of Poisson type.* Soviet J. Contemporary Math. Analysis, Armenian Academy of Sciences, 1979, Vol. 14, B. 1, pp. 1-18.

169. Kutoyants Yu.A., *Parameter Estimation for Stochastic processes.* Berlin, Helderman, 1983.

170. Kushner H.J., Huang H., *Rates of convergence for stochastic approximation.* SIAM J. Contr. Optim., 1979, Vol. 17, pp. 607-617.

171. Kushner H.J., *Diffusion approximations to output processes of nonlinear systems with wide band inputs, and application.* IEEE Trans. Inform. Theory, 1980, IT-26, pp. 715-725.

172. Kushner H.J., Huang H., *On the weak convergence of a sequence of general stochastic difference equations to a diffusion.* SIAM J. Appl. Math., 1981, Vol. 40, N. 3, pp. 528-541.

173. Lavrent'ev V.V., *On the weak convergence of Hilbert space-valued semimartingales to stochastically continous processes with conditionally independent increments.* Russian Math. Surveys, 1983, Vol. 38, N. 1, pp. 149-150.

174. Lavrent'ev V.V., *A functional central limit theorem for semimartingales with values in Hilbert space.* Soviet Math. Dokl., 1983, Vol. 27, N. 3, pp. 548-552.

175. Lebedev V.A., *On relative compactness of families of distributions of semimartingales.* Theory Probab. Appl., 1981, Vol. 26, N. 1. pp. 140-148.

176. Lebedev V.A., *On the weak compactness of families of distributions of general semimartingales.* Theory Probab. Appl., 1982, Vol. 27, N. 1, pp. 15-23.

177. Levy P., *Théorie de l'addition des variables aléatoires.* Paris, Gauthiers-Villars, 1948.

178. Lenglart E., *Relation de domination entre deux processus.* Ann. Inst. H. Poincaré, Sect. B, 1977, Vol. XIII, N. 2, pp. 171-179.

179. Lepingle D., *Sur la répresentation des sauts d'une martingale.* Séminaire de Probabilités XI, Berlin, Springer, 1977, pp. 418-434 (Lect. Notes Math., Vol. 581).

180. Letta G., *Compacité et convergence des pois d'une suite de processus de comptage.* Repdicontl. Accademia Nazionale delle Scienze detta del XL Memorie di Matematica, 1983, Vol. 7, N. 101, pp. 193-200.

181. Liese F., Vom Scheidt J., *A limit theorem for sequences of weakly dependent stochastic processes*. Serdika Bulg. Mat. Opisanije, 1973,Vol. 10. pp. 18-30.
182. Lindvall T., *Weak convergence of probability measures and random functions in the function space* D [0, ∞). J. Appl. Probab., 1973, Vol. 10, pp. 109-121.
183. Linnik Yu.V., Ostrovskii I.V., *Decomposition of Random Variables and Vectors*. Providence, A.M.S., 1977.
184. Liptser R.Sh., *On the representation of local martingales*. Theory Probab. Appl., 1976, Vol. 21, N. 4, pp. 698-705.
185. Liptser R.Sh., *Gaussian martingales and a generalization of the Kalman-Bucy filter*. Theory Probab. Appl., 1975, Vol. 20, N. 2, pp. 285-301.
186. Liptser R.S., *On functional limit theorem for Markov process with finite state space*, in: "Statistics and Control of Stochastic processes". Eds. N.V. Krylov, R.Sh. Liptser, A.A. Novikov. Berlin, Springer, 1985, pp. 305-316.
187. Liptser R.Sh., *A strong law of large numbers for local martingales*. Stochastics, 1980, Vol. 3, N. 3, pp. 217-228.
188. Liptser R.Sh., Shiryayev A.N., *Statistics of Random Processes*. Moscow, 1974, (Russian).
189. Liptser R.Sh., Shiryayev A.N., *Statistics of Random Processes, I: General Theory*. New York, Springer, 1977.
190. Liptser R.Sh., Shiryayev A.N., *Statistics of Random Processes, II: Applications*. New York, Springer, 1978.
191. Liptser R.Sh., Shiryayev A.N., *A functional central limit theorem for semimartingales*. Theory Probab. Appl., 1980, Vol. 25, N. 4, pp. 667-688.
192. Liptser R.Sh., Shiryayev A.N., *On necessary and sufficient conditions in the functional central limit theorem for semimartingales*. Theory Probab. Appl., 1981, Vol. 26, N. 1, pp. 130-135.
193. Lipcer R.S., Sirjaev A.N., *On weak convergence of semimartingales to stochastically continuous processes with independent and conditionally independent increments*. Math. USSR Sbornik, 1983, Vol. 44, N. 3, pp. 299-323.
194. Liptser R.Sh., Shiryayev A.N., *On the rate of convergence in the central limit theorem for semimartingales*. Theory Probab. Appl., 1982, Vol. 27, N. 1, pp. 1-13.
195. Liptser R.Sh., Shiryayev A.N., *On a problem of necessary and sufficient conditions in the functional central limit theorem for local martingales*. Z. Wahrsch. verw. Geb., 1982, Bd. 59, pp. 311-318.
196. Liptser R.Sh., Shiryayev A.N., *Weak convergence for a sequence of semimartingales to a process of diffusion type*. Math. USSR Sbornik, 1984,

Vol. 49, N. 1, pp. 171-195.

197. Liptser R.Sh., Shiryayev A.N., *On the invariance principle for semimartingales: the "non-classical" case*. Theory Probab. Appl., 1983, Vol. 28, N. 1, pp. 1-34.

198. Lin'kov Ju.N., *The asymptotic nomality of stochastic integrals and the statistics of discontinuous processes*. Theory Probab. Math. Statist., 1981, N. 23, pp. 97-107.

199. Lin'kov Yu.N., *Asymptotic normality of locally square-integrable martingales in a series scheme*. Theory Probab. Math. Statist., 1983, N. 27, pp. 95-103.

200. Lifshits B.A., *On the central limit theorem for Markov chains*. Theory Probab. Appl., 1978, Vol. 23, N. 2, pp. 279-297.

201. Loynes R.M., *The central limit theorem for backwards martingales*. Z. Wahrsch. verw. Geb., 1969, Bd. 13, pp. 1-8.

202. Loynes R.M., *A criterion for tightness for a sequence of martingales*. Ann. Probab., 1976, Vol. 4, pp. 859-862.

203. Loéve M., *Probability Theory, 4th ed.* New York, Springer, 1977-1978.

204. Mc Leish D.L., *Dependent central limit theorems and invariance principles*. Ann. Probab., 1974, Vol. 2, pp. 620-628.

205. McLeish D.L., *Invariance principles for dependent variables*. Z. Wahrsch. verw. Geb., 1975, Bd. 32, pp. 165-178.

206. McLeish D.L., *An extended martingale principle*. Ann. Probab., 1978, Vol. 6, N. 1, pp. 144-150.

207. Macys J., *Sur la convergence des repartitions de sommes de variables aléatoires independantes vers les lois de la classe* I_0 *de Linnik*. C. R. Acad. Sci., Paris, 1968, Vol. 267, pp. 316-317.

208. Machis Yu.Yu., *Non-classical limit theorems*. Theory Probab. Appl., 1971, Vol. 16, N. 1, pp. 175-182.

209. Mackevicius V., *Weak convergence of random processes in spaces* $D_{[0,\infty)}(\aleph)$. Lithuanian Math. Transactions, 1974, Vol. 14, N. 4, pp. 620-623.

210. Meyer P.-A., *A decomposition theorem for supermartingales*. Illinois J. Math., 1962, Vol. 6, pp. 193-205.

211. Meyer P.-A., *Decomposition of supermartingales: the uniqueness theorem*. Illinois J. Math., 1963, Vol. 7, pp. 1-17.

212. Meyer P.-A., *Multiplicative decomposition of positive supermartingales, Markoff processes and potential theory*. Ed. J. Chover, New York, Wiley, 1967, pp. 117-139.

213. Meyer P.-A., *Integrales stochastiques. I-IV*. Séminaire de Probabilités I, Berlin, Springer, 1967, pp. 72-162 (Lect. Notes Math., Vol. 39).

214. Meyer P.-A., *Sur un probleme de filtration*. Séminaire de Probabilités VI,

Berlin, Springer, 1972, pp. 223-247 (Lect. Notes Math., Vol. 258).
215. Meyer P.-A., *Martingales and Stochastic Integrals*. Berlin, Springer, 1972, (Lect. Notes Math., Vol. 284).
216. Meyer P.-A., *Un cours sur les integrales stochastiques*. Séminaire de Probabilités X, Berlin, Springer, 1976, pp. 246-400 (Lect. Notes Math., Vol. 511).
217. Meyer P.-A., *Probabilités et potentiel*. Paris, Hermann, 1966.
218. Meyer P.-A., Zheng W.A., *Tightness criteria for laws of semimartingales*. Ann. Inst. H. Poincaré, 1984, Vol. 20, N. 4, pp. 353-372.
219. Métivier M., *Semimartingales*. Berlin, de Gruyter, 1982.
220. Métivier M., *Une condition suffisante de compacité faible pour une suite de processus. Cas des semimartingales*. Bolaiseau, Centre de Mathématique Appliqués, École Polytechnique, 1983 (Rappt. Interne, N. 61).
221. Métivier M., *Weak convergence of sequences of semimartingales*. Depart. Statist. Univer. of North Carolina, Technical Report N. 49, 1983, 38p.
222. Métivier M., Pellaumail J., *A basic course on stochastic integration*. Sém. Prob. Rennes, Paris, Hermann, 1978, 77, Vol. 1, pp. 1-56.
223. Meshalkin L.D., *Limit theorems for Markov chains with a finite number of states*. Theory Probab. Appl., 1958, Vol. III, N. 4, pp. 335-357.
224. Môri T., *Non-classical central limit theorem for martingales*. Ann. Univ. Sci., Budapest, Sec. Math., 1981, Vol. 24, pp. 113-121.
225. Môri T., *On the rate of convergence in the martingale central limit theorem*. Studia Scientiarum Mathematicarum Hungarica, 1977, Vol. 12, pp. 413-417.
226. Morkvenas R., *Weak convergence of random processes to the solution of a martingale problem*. Lithuanian Math. Transections, 1975, Vol. 15, N. 2, pp. 247-253.
227. Nagaev S.V., *Some limit theorems for stationary Markov chains*. Theory Probab. Appl., 1957, Vol. 2, N. 1, pp. 378-406.
228. Nagaev S.V., *The central limit theorem for Markov processes with discrete time*. Izvestia Academy Uzbek SSR (mathematical series), 1962, N. 2, pp. 12-20 (Russian).
229. Nadaraia E.A., *Application of the central limit theorem for martingales to the study of the limit distribution of quadratic deviation of the kernel-type estimator of a density function*. Bulletin Academy Georgian SSR, Vol. 113, N. 3, pp. 253-256 (Russian).
230. Nakata T., *On the rate of convergence in mean central limit theorems for martingale differences*. Rep. Statist. Appl. Res. Un. Japan Sc. Engrs., 1976, Vol. 23, pp. 10-15.

231. Neveu J., *Martingales à temps discret*. Paris, Masson, 1972.
232. Norman M.F., *Diffusion approximation of non-Markovian processes*. Ann. Probab., 1975, Vol. 3, pp. 358-364.
233. Newman C.M., Wright A.L., *On invariance principles for certain dependent sequences*. Ann. Probab., 1981, Vol. 9, N. 4, pp. 671-675.
234. O'Brien G.L., *Limit theorems for sums of chaindependent processes*. J. Appl. Probab., 1974, Vol. 11, pp. 582-587.
235. Orey S., F-*processes*. Proc. Fifth Berkeley Symp., Vol. 2, Berkeley, Univ. of California Press, 1965, pp. 301-313.
236. Papanicolaou G.C., Stroock P.W., Varadhan S.R.S., *A martingale approach to some limit theorems*. Proc. Duke University Conference on Turbulence, 1975.
237. Papanicolaou G.C., Kohler W., *Asymptotical theory of mixing stochastic ordinary differential equations*. Comm. Pure and Appl. Math., 1974, Vol. 27, pp. 641-668.
238. Peligrad M., *An invariance principle for dependent random variables*. Z. Wahrsch. verw. Geb., 1981, Bd. 57, N. 4, pp. 495-507.
239. Peligrad M., *A criterion for tightness for a class of dependent random variables*. Acta Math. Sci. Hungar., 1982, Vol. 39, N. 4, pp. 311-314.
240. Petrov V.V., *Sums of Independent Random Variables*. Berlin, Springer, 1975.
241. Pyke R., Schanffle R., *Limit theorems for Markov renewal processes*. Ann. Math. Stat., 1964, Vol. 35, pp. 1746-1764.
242. Pisanec S.I., *Limit theorems for a sequence of diffusion type processes*. Theory of Random Processes, 1982, Vol. 10, pp. 73-80.
243. Platen E., Rebolledo R., *Weak convergence of semimartingales and discretisation methods*. Preprint, Akademie der Wissenschaften der DDR, Institut für Mathematik, 1981.
244. Popesku Gh., *A functional central limit theorem for a class of Markov chains*. Rev. Roum. math. pures et appl., Vol. 21, N. 6, pp. 757-750.
245. Pragarauskas H., *Approximation of controlled solutions of Ito's equation by controlled Markov chains*. Liethuanian Math. Journal, 1983, Vol. 23, N. 1, pp. 98-108.
246. Prakasa Rao R.M., *On central limit theorem, invariance principles and rate of convergence for backward martingale arrays*. Liet. Matem. Rink., 1979, Vol. 19, N. 4, pp. 153-165.
247. Prohorov Yu.V., *Probability distributions in functional spaces*. Uspekhi Matem. Nauk, 1953, Vol. 8, N. 3, pp. 165-167 (Russian).
248. Prokhorov Yu.V., *Convergence of random processes and limit theorems in probability theory*. Theory Probab. Appl. , 1956, Vol. 1, N. 2, pp. 157-214.
249. Prokhorov Yu.V., Rozanov Yu.A., *Probability Theory: Basic concepts, Limit*

theorems, Random processes. Berlin, Springer, 1969.

250. Rao K.M., *Quasimartingales*. Math. Scand., 1969, Vol. 24, pp. 79-92.

251. Rebolledo R., *Remarques sur la convergence en loi de martingales vers des martingales continues*. C. R. Acad. Sci. Paris, 1977, Vol. 285, pp. 517-520.

252. Rebolledo R., *La méthode des martingales appliquée à l'étude de la convergence en loi de processus*. Bull. de la Société Mathématique de France, 1979, Mém. 62, pp. 1-125.

253. Rebolledo R., *Sur le théorème de la limite central pour les martingales*. C. R. Acad. Sci. Paris, 1979, Ser. A288, pp. 879-882.

254. Rebolledo R., *Central limit theorems for local martingales*. Z. Wahrsch. verw. Geb., 1980, Bd. 51, pp. 269-286.

255. Rebolledo R., *The central limit theorem for semimartingales: necessary and sufficient conditions*. Z. Wahrsch. verw. Geb., 1981, Bd. 54, pp. 260-278.

256. Rényi A., *On mixing sequences of sets*. Acta Math. Acad. Sci. Hungar. 1968, Vol. 9, pp. 215-228.

257. Rényi A., *On stable sequences of events*. Indian J. of Statistics, Ser. A., 1963, Vol. 25, N. 3, pp. 293-302.

258. Rychlik Z., Szyszkowski I., *On the rate of convergence for distributions of integral type functionals*. Séminaire de Probabilités XVIII, Berlin, Springer, 1984, pp. 255-275 (Lect. Notes Math., Vol. 1080).

259. Rozanov Yu.A., *A central limit theorem for additive random functions*. Theory Probab. Appl., 1960, Vol. 5, N. 2, pp. 243-246.

260. Rozanov Yu.A., *Stationary Random Processes*. San Francisco, Holden-Day, 1967.

261. Rosen B., *On the central limit theorem for sums of dependent random variables*. Z. Wahrsch. Geb., 1967, Bd. 7, pp. 48-82.

262. Rosenblatt M., *A central limit theorem and a strong mixing condition*. Proc. Nat. Acad. Sci. USA, 1956, Vol. 42, N. 1, pp. 43-47.

263. Rosenblatt M., *A strong mixing condition and central limit theorem on compact groups*. J. Math. Mech., 1967, Vol. 17, N. 2, pp. 189-198.

264. Rosenkranz W.A., *Limit theorems for solution to a class of stochastic differential equations*. Indiana Univ. Math. J., 1975, Vol. 24, pp. 613-625.

265. Rotar' V.I., *An extension of the Lindeberg-Feller theorem*. Math. Notes, 1975, Vol. 18, N. 1, pp. 660-663.

266. Rotar' V.I., *Some remarks on summing independent variables in the non-classical case*. Theory Probab. Appl., 1976, Vol. 21, N. 1, pp. 130-137.

267. Rotar' V.I., *Non-classical estimates of the rate of convergence in the multi-dimensional central limit theorem. I; II*. Theory Probab. Appl., 1977, Vol. 22,

N. 4, pp. 755-772; 1978, Vol. 23, N. 1, pp. 50-62.

268. Rotar' V.I., *Non-classical estimates of the error of approximation in the central limit theorem.* Math. Notes, 1978, Vol. 23, N. 1, pp. 77-83.

269. Rotar' V.I., *Limit theorems for polylinear forms.* J. Multivar. Anal., 1979, Vol. 9, N. 4, pp. 511-530.

270. Rotar' V.I., *On summation of independent variables in a non-classical situation.* Russian Math. Surveys, 1982, Vol. 37, N. 6, pp. 151-175.

271. Rootzén H., *Fluctuations of sequences in distribution.* Ann. Probab., 1976, Vol. 4, pp. 456-463.

272. Rootzén H., *On the functional limit theorem for martingales.* Z. Wahrsch. verw. Geb., 1977, Bd. 38, pp. 199-210.

273. Rootzén H., *A note on convergence to mixtures of normal distributions.* Z. Wahrsch. verw. Geb., 1977, Bd. 38, pp. 211-216.

274. Rootzén H., *On the functional limit theorem for martingales.* Z. Wahrsch. verw. Geb., 1980, Bd. 51, pp. 579-593.

275. Rootzén H., *Limit distributions for the error in approximations of stochastic integrals.* Ann. Probab., 1983, Vol. 8, N. 2, pp. 241-251.

276. Sen P.K., *Weak convergence of generalized U-statistics.* Ann. Probab., 1974, Vol. 2, pp. 90-102.

277. Sen P.K., Tsong Y., *On the functional limit theorems for certain continuous time parameter stochastic processes.* J. Multivar. Anal., 1980, Vol. 10, N. 3, pp. 371-378.

278. Serfling R.J., *A general Poisson approximation theorem.* Ann. Probab., Vol. 3, pp. 726-731.

279. Sil'vestrov D.S., *Limit Theorems for Composite Random Functions.* Kiev, Visca Skola, 1974, (Russian).

280. Sil'vestrov D.S., Tursunov G.T., *General limit theorems for sums of controlled random variables, I.* Theory Probab. and Math. Statist., 1979, N. 17, pp. 131-146.

281. Sil'vestrov D.S., Tursunov G.T., *General limit theorems for sums of controlled random variables, II.* Theory Probab. and Math. Statist., 1980, N. 20, pp. 131-141.

282. Skorohod A.V., Slobodenuk N.P., *Limit Theorems for Random Walks.* Kiev, Naukova Dumka, 1970, 303 p. (Russian)

283. Scott P.J., *An invariance principle for reversed martingales.* Z. Wahrsch. verw. Geb., 1971, Bd. 20, pp. 9-27.

284. Skott D.J., *Central limit theorems for martingales with stationary increments using Skorokhod representation approach.* Adv. Appl. Probab., 1973, Vol. 5,

pp. 119-137.

285. Snell J.L., *Application of martingale system theorems*. Trans. AMS, 1952, Vol. 73, pp. 292-312.

286. Smorodinskii A.V., *Asymptotic distribution of the queue length of a service system*. Automatika i Telemekhanika, 1986, N. 2, pp. 92-92. (Russian)

287. Statulyavichus V.A., *Some new results for sums of weak dependent random variables*.Theory Probab. Appl., 1960, Vol. 5, N. 2, pp. 233-234.

288. Statulyavichus V.A., *Limit theorems for sums of random variables forming a Markov chain. I-III*. Liet. Matem. Rink., 1969, Vol. 9, N. 2, pp. 345-361; Vol. 9, N. 3, pp. 635-672; 1970, Vol. 10, N. 1, pp. 161-169 (Russian).

289. Stout W.F., *Almost Sure Convergence*. New York, Wiley, 1973.

290. Stout W.F., *On convergence of ϕ-mixing sequences of random variables*. Z. Wahrsch. verw. Geb., 1974, Bd. 31, N. 7, pp. 69-70.

291. Stone C., *Weak convergence of stochastic processes defined on semi-infinite time intervals*. Proc. Amer. Math. Soc., 1963, Vol. 14, pp. 694-696.

292. Straf M.L., *Weak convergence of stochastic processes with several parameters*. Proc. 6th Berkeley Symp. Math. Statist. Probab., Vol. 2, Berkeley, Univ. of California Press, 1972, pp. 187-221.

293. Stricker C., *Quasimartingales et variations*. Séminaire de Probabilités XV, Berlin, Springer, 1981, pp. 493-498 (Lect. Notes Math., Vol. 850).

294. Stricker C., *Quasimartingales, martingales locales, semimartingales et filtration naturelle*. Z. Wahrsch. verw. Geb., 1977, Bd. 30, pp. 55-63.

295. Stricker C., *Semimartingales gaussiennes - Applications au problème de l'innovation*. Strasbourg, Publications de L'IRMA, 1982.

296. Stricker C., *Semimartingales gaussiennes - Application au problème de l'innovation*. Z. Wahrsch. verw. Geb., 1983, Bd. 64, pp. 303-312.

297. Stroock D.W., Varadan S.R.S., *Diffusion process with continuous coefficients, I*. Commun. Pure and Appl. Math., 1969, Vol. 22, pp. 345-400.

298. Stroock D.W., Varadan S.R.S., *Multidimensional Diffusion Processes*. New York, Springer, 1979.

299. Touati A., *Théorèmes de limite central fonctionelle pour les processus de Markov*. Ann. Inst. H. Poincaré (B), 1983, Vol. XIX, pp. 43-59.

300. Whitt W., *Some usefull functions for functional limit theorems*. Math. Operations Res., 1980, Vol. 5, pp. 67-85.

301. Feigin P.D., *Stable convergence of semimartingales*. Stoch. Proc. Appl., 1985, Vol. 19, N. 1, pp. 125-134.

302. Fernique X., *Regularité des trajectoires des fonctions aleatoires Gaussiennes*. Ecole d'Eté de Probabilités de Saint-Flour IV, Berlin, Springer, 1975, pp. 2-96. (Lect. Notes Math., Vol. 480.)

303. Fisk D.L., *Quasimartingales*. Trans. AMS, 1965, Vol. 20, pp. 369-389.
304. Friedman D., *Some invariance principle for functional of Markov chain*. Ann. Math. Statist., 1967, Vol. 38, pp. 1-7.
305. Friedman D., *The Poisson approximation for dependent event*. Ann. Probab., 1974, Vol. 2, pp. 256-269.
306. Formanov S.K., *On invariance principles for homogeneous Markov chains*. Soviet Math. Dokl., 1975, Vol. 16, N. 2, pp. 301-304.
307. Hadjiev D.I., *On the filtering of semimartingales in case of observations of point processes*. Theory Probab. Appl., 1978, Vol. 23, N. 1, pp. 169-178.
308. Haussmann U.G., *On the integral representation of functional of Ito processes*. Stochastics, 1979, Vol. 3, pp. 17-27.
309. Heyde C.C., Brown B.M., *On the departure from normality of a certain class of martingales*. Ann. Math. Statist., 1970, Vol. 41, pp. 2161-2165.
310. Helland I.S., *On weak convergence to Brownian motion*. Z. Wahrsch. verw. Geb., 1980, Bd. 52, pp. 251-265.
311. Helland I.S., *Minimal conditions for weak convergence to a diffusion process on the line*. Ann. Probab., 1981, Vol. 9, pp. 429-452.
312. Helland I.S., *Central limit theorems for martingales with discrete or continuous time*. Scand. J. Statist., 1982, Vol. 9, pp. 79-84.
313. Helland I.S., *Convergence to diffusions with regular boundaries*. Stoch. Proc. Appl., 1982, Vol. 12, pp. 27-58.
314. Hennequin P.L., Tortrat A., *Théorie des probabilités et quelques applications*. Paris, Masson, 1965.
315. Hertz E., *On convergence rates in the central limit theorem*. Ann. Math. Statist., 1969, Vol. 40, pp. 475-479.
316. Haüsler E., *On the rate of convergence in the invariance principle for real-valued functions of Doeblin processes*. J. Multivar. Anal., 1984, Vol. 15, N. 1, pp. 73-90.
317. Holewiyn P.J., Meilijson I., *Note on the central limit theorem for stationary processes*. Séminaire de Probabilités XVI, Berlin, Springer, 1983, pp. 240-242 (Lect. Notes Math., Vol. 986).
318. Hall P., *Martingale invariance principles*. Ann. Probab., 1977, Vol. 5, pp. 875-887.
319. Hall P., *On the invariance principle for U-statistics*. Stoch. Proc. Appl., 1979, Vol. 9, pp. 175-187.
320. Hall P., *On the Skorokhod representation approach to martingale invariance principles*. Ann. Probab., 1979, Vol. 7, pp. 371-376.
321. Hall P., Heyde C.C., *Martingale Limit Theory and Its Application*. New York,

Academic Press, 1980.
322. Hall P., Heyde C.C., *Rates of convergence in the martingale central limit theorem*. Ann. Probab., 1981, Vol. 9, N. 3, pp. 395-404.
323. Chow Y.S., Teicher H., *Probability Theory*. New York, Springer, 1978.
324. Chen L.H.Y., *Martingale convergence via square function*. Proc. Amer. Math. Soc., 1981, Vol. 83, N. 1, pp. 125-127.
325. Chikin D.O., *On martingale methods of proving the central limit theorem for dependent random variables*. Moscow, Institute of Control Problems, 1984 (Russian).
326. Chou C.S., Meyer P.-A., *Sur la représentation des martingales comme intégrales stochastiques dans les processus ponctuels*. Séminaire de Probabilités IX, Berlin, Springer, 1975, pp. 117-131 (Lect. Notes Math., Vol. 465).
327. Chou C.S., *Le processus des sauts d'une martingale locale*. Séminaire de Probabilités XI, Berlin, Springer, 1977, pp. 356-361 (Lect. Notes Math., Vol. 581).
328. Csörgo M., *On a strong law of large numbers and the central limit theorem for martingales*. Trans. Amer. Math. Soc., 1968, Vol. 131, pp. 259-275.
329. Csörgo M., Fishler R., *Departure from independence: the strong law, standard and random-sum central limit theorem*. Acta Math. Acad. Sci. Hungar., 1970, Vol. 21, pp. 105-114.
330. Shiryayev A.N., *Optimal Stopping Rules*. New York, Springer, 1978.
331. Shiryayev A.N., *Absolute continuity and singularity of probability measures in functional spaces*. Proc. of the Intern. Congress of Mathematicians, Helsinki, Vol. 1, 1978, pp. 209-225.
332. Shiryayev A.N., *Probability*. Moscow, Nauka, 1980 (Russian).
333. Shiryayev A.N., *Martingales: Recent Developments, Results and Applications*. International Statistical Review, 1981, Vol. 49, pp. 199-233.
334. Shiryayev A.N., *Probability*. New York, Springer, 1984.
335. Elliott R. J., *Stochastic Calculus and Applications*. New York, Springer, 1982, 302 p.
336. Emery M., *Covariance des semimartingales gaussiennes*. C. R. Acad. Sci. Paris, 1982, Vol. 295, Sér. I, pp. 703-705.
337. Engelbert H.J., Hess J., *Integral representation with respect to stopped martingales*. Jena, Friedrich-Schiller Univ., 1979, Forschungsergebnisse N. 5.
338. Engelbert H.J., Hess J., *Stochastic integrals of continuous local martingales. I; II*. Jena, Friedrich-Schiller Univ., 1980, Forschungsergebnisse N. 2; N. 4.
339. Erickson R.V., Quine M.P., Weber N.C., *Explicit bounds for the departure from normality of sums of dependent random variables*. Acta Math. Acad. Sci. Hungar., 1978, Vol. 34, pp. 27-32.

340. Erickson R.V., *Lipschitz smoothness and convergence with applications to the central limit theorem for summation processes*. Ann. Probab., 1981, Vol. 9, N. 5, pp. 831-851.

341. Janson S., Wichura M.J., *Invariance principles for stochastic area and related stochastic integrals*. Stoch. Proc. Appl., 1983, Vol. 16, pp. 71-84.

Additional references to the English edition.

342. Anulova S.V., Liptser P. Sh., *Diffusion approximation for processes with normal reflection*. Submitted to Theory Probab. Appl.

343. Bhattacharya R.N., *On the functional central limit theorem and the law of the iterated logarithm for Markov processes*. Z. Wahrsch. verw Geb., 1982, Vol. 60, pp. 185-201.

344. Freidlin M.I., Wentzell A. D., *Random Perturbations of Dynamical Systems*. New York, Springer, 1984.

345. Grigelionis B., Mikulevicius R., *On weak convergence of random processes with boundary conditions*. Nonlinear Filtering and Stochastic Control, Proceedings, Cortona 1981, Berlin, Springer 1983, pp. 260-275 (Lect. Notes Math., Vol. 972).

346. Grigelionis B., Mikulevicius R., *On the diffusion approximations in queueing theory*. Proceedings Conf. on Teletrafic, Moscow, 1984.

347. Doss H., *Liens entre équations différentielles stochastiques et ordinaires*. Ann. Inst. H. Poincaré, 1977, Vol. 13, pp. 99-125.

348. Gyöngy I., *On the approximation of stochastic differential equations*. Stochastics, 1988, Vol. 23, pp. 331-352.

349. Dürr D., Goldstein S., *Remarks on the central limit theorem for weakly dependent random variables*. Proc. 1st BIBOS Conf., Bielefeld 1984, Berlin, Springer, 1986, pp. 104-118 (Lect. Notes Math., Vol. 1158).

350. Ibragimov I. A., *A central limit theorem for a class of dependent random variables*. Theory Probab. Appl., 1963, Vol. 8, pp. 83-89.

351. Ikeda N., Watanabe S., *Stochastic Differential Equations and Diffusion Processes*. North-Holland, Amsterdam, 1981.

352. Kesten H., Papanicolau G. C., *A limit theorem for turbulent diffusion*. Commun. Math. Phys. 1979, Vol. 65, pp. 97-128.

353. Krasnosel'skii M. A., Pokrovskii A. V., *Systems with Hysteresis*. Nauka, Moscow, 1983 (Russian).

354. Mackevicius V., *Stability of solutions of symmetric stochastic differential*

equations. Lithuanian Math. Journal, 1985, Vol. 25, No. 4, pp. 343-352.

355. Protter P., *Semimartingales and measures preserving flots*. Ann. Inst. H. Poincaré, 1986, Vol. 22, No. 2, pp. 127-147.

356. Rosén B., *On the central limit theorem for sums of dependent random variables*. Z. Wahrsch. verw. Geb., 1967, Bd. 7, pp. 48-82.

357. De Sam Lazaro J., Meyer P.-A., *Méthodes de martingales et théorie des flots*. Z. Wahrsch. verw. Geb., 1971, Bd. 18, pp. 116-140.

358. De Sam Lazaro J., Meyer P.-A., *Questions de théorie des flots*. Séminaire de Probabilités IX, Berlin, Springer, 1975, pp. 213-224 (Lect. Notes Math., Vol. 465).

359. Serfling R.J., *Contributions to central limit theory for dependent variables*. Ann. Math. Stat., 1968, Vol. 39, pp. 1158-1175.

360. Skorokhod A.V., *Stochastic equations for diffusion processes in a bounded region*. Theory Probab. Appl., 1961, Vol. 6, N.3, pp. 264-274; 1962, Vol. 7, N.1, pp. 3-23.

361. Tanaka H., *Stochastic differential equations with reflecting boundary condition in convex regions*. Hiroshima Mathematical Journal, 1979, Vol. 9, N.1, pp. 163-177.

362. Chikin D.O., *Functional limit theorem for stationary processes*. Submitted to Theory Probab. Appl.

363. Khas'minskii R. Z., *On stochastic processes defined by differential equations with a small parameter*. Theory Probab. Appl., 1966, Vol. 11, N.2, pp. 211-228.

364. Wentzell A. D., *Limit Theorems on Large Deviations for Markov Random Prcesses*. Nauka, Moscow, 1986 (Russian; to be translated into English by Reidel, Dordrecht, 1989).

365. Varadhan S. R. S., *Large Daviations and Applications*. CBMS 46, SIAM, Philadelphia, 1984.

366. Dacunha - Castelle D., Duflo M., *Probabilités et Statistiques*, I, Masson, Paris, 1982.

367. Dacunha - Castelle D., Duflo M., *Probabilités et Statistiques*, II, Masson, Paris, 1983.

INDEX

GPSR Compliance

The European Union's (EU) General Product Safety Regulation (GPSR) is a set of rules that requires consumer products to be safe and our obligations to ensure this.

If you have any concerns about our products, you can contact us on ProductSafety@springernature.com

In case Publisher is established outside the EU, the EU authorized representative is:

Springer Nature Customer Service Center GmbH
Europaplatz 3
69115 Heidelberg, Germany

Batch number: 10823769

Printed by Printforce, the Netherlands